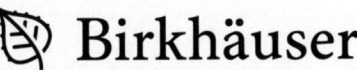

Fog and Dew Observations and Modeling

Edited by
Ismail Gultepe
Jörg Bendix
Otto Klemm
Werner Eugster
Jan Cermak

Previously published in *Pure and Applied Geophysics* (PAGEOPH), Volume 169, No. 5–6, 2012

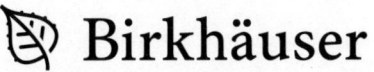 Birkhäuser

Editors

Ismail Gultepe
Cloud Physics and Severe
Weather Research Section
Environment Canada
Dufferin St. 4905
M3H 5T4 Toronto Ontario
Canada

Jörg Bendix
Philipps University Marburg
Faculty of Geography
Deutschhausstr. 10
35032 Marburg
Germany

Otto Klemm
University of Münster
Climatology Working Group
Robert-Koch-Str. 28
48149 Münster
Germany

Werner Eugster
ETH Zurich
Institute of Agricultural Sciences
Universitätstr. 2
8092 Zurich
Switzerland

Jan Cermak
Ruhr-Universität Bochum
Department of Geography
Universitätsstr. 150
44801 Bochum
Germany

ISBN 978-3-0348-0456-1   e-ISBN 978-3-0348-0457-8
DOI 10.1007/978-3-0348-0457-8

Library of Congress Control Number: 2012939391

Cover illustration: Photography by Ismail Gultepe taken at the port of Lunenburg, Nova Scotia, Eastern Canada in June 2006 (visibility less than 100 m).

Cover design: deblik, Berlin.

Printed on acid-free paper

Springer Basel AG is part of Springer Science+Business Media

www.birkhauser-science.com

Contents

Pure Appl. Geophys. 169 (2012), 765–766
© 2012 Springer Basel AG
DOI 10.1007/s00024-012-0471-y

Fog and Dew Observations and Modeling: Introduction

Ismail Gultepe[1]

This special issue of the Journal of Pure and Applied Geophysics contains 28 papers related to fog and dew physics and chemistry, fog hazards and low visibility, dew and fog climatology, fog remote sensing, fog modeling and forecasting, fog and dew interactions with vegetation, and condensation over the desert regions. The results discussed in this special issue are generated by research efforts conducted all over the world. The papers relate to presentations that were given during the 5th International Conference on Fog, Fog Collection and Dew, which took place in Münster, Germany, during 25–30 July 2010.

Irrespective of the knowledge gained on fog and dew, which is documented in this special issue, our understanding of the physics of fog and dew still remains incomplete due to all the complex processes influencing fog and dew formation, development, and dissipation along a spatio-temporal scale continuum (e.g., nucleation, microphysics and radiation processes, turbulence and turbulent fluxes, boundary layer processes and meso- to macroscale circulation processes, all influenced by the great variability of the soil–vegetation–atmosphere interface).

To cope with open questions, new technologies regarding instrumentation, remote sensing and numerical modeling must be developed, tested and applied. Respective fog research is presented in this special issue. In particular, the accurate forecasting and nowcasting of fog using numerical models and remote sensing techniques remains difficult and related issues need to be further studied in detail. Presently, typical numerical forecast models lack sufficient resolution, appropriate physical parameterizations to represent

fog and sophisticated approaches in estimating visibility. Simple approaches, such as using the liquid water content to calculate visibility, lead to uncertainties of more than 50% in many conditions. This special issue points out that model improvements have been realized, especially by using 3D fog models; however, they are currently not practical to be used operationally because of their time consuming numerical algorithms. On the fog hazard side, better forecasts are expected to help mitigate the financial losses and would reduce injuries caused by fog hazards in air, sea and road traffic. For example, an average 60 people die every year in Canada, which is comparable to impacts resulting from other severe weather events.

There are important feedback mechanisms between the fog physics and fog chemistry. The number of cloud condensation nuclei and their chemical composition impact the number and size distribution, and eventually the visibility, during hazy and foggy conditions. This has become mostly evident in urban fog studies and in decadal trend analyzes of the visibility on various continents. Global Change processes, including regional air pollution issues, exhibit a strong influence on fog. However, we are just starting to develop understanding for these processes, and the papers addressing this issue all show a huge natural temporal variability in time series of fog frequency and visibility.

From an analysis perspective, the modeling issue of fog needs to be further developed because surface observations are too sparsely distributed to adequately capture the temporal and spatial variability of fog. Similarly, satellite imagery, which could provide the necessary information on spatial variability, has serious limitations with temporal resolution and also with spatial coverage in the presence of mid- and high-level clouds. Surface observations in principle can easily detect the fog events, but there are not

---

[1] Environment Canada, Cloud Physics and Severe Weather Research Section, Toronto, ON M3H5T4, Canada. E-mail: ismail.gultepe@ec.gc.ca

many surface sites that include the necessary novel sensors such as liquid water content sensors, fog droplet spectrometers, cloud radars, profiling microwave radiometers, and ceilometers, for detecting fog. Moreover, fog is a discontinuous phenomenon in time and space that requires the detailed analysis of boundary layer structure, atmospheric stability, and state of turbulence. All these are important variables for better understanding the meteorology of fog. Integrated methods are however needed to combine all key variables in a consistent mechanistic way for nowcasting applications, for example, predictions over less than 3 h. When observations are integrated with numerical model output, better methods for fog nowcasting and forecasting can be developed, but the lack of surface observations may still result in important problems for fog prediction.

Dew typically provides less liquid water than fog, if collection efficiencies are expressed per unit sampling area. However, dew plays important roles in the boundary layer between the surface (vegetation or soil) and the atmosphere. The soil micro-hydrology is absolutely dependent on dew in some desert environments and can become an important component in the hydrology of desert ecosystems. Leaf surfaces, as wetted by dew, may be of key importance for the exchange of nutrients and pollutants of plants, since many gaseous species are water soluble and hence deposit to wet vegetation at much higher rates than under absence of dew. In some cases, dew can even be successfully collected for drinking water production. The physics and chemistry of dew are challenging issues in analytical chemistry and micrometeorology.

This special issue focuses on warm fog and dew conditions, but in reality freezing fog and ice fog can occur often in cold climates, particularly in the high Arctic and Antarctic regions, and can create adverse weather conditions for aviation and marine applications that need to be addressed seriously in the near future as air traffic and marine transportation in these areas increases.

We wish to thank the scientific contributors to this topical issue and the members of the science team of the 5th International Conference on Fog, Fog Collection, and Dew around the convener Dr. Otto Klemm (University of Münster, Germany). We also gratefully acknowledge funding support for Editor I. Gultepe by Environment Canada (EC) and the Canadian National Search and Rescue (SAR) Secretariat. Technical support for the data collection was provided by the Cloud Physics and Severe Weather Research Section of the Science and Technology Branch of EC, Toronto, Ontario. In addition, our sincere thanks go to Drs. Renata Dmowska and Brian Mitchell, Editors in chief for topical and regular issues at Pure and Applied Geophysics, inviting our working group to prepare this special issue, handling the review process and providing invaluable advice during the preparation of this special issue.

The editor is indebted to many reviewers who took their valuable time and effort to carefully review the manuscripts and make suggestions for improvements.

June 20 2011
Guest Editors
Dr. Ismail Gultepe,
Environment Canada,
Cloud Physics and Severe Weather Research
Section, Toronto, ON M3H5T4, Canada

Dr. Jörg Bendix,
Laboratory for Climatology and
Remote Sensing, Faculty of Geography,
University of Marburg, 35032 Marburg, Germany

Dr. Otto Klemm,
University of Münster,
Climatology Working Group,
48149 Münster, Germany

Dr. Werner Eugster,
ETH Zurichm,
Institute of Agricultural Sciences,
8092 Zurich Switzerland

Dr. Jan Cermak,
ETH Zurich,
Institute for Atmospheric and Climate Science,
8092 Zurich, Switzerland

(Published online: April 10, 2012)

Pure Appl. Geophys. 169 (2012), 767
© 2012 Springer Basel AG
DOI 10.1007/s00024-012-0472-x

Memorial Note

I. GULTEPE[1]

This special issue on Fog and Dew Observations and Modeling is dedicated to the memory of Dr. Mehmet Karadeniz, a manager at the Turkish State Water Company (DSI), who passed away prematurely in 1999. He was born on 23 January 1948, in Elazig, Turkey. I worked with him for a short time period (1979–1980) during which he was a supervisor for my research on snow melting processes that resulted in my current research interest on cold precipitation processes, including snow precipitation, freezing fog, and ice nucleation processes.

Dr. Karadeniz graduated from the Civil Engineering Department of the Middle East Technical University (METU), Ankara, Turkey in 1972 with a Master of Engineering Degree (ME). He studied water resources and he became an expert on hydrometeorological applications related to water management, snow melting and flood control issues. He later became an assistant researcher at the METU (1973). After that, he received a Ph.D. in 1978 also on water resources from the Faculty of Agriculture of the Ankara University (AU). He then became a Hydrology Branch Manager at the Surveying and Planning Department of the Turkish State Water Company (DSI), Ankara, Turkey. He left DSI in 1983, when he started his own company that focused on hydrological applications, while also working for the Scientific and Technological Research Council of Turkey (TUBITAK). In 1985, he received his Associate Prof. title from Ankara University. During his tenure at DSI, he was a member of the UNESCO International Hydrology Program team and represented Turkey in various aspects of hydrological applications. He passed away prematurely in 1999 in Ankara.

The passing of Dr. M. Karadeniz represents a great loss for applied hydrometeorological sciences; we have lost an exceptional colleague, a supervisor, and a good friend. He will be remembered by many colleagues and students for his great help to others, enthusiasm, intellectual mind, and kind personality.

(Published online: April 10, 2012)

[1] Environment Canada, Cloud Physics and Severe Weather Section, 4905 Dufferin St., Toronto, ON M3H 5T4, Canada.
E-mail: ismail.gultepe@ec.gc.ca

Pure Appl. Geophys. 169 (2012), 769–781
© 2011 Springer Basel AG
DOI 10.1007/s00024-011-0334-y

Use of a Sodar to Improve the Forecast of Fogs and Low Clouds on Airports

ALAIN DABAS,[1] SAMUEL REMY,[2] and THIERRY BERGOT[1]

Abstract—A sodar was deployed at Roissy–Charles de Gaulle airport near Paris, France, in 2008 with the aim of improving the forecast of low visibility conditions there. During the winter of 2008–2009, an experiment was conducted that showed that the sodar can effectively detect and locate the top of fog layers which is signaled by a strong peak of acoustic reflectivity. The peak is generated by turbulence activity in the inversion layer that contrasts sharply with the low reflectivity recorded in the fog layer below. A specific version of the 1D-forecast model deployed at Roissy for low visibility conditions (COBEL-ISBA) was developed in which fogs' thicknesses are initialized by the sodar measurements rather than the information derived from the down-welling IR fluxes observed on the site. It was tested on data archived during the winters of 2008–2009 and 2009–2010 and compared to the version of the model presently operational. The results show a significant improvement—dissipation times of fogs are better predicted.

1. Introduction

The prediction of low visibility conditions is a formidable challenge for operational weather forecast services. Adverse visibility conditions can strongly reduce the efficiency of a terminal area traffic flow. At Paris—Charles de Gaulle (CDG) international airport, Low Visibility Procedures (LVP) are applied when visibility is under 600 m (2,000 feet) or the ceiling is below 60 m (200 feet). The application of LVP reduces airport efficiency for takeoffs and landings by a factor of two. Costly delays and flight cancellations ensue. CDG is the largest airport in France and was number 2 in 2009 in Europe for the number of passengers. It is a hub for AirFrance-KLM and, as such, receives many passengers connecting

from one flight to another. It is strongly affected by fogs during a significant amount of time every year. Owing to its importance at the national and European level, CDG was equipped with a specific system for the prediction of low-visibility conditions. The system is based on the 1D COBEL-ISBA numerical model and a dedicated observation package (see BERGOT *et al.* 2005 and BERGOT, 2007 for a detailed description). The high resolution 1D COBEL-ISBA model was developed in collaboration between the Laboratoire d'Aérologie (Université Paul Sabatier/CNRS, Toulouse, France) and the Centre National de Recherches Météorologique (Météo-France/CNRS, Toulouse, France). A detailed description of the model can be found in BERGOT and GUEDALIA, (1994). The COBEL equations are solved on a high-resolution vertical grid. Near the surface, in the region of significance for fog, numerical computations are made on a very fine mesh grid (20 vertical levels in the first 200 m, with a first level at 0.5 m). The physical package includes a parameterization of boundary layer turbulent mixing for stable, neutral and unstable conditions, a microphysical parameterization adapted for fogs and low clouds and a detailed radiation transfer scheme. The COBEL atmospheric model is coupled with the ISBA seven layer surface scheme (NOILHAN and PLANTON, 1989).

Accurate short-term forecasting of fogs and low clouds strongly depends on the accuracy of initial conditions (REMY and BERGOT, 2009). Specific observations are made at CDG in order to improve the description of the surface boundary layer. Atmospheric temperature and humidity profiles, as well as short- and long-wave radiation fluxes, are performed on a 30 m high meteorological tower (levels of measurements: 1, 2, 5, 10 and 30 m). These observations are used in a local assimilation scheme (see BERGOT *et al.* 2005 for details, only a brief

---
[1] CNRM-GAME, 42 Av Coriolis, 31057 Toulouse Cedex, France. E-mail: alain.dabas@meteo.fr
[2] Météo-France Direction Ile de France Centre, 2 Av Rapp, 75340 Paris Cedex07, France.

description will be given hereafter). The assimilation process is done in three steps:

- assimilation of atmospheric profiles: local observations, forecast from the French numerical weather prediction (NWP) model ALADIN and the guess profiles are mixed following a classical BLUE equation (Best Linear Unbiased Estimator). The error statistics are imposed in order that the initial profiles are close to observations near the surface and get closer to the NWP forecast at the top of the model.
- Assimilation of fog and low cloud layers: the atmospheric profiles inside the fog layer are adjusted following the hypothesis that the cloudy layer is well-mixed. The fog depth is determined using an iterative method that minimizes the model error on radiation fluxes.
- Assimilation of soil profiles: the temperature and moisture profiles inside the soil are linearly interpolated from in situ measurements.

The second step of assimilation, i.e. the initialization of the fog layer, is very important for an accurate forecast of the dissipation of the fog layer. When the fog layer is between the two levels of the down-welling IR radiation measurements (at Roissy CDG, the radiation sensors are at ground and at 45 m above the surface), the minimization of radiation fluxes divergence between the two levels allows an accurate initialization of the fog layer (see BERGOT et al. 2005 for details). However, when the top of the fog layer is above the highest sensor, the divergence between the two sensors is null and the estimation of the fog layer height is then more difficult. The errors in the measurements of radiation fluxes due, in particular, to the deposition of droplets on the sensor window lead to errors in the fog height that can be up to several tens of meters. Another source of error is the possible presence of a cloud layer above the fog that produces a stronger down-welling IR flux that the model interprets as a fog layer thicker than it is in reality. The goal of this article is to test how the use of sodar measurements can help the initialization of fog layers (instead of the use of radiation profiles). The first section is devoted to the sodar and the possibility of using this instrument for detecting the top of fog layers. Results of an experiment conducted in the winter of 2008–2009 at Roissy

CDG are presented. Then, in Sect. 4, the impact of sodar detected fog top data on the forecast of LVP conditions by COBEL-ISBA at CDG is studied.

2. Sodar

2.1. The Sodar Technology

Acoustic sounders, or sodars, have been widely used since the 1970s for observing the atmospheric boundary layer (see review paper by COULTER and KALLISTRATOVA, 2004). Benefiting from several decades of development, sodar technology has reached a high degree of maturity. Today, fully automatic sodars, able to operate unattended for long time periods with minimal maintenance, are available off the shelf from several manufacturers at prices below 100 k€.

A sodar probes the atmosphere by emitting pulsed sound waves. While it propagates, the sound wave is backscattered by turbulent temperature heterogeneities. The small fraction of the emitted sound power thus scattered back to the sodar is detected by microphones and analyzed. The strength P_r of the backscattered sound is given by the sodar equation (LITTLE, 1969)

$$P_r(r) = PA \frac{c\tau}{2} \frac{\sigma(r)L(r)}{r^2} \qquad (1)$$

It is proportional the power P (W) of the emitted sound, the pulse duration τ (s), the receiver "efficiency" A (m^2) and the celerity of sound c (ms^{-1}). It is a function of the range r (m) through the $1/r^2$ term, the round-trip transmission factor $L(r)$ (which decreases with the range r) and the backscattering coefficient $\sigma(r)$ of the atmosphere. This latter parameter characterizes the "reflective" power of the atmosphere; it is related to its thermodynamic properties

$$\sigma(r) = 0.0039 \, k^{\frac{1}{3}} C_\tau^2(r)/T^2(r) \qquad (2)$$

Here, $k = 2\pi/\lambda$ is the wavenumber (λ (m) is the wavelength—typically of the order of 10 cm), $C_\tau^2(r)$ (K^2 m$^{2/3}$) is the structure coefficient of the temperature turbulence, and $T(r)$ (K) is the absolute temperature.

The most common application of a sodar is the measurement of highly resolved, vertical profiles of wind within the first several hundreds of meters of the atmosphere. This application is based on the estimation of the frequency Doppler shift δf (Hz) between the outgoing and the backscattered sound waves. It is proportional to the wind velocity component v_r along the sodar beam

$$\delta f = -\frac{2v_r}{\lambda} \qquad (3)$$

Combining at least 3 beam directions (with 3 different antennas, or a phased-array antenna), it is then possible to retrieve vertical profiles of the 3D wind vector. In comparison to other wind profiling systems (lidar or UHF radar), a sodar offers the advantage of a fine time and space resolution (a few minutes and several tens of meters in the vertical direction) but limited ranges (the transmission factor $L(r)$ decreases rapidly with the range).

In the present article, the sodar is used for real-time detection of the top of fog layers. Although several papers report sodar observations of stable boundary layers—see for instance GUÉDALIA et al. 1980, or ANDERSON, 2003—this is an unusual application (to our knowledge). An enhanced turbulent activity is expected at the top of the fog layers due to radiative cooling and wind shear. Therefore, sodar backscatter at the top of a fog layer is stronger than above or below. It must be noted here that the fine time and space resolutions of a sodar are particularly well suited to the observation of fogs which are thin (the typical thicknesses of fogs is several tens to several hundreds of meters) and evolve at time scales of several minutes.

2.2. Instrumental Set-up

The ability of a sodar to detect the top of fog layers was tested during an experiment conducted by the Groupe d'Étude de l'Atmosphère Météorologique (a research laboratory operated by the French weather service Météo-France and the Centre National de la Recherche Scientifique, the main research operator in France) during the winter of 2008–2009. It took place at Roissy Charles de Gaulle (CDG) airport.

The experimental set-up was composed of a sodar PCS.2000-64 from METEK (see details below), a tethered balloon, and a ceilometer. It was complemented by various observation systems operated routinely at the airport (including 12 visibility sensors DF320 from DEGREANE, and 4 ceilometers LD-WHX05 from IMPULSPHYSICS).

The sodar was deployed in the summer of 2008 (see Fig. 1) at a location (see Fig. 2) carefully selected after measurements of ambient acoustic noise at different possible places in the airport area. The site is in the western, restricted area of the airport, and in the axis of the two runways 09R and 09L forming the northern "doublet" of CDG. The distance to the runways is 700 and 1,950 m respectively. There, the ambient noise can peak high above 100 dBa when an aircraft lands or takes-off, which happens once every 90 s during rush hours, but it is otherwise surprisingly low—below 45 dBa. The calm periods between aircraft noise peaks last long enough (about 70 s) to allow the sodar to make reliable measurements (the sodar software is able to discard noisy signals from the wind and the reflectivity retrieval process). The main characteristics of the sodar are listed in Table 1, while Table 2 shows the instrument parameters set by METEK for the experiment. The antenna was a phased array of $8 \times 8 = 64$ transducers. It was divided into 4 sub-panels. Independent 90° phase delays can be added to each sub-panel so that the sodar can emit beams in up to 5 different directions, one of them the vertical (all sub-panels are in phase), the other four pointing ~20° off-zenith and at 90° azimuth increments. In practice, only 3 directions were used (vertical and two of the four titled directions, the last two being redundant and affected by echoes from nearby towers). The sodar was parameterized so as to achieve the best possible vertical resolution (10 m) and a first range gate as low as possible (bottom altitude at 15 m). The integration time was set to 10 min as no major fog evolutions were expected during so short a time period. As for the maximum altitude (350 m), the requirement was based on a climatology study according to which most fogs at CDG (75%) do not exceed 100 m in thickness. It was raised to 460 m after the experiment (from October 2009) as sodar data showed the study had probably underestimated the occurrence of thick fogs.

The tethered balloon system was aimed at measuring vertical profiles of temperature and humidity at the high repetition rate of 1 or 2 per hour. It served as the reference for the altitude of fog layer tops that are marked by a thermal inversion and a sudden drop in the relative humidity. For safety reasons, the tethered balloon could not be deployed close to the sodar, but was located 7 km away to north of the airport. Another constraint was that the maximum balloon altitude was limited to about 300 m above the surface. As the topography around the airport is rather flat and homogeneous (the terrain slope is $\sim 0.13°$, the sodar being only 16 m below the balloon sounding system), it was expected that the fogs at CDG would be rather horizontally homogeneous, in particular when caused by radiative cooling, but differences between balloon and sodar fog top altitudes could be partly due to the distance between the two instruments.

The tethered balloon system was composed of a winch TTW111 (see Fig. 3), an SPS220 sonde receiver and a DIGICORA III processing system, all from Vaisala. The sondes were standard RS92SGP from the same manufacturer, but placed in a home-made ventilating shield that was developed in order to compensate for the lack of natural ventilation of the sensing elements (in standard radiosounding, the ascending speed of ~ 5 ms^{-1} naturally ventilates the sensors; with the tethered balloon, the ascending speed was only ~ 0.5 ms^{-1}). Vertical profiles of temperature and relative humidity were measured at the rate of 1 or 2 profiles per hour by letting the helium inflated balloon go up to the maximum altitude and then bringing it down to the surface with the winch. Ascents and descents lasted 15–30 min depending on winch speed. The sounding system was activated every time a fog event was predicted by the local station of Météo-France (the French weather service) or when an unexpected fog was observed at the airport. The purpose was to start the measurements 1 or 2 h before the formation of fog and to make vertical profiles of the atmosphere at regular time intervals until it disappeared. The sonde in operation was replaced every 2 to 3 h as the first nights of operation showed a saturation effect on the humidity sensor after a few hours. An example of a set of temperature and humidity profiles measured during a fog event is given in Fig. 4. The top of the fog layer is indicated by the sharp temperature increase and the correlated drop of humidity. In this

Figure 1
Picture of the sodar METEK PCS2000-64 deployed at Roissy CDG airport. The sodar antenna is *inside the white shield* in the middle of the picture. The shield is made of 4 vertical panels aimed at protecting the receiver from ambient noise .

Figure 2

Satellite view of Roissy Charles de Gaulle airport with the position of the sodar west of the two runways forming the northern "doublet" of the airport. The distance to runway 09R is 700 m, and 1,950 m to runway 09L

Table 1

Main characteristics of the sodar PCS.2000-64 from METEK

Overall characteristics	
Frequency	~1,600 Hz
Wavelength	~21 cm
Emitted power	118 dbA and 50 W
Power consumption	
Without heater	250 W
With heater	550 W
Antenna	
Number of transducers	$8 \times 8 = 64$
Dimension	110×110 m
Weight	136 kg
Number of beam directions	Up to 5
Beam angles (nadir)	$0 \pm 22°$
Beam angles (azimuth)	0, 90, 180 and 270°
Measurement range	
Max horiz. velocity	30 ms^{-1}
Max vertical velocity	± 10 ms^{-1}
Number of range gates	40 max
Vertical resolution	5–100 m
Integration time	10–1,800 s
Precision	
Horizontal velocity	10%
Vertical velocity	5%
Direction	5°
Operating conditions	
Temperature	−30 to 45°C
Humidity	5–100%

Table 2

Sodar settings during the experiment conducted in the winter 2008–2009

Sodar settings	
Number of frequencies	1
Number of beam directions	3
Base altitude of lowest range gate	15 m
Nbre of range gates	34
Top altitude of highest range gate	360 m
Vertical resolution	10 m
Integration time	10 min

particular case (7th of December, 2008, at 12:20 UTC), its altitude was 250 m above sea-level.

The ceilometer was a CT25K from Vaisala. It was deployed at the local station of Météo-France (indicated by CDM95 in Fig. 2). It is a small backscatter lidar that operates in the near IR at 0.9 μm. It produces two types of messages. The most common contains the height of cloud bottoms. During the experiment, we were mostly interested by the second type, which contains lidar backscatter signals. As shown in Fig. 5, the presence of a fog is signaled by a strong return at the surface (the small water droplets in the fog strongly reflect the light emitted by the lidar) followed

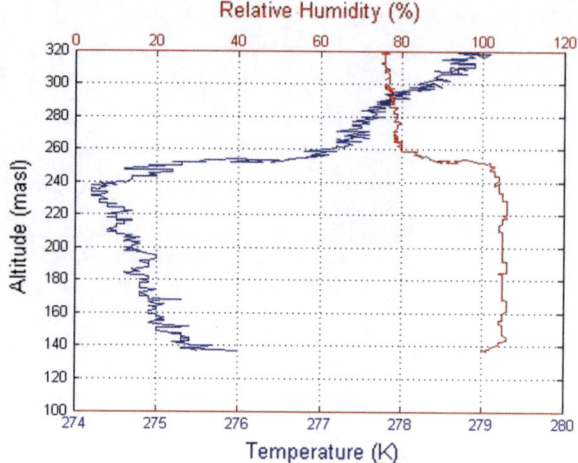

Figure 4
Temperature (*blue*) and relative humidity (*red*) profiles measured by the tethered balloon on the 7th of December 2008 at 12:20 UTC. The top of the fog layer (*black dashed line*) appears very clearly at the altitude of 250 m with the sharp increase of temperature and the drop of the relative humidity. Note that the humidity sensor is oversaturated in the fog layer and measures a relative humidity of ∼ 105%

Figure 3
Picture of the sounding system used during the experiment of 2008–2009. A standard radiosounding balloon with a radiosonde RS92SGP mounted in a home-made ventilating shield is tied to the electric winch. During fogs, the balloon and the sonde were allowed to go up and brought back one or two times every hour

by a steep attenuation (the strong scattering attenuates the laser beam very steeply). It must be stressed here that the thickness of the return is in no way related to the thickness of the fog; it is a function of the laser pulse length (15 m in the present case) and the multiple scattering effect (many photons captured by the ceilometers were scattered several times by several particles—see for instance Bissonnette *et al.* (1995), for details on multiple scattering effects on lidar returns). During the experiment, we used ceilometer returns in order to determine precisely the time

of fog formation and dissipation (see the following section).

3. Data Analysis

Eleven observation periods were carried out during the campaign. During seven of them, a fog event was actually observed. For the other four, the fog prediction turned out to be a false alarm.

Figure 5 is a typical example of the observations acquired by the various sensors during a fog event. The particular event portrayed by the figure lasted a little more than 24 h, from 10:00 UTC on the 7th of December 2008 to 10:00 UTC on the 8th of December. The top graph contains the visibility (blue line) and cloud ceiling (red line) measured by airport sensors close to the sodar. At 10 UTC on the 7th of December, the visibility drops below 1,000 m indicating the sudden formation of a fog at CDG. The suddenness of the formation is clear on ceilometer data (bottom graph) where a strong lidar reflectivity appears almost instantly at the surface. The fog lasts until 10:00 UTC the next day, but a temporal dissipation at the surface occurs between 18:00 UTC and 24:00 UTC on the 7th

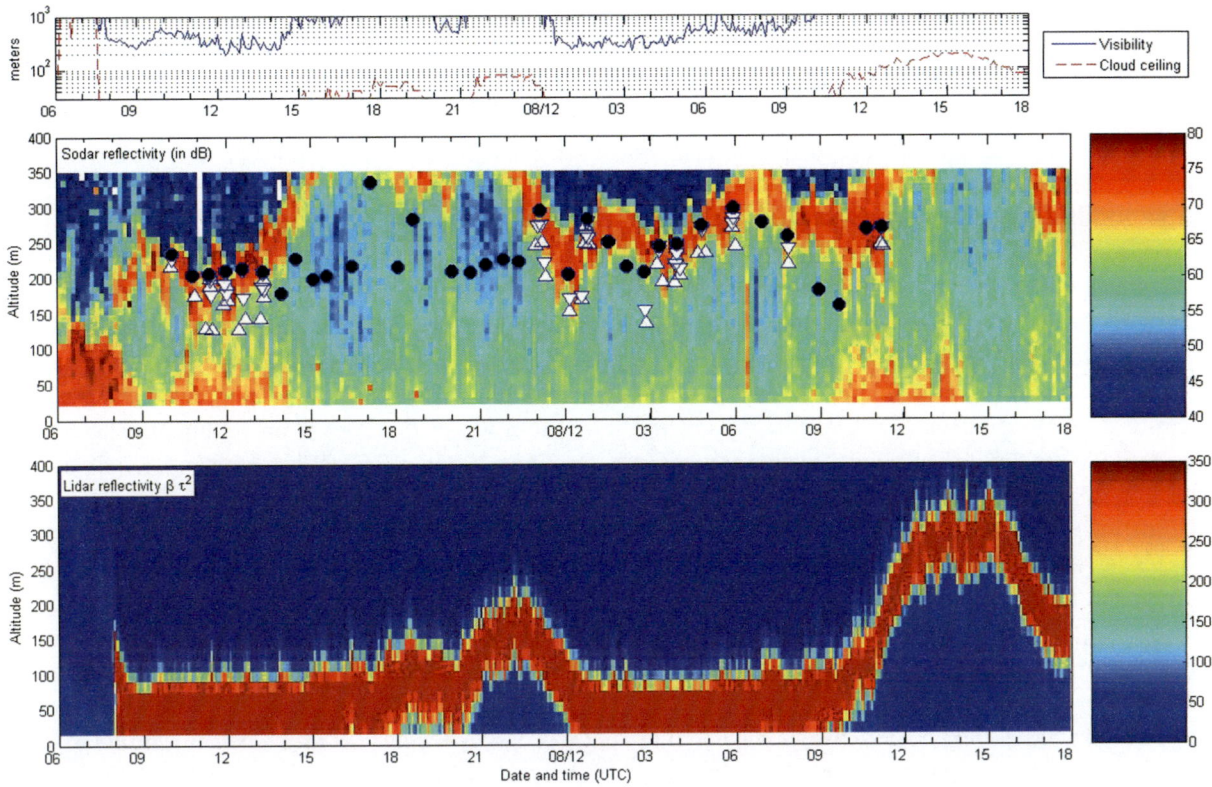

Figure 5

Visibility, cloud ceiling (*top*), sodar reflectivity (*middle*) and lidar reflectivity (*bottom*) measured from 6:00 UTC on 7th of December 2008 to 18:00 UTC on 8th of December 2008. Strong lidar reflectivities at the surface indicate the presence of fog. The fog event started suddenly at 8:00 UTC on December 7th. The dissipation occurred the next day at about 10:00 UTC. On the evening of the 7th, the fog dissipated temporally at the surface between 18:00 UTC and 24:00 UTC. Then the visibility reached over 1,000 m, but the cloud ceiling remained very low. Low visibility procedures were activated when the visibility was below 600 m or the cloud ceiling below 200 ft, thus remained in effect during the whole period. On the sodar plot (*middle graph*), a thin, *red layer* of strong acoustic reflectivities can be seen during the whole event. It is tightly correlated to the temperature inversions detected on the balloon soundings and shown with Δ (*bottom of inversions*) and ∇ (*top of inversions*). The *black spots* show the maximum altitude reached by the balloon. No inversion was detected on soundings with a maximum altitude below the red sodar reflectivity layer

of December. During this period, the visibility exceeds 1,000 m, but the cloud ceiling remains very low. Low visibility procedures were thus maintained at the airport throughout the whole period. Sodar reflectivities are displayed in the middle graph in dB units. At the time the fog forms, the strong reflectivity above the surface drops, and a thin layer of strong acoustic reflectivity (red) appears at the height of about 200 m. The layer remains during the whole fog period. Its height varies, ascends beyond the sodar maximum range (360 m) during the afternoon of 7 December, and then descends and stabilizes at about 250 m until dissipation. This layer contrasts sharply with the low reflectivity ($\lesssim$55 db) at the lower heights. The

temperature inversions detected manually (that is by a visual inspection) on the balloon profiles are shown with white triangles pointing up (bottom of the inversion) or down (top of the inversion). The gradient at the inversion is typically of the order of 0.01 km^{-1}. The inversions are tightly correlated with the peak sodar reflectivities. This and the absence of temperature inversions on the profiles that did not reach the red layer (the maximum altitude of balloon profiles are indicated with black spots) suggest that strong sodar reflectivities during fogs are indeed a good signature of the presence of a fog top temperature inversion.

The correlation between the altitude of peak sodar reflectivity during fogs and the altitude of detected inversions was extended to the 7 fog events documented during the experiment. The result is summarized with a scatter plot in Fig. 6. The altitudes of the base and the top of inversions are distinguished by the use of two different markers (Δ: base, and ∇: top). The correlation coefficients are nearly 70% for both, with inversion tops 8 m above sodar peaks on the average, and inversion bottoms −27 m below. This confirms that there is a good correlation between peak sodar reflectivities and the tops of fog layers. Differences between the altitudes of both may reach several tens of meters on some occasions. Their significance is hard to assess. They are partly due to the difficulty of detecting inversion layers on some balloon profiles (the thermal inversion is sometime very small, less than 0.5 K), and could also be due to the distance between the sodar and the balloon. On some occasions, it appeared, for instance, that the fog was present at the balloon site several tens of minutes before the sodar, and vice versa. Considering these sources of uncertainty, the correlation coefficient reached here seems good enough to conclude that the sodar provides a reliable estimation of the top level of fog layers.

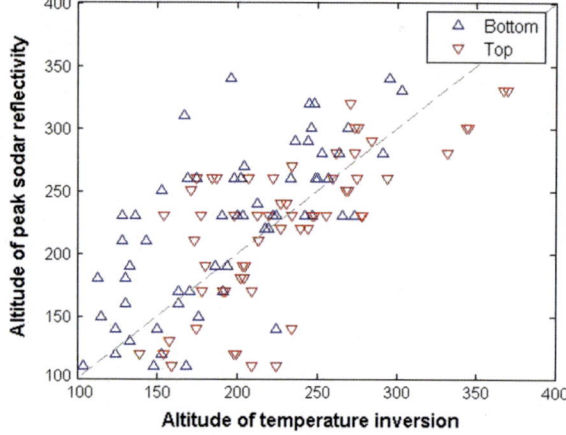

Figure 6
Scatter plot of peak sodar reflectivity altitudes versus temperature inversion altitudes. Top and bottom altitudes of temperature inversions are distinguished with *triangles* pointing *up* (*bottom*) and *down* (*top*). The number of temperature inversion tops is 60. On the average, they are 8 m above the sodar reflectivity peak (230 vs. 222 m). The correlation coefficient between inversions tops and sodar reflectivity peaks is 69.97%. As far as inversion bottoms are concerned (61 values), they are −27 m below sodar reflectivity peaks on the average (195 against 223 m) with a correlation coefficient of 69.44%

4. The Impact of Sodar Data on The COBEL-ISBA Forecasting System

The sodar data was used in the COBEL-ISBA forecasting system in order to check whether the forecasts of LVP conditions were improved or not.

4.1. Experimental setup

When fog was present at initialization time, the height of the fog layer provided by the sodar was used directly to initialize the liquid water content, with a value of 0.2 g/kg for the liquid water mixing ratio within the cloud. When stratus (i.e. a cloud with a base that is not touching the ground) was present, observations of the radiative fluxes were used as in the operational setup. As shown above, the sodar provided reliable estimates of the height of the top of a fog layer; however it has not been tested yet for stratus clouds.

The period of study covered 9 months from two successive winters: from 1 November 2008 to 28 February 2009 and from 1 October 2009 to 28 February 2010. LVP conditions were observed during 470 h during these two periods; they occurred more frequently during the early morning hours (local time is UTC plus 1 h during winter), as shown by Fig. 7. Overall, half of the LVP conditions were caused by fogs. Simulations were carried out at 1 h intervals: the model was run 6,492 times in total. The sodar observations were available for every simulation with fog at initialization time except one: the sodar was very reliable during the period of study (the average data availability has been larger than 95% until now, that is, after more than 2 years of operation).

The quality of the forecasts of LVP conditions was assessed in terms of Hit Ratio (HR) and pseudo-False Alarm Ratio (pseudo-FAR). When studying rare events, such as fog and LVP conditions, the pseudo-FAR is convenient because it removes the impact of the "no–no good forecasts" (no LVP forecast and no LVP observed), which mostly dominate the data sample and hide the true merits of the LVP forecast system. If a is the number of events forecasted and observed, b the number of events forecasted but not observed, and c the number

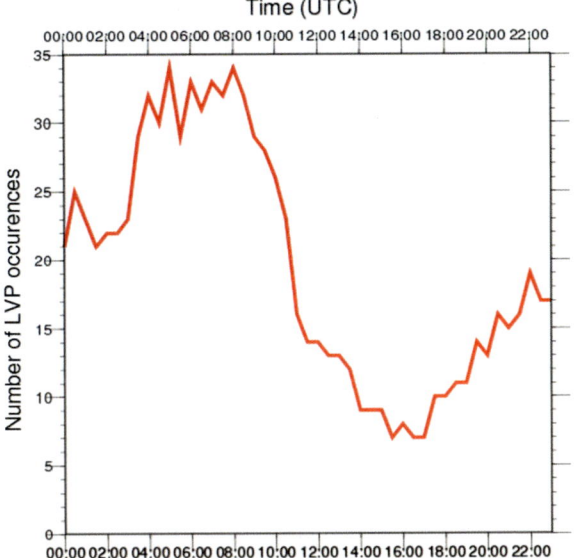

Figure 7

Number of occurrences of LVP conditions, depending on time of the day, from 1 November 2008 to 28 February 2009 and from 1 October 2009 to 28 February 2010

of events observed but not forecasted, HR and pseudo-FAR are then defined as follows:

$$HR = \frac{a}{a+c}; \quad pseudoFAR = \frac{b}{a+b} \qquad (4)$$

5. Results

Figure 8 shows the HR and pseudo-FAR of LVP conditions versus forecast time for simulations with the operational setup and with the use of sodar estimates of the thickness of the fog layer. A Student test for correlated samples was carried out to assess the significance level of the differences between the two experiments: its results are also shown in Fig. 8. The HR was higher when using sodar data as compared to the operational setup, with an increasing improvement for higher forecast times. This increase in the HR has a 95% significance level or above for forecast times larger than 1 h. The HR of LVP conditions for all simulations and forecast times was 0.585 with the sodar against 0.556 with the operational setup. The pseudo FAR was slightly degraded for forecast times below 2 h and slightly improved for higher forecast times,

with a significance level over 95% for forecast times larger than 1 h. The overall pseudo FAR was 0.439 with the sodar against 0.443 with operational setup.

These statistics cover many cases for which the sodar data were not used, such as LVP due to stratus, or due to fogs that appear after initialization time. Figure 9 shows the HR and Pseudo-FAR of LVP conditions computed for simulations with a fog present at initialization time. A Student test was also carried out to check the significance level of the differences between the two experiments. The figure gives a better assessment of the impact of the sodar on fog predictions. A total of 263 h of LVP conditions correspond to these situations. The HR is improved for forecast times higher than 3 h, with an improvement larger for forecast times of 5 h and beyond. The pseudo-FAR follow the same pattern, with an important improvement for forecast times larger than 5 h. The differences were significant for forecast hours larger than 1 h for the HR and 3.5 h for the pseudo-FAR. It can thus be concluded that the sodar clearly improves the initialization of fogs in COBEL.

Figure 10 shows the compared statistics of both experiments for the forecast of the end time of LVP conditions, for simulations with fog at initialization time. The standard deviation of the forecast errors is slightly reduced when using the sodar, while the bias is slightly increased. 43% of simulations show an error smaller than 30 min when using the sodar, up to 40% with the operational setup. The bias difference between the two experiments was assessed to be significant with a level close to 100%, using a correlated samples Student test.

The initial fog thickness given by the sodar and the operational setup were compared in Fig. 11. When the operational setup gives values below 100 m, the sodar data are larger in nearly 80% of the cases. This could explain the better HR scores when using the sodar, as a thicker fog at initialization time lasts longer during the simulation. On the other hand, the validity of sodar data when the fog is thin is not very well known. The turbulence activity is generally stronger in the first tens of meters of the atmosphere, so the detection of a reflectivity peak due to the inversion at the top of

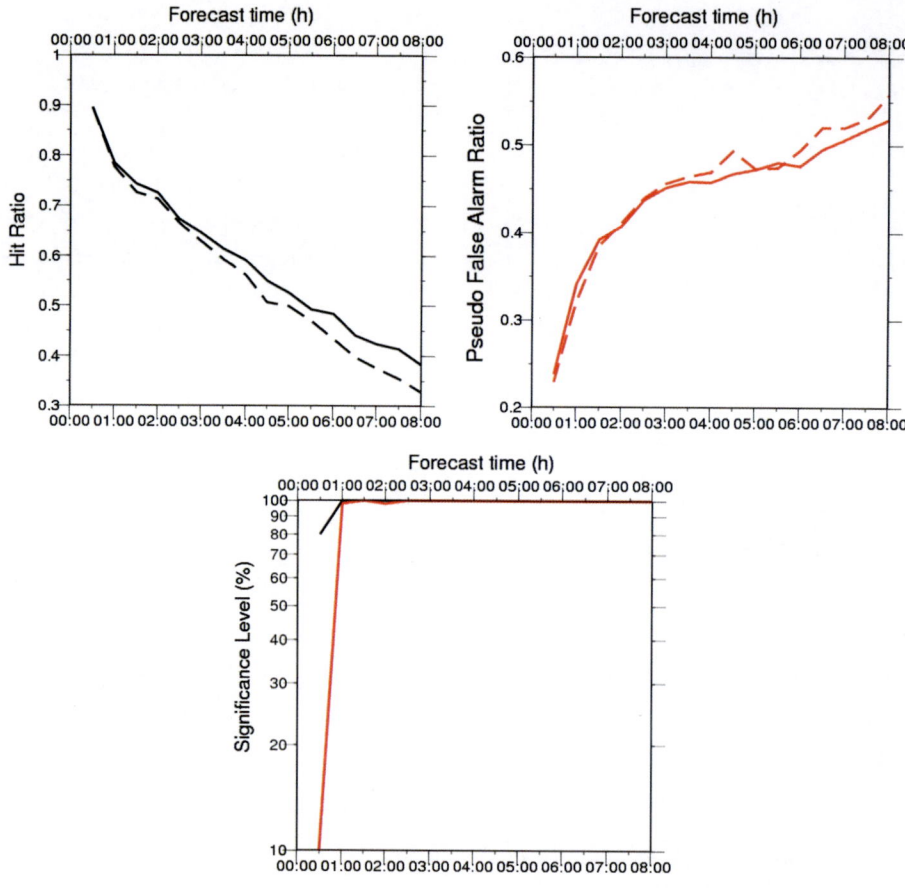

Figure 8

Hit Ratio (*top left*) and Pseudo-False Alarm Ratio (*top right*) versus forecast time. Simulations with the operational setup (*dashed line*) and using the sodar (*continuous line*), for the winters 2008–2009 and 2009–2010. The significance level for the difference between the operational setup and using the sodar is given in the *bottom*, for the Hit Ratio (*black*) and the Pseudo-False Alarm Ratio (*red*)

the fog is more difficult to detect. On the contrary, when the operational setup gives values between 100 and 300 m, no clear tendency can be drawn. For values above 300 m, the sodar data are always equal or lower, partly because of the range limit, but also partly because the operational set-up is sometimes misled by the presence of cloud above the fog layer.

6. Conclusions

The experiment conducted at CDG during the winter of 2008–2009 has proven that a sodar can operate efficiently at an airport to provide reliable measurements of the thickness of fog layers. Using sodar information (instead of down-welling IR fluxes) in the COBEL-ISBA model improved its hits and misses on low-visibility forecasts. In particular, the time of fog dissipation is better predicted. Part of the improvement is due to the enhanced data availability—nearly 1/3 of IR flux measurements are not available at the time of COBEL run while the average monthly availability of sodar data is better than 95%—but other problems, such as the presence of a stratus layer above the fog that led to errors in the estimates of the fog thickness using IR flux measurements and simulations, may be solved by the sodar. Future works will focus on a detailed understanding of the results that have been obtained;

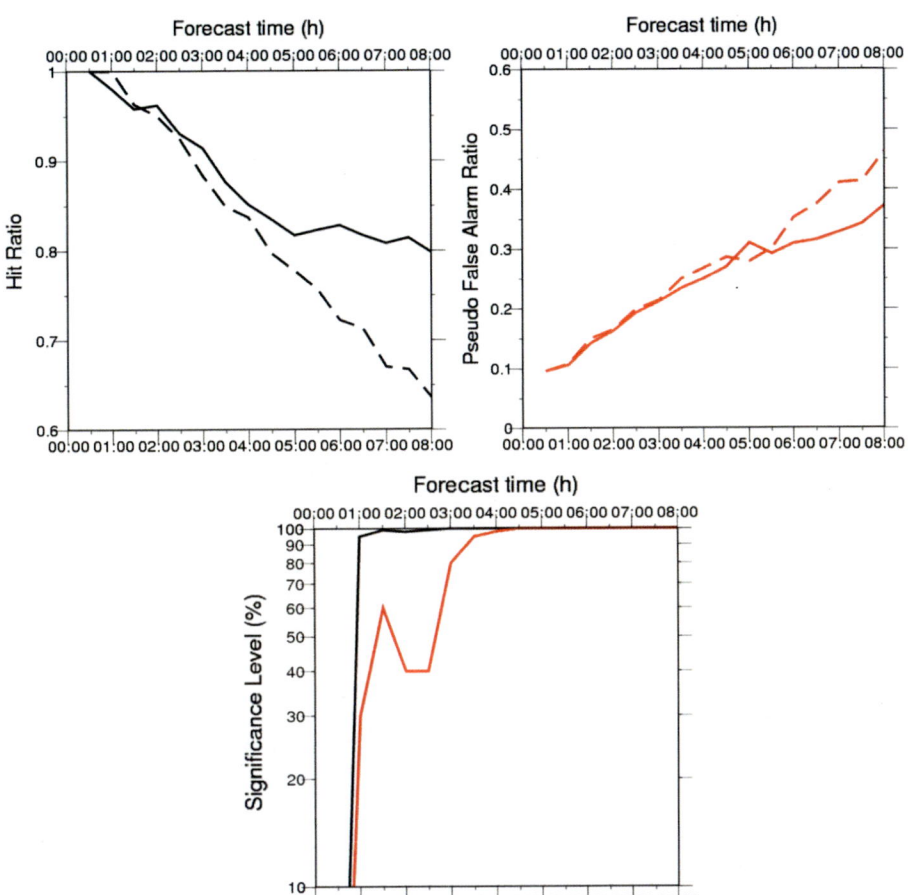

Figure 9
Same as Fig. 8 for simulations with fog at initialization time

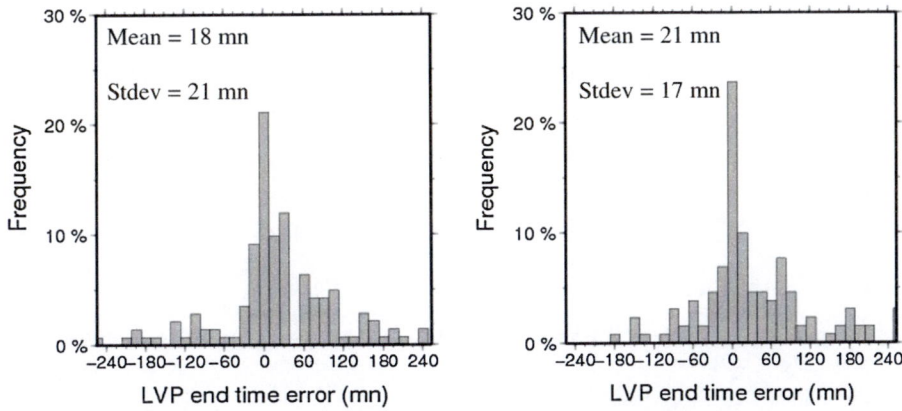

Figure 10
Frequency distribution histogram of the error of the predicted end time of LVP conditions, in minutes. The operational setup is on the *left*, simulations using the sodar data are on the *right*. Positive values correspond to a forecast of onset or end time that is too late. Errors larger than 240 min are grouped in the 240 min column. The mean and standard deviation of errors smaller than 240 min are indicated. The statistics were computed only for simulations with fog at initialization time

Reprinted from the journal

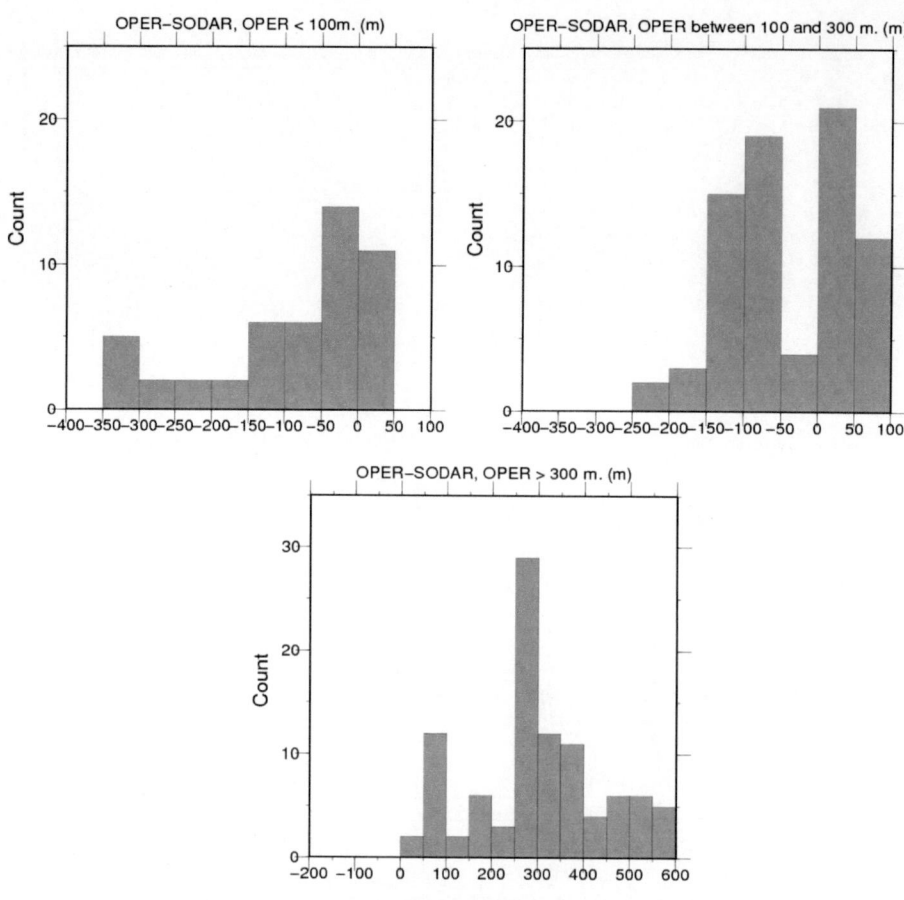

Figure 11
Fog thickness estimates using IR observations and simulations minus estimates by the sodar for thickness (as provided by the operational setup) below 100 m (*upper left*), between 100 and 300 m (*upper right*) and above 300 m (*bottom*)

in particular, they will aim at providing an explanation as to why some sodar estimations differ significantly from estimates using IR flux observations.

Acknowledgment

The work presented in this paper was financed by Météo-France. It is the fruit of the work of many people, in particular the people who spent long hours on cold nights operating the tethered balloon. We thank J.M. Donier, T. Douffet, F. Lavie, M. Olivier, O. Garrouste, O. Legain, A. Minisini, E. Moulin, G. Bouhours, D. Suquia, and J. Barrié. We also thank the staff of the local station of Météo-France (the Centre Départemental de la Météorologie 95) for their help.

REFERENCES

ANDERSON, PS. (2003), *Fine-scale structure observed in a stable atmospheric boundary layer by sodar and kite-borne tether-sonde*, Boundary Layer Meteorology *107*, 323–350.

BERGOT T. (2007), *Quality Assessment of the Cobel-Isba numerical forecast system of fog and low clouds*, Pure appl. Geophys. *164*, 1265–1282.

BERGOT, T. and GUEDALIA, D. (1994), *Numerical forecasting of radiation fog. Part I : Numerical model and sensitivity tests*, Mon. Wea. Rev. *122*, 1218–1230.

BERGOT, T., CARRER, D., NOILHAN, J. and BOUGEAULT, P. (2005), *Improved site-specific numerical prediction of fog and low clouds : a feasibility study*, Weather and Forecasting *20*, 627–646.

BISSONNETTE, L. R., BRUSCAGLIONI, P., ISMAELLI, A., ZACCANTI, G., COHEN, A., BENAYAHU, Y., KLEIMAN, M., EGERT, S., FLESIA, C., SCHWENDIMANN, P., STARKOV, A.V., NOORMOHAMMADIAN, M., OPPEL, U.G., WINKER, M., ZEGE, E.P., KATSEV, I.L. and POLONSKY, I.N. (1995), *LIDAR multiple scattering from clouds*, Applied Physics B. *60*, 355–362.

COULTER R.L. and KALLISTRATOVA, M. A. (2004) *Two decades of progress in sodar techniques: a review of 11 ISARS proceedings*, Meteorol Atmos Phys. *85*, 3–19.

GUÉDALIA, D., NTSILA, A., DRUILHET, A. and FONTAN, J. (1980), *Monitoring of the atmospheric stability above urban and suburban site using sodar and radon measurements*, Journal of Applied Meotorology *19*, 839–848.

LITTLE, C.G. (1969), *Acoustic methods for the remote probing of the lower atmosphere*, Proc. IEEE. *57*, 571–578.

NOILHAN, J. and PLANTON, S. (1989), *A simple parametrization of land surface processes for meteorological models*, Monthly Weather Review A., *117*, 536–549.

REMY, S. and BERGOT, T. (2009), *Assessing the impact of observations on a local numerical fog prediction system*, Quarterly Journal of the Royal Meteorology *135*, 1248–1265.

(Received December 1, 2010, revised March 25, 2011, accepted April 8, 2011, Published online June 29, 2011)

Pure Appl. Geophys. 169 (2012), 783–791
© 2011 Her Majesty the Queen in Right of Canada
DOI 10.1007/s00024-011-0349-4

FTS (Fog To Snow) Conversion Process During the SNOW-V10 Project

I. Gultepe[1] and B. Zhou[1]

Abstract—The objective of this work is to understand how winter fog which occurred on Whistler Mountain on 3–4 March 2010 developed into a snow event by the means of the FTS (Fog To Snow) process. This event was documented using data collected during the Science of Nowcasting Winter Weather for Vancouver 2010 (SNOW-V10) project that was supported by the Fog Remote Sensing and Modelling (FRAM) project. The FTS resulted in a snow event at about 1,850 m altitude where the RND (Round-house) meteorological station was located. For both days, there was no large scale system that affected local fog formation and its development into snow. Patchy fog occurred in the early hours of both days and was based below 1,500 m. Clear skies at night likely resulted in cooling, the valley temperature (T) was about −1°C in the early morning, and snow was on the ground. Winds were relatively calm (<1 m s^{-1}). At the RND site, T was about −3°C. Weather at RND was clear and sunny till noon. When fog moved over the mountain peak/near RND, light snow started and lasted for about 4–5 h and was not detected by precipitation sensors except the Ground Cloud Imaging Probe (GCIP) and Laser Precipitation Sensor (LPM). In this work, the FTS process is conceptually summarized. Because clear weather conditions over the high mountain tops can become hazardous with low visibilities and significant snow amounts (<1.0 mm h^{-1}), such events are important and need to be predicted.

Key words: Fog forecasting, fog measurements, light precipitation, fog models, snow.

1. Introduction

Mountain weather conditions change quickly because of variability over the short time and space scales of the various boundary layer processes such as radiative heating, turbulence and lifting over sloped surfaces. Time and space scales for changes in microphysical, dynamical, and radiative processes over mountain surfaces can be much shorter than those associated with a flat terrain. Warm fog usually forms when relative humidity with respect to water (RH$_w$) reaches ~100%; reaching this point can be affected by various conditions, e.g. radiative cooling, lifting, turbulence and evaporative mixing [Pagowski *et al.*, 2004; Gultepe *et al.* 2010; Gultepe and Issac 2010]. The studies by Hansen *et al.* [2007] and Gultepe *et al.* [2009] stated that fog occurrence is very common in Canada and can reach up to 20–30% of time depending on time, season, and location, but these numbers do not represent fog events specific to mountain regions.

Usually, fog is considered as a nowcasting event because of its occurrence in short scales of time and space. Forecasting fog using large scale 2D models [Teixeira, 1999] can be very difficult, so a 3D fog model must be considered [Gultepe *et al.* 2007]. This latter study stated that 1D numerical models usually cannot predict the physical parameters of fog accurately because of limited knowledge on dynamical processes (e.g. Turbulence) in the boundary layer and the 3D structure of the fog layer. Fog events over mountain regions can play a much different role than those occurring over flat surfaces. Over the flat surfaces, fog usually reduces the visibility and dissipates with the effect of increasing solar radiation [Duynkerke, 1991; Brown, 1980, Brown and Roach, 1976]. Over mountain slopes, however, solar radiation increases dynamical activity, causing fog lifting and then eventually precipitation formation at high levels [Gultepe and Isaac, 2010] where temperature often goes down to 0°C during winter.

The fog events over the Swiss Alps have been studied by [Müller *et al.* 2007] using a 1D fog model initialized by a 3D forecasting model with detailed microphysical and dynamical processes. They emphasized the importance of 3D high resolution fog models for mountain fog forecasting. [Bott *et al.* 1990], and [Bergot and Guedalia, 1994] also studied

---
[1] Cloud Physics and Severe Weather Research Section, Environment Canada, Toronto, ON M3H5T4, Canada. E-mail: Ismail.gultepe@ec.gc.ca

fog using detailed microphysical processes based on 1D fog models, and also emphasized the importance of 3D fog modelling applications. The work by [DUYNKERKE *et al.* 1999] studied 1D and 3D model simulations of fog and compared the results with in situ observations, suggesting that 3D model simulations of fog will likely be beneficial for fog prediction over inhomogeneous surface conditions.

Fog over mountain basins occurs in the morning hours due to a cooling process that can be related to advection, radiation, or a slow adiabatic cooling. When a large scale process doesn't exist, fog likely is not related to clouds because it forms usually at the lower levels of the mountains with sloping surfaces.

The objective of this work is to understand how winter fog at low levels (below 800 m) of Whistler Mountain range during the period 3–4 March 2010 developed into a snow event at higher elevations by means of the FTS (Fog To Snow) process. This process has been documented for the first time in the literature to explain how fog at the low levels (1 km) of a mountain during a clear night can lead to snow precipitation at higher elevations, e.g. >2 km.

The FTS event occurred during the Science of Nowcasting Winter Weather for Vancouver 2010 [SNOW-V10, ISAAC *et al.* 2010] project that was supported by the Fog Remote Sensing and Modelling [FRAM; GULTEPE *et al.* 2009] project. The FTS resulted in a snow event above 1,850 m altitude where the RND (Roundhouse) meteorological station, with several precipitation and visibility sensors, was located. For both days, there was no large scale system that affected local fog formation and its development into snow.

2. *Observations*

During the SNOW-V10 and FRAM projects, several instruments at the RND site [GULTEPE *et al.* 2009; 2010] were used to measure visibility (Vis), surface precipitation rate (PR), wind speed (Uh) and direction, radiative fluxes, and particle size and shape. The deployed sensors are shown in Fig. 1, which also includes a Vaisala FD12P for visibility, a YES Inc. TPS (Total Precipitation Sensor) PR, a DMT (Droplet Measurement Technologies) GCIP

Figure 1

Pictures taken during the FTS process: **a** low level fog at VOL station (early morning, *top box*), **b** elevated fog taken from the RND station (*second from top*), **c** clear conditions at RND (late afternoon, *3rd from top*), and **d** snowing conditions at RND station (evening, *bottom box*)

(Ground Cloud Imaging Probe) for particle shape and size spectra, 2D and 3D-Yonge sensors for wind speed and direction, both HMP45C and HMP45C-

232 sensors for T and RH$_w$, an LPM (Laser Precipitation Monitor) for PR and Vis, a CRN1 (Kipp & Zonen Inc.) for IR + SW broadband radiative fluxes), and an SPN1 (Biral Inc.) for direct and diffuse radiative fluxes. In case of instrument failure, another sensor measurement can usually be used for measurement of a given parameter.

The GCIP measurements of particle shapes and sizes from 7.5 µm up to 960 micron over 62 bins (64 diodes) with 15 µm resolution were used for snow/fog particle detection together with a fog device (DMT-FMD). A 7.5 micron particle travelling directly over the diodes' centre in the receiver could meet the 50% shadowing criteria and thus be detected. The GCIP instrument, which is based on a DMT CIP probe that is usually mounted on an aircraft, is adapted for ground-based applications. The heated inlet is used to keep the surfaces clean and primary suction is provided by 2–6 inches ports giving a flow rate of 25 m s^{-1} at the measurement area. Three distinct zones of heat control are provided and the entire horn exterior is insulated to reduce thermal losses. The flow cross-sectional diameter varies from 12" at the inlet to 4" between the CIP arms.

The measurements of the cloud base height was made by a VAISALA ceilometer (Model C31). Both T and RH$_w$ were also obtained using the sensors mounted on a Whistler gondola travelling along Whistler mountain slope. These parameters were used to obtain T and RH$_w$ profiles. Distance between the town center and the RND site was about 6 km, data were collected each minute, and vertical height difference between two locations was about 1,100 m. The gondola location is referenced using a GPS system.

3. Analysis and Results

The intense fog event (<500 m) occurred in the early hours of March 3 and 4, 2010, below the RND (1.9 km) and VOL (1.6 km) sites, covering the town of Whistler below the 800 m level before noon (Fig. 1a). The fog occurred during early morning, likely due to adiabatic and/or radiative cooling. Then, the fog lifted slowly to higher levels after solar radiative heating occurred when the sun came out.

Note that some effects on fog lifting to higher levels can also be due to upslope winds resulting from the radiative heating processes. It is assumed that radiative warming was a leading force for lifting the fog to higher levels (as observed by first author). Here, the results from only the March 4 event will be summarized in detail for the FTS process [GULTEPE and ISAAC, 2010]; the March 3 event occurred similarly. Figures 1a–d show the time development of the FTS process. Clear skies at night likely resulted in radiative cooling, and the valley temperature (T) at the VOL station and town center on March 4 (March 3) was about −1°C (+3°C) in the early morning, with snow on the ground on both days. The foggy air below VOL (Fig. 1a) lifted to higher elevations (Fig. 1b) around local noon. Weather at RND was clear and sunny until usually shown as 4 p.m. (Fig. 1c). When shortwave (SW) radiation provided additional heat at the low levels and sloping surfaces, the fog started to lift over a 3–4 h time period, gaining additional moisture from melting snow. Then, the stratiform fog layers were eventually converted to altostratus. In the afternoon, at about 07:00 p.m., the entire valley at higher levels was filled with a fog/cloud mixture (Fig. 1d). The snow occurred at about 07:00 p.m. local time at the RND site.

The time series of cloud (fog) ceiling height from the Vaisala C31 ceilometer is used for detecting fog. Figure 2a shows the measured fog ceiling levels at the various stations along the Whistler Mountain slope on March 4. The green and brown dots represent the measurements of the C31 ceilometer and the PMPR (Profiling Microwave Radiometer, Radiometrics Inc.). The profiles of RH$_w$ measured by the sensors mounted on a Gondola along the Whistler Mountain slope are shown in Fig. 2b. The low level cloud base height was at about 2 km after 16:00 UTC, indicating the precipitation at the RND site (1.9 km) After 12:00 UTC, cloud base was about 1.2 km above the sea level. The ceilometer and PMWR were located at about 600 m above sea level. Figure 2b shows that RH$_w$ was about 90% at the RND site after 0000 UTC (04:00 p.m. LST on April 4), leading to cloud formation. Figure 2c shows that at about 04:00 p.m. LST, T was about −4°C at the RND site and 4°C at the VOT site, where the C31 ceilometer and PMWR were located.

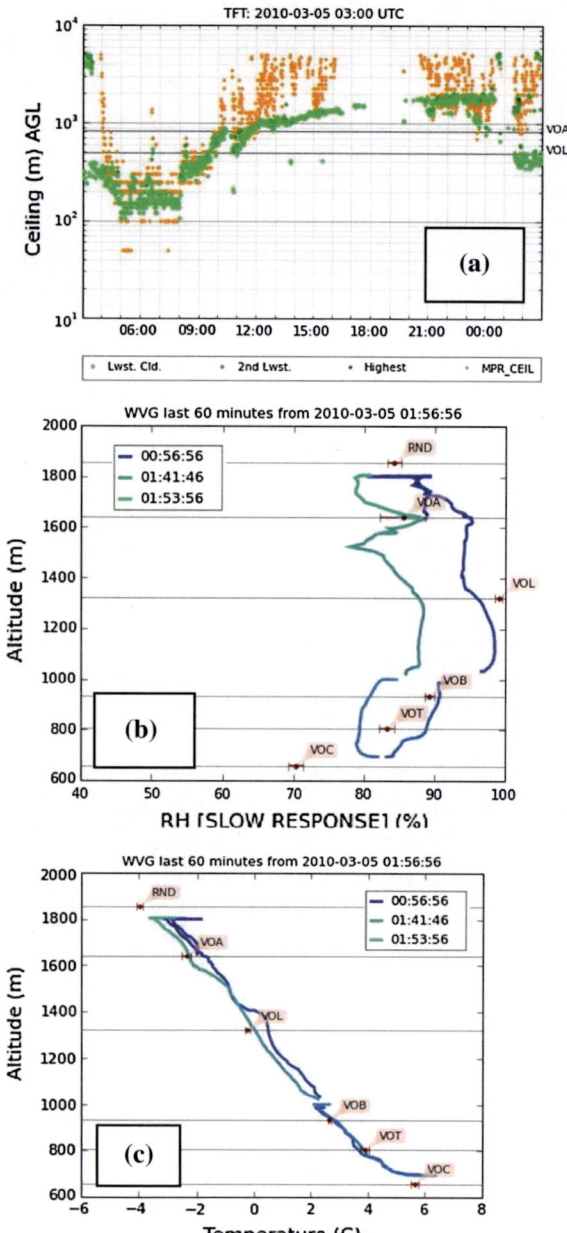

Figure 2

Times series of fog base height measured at the Whistler Mountain base (VOT at 800 m ASL) on March 4. The *green dots* represent VAISALA ceilometer readings and *brown dots* represent those of the MPR (Microwave Profiling Radiometer, Radiometrics Inc.) (**a**). The profiles of RH and T measured by the sensors mounted on a Gondola along the Whistler Mountain slope are shown in (**b**) and (**c**), respectively. Times are given in UTC

Time and height cross-section of T, RH_w, and vapour density (qv) measured with Radiometrics Inc. PMWR are shown in Fig. 3. The T values were about

-5 to $-8°C$ at the levels close to the RND site, and they were about $3-4$ C at the VOT site near the town center. The RH_w plot shows that from 1800 UTC till 0000 UTC, the low levels became subsaturated and, for the same time period, high levels (e.g. RND site) became more saturated. The RH_w of $\sim 100\%$ is seen up to 5 km, which likely results in snowing at the levels of about 2 km. The qv values at the surface shown in Fig. 3 decreased from about 5 to 3 g m^{-3} during FTS processes, supporting the FTS hypothesis.

The results obtained using measurements from the RND and VOL sites (see Fig. 2c) and GEM model simulations obtained during the project are shown in Fig. 4. In Fig. 4, the models used are based on the Canadian GEM model [GULTEPE and MILBRAND, 2007] with different spatial resolutions (e.g. 1 and 2.5 km in GEM-LAM, and 15 km in GEM-REG). Time resolution used is 1 min. The winds on this day were relatively calm (<1 m s^{-1}, not shown) and T was about -2 ($-6°C$) at 15 UTC at VOL (RND) (Fig. 4a, c). Before 1500 UTC, the RH_w was 100% at the VOL (Fig. 4b) and it was about 90% at the RND site (Fig. 4d). Significant differences between measurements and simulations for some time periods suggest that model parameterizations need to be improved to account topographical variability and boundary layer physics.

Figure 5a, b show the time series of Vis and RH_w, and PR from the FD12P and TPS sensors, respectively, observed at the RND site on March 4 2010. After 0000 UTC, PR for maximum snow amount was about 0.7 mm h^{-1} and lasted till midnight (LST). The FD12P Vis was about 50 km before the snow began, showing very low PR values (Fig. 5b). When TPS PR was ~ 0.7 mm h^{-1}, Vis decreased down to 10 km. Vis ~ 100 m at VOL during the fog event at the low levels (not shown).

During the FTS process at the RND site, snow was not detected by the weighing precipitation sensors except with the Ground Cloud Imaging Probe (GCIP; Fig. 6) and disdrometers [Laser Precipitation Sensor (LPM, Fig. 7) and OTT]. Figure 6 shows that precipitation size particles were about 10–20 μm at early after noon and about 500 μm in late afternoon. The maximum size of particles is about 1000 micron. Many small particles existed with sizes less than 100 μm at about local noon at the RND site. The unique data set from the GCIP provided more

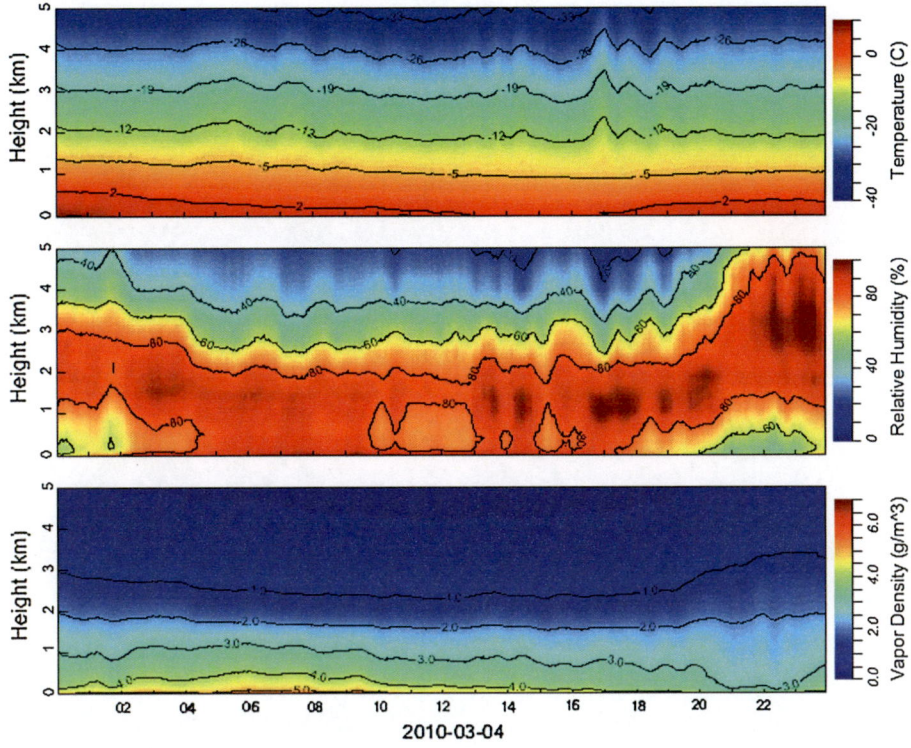

Figure 3
Time (UTC)–Height(km) cross-section of temperature (T), relative humidity with respect to water (RH_w), and vapor density (q_v) for the April 4 case. After 2200 UTC, RH_w increases significantly where snow precipitation amount was observed at the RND site

information on light precipitation estimation at the cold temperatures over the mountain regions but it is not given here.

The LPM sensor at the RND site provided detailed measurements of precipitation rate, size, fall speed, Vis, and accumulated precipitation amount. Figure 7 shows the snow particle fall velocities versus particle size together with a fit to droplets' fall velocities as a function of their size obtained theoretically for liquid droplets. Any deviation from the fit suggests the existence of snow particles and small ice crystals. In this case, many snow particles had fall velocities of less than about 2.0 m s^{-1}, and counts were maximum at about 200. The red vertical line in the bottom panel indicates the location of the particle concentration measurements made in box a. After 0300 LST, Vis values were about 10 km, which matched with the snow occurrence and Vis measurements obtained from FD12P (Fig. 5).

Both GCIP and LPM provided results that indicated occurrence of the snow at the RND site (at

6 p.m. LST) after the fog event occurred in early morning at the VOL site (at about 9 a.m. LST).

4. Discussion

The observations collected at the RND site during the SNOW-V10 project were very extensive and covered many parameters related to remote sensing, microphysics, dynamics, radiation, and aerosols. The lower stations, i.e. VOL, VOA, and VOT sites, together with the Gondola T and RH_w measurements, provided observations for better understanding the FTS process.

The models over small scales such as over the mountain slopes need better measurements of dynamical and microphysical processes to improve their predictability. The weather processes over the mountains usually show nonlinear characteristics; therefore, initialization of the models over a limited area from a large scale model cannot always be a

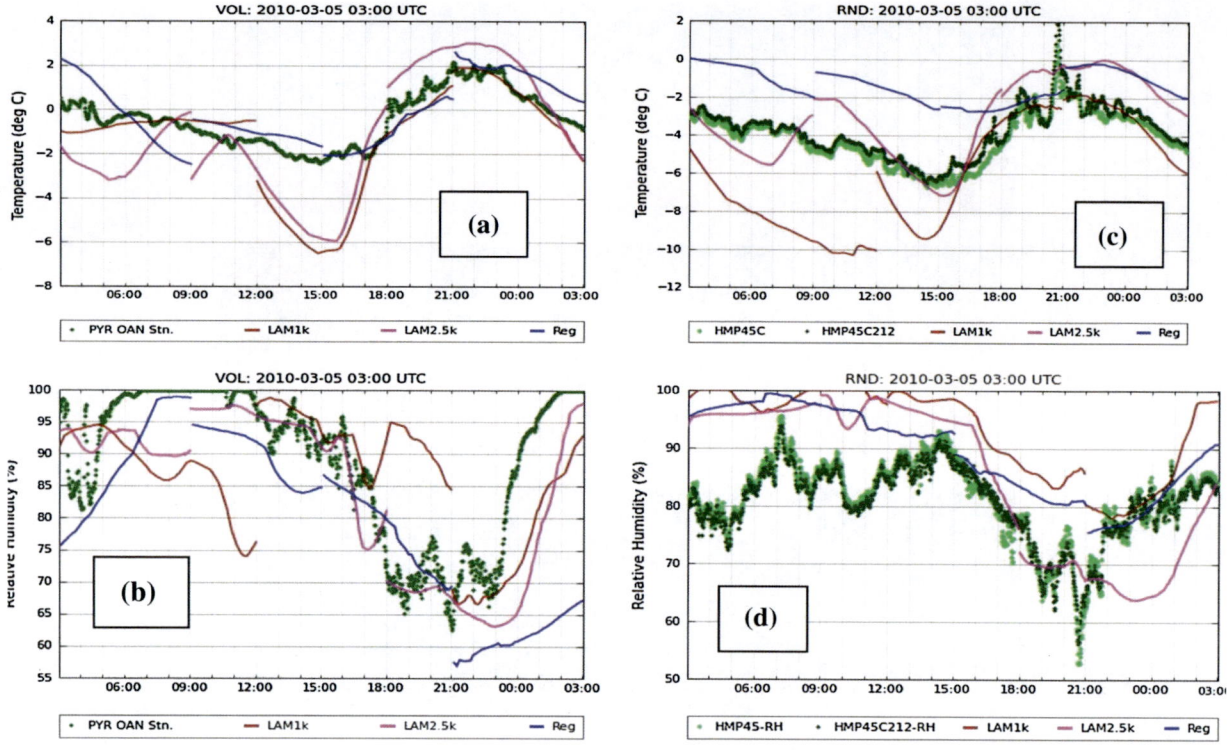

Figure 4

Time series of T and RH$_w$ for VOL (**a** and **b**) and RND (**c** and **d**) stations, respectively. The *solid colour lines* are predicted parameters obtained from the Canadian GEM model simulations with 1, 5, and 15 km resolutions. The *green lines* are for measured values from instruments

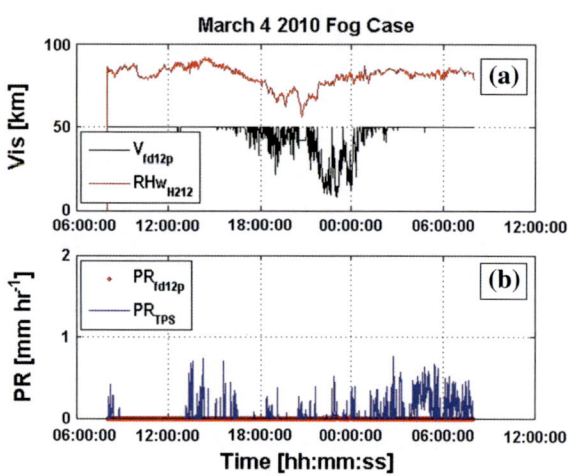

Figure 5

Time series of Vis and PR from FD12P, RH$_w$, and PR from the TPS sensor. Note that FD12P did not measure any precipitation amount because of its threshold value. Time is given in UTC. Significant snow amount is seen after 5 p.m. LST(=UTC-8 h)

good idea because of their large scale resolutions. The FTS process needs to be better documented because snow resulting from fog is usually not

predicted by the models and forecasters, and can present very adverse conditions during search and rescue operations, as well as sporting activities, over the mountain ranges.

As a part of both the SNOW-V10 and FRAM projects, this work provided a starting point for the better understanding of the FTS process. It is hoped that more detailed research and application work will be conducted in the future. Clearly, the FTS process can be considered by the forecasters for short term predictions based on observations, but forecasting models will be inadequate until a detailed regional 3D cloud-fog model is developed. One main reason is that model boundary layer microphysics cannot handle radiation effects on the particle growth equation. If this is not accounted for in the calculations, then lifting of a fog layer over a mountain slope because of radiative heating/lifting processes cannot be modelled accurately. In fact, forecasting models can resolve the scales down to 1–2 km, but the entire process occurred over 2 km height level and a ~10 km sloped surface; therefore, a 3D cloud-fog

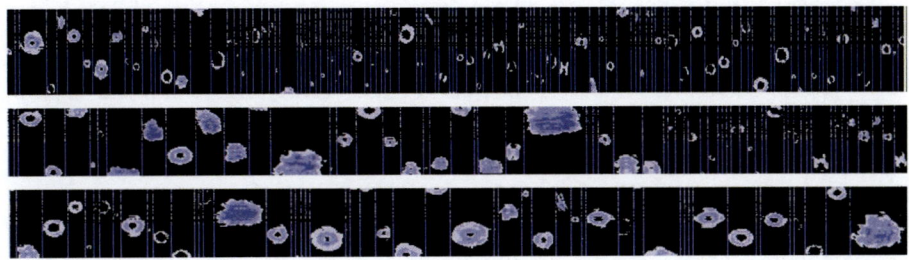

Figure 6
The GCIP probe measurements of the snow particle shape. The *boxes* from top to bottom are taken at 125704, 135809, and 170913 LST, respectively

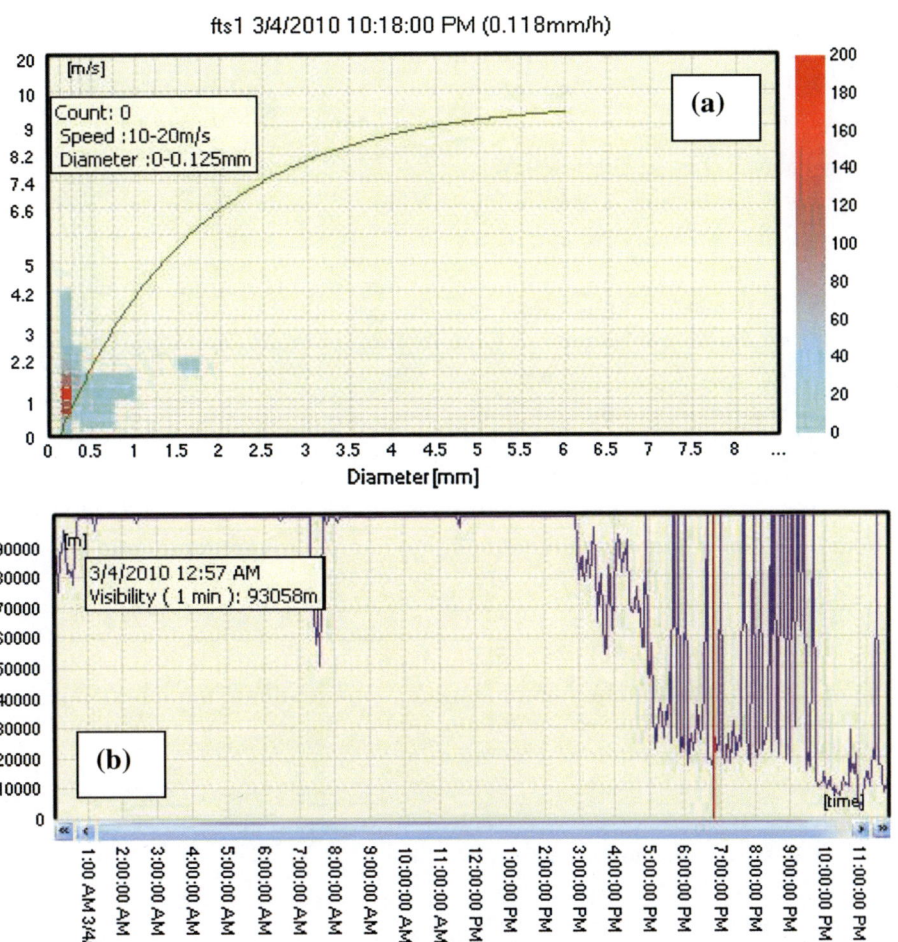

Figure 7
The LPM measured snow particle counts as a function of particle fall speed and size. Counts are shown in *coloured bars*. The LPM snow visibility (km, *bottom panel*) time series (LST) is given in the *bottom panel*

model developed for mountainous regions is needed to resolve the FTS process.

Because of quick changes in the weather parameters in mountain environments, a new set of the sensors also needs to be developed. Parameters from regular meteorological stations are insufficient for accurately predicting occurrences of the FTS process.

25

5. *Conclusions*

During the FTS process, observed snow precipitation rates at high elevations after noon was not predicted with model simulations; this was likely due to issues related to microphysical parameterizations, and resolution and subgrid scale variability. Snow PR was detected by some optical sensors, but most of the weighing sensors missed this precipitation type because of their higher sensitivity threshold, such as 0.5 mm h^{-1} and wind effects. The optical probes performed better than weighing gauges when light snow occurred.

The following conclusions can be drawn from this work:

- Visibility was low (300–500 m) at the low levels early in the morning but it became worse during the FTS process at high elevations (~ 2 km) after 3 p.m. local time.
- Forecast of the FTS process is important to meteorologists because it lowers mountain Vis down to a few km from a clear sky condition and increases snow precipitation rates up to 0.2–0.5 mm h^{-1}; meanwhile it is not predicted by the models and weighing gauges due the problems given above.
- Cloud base decreased gradually at the high elevations and it was not detected accurately by the model predictions.
- Solar radiation (indicated by the pictures taken) played an important role over the sloped surfaces by increasing radiative heating at low levels, resulting in a moisture source for high level snow formation.
- GCIP sensor measurements were effective at measuring the light snow amounts, e.g. snow particles, and their spectra; therefore, its measurements can be used to parameterize snow particle spectra during light precipitation events (<0.5 mm h^{-1}).

The results of this work, in the future, will be further used to detect light precipitation amounts and compare the measurements obtained from the various weighing precipitation gauges and optical sensors. The results based on observations, in the future, will likely be used in a 3D fog model simulation to assess FTS processes. Clearly, not all precipitation sensors measure light precipitation amounts accurately; optical sensors can work better than weighing sensors for light snow measurements over various geographical locations and environmental conditions.

Acknowledgments

The author would like to thank to science team of Environment Canada (especially Drs. G. Isaac and P. Joe) and other SNOW-V10 and FRAM project participants who helped to this work having a success. This project was funded by the DND (Department of National Defence) National Search and Rescue Secretariat (NSS) Office and Environment Canada (EC).

References

BERGOT, T. and GUÉDALIA, D. (1994), *Numerical forecasting of radiation fog. Part I: numerical model and sensitivity tests.* Monthly Weather Review *122*, 1218–1230.

BOTT, A., SIEVERS, U., and ZDUNKOWSKI, W. (1990), *A radiation fog model with a detailed treatment of the interaction between radiative transfer and fog microphysics,* J. Atmos. Sci. *47*, 2153–2166.

BROWN, R. (1980), *A numerical study of radiation fog with an explicit formulation of the microphysics,* Quart. J. Roy. Meteor. Soc. *106*, 781–802.

BROWN, R. and ROACH, W.T. (1976), *The physics of radiation fog: II—a numerical study,* Quart. J. Roy. Meteor. Soc. *102*, 335–354.

DUYNKERKE, P.G., (1991) *Radiation fog: a comparison of model simulation with detailed observations,* Monthly Weather Review *119*, 324–341.

DUYNKERKE, P.G., JONKER, P.J., CHLOND, A., VAN ZANTEN, M.C., CUXART, J., CLARK, P., SANCHEZ, E., MARTIN, G., LENDERINK, G., and TEIXEIRA, J. (1999), *Intercomparison of three- and one-dimensional model simulations and aircraft observations of stratocumulus,* Bound. Layer Meteorol. *92*, 453–487.

GULTEPE, I., R. TARDIF, S.C. MICHAELIDES, J. CERMAK, A. BOTT, J. BENDIX, M. MÜLLER, M. PAGOWSKI, B. HANSEN, G. ELLROD, W. JACOBS, G. TOTH, S.G. COBER, (2007), *Fog research: a review of past achievements and future perspectives.* J. of Pure and Applied Geophy., special issue on fog, edited by I. Gultepe. Vol. 164, 1121–1159.

GULTEPE, I., AND J. MILBRANDT, (2007), *Microphysical observations and mesoscale model simulation of a warm fog case during FRAM project.* J. of Pure and Applied Geophy., Special issue on fog, edited by I. Gultepe. Vol. 164, 1161–1178.

GULTEPE, I., G. PEARSON J. A. MILBRANDT, B. HANSEN, S. PLATNICK, P. TAYLOR, M. GORDON, J. P. OAKLEY, and S.G. COBER, (2009), *The fog remote sensing and modeling (FRAM) field project.* Bull. Of Amer. Meteor. Soc., v.90, 341–359.

GULTEPE, I., G. A. ISAAC and R. RASMUSSEN, (2010), GCIP measurements of precipitation and fog during SNOW-V10 and

FRAM projects. The 13th Conference on Cloud Physics, 28 June–2 July 2010, Portland, Oregon, USA. CD version.

GULTEPE, I. and G.A. ISAAC, (2010), FTS (Fog To Snow) conversion process during the SNOW-V10 Project 2010, 5th International Conference on Fog, Fog Collection and Dew, 26-30 July 2010, Munster, Germany, 223-226.

HANSEN, B., GULTEPE, I., KING, P., TOTH, G., and MOONEY, C. (2007), Visualization of seasonal-diurnal climatology of visibility in fog and precipitation at Canadian airports, AMS Annual Meeting, 16th Conference on Applied Climatology, San Antonio, Texas, 14–18 January 2007, In CD volume.

ISAAC, G., P. JOE, J. MAILHOT, M. E. BAILEY, S. BELAIR, F. S. BOUDALA, M. BRUGMAN, E. CAMPOS, R. L. CARPENTER, S. G. COBER, B. DENIS, C. DOYLE, D. E. FORSYTH, I. GULTEPE, T. HAIDEN, L. HUANG, J. A. MILBRANDT, R. MO, R. M. RASMUSSEN, T. SMITH, R. E. STEWART, and D. WANG, (2010), Nowcasting winter weather in complex terrain—experiences from SNOW-V10 AMS Mountain Meteorology, San Diego, CA, USA, CD version.

MÜLLER, M.D., SCHMUTZ, C., and PARLOW, E. (2007), A one-dimensional ensemble forecast and assimilation system for fog prediction, J. Pure Appl. Geophys., 164, (Special issue), 1241–1264.

PAGOWSKI, M., GULTEPE, I., AND KING, P. (2004), Analysis and Modeling of an Extremely Dense Fog Event in Southern Ontario, J. Appl. Meteor. 43, 3–16.

TEIXEIRA, J. (1999), Simulation of fog with the ECMWF prognostic cloud scheme, Quart. J. Roy. Meteor. Soc, 125, 529–553.

(Received November 19, 2010, revised April 28, 2011, accepted May 3, 2011, Published online July 26, 2011)

Reprinted from the journal

Pure Appl. Geophys. 169 (2012), 793–807
© 2011 Springer Basel AG
DOI 10.1007/s00024-011-0332-0

Simulating Z–LWC Relations in Natural Fogs with Radiative Transfer Calculations for Future Application to a Cloud Radar Profiler

F. Maier,[1] J. Bendix,[1] and B. Thies[1]

Abstract—The vertical distribution of liquid water content (LWC) in natural fog and low stratus is a crucial variable in many applications, e.g. the development of satellite based retrievals of ground fog. Unfortunately, there is very little data concerning fog LWC-profiles, mainly due to the lack of suitable operational instrumentation. A novel ground-based 94 GHz FMCW cloud radar could fill this gap if radar reflectivity Z could be converted to LWC by using appropriate Z–LWC relations. However, this relation strongly depends on drop size distribution (DSD) and is hardly known for natural fog types. In this sensitivity study, the influence of the DSD on the Z–LWC relation in different types and life cycle stages of natural fogs is analyzed using a radiative transfer code (RTC) and published fog drop size distributions. It could be shown that there is a direct but nonlinear relationship between LWC and radar reflectivity. The proportionality factor of the Z–LWC equation in particular reveals specific ranges for the different life cycle stages. If a proper classification of fog life cycle in the field is possible, the results could be used to properly convert Z to LWC.

Key words: Fog, low stratus, cloud radar, radiative transfer, modified gamma distribution.

1. Introduction

Fog defined as horizontal visibility <1 km (WMO 1992) is a major obstruction for air, land and sea traffic, but can also have severe impacts on air quality (smog) if air pollutants are present. Unfortunately, the spatio-temporal observation with an operational (human observer, transmissiometer, scatterometer) network is difficult due to the sparse density of observations, particularly in complex terrain such as the lower mountain ranges in Germany (e.g. SCHULZE-NEUHOFF 1976). To overcome this problem, different methods for operational fog forecasting and

nowcasting have been developed over recent decades (for an overview, see GULTEPE *et al.* 2007; JACOBS *et al.* 2008).

From the numerical modeling point of view, accurate fog forecasting still remains a challenge. One reason for the problem in accurately forecasting fog is the difficulty in representing the microphysical processes involved, which are still not completely understood (GULTEPE *et al.* 2007). Several studies suggest that a better understanding of fog microphysics is needed to develop more accurate forecasting models (TARDIF 2007; GULTEPE and MILBRANDT 2007; PAGOWSKI *et al.* 2004). The life cycle of fog (formation, development and dissipation) is directly related to microphysical processes that are not represented accurately enough in models. SIEBERT *et al.* (1992a, b) and von GLASOW and BOTT (1999) developed models that explicitly resolve the evolution of the droplet size distribution and cloud condensation nuclei. In a comparative study, TERRADELLAS and BERGOT (2007) showed that 1-D models can simulate the major features of the fog cycle. Their results also indicate a high sensibility to the model's physical parametrization and vertical resolution. However, such 1-D model approaches are computationally very intensive. Parametrized versions of the detailed 1-D fog microphysics models can be incorporated in 3-D models, resulting in more precise forecast results (GULTEPE *et al.* 2006; GULTEPE and MILBRANDT 2007; PAGOWSKI *et al.* 2004).

Another major challenge in nowcasting the spatio-temporal fog dynamics is its detection based on weather satellite data. Here, particularly the ability to distinguish low stratus and ground fog is not yet conclusively established. First approaches for ground fog detection using TERRA-MODIS and Meteosat Second Generation SEVIRI daytime data have been

---

[1] Faculty of Geography, University of Marburg, Marburg, Germany. E-mail: frank.maier@staff.uni-marburg.de

proposed by BENDIX *et al.* (2005) and CERMAK and BENDIX (2010). However, these techniques require information on the vertical structure of liquid water content (LWC) within fog layers, which are not directly available from optical satellite data. For this reason, the approaches mentioned assume a sub-adiabatic LWC profile, which might not coincide with real conditions. This uncertainty is mainly due to a lack of data on the vertical structure of fog for different fog types and life cycle stages. In situ airborne measurements, as frequently conducted for higher clouds, are not permitted and not even possible during fog situations. Balloon-borne systems with suitable sensors for temporally continuous profile measurements are not available. Passive microwave profilers might bridge this gap, but their poor vertical resolution of >500 m partly exceeds the vertical extent of fog layers (e.g. CERMAK *et al.* 2006, RUF-FIEUX *et al.* 2006). In this context, millimeter-wave cloud profiling radars operating at 35 or 94 GHz are well suited for continuous cloud observations, since the attenuation of the beam by oxygen and water vapor absorption is minimal near these frequencies. Furthermore, these radars are more sensitive for cloud particles with diameters of a few micrometers to precipitating drops. The sensitivity to small hydrometeors, the high spatial resolution and the ability to provide information on multiple cloud layers from millimeter-wavelength radars make them ideal for continuously monitoring the vertical distribution of clouds and their microphysical properties (e.g. CLOTHIAUX *et al.* 1995, KOLLIAS *et al.* 2007). Especially cloud radars that rely on the frequency modulated continuous wave technique (FMCW) are highly suitable for monitoring fog and low stratus layers because of their small near field. They enable the detection of clouds as low as about 30 m (CERMAK *et al.* 2006; NOWAK *et al.* 2008; BENNETT *et al.* 2009a), which contrasts starkly with existing pulse cloud radars with larger near fields reaching up to about 500 m (e.g. LÖHNERT *et al.* 2001).

Simple techniques use empirically derived relationships linking both parameters to retrieve LWC profiles from radar reflectivity Z (e.g. SAUVAGEOT and OMAR 1987; LIAO and SASSEN 1994, FOX and ILLINGWORTH 1997, BOERS *et al.* 2000). The main advantage of these approaches lies in their simple application.

However, the Rayleigh approximation applies since the radar wavelength is much longer compared to the size of the observed cloud droplets. Consequently, the radar reflectivity is proportional to the sixth moment of the droplet spectrum, whereas the LWC is proportional to the third moment of the droplet spectrum. A variation of the droplet spectrum thus strongly influences the relationship between Z and LWC. Using only one fixed relationship to retrieve the LWC from Z without considering the droplet spectrum leads to high uncertainties in the calculated LWC profile (e.g. LÖHNERT *et al.* 2001). KHAIN *et al.* (2008) emphasize the ambiguous character of Z–LWC relationships in low level water clouds, which is responsible for the low accuracy of retrieval algorithms. According to the authors, this is partially related to the fact that empirical Z–LWC relationships are often derived without a corresponding understanding of the microphysical processes within the cloud. The results of their numerical simulations indicate the significant importance of microphysical processes for the retrieval of LWC from radar reflectivity. Additional information about droplet size distributions and specific regimes of cloud formation has to be taken into account in order to derive proper Z–LWC relationships.

To minimize the uncertainties caused by the unknown droplet spectrum, more sophisticated techniques use different sensors in combination to incorporate additional information about the liquid water path and the backscatter coefficient measured by lidar ceilometers (e.g. FRISCH *et al.* 1995, 1998, DONOVAN and VAN LAMMEREN 2001; MCFARLANE *et al.* 2002; LÖHNERT *et al.* 2001, 2004). However, these techniques also rely on assumptions about the droplet spectrum (e.g. lognormal or gamma) and/or the droplet concentration. Concerning the high temporal dynamic of low level stratiform clouds, such assumptions are not suitable for a proper retrieval of the LWC from radar reflectivity. Especially for fog and low stratus clouds, several authors have identified different evolutionary stages with strong influences on the drop size spectrum and thus the relationship between Z and LWC (e.g. Pilié *et al.* 1975b, MEYER *et al.* 1986, WELCH and RAVICHANDRAN 1986, WENDISCH *et al.* 1998).

The aim of the present study is to investigate the influence of the drop size spectrum on the relationship between Z and LWC by means of radiative transfer calculations. To this end, characteristic drop size distributions of natural fogs, as measured during various field studies, are taken from the literature. The results are used to test the sensitivity of the Z–LWC relation in natural fogs regarding variations in drop size distributions as reported for different fog situations and life cycle stages. In this context, the study aims to investigate three questions: (1) the influence of the drop size spectrum on the relationship between Z and LWC, (2) the different fog types with regard to their specific influence on the relationship between Z and LWC and (3) the relevance of the fog life cycle for the Z–LWC relation.

The results of the study provide valuable information for the development of a Z–LWC retrieval technique for a novel 94 GHz FMCW cloud radar profiler (HUGGARD et al. 2008; Bennett et al. 2009a). In a next step, the retrieved LWC profiles from the radar reflectivity might be used to optimize satellite retrievals for ground fog. From the numerical modeling point of view, the investigation of the fog life cycle and its relevance for the droplet size distribution and the resulting Z–LWC relationship can contribute to a better understanding and parametrization of the microphysical processes shaping the microstructure of fog. The paper is structured as follows: the theoretical background of the study is described in Sect. 2. This concerns the radar characteristics, the relationship between the droplet spectrum, the liquid water content and the radar reflectivity, the adaptation of the radiative transfer code and the fog life cycles. The results of the sensitivity study are presented in Sect. 3 and discussed in Sect. 4. Finally, some conclusions and an outlook are given in Sect. 5.

2. Materials and Methods

As mentioned in the introduction, the current study is focused on the development of an appropriate Z–LWC retrieval for a novel 94 GHz cloud radar profiler. The instrument is a 94 GHz frequency modulated continuous wave radar that is developed and crafted by Rutherford Appleton Laboratory,

Great Britain. Because of its high frequency, it is highly sensitive to cloud droplets. It delivers vertical profiles of Z every 15 s in a vertical resolution of 4 m up to 2 km above ground. Because of the FMCW technique, continuous measurements can be provided from 30 m above the ground. For further details on the instrument see also BENNETT et al. (2009b) and HUGGARD et al. (2008).

Some theoretical considerations on the drop size spectra of natural fog and their statistical representation in a distribution function are needed to develop appropriate Z–LWC relations for different fog situations.

Generally, the drop size spectrum is defined such that $n(r)dr$ is the total number of particles within a given radius range of interest $r_1 \leq r \leq r_2$ per unit volume of air. Hence,

$$n(r) = \int_{r_1}^{r_2} n(r)dr \qquad (1)$$

is the total concentration of drops per unit volume of air with radii less than r_2 (FLATAU et al. 1989). In order to describe the drop size spectrum, a suitable probability density function $f(r)$ has to be selected, which is defined by the intervals r_1 and r_2 (FLATAU et al. 1989).

The drop size spectrum can then be represented by

$$n(r) = N_t f(r) \qquad (2)$$

where N_t is the total number of drops per unit volume of air (cm^{-3}).

$n(r)$ is a continuous and integrable function within the radius range of the fog drops for a given volume of air. Thus, its unit is (μm^{-1} cm^{-3}).

The suitable probability density function for describing the drop size distribution of low stratus clouds and fog is commonly characterized in literature by the so-called modified gamma distribution (DEIRMENDJIAN 1964).

$$f(r) = ar^{\alpha} \exp(-br^{\gamma}) \qquad (3)$$

which is the derivative of the ordinary gamma-distribution to which it reduces when $\gamma = 1$. Its probability density function is defined by the scale parameter b and the shape parameter p.

$$f(x) = \frac{b^p}{\Gamma(p)} x^{p-1} \exp(-bx), \quad p, b \geq 0 \text{ and } p, b \in \mathbb{R}$$

(4)

with the gamma-function

$$\Gamma(x) = \int_0^\infty t^{x-1} \exp(-t) dt \quad \text{for } x > 0 \qquad (5)$$

which is an extension for the faculty of real and complex numbers. It has the following features for positive integer numbers.

$$\Gamma(x) = (x - 1)! \qquad (6)$$

and (FLATAU et al. 1989).

$$\Gamma(x + 1) = x \Gamma(x) \qquad (7)$$

The four factors a, α, b and γ in Eq. 3 are positive and real numbers. Additionally, a is an integer and the intercept of Eq. 3. The latter three parameters completely determine the shape of the distribution. The constant α is called the shape parameter and b the slope or gradient. The function is normalized by a to obtain the total number of drops per unit volume N_t. This guarantees that the integral of the size distribution over all radii is N_t.

The four parameters are not independent of each other and are related to quantities of the frequency distribution. An integration over the range of radii $r_1 \leq r \leq r_2$ by the drop size spectrum shows that a is only related to N_t (DEIRMENDJIAN 1969).

$$N_t = \int_{r_1}^{r_2} ar^\alpha \exp(-br^\gamma) dr \qquad (8)$$

$$N_t = \frac{a}{\gamma} b^{-\left(\frac{\alpha+1}{\gamma}\right)} \Gamma\left(\frac{\alpha+1}{\gamma}\right) \qquad (9)$$

The latter (N_t) is, therefore, determined by the zeroth moment of the drop size spectrum. The following general solution for the modified gamma distribution is used for fog by Tampieri and TOMASI (1976) to compute the integral of the product of the modified gamma distribution multiplied with the power r^k over the entire radii interval.

$$I_k = a \int_{r_1}^{r_2} r^{\alpha+k} \exp(-br^\gamma) dr \qquad (10)$$

$$I_k = \frac{a}{\gamma} b^{-\left(\frac{\alpha+k+1}{\gamma}\right)} \Gamma\left(\frac{\alpha+k+1}{\gamma}\right) \qquad (11)$$

These equations are based on the calculations of the integrals from GRADSHTEYN and RYSHIK (1980).

$$\int_0^\infty r^\alpha \exp(-br) dr = a! b^{-\alpha-1} = b^{-(\alpha+1)} \Gamma(\alpha + 1)$$

(12)

A differentiation of the drop size spectrum with respect to r leads to

$$\frac{d}{dr} n(r) = ar^{\alpha-1}(\alpha - \gamma br^\gamma) \exp(-br^\gamma) \qquad (13)$$

that has three zeros. One is at $\alpha = 0$ and one at $\alpha = \infty$. One finds the third zero by setting the second term from Eq. 13 equal to 0. That is also the absolute maximum of the equation and thus the mode radius r_c (μm).

$$r_c = \left(\frac{\alpha}{\gamma b}\right)^{\frac{1}{\gamma}} \qquad (14)$$

The latter (r_c) represents the radius size of maximum frequency in the drop size distribution.

The modified gamma distribution is often also expressed by substituting b by

$$b = \frac{\alpha}{\gamma r_c^\gamma} \qquad (15)$$

in the literature which yields to (TAMPIERI and TOMASI 1976):

$$n(r) = ar^\alpha \exp\left(-\frac{\alpha}{\gamma}\left(\frac{r}{r_c}\right)^\gamma\right) \qquad (16)$$

As can be seen, the parameter b is characterized by mode radius r_c if α and γ are known.

With regard to the frequency of the radar at 94 GHz (= wavelength of 3.2 mm), the dominant scattering process is described by the Rayleigh-approximation. Therefore, the radar reflectivity Z can be written as the sixth moment of the continuous drop size spectrum (DANNE 1996)

$$Z = \int_{r_1}^{r_2} n(r) r^6 dr \qquad (17)$$

Inserting the drop size spectrum yields

$$Z = a \int_{r_1}^{r_2} r^{\alpha+6} \exp(-br^\gamma)dr \qquad (18)$$

An integration over the range of radii $r_1 \le r \le r_2$ with respect to Eq. 10 yields

$$Z = \frac{a}{\gamma} b^{-\left(\frac{\alpha+7}{\gamma}\right)} \Gamma\left(\frac{\alpha+7}{\gamma}\right) \qquad (19)$$

The unit of radar reflectivity Z is either $\mathrm{mm}^6\,\mathrm{m}^{-3}$ or in a logarithmical unit according to

$$dBz = 10 \log_{10} \frac{z}{z_0} \qquad (20)$$

with $Z_0 = 1\ \mathrm{mm}^6\,\mathrm{m}^{-3}$ (DANNE 1996).

The determination of Z–LWC relations is directly based on the drop size distribution, because the liquid water content LWC ($\mathrm{g\,m}^{-3}$) can be derived from the drop size spectrum and the mass of drops:

$$LWC = \int_{r_1}^{r_2} n(r)m(r)dr \qquad (21)$$

with

$$m(r) = \frac{4}{3}\pi r^3 \rho_{H_2O} \qquad (22)$$

where ρH_2O is the water density $1\ \mathrm{g\,cm}^{-3}$ (DANNE 1996).

As can be seen, the LWC is proportional to the third moment of the drop size spectrum. The integration of the LWC

$$LWC = \frac{4}{3}\rho_{H_2O}\pi a \int_{r_1}^{r_2} r^{\alpha+3} \exp(-br^\gamma)dr \qquad (23)$$

yields with the usage of Eq. 10

$$LWC = \frac{4}{3}\pi\rho_{H_2O}\frac{a}{\gamma}b^{-\left(\frac{\alpha+4}{\gamma}\right)}\Gamma\left(\frac{\alpha+4}{\gamma}\right) \times 10^{-6} \qquad (24)$$

Consequently, the LWC–Z relation can be derived by dividing Eq. 24 by 19.

$$\frac{LWC}{Z} = \frac{4}{3}\pi\rho_{H_2O}\frac{\Gamma\left(\frac{\alpha+4}{\gamma}\right)}{\Gamma\left(\frac{\alpha+7}{\gamma}\right)}b^{\frac{3}{\gamma}} \times 10^{-6} \qquad (25)$$

The right part of the equation is defined as a proportionality factor Ω ($\mathrm{g\,mm}^{-6}$) which enables a linear relationship between $Z\,(\mathrm{mm}^6\,\mathrm{m}^{-3})$ and LWC ($\mathrm{g\,m}^{-3}$).

$$LWC = \Omega \times Z \qquad (26)$$

If the factor Ω can be determined for natural fog, it is possible to directly convert Z to LWC. Unfortunately, a universal value of the factor is not known and it is hypothesized that it varies with fog type and life cycle stage.

To investigate the behavior of Ω in different natural fogs, the radiative transfer code QuickBeam (HAYNES et al. 2007) was adapted to the ground-based 94 GHz FMCW cloud radar. Major modifications included the increase of the spatial resolution to 4 m and the selection of the vertical path length of 2 km. According to Petterssen's classification of fog based on temperature, there are three types of fog (PETTERSSEN 1956). As our radar measurements are performed in central Europe mainly in spring and autumn, only liquid fog, defined by temperatures $T > -10°C$, is of importance; mixed phase fog ($-30°C > T > -10°C$) and ice fog ($T < -30°C$) are negligible. A major modification was the replacement of default drop size distributions by the freely adjustable modified gamma distribution for fog according to DEIRMENDJIAN (1964).

With the adapted radiative transfer code, a sensitivity study on the Z–LWC relation was conducted based on fog drop size distributions as taken from literature. In this study, the modified gamma distribution varied within typical ranges and the resulting radar reflectivity and liquid water content (and so, the Z–LWC relation) were calculated by the RTC. To find typical ranges of the drop size spectrum for different fog types and evolutionary stages, a literature survey was conducted to find respective field data. The results are displayed in Table 1. However, even more parameter sets were found, not all could be included because the level of information on fog occurrence was insufficient (e.g. KALASHNIKOVA et al. 2002; KIM et al. 2001) or because they were only recalculated from existing values (HESS et al. 1998; AWAN et al. 2008).

Furthermore, we depicted the four parameter sets of HARRIS (1995) and not the original ones of SHETTLE and FENN (1979), because the probability density function of the advection fog examples did not always result in 1.

Table 1

Parameter sets of the modified gamma distribution (a, α, b, γ) for the drop size spectrum

Fog type	DSD-model	a	α	γ	r_c (μm^{-1})	b	N_t (cm^{-3})	LWC (g m^{-3})	Source
Radiation fog	1	4.3239E0	4	0.70	2.13	*3.36586*	249.93	6.137E-2	T&T
	2	7.4438E-3	4	1.23	4.98	*0.4514*	76.29	9.899E-2	T&T
	3	3.0410E-4	4	1.77	8.06	*0.05622*	35.85	1.379E-1	T&T
	4	7.5475E-6	5	1.62	12.22	*0.05351*	15.90	2.061E-1	T&T
	5	*2.3730E-2*	6	1.00	4.00	1.5	99.87	6.247E-2	H
	6	*3.0375E0*	6	1.00	2.00	3.0	199.83	1.563E-2	H
Advection fog	1	4.9763E-1	4	0.85	2.75	*1.99163*	184.55	6.795E-2	T&T
	2	4.2028E-3	5	1.17	5.04	*0.64408*	78.75	9.534E-2	T&T
	3	8.1656E-3	3	1.05	6.20	*0.42065*	37.73	1.537E-1	T&T
	4	3.0861E-5	6	1.47	8.10	*0.18852*	36.13	1.348E-1	T&T
	5	8.9728E-8	5	3.09	21.16	*1.29768E-4*	6.50	3.059E-1	T&T
	6	1.3500E-3	3	1.00	10.00	0.3	20.00	3.724E-1	H
	7	3.4752E-3	3	1.00	7.89	0.38	18.99	1.740E-1	H
Evolutionary stage of valley fog									
Ground fog	1	1.3823E0	1	0.89	0.86	*1.28501*	509.99	5.141E-2	T&T
Ground fog	2	2.0338E-1	1	2.04	2.22	*0.09634*	309.79	5.146E-2	T&T
Formation stage	1	5.6434E-2	3	1.26	3.49	*0.49294*	138.04	7.548E-2	T&T
Formation stage	2	2.1192E-4	5	1.40	7.50	*0.21269*	39.32	1.325E-1	T&T
Formation stage	3	4.0633E-5	5	1.69	9.07	*0.07124*	29.26	1.504E-1	T&T
Formation stage	4	5.1199E-5	4	2.17	10.82	*0.0105*	21.73	1.738E-1	T&T
Mature fog	1	4.0001E-2	1	1.93	4.99	*0.02329*	59.02	1.236E-1	T&T
Mature fog	2	1.5416E-3	4	1.14	7.12	*0.37439*	35.67	1.484E-1	T&T
Mature fog	3	6.1487E-5	5	1.47	9.00	*0.13456*	28.00	1.565E-1	T&T
Mature fog	4	1.3900E-2	1	4.39	9.10	*1.40392E-5*	34.68	1.450E-1	T&T
Mature fog	5	1.4085E-3	2	3.41	11.18	*1.55987E-4*	22.08	1.754E-1	T&T
Dissipation stage	1	3.3437E-3	4	0.94	6.92	*0.69061*	32.51	1.623E-1	T&T
Dissipation stage	2	8.0977E-5	4	2.00	10.10	*0.01961*	24.14	1.662E-1	T&T

T&T is taken from Tomasi and Tampieri (1976), H from Harris (1995). N_t is the total number of drops per unit volume, r_c the mode radius and LWC the liquid water content. Numbers in italic are calculated by the radiative transfer code

It should be stressed that the selected fog types follow the classification of Tampieri and Tomasi, which were referred to by EADIE *et al.* (1971). The life cycle stages of fog are taken from Pilié's fog life cycle (PILIÉ *et al.* 1972, 1975a, b) and are: (a) ground fog, (b) formation stage, (c) mature fog and (d) dissipation stage.

In their bulk fog study, PILIÉ *et al.* (1975a, b) found typical patterns of fog evolution in the Chemung River Valley, which can be accepted as generally valid. Ordinarily, preliminary plane ground fog (defined by visibility of less than 4,000 m) appears before a deep fog event. It is characterized by a large total number of droplets ($N_t = 100$–200 cm^{-3}) with a low mean radius (2 μm $< r <$ 4 μm), which results in a positive skewness of the drop size spectrum. In the so-called formation stage the visibility decreases as well as total number concentration down to 2 cm^{-3}. Therefore, the mean radius rises up to 6–12 μm, as

does the LWC. The curve of the drop size distribution is typically almost parallel to the x axis, which stretches from 3 to about 25 μm. The very small droplets disappear during this stage. The visibility minimum after the first quarter of the fog life cycle delineates the formation stage of the mature fog stage. Its features are the reappearance of very small droplets (2 μm $< r < $ 4 μm) and an enlargement of both the total number concentration (12 cm$^{-3} < N_t < 25$ cm^{-3}) and the LWC to between 0.05 and 0.15 mg m^{-3}. Both fluctuate synchronously with visibility. Consequently, a bimodal drop size distribution can be recorded with a mode radius from 2 to 3 μm and a mode from 6 to 12 μm. As fog thins out to a visibility above 1,000 m after three-fourths of the life cycle, the visibility rises continuously. This last evolutionary step is called the dissipation stage. Here, LWC, mean radius and total number concentration decrease continuously. However, the fog does not always disperse equally. In three out of eight examples measured, the drop size distribution could be described by a bimodal curve with a mode radius of 2–3 µm and one with 6–12 µm, another three out of eight measurements showed a curve with positive skewness. In the last two measurements the drop sizes seemed to be distributed randomly.

As is shown in Table 1, α is the only integer number and is in the range of 1–6. The other parameters of the modified gamma distribution are real numbers. The values of b are between 0.05351 and 3.36586. The minimum of γ is 0.7 and the maximum 2.17. The normalization factor a of the modified gamma distribution does not have to be taken into account, because it does not have an influence on the shape of the distribution but only on the slope. The mode radius r_c lies in the region between 0.86 and 12.22 µm. The total number of drops per unit volume of air is greater than 6.5 but smaller than 509.99 $\mu m^{-1}\ cm^{-3}$. The values of the measured LWC range from 0.016 to 0.37 g m^{-3}.

The values presented in Table 1 are taken as minimum and maximum boundaries for the sensitivity study. The parameters were iterated between the minimum and maximum values with the RTC, and the resulting Z and LWC were calculated. The increment for α is 1, for γ and Z 0.1 and for b 0.05. All different cases of the modified gamma distribution which could be taken into account were considered in four intertwined loops. The limits for Z were −25 and −85 dBz, which corresponds to the dynamic range of the radar.

The first parameter calculated by the RTC was the normalization factor a (g mm^{-6}) of the modified gamma distribution by resolving Eq. 19 for:

$$a = \frac{10^{\frac{z}{10}} \cdot \gamma \cdot 10^{12}}{b^{-\left(\frac{\alpha+7}{\gamma}\right)} \Gamma\left(\frac{\alpha+7}{\gamma}\right)} \quad (27)$$

Afterwards, the mode radius r_c was calculated by means of Eq. 14. The total number of drops per unit volume of air N_t was computed by Eq. 9. The LWC was determined by Eq. 24.

In order to investigate different fog types in regards to their specific influence on the relationship between Z and LWC, the fogs were classified into

two main groups, radiation and advection fog, after having been calculated by the RTC.

Furthermore, the calculated dataset was subdivided into four different stages of fog evolution according to the life cycles of valley fog identified by PILIÉ et al. (1972, 1975a, b).

This was done to investigate the relevance of the fog life cycle for the Z–LWC relationship. The resulting parameter sets are shown in Table 2.

For a better comparability of the parameter sets, the normalization factor a was recalculated for a probability function equal to 1. As Tomasi's and Tampieri's parameter sets were expressed in terms of r_c, the corresponding b had to be calculated by the usage of Eq. 14. The same equation was used for calculating r_c for the given parameter b in Harris' fog parameter sets.

3. Results

Figure 1 illustrates the absolute frequency distribution of fog traits as extracted from the literature, related to radar reflectivity Z. The absolute number on the y axis is the frequency in terms of the parameter of the x axis. The population of parameter sets from the RTC equals 190988. It can be shown that fog is generally characterized by small LWCs with a mode at 0.016 g m^{-3} and a tendency to smaller droplet number concentrations. Also, the mode radius is small, mostly below 6 µm. The parameter of the modified gamma distribution reveal a wide range of values, where b and γ show a mode at lower values (0.2 or 0.8), mainly oriented on the shape of the LWC and mode radius histograms, while the parameter α is monotonously increased in frequency with increasing value, inversely shaped to the histogram of number concentration. The consequences for the theoretically expectable radar reflectivity point to a bell-shaped histogram with a mode at −52.2 dBz and very low values of the Z–LWC relation (mode at 4231.18295).

To derive a Z–LWC relation which can be used for the cloud radar, it is necessary to know how sensitive the relation is to variations in the three parameters. In order to address this uncertainty, the Z–LWC relationship factor Ω was plotted as a function of the three parameters of the modified gamma

Table 2

Ranges of the characteristic coefficients of fog. Fog events (all fogs) are separated by type (radiation fog, advection fog and valley fog) and by time (ground fog, formation stage, mature fog and dissipation stage)

	All fog events	Radiation fog	Advection fog	Valley fog	Ground fog	Formation stage	Mature fog	Dissipation stage
α_{min}	1	4	3	1	1	3	1	4
α_{max}	6	6	6	5	1	5	5	4
b_{min}	0.05351	0.05351	0.18852	0.0105	0.09634	0.0105	0.02329	0.01961
b_{max}	3.36586	3.36586	1.99163	1.28501	1.28501	0.49294	0.37439	0.69061
γ_{min}	0.7	0.7	0.85	0.89	0.89	1.26	1.14	0.94
γ_{max}	2.17	1.77	1.47	2.17	2.04	2.17	1.93	2
$r_{c\ min}$ (µm)	0.86	2.13	2.75	0.86	0.86	3.49	4.99	6.92
$r_{c\ max}$ (µm)	12.22	12.22	10	10.82	2.22	10.82	9.00	10.1
$N_{t\ min}$ (cm^{-3})	6.5	15.9	20	21.73	309.79	21.73	28	24.14
$N_{t\ max}$ (cm^{-3})	509.99	249.93	184.55	509.99	509.99	138.04	59.02	32.51
LWC$_{min}$ (g m^{-3})	0.016	0.016	0.06795	0.05141	0.05141	0.07548	0.1236	0.1623
LWC$_{max}$ (g m^{-3})	0.37	0.2061	0.1537	0.1754	0.05146	0.1738	0.1565	0.1662
Counts	190,988	60,456	8,704	26,022	774	3,226	1,200	84

Differentiators are the three parameters of the drop size spectrum (α, b, γ), the mode radius, the total numbers of drops per unit volume and the LWC. The counts of fitting parameter sets are also itemized as well

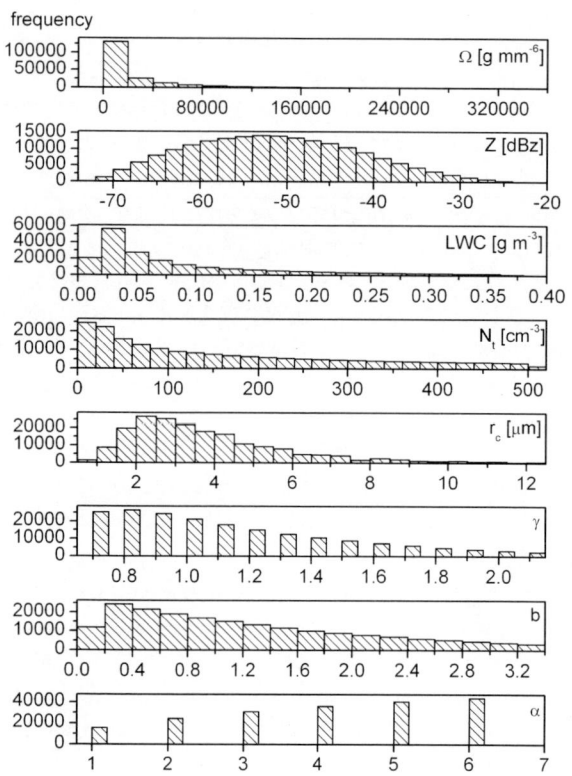

Figure 1
Histogram of the characteristics for all (population = 190,988) computed fog types in the sensitivity study

distribution in Fig. 2. Maximum values of Ω (3.2E + 05 g mm^{-6}) can only be found for high values of both α and γ accompanied by smaller values of b. Minimum values of Ω (0 g mm^{-6}) occur in the entire range of α together with smaller values of γ almost independently of b. Ω remains fairly constant along the b and α axis at constant γ values. For fixed b values, Ω grows with increasing γ. Assuming fixed α, high values of γ and b result in high values of Ω, whereas high values of γ have priority. In total, the impact of α on Ω is higher than the impact of γ and b, as a maximum value of alpha is a necessary condition for a maximum value of Ω. Further, γ has priority to b as Ω grows with rising γ at constant values for α and b, whereas Ω stays nearly constant for fixed values for α and γ and variable b.

A second question is if the fog type has any specific impact on the Z–LWC relation when the main fog types considered are radiation and advection fog. To answer this question, Z and LWC for all simulations are plotted for both fog types, whereby more information is available (Table 2; Fig. 3) for radiation fog. There are 60,456 calculated parameter sets for radiation and 8,704 for advection fog. It should be emphasized that the denser the plotted surface in Fig. 3 is, the more Z–LWC combinations lie in this range. The range of the Z–LWC-interaction factor Ω is according to Table 3 almost five times higher for radiation fog (142584.59338) than for advection fog (21110.49235), whereby the last range lies completely in the first one. For radiation fog it ranges from 253.43537 to 142838.02875 g mm^{-6}

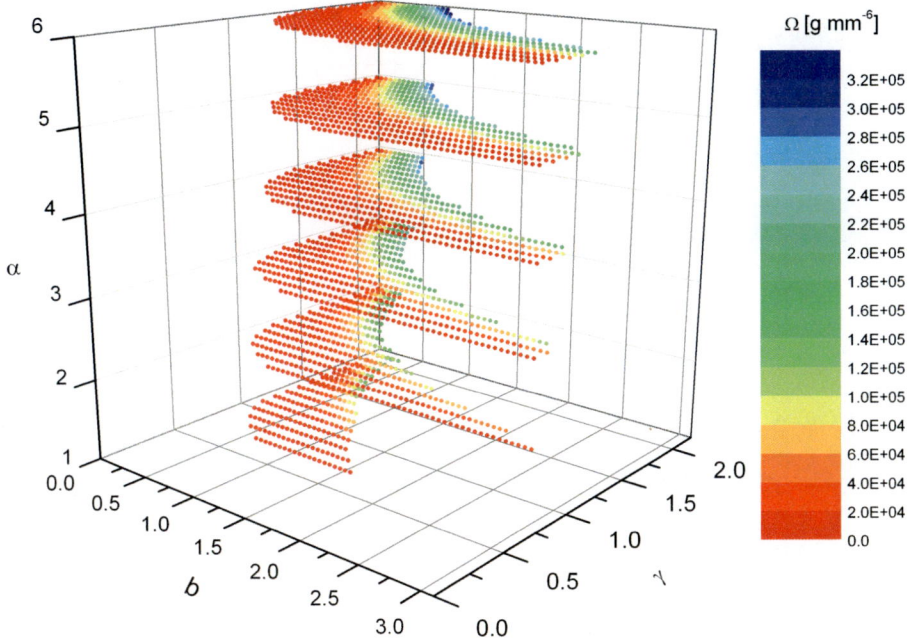

Figure 2
Influence of the three parameters of the drop size spectrum γ, b, α on the LWC–Z relationship Ω for all computed fog types

and for advection fog from 521.57568 to 21632.06803 g mm^{-6}. It can be stated that the values of Ω are widely spread.

Figure 3 also reveals that the range of the radar reflectivity for radiation fog has a broader range of Z—between −69.5 to −30.9 dBz—in comparison with advection fog, with Z ranging from −55.0 to −35.4 dBz. The same holds for the LWC which shows

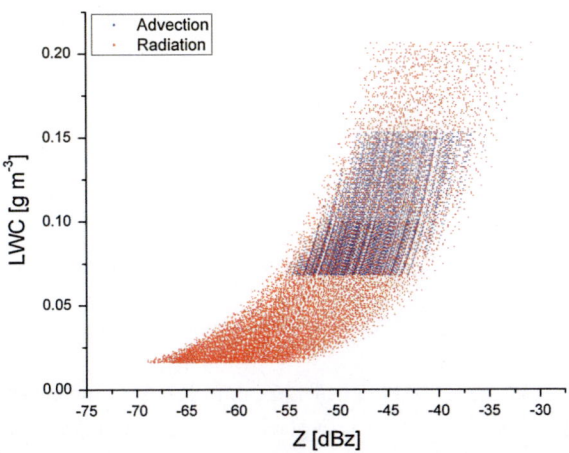

Figure 3
Derived values of LWC from radar reflectivity dependent on fog type. The plots for radiation fog were thinned out by a factor of 5

a broader range for radiation fog (0.016–0.206 g m^{-3}) than for advection fog (0.068–0.153 g m^{-3}). However, looking at the point density in Fig. 3 makes it clear that radiation fog is most frequently characterized by smaller LWC values <0.06 g m^{-3}, which is approximately the lower boundary of advection fog.

The last question is the relevance of the fog life cycle for the Z–LWC relation. Its impact on the Z–LWC interaction is presented in Fig. 4. Generally, the radar reflectivity shifts to higher values as a function of time in the course of the fog life cycle. With this, all four life cycle stages show a characteristic Z–LWC relation.

In the first three evolutionary stages (ground fog, formation stage, mature fog) the LWC values corresponding to a given radar reflectivity are spread over a wide range. Particularly in the formation stage, no unique LWC values can be derived from radar reflectivity. However, the dissipation stage features a narrow range of LWC values corresponding to a given radar reflectivity.

As can be seen in Table 3, the range of the Z–LWC relation factor Ω is initially very wide for a whole valley fog event (92954.42255). A separation into the four characteristic states of a fog event results

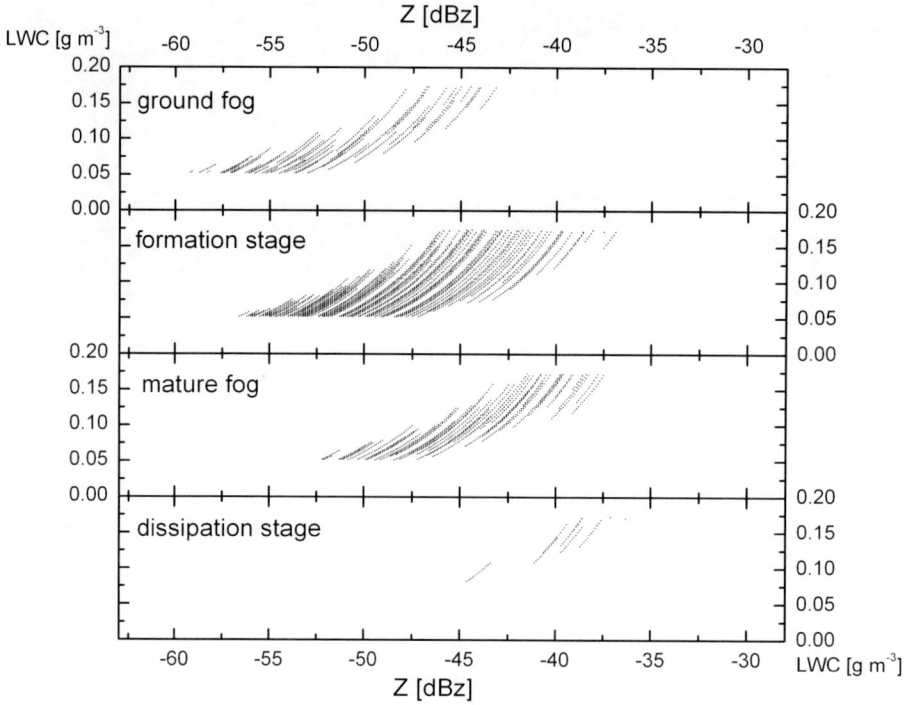

Figure 4
Derived values of the LWC from Z relative to the evolutionary stage of fog. The temporal stages are listed in chronological order

Table 3

Characteristic numbers (minimum, maximum, mode and range) of the Z–LWC relationship factor Ω depending on the type and evolutionary stage of fog

	Ω_{min} (g mm^{-6})	Ω_{max} (g mm^{-6})	Ω_{mode} (g mm^{-6})	Ω_{delta} (g mm^{-6})
All fog events	43.95566	336376.67707	4231.18295	336332.72141
Radiation fog	253.43537	142838.02875	4233.16494	142584.59338
Advection fog	521.57568	21632.06803	5248.0746	21110.49235
Valley fog	227.41854	93181.84109	6359.51148	92954.42255
Ground fog	3573.75354	43750.77436	8400.81409	40177.02082
Formation stage	836.98268	23594.8071	6214.75132	22757.82442
Mature fog	958.0055	8669.36576	2054.87482	7711.36026
Dissipation stage	725.18518	2359.59368	–	1634.4085

in smaller ranges of Ω. The Ω-intervals lessen with advancing time. While the Ω-range amounts to 40177.02082 during ground fog, the dissipation stage accounts only to 1634.4085.

The same applies to the mode of Ω, which decreases with ongoing fog development (Fig. 5).

The mode of Ω is 8400.81409 for ground fog, 6214.75132 for the formation stage and 2054.87482 for mature fog. There is no clear mode value for Ω for the dissipation stage. However, the histogram in Fig. 5 suggests a shift of its Ω-mode to relatively low

values in comparison to the Ω-modes in the preliminary stages.

4. Discussion

The development of a method for deriving LWC-profiles from radar reflectivity raised three questions.

The first one was how the drop size spectrum influences the relationship between radar reflectivity and LWC. Since there are three factors which

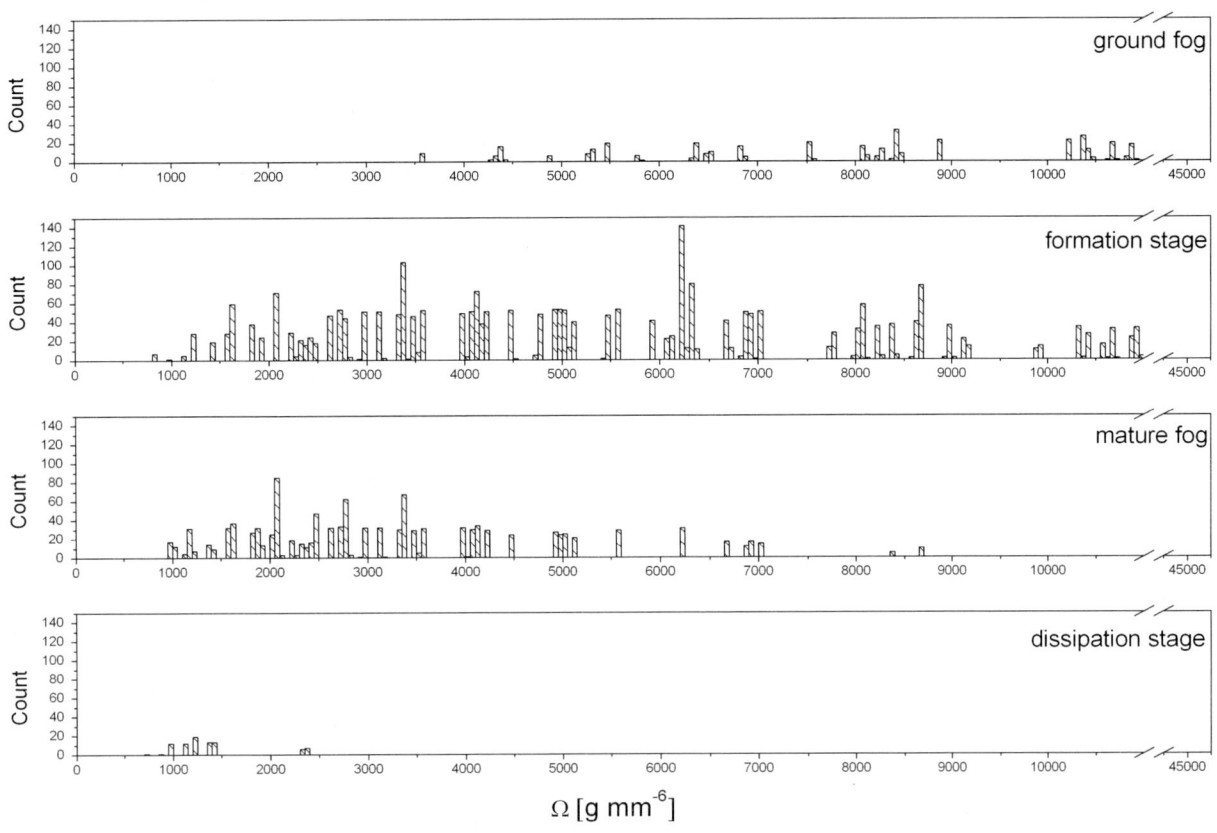

Figure 5
Histogram of Ω dependent on evolutionary stage of fog

determine its shape by means of the modified gamma distribution, their frequency was evaluated based on the measurement examples by Tomasi and Tampieri (1976). The first factor α showed a reverse behavior to the total number concentration N_t (Fig. 1). The two other factors b and γ determine, in concert with α, the mode radius of fog drops by Eq. 14. Thus, the radar reflectivity is not only influenced by the drop radius (represented by b and γ) but also by the total number concentration (represented by α). Figure 2 shows that both α and γ have a higher influence on the Z–LWC relationship than b because an individual change of their value resulted in a higher change of Ω compared to b. It can be observed that both α and γ have an important role on the Z–LWC interaction in relation to the mode radius r_c as computed by Eq. 14.

The second question was if the fog type influences the Z–LWC-interaction. Table 3 and Fig. 3 indicate not only that the range of the Z–LWC interaction factor Ω is much larger for radiation fog than for

advection fog, but also that the Ω values of advection fog lie completely in the range of radiation fog. This behavior is related to the total number of droplets and the mode radius, which can be higher in radiation fog (Table 2). Unfortunately, the intersection of the Ω values of both fog types does not allow a strict classification of fog specific Ω values. Although Fig. 3 reveals a strong accommodation of Ω values for each fog type, advection fog has a strong concentration of Ω values from about -55 to -43 dBz. This results in a LWC-range of 0.06795–0.1537 g m^3. Radiation fog shows a strong concentration of Ω values from -68 to -50 dBz, which results in a LWC-range from 0.016 to 0.2061 g m^{-3}. Comparing these ranges makes it apparent that the type of fog plays a very important role on the Z–LWC interaction.

The third question concerns the relevance of the fog life cycle for the Z–LWC relationship. To investigate this question, the parameter sets of valley fog were classified by valley fog's four development stages (Table 2). Regarding the fog event as a whole a wide range of 92954.4225 g mm^{-6} was found for the relationship factor Ω (Table 3). Evidently it is difficult to derive certain LWC-values from a given radar reflectivity without further specifications. Distinguishing the stage specific parameter sets from Table 2 for valley fog delivered narrower Ω ranges from Eq. 26. The Ω-interval decreases continuously from 40177.02082 for ground fog to 1634.4085 for the dissipation stage (Table 3). As can be seen in Fig. 4, the LWC–Z relationship could be specified for all four fog development stages, although the method is still in need of further improvement.

The radar reflectivity increases during the fog life cycle (Fig. 4). During ground fog the Z varies from -59.2 to -43.2 dBz. In the formation stage, it varies from -56.6 to -36.9 dBz. While Z ranges from -52.2 to -37.5 dBz during the mature stage, it varies from -44.6 to -36.3 dBz in the dissipation stage. A possible cause for the shift of Z to higher values with ongoing fog development might be the increasing drop radii. According to Eq. 18, bigger radii result in higher radar reflectivity. According to Eq. 23, bigger radii also result in higher LWC values. As the LWC is proportional to the drop size spectrum (a function of the drop radius) by the third moment and the radar reflectivity is proportional to DSD by the sixth moment, an increase of the radius results in smaller Ω values being the quotient, with LWC as the numerator and Z as the divisor. This also coincides with Table 3, where the mode of Ω decreases with advancing fog life cycle from 8400.81409 for ground fog to 2054.87482 for mature fog.

From the increasing radar reflectivity and the decreasing proportionality factor Ω with fog development, it can be concluded that the LWC stays constant or grows as well. This matches with Pilié's observations (PILIÉ *et al.* 1975a, b) of increasing LWC during fog events.

The narrower ranges of Ω in Table 3 for the consecutive fog stages imply clearer Z–LWC ratios as both factors determine the quotient Ω.

With regard to the three questions, there is a direct link between radar reflectivity and LWC even though it is not a direct one. For a more precise relationship between these two key figures, a fog differentiation by type and evolutionary stage in its life cycle promises good improvements.

5. Conclusion

The main purpose of the work was to show if it is possible to derive LWC-profiles from measured radar reflectivity of a new 94 GHz FMCW cloud radar. In order to answer this question, a radiative transfer model was developed considering the radar reflectivity as well as the LWC-profiles. Since both factors depend on the drop size spectrum, its influence on both factors was investigated. It could be shown that the parameters of the modified gamma distribution have a strong effect on their relationship, expressed by the proportionality factor Ω. Hence, the parameters of the modified gamma distribution (α, γ, b) were analyzed as a function of the fog type and of the fog life cycle. Existing parameter sets of α, γ and b were taken from the relevant literature. The eligible parameter sets were separated into characteristic value ranges corresponding to the respective fog evolutionary stages. Next, new parameter sets for the drop size spectrum within these ranges were recalculated to obtain accommodated sets for specific fog cases and for error minimization. Both the fog type and the classification by time permitted a reduction of the contemplable parameter sets which fell within a narrower range of the Z–LWC-relation factor Ω. The results indicate a direct but nonlinear relation between LWC and radar reflectivity. Particularly, the Z–LWC-relation factor Ω reveals characteristic ranges for the different life cycle stages. If a proper classification of the respective development stages in the field can be accomplished, it should be possible to apply appropriate Z–LWC relationships to calculate the LWC from the radar reflectivity for the respective fog life cycle stage. Microwave cloud radar profilers with the frequency modulated continuous wave technique are very well suited for monitoring fog and low stratus clouds. Because of the novelty of these radar devices, no published Z–LWC-relationships for fog and low stratus relying on direct radar and in situ measurements are available. GULTEPE *et al.* (2009) derived Z-values from LWC- and r_{eff} values that were measured with a fog measuring device during the

FRAM-project. In their study they concluded that it is possible to obtain LWC and visibility from Z if r_{eff} is assumed to be constant. However, assuming a constant r_{eff} might not coincide with reality for most fog events. The results of this study indicate that r_{eff} cannot be taken as a constant (Fig. 1). According to GULTEPE et al. (2009), the main reason for discontinuous r_{eff} values as well as varying Z–LWC-relationships is the large variability in fog drop size spectra.

Generally, the results imply that satellite and model diagnosing and forecasting applications relying on correct LWC profiles should be generally adapted to the life cycle stage of fog and fog type. Most fog diagnosing and nowcasting applications based on satellite information are reliant on passive instruments onboard operational polar-orbiting and geostationary weather satellite imagery (GULTEPE et al. 2007). Since radiation in the wavebands of these instruments does not normally completely penetrate opaque cloud layers, it is only possible to observe the upper part of a potential fog layer, making the distinction between low stratus and ground fog a difficult task. As stated in the introduction, novel approaches were recently developed to overcome this problem (BENDIX et al. 2005, CERMAK and BENDIX 2010). The major initialization parameter of these approaches is the LWC profile in a pixel. Because of the two-dimensional nature of the images, only the columnar liquid water path (LWP) can be retrieved (NAUß et al. 2005), which is not bijectively related to the LWC value or its profile. Different LWC profiles can occur, depending on LWP and geometrical depth of the fog layer. Because of hitherto lacking information, it is assumed in the current ground fog detection schemes that LWP can be related to the respective LWC by a sub-adiabatic model of cloud microphysics (CERMAK and BENDIX 2010). It is obvious that this simplification must impose inaccuracies in the processing. A comprehensive investigation of the relation between LWC profiles and integrated LWP under different weather situations using the cloud radar-based Z–LWC relationships will be a clear step forward for a realistic initialization of the process. However, this will also require a proper estimate of fog type and life cycle in the satellite images or by ancillary data (e.g. time of day). With

regard to ground fog forecast models, proper LWC profiles are important for validation and model improvements (GULTEPE and MILBRANDT 2007). Most models are initialized with standard meteorological parameters with poor vertical resolution, especially close to the ground (e.g. temperature, humidity etc.). Some studies show that the correct prediction of LWC as dependent on the vertical resolution of the models is hardly possible, thus (negatively) influencing important forecast parameters like ground visibility (TARDIF 2007). Continuously measured LWC profiles based on the Z–LWC relation could particularly improve 1D model results by initializing with LWC profiles with a high vertical resolution. This could also alleviate the common problem of fog modeling that reliable vertical humidity profiles used for data assimilation are often not available (MÜLLER et al. 2007). With regard to straightforward, partly statistical fog models, it is obvious that those which rely directly on columnar LWC data (as e.g. REUDENBACH 1998) will benefit from cloud radar-based retrievals of LWC profiles.

Acknowledgments

The authors thank the University of Marburg for granting a PhD scholarship for F. Maier (03-06-2010) and the German Research Foundation DFG for funding the project (BE1780/14-1; TH1531/1-1). The work is also part of the COST action EG Climet.

REFERENCES

AWAN, M.S., LEITGELB, E., MUHAMMAD, S.S., MARZUKI, F.N., KHAN, M.S., and CAPSONI, C. (2008), *Distribution function for continental and maritime fog for environments for optical wireless communication*, doi:10.1109/CSNDSP.2008.4610728, 260–264.

BENDIX, J., THIES, B., CERMAK, J., and NAUSS, T. (2005), *Ground fog detection from space based on MODIS daytime data—a feasibility study*, Weather and Forecasting *20*, 989–1005.

BENNETT, A.J., GAFFARD, C., OAKLEY, T., and MOYNA, B.P. (2009), Cloud radar—initial measurements from the 94 GHz FMCW cloud radar, Proceedings of the 8th International Symposium on Tropospheric Profiling 19–23 October, Delft.

BENNETT, A., NASH, J., GAFFARD, C., MOYNA, B., OLDFIELD, M., HUGGARD, P., AND WRENCH, C. (2009), Observations from the UK Met Office 94 GHz FMCW cloud radar, Proceedings of the 8th International Symposium on Tropospheric Profiling 19–23 October, Delft.

Reprinted from the journal

BOERS, R., RUSSCHENBERG, H., ERKELENS, J., VENEMA, V., VAN LAMMEREN, A., APITULE, A., PITULEY, A., and JONGEN, S. (2000), *Ground-based remote sensing of stratocumulus properties during CLARA, 1996*, Journal of Applied Meteorology *39*, 169–181.

CERMAK, J., SCHNEEBELI, M., NOWAK, D., VUILLEUMIER, L., and BENDIX, J. (2006), *Characterization of low clouds with satellite and ground-based remote sensing systems*, Meteorologische Zeitschrift *15*, 65–72.

CERMAK, J., and BENDIX, J. (2010), *Detecting ground fog from space—a microphysics-based approach*, International Journal of Remote Sensing, in press.

CLOTHIAUX, E.E., MILLER, M.A., ALBRECHT, B.A., ACKERMANN, T.P., VERLINDE, J., BABB, D.M., PETERS, R.M., and SYRETT, W.J. (1995), *An evaluation of a 94 GHz radar for remote sensing of cloud properties*, Journal of Atmospheric and Oceanic Technology *12*, 201–229.

DANNE, O. (1996), Messungen physikalischer Eigenschaften mit stratiformer Bewölkung mit einem 94 GHz-Wolkenradar, Dissertation, University of Hannover.

DEIRMENDJIAN, D. (1964), *Scattering + polarization properties of water clouds + hazes in the visible + infrared*, Applied Optics *3*, 187–196.

DEIRMENDJIAN, D., Electromagnetic scattering on spherical polydispersions (Elsevier Scientific Publishing, New York 1969).

DONOVAN, D.P., and VAN LAMMEREN, A.C.A.P. (2001), *Cloud effective particle size and water content profile retrievals using combined lidar and radar observations 1. Theory and examples*, Journal of Geophysical Research-Atmospheres *106*, 27425–27448.

EADIE, W.J., KOCMOND, W.C., LEONARD, R.P., MACK, E.J., and PILIÉ, R.J. Investigation of warm fog properties and fog modification concepts (NASA Contractor Report 1731, Washington D.C. 1971).

FLATAU, P.J., TRIPOLI, G.J., VERLINDE, J., and COTTON, W.R. (1989), *The CSU-RAMS cloud microphysics module: general theory and code documentation, Colorado State University*, Atmospheric Science Paper *45*.

FRISCH, A.S., FAIRALL, C.W., and SNIDER, J.B. (1995), *Measurement of stratus cloud and drizzle parameters in ASTEX with a K-alpha-band Doppler radar and a microwave radiometer*, Journal of the Atmospheric Sciences, *52*, 2788–2799.

FRISCH, A.S., FAIRALL, C.W., FEINGOLD, G., UTAL, T., and SNIDER, J.B. (1998), *On cloud radar microwave radiometer measurements of stratus cloud liquid water profiles*, Journal of Geophysical Research *103*, 23195–23197.

FOX, N.I., and ILLINGWORTH, A.J. (1997), *The retrieval of stratocumulus cloud properties by ground-based cloud radar*, Journal of Applied Meteorology *36*, 485–492.

GRADSHTEYN, I.S., and RYZHIK, I.M., Table of integrals, series, and products (Academic Press Inc, New York 1980).

GULTEPE, I., MULLER, M.D., and BOYBEYI, Z. (2006), *A new visibility parameterization for warm-fog applications in numerical weather prediction models*, Journal of Applied Meteorology and Climatology *45*, 1469–1480.

GULTEPE, I., and MILBRANDT, J.A. (2007), *Microphysical observations and mesoscale model simulation of a warm fog case during FRAM project*. Pure and Applied Geophysics *164*, 1161–1178.

GULTEPE, I., TARDIF, R., MICHAELIDES, S.C., CERMAK, J., BOTT, A., BENDIX, J., MUELLER, M.D., PAGOWSKI, M., HANSEN, B., ELLROD, G., JACOBS, W., TOTH, G., and COBER, S.G. (2007), *Fog reasearch: A review of past achievements and future perspectives*, Pure and Applied Geophysics *164*, 1121–1159.

GULTEPE, I., PEARSON, G., MILBRANDT, J.A., HANSEN, B., PLATNICK, S., TAYLOR, P., GORDON, M., OAKLEY, J.P., and COBER, S.G. (2009), *The fog remote sensing and modeling field project*, Bulletin of the American Meteorological Society *90*, 341–359.

HAYNES, J.M., MARCHAND, R.T., LUO, Z., BODAS-SALCEDO, A., and STEPHENS, G.L. (2007), *A multipurpose radar simulation package: QuickBeam*, Bulletin of the American Meteorological Society *88*, 1723–1727.

HARRIS, D. (1995), *The attenuation of electromagnetic-waves due to atmospheric fog*, International Journal of Infrared and Millimeter Waves *16*, 1091–1108.

HESS, M., KOEPKE, P., and SCHULT, I. (1998), *Optical properties of aerosols and clouds: The software package OPAC*, Bulletin of the American Meteorological Society *5*, 831–844.

HUGGARD, P.G., OLDFIELD, M.L., MOYNA, B.P., ELLISON, B.N., MATHESON, D.N., BENNETT, A.J., GAFFARD, C., OAKLEY, T., and NASH, J. (2008), 94 GHz FMCW cloud radar, Proceedings of the SPIE symposium on millimetre wave and terahertz sensors and technology, 15–18 September 2008, Cardfiff, 6 pp.

JACOBS, W., NIETOSVAARA, V., BOTT, A. BENDIX J., CERMAK, J., MICHAELIDES, S., and GULTEPE, I. (eds) (2008), Short range forecasting methods of fog, visibility and low clouds, COST Action 722 final report, Brussels, Office for official publications of the European Communities, 489 pp.

KALASHNIKOVA, O.V., HEINZ, A.W., and MAYHEW, L.M. (2002), *Wavelength and altitude dependence of laser beam propagation in dense fog*, doi:10.1117/12.464103, 278–287.

KIM, I.I., MCARTHUR, B., and KOREVAAR, E. (2001), *Comparison of laser beam propagation at 785 nm and 1,550 nm in fog and haze for optical wireless communications*, doi:10.1117/12.417512, 12 pp.

KHAIN, A., PINSKY, M., MAGARITZ, L., KRASNOV, O., and RUSSCHENBERG, H.W.J. (2008), *Combined observational and model investigations of the Z–LWC relationship in stratocumulus clouds*, Journal of Applied Meteorology and Climatology *47*, 591–606.

KOLLIAS, P., CLOTHIAUX, E.E., MILLER, M.A., ALBRECHT, B.A., STEPHENS, G.L., and ACKERMAN, T.P. (2007), *Millimeter-wavelength radars—New frontier in atmospheric cloud and precipitation research*, Bulletin of the American Meteorological Society *88*, 1608–1624.

LIAO, L., and SASSEN, K. (1994), *Investigation of relationships between Ka-band radar reflectivity and ice and liquid water contents*, Atmospheric Research *34*, 231–248.

LÖHNERT, U., CREWELL, S., SIMMER, C., and MACKE, A. (2001), *Profiling cloud liquid water by combining active and passive microwave measurements with cloud model statistics*, Journal of Atmospheric and Oceanic Technology *18*, 1354–1366.

LÖHNERT, U., CREWELL, S., and SIMMER, C. (2004), *An integrated approach toward retrieving physically consistent profiles of temperature, humidity, and cloud liquid water*, Journal of Applied Meteorology *43*, 1295–1307.

MCFARLANE, S.A., EVANS, K.F., and ACKERMAN, A.S. (2002), *A Bayesian algorithm for the retrieval of liquid water cloud properties from microwave radiometer and millimeter radar data*, Journal of Geophysical Research-Atmospheres *107*, 4317–4338.

MEYER, M.B., LALA, G.G., and JIUSTO, J.E. (1986), *FOG-82–A cooperative field study of radiation fog*, Bulletin of the American Meteorological Society *67*, 825–832.

MÜLLER, M.D., SCHMUTZ, C., and PARLOW, E. (2007), *A one-dimensional ensemble forecast and assimilation system for fog prediction*, Pure and Applied Geophysics *164*, 1241–1264.

NAUß, T., KOKHANOVSKY, A., NAKAJIMA T.Y., REUDENBACH, C., and BENDIX, J. (2005), *The intercomparison of selected cloud retrieval algorithms*, Atmospheric Research *78*, 46–78.

NOWAK, D., RUFFIEUX, D., AGNEW, J.L., and VUILLEUMIER, L. (2008), *Detection of fog and low cloud boundaries with ground-based remote sensing systems*, Journal of Atmospheric and Oceanic Technology *25*, 1357–1368.

PAGOWSKI, M., GULTEPE, I., and KING, P. (2004), *Analysis and modeling of an extremely dense fog event in southern Ontario*, Journal of Applied Meteorology *43*, 3–16.

PETTERSSEN, S., Weather analysis and forecasting (McGraw-Hill, New York 1956).

PILIÉ, R.J., EADIE, W., MACK, E.J., ROGERS, C., and KOCMOND, W.C. (1972), Project Fog Drops. Part 1: Investigations of warm fog properties, Washington D.C., NASA Contractor Report 2078, 149 pp.

PILIÉ, R. J., MACK, E. J., KOCMOND, W. C., ROGERS, C. W., and EADIE, W.J. (1975a), *The life-cycle of valley fog. 1. Micrometeorological characteristics*, Journal of Applied Meteorology *14*, 347–363.

PILIÉ, R. J., MACK, E. J., KOCMOND, W. C., EADIE, W. J., and ROGERS, C. W. (1975b), *The life-cycle of valley fog. 2. Fog microphysics*, Journal of Applied Meteorology *14*, 364–374.

REUDENBACH, C., and BENDIX J. (1998), *Experiments with a straightforward model for the spatial forecast of fog/low stratus clearance based on multi-source data*, Meteorological Applications *5*, 205–216.

RUFFIEUX, D. NASH, J., JEANNET, P., and AGNEW, J.L. (2006), *The COST 720 temperature, humidity, and cloud profiling campaign: TUC*, Meteorologische Zeitschrift *15*, 5–10.

SAUVAGEOT, H., and OMAR, J. (1987), *Radar reflectivity of cumulus clouds*, Journal of Atmospheric and Oceanic Technology *4*, 264–272.

SCHULZE-NEUHOFF, H. (1976), *Nebelfeinanalyse mittels zusätzlicher 420 Klimastationen*, Meteorologische Rundschau *29*, 75–84.

SHETTLE, E.P., and FENN, R.W. (1979), *Models for the aerosols of the lower atmosphere and the effects of humidity variations on their optical properties*, Environmental Research Paper 676.

SIEBERT, J., BOTT, A., and ZDUNKOWSKI, W. (1992a), *Influence of a vegetation-soil model on the simulation of radiation fog*, Contributions to Atmospheric Physics/Beiträge zur Physik der Atmosphäre *65*, 93–106.

SIEBERT, J., BOTT, A., and ZDUNKOWSKI, W. (1992b), *A one-dimensional simulation of the interaction between land surface processes and the atmosphere*, Boundary-Layer Meteorology *59*, 1–34.

TAMPIERI, F., and TOMASI, C. (1976), *Size distribution models of fog and cloud droplets in terms of the modified gamma function*, Tellus *28*, 333–347.

TARDIF, R. (2007), *The impact of vertical resolution in the explicit numerical forecasting of radiation fog. A case study*. Pure and Applied Geophysics *164*, 1221–1240.

TERRADELLAS, E., and BERGOT, T. (2007), *Comparison between two single-column models designed for short-term fog and low-clouds forecasting*. Física de la Tierra, *19*, 189–203.

TOMASI, C., and TAMPIERI, F. (1976), *Features of the proportionality coefficient in the relationship between visibility and liquid water content in haze and fog*, Atmosphere *14*, 61–76.

WELCH, R.M., and RAVICHANDRAN, M.G. (1986), *Prediction of quasi-periodic oscillations in radiation fogs. 1. Comparison of simple similarity approaches*, Journal of the Atmospheric Sciences *43*, 633–651.

VON GLASOW, R. and BOTT, A. (1999), *Interaction of radiation fog with tall vegetation*, Atmospheric Environment *33*, 1333–1346.

WENDISCH, M., MERTES, S., HEINTZENBERG, J., WIEDENSOHLER, A., SCHELL, D., WOBROCK, W., FRANK, G., MARTINSSON, B.G., FUZZI, S., ORSI, G., KOS, G., and BERNER, A. (1998), *Drop size distribution and LWC in Po Valley fog*, Contributions to Atmospheric Physics/Beiträge zur Physik der Atmosphäre *71*, 87–100.

WMO, International Meteorological Vocabulary 182 (WMO, Geneva 1992).

(Received November 18, 2010, revised March 7, 2011, accepted April 9, 2011, Published online May 19, 2011)

Pure Appl. Geophys. 169 (2012), 809–819
© 2011 Springer Basel AG
DOI 10.1007/s00024-011-0343-x

Summary of a 4-Year Fog Field Study in Northern Nanjing, Part 1: Fog Boundary Layer

D. Y. Liu,[1] S. J. Niu,[1] J. Yang,[1] L. J. Zhao,[1] J. J. Lü,[1] and C. S. Lu[1]

Abstract—Comprehensive fog field observations were conducted during the winters of 2006–2009 at the Nanjing University of Information Science and Technology to study the macro and micro-physical structures and the physical–chemical processes of dense fogs in the area. The observations included features of the fog boundary layer, characteristics of fog water, the particle spectrum, the chemical composition of atmospheric aerosols, radiation and heat components, turbulence, meteorological elements (air temperature, pressure, wind speed, wind direction), and environmental monitoring. The fogs observed were divided into four types: radiation fog, advection–radiation fog, advection fog, and precipitation fog, according to the mechanisms and primary factors of the fog processes. Fog boundary-layer structures of different types and their corresponding characteristics were then studied. Fog boundary-layer features, temperature structures, wind fields, and fog maintenance are discussed. The results show that radiation fog had remarkable diurnal variation and formed mostly at sunset or midnight, and lifted after sunrise or at noon, and that advection–radiation fog and advection fog were of very long duration. Extremely dense fogs occurred only in radiation-related cases. Inversion in radiation fog was short-lived, disappearing 1 or 2 hours after sunrise or at noon, faster than that in advection–radiation fog. When wind direction reversed from easterly to westerly or from southerly to northerly, the fog became an extremely dense fog. Low-level jet at times impeded fog development, whereas at other times it encouraged fog continuance. The deep inversion was merely an essential condition for a thick fog layer; sufficient vapor supply was advantageous to the formation and maintenance of a deep fog layer.

Key words: Fog observation, fog boundary layer, inversion, dense fog, low-level jet.

1. Introduction

The effect of fog on human life was recognized long ago, but its effect on society has significantly

increased in recent decades because of increasing air, marine, and road traffic. The total economic loss related to fog is comparable with that for tornadoes, even comparable to that for hurricanes or winter storms in some situations (Gultepe et al., 2007, 2009). Fog can form under specific weather background conditions and is also related to local conditions, for example terrain and ecological environments, so fog structures and evolution can differ greatly.

Observations of fog boundary-layer structure have been undertaken in many field campaigns to study different kinds of fog (Meyer 1986; Fitzjarrald and Lala 1989; Huang et al., 1992; Fuzzi et al., 1992, 1998; Zhou et al., 2007; Van Der Velde and Steeneveld, 2010). Early studies revealed that the balance between radiative cooling and turbulent mixing was the primary factor in radiation fog development (Roach et al., 1976; Nakanishi 2000; Terradellas et al., 2008). Some studies focused on the radiation fogs formed in valleys, and showed that the mountain wind, urban heat island, and water surface were important in the life cycle of valley fogs (Pilie et al., 1979).

Long fog duration (Huang et al., 1998) and deep fog layers (Holets and Swanson, 1981) have been studied by many authors. Taylor (1917) and Pilie et al. (1979) noted that cooling from below cannot produce deep fog layers. Rodhe (1962) suggested two or more processes, including turbulence and radiation, are usually involved in the formation of deep fog. Tomine et al. (1991) gave six examples confirming that rapid fog thickening is accompanied by a rapid wind change.

Double fog-layer structures (Li et al., 1999; Pu et al., 2001) have been investigated in many field campaigns, to understand its formation mechanism. Fitzjarrald and Lala (1989) observed a two-layer

---
[1] Key Laboratory of Meteorological Disaster of Ministry of Education, School of Atmospheric Physics, Nanjing University of Information Science and Technology, Nanjing, China. E-mail: niusj@nuist.edu.cn

fog case in the Hudson River valley. In this case, the cool layer grew from 10 to 60 m and a deeper layer above reached 180 m. LI *et al.* (1999) and PU *et al.* (2001) found that a double fog layer was formed at the upper level and the ground; afterwards, the ground fog layer developed upward and the upper fog layer that was caused by vapor accumulation at the inversion base developed downward and upward; thereafter the two combined to form a very deep, dense fog.

Recently, field experiments have been conducted to investigate dynamic, thermodynamic, micro-physical, and radiative processes in Beijing, China (ZHANG *et al.,* 2005), in Canada (GULTEPE *et al.,* 2009), and in Paris, France (HAEFFELIN *et al.,* 2010). ZHANG *et al.* (2005) detected a signal foretelling potential strong disturbances in the lower boundary layer approximately 10 h before fog onset and observed a significant rise of both mean and disturbance kinetic energies. They considered this strong signal to be very meaningful in monitoring and predicting fog occurrence and development.

In an effort to gain insight into the mechanisms of triggering, formation, maintenance, and dissipation of fog and extremely dense fog, four field campaigns of the Fog Monitoring and Early Warning and Disaster Damage Assessment Study in the Yangtze River Delta (YRDFOG) project were conducted at the meteorological observation station of Nanjing University of Information Science and Technology (NUIST; 32°12′N and 118°42′E; 25 m above the sea level) in the winters of 2006–2009 (LU *et al.,* 2008, 2010a, b; PU *et al.,* 2008a, b; YAN *et al.,* 2009; LIU *et al.,* 2010, 2011a, b; NIU *et al.,* 2010a, b; LI *et al.,* 2011; YANG *et al.,* 2011). TONG *et al.* (2009) reported fog climatology of Nanjing which showed the seasonal characteristics with the maximum frequency in winter and autumn to be more than 60% of the annual sum, followed by spring. These in-situ observations include the structures of fog boundary layer (FBL; ZHOU and FERRIER, 2008) micro-physical data, turbulence, radiation, and thermal equilibrium components. The descriptions and discussions of these observations are presented in two parts: Part I, presented here, covers the investigations of fog boundary features, whereas Part II, to be presented in a separate paper,

focuses on investigation of the micro-physical characteristics of these observations.

The main objectives of this paper are to summarize the preliminary results of the FBL features in the YRDFOG field experiment project, namely:

1. to understand different FBL features of four types of fog;
2. to gain insight into mechanisms of long-lasting and deep fog layers; and
3. to understand explosive development mechanisms of fog layer thickness.

A brief introduction of instruments and data is given in the section "Observations and data". The "Results" section focuses on fog classification, fog maintenance, fog-layer features, temperature structures, and wind fields. This is followed by the "Discussion" and "Conclusions" sections.

2. *Observations and Data*

The observational site is located on flat ground with no tall buildings or tall trees within 200 m. The instruments included a fog-measuring device (FMD; FM-100), a visibility meter, an automatic weather station, acoustic Doppler radar, and a tether-sonde system (Table 1).

The tether-sonde system (DigiCORA III by Vaisala, Finland) was used to monitor the vertical distributions of temperature, pressure, humidity, and wind in the atmosphere (LUI *et al.,* 2011a). Precision of temperature, relative humidity (RH), air pressure, wind speed, and wind direction sensors was 0.1°C, 0.1%, 0.1 h Pa, 0.1 m/s, and 1°, respectively. The acoustic radar also detected 3D boundary-layer winds, with an effective probed height reaching ~ 500 m above ground and a wind profiler that provided a complete set of wind directions/velocities at 15-min intervals (LUI *et al.,* 2011a). The droplet measurement technology (DMT) FMD (FM-100) continuously measured droplet number concentration, fog droplet spectrum, liquid water content (LWC), and N_d, at 1-Hz sampling rate with a diameter within 2–50 μm and maximum number concentration limit $10^4/cm^3$ (LUI *et al.,* 2011a). The ground weather elements and visibility were collected at 1-min

Table 1

List of instruments used during the YRDFOG project

Instrument	Measurement	Resolution time	Precision
Tether-sonde system (DigiCORA III; Vaisala, Finland)	T, P, RH, wind profiles	1–3 s	T, 0.1°C; P, 0.1 hPa; RH, 0.1%; wind, 0.1 m/s
Visibility meter (ZQZ-DN; Jiangsu Province Radio Science Institute, China)	Visibility	1 min	10%, <1,000 m; 20%, >1,000 m
Fog-measuring device (FM-100; DMT, USA)	Fog droplet spectrum, LWC, N_d	1 s	Range 2–50 μm
Ultrasonic anemometer (CSAT3; Campbell, USA)	3D wind velocity, turbulence, and ultrasonic virtual temperature	0.1 s	u, v, 1 mm s^{-1}; w, 0.5 mm s^{-1}; UVT, 0.002°C
Temperature and RH probe (HMP45C; Campbell, USA)	T, RH	15 s	T, 0.2–0.5 ±2%; RH (0–90% RH) ±3%; RH (90–100% RH)
Heat flux transducer (HFP01-L; Campbell, USA)	Heat flux	<18 s	Within −15 to +5% in most common soils
Water content reflectometers (S616-L; Campbell, USA)	Soil water content		0.05%
Radiation monitor (CNR-1; Kipp & Zonen, The Netherlands)	Incoming and outgoing short-wave and long-wave radiation	<18 s	
CO$_2$/H$_2$O analyzer (Li7500; LiCor, USA)	CO$_2$ and water vapor fluxes	0.1 s	
Automatic weather station (EnviroStation^{TM}; ICT, Australia)	T, P, RH, wind	1 min	1–3%
Acoustic radar (MFAS; Scintec Corporation German)	3D vertical winds	15 min	Effective height: 500 m in fog

T, P, LWC, N_d, and RH stand for temperature, atmospheric pressure, liquid water content, droplet number concentration, and relative humidity, respectively. After Lui *et al.*, (2011a)

intervals, by use of an automatic weather station (EnviroStation^{TM}; ICT, Australia) and visibility meter (ZQZ-DN; Jiangsu Province Radio Science Institute, China), respectively (Lui *et al.*, 2011a).

From 2006 to 2009, the total number of fog cases with visibility <1,000 m was 29. A fog layer is defined by RH = 97% in a vertical profile. Among these 29 cases, full FBL data were acquired for 16. On the basis of the FBL data, we carried out analysis and present the results in this paper (Table 2).

3. Results

3.1. Fog Classification and Types

There are different kinds of fog classification in the literature. Gultepe *et al.* (2007) summarized fogs into radiation fog, high-inversion fog, advection–radiation fog, advection fog, and steam fog. The first three types are associated with radiative cooling over land.

Many factors, for example radiation cooling, warm/cold advection, and precipitation, can affect fog processes. Some fog cases are affected by several factors, but the classification is based on the primary factor that affects fog formation, maturation, and dissipation. According to the mechanisms of fog processes, 16 fog episodes observed during this project are divided into four types: radiation fog, advection–radiation fog, advection fog, and precipitation fog.

Radiation fog is most likely to occur over land (Taylor, 1917; Li, 2001), and usually forms near the surface under clear sky in stagnant air in association with an anticyclone (Gultepe *et al.*, 2007). So the primary factor for radiation fog is radiative cooling. Two factors affect advection–radiation fog—radiative cooling and advection. It is mostly a coastal phenomenon (Gultepe *et al.*, 2007, 2009; Niu *et al.*, 2010a, b), but may appear over land. Advection fog has advection as its dominant factor (Liu *et al.*, 2011b). Precipitation fog forms as raindrops fall into drier air below the cloud; the liquid droplets evaporate to become water vapor, the water vapor cools, and at the dew point it condenses and becomes fog (Tardif and Rasmussen, 2009).

Table 2

Macro-structure characteristics of fog processes in Nanjing

Fog types	Data	Time 1 (LST)	Time 2 (LST)	Duration (min)	Min vis. (m)	PTDF (L < 50 m)	FWD	FWS (m/s)	DWD	WSR (m/s)	WDPDF	T(°C) Ave	T(°C) Range	Height (m) Ave	Height (m) Range	T(°C) Ave	T(°C) Range
Precipitation fog	06–07 December 2006	16:15	07:09	894	51	/	NW	/	NW	/	/	9.7	7.7 to 11.7	331	261–428	12.8	12–13.7
Radiation fog	09–10 December 2006 −1.4–1.0 −0.72	21:20	09:30	730	290	/	/	/	/	/		–	0.2	–	1.8 to 3.0	426	262–483
Radiation fog	11–12 December 2006	23:00	17:00	1080	15	0426–1020 12th	/	/	/	/		4.1	–1.8 to 6.5	347	192–527	4.9	3.8–6.3
Radiation fog	13–14 December 2006	22:00	12:00	840	15	0837–1011 14th	/	/	/	/		5.7	4.8 to 7	526	258–738	5.7	4.6–7.5
Advection–radiation fog	24–27 December 2006	22:00	14:14	3840	15	0041 25th –1236 26th 1932–2310 26th	NW	/	NW NE	/	NE–SW/ NE–NW	5.0	0.6 to 7.0	447	175–653	7.7	5.1–9.7
Radiation fog	03–04 December 2007	18:12	09:11	900	15	0630–0655 4th	SSE	0.4	N/S	0–2.2	N	3.5	0.6 to 5.5	144	70–220	4.5	4–4.7
Radiation fog	10–11 December 2007	22:31	12:30	840	15	0735–0935 11th	NW	0.5	N/S	0–3.0	S–N	3.8	1.8 to 7.7	131	103–150	5.4	5.2–5.8
Radiation fog	13–14 December 2007	19:50	11:21	930	15	0548–0958 14th	NNE	0.3	SE/ SW/ N	0–1.0	SW–NW	1.0	–1.1 to 4.5	223	30–623	3.0	0.2–4.4
Radiation fog	14–15 December 2007	20:48	12:00	902	15	0450–0612 15th	NE	0.5	S/NW	0–2.7	SW	2.2	–0.3 to 7.2	106	55–152	5.7	4.3–6.5
Radiation fog	18 December 2007	02:30	11:30	540	15	0830–1015	NW	0.4	SW/ NW	0–2.0	NW	4.2	2.9 to 6.6	768	739–840	3.1	2.1–4.0
Advection–radiation fog	18–19 December 2007	16:06	12:30	1230	15	0143–0400 19th 0445–0925 19th	NW	0.4	S N	0–5.3	S–NW	3.8	–0.2 to 8.6	477	200–650	5.9	4.3–7.5
Advection fog	20–21 December 2007	17:47	19:07	1520	51	/	S	0.6	W/N	0–3.3	/	7.7	6.7 to 9.7	848	664–1050	6.7	5.6–7.7
Radiation fog	22–23 December 2007	22:10	00:38	149	66	/	NW	0.2	W–S	0–0.6	/	7.6	7.3 to 9.1	124	50–180	9.4	8.6–10.0
Radiation fog	23 December 2007	01:16	05:30	255	15	0146–0308	NE	0.3	NE	0–0.8	NE	5.6	5.4 to 7.7	427	110–780	7.0	5.1–8.6
Advection fog	01–02 December 2009	19:00	11:20	980	51	/	NE	2.0	/	0.4–3.9	/	3.9	3 to 6.1	555	260–690	4.9	2.9–5.9

Formation time (Time1); dissipation time (Time 2); persistence time of dense fog (PTDF); formation wind direction (FWD); formation wind speed (FWS); dominant wind direction (DWD); wind speed range (WSR); wind direction pre-dense fog (WDPDF)

[a] Only lists cases with boundary-layer data

During the winter fog observations in 2006–2009, 10 radiation fog cases, three advection fog cases, two advection–radiation fog cases, and one precipitation fog case were obtained (Table 2). Note that Table 2 only lists cases with boundary-layer data collected.

3.2. Formation, Maintenance, and Dissipation Time

Radiation fog has remarkable diurnal variation. It forms mostly at sunset or midnight, and lifts after sunrise or at noon, with duration mostly between 9 and 18 h (Table 2). Advection–radiation fog and advection fog tends to have very long duration, between 16 and 64 h; the longest case persisted for more than 64 h during 24–27 December, 2007. These fogs have formation and dissipation times similar to those of radiation fog, and slight diurnal variation that weakens or disappears at noon.

Extremely dense fog occurs only in radiation-related cases, namely, radiation fog or advection–radiation fog. However, the lowest visibility of advection fog was more than 50 m (Table 2). The persistence time of extremely dense fog is mostly between 1 and 6 h, with the shortest not much longer than half an hour and the longest nearly 40 h during 24–27 December, 2006. Such a difference can be caused by micro-physical processes, which will be analyzed in detail in Part II. The extremely dense fogs appeared mostly in the morning, 1 or 2 h before or after sunrise; at other times they form at midnight. They are related to radiative cooling or evaporation (Table 2).

3.3. Fog-Layer Features

3.3.1 General Features

Table 2 shows that the advection fog and advection–radiation fog have the thickest layers. The maximum fog-layer altitude for the five cases is more than 650 m, with an average of >420 m. The observed radiation fog also has a thick layer. For about 60% of these, the maximum thickness reaches 500–800 m; for the others, it reaches 100–200 m. The layer thickness of precipitation fog is related to the cold air layer thickness above the ground.

3.3.2 Double Fog Layers

Double fog layers were detected in three cases (13–14 December, 2006; 10–11 December, 2007; and 13–14 December, 2007). In the first case, there was a thick fog layer (maximum 738 m) during the fog development process; afterward a double fog-layer structure occurred, which was caused by the upper-layer sinking motion during fog-top decline. There were double inversions in the second case, which caused a double fog-layer structure; the lower-level jet between the fog layers separated the two layers (Lui *et al.*, 2011a). In the third case, the surface fog layer was formed by radiative cooling and was followed by a cloud layer caused by low-level cold air advection. The cloud layer developed downward and the surface fog layer developed upward; the two combined to form a very deep dense fog (Fig. 1).

3.4. Fog Temperatures

3.4.1 General Features

Table 2 indicates that in most cases Nanjing fogs are warm fogs for which the temperature (T) in the fog body is higher than 0°C during the whole fog life cycle. The unique precipitation fog case has the highest temperature in its body, 12.0–13.7°C at the fog top and 7.7–11.7°C on the ground. The radiation fog has a lower fog-top temperature between −1.4 and 10.0°C, and a temperature on the ground between −1.8 and 9.1°C. The fog-top temperatures of advection fog and advection–radiation fog change little between 5.1 and 9.7°C, and their temperatures on the ground are 0.6–7.0°C during their whole lifetime. All of these fogs have the common characteristics that the temperature at the fog top is higher than that on the ground; for the advection–radiation fog, especially, the difference is 3–4°C. These results clearly show that these four kinds of fog all form in the inversion.

3.4.2 Relationship Between Fog Top and Inversion

Studies have shown that the low-level inversion formed shortly before the radiation fog appeared, i.e., the fog formed in the inversion (Pilie *et al.*, 1979). As the surface became radiatively shielded by the

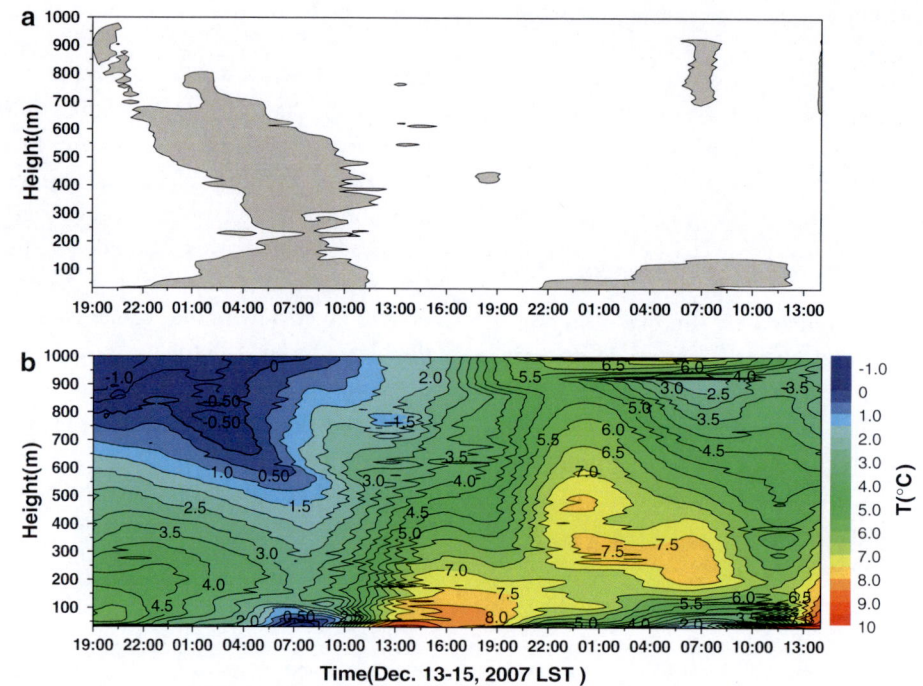

Figure 1
Time–height cross-section of relative humidity (**a** RH, shaded area with RH ≥ 97% indicates the fog layer) and temperature (**b** *T*) observed by the tether-sonder system during the fog event of 13–15 December, 2007

developing fog, the radiation inversion migrated to the top of the fog, accompanied by a moisture adiabatic lapse rate in the fog. Most cases observed at the NUIST had this feature. The surface inversion during radiation fog was formed as a result of radiative cooling, and then rose after sunset. Ten radiation fogs and advection–radiation fogs formed under the same conditions; however, the inversion in radiation fog lasts for only a short time, disappears 1 h or 2 after sunrise or at noon, and is shorter than that in advection–radiation fog.

Subsidence inversion was also measured in this project. This was caused by a systematic upper-layer sinking motion above the warm advection. It played dual roles in the fog on 14 December, 2006—it encouraged the explosive development of the fog top and accelerated the fall of the fog top in the disappearing stage (Lu *et al.*, 2010a, b). Liu *et al.* (2011b) suggested another role in the fog during 1–2 December, 2009—the fog top was related to the subsidence inversion intensity, and would drop when the subsidence inversion weakened with the sinking

motion, and would rise as the subsidence inversion strengthened again.

3.5. Wind Fields

3.5.1 FBL Low-Level Wind

In the cases of radiation fog and advection–radiation fog, the low-level wind was always light before fog formation; the maximum surface wind speed was no more than 0.6 m/s, and the wind direction was mostly northwesterly or southeasterly (Table 2). The wind direction varied substantially during the whole process of all these fogs, but the dominant wind directions that persisted for the longest times were from the northwest or southeast. Speed mostly dis not exceed 2 m/s, yet the instantaneous wind speed could exceed 3 m/s or even reach 5.3 m/s, because of advection (Table 2).

Similar to fog formation, the surface wind direction changed substantially before extremely dense fog, especially in cases during which extremely dense fogs persist for more than 2 h. When the wind reverses its direction from easterly to westerly or from southerly to northerly, the fog develops into extremely dense fog.

The lowest visibility was more than 50 m in all three observed cases of advection fog, which did not form extremely dense fog. The formation wind direction was southerly or easterly. During 1–2 December, 2009, the formation wind speed was 2 m/s; it varied between 0.4 and 3.9 m/s but was mostly 2 m/s during the whole episode.

A distinct feature of radiation fog and advection–radiation fog is that the formation wind direction is easterly or southerly if the onset of the fog is before or after sunset but is northerly in the midnight cases.

3.5.2 FBL Wind Aloft

The vertical wind direction and speed were obtained by the tether-sonde system and partly by the acoustic radar. The acoustic radar can detect horizontal and vertical wind speeds; however, it has a limited measuring height for capturing the whole fog layer.

The wind direction, inversion, and fog layer were observed to have a special relationship in an advection fog during 1–2 December, 2009 (Fig. 2). There were two inversion layers beneath 1,000 m: the bottom of the lower inversion was located at the shear zone of the easterly and southeasterly, and the top was located within the southeasterly. Additionally, the fog top was above the top of the lower inversion but below the bottom of the upper inversion.

Wind direction variation with height and time were detected by the acoustic radar data. The time–height cross-section of the horizontal and vertical winds from pre-fog until its dispersal, as probed by the acoustic radar, during 10–11 December, 2007, are shown in Fig. 3. It indicates that during the whole process the wind direction changed clockwise with height at any given time and with time at any given height, illustrating that the fog was affected by warm advection. Furthermore, the vertical wind changed during the mature phase, being upward flow for a time then downward flow for another time from the surface layer to 200 m, with an approximately 1–2 h quasi-period (Fig. 3). The changes were caused by a low-level jet.

3.5.3 Low-Level Jet

Low-level jet sometimes impeded the development of fog whereas at other times it helped to sustain fog. In the case of advection fog during 1–2 December, 2009, there was a low-level jet at the top of the lower inversion (Fig. 4). The air masses near the jet axis exchanged constantly, mixing the cold lower-layer air with the warm upper-layer air and leading to an increase of cooling area height and of the top/bottom heights of the lower inversion. Simultaneously, the vapor was transported downward from the warm, moist upper-layer air via the jet, also raising the bottom of the high-humid area and causing upward

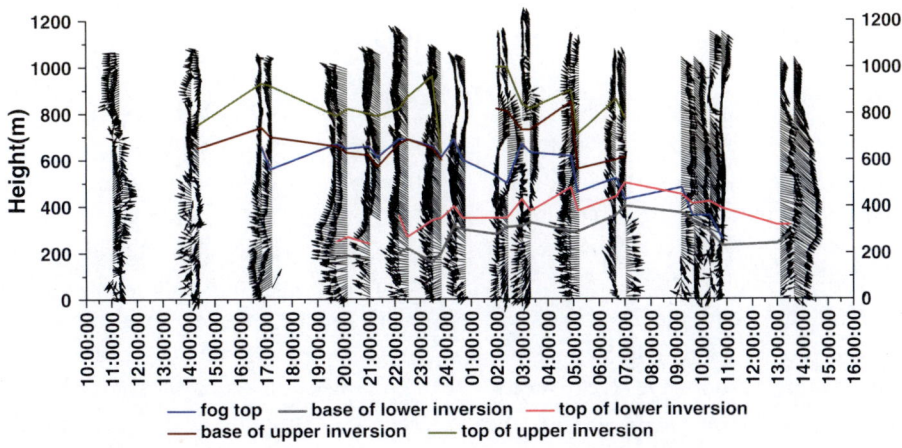

Figure 2
Time–height cross-section of wind vector/fog top/and inversions observed by the tether-sonder system during the fog of 1–2 December, 2009 (LST)

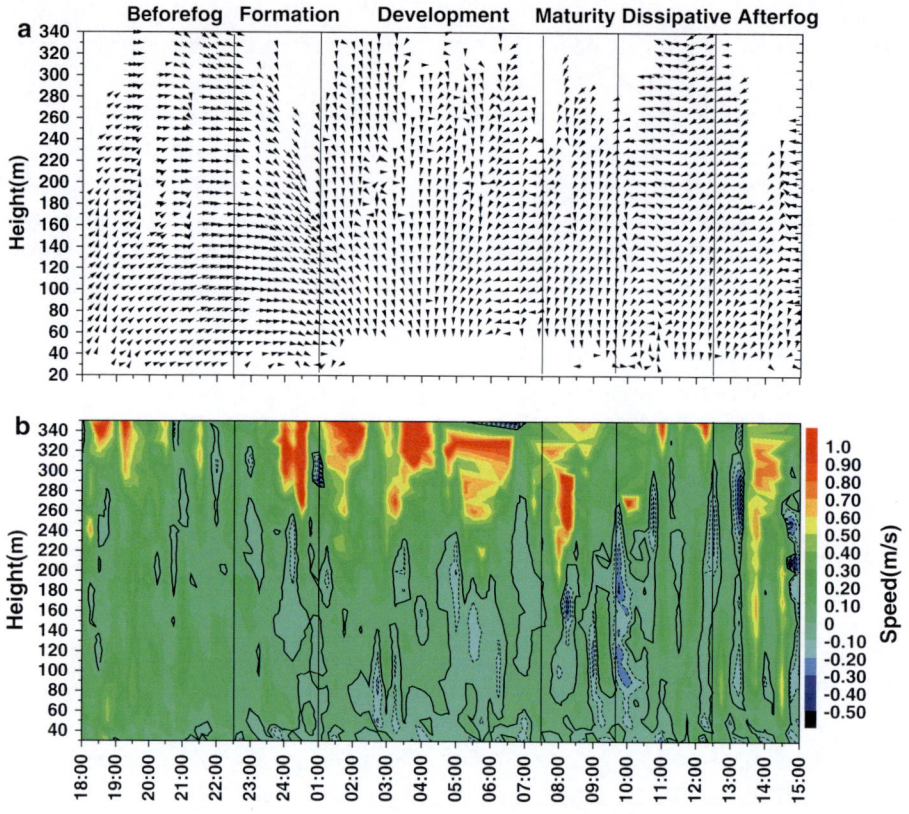

Figure 3

Time–height cross-section of horizontal wind and vertical speed observed by the acoustic radar during 10–11 December, 2007 LST (**a** wind direction, with invalid value removed; **b** vertical speed, with positive value for updraft, negative value (*dotted line*) for downdraft, and *solid black curves* for zero contour). After Li *et al.* (2011)

movement of the low-humid area; in this way sufficient moisture was provided to maintain the fog.

4. Discussion

As described in the sections "Formation, maintenance, and dissipation time" and "Fog-layer features", long, persistent fogs and deep fogs occurred not only in advection and advection–radiation fogs but also in radiation fogs. Deep and long persistent advection-related fogs are usually related to strong inversion or double-inversion structures, and the inversions are mainly affected by cold/warm–moist advections. On the other hand, long-lasting radiation fogs are related to precipitation before fog formation and to surface cold advection and turbulence.

ZHOU and FERRIER (2008) have a special opinion on the persistence of radiation fog—they suggested

that the primary reason for the persistence of radiation fog, or at least one reason but not the least reason, is the role of turbulence in radiation fog. According to a theoretical study, they recommended the persistence condition for a steady fog be characterized by either the critical turbulent exchange coefficient or the characteristic depth of the FBL. If the turbulence intensity inside a fog is smaller than the turbulence threshold, the fog persists, whereas if the turbulence intensity exceeds the turbulence threshold or the characteristic depth of the FBL dominates the entire fog bank, then the balance will be destroyed, leading to dissipation of the existing fog. NAKANISHI (2000) carried out a numerical experiment on radiation fog to investigate and validate fog structure and the mechanisms behind its formation. He indicated that before formation, the atmospheric stratification was stable, with an inversion layer immediately above the surface; during

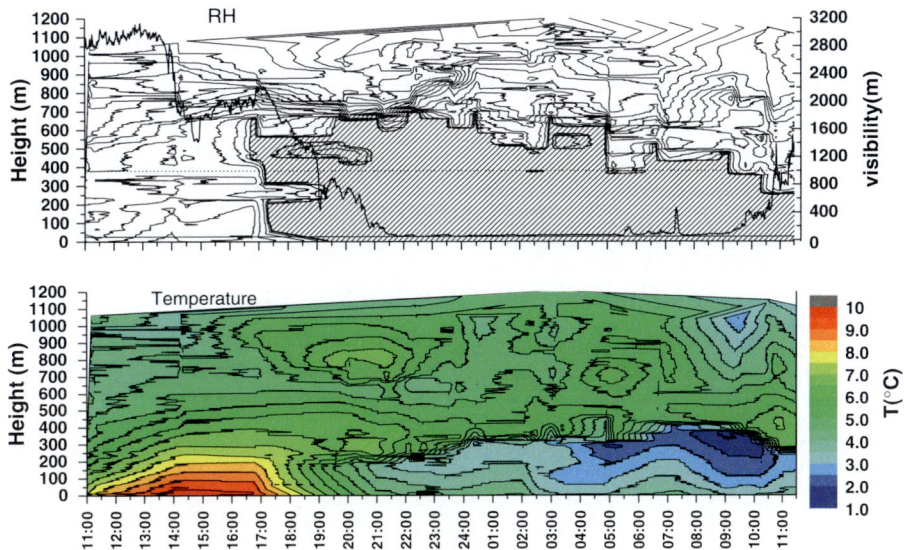

Figure 4

Time–height cross-section of relative humidity (RH, shaded area with RH ≥ 97% indicates the fog layer; *dotted line* indicates visibility of 1,000 m) and air temperature observed by the tether-sonder system during the fog of 1–2 December, 2009 LST

development the stratification became unstable, leading to the occurrence of a mixed layer. It is noted that the turbulent structure varies in the formation, development, and dissipation phases of radiation fog.

Previous analysis illustrated that cold/warm–moist advections help to form deep fog layers. However, warm advection alone was adverse to the development of fog thickness. Two cases during 13–15 December, 2007 can be used for comparison (Fig. 1): the temperature, wind field, moisture, and fog-layer thickness structures of the two cases were in substantial agreement below 100 m but very different above 100 m. This is because there was a cold advection on 13–14 December, 2007, which formed low cloud; in contrast, there was a deep warm temperature area on 14–15 December, 2007, which led to a deep inversion layer from the ground to 700 m. The warm temperature area prevented the fog from developing upward, and the maximum fog-layer thickness was merely 150 m as a result.

Note that in the case on 13–14 December, 2007 the extremely dense fog lasted more than 4 h, but on 14–15 December, 2007 it lasted just an hour or so after the appearance of an extremely dense fog. It is thus evident that the deep inversion was merely an essential condition for a thick fog layer; sufficient vapor supply was advantageous to the formation and maintenance of the deep fog layer.

5. Conclusions

This paper provides an overview on the FBL features of the YRDFOG project in Nanjing. The project's field observations were conducted at the meteorological observation station of the NUIST during the winters of 2006–2009 in an effort to reveal FBL structures and macro/micro-physical characteristics and to gain insight into the mechanisms of triggering, formation, maintenance, and dissipation of fog (including extremely dense fog). Detailed FBL observations were obtained using a FMD (FM-100), a visibility meter, an automatic weather station, acoustic Doppler radar, and a tether-sonde system. The important conclusions we obtained are as follows:

1. Four types of fog were observed: radiation fog, advection–radiation fog, advection fog, and precipitation fog, according to the mechanisms and primary factors of fog processes.
2. The radiation fog had a remarkable diurnal variation. It formed mostly at sunset or midnight, lifted after sunrise or at noon, and its duration was mostly between 9 and 18 h. Advection–radiation fog and advection fog had very long durations. Extremely dense fogs occurred only in radiation–related cases, namely, radiation fog or

advection–radiation fog, and were mostly between 1 and 6 h.

3. The inversion in radiation fog was short-lived, disappearing 1 h or 2 after sunrise or at noon, shorter than that in advection–radiation fog.

4. When the wind direction reversed its direction from easterly to westerly or from southerly to northerly, the fog developed into extremely dense fog.

5. Low-level jet sometimes impeded the development of fog; at other times it encouraged its persistence.

6. Deep inversion was essential for a thick fog layer; sufficient vapor supply was advantageous to the formation and maintenance of a deep fog layer.

Acknowledgments

Funding for this work was jointly provided by the National Natural Science Foundation of China (grant no. 40775012), the Natural Science Fund for Universities in Jiangsu Province (grant no. 08KJA170002), the Meteorology Fund of the Ministry of Science and Technology (grant nos GYHY (QX) 2007-6-26, GYHY200906012], Jiangsu Province Qinglan Project of "Cloud fog precipitation and Aerosol Research", and the Graduate Student Innovation Plan in Universities of Jiangsu Province (grant no. CX10B_292Z).

REFERENCES

FITZJARRALD, D. R. and LALA, G. G. (1989), *Hudson Valley Fog Environments,* Journal of Applied Meteorology *28,* 1303-1328.

FUZZI, S., FACCHINI, M. C., ORSI,G., LIND, J. A., WOBROCK, W., KESSEL, M., MASER, R., JAESCHKE, W., ENDERLE, K. H., ARENDS, B. G., BERNER, A., SOLLY, I., KRUISZ, C., REISCHL, G., PAHL, S., KAMINSKI, U., WINKLER, P., OGREN, J. A., NOONE, K. J., HALLBERG, A., FIERLINGEROBERLINNINGER, H., PUXBAUM, H., MARZORATI, A., HANSSON, H. C., WIEDENSOHLER, A., SVENNINGSSON, I. B., MARTINSSON, B. G., SCHELL, D., AND GEORGII, H. W. (1992), *The Po Valley fog experiment 1989,* Tellus B *44*(5), 448-468.

FUZZI, S., LAJ, P., RICCI, L., ORSI, G., HEINTZENBERG, J., WENDISCH, M., YUEKIEWICZ, B., MERTES, S., ORSINI, D., SCHWANZ, M., WIEDENSOHLER, A., STRATMANN, F., BERG, O.H., SWIETLICKI, E., FRANK, G., MARTINSSON, B.G., GUNTHER, A., DIERSSEN, J.P., SCHELL, D., JAESCHKE, W., BERNER, A., DUSEK, U., GALAMBOS, Z., KRUISZ, C., MESFIN, N.S., WOBROCK, W., ARENDS, B., AND BRINK, H.T. (1998), *Overview of the Po Valley Fog Experiment 1994(CHEMDROP),* Contributions to Atmospheric Physics *71:* 3-19.

GULTEPE, I., TARDIF, R., MICHAELIDES, S. C., CERMAK, J., BOTT, A., BENDIX, J., MULLER, M. D., PAGOWSKI, M., HANSEN, B., ELLROD,

G., JACOBS, W., TOTH, G., and COBER, S. G. (2007), *Fog Research: A Review of Past Achievements and Future Perspectives,* Pure & Applied Geophysics *164,* 1121-1159.

GULTEPE, I., PEARSON, G., MILBRANDT, J. A., HANSEN, B., PLATNICK, S., TAYLOR, P., GORDON, M., OAKLEY, J. P., and COBER, S. (2009), *the Fog Remote Sensing and Modeling Field Project,* Bulletin of the American Meteorological Society *90,* 341-359.

HAEFFELIN, M., BERGOT, T., ELIAS, T., TARDIF, R., CARRER, D., CHAZETTE, P., COLOMB, M., DROBINSKI, P., DUPONT, E., DUPONT, J. C., GOMES, L., MUSSON-GENON, L., PIETRAS, C., PLANA-FATTORI, A., PROTAT, A., RANGOGNIO, J., RAUT, J. C., REMY, S., RICHARD, D., SCIARE, J., and ZHANG, X. (2010), *PARISFOG: Shedding New Light on Fog Physical Processes,* Bulletin of the American Meteorological Society *91(6),* 767–783.

HOLETS, S. and SWANSON R. N. (1981), *High-Inversion Fog Episodes in Central California,* Journal of applied meteorology *20(8),* 890-899.

HUANG, J.P., ZHU, S.W., and ZHU, B. (1998), *Characteristics of the atmospheric boundary layer during radiation fog,* Journal of Nanjing Institute of Meteorology *21,* 258-265. (In Chinese)

HUANG, Y. S, XU, W.R., LI, Z.H., FAN, L., and HUANG W.J. (1992), *an observation and analysis on the radiation fog in Xishuangbanna,* Acta Meteorologica Sinica *50,* 112-117. (In Chinese)

LI, Z.H., HUANG, J.P., ZHOU, Y.Q., and ZHU, S.W. (1999), *Physical structures of the five-day sustained fog around Nanjing in 1996,* Acta Meteorologica Sinica *57,* 622-631. (In Chinese)

LI, Z.H. (2001), *Study of fog in China over the past 40 years,* Acta Meteorologica Sinica *59,* 616-624. (In Chinese)

LI, Z.H., LIU, D.Y., and YANG, J. (2011), *The microphysical processes and macroscopic conditions of the radiation fog droplet spectrum broadening,* Chinese Journal of Atmospheric Sciences (In Chinese) *35*(1), in press.

LIU, D.Y, PU, M.J, YANG, J., ZHANG, G.Z., YAN, W.L., and LI, Z.H. (2010), *Microphysical Structure and Evolution of a Four-Day Persistent Fog Event in Nanjing in December 2006,* Acta Meteorologica Sinica *24,* 104-115.

LIU D.Y., YANG, J., NIU, S.J., and LI, Z.H. (2011a), *On the Evolution and Structures of a Radiation Fog Event in Nanjing,* Adv. Atmos. Sci., *28*(1), 1-15, doi:10.1007/s00376-010-0017-0.

LIU,D.Y., NIU, S.J., PU, M.J., YANG, J., and LI, Z.H. (2011b), *On the physical process of advection fog influenced by cold & warm advection,* Chinese Journal of Geophysics, *in press*(In Chinese)

LU, C.S., NIU, S.J., YANG, J., and WANG, W.W. (2008), *An observational study on physical mechanism and boundary layer structure of winter advection fog in Nanjing,* Journal of Nanjing Institute of Meteorology *31,* 520-529. (In Chinese)

LU, C.S., NIU, S.J., YANG, J., LIU, X., and ZHAO, L.J. (2010a), *Jump features and causes of macro and microphysical structures fog a winter fog in Nanjing,* Chinese Journal of Atomospheric Sciences *34,* 681-690. (In Chinese)

LU, C.S. NIU,S.J., TANG, L.L., LV, J.J., ZHAO, L.J., and ZHU, B. (2010b). *Chemical composition of fog water in Nanjing area of China and its related fog microphysics.* Atmospheric Research *97*(1-2): 47-69.

MEYER, M. B., LALA G. G., and JIUSTO J.E. (1986), *Fog-82: A Cooperative Field Study of Radiation Fog,* Bulletin of the American Meteorological Society *67*(7), 825–832.

NAKANISHI, M. (2000), *Large-Eddy simulation of radiation fog,* Boundary-Layer Meteorology *94,*461-493.

NIU, S.J., LU, C.S., YU, H.Y., ZHAO, L.J., and ZHAO, L.J. (2010a), *Fog research in China: an overview*, Adv. Atmos. Sci., *27*(3), 639-662.

NIU, S.J., LU, C.S., ZHAO, L.J., LU, J.J, and YANG, J. (2010b), *Analysis of the microphysical structure of heavy fog using a droplet spectrometer: a case study,* Adv. Atmos. Sci., *27*(6), 1259-1275,doi:10.1007/s00376-010-8192-6.

PILIE, R.J., MACK, E.J., ROGERS, C. W., KATZ, U., and KOCMONDET, W. C. (1979), *The Formation of Marine Fog and the Development of Fog-Stratus Systems along the California Coast,* Journal of Applied Meteorology *18*(10), 1275–1286.

PU, M.J., LI, L.F., LI, Z.H., and HUANG, J.P. (2001), *Study of physical process of the Xishuagnbanna valley fog in winter,* Scientia Meteorologica Sinica *21*, 425-432. (In Chinese)

PU, M.J., YAN, W.L., SHANG, Z.T., YANG, J., and LI, Z.H. (2008a), *Study on the physical characteristics of burst reinforcement during the winter fog of Nanjing*, Plateau Meteorology *27*, 1-8. (In Chinese)

PU, M.J., ZHANG, G.Z., YAN, W.L., and LI, Z.H. (2008b), *Features of a rare advection-radiation fog event,* Science in China Series D: Earth Sciences *51,* 1044-1052.

ROACH, W.T., BROWN, R., CAUGHEY, S. J., GARLAND, J. A., and READINGS, C. J. (1976), *The physics of radiation fog: I - a field study,* Quarterly Journal of the Royal Meteorological Society *102*(432), 313-333.

RODHE, B. (1962), *The effect of turbulence of fog formation,* Tellus *14,* 49-86.

TAYLOR, G.I. (1917), *The formation of fog and mist,* Quart. J. Roy. Meteor. Soc *43,* 241–268.

TARDIF, R., and RASMUSSEN, R.M. (2009), *Evaporation of non-equilibrium raindrops as a fog formation mechanism,* Journal of the Atmospheric Sciences *49,* 1247-1267.

TERRADELLAS, E., FERRERES, E., and SOLER, M.R. (2008). *Analysis of turbulence in fog episodes.* Adv. Sci. Res. *2:* 31-34.

TOMINE, K., MCHIMOTO, K., HIKIJI, I., ABE, S., and TAKEUCHI, N. (1991), *The Vertical Structure of Fog Observed with a Lidar System at Misawa Airbase, Japan,* Journal of Applied Meteorology *30*(8), 1088–1096.

TONG, Y.Q., YIN, Y., XU, X.Z., XU, M.L., and XIANG, Y. (2009), *Climatological characteristics of fog in Nanjing,* Journal of Nanjing Institute of Meteorology *32,* 115-120.

VAN DER VELDE I.R. and STEENEVELD G.J. (2010). *Modeling and forecasting the onset and duration of severe radiation fog under frost conditions.*Monthly Weather Review.

YAN, W.L., PU, M.J., WANG, W.W., YANG, J., and LIU, D.Y. (2009), *A study on a rare radiation-advection fog (I): the analysis of physical process of genesis and dissipation,* Scientia Meteorologica Sinica *29*, 9-16. (In Chinese)

YANG, J., WANG, L., LIU, D.Y., and LI, Z.H. (2011), *Boundary layer structure and physical mechanisms in the evolution of a very deep dense fog,* Acta Meteorologica Sinica, *in press.* (In Chinese)

ZHANG, G.Z, BIAN, L.G, WANG, J.Z., YANG, Y.Q., YAO, W.Q., and XU, X.D. (2005), *The boundary layer characteristics in the heavy fog formation process over Beijing and its adjacent areas,* Science in China Series D (Earth Sciences) *48,* suppl. 2, 88-101.

ZHOU B.B., DU J., FERRIER B., MCQUEEN J., and DIMEGO G. (2007). *Numerical forecast of fog-central solutions.* 18th Conference on Numerical Weather Prediction, AMS, 25-29 June 2007, Park City, UT.

ZHOU B.B. and FERRIER B.S. (2008), *Asymptotic Analysis of Equilibrium in Radiation Fog,* Journal of Applied Meteorology & Climatology, *47*(6), 1704-1722

(Received November 14, 2010, revised May 2, 2011, accepted May 7, 2011, Published online May 31, 2011)

Reprinted from the journal

Pure Appl. Geophys. 169 (2012), 821–833
© 2011 Springer Basel AG
DOI 10.1007/s00024-011-0270-x

The Relation Between Humidity and Liquid Water Content in Fog: An Experimental Approach

STEFAN GEORG GONSER,[1] OTTO KLEMM,[2] FRANK GRIESSBAUM,[2] SHIH-CHIEH CHANG,[3] HOU-SEN CHU,[3]
and YUE-JOE HSIA[3]

Abstract—Microphysical measurements of orographic fog were performed above a montane cloud forest in northeastern Taiwan (Chilan mountain site). The measured parameters include droplet size distribution (DSD), absolute humidity (AH), relative humidity (RH), air temperature, wind speed and direction, visibility, and solar short wave radiation. The scope of this work was to study the short term variations of DSD, temperature, and RH, with a temporal resolution of 3 Hz. The results show that orographic fog is randomly composed of various air volumes that are intrinsically rather homogeneous, but exhibit clear differences between each other with respect to their size, RH, LWC, and DSD. Three general types of air volumes have been identified via the recorded DSD. A statistical analysis of the characteristics of these volumes yielded large variabilities in persistence, RH, and LWC. Further, the data revealed an inverse relation between RH and LWC. In principle, this finding can be explained by the condensational growth theory for droplets containing soluble or insoluble material. Droplets with greater diameters can exist at lower ambient RH than smaller ones. However, condensational growth alone is not capable to explain the large observed differences in DSD and RH because the respective growth speeds are too slow to explain the observed phenomena. Other mechanisms play key roles as well. Possible processes leading to the large observed differences in RH and DSD include turbulence induced collision and coalescence, and heterogeneous mixing. More analyses including fog droplet chemistry and dynamic microphysical modeling are required to further study these processes. To our knowledge, this is the first experimental field observation of the anti-correlation between RH and LWC in fog.

Key words: Fog, orographic fog, humidity, liquid water content, droplet size distribution, fog microphysics.

1. Introduction

Formation of clouds and fog in the atmosphere occurs in an air mass after a sufficiently high level of humidity has been reached. Adiabatic cooling, isobaric cooling, and mixing of air masses with high relative humidity (RH) but different temperature are the main processes leading to condensation of water vapor on aerosol particles and activation of cloud condensation nuclei (CCN). Several mechanisms have been proposed to allow rapid droplet growth, yielding droplet diameters greater than 10 μm within timescales of less than 30 min. The most prominent ones are influence of turbulence on collision and coalescence, and heterogeneous mixing (ROGERS and YAU, 1989; WALLACE and HOBBS, 2006).

The activation of CCN requires supersaturation (above 100% RH) in the direct vicinity of the particles. After activation of CCN has occurred, the condensation process leads to a decrease of the gas phase humidity content, an increase of the liquid water content (LWC), and to a modification of the droplet size distribution (DSD). However, these processes do not occur homogeneously throughout the entire cloudy or foggy air mass. In contrast, clouds and fogs are heterogeneous systems. For RH and LWC, variations within clouds and fogs have been observed to occur (e.g. GERBER, 1991; PRUPPACHER and KLETT, 1997; GARCIA-GARCIA *et al.*, 2002). The scope of this work is to observe rapid humidity fluctuations and their effects on DSD in orographic fog above a subtropical montane cloud forest in northeastern Taiwan.

Many studies of fog microphysics have been carried out in the past. However, most of these dealt with radiation and advection type of fogs (for a summary see GULTEPE *et al.* (2007)), which represent the most

---

[1] BayCEER, University of Bayreuth, 95440 Bayreuth, Germany. E-mail: Stefan.Gonser@uni-bayreuth.de

[2] Institute of Landscape Ecology, University of Muenster, 48149 Munster, Germany.

[3] Institute of Natural Resources, National Dong Hwa University, 974 Hualien, Taiwan.

widespread type of fog in populated areas. Besides the variation of humidity, also temporal droplet spectra evolutions in radiation fogs have been described, among others, by PILIÉ *et al.* (1975) and GERBER (1991). Since this fog type is characterized by low wind speeds, the evolution of the DSD can be observed at temporal scales on the order of several minutes to a few hours at a fixed measurement position.

For orographic fog, the situation is different. Only very few analyses have been performed up to now (e.g. EUGSTER *et al.*, 2006; BEIDERWIEDEN *et al.*, 2008), of which the main focus was not on microphysics. The apparent lack of investigations probably is based on the fact that the occurrence of this fog type is generally limited to hardly accessible and scarcely populated montane regions.

Orographic fog can be regarded as a highly heterogeneous system, composed of various air volumes with different characteristics. The advection of air volumes with changing fog density is clearly observable even with the naked eye. Because of the permanent advection of air with differing temperature, humidity, aerosol number, size, and chemical composition, any measurement techniques should allow for a high temporal resolution, preferably of less than one second. Only then can the rapid changes of fog microphysical parameters be appropriately studied.

For the present study, a set of instruments was installed in an environment with a daily occurrence of orographic fog. Data of water vapor density, temperature and DSD were sampled with a temporal resolution of 3 Hz. Any similar short-time dynamics of DSD in conjunction with humidity and temperature fluctuations have not, to our knowledge, been studied yet. For the study site in mountainous Taiwan, it has been shown that vaporization of water from the forest canopy into the foggy boundary layer air frequently occurs (BEIDERWIEDEN *et al.*, 2008). The occurrence of this process can only be explained through the heterogeneous nature of the boundary layer air. It is our goal to further develop an understanding of the variability of microphysical properties and processes in fog. Our focus is on the relationship between humidity, LWC, and DSD, and on the temporal and spatial scales in which the respective variations occur.

2. Instrumentation and Methods

2.1. Measurement Site

The present study was performed at the Chilan Nature Reserve above a subtropical montane cloud forest in north-eastern Taiwan. The Chilan Research Site (CLM) is located at 24°35′N, 121°24′E at 1,650 m above mean sea level on the eastern slope of the Syue-Shan Mountain Range. The instrumentation was installed on a 22 m high tower, approximately 10 m above the canopy of a homogeneous yellow cypress (*Chamaecyparis obtusa* var. formosana) seed regenerated forest with a slope of 14° towards the southeast (CHANG *et al.*, 2006). The mean annual temperature at the research site is 13°C and the annual rainfall is between 2,000 and 5,000 mm (CHANG *et al.*, 2002). Fog occurs almost every day with monthly averaged durations between 2.7 and 14.2 h day^{-1} in July and November, respectively (CHANG *et al.*, 2006). The wind regime is dominated by a diurnal variation. During day time, southeasterly directions from the valley prevail, whereas at night the wind turns to the opposite direction and blows downslope from the northwesterly sector. Fog normally appears over noon to afternoon and begins to dissipate in the early evening when the wind direction begins to turn towards the NW. These factors render the Chilan Research Site (CLM) a perfect location for studying microphysical characteristics of orographic fog.

2.2. Instrumentation

From 7 July through 5 August 2009, the following array of instruments was installed at the CLM site to analyze the dynamics of humidity and the DSD in fog: a FM-100 Fog Monitor (Droplet Measurement Technologies, Boulder, CO, USA), a LI-7500 Open Path CO_2/H_2O Analyzer (LI-COR Biosciences, Lincoln, NE, USA), a R3-50 Sonic Anemometer (Gill Instruments Ltd., Lymington, Hampshire, UK), a TR-1050 Temperature Recorder (RBR Ltd., Ottawa, Ontario, CDN), a Mira Visibility Sensor (Aanderaa Data Instuments, Bergen, N) and a CNR 1 Net Radiometer (Kipp & Zonen, Delft, NL). All sensors were mounted in immediate vicinity to each other (maximum distance 0.8 m); solely, the CNR 1 was positioned on a nearby tower at an approximate horizontal distance of 30 m. The Fog Monitor (FM-100) is an optical spectrometer which samples droplet-laden ambient air and measures the droplet size distribution in the range of 2–50 μm diameters. The liquid water content was derived using the DSD and the air flow rate through the instrument's sample area. Absolute humidity (AH) was recorded in g m^{-3} by the LI-7500 (for detail see below). The three-dimensional wind speed and direction was conducted by the sonic anemometer with a resolution of 0.01 m s^{-1}. Wind, AH and DSD were sampled with a frequency of 10 Hz. The TR-1050 temperature recorder can be polled with a frequency of 3 Hz and exhibits an accuracy in laboratory conditions of 0.002 K. Its thermistor is embedded in a cylindrical housing made of stainless steel (with 0.89 mm diameter and a wall thickness of 0.18 mm). The TR-1050's temperature response time is slightly reduced in comparison to the polling frequency, since the housing does not adapt instantaneously to short term temperature fluctuations. During data post processing, all sample intervals where reduced to the temperature recorder's polling frequency of 3 Hz. From the recorded AH and ambient temperature, RH was calculated by means of the Magnus formulation. Because of general electricity blackouts, absence of fog, and technical problems with the measurement equipment; in total, three complete days of optimum quality data could be recorded.

2.3. Quality Assurance

The precise determination of the gas phase humidity content in air is critical for this study. The measurement of the relative humidity in the range of around 100% is a challenge. Specific instruments are needed. GERBER (1980) developed a hygrometer, solely to measure humidity in the range of 95–105% RH. This is the only sensor capable of directly measuring the supersaturation in the field with sufficient accuracy for cloud physical analysis up to now. Unfortunately, his hygrometer was only built once and has never been developed beyond a prototypical state. Therefore, we approach the issue through the measurement of the water vapor

concentration through infrared light absorption (2.59 μm wavelength). The LI-7500 is a suitable instrument that is commonly applied in micrometeorology (BALDOCCHI et al., 2001). The open path length is 12.5 cm. Erroneous readings may originate from aerosol particles in the sampling path or deposited on the sensor windows. To mitigate these interfering effects, the instrument operates with a reference beam at a wavelength (2.4 μm), located very close to the measurement beam. The water vapor concentration is deduced from the signal ratio of the measurement beam versus the reference beam. Since the reference beam is not located in the absorption band of water vapor, and its wavelength is still very close to the measurement beam, it is assumed that contaminations on the sensor windows and aerosols in the measurement path will affect both beams in the same way. This leads to the assumption that the measured water vapor concentrations are not affected by the abovementioned effects. Unfortunately, this presumption is not unrestrictedly valid, as it has been observed that water droplets or ice on the optical windows will cause biases in concentration readings (WEISENSEE and LEPS, 2002). Such contaminations are indicated by the instrument's automatic gain control (AGC) operational parameter. For our sensor, it increases from a value of 50% for clean windows up to 100% for heavily contaminated ones. Because of the necessity of high accuracy in readings of AH, all data with an AGC value above 50% was discarded.

The presence of droplets in the measurement path will also cause biases in the sensors readings. CERNI (1994) described an infrared hygrometer very similar to the LI-7500. He studied the potential interference by applying a radiation model in combination with laboratory studies. His analysis showed that fog and clouds only have a minimal influence on the humidity readings as long as their DSD are broad enough. Narrow distributions yielded noticeable errors. In order to assess this error for the LI-7500, the extinct portion of the measurement and reference infrared beams due to typical DSD from our experimental data set was calculated by means of the light scattering calculator MieCalc (MICHEL, 2007). Our results are similar to those of CERNI (1994). Fog droplets in the sampling path generally exhibit a stronger attenuating effect on the measurement beam than on the reference beam. This results in a slight overestimation of AH readings when fog is present in the sample path. The expected maximum error lies on the order of 0.2%. The overestimation is particularly discernible when a narrow DSD is present. Broader spectra yield less intense overestimations. When the type of DSD changes from narrow to broad (see Sect. 3 for details), the error from these LI-7500 artifacts lies in the range of 0.09–0.1%. Our recorded AH data show a decrease on the scale of 0.7–1.0% during transitions from narrow DSD to broader ones. Therefore, the variations as observed in the field are roughly one order of magnitude larger than the instrumental error. The error associated to droplets in the measurement path is, therefore, sufficiently small to be neglected in the following.

The AH measurement data yields a relatively poor absolute accuracy. This is due to the zero and gain drift with ambient temperature, the cross-sensitivity towards ambient CO_2 concentrations, and the signal noise of the LI-7500. The uncertainties related to temperature drift and CO_2 variations have a notable effect on the absolute AH readings, especially when temporal scales of several hours are considered. In this study, the focus lies on short-term variations at temporal scales between seconds and a few minutes. For such short term variations the sensor's accuracy is expected to be sufficient, since both temperature drift and CO_2 concentrations exhibit a rather slow variation with time. In the following, only the relative fluctuations of AH and RH will be considered. The absolute readings are not discussed.

To exclude radiation effects on the temperature readings, a spherical radiation shield with 50% shading effect was mounted around the TR-1050 Temperature Recorder. In addition, all periods with short wave radiation intensity exceeding 100 W m^{-2} where discarded. This threshold is based on laboratory experiments performed prior to the field campaign. Radiation intensities of less than 100 W m^{-2} resulted in an acceptable accuracy of 0.01 K in the field.

Furthermore, a wind filter was applied to the data, in order to exclude all directions which diverged more than 30° from the orientation of the FM-100 nozzle. This excluded preferential sampling of small fog droplets and yielded correct DSD.

The FM-100 principle of operation is based on forward scattering of the internal laser beam on droplets of different sizes. The scattering intensity is linked to the droplet diameter via the Mie-theory (MIE, 1908), and measured droplet concentrations can be assigned to up to 40 size classes. The Mie-curve for pure spherical water droplets does not show an even progression but is characterized by irregularities (Fig. 1). The irregularities make it impossible, in most cases, to allocate a single scattering cross section to an exact particle diameter, since several particle diameters apply to the same scattering cross section. Therefore the theoretically maximum possible size resolution of the probe is limited by the Mie-curve. Considering the irregularities, we adapted 23 droplet size classes (grey shaded rectangles in Fig. 1) to the Mie-curve in order to achieve the best possible results for DSD measurements. Nevertheless, droplet concentrations in some size classes (e.g. the third size class in Fig. 1) are overestimated, since a few droplets from adjacent classes will be included. During data post processing these outliers in the measured DSD where flattened by applying a Savitzky–Golay smoothing filter (SAVITZKY and GOLAY, 1964). This is justified by the application of a computer model, simulating the FM-100. The model is based on the Mie-curve, provided by the manufacturer of the probe, and the spacing of our chosen 23 droplet size classes (Fig. 1). By feeding the model with artificially created DSD, it simulates the output of the FM-100. The

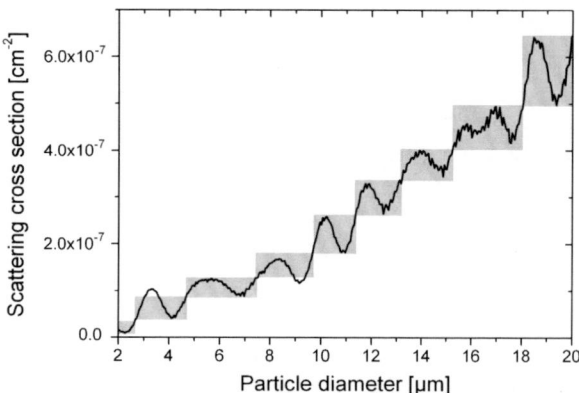

Figure 1
Segment of the Mie-curve for spherical particles (*black line*) of pure water, refractive index is 1.33 and wavelength of incident light is 680 nm (courtesy of D. Baumgardner). *Grey shaded rectangles* represent the spacing of our selected droplet size classes

resulting DSD do not show the identical spectrum as the input data does, but are characterized by positive and negative outliers. These outliers resemble those observed in our experimental dataset (not shown in detail). After application of the Savitzky-Golay smoothing filter, the simulated and subsequently smoothed DSD were virtually identical to those of the input spectra. Eventually, the adaption of our 23 size classes and the subsequent smoothing of the spectra results in a significant improvement of the FM-100 readings.

2.4. Equations

In order to derive RH from temperature and AH measurements, the Magnus formulation to calculate the saturation vapor pressure e_s (Pa) is applied. A formulation, originally stated by SONNTAG (1990), from a review paper by ALDUCHOV and ESKRIDGE (1996) is utilized.

$$e_s = \exp(-6096.94T^{-1} + 16.64 - 2.71 \times 10^{-2}T + 1.67 \times 10^{-5}T^2 + 2.43 \ln T) \tag{1}$$

where T (K) is the ambient temperature measured with the TR-1050 temperature recorder.

The actual vapor pressure e (Pa) was derived by applying the ideal gas equation

$$e = \frac{m_v}{V}R_v T \tag{2}$$

where $\frac{m_v}{V}$ is the mass of vapor per air volume, hence the reading of the LI-7500 (kg m^{-3}) and R_v is the gas constant for water vapor 461 (J K^{-1} kg^{-1}). Finally RH in units % was calculated as the ratio of e and e_s multiplied with 100.

The derived RH yields a relatively poor absolute accuracy; a maximum error of approximately ±5% RH can be expected. This is due to the uncertainties related to the LI-7500 mentioned in Sect. 2.3.

In order to derive LWC from spectrometer readings the following formulation was used:

$$\text{LWC} = \frac{\pi \rho_w}{6} \sum_{i=1}^{m} c_i d_i^3 \tag{3}$$

The density of water ρ_w was set to 1×10^6 (g m^{-3}). The temperature dependence of water density was not considered, since the

temperature showed only slight changes throughout the fog events. Further c_i is the droplet number concentration in air (m^{-3}), and d_i the average droplet diameter (m) in size class i, respectively.

The droplet concentration c_i is derived in the following way,

$$c_i = \frac{n_i}{V_s} \tag{4}$$

Here n_i represents the total number of droplets in size class i per sample interval, and V_s is the sampling volume (1.24 × 10^{-16} m^3).

The recorded DSD show a great variety regarding their respective shapes. In order to allocate the measured droplet spectra to three general DSD-types, a simple method, based on two ratios (R_1 and R_2) was applied during data post processing.

$$R_1 = \frac{\sum_{i=1}^{2} c_i}{\sum_{i=3}^{23} c_i} \quad R_2 = \frac{\sum_{i=3}^{4} c_i}{\sum_{i=5}^{23} c_i} \tag{5}$$

R_1 and R_2 represent droplet concentration ratios between different size ranges of a complete DSD. The ratios compare the amount of smaller droplets to the amount of larger ones. Hence a combination of R_1 and R_2 represents the recorded DSD in a rough manner, and can therefore be used to allocate each DSD to one of three general DSD-types. In order to assign the recorded droplet spectra to the three general DSD-types the following threshold values were used:

DSD-type 1: Periods with R_1 and R_2 greater than 0.5.

DSD-type 2: Periods where neither type 1 nor type 3 is established.

DSD-type 3: Periods with R_1 and R_2 equal or smaller than 0.5.

Condensation or vaporization of water will change the ambient temperature by the release or absorption of latent heat. To estimate the influence of this heat source/sink, the air temperature change (ΔT_{lh}). on basis of the variation of the LWC, was calculated.

$$\Delta T_{lh} = \frac{l_v \Delta m}{c_p \rho} \tag{6}$$

where l_v is the specific latent heat of vaporization or condensation of water, l_v is a linear function of temperature and varies from 2.5×10^6 to 2.43×10^6 J kg^{-1}, for 0 and 30°C, respectively (SCHMIDT and GRIGULL, 1979). Further, Δm (kg m^{-3}) is the change in the mass of available water, hence the change in LWC, ρ (kg m^{-3}) is the density of moist air, and c_p is the specific heat at constant pressure for moist air (ELTING, 2008).

3. Results

Data with good quality was acquired on 3 days. Two fog events on 8 and 9 July persisted for 6 and 5 h, respectively. Both days were characterized by the typical diurnal variation at the measurement site: fog appeared during the afternoons and disappeared in the early evening hours. A third complete event was recorded on 12 July. It differed from the others with respect to its persistence: it lasted from 13:00 hours for 11 h, until midnight.

During all three events, fog was characterized by three general types of DSD (Fig. 2). The distinction of the three DSD-types was performed by applying Eq. 5. DSD-type 1 was characterized by narrow spectra with maximum number concentration in the 2.33 μm class and a continuous decrease towards larger diameters, type 2 exhibited slightly broader spectra with a plateau or peak around 10 μm diameter, and type 3 showed broad spectra with the concentration peaking around 15 μm diameter. The three DSD-types did also show distinct differences with respect to their LWC distribution within the droplet size spectra. The total LWC increased from type 1 throughout type 3. Occurrence of the three

general DSD-types was strongly related to RH and the general evolutional state of the fog. At the onset of a fog event, the DSD were largely dominated by small droplets (DSD-type 1). During these stages, only short periods with broader distributions occurred. Such intermissions were mostly present for less than a minute each and were associated with a decrease in humidity, both RH and AH, and rising temperature. Later on the spectra tended to become broader, and DSD-type 3 was dominating. Whether DSD-type 3 dominated also during the depletion phase of the fog can unfortunately not be deduced from our dataset. This is due to the reversion of the wind direction at the end of each fog event, and the consequent violation of our wind direction quality criterion. DSD-type 2 may be regarded as a transitional spectrum type, as it occurred in between the two other types and was mainly characterized by short persistence.

In order to obtain a general view on the characteristics of the heterogeneously composed orographic fog at the CLM site, we determined the median values for RH, LWC, temperature, and persistence of every single period with the same DSD-type. The results are presented in Table 1 for all three fog events. In addition to the medians, the standard deviations are presented to characterize the variations within the different DSD-types. Considering LWC, RH, and temperature, a straightforward development from DSD-type 1 throughout type 3 can be observed. It is very remarkable that the LWC shows a clear inverse behavior to RH. The LWC increases with DSD-type number, while RH decreases. Remarkable are also the high standard deviations of LWC and RH

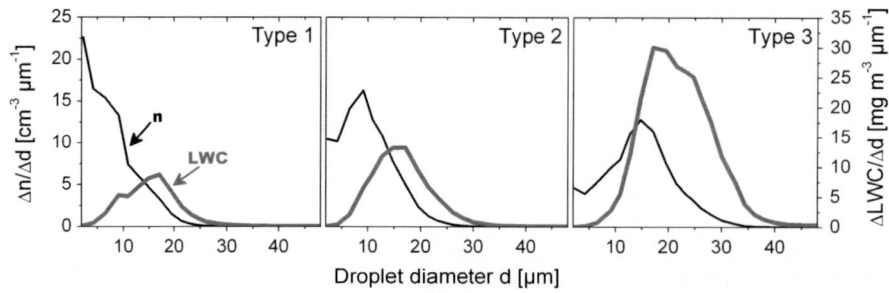

Figure 2
Averaged DSD (*n*, *left* y-axis) and LWC (*right* y-axis) distributions for the three general DSD-types, for the whole fog event on 12 of July 2009

Table. 1

Median values (Med) and standard deviations (Std) for LWC (g/m³), RH (%), T (°C), duration (s) and count of the three DSD-types on 8, 9 and 12 of July 2009

Type	Med LWC	Std LWC	Med RH	Std RH	Med T	Std T	Med duration	Std duration	Count
8 July 2009									
1	0.027	0.028	99.84	0.74	18.95	0.27	6.7	20.1	287
2	0.097	0.036	99.31	0.92	19.00	0.22	2.3	3.2	77
3	0.541	0.331	97.84	2.10	18.42	0.39	5.7	206.9	41
9 July 2009									
1	0.099	0.050	100.12	0.91	18.39	0.32	4.8	23.1	308
2	0.150	0.048	99.86	0.96	18.37	0.32	2.0	3.1	300
3	0.218	0.129	99.34	1.42	18.15	0.28	6.3	62.5	280
12 July 2009									
1	0.053	0.031	99.30	1.06	18.51	0.51	4.0	7.2	333
2	0.087	0.038	98.61	1.39	18.35	0.53	3.3	9.0	556
3	0.175	0.108	96.66	2.20	17.84	0.53	6.7	44.1	603

in DSD-type 3. Apparently, this DSD-type existed during various RH conditions with a large variability of LWC. Regarding the median durations, DSD-type 2 exhibits the shortest time periods and also the smallest standard deviations, confirming that type 2 can be regarded as a transition between DSD-type 1 and type 3.

As a case study, two periods, each of 1 min duration, are presented in the following. Figure 3 shows an example for a transition from DSD-type 1 to type 3 on 9 July, shortly before sunset. At 18:00:59 hours, a rapid change of the microphysical conditions occurred. Before that time, the fog was dominated by small droplets with a narrow DSD. The contour plot in the upper part of Fig. 3 shows that the transition occurred almost instantaneously. With broadening of the droplet spectra the LWC increased. Simultaneously, the RH decreased in the order of one percent.

Note that the absolute values of RH have to be considered with extreme caution; because of this, we focus on the variations (ΔRH) of its 500 digit moving average, rather than on the absolute magnitudes.

The observed decrease of AH around 18:00:59 hours implies two effects: First, the decrease of ΔRH was not simply a temperature effect as both the relative and absolute humidity in the gas phase decreased. Secondly, it indicates that the present water vapor experienced a change of phase and condensed onto the fog droplets, leading to a broadening of the DSD. This is confirmed by the behavior of the total water content

(TWC) per m³ air. Its concentration, which is the sum of AH and LWC, is rather constant over the entire period since LWC and AH vary in the same scale but in opposite directions. All these data indicate that the observed processes occurred within a single air mass of stationary total water (gas plus liquid phase) content.

In the bottom part of Fig. 3, two temperatures are depicted. The black line represents the temperature T_{sa} as recorded with the sonic anemometer and corrected for humidity effects. Despite the correction, T_{sa} still showed a deviation of one Kelvin in comparison to the readings of the TR-1050 temperature recorder. Despite this uncertainty in absolute temperature reading we show T_{sa} in Figs. 3 and 4 because the sonic anemometer showed immediate response to short term fluctuations. The bold gray temperature curve in Fig. 3 represents the calculated change in air temperature (ΔT_{lh}), resulting from the release or absorption of latent heat by condensation or vaporization of water, respectively. As ΔT_{lh} only represents the potential variations of air temperature, its first values in Figs. 3 and 4 are set equal to zero. The scale of variation of ΔT_{lh} and T_{sa} does not match exactly, but their order of magnitude is identical, hence ΔT_{lh} can be regarded as a plausible value. By considering the course of the two temperature curves it becomes apparent that they exhibit a similar behavior. Most of the variations in ΔT_{lh} are also evident in T_{sa}. Particularly during the transition from one DSD-type to another, the two temperatures show a remarkably

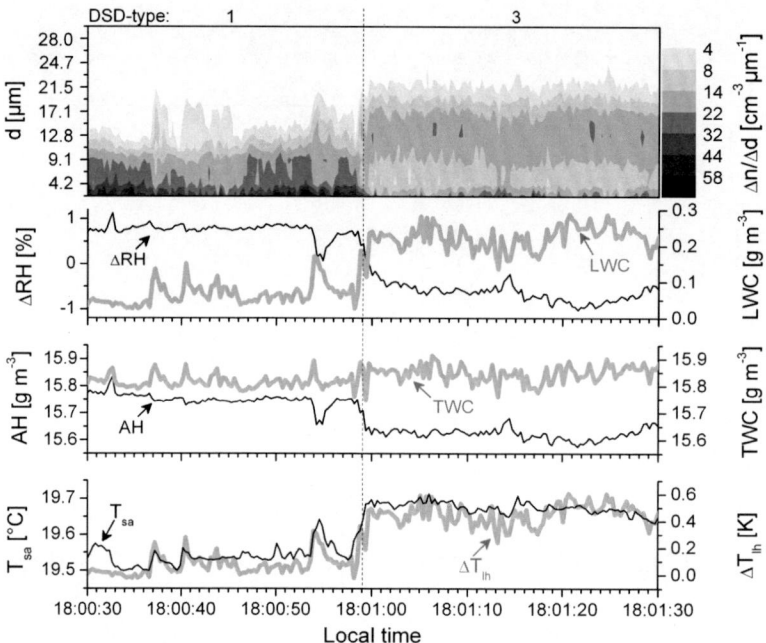

Figure 3
Time series of DSD-type, droplet size distribution ($\Delta n/\Delta d$), relative humidity fluctuations (ΔRH), liquid water content (LWC), absolute humidity (AH), total water content (TWC), temperature (T_{sa}), and change in air temperature due to condensation/vaporization (ΔT_{lh}) for 1 min on 9 of July 2009

simultaneous increase. The parallel behavior of the two temperatures suggests that the change in air temperature was solely driven by the change of latent heat. The behavior of ΔT_{lh} confirms the suggestion that the period depicted in Fig. 3 represents a single undisturbed air volume.

With the development of the humidity parameters and the temperatures, we derived two strong arguments supporting the hypothesis that the changes as shown in Fig. 3 occurred within one single air mass. The shift of the DSD from type 1 to type 3 occurred abruptly, though through an unknown mechanism.

On the other hand, the measured CO_2 concentrations (not shown) exhibit a course similar to that of LWC and T_{sa}, i.e., a clear increase in concentrations at the moment when the DSD transition occurs. The CO_2 concentrations should be generally independent from the humidity or temperature parameters. Therefore, the increase of CO_2 at 18:00:59 hours disproves the hypothesis of a single undisturbed air mass to have been observed in Fig. 3. Following this argument, the DSD-shift must have been induced by the advection of a new air volume with different

DSD. However, the total water content and internal energy parameters are almost identical.

Overall, our data was characterized by clear and persistent DSD transitions as exemplified in Fig. 3. Occasional intermissions with other DSD-types occurred, generally with a persistence of a few seconds. In Fig. 4, a period of 1 min duration of 12 July 2009, is presented. The major difference to the prior example is that the fog was already in an advanced evolutional state, with generally high LWC, lower droplet concentrations, and lower humidities (RH and AH). In this state the fog is usually dominated by continuous DSD with large droplets (DSD-type 3), only interrupted by short periods with smaller droplets. Until 21:24:14 hours, DSD-type 3 was present and the LWC showed a relatively constant behavior. After 21:24:14 hours, several fast changes of the DSD-types appeared. This behavior also appears in LWC. The short intermissions exhibited persistence between 1 and 5 s.

It is striking in this example that ΔRH did not show the clear contrasting behavior to LWC like in the previous example. Nor did AH and ΔRH show

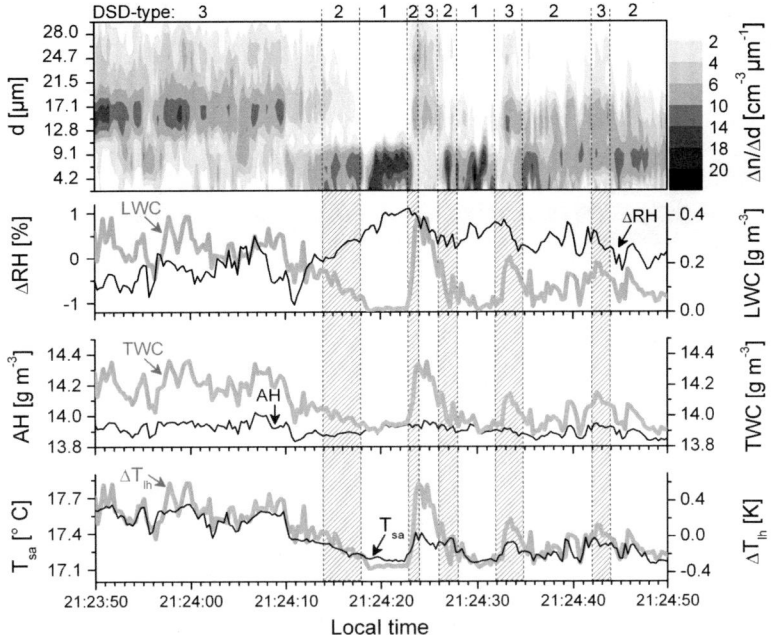

Figure 4
As Fig. 3, but for 12 of July 2009

synchronous behavior. A further remarkable difference is that the TWC was not constant. In this case, the variations of AH are smaller than those of LWC, it is the latter that dominates the fluctuations of TWC.

The temperatures in the bottom part of Fig. 4 show, once again, a parallel behavior. This indicates that the variations in air temperature are largely influenced by the change in latent heat. The intense variations of TWC and the fast fluctuations of the DSD in Fig. 4 indicate that this period is characterized by the continuous advection of fog volumes with different properties.

Our air volume analysis at the beginning of Sect. 3 indicated that RH and LWC exhibit an inverse relation. To strengthen this argument we present the correlation between RH and LWC for the complete 3 Hz dataset, for every fog event separately. Again, the absolute values of RH must be considered with caution, especially due to the hygrometers drift with temperature during entire fog events and between measurement days.

The three recorded fogs showed temperature variations between 1.5 and 2.5 K, over the entire duration of the events. Temperature changes of this scale yield a noticeable drift in the hygrometer readings. To account for this AH-drift with temperature, the correlation between RH and LWC is calculated from the deviations of their 500 digit moving averages. An anti-correlation is evident from the analyses of the three recorded fog events. In Fig. 5, the scatter plot for the ΔRH–ΔLWC anti-correlation on 9 of July is depicted. The Pearson correlation coefficient of -0.62 denotes a clear negative-correlation between the two parameters. The other two fog events showed this anti-correlation as well, although their correlation coefficients were less clear, -0.34 and -0.49 on 8 and 12 July, respectively. The inverse relation between RH and LWC over entire fog events is in good accordance with the exemplary datasets presented in Figs. 3 and 4.

4. Discussion

Orographic fog at the Chilan Research Site is clearly composed of various volumes of air masses that are intrinsically homogeneous but exhibit clear differences between each other with respect to their size, RH, and DSD. Such distinct volumes could

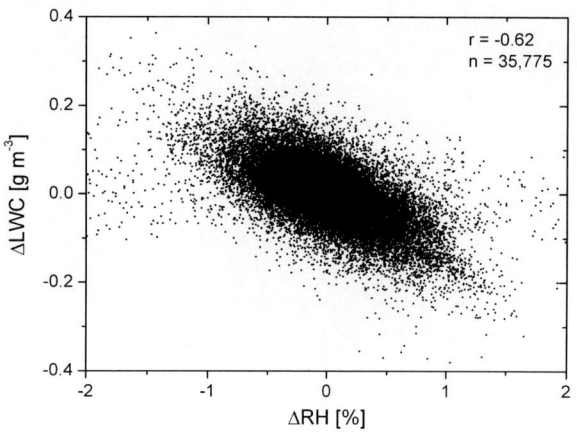

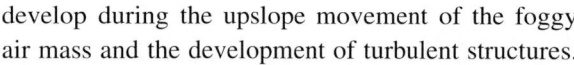

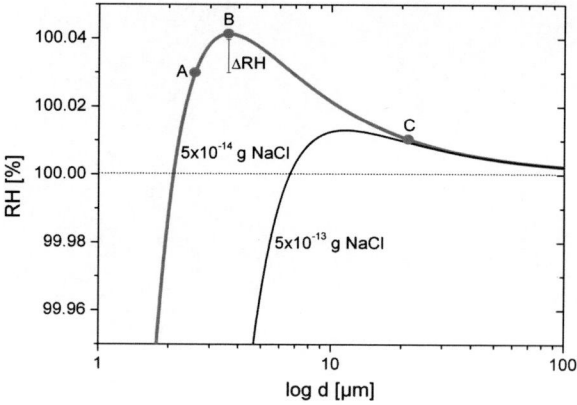

Figure 5
Negative-correlation between variations of RH and LWC for the fog event on 9 of July 2009. The Pearson Correlation Coefficient (r) of -0.62 denotes the negative-correlation between RH and LWC

Figure 6
Köhler-curves for NaCl nuclei with differing masses. Point A denotes the equilibrium state of a non activated droplet, at point B the droplet reached the deflection point of the Köhler-curve—as humidity was elevated sufficiently (ΔRH) for the droplet activation. Subsequently it will grow until it finally reaches point C, where it is in equilibrium at lower ambient RH but exhibits a greater diameter

develop during the upslope movement of the foggy air mass and the development of turbulent structures.

The variabilities in humidity and DSD between air volumes can, in principal, be explained by the Köhler-theory for diffusional droplet growth (KÖHLER, 1921). Fog droplets are never composed of pure water only, but contain soluble and/or insoluble material. The Köhler-theory describes the growth of droplets due to condensation; hence, the diameter of a solution droplet as a function of RH (or supersaturation). For further detail see PUPPACHER and KLETT (1997).

To apply the Köhler-theory to data measured in the field, knowledge of the chemical composition of the droplets is crucial. Unfortunately, no chemical analysis of the fog water or interstitial aerosols was performed for our study. Therefore we will assume a fixed solved mass of NaCl per droplet for the following thought experiment. In Fig. 6, two equilibrium curves for different sodium chloride nuclei are depicted. Equilibrium states of both particles show the typical Köhler-curve shape, with a steep increase at smaller diameters and a gentle decrease for larger diameters. To further understand the process of droplet growth by condensation, let us assume a hypothetical droplet of diameter 2.5 μm and 5×10^{-14} g NaCl solved in it (marked as point A in Fig. 6). This droplet is in equilibrium with its

surroundings at a RH of 100.03%. If the RH is further elevated due to cooling of the ambient temperature or the increase of AH, the droplet will no longer be in equilibrium, but will grow along the gray curve, keeping its equilibrium condition fulfilled. If the RH is being raised sufficiently (denoted as ΔRH in Fig. 6), the droplet reaches the inflection point of the Köhler-curve (denoted by B). Now, the droplet experiences an intense growth without having the need of a further elevation of RH. Such a droplet is said to be activated. During the growth process water in the vapor phase condenses onto the droplet. As a result, the temperature rises and the available vapor (AH) will decrease. Both factors lead to a decrease of RH. The process proceeds until the droplet is again in equilibrium with its surroundings (for example at diameter 22 μm, point C). Eventually the droplet did gain mass (about a factor of 1,000 in comparison to point A) and is able to exist in an environment of lower RH than it was at point A.

The growth process of an activated droplet does not occur promptly but takes some time. The growth speed depends on the droplet size and the mass and nature of the solved material. For example, a diameter change from 4 to 20 μm of a droplet containing 10^{-13} g NaCl due to pure condensational growth would take approximately 30 min (BEST, 1951). The black equilibrium curve in Fig. 6 represents a droplet with 10^{-13} g of dissolved NaCl. The higher mass yields an activation of the droplet at lower ambient RH, greater equilibrium diameters at the same RH, and a faster growth speed, in comparison to a droplet containing less NaCl.

Up to this point we considered single droplets. However, fogs and clouds are composed of droplet populations with a large variety of diameters and chemical compositions. By reducing RH, the non-activated droplets will shrink or even evaporate completely, while the activated droplets experience a further growth due to the extra vapor produced by the evaporation of the smaller droplets. Thus the droplets with different diameters and solutions compete for the excess water vapor. Smaller activated droplets grow faster in comparison to larger ones, since the rate of growth of a droplet is inversely proportional to its diameter (WALLACE and HOBBS, 2006). Eventually, all activated droplets of different diameters will approach the same size.

Although the described mechanisms are plausible, the increase of droplet diameters solely by diffusional growth is not a valid explanation for the large observed variabilities of LWC and RH (Figs. 3, 4) in our field study. The observed DSD-transitions occurred almost instantaneously, whereas growth due to condensation would take too much time. If condensational growth would be the only mechanism involved, the differences in DSD between the single air volumes would be less pronounced. Other possible mechanisms must be sought to explain the observed phenomena. Potential mechanisms yielding faster growth include (1) the influence of turbulence on collision and coalescence, and (2) entrainment of air volumes of different composition and history into the cloud (heterogeneous mixing) (ROGERS and YAU, 1989; WALLACE and HOBBS, 2006).

(1) Droplet growth by collision and coalescence, as induced by gravitational movement, is relevant when collector drops with diameters of 40 μm or larger are present (ROGERS and YAU, 1989). Droplets smaller than this, as were mostly observed in our study (see Fig. 2), will rather be deflected around each other by their streamlines, instead of experiencing a collision. Only when turbulence is present, collision and coalescence will include smaller droplets. Under these conditions, droplets will experience acceleration and thus overlapping streamlines. Turbulence occurs frequently in orographic fog above montane cloud forests, as the forest canopy is characterized by a high roughness-length, and because a constant wind field is present. Consequently, turbulence-induced collision and coalescence is a possible mechanism contributing to a fast broadening of DSD above montane cloud forests.

(2) Entrainment of air volumes with different composition and history is also a possible reason for the observed LWC differences. Some of our time series show large variabilities in humidity. When, for example, sub-saturated air volumes are mixed into the fog, most droplets will experience evaporation, yielding an essentially water-saturated air volume with only a few or even no droplets dispersed in it. This volume may mix with the unaffected part of the fog and consequently reduce the droplet concentration by dilution, without affecting their sizes. As a consequence, the activated droplets in the mixed air mass will grow much faster because there are fewer droplets competing for the available water vapor.

5. Summary and Conclusions

Microphysical properties of orographic fog have been recorded with a temporal resolution of 3 Hz above a montane cloud forest in North Eastern Taiwan. The measured data includes DSD, AH, RH, air temperature, wind speed and direction, visibility and solar short wave radiation. The formation of orographic fog is characterized by moist air forced to higher elevations, its subsequent cooling by adiabatic expansion and the resulting condensation of the present water vapor. This type of fog is generally composed of randomly distributed air volumes, differing in size, temperature, humidity, DSD, LWC, aerosol number and size, and chemical composition. Three general types of air volumes could be distinguished via the recorded DSD. A statistical analysis of these different volumes was performed with focus on their median RH, LWC, temperature and persistence. The analysis yielded intense variations between the air volumes with respect to RH and LWC. By considering entire fog events, these large differences resulted in a clear inverse relation between RH and LWC. This finding is not new regarding the current state of knowledge of fog microphysics, but is, to our knowledge, the first time to have been observed in the field.

Theory on droplet growth by condensation in fogs and clouds can, in principal, explain the observed variability. However, kinetic considerations imply that the rapidly occurring differences in DSD and RH can not have happened exclusively through condensation and evaporation processes. The involved diffusion is too slow. Certainly other mechanisms play a dominating role. Possible processes leading to the large observed differences in RH and DSD include turbulence effects on collision and coalescence, and entrainment of dryer air volumes into the cloud (heterogeneous mixing). To which degree these mechanisms account for the large observed variabilities in RH and DSD, and consequently LWC, remains unclear so far. More analyses including fog droplet chemistry and dynamic microphysical modeling are required to further study these processes.

Acknowledgments

This study was supported by the German Academic Exchange Service DAAD through the PPP program. We thank D. Baumgardner, H.M. Chung, T. El-Madany, Y.T. Jian, J.Y. Jiang, C.W. Lai, M.C. Li, J.Y. Lin, T.Y. Lin, P. Sulmann, C.P. Wang, T. Wolf, and C.C. Wu for help and advice in the field experiment and during data analysis. We gratefully acknowledge L. Harris for language editing of the manuscript.

REFERENCES

ALDUCHOV, O.A., and ESKRIDGE, R.E. (1996), *Improved magnus form approximation of saturation vapor pressure*, Journal of Applied Meteorology 35, 601–609.

BALDOCCHI, D., FALGE, E., GU, L.H., OLSON, R., HOLLINGER, D., RUNNING, S., ANTHONI, P., BERNHOFER, C., DAVIS, K., EVANS, R., FUENTES, J., GOLDSTEIN, A., KATUL, G., LAW, B., LEE, X.H., MALHI, Y., MEYERS, T., MUNGER, W., OECHEL, W., U, K.T.P., PILEGAARD, K., SCHMID, H.P., VALENTINI, R., VERMA, S., VESALA, T., WILSON, K., and WOFSY, S. (2001), *FLUXNET: a new tool to study the temporal and spatial variability of ecosystem-scale carbon dioxide, water vapor, and energy flux densities*, Bulletin of the American Meteorological Society 82, 2415–2434.

BEIDERWIEDEN, E., WOLFF, V., HSIA, Y.J., and KLEMM, O. (2008), *It goes both ways: measurements of simultaneous evapotranspiration and fog droplet deposition at a montane cloud forest*, Hydrological Processes 22, 4181–4189.

BEST, A.C. (1951), *The Size of Cloud Droplets in Layer-Type Cloud*, Quarterly Journal of the Royal Meteorological Society 77, 241–248.

CERNI, T.A. (1994), *An Infrared Hygrometer for Atmospheric Research and Routine Monitoring*, Journal of Atmospheric and Oceanic Technology 11, 445–462.

CHANG, S.C., LAI, I.L., and WU, J.T. (2002), *Estimation of fog deposition on epiphytic bryophytes in a subtropical montane forest ecosystem in northeastern Taiwan*, Atmospheric Research 64, 159–167.

CHANG, S.C., YEH, C.F., WU, M.J., HSIA, Y.J., and WU, J.T. (2006), *Quantifying fog water deposition by in situ exposure experiments in a mountainous coniferous forest in Taiwan*, Forest Ecology and Management 224, 11–18.

ELTING, D., *Theoretische Meteorologie: Eine Einführung* (Berlin: Springer, 2008).

EUGSTER, W., BURKARD, R., HOLWERDA, F., SCATENA, F.N., and BRUIJNZEEL, L.A.S. (2006), *Characteristics of fog and fogwater fluxes in a Puerto Rican elfin cloud forest*, Agricultural and Forest Meteorology 139, 288–306.

GARCIA-GARCIA, F., VIRAFUENTES, U., and MONTERO-MARTINEZ, G. (2002), *Fine-scale measurements of fog-droplet concentrations: a preliminary assessment*, Atmospheric Research 64, 179–189.

GERBER, H.E. (1980), *A Saturation Hygrometer for the Measurement of Relative-Humidity between 95-Percent and 105-Percent*, Journal of Applied Meteorology 19, 1196–1208.

GERBER, H.E. (1991), *Supersaturation and Droplet Spectral Evolution in Fog*, Journal of the Atmospheric Sciences 48, 2569–2588.

GULTEPE, I., TARDIF, R., MICHAELIDES, S.C., CERMAK, J., BOTT, A., BENDIX, J., MULLER, M.D., PAGOWSKI, M., HANSEN, B., ELLROD, G., JACOBS, W., TOTH, G., and COBER, S.G. (2007), *Fog research: A review of past achievements and future perspectives*, Pure and Applied Geophysics 164, 1121–1159.

KÖHLER, H. (1921), *Zur Kondensation des Wasserdampfes in der Atmosphäre*, Geofys. Publ. 2, 3–15.

MICHEL, B. (2007), *MieCalc—freely configurable program for light scattering calculations (Mie theory)*, http://www.lightscattering.de/MieCalc/eindex.html.

MIE, G. (1908), *Beiträge der Optik trüber Medien, speziell kolloidaler Metallösungen*, Annalen der Physik 4, 377–445.

PILIÉ, R.J., MACK, E.J., KOCMOND, W.C., EADIE, W.J., and ROGERS, C.W. (1975), *Life-Cycle of Valley Fog .2. Fog Microphysics*, Journal of Applied Meteorology 14, 364–374.

PRUPPACHER, H.R., and KLETT, J.D., *Microphysics of clouds and precipitation*, (Boston: Kluwer Academic Publishers, 1997).

ROGERS, R. R., and YAU, M. K., *A short course in cloud physics*, (Oxford: Pergamon Press, 1989).

SAVITZKY, A., and GOLAY, M.J.E. (1964), *Smoothing + Differentiation of Data by Simplified Least Squares Procedures*, Analytical Chemistry 36, 1627–1639.

SCHMIDT, E., and GRIGULL, U., *Properties of water and steam in SI-units : thermodynamische Eigenschaften von Wasser und Wasserdampf : 0-800 p0 sC, 0-1000 bar*, (Berlin: Springer, 1979).

SONNTAG, D. (1990), *Important new values of the physical constants of 1986, vapor pressure formulations based in ITS-90, and psychrometer formulae*, Zeitschrift Fur Meteorologie *70*, 340–344.

WALLACE, J.M., and HOBBS, P.V., *Atmospheric science : an introductory survey*, (Boston: Elsevier Academic Press, 2006).

WEISENSEE, U., and LEPS, J.-P. (2002), *Fast-Response, open path optical Hygrometer for long-term measurements—Experiences, results, future requirements*, Deutscher Wetterdienst, Meteorologisches Observatorium Lindenberg.

(Received June 21, 2010, revised December 10, 2010, accepted December 24, 2010, Published online February 13, 2011)

Pure Appl. Geophys. 169 (2012), 835–845
© 2011 Springer Basel AG
DOI 10.1007/s00024-011-0330-2

Leaf Surface Wettability and Implications for Drop Shedding and Evaporation from Forest Canopies

W. Konrad,[1] M. Ebner,[1] C. Traiser,[1] and A. Roth-Nebelsick[2]

Abstract—Wettability and retention capacity of leaf surfaces are parameters that contribute to interception of rain, fog or dew by forest canopies. Contrary to common expectation, hydrophobicity or wettability of a leaf do not dictate the stickiness of drops to leaves. Crucial for the adhesion of drops is the contact angle hysteresis, the difference between leading edge contact angle and trailing edge contact angle for a running drop. Other parameters that are dependent on the static contact angle are the maximum volume of drops that can stick to the surface and the persistence of an adhering drop with respect to evaporation. Adaption of contact angle and contact angle hysteresis allow one to pursue different strategies of drop control, for example efficient water shedding or maximum retention of adhering water. Efficient water shedding is achieved if contact angle hysteresis is low. Retention of (isolated) large drops requires a high contact angle hysteresis and a static contact angle of 65.5°, while maximum retention by optimum spacing of drops necessitates a high contact angle hysteresis and a static contact angle of 111.6°. Maximum persistence with respect to evaporation is obtained if the static contact angle amounts to 77.5°, together with a high contact angle hysteresis. It is to be expected that knowledge of these parameters can contribute to the capacity of a forest to intercept water.

Key words: Wettability, canopy, contact angle, contact angle hysteresis, hydrology, interception.

1. Introduction

A significant amount of water that precipitates within canopies as rain, fog or dew is intercepted by the forest canopy. Interception comprises various processes that occur after the water has come into contact with plant surfaces: evaporation of water retained inside the canopy (interception loss), down

drip off the canopy (drip) and down flow along the stems (stemflow). Canopy interception is a process of considerable importance for the hydrological cycle since annual interception losses in forests can amount to more than a quarter of total rainfall (HÖRMAN et al. 1996; DINGMAN 2002). Determination of hydrological input by fog is partially very difficult due to the exact determination of interception. The actual rate of interception loss is dependent on various factors, such as forest structure, fog or rain intensity and meteorological parameters. Precise knowledge on interception losses are of substantial importance for predicting and modeling hydrological processes, such as effects of woodland, climate or land cover on water resources (GASH 1979; CALDER 1990; ABOAL et al. 1999; MUZYLO et al. 2009).

The maximum amount of water that can be temporarily held by a canopy is mainly distributed between bark and leaves (HERWITZ 1985). For the leaves, the amount of stored water is a function of leaf area index (LAI), but with substantial interspecific differences (ASTON 1979). Besides leaf area, wettability of the leaves is expected to be important. Wettability describes the behaviour of water after coming in contact with a surface. Water repellent surfaces are hydrophobic, and droplets upon these surfaces develop spherical forms with contact angles of >90° (CALIES and QUÉRÉ 2005). On hydrophilic surfaces, droplets attain contact angles of <90°. Complete wetting leads to the spreading of drops into films. Water repellency differs substantially between upper and lower sides of leaves, between species and between forest types (HOLDER 2007).

Wettability and the resulting contact angle also influences the gliding angle, that is, the angle of inclination of an object that leads to the rolling off of drops lying upon an object or being attached to the

[1] Department of Geosciences, University of Tübingen, Sigwartstrasse 10, 72076 Tübingen, Germany. E-mail: wilfried.konrad@uni-tuebingen.de

[2] State Museum of Natural History, Rosenstein 1, 70191 Stuttgart, Germany.

underside of the object. Wettability of a leaf depends on the chemical nature of the leaf waxes, as well as on structures of the leaf surface, including papillae and trichomes (BARTHLOTT and NEINHUIS 1997; SHIRTCLIFFE et al. 2009). It is intuitively expected that the more water repellent a surface is, the lower is the gliding angle. This would mean that a low inclination would be sufficient to remove drops from the object, and species with water repellent leaves would, therefore, show a lower storage capacity for rain interception than a species with more hydrophilic leaves.

However, the gliding angle is not only dependent on wettability. Rather, contact angle hysteresis governs the gliding behavior of droplets on a surface (QUÉRÉ 2008), that is, the difference between the developing contact angle when a droplet moves forwards (advancing contact angle) or backwards (receding contact angle). For example, petals can be very sticky with respect to droplet behavior despite their low wettability (FENG et al. 2008). In this contribution we consider the interrelationship between wettability and contact angle hysteresis and gliding behaviour of drops attached to surfaces. In particular, we will address the following questions: which contact angle leads to (1) a maximum storage capacity of a surface with respect to sitting or hanging drops, and (2) a maximum persistence of drops with respect to evaporation (under a given humidity and temperature.)

2. Basic Properties of Droplets Attached to a Plane

In this section we derive volumes and areas of droplets attached to a horizontal or inclined plane and the surface and gravitational forces acting on them.

2.1. Droplet Hanging Down from a Horizontal Plane

We consider a droplet hanging down from a horizontal plane under its own weight (Fig. 1). Assuming that the droplet is shaped as a segment of a sphere of radius R and that the contact angle formed between droplet and substrate is θ the volume of the spherical segment amounts to

$$
\begin{aligned}
V &= \frac{\pi R^3}{3}(1 - \cos\theta)^2(2 + \cos\theta) \\
&= \frac{\pi s^3}{3}\frac{(1 - \cos\theta)^2(2 + \cos\theta)}{\sin^3\theta}
\end{aligned}
\tag{1}
$$

The second version is a result of the substitution $s = R\sin\theta$, where s denotes the radius of the circle which forms the contact line where water, air and the solid of the plane meet. The surface area of the droplet is the sum $M + S$, where

$$
S = \pi R^2 \sin^2\theta = \pi s^2
\tag{2}
$$

$$
M = 2\pi R^2(1 - \cos\theta) = \frac{2\pi s^2}{1 + \cos\theta}
\tag{3}
$$

S denotes the attachment area between drop and plane and M denotes that part of the droplet surface which is in contact with air.

As long as the droplet is pending the force due to its weight is compensated by the force which originates from the surface tension of the water/air interface at the contact line (see Fig. 1). The infinitesimal force arising from an infinitesimal element of this circle can be decomposed into a horizontally and a vertically oriented component. Integrating along the contact line (i.e. the circle) the integral of the horizontal force component vanishes because of the axial symmetry of the situation while the vertical component of the tension force adds up to

$$
F_\sigma = 2\pi\sigma s \sin\theta
\tag{4}
$$

where σ denotes the surface tension between water and air. For a droplet with maximum volume F_σ should be balanced by the gravitational force

$$
F_g = \rho g V
\tag{5}
$$

caused by the droplet's weight. Exploiting expression (1) and then solving $F_g = F_\sigma$ for the radius s_m of the contact circle corresponding to a maximum volume droplet we arrive at

$$
s_m = \frac{l(1 + \cos\theta)}{\sqrt{2 + \cos\theta}}
\tag{6}
$$

The quantity

$$
l := \sqrt{\frac{6\sigma}{\rho g}} \approx 6.60 \times 10^{-3}\,\text{m}
\tag{7}
$$

(a)

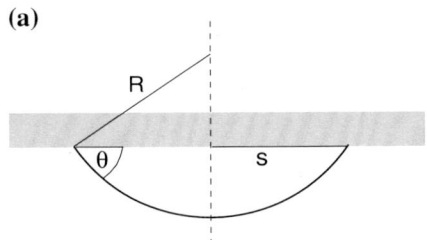

(b)

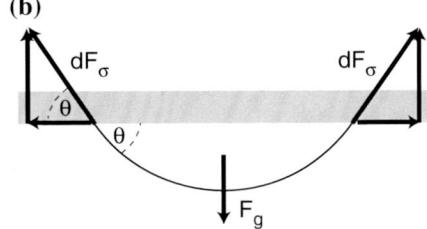

Figure 1

a Droplet hanging down from a horizontal plane. **b** Forces acting on the droplet. F_g denotes the force of gravity, F_σ the force due to the surface tension of the water/air interface at the contact line. Notice that the vectors $\mathbf{dF}_\sigma$ represent infinitesimal forces contributed by two infinitesimal elements of the contact line. Because these are diametrically located, the horizontal component of F_σ vanishes if integrated around the circular contact line. The vertical component of F_σ may surmount F_g, in which case the droplet remains attached to the horizontal plane. For $F_\sigma < F_g$ the droplet falls down

depends merely on the natural constants surface tension (between air and water) $\sigma \approx 72 \times 10^{-3}$ N/m, the gravitational acceleration $g \approx 10$ m/s^2 and the density of water $\rho \approx 10^3$ kg/m. Thus, it represents a characteristic length of the problem. (In the literature the expression $\sqrt{\sigma/(\rho g)}$ is known as the capillary constant.) From $s = R \sin \theta$ and (6) we find

$$R_m = \frac{l(1 + \cos \theta)}{\sin \theta \sqrt{2 + \cos \theta}} \tag{8}$$

and (upon insertion of (6) into expression (1)) we obtain the maximum volume V_m of the droplet:

$$V_m = \frac{\pi l^3}{3} \frac{(1 + \cos \theta) \sin \theta}{\sqrt{2 + \cos \theta}} \tag{9}$$

Notice that although $s_m(\theta)$, $R_m(\theta)$ and $V_m(\theta)$ represent a droplet which is—due to its weight—on the verge of falling down these functions depend still on the contact angle θ (see Fig. 2). Hence, we may calculate the contact angle(s) where they attain their maxima. Interestingly, the maxima of $s_m(\theta)$ and $V_m(\theta)$ do *not* coincide. We, rather, find:

$$\theta_{s_m} = 0° : \quad s_m(\theta_{s_m}) = \frac{2\sqrt{3}}{3} l \approx 7.62 \times 10^{-3} \text{ m} \tag{10}$$

$$\theta_{V_m} = \arccos(\sqrt{2} - 1) \approx 65.5° :$$
$$s_m(\theta_{V_m}) = \left(\sqrt{2} \left(\sqrt{2} - 1 \right) \sqrt{\sqrt{2} + 1} \right) l \tag{11}$$
$$\approx 6.01 \times 10^{-3} \text{ m}$$

$$V_m(\theta_{V_m}) = \frac{2\pi(\sqrt{2} - 1)}{3} l^3 \approx 250 \times 10^{-9} \text{ m}^3 \tag{12}$$

2.2. Droplet and Inclined Plane

In conjunction with horizontally oriented planes, only hanging droplets are interesting, because the vertically directed force of gravitation has no horizontal component which could push around droplets attached to the upper side of the plane. If the plane is inclined with respect to the horizontal the situation changes. Hence, we consider now both hanging and sitting droplets.

2.2.1 Forces Acting on a Hanging Droplet, Conditions for Detachment and Downslide

Consider a droplet of given volume V hanging down from a plane which is inclined against the horizontal by an angle α (Fig. 3). If α is increased, the shape of the droplet deviates—at first slightly, then increasingly—from being a segment of a sphere. The droplet as a whole, however, does not move. Eventually—when a critical angle of inclination has been reached—the droplet either starts to slide down or it detaches from the plane and falls down. Similarly, as in Sect. 2.2.1 (Eq. 6), we would like to find a relation between the "system defining" variables like s, θ and α which characterise the onset of slide or detachment.

Experimentally, it has been found that at the critical inclination the contact angle θ assumes along the "upstream" segment of the contact circle the receding value θ_r. At the "downstream" segment of the contact circle, however, the advancing contact angle θ_a is realised.

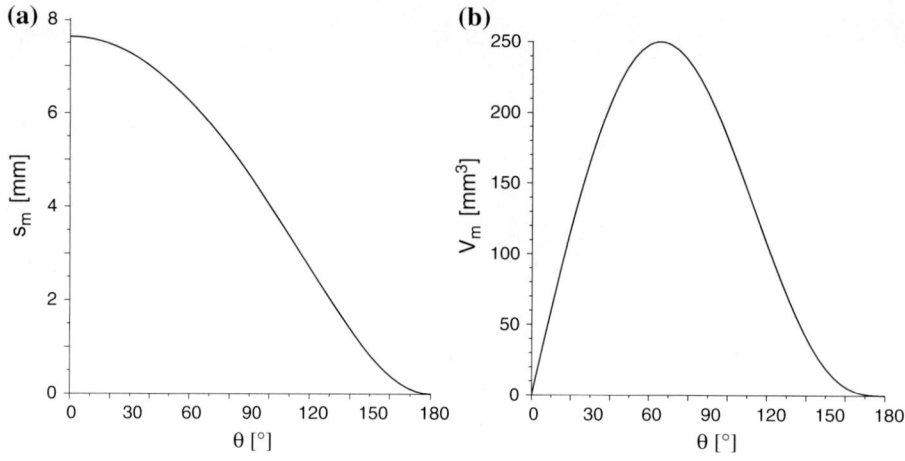

Figure 2

a Contact circle radius $s_m(\theta)$ related to the maximum volume of a droplet as a function of contact angle θ. The maximum of $s_m(\theta)$ is located at $\theta_{s_m} = 0°$ (see (10)). **b** Volume $V_m(\theta)$ related to the maximum volume of a droplet as a function of contact angle θ. The maximum of $V_m(\theta)$ is located at $\theta_{V_m} \approx 65.5°$ (see (12))

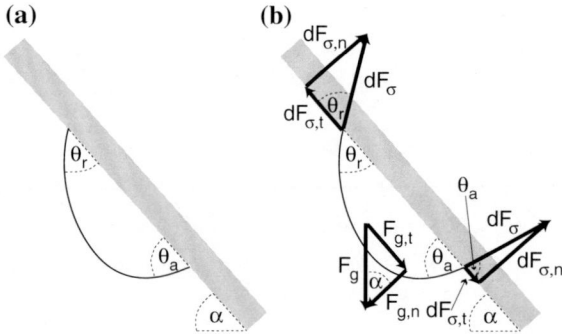

Figure 3

a Droplet hanging down from a plane which is inclined against the horizontal by an angle α. θ_r and θ_a denote the receding and advancing contact angle, respectively. **b** Forces acting on a droplet hanging down from a plane which is inclined against the horizontal by an angle α. θ_r and θ_a denote the receding and advancing contact angle, respectively. $\mathbf{F_g}$ denotes the force of gravity, $\mathbf{F_\sigma}$ the force due to the surface tension of the water/air interface at the contact line. Subscripts n and t denote normal and tangential components with respect to the plane. Because the infinitesimal force vectors $\mathbf{dF_\sigma}$ are oriented tangentially to the water/air interface, the inequality $\theta_r \neq \theta_a$ implies that the magnitude of their (infinitesimal) components parallel and normal to the plane vary along the contact line

Decomposition of $\mathbf{F_g}$ and $\mathbf{F_\sigma}$ into components parallel and normal to the inclined plane (Fig. 3) leads to two conditions:

(a) Downslide of the droplet sets in when the tangential component of gravity surmounts the tangential component of the surface tension

$$F_{g,t} \equiv \rho g V \sin \alpha \geq k\sigma w(\cos \theta_r - \cos \theta_a) \equiv F_{\sigma,t} \quad (13)$$

w denotes denotes the maximum halfwidth of the droplet. The detailed theory of a sliding droplet is rather involved (see EXTRAND and KUMAGAI 1995; PETRISSANS and CSCAPO 2003; PODGORSKI *et al.* 2001; DIMITRAKOPOULOS and HIGDON 1998, 1999; GLASNER 2007). Most authors agree on the structure of Eq. 13 but disagree on the value of the numerical constant k to which various values in the range $k = \pi/4...2$ have been assigned. Moreover, k depends also on the shape of the droplet.

(b) Detachment of the droplet from the plane occurs if the normal gravity component is greater than the normal component of the surface tension

$$F_{g,n} \equiv \rho g V \cos \alpha \geq \frac{\pi}{2} k\sigma w(\sin \theta_r + \sin \theta_a) \equiv F_{\sigma,n}$$

$$(14)$$

As they stand, Eqs. 13 and 14 are of limited use because they contain the droplet volume V which is difficult to obtain experimentally. If, however, the shape of the droplet deviates not too much from a segment of a sphere, the contact line is a circle with radius s and we can use Eq. 1 to express V in terms of s and the contact angle θ which is in this context defined as the arithmetic mean of the receding and the advancing contact angle:

$$\theta = \frac{\theta_a + \theta_r}{2} \qquad (15)$$

Moreover, if the droplet is quasi-spherically shaped, we may conclude $w = s$ and $k = 2$ (see PETRISSANS and CSCAPO 2003), which we shall assume in the sequel. Then, insertion of (1) into (13) and (14) allows us to solve these expressions for the contact circle radii s at which the equals signs in (13) and (14) are realised. Employing the definition

$$\chi := \frac{\theta_a - \theta_r}{2} \qquad (16)$$

and the relations

$$\cos x - \cos y = -2 \sin \frac{x+y}{2} \sin \frac{x-y}{2}$$
$$\sin x - \sin y = 2 \sin \frac{x+y}{2} \cos \frac{x-y}{2} \qquad (17)$$

conditions (a) and (b) can be stated as follows:

(a) Downslide sets in if $s > s_\sigma$, where

$$s_\sigma := \frac{l(1 + \cos \theta)}{\sqrt{2 + \cos \theta}} \sqrt{\frac{2 \sin \chi}{\pi \sin \alpha}} \qquad (18)$$

(b) Detachment occurs if $s > s_g$, where

$$s_g := \frac{l(1 + \cos \theta)}{\sqrt{2 + \cos \theta}} \sqrt{\frac{\cos \chi}{\cos \alpha}} \qquad (19)$$

Meaningful ranges of θ and χ are: $0 \leq \theta \leq \pi$ and $0 \leq \chi \leq \pi/2$. The latter is equivalent to $0 \leq \theta_a - \theta_r \leq \pi$.

Whether a droplet of contact radius s detaches itself from or slides down along the inclined plane depends on the relation between s, s_σ and s_g: for $s > s_g > s_\sigma$, sliding sets in, for $s > s_\sigma > s_g$, however, detachment occurs.

The relation between s_σ and s_g reduces via (18) and (19) to a relation between α and χ:

$$s_\sigma < s_g \iff \tan \alpha > \frac{2}{\pi} \tan \chi \qquad (20)$$
$$\text{(downslide if } s > s_g)$$

$$s_\sigma > s_g \iff \tan \alpha < \frac{2}{\pi} \tan \chi \qquad (21)$$
$$\text{(free fall if } s > s_\sigma)$$

2.2.2 Critical Volume of a Hanging Droplet, Contact Angle and Gliding Angle

Combination of (18) and (19) with (1) results in an expression for the critical droplet volume:

$$V_c := \left\{ \frac{\pi l^3}{3} \frac{(1 + \cos \theta) \sin \theta}{\sqrt{2 + \cos \theta}} \right\}$$
$$\times \begin{cases} \sqrt{\left(\frac{2}{\pi} \frac{\sin \chi}{\sin \alpha}\right)^3} & \text{if} \quad \tan \alpha > \frac{2}{\pi} \tan \chi \quad \text{(downslide if } V > V_c) \\ \sqrt{\left(\frac{\cos \chi}{\cos \alpha}\right)^3} & \text{if} \quad \tan \alpha < \frac{2}{\pi} \tan \chi \quad \text{(free fall if } V > V_c) \end{cases}$$
$$(22)$$

The condition that a pending droplet starts to move (either by downslide or by detachment) can be stated as $V > V_c$. Critical droplet volumes $V_c(\alpha)$ for a few combinations of θ and χ are depicted in Fig. 4. This figure illustrates also that the curves $V_c(\alpha)$ exhibit maxima at $\alpha_m = \arctan\left(\frac{2}{\pi} \tan \chi\right)$.

Notice, that the expression in braces equals the volume V_m (defined in (9)) which emerged in the context of droplets hanging down from a horizontal plane. In order to clarify the role of χ with respect to droplet motion we divide the equation $V = V_c$, which separates immobile ($V < V_c$) and falling or sliding ($V > V_c$) droplets, by V_m (see (9)). It then becomes feasible to reformulate the conditions for droplet (im-)mobility in terms of relations between the inclination angle α and the half-difference of advancing and receding contact angle χ: depending on the value of V/V_m, the pair of curves

$$\alpha = \begin{cases} \arcsin\left(\frac{2}{\pi} \left(\frac{V_m}{V}\right)^{\frac{2}{3}} \sin \chi\right) & \text{if} \quad \tan \alpha > \frac{2}{\pi} \tan \chi \\ \arccos\left(\left(\frac{V_m}{V}\right)^{\frac{2}{3}} \cos \chi\right) & \text{if} \quad \tan \alpha < \frac{2}{\pi} \tan \chi \end{cases}$$
$$(23)$$

divides the (χ, α)-plane into either (a) three sections, if V is in the interval $0 < V \leq (2/\pi)^{3/2} V_m \approx 0.5 V_m$ (Fig. 5a), or (b) two sections, if $(2/\pi)^{3/2} V_m < V < V_m$ applies (Fig. 5b). (χ, α)-pairs lying between the two curves indicate that a droplet of volume V remains immobile. Droplets (of this same volume) characterised by a (χ, α)-pair above/left of the upper curve slide down, droplets below/right of the lower curve detach from the plane and fall down.

The important conclusion is, that droplets of a given volume that fall into category (a) can be made to adhere

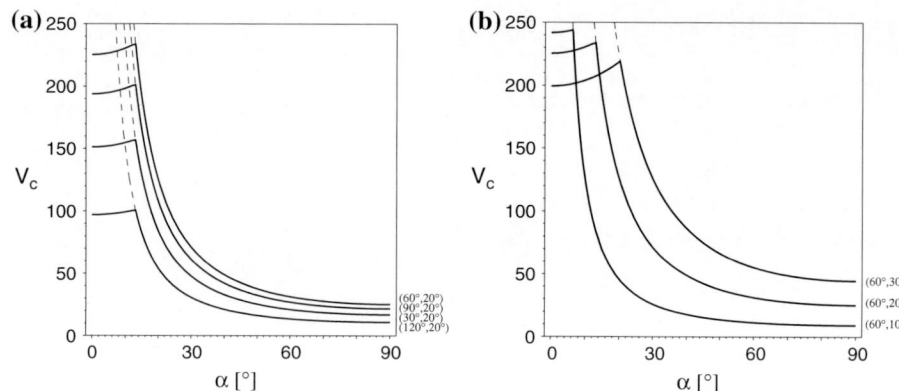

Figure 4

Critical volume V_c of a hanging droplet (*solid lines*) and of a sitting droplet (*broken lines*) as a function of the inclination angle α. Droplets characterised by (V, α)-pairs lying below a curve (V denotes the volume of the droplet) remain immobile, whereas droplets characterised by a (V, α)-pair above this curve either detach from the leaf and fall down (curve segments to the *left* of the cusps) or slide down while keeping contact with the inclined leaf (curve segments to the right of the cusps). **a** The different curves are generated by insertion of the values $(\theta, \chi) = (60°, 20°)$, $(90°, 20°)$, $(30°, 20°)$ and $(120°, 20°)$ (from *top* to *bottom*) into expression (22). **b** The different curves are generated by insertion of the values $(\theta, \chi) = (60°, 30°)$, $(60°, 20°)$ and $(60°, 10°)$ (from *top* to *bottom*) into expression (22)

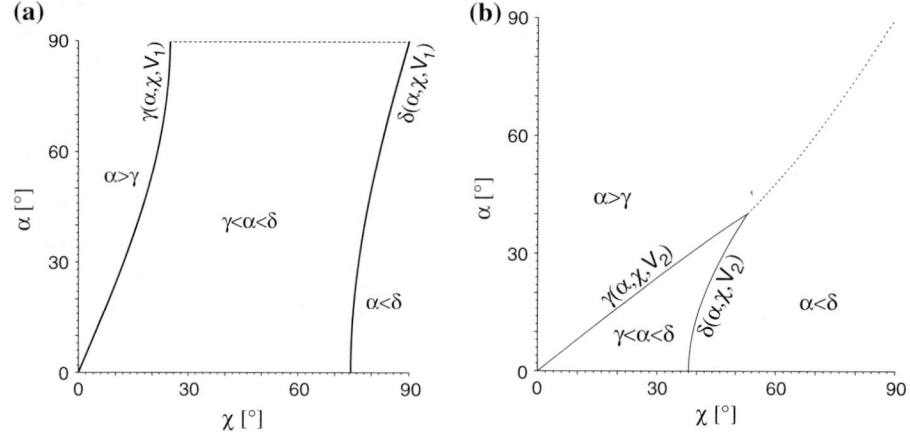

Figure 5

Areas of moving ($\alpha > \gamma$ and $\alpha < \delta$) and immobile ($\gamma < \alpha < \delta$) droplets hanging from a plane which is inclined against the horizontal by an angle α. χ denotes the half-difference of advancing and receding contact angle. $\gamma(\alpha, \chi, V)$ and $\delta(\alpha, \chi, V)$ denote gliding angle and "detachment angle" defined in Eqs. 24 and 25. Immobile droplets are related to (χ, α)-pairs lying within the tetragon. Droplets characterised by (χ, α)-pairs outside the tetragon either slide down (area denoted $\alpha > \gamma$) or detach from the plane and fall down (area denoted $\alpha < \delta$). **a** For $V_1 = 0.14V_m < (2/\pi)^{3/2}V_m$ the curves $\gamma(\alpha, \chi, V_1)$ and $\delta(\alpha, \chi, V_1)$ form a "tetragon of immobility" which covers the complete interval $0 < \alpha < \pi/2$. Hence, by adjusting χ appropriately, a droplet of this volume can be made to adhere to an arbitrarily inclined plane. **b** For $V_2 = 0.7V_m > (2/\pi)^{3/2}V_m$ the curves $\gamma(\alpha, \chi, V_2)$ and $\delta(\alpha, \chi, V_2)$ intersect and form a "triangle of immobility" which covers only part of the interval $0 < \alpha < \pi/2$. Hence, droplets of this volume cannot be kept immobile for every inclination $0 < \alpha < \pi/2$, even if χ can be arbitrarily chosen. The *broken curve* indicates whether a droplet slides down ($\alpha > \gamma$) or detaches ($\alpha < \delta$) from the plane

to an arbitrarily inclined plane if χ can be adjusted appropriately, whereas droplets belonging to category (b) cannot be kept immobile for every inclination $0 < \alpha < \pi/2$, even if χ can be arbitrarily chosen.

It is thus justified to identify the upper expression of (23) with the gliding angle γ (the angle of

inclination of an object that leads to the rolling off of drops lying upon the object) and its lower counterpart with a "detachment angle" δ (the angle of inclination of an object that leads to the detachment of drops from the object). Upon insertion of (9) expression (23) transforms to

$$\gamma := \arcsin\left(\sqrt[3]{\frac{8}{9\pi}\left(\frac{l^3}{V}\right)^2} \ \frac{\sin^2\theta\sin\chi}{\sqrt{[3]}(1-\cos\theta)^2(2+\cos\theta)} \right) \tag{24}$$

$$\delta := \arcsin\left(\sqrt[3]{\frac{\pi^2}{9}\left(\frac{l^3}{V}\right)^2} \ \frac{\sin^2\theta\cos\chi}{\sqrt{[3]}(1-\cos\theta)^2(2+\cos\theta)} \right) \tag{25}$$

where it is implicit that the condition $V \leq V_m$ should be fulfilled, if (24) and (25) are applied to hanging droplets. If so, the condition that a droplets starts to slide down can be expressed by the statement $\alpha > \gamma$, and the condition that a droplets starts to detach from the plane is equivalent to $\alpha < \delta$. Droplet immobility is realised where $\gamma < \alpha < \delta$ applies (cf. Fig. 5). If $\delta > \gamma$ is valid (as in the upper, right part of Fig. 5b) immobile droplets cannot exist.

2.2.3 Sitting Droplet, Conditions for Downslide

A droplet sitting on the upper surface of an inclined leaf cannot choose between "downslide" and "free fall" (although free fall may eventually become an option, if the droplet reaches the leaf margin). Thus, the lower line of (22) and "detachment condition" (25) become void. The condition that a droplet starts to move can be expressed either by the statement $V > V_c$, where

$$V_c = \left\{\frac{\pi l^3}{3}\frac{(1+\cos\theta)\sin\theta}{\sqrt{2+\cos\theta}}\right\}\sqrt{\left(\frac{2\sin\chi}{\pi\sin\alpha}\right)^3} \tag{26}$$

or by the condition $\gamma > \alpha$ (see (24)) which is in this case valid without the restriction $V \leq V_m$.

Similarly, as in the case of the hanging droplet, an increase of droplet volume brings about a drastic decrease of the (χ, α)-area wherein droplets remain sessile (Eq. 24; Fig. 6).

The main result of this section concerns the dependence of the critical volume $V_c(\theta, \chi, \alpha)$ on the contact angle θ, the half-difference of advancing and receding contact angle χ and the inclination α of the plane to which the hanging or sitting droplet is attached (cf. Figs. 4, 5):

- The critical volume shows a maximum with respect to θ at $\theta_{V_m} = \arccos(\sqrt{2}-1) \approx 65.5°$

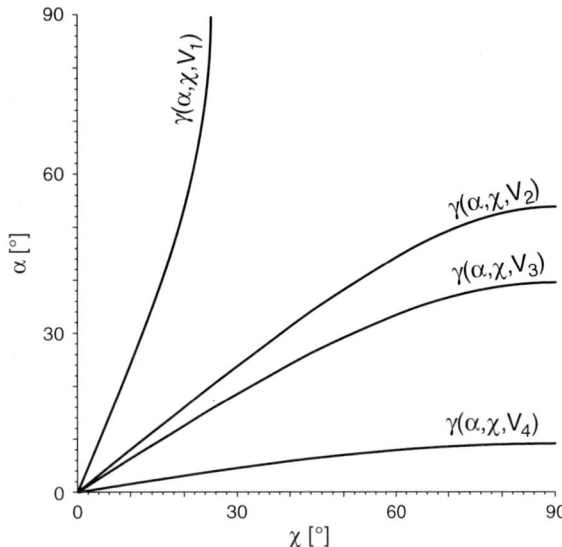

Figure 6

Areas of moving and immobile droplets sitting on a plane which is inclined against the horizontal by an angle α. χ denotes the half-difference of advancing and receding contact angle. The curves $\gamma(\alpha, \chi, V_k)$ ($k = 1, ..., 4$) represent the gliding angle defined in Eq. 24 for the droplet volume V_k. Immobile droplets are related to (χ, α)-pairs lying above/to the left of a curve related to a given V_k, droplets characterised by (χ, α)-pairs below/to the right of the same curve slide along the plane. For $V_1 = 0.14V_m < (2/\pi)^{3/2}V_m$ the curve covers the complete interval $0 < \alpha < \pi/2$. Hence, by adjusting χ appropriately, a droplet of this volume can be made to adhere to an arbitrarily inclined plane. The curves $\gamma(\alpha, \chi, V_k)$ related to $V_2 = 0.7V_m > (2/\pi)^{3/2}V_m$, $V_3 = V_m$ and $V_4 = 8\ V_m$, however, cover only part of the interval $0 < \alpha < \pi/2$. Hence, droplets of these volumes cannot be kept immobile for every inclination $0 < \alpha < \pi/2$, even if χ can be arbitrarily chosen

(for constant values of χ and α, Fig. 2b illustrates—apart from a constant factor—the θ-dependence of $V_c(\theta, \chi, \alpha)$).

- For sitting droplets (and for hanging droplets, provided $\tan\alpha > \frac{2}{\pi}\tan\chi$) the critical volume increases with increasing χ.
- For sitting droplets (and for hanging droplets, provided $\tan\alpha > \frac{2}{\pi}\tan\chi$) the critical volume decreases with increasing α.

Consequently, if a leaf surface wants to dispose of hanging or sitting water droplets for a given leaf inclination α, it has two options: (1) it may minimise contact angle hysteresis (i.e. aiming for $\chi \to 0$), or (2) it can try to produce a contact angle which either much smaller or much larger than the value $\theta_{V_m} \approx 65.5°$ for which V_c attains its maximum. Minimising

χ is probably the more promising choice, since it has—according to Fig. 4 or Eq. 22—the effect of shifting the maxima of the V_c-curves (which exist for hanging droplets only) to smaller values of α, thereby enhancing the tendency to detach or slip down of a droplet.

Contrariwise, if a leaf wants to store much water it should try to arrange for (1) a large angle χ and (2) a contact angle close to $\theta_{V_m} \approx 65.5°$.

3. Maximum Storage Capacity and Optimum Spacing of Droplets

Consider a leaf that wants to store as much water as possible in the form of droplets attached to its lower or upper surface. Two questions arise: (1) which geometric pattern should the droplets form, and, (2) exists an optimum contact angle?

Obviously, the optimum pattern is realised by partitioning the leaf surface into hexagons and "inscribing" into each of them one droplet. If the contact angle is in the range $0 \leq \theta \leq \pi/2$ the contact circle represents the greatest lateral extension of the droplet, hence the radius of the hexagon's incircle should equal s. For $\pi/2 \leq \theta \leq \pi$, however, $R > s$, thus the incircle radius should equal R. The respective areas amount to

$$A_{\text{hex}} = \begin{cases} 2\sqrt{3}s^2 & \text{if} \quad 0 \leq \theta \leq \pi/2 \\ 2\sqrt{3}R^2 & \text{if} \quad \pi/2 \leq \theta \leq \pi \end{cases} \quad (27)$$

In order to calculate the contact angle related to maximum storage capacity, we form the quantity μ :
= [maximum water volume stored in one droplet]/ [leaf area required for one droplet].

3.1. Horizontal Plane

Recalling that the volume of hanging droplets exhibits a maximum value V_m which depends, according to (9), on the contact angle θ, we find from (6), (8), (9) and (27)

$$\mu_m := \begin{cases} \left(\frac{l\pi\sqrt{3}}{18}\right) \frac{\sin\theta\sqrt{2+\cos\theta}}{1+\cos\theta} & \text{if} \quad 0 \leq \theta \leq \pi/2 \\ \left(\frac{l\pi\sqrt{3}}{18}\right) \frac{\sin^3\theta\sqrt{2+\cos\theta}}{1+\cos\theta} & \text{if} \quad \pi/2 \leq \theta \leq \pi \end{cases} \quad (28)$$

It appears that μ_m features a maximum with respect to θ which is located at

$$\theta_{\mu_m} = \pi - \arctan\left(\frac{\sqrt{10\sqrt{10}+26}}{3}\right) \approx 111.6°, \quad (29)$$

i.e. in the hydrophobic range of contact angles. Insertion into (28) produces

$$\mu_{m,\max} = \left(\frac{l\pi\sqrt[4]{1125}}{2250}\right)\sqrt{\left(\sqrt{10}-1\right)^3\left(\sqrt{5}+\sqrt{2}\right)}$$

$$\approx 0.49l \quad (30)$$

Since droplets sitting on a horizontal plane cannot be detached by gravitation, their volume is—in principle—unlimited; that is, both s and R can be increased (almost) indefinitely. Forming $\mu = V/A_{\text{hex}}$ from expressions (1) and (27) results in

$$\mu = \begin{cases} s\left(\frac{\pi\sqrt{3}}{18}\right) \frac{(1-\cos\theta)^2(2+\cos\theta)}{\sin^3\theta} & \text{if} \quad 0 \leq \theta \leq \pi/2 \\ R\left(\frac{\pi\sqrt{3}}{18}\right)(1-\cos\theta)^2(2+\cos\theta) & \text{if} \quad \pi/2 \leq \theta \leq \pi \end{cases}$$

$$\quad (31)$$

The absence of θ-dependent upper limits for s and R effects the θ-dependence of μ (compared to μ_m): other than (28), expression (31) exhibits no maximum with respect to θ.

3.2. Inclined Plane

Assuming that the shape of droplets attached to an inclined plane deviates not much from a segment of a sphere, expression (31) is valid, provided that $V < V_c$ (where V and V_c are defined in (1) and (22), respectively).

In the limit $V = V_c$, expression (27) assumes upon use of (22) (31) (respectively, (26)), (18) (19) and (27) the form

$$\mu_c := \frac{V_c}{A_{\text{hex}}} = \mu_m(\theta)$$

$$\times \begin{cases} \sqrt{\left(\frac{2}{\pi}\frac{\sin\chi}{\sin\alpha}\right)^3} & \text{if} \quad \tan\alpha > \frac{2}{\pi}\tan\chi \quad \text{(downslide if } V > V_c) \\ \sqrt{\left(\frac{\cos\chi}{\cos\alpha}\right)^3} & \text{if} \quad \tan\alpha < \frac{2}{\pi}\tan\chi \quad \text{(free fall if } V > V_c) \end{cases}$$

$$\quad (32)$$

Notice that the θ-dependence of μ_c is contained in $\mu_m(\theta)$ (as given in (28)). Therefore, it shares with $\mu_m(\theta)$ the maximum described in (29). As it stands, expression (32) is valid for hanging droplets of critical (maximum) volume V_c. Sitting droplets of volume V_c are represented by the upper line of (32), only.

4. Lifetime of a Droplet

In Sect. 3 we explored the maximum storage capacity of a leaf. In this section we evaluate (1) the lifetime of a droplet subjected to evaporation, and (2) whether its lifetime depends on its radius and on the leaf contact angle.

4.1. Horizontal Plane

A droplet which is attached to a leaf loses water by evaporation through the droplet's water/air interface M (see (3)). This particle loss leads to a decrease of the droplet volume V, according to

$$-\frac{dV}{dt} = j_M M \qquad (33)$$

For what follows, we assume that the evaporation flux j_M is a constant with respect to t and θ. Applying textbook thermodynamics (e.g. REIF 1974; ATKINS 1998), j_M can—under these assumptions—be expressed as

$$j_M = D_{wv} V_{wv} \left(\frac{c_{sat} - c_{wv}}{b} \right) \qquad (34)$$

D_{wv} denotes the diffusional constant of water vapour in air, V_{wv} the molar volume of water vapour, b is the thickness of the boundary layer surrounding the leaf. c_{wv} denotes the atmospheric molar concentration of water vapour, and c_{sat} the saturation value related to it.

Employing the differentiation rule $dV(s)/dt = (dV/ds)(ds/dt)$ and expressions (1)–(3), Eq. 33 becomes, after a few rearrangements, a simple differential equation for $s(t)$

$$-\frac{ds}{dt} = \frac{2j_M(1 + \cos\theta)}{(2 + \cos\theta)\sin\theta} \qquad (35)$$

with the solution

$$s(t) = s_0 - \frac{2j_M(1 + \cos\theta)}{(2 + \cos\theta)\sin\theta} t \qquad (36)$$

where s_0 denotes the radius of the contact line at time $t = 0$ when evaporation and/or absorption set in

(cf. Fig. 1). The lifetime τ of the droplet is obtained by letting $s = 0$ in (36) and solving for t,

$$\tau = \frac{(2 + \cos\theta)\sin\theta}{2j_M(1 + \cos\theta)} s_0 \qquad (37)$$

If the initial radius of the contact line attains the value s_m (corresponding to the maximum volume $V_m(\theta)$ of a hanging droplet), the droplet's lifetime becomes

$$\tau_m = \frac{l\sin\theta\sqrt{2 + \cos\theta}}{2j_M} \qquad (38)$$

It is interesting to calculate the contact angle θ_τ for which the lifetime τ_m becomes a maximum. It turns out that it depends neither on l nor on j_M. Its value amounts to

$$\theta_\tau := \arccos\left(\frac{\sqrt{7} - 2}{3}\right) \approx 77.5° \qquad (39)$$

which is in the hydrophilic range of contact angles. Upon insertion of θ_τ into τ_m we find

$$\tau_{m,max} = \frac{l}{j_M} \frac{\sqrt{10 + 7\sqrt{7}}}{3\sqrt{6}} \approx 0.73 \frac{l}{j_M} \qquad (40)$$

An explicit expression for $V(t)$ results from insertion of (36) into (1):

$$V(t) = \frac{\pi}{3}(2 + \cos\theta)\sqrt{1 - \cos\theta}$$
$$\times \left\{ \frac{s_0}{\sqrt{1 + \cos\theta}} - \frac{2j_M}{(2 + \cos\theta)\sqrt{1 - \cos\theta}} t \right\}^3 \qquad (41)$$

4.2. Inclined Plane

If the plane is inclined, the basic Eq. 33 remains valid both for hanging and sitting droplets. Treating the droplets as quasi-spherical segments, the same reasoning as above applies with the same results regarding the t-dependance of s and V and the droplet lifetime (Eqs. 36, 37 and 41, respectively).

In the case of hanging droplets we have to take into account that the initial value s_0 of the contact circle should—due to the inclination of the leaf—fulfill the conditions $s_0 < s_\sigma$ and $s_0 < s_g$ [see (18) and (19)].

W. Konrad et al.   Pure Appl. Geophys.

If a hanging droplet is initially of maximum volume (i.e. $s_0 = s_\sigma$ or $s_0 = s_g$, depending on the values of the inclination angle α and on the half-difference of advancing and receding contact angle χ), τ simplifies to

$$\tau_c = \left\{ \frac{l \sin\theta \sqrt{2 + \cos\theta}}{2 j_M} \right\} \\ \times \begin{cases} \sqrt{\frac{2}{\pi} \frac{\sin\chi}{\sin\alpha}} & \text{if} \quad \tan\alpha > \frac{2}{\pi}\tan\chi \\ \sqrt{\frac{\cos\chi}{\cos\alpha}} & \text{if} \quad \tan\alpha < \frac{2}{\pi}\tan\chi \end{cases} \quad (42)$$

where the expression in braces is just τ_m, the lifetime of a droplet hanging from a horizontal plane.

5. Discussion

The easiness with which water drops roll off a surface is mainly dependent on the contact angle hysteresis of a surface, and not on the static contact angle. The smaller the contact angle hysteresis the easier a drop will roll off the surface. Thus, hydrophobicity does not necessarily lead to a low stickiness of drops to the surface. To determine the stickiness of drops to surfaces it is not sufficient to measure the static contact angle θ. It is necessary to also measure χ. Potential benefits of efficient shedding of drops for a leaf are frequently discussed with respect to self cleaning. Pathogens, such as fungal spores, are easily washed off then, or are prevented from germination on the constantly dry surface (BARTHLOTT and NEINHUIS 1997). If a plant species pursues the strategy to efficiently get rid of water drops upon its surface, it should minimize χ. This quantity is usually influenced by the often heterogeneous nature of surfaces, with respect to surface micro/nanostructure and/or local contact angle variations (QUÉRÉ 2008). In leaves, these surface effects will be mostly created by cuticle wax structures and/or trichomes.

Another benefit has to do with photosynthesis since the development of a water film above stomata, the gas exchange pores of a leaf, is deleterious for photosynthesis. To keep stomata free from a water film, it is beneficial—besides to get rid of the drops by a low χ—to reduce the contact area between a droplet and the surface. Consequently, high water repellency, together with low stickiness for drops, is expected for species in foggy habitats and especially for stomatous leaf sides (frequently the abaxial side). HOLDER (2007) found that for different forest species θ tended to be unexpectedly low for many cloud forest species. He discussed this result as being caused by the high erosion of cuticle waxes due to intense precipitation. However, since χ was not determined in Holder's study, leaves may pursue also other strategies by their surface properties than easy shedding of spherical drops. For example, it may be speculated that fog harvesting species are interested in collecting large drops before they roll off the leaf and fall to the ground. Fog drip can be an important source of water input (FENG et al. 2008). The shedding of large drops is expected to be more efficient in wetting the soil around the plant than the shedding of many small or very small droplets because large drops have a higher potential for throughfall. Furthermore, due to their low surface-to-volume ratio, they will resist evaporation much more strongly than small droplets. Maximum drop size can be achieved with a $\theta \approx 65.5°$. A good ability for water retention may have also other beneficial aspects. There are various plants that are able to absorb water via the leaves. This was unambiguously demonstrated for Californian redwoods (BURGESS and DAWSON 2004), but also indicated for other species (BRESHEARS et al. 2008). If water absorption by leaves represents an important water source for a species, then it should be expected that both a good wetting behavior (low θ and a high χ) leading to a high persistence of the water on the leaf is beneficial. This would then prolong the water amount and the time interval available for absorption. The highest leaf storage capacity of a leaf would be obtained with a contact angle of $\theta \approx 111.6°$ and a high χ. For maximum persistence of a drop under given meteorological conditions, the contact angle should amount to $\theta \approx 77.5°$.

Furthermore, nutrient input via aerosols are important for many ecosystems, and it was shown that hygroscopic mineral salts can be found upon leaf surfaces (BURKHARDT 2010). Absorption not only of water but also of minerals dissolved in leaf surface water may, therefore, be of substantial importance for species living on nutrient-poor soils (BURKHARDT 2010) and a good water retention should be also beneficial in these cases.

Reprinted from the journal   80

Leaf surface properties do not only affect the hydrological cycle via canopy interception but can also be important for the plant. The ecophysiological interrelationships between leaf surface and drops can be manyfold, and may additionally be changed by external factors, such as dust particles, abrasion or insect-mediated structural changes. Furthermore, the mechanical impacts of wind currents or animal activities upon leaves will frequently be sufficient to lead to detachment of drops that otherwise would be attached quite firmly to the leaf surface. The consequences and impacts of the behaviour of deposited drops upon leaf surfaces are complex for a considered species and depend on its habitat, ecological niche and other ecophysiological traits. Often, an interrelationship to any vital function cannot be provided. The ability of hydrophobic petals, for example, to retain a water droplet firmly, even if upside down (Petal Effect) (FENG et al. 2008), can presently not be explained as being of any adaptive value. Considering the various complex interrelationships of leaves with their environment, the interaction of leaf surfaces with water is a fascinating and important topic that deserves further attention.

Acknowledgments

This work was supported by the German Federal Ministry of Education and Research by a grant to A. R. -N., within the project 02WT0906.

REFERENCES

ABOAL, J.R., JIMÉNEZ, M.S., MORALES, D., HERNÁNDEZ, J.M. (1999), *Rainfall interception in laurel forest in the Canary Islands*. Agricultural and Forest Meteorology 97, 73–86.

ASTON, A.R. (1979), *Rainfall interception by eight small trees*. Journal of Hydrology 42, 383–396.

ATKINS, P., Physical Chemistry (W.H. Freeman, New York 1998).

BARTHLOTT, W., NEINHUIS, C. (1997), *Purity of the sacred lotus, or escape from contamination in biological surfaces*. Planta 202, 1–8.

BRESHEARS, D., McDOWELL, N.G., GODDARD, K. L., DAYEM, K. E., MARTENS, S. N., MEYER, C. W., BROWN, K. M. (2008), *Foliar absorption of intercepted rainfall improves woody plant water status most during drought*. Ecology 89, 41–47.

BURGESS, S.S.O., DAWSON, T.E., (2004), *The contribution of fog to the water relations of Sequoia sempervirens (D. Don): foliar uptake and prevention of dehydration*. Plant, Cell and Environment 27, 1023–1034.

BURKHARDT, J. (2010), *Hygroscopic particles on leaves: nutrients or desiccants?* Ecological Monographs, 80, 369–399.

CALDER, I. (1990), Evaporation in the uplands. John Wiley and sons Inc., Chichester. 149 pp.

CALIES, M., QUÉRÉ, D. (2005), *On water repellency*. Soft Matter 1, 55–61.

DIMITRAKOPOULOS, P., and HIGDON, J.J.L. (1998), *On the displacement of three-dimensional fluid droplets from solid surfaces in low-Reynolds-number shear flows*, J. Fluid Mech. 377, 189–222.

DIMITRAKOPOULOS, P., and HIGDON, J.J.L. (1999), *On the gravitational displacement of three-dimensional fluid droplets from inclined solid surfaces*, J. Fluid Mech. 395, 181–209.

DINGMAN, S., Physical hydrology (Prentice Hall, Upper Saddle River 2002).

EXTRAND, C.W., and KUMAGAI, Y., (1995), *Liquid drops on a inclined plane: the relation between contact angles, drop shape, and retentive force*. Journal of Colloid and Interface Science 170, 515–521.

FENG, L., ZHANG, Y.A., XI, J.M., ZHU, Y., WANG, N., XIA, F., JIANG, L. (2008), *Petal effect: Two major examples of the Cassie-Baxter model are the Petal effect and Lotus effect. A superhydrophobic state with high adhesive force*. Langmuir 24, 4114–4119.

GASH, J.H.C. (1979), *An analytical model of rainfall interception by forests*. Journal of Hydrology 105, 43–55.

GLASNER, K.B. (2007), *The dynamics of pendant droplets on a one-dimensional surface*, Physics od Fluids 19, 102–104.

HERWITZ, S.R. (1985), *Interception storage capacities of tropical rainforest canopy trees*. Journal of Hydrology 77, 237–252.

HÖRMAN, G., BRANDING, A., CLEMEN, T., HERBST, M., HINRICHS, A. (1996), *Calculation and simulation of wind controlled canopy interception of beech forest in Northern Germany*. Agricultural and Forest Meteorology 79, 131–148.

HOLDER, C.D. (2007), *Leaf water repellency of species in Guatemala and Colorado (USA) and its significance to forest hydrology studies*. Journal of Hydrology 336, 147–154.

MUZYLO, A., LLORENS, P., VALENTE, F., KEIZER, J.J., DOMINGO, F., GASH, J.H.C. (2009), *A review of rainfall interception modelling*. Journal of Hydrology 370, 191–206.

PETRISSANS, M., and CSCAPO, E. (2003), *Retention of glycerol sessile drop on MDF wood material*, Holz als Roh- und Werkstoff, 61, 12–116.

PODGORSKI, T., FLESSELLES, J.-M., and LIMAT, L. (2001), *Corners, Cusps, and Pearls in Running Drops*, Phys. Rev. Lett. 87, 036102.

QURÉRÉ, D. (2008), *Wetting and roughness*. Annual Review of Materials Research 38, 71–99.

REIF, F., Fundamentals of statistical and thermal physics (McGraw-Hill, New York 1974).

SHIRTCLIFFE, N.J., McHALE, G., NEWTON, M.I. (2009), *Learning from superhydrophobic plants: The use of hydrophilic areas on superhydrophobic surfaces for droplet control*. Langmuir 25, 14121–14128.

(Received November 15, 2010, revised February 26, 2011, accepted March 31, 2011, Published online July 9, 2011)

Pure Appl. Geophys. 169 (2012), 847–857
© 2011 Springer Basel AG
DOI 10.1007/s00024-011-0328-9

A Method for Direct Assessment of the "Non Rainfall" Atmospheric Water Cycle: Input and Evaporation From the Soil

KUDZAI FARAI KASEKE,[1,4,6] ANTHONY J. MILLS,[2] ROGER BROWN,[3] KAREN J. ESLER,[1] JOHANNES. R. HENSCHEL,[4] and MARY K. SEELY[5]

Abstract—"Non rainfall" atmospheric water (dew, fog, vapour adsorption) supplies a small amount of water to the soil surface that may be important for arid soil micro-hydrology and ecology. Research into the direct effects of this water on soil is, however, lacking due to instrument and technical constraints. We report on the design, development, construction and findings of an automated microlysimeter instrument to directly measure this soil water cycle in Stellenbosch, South Africa during winter. Performance of the microlysimeter was satisfactory and results obtained were compared to literature and fell within the expected range. "Non rainfall" atmospheric water input into bare soil (river sand) was between 0.88 and 1.10 mm per night while evaporation was between 1.39 and 2.71 mm per day. The study also attempted to differentiate the composition of "non rainfall" atmospheric water and results showed that vapour adsorption contributed the bulk of this input.

Key words: "Non rainfall" atmospheric water, dew, vapour adsorption, microlysimeter.

1. Introduction

"Non rainfall" atmospheric water (fog, dew, vapour adsorption) supplies a small but critical amount of water to arid zones (MALEK *et al.*, 1999;

KIDRON, 2000), promoting biological decomposition and nutrient recycling in the upper few centimetres of the soil profile (WHITFORD, 2002). The ecological significance of this water input is highlighted by the fact that in some arid environments, dew and fog input equals or exceeds annual rainfall and is the sole source of liquid water for plants (AGAM and BERLINER, 2006).

Despite its acknowledged importance, there is no international standard for measuring this input directly into the soil (ZANGVIL, 1996; BROWN *et al.*, 2008). This is primarily because of the small fluxes involved which pose difficulties for measurement (SCOTT, 1962; JACOBS *et al.*, 2002; NINARI and BERLINER, 2002). MONTEITH and UNSWORTH (1990), estimated the theoretical dew maximum at 0.40 mm per night. Thus, instrumentation with a minimum resolution of less than 0.10 mm would be ideal for such studies (AGAM and BERLINER, 2006). However, achieving this degree of accuracy in a highly mobile soil environment is difficult given the potential interferences from inter alia wind, animals and drifting sand (BROWN *et al.*, 2008).

Lack of agreement on the very definition of "non rainfall" atmospheric water and its vectors has further compounded the problem of lack of appropriate instrumentation (NOFFSINGER, 1965; ZANGVIL, 1996). Several methods and instruments have been developed to measure this water, e.g., fog collector (SCHEMENAUER and CERECEDA, 1994), cloth plate method and Duvdevani blocks (KIDRON, 2000). These methods are useful in that they provide proxy measurements for inter-site comparisons (BERKOWICZ *et al.*, 2001), but are limited in their application because they rely on artificial collecting surfaces which differ from the natural receiving substrate

[1] Department of Conservation Ecology and Entomology, Stellenbosch University, P. Bag X1, Matieland 7602, South Africa. E-mail: faraikaseke@gmail.com; kje@sun.ac.za
[2] Department of Soil Science, Stellenbosch University, P. Bag X1, Matieland 7602, South Africa. E-mail: mills@sun.ac.za
[3] Climate Analysis Systems Group, University of Cape Town, P. Bag X3, Rondebosch 7701, South Africa. E-mail: brown.roger@gmail.com
[4] Gobabeb Research Centre, P. O. Box 953, Walvis Bay, Namibia. E-mail: joh.henschel@gobabeb.org
[5] Desert Research Foundation of Namibia, 7 Rossini Street, Windhoek, Namibia. E-mail: Mary.seely@drfn.org
[6] 26 Gobvu Road, Zengeza 3, Chitungwiza, Zimbabwe.

(soil) surface. They, therefore, tend to under- and/or over-estimate "non rainfall" atmospheric water input direct into the soil (BERKOWICZ *et al.*, 2001; NINARI and BERLINER, 2002; HEUSINKVELD *et al.*, 2006).

Human reliance for observation is an additional handicap that makes these traditional methods of instrumentation unsuitable for studies in remote or sparsely populated areas (JACOBS *et al.*, 2002). There is, therefore, a need for the development of automated instrumentation for "non rainfall" atmospheric water studies (HEUSINKVELD *et al.*, 2006) and, according to NINARI and BERLINER (2002), the loadcell microlysimeter method is the most promising approach. The microlysimeter method is an in situ method that directly measures mass loss or gain (HEUSINKVELD *et al.*, 2006) and residence time of water derived from "non rainfall" atmospheric sources (BROWN *et al.*, 2008) in a soil sample.

1.1. Microlysimeter Design and Specifications

The microlysimeter method has been applied to evaporation studies from the soil surface of irrigated crops (STORLIE and ECK, 1996). However, design specifications were meant for a high evaporation flux after irrigation (BOAST and ROBERTSON, 1982), whilst "non rainfall" atmospheric water cycles involve a much smaller latent heat flux (NINARI and BERLINER, 2002). Design and specifications of the microlysimeter method for high evaporation fluxes are thus unsuitable for "non rainfall" atmospheric water studies and consequently require adjustments. Figure 1 shows a schematic of the microlysimeter developed for this study which is based on the automated loadcell microlysimeter design by HEUSINKVELD *et al.* (2006) and BROWN *et al.* (2008).

We used a Tedea-Huntleigh 1004 (1,500 g) aluminium single point low capacity loadcell, dimensions (110 × 10 × 33) mm. Excitation with 10 V DC is stated as 7.69 mV which translates to 0.0051 mV g^{-1}. Temperature variation of the loadcell can cause significant errors to loadcell output (STORLIE and ECK, 1996). However, these temperature effects on loadcell output can be buffered by installing the loadcell below ground. This measure significantly reduced temperature variation on the loadcell but we still needed to account for the effect

of the remaining temperature variation on loadcell output. We thus housed the loadcell in a chamber with an aluminium base plate (10 mm thick) and sides/top of polyvinylchloride (PVC) (5 mm thick). The PVC acted as an additional heat buffer to the loadcell from lateral and vertical heat flow from the surface and surrounding soil. Assuming 100% thermal conductivity by the aluminium base plate, loadcell temperature variation should be similar to that of the underlying soil. Based on these assumptions we calculated and compensated for the temperature error effect on loadcell output (STORLIE and ECK, 1996). A small depression was carved onto the base plate to allow the loadcell to bend, and an overload screw was inserted on the loading end of the loadcell through the base plate to protect against overload and damage to the loadcell (HEUSINKVELD *et al.*, 2006).

Vertical heat conduction towards the loadcell from the surface was further reduced by attaching the soil-sample dish to the loadcell via a 180 mm non-conductive plastic pipe (HEUSINKVELD *et al.*, 2006). This connecting pipe slotted into a plastic holder on the loading end of the loadcell, further reducing heat conduction from the surface (Fig. 1). The connecting pipe also reduced dead weight exerted on the loadcell by microlysimeter components keeping the unit within loadcell capacity. The connecting tube encased the connecting pipe and limited any lateral movement due to air turbulence affecting the sample dish at the surface. The connecting pipe also slotted into a cap on the underside of the sample dish, that was designed with a 5 mm protruding ridge (overflow protector) at the bottom. The purpose of the ridge was to divert possible overflow from the sample dish from running down the connecting pipe directly into the loadcell compartment, thereby ensuring a dry environment in the loadcell compartment.

Theoretically, microlysimeters provide an absolute reference for latent heat fluxes provided that the soil and heat balance of the microlysimeter are similar to those of the surrounding area (NINARI and BERLINER, 2002). Microlysimeter casing affects thermal transfer between surrounding soil and the sample in the microlysimeter (EVETT *et al.*, 1995). Sample dishes constructed from high thermal conducting materials do result in lower soil sample surface

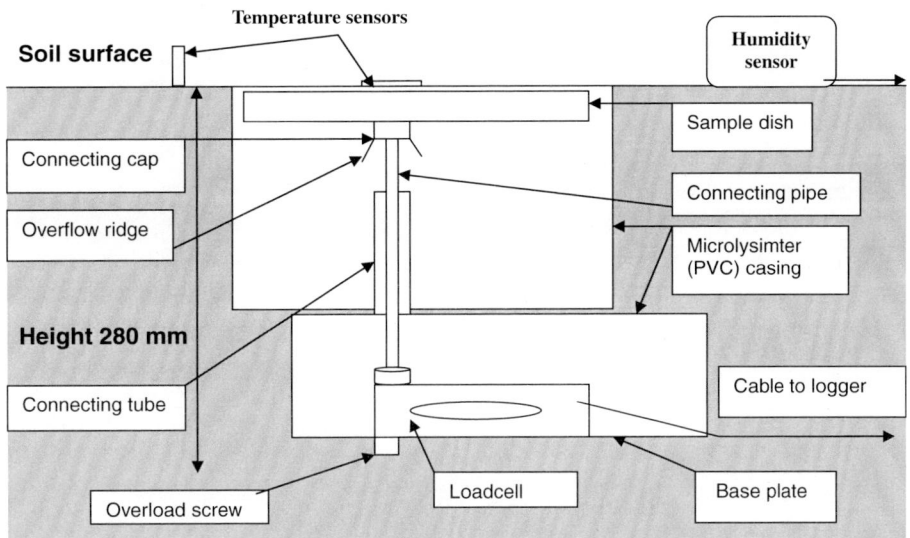

Figure 1
A schematic illustration of the automated loadcell microlysimter, designed following HEUSINKVELD et al., (2006) and BROWN et al., (2008)

temperature in the microlysimeter compared to adjacent soils (EVETT et al., 1995). During the day, these materials conduct heat away from the surface of the sample and the reverse occurs at night. This energy reduction may result in lower evaporation estimates from metal microlysimeters during the day than actual field evaporation (EVETT et al., 1995). It is, therefore, impossible to eliminate material influences on the sample, but selection of low thermal conducting materials could minimise the influence of the microlysimeter on the soil sample (STORLIE and ECK, 1996). According to WALKER (1983), there is no significant temperature difference between soil samples in plastic sample dishes and the adjacent soil. EVETT et al. (1995) recommended the use of PVC which has a similar thermal conductivity to that of a dry mineral soil to construct sample dishes. Sample dishes used for this study were constructed from 2 mm transparent polypropylene to minimise material influences on sample temperature (EVETT et al., 1995; HEUSINKVELD et al., 2006).

Insufficient depth and surface of the sample dish can significantly affect the soil heat balance through distortion of water and temperature profiles in the sample (NINARI and BERLINER, 2002). However, according to JACOBS et al. (1999), the daily moisture cycle is confined to the upper 2–3 cm of the soil profile and there is no significant difference between

sample dish depths ranging between 30 and 70 mm. We thus adopted a 35 mm sample dish depth, as a shallower depth would block vapour transport to deeper soil layers and lead to possible under- and or over-estimation of "non rainfall" atmospheric water input (HEUSINKVELD et al., 2006). A larger depth would exert more dead weight onto the loadcell and decrease loadcell sensitivity. In theory, small sample dishes exacerbate edge effects. However, DAAMEN et al. (1993) found no evidence to support this after testing dish sizes ranging from 50 to 210 mm in diameter. Larger sample dish sizes passively amplify the signal (BROWN et al., 2008) and reduce the aerodynamic edge effects (HEUSINKVELD et al., 2006) provided the dead weight is kept within loadcell capacity. We settled for a 140 mm diameter plastic soil-sample dish for this study.

FRANCIS et al. (2007) pointed to the possibility of moisture recharge from below the soil surface (capillary rise and overnight distillation processes). It was thus necessary to physically isolate the microlysimeter sample from below ground moisture recharge by using a non-porous sample dish. Physical isolation of the sample ensures that water input into the sample is from the atmosphere and not from below ground (BROWN et al., 2008). HEUSINKVELD et al. (2006) noted the need for automated instrumentation for "non rainfall" atmospheric water studies, given the handicap of traditional methods which rely on human observation and are unsuitable for remote areas (JACOBS *et al.*, 2002). Our microlysimeter was therefore fully automated, capable of repeated measurements and solar powered. After compensating for temperature error effects on loadcell output and isolating the sample from below ground moisture recharge, it was assumed that the remaining weight change would reflect "non rainfall" atmospheric water input and evaporation from the soil sample.

1.2. Vector Definitions

"Non rainfall" atmospheric water in arid ecosystems is supplied via three vectors namely: fog, dew and vapour adsorption (AGAM and BERLINER, 2006). However, for the duration of this study, no fog was observed. It is, nonetheless, important to define dew and vapour adsorption as used in this particular study.

Dew is the natural condensation of water vapour into liquid droplets on a sufficiently cooled substrate surface (STONE, 1963; BEYSENS, 1995; AWANOU and HAZOUME, 1997; MALEK *et al.*, 1999). It is a phase transition on the soil–plant–atmosphere interface affecting energy balance. Dew formation is dependent on the receiving substrate characteristics (BEYSENS, 1995), occurring when the substrate surface equals or falls below ambient dew point temperature but is above freezing point, otherwise frost forms (ASHBEL, 1949; BEYSENS, 1995; AGAM and BERLINER, 2006).

Vapour adsorption is a reversible interfacial physical process resulting from differential forces of attraction and repulsion between vapour molecules and soil particles (AGAM and BERLINER, 2006). It occurs as a result of vapour movement from the atmosphere into the soil due to the establishment of a vapour gradient between the two. Although there are two types of adsorption—physical and chemical— physical adsorption is dominant in the soil because the energy available under natural conditions cannot support chemical adsorption (HILLEL, 1998). This study makes no attempt to distinguish between osmotic effects due to salinity and strict vapour adsorption. Instead, their combined effect is collectively referred to as vapour adsorption or hygroscopic uptake.

In theory, conditions conducive for one input vector preclude the others from occurring concurrently (BROWN *et al.*, 2008). Dew formation occurs when the receiving substrate surface temperature equals or falls below ambient dew point (BEYSENS, 1995), and vapour adsorption occurs when a vapour gradient is established between the atmosphere and the soil, independent of dew point temperature (BROWN *et al.*, 2008). Therefore, the receiving substrate (soil) surface temperature can be used to distinguish between dew and vapour adsorption input (AGAM and BERLINER, 2006). According to our study, input that occurred when soil surface temperature was below ambient dew temperature was classified as dew and if temperature was above ambient dew point this was classified as vapour adsorption.

2. Materials and Methods

The microlysimeters were calibrated in the laboratory and the effect of temperature variation on loadcell output was calculated (Fig. 2). The temperature response graph was calculated by exposing the microlysimeters to varying temperature regimes (range 20.7–26.3°C) whilst loaded with a standard 500 g mass. Calibration of the microlysimeter unit showed that for the dimensions of the loading dish used in this study, the minimum resolution of the loadcell was 0.038 g equivalent to a depth of 0.0026 mm of water.

After laboratory calibrations, field trials were conducted from 11 to 23 July 2008 at Stellenbosch University farm (S33 56.577 E18 51.152).

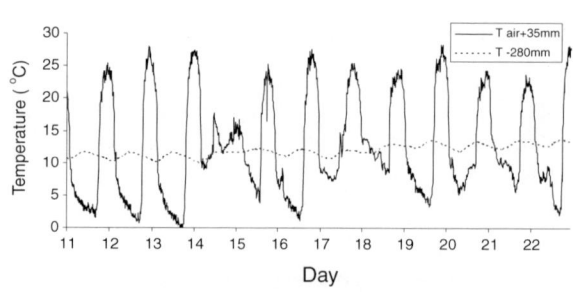

Figure 2
Temperature profiles 35 mm above ground and 280 mm below ground, Stellenbosch, July 2008. *T*-280 mm was taken as the loadcell temperature profile

Stellenbosch experiences a Mediterranean climate and is located about 50 km east of Cape Town, South Africa, 119 m a.s.l. Summers are dry and sunny averaging 30°C, while winters are warm averaging 18°C (midday) and 6°C at night; rainfall ranges between 600 and 800 mm year^{-1}.

The microlysimeters were positioned in the soil, with the sample dishes—filled with river sand (<2 mm)—flush with the surface. Dallas Semiconductor DS18820+ temperature sensors (resolution 0.1°C) were placed at loadcell depth (Fig. 1), the surface of the soil sample and 35 mm above ground level. A maxi control temperature humidity combo sensor was used to monitor relative humidity at ground level and sampling interval for all instruments was set at 10 min for the entire test period. These data allowed calculation of dew point at the soil surface using the equation by BERRY (1945):

$$\text{Dew point} = ((0.66077 - \log EW) \times 237.3/(\log EW - 8.16077))$$

where

$$\log EW = 0.66077 + (7.5 \times T/(237.3 + T)) + \log_{10}(RH) - 2$$

where T is atmospheric temperature (°C) and RH relative humidity (%).

The microlysimeter unit was solar powered using a single 12 V 26 Ah battery recharged by a 10 W photovoltaic cell connected to our specially designed data logger. The power pack and logger were placed above ground to prevent flooding and south of the microlysimeter to prevent shadow casting onto the samples which would influence the temperature regime of the sample.

3. Results and Discussion

Different temperatures affected loadcell performance (output) in the laboratory. An increase in loadcell temperature of +0.1°C resulted in a corresponding mass change equivalent to −0.00185 mm of water. This relationship followed the linear regression ($y = -0.0185x$, $R^2 = 0.9811$, $p < 0.01$). Temperature variation of the loadcell can cause significant errors to loadcell output (STORLIE and ECK, 1996) and given the small fluxes involved in "non rainfall" atmospheric water measurements it is therefore important to compensate for the effect of temperature variation on loadcell output.

Temperature profiles of the air 35 mm above the soil surface exhibited high variation with a diurnal range of over 31°C, while the soil temperature at loadcell depth (280 mm) had a diurnal range of less than 2°C (Fig. 2). This is in agreement with WHITFORD (2002), who states that temperature variation at depths 150–450 mm is less than 3°C. However, temperature variation at loadcell depth over the test period was 3.6°C and this could have been related to the water status of the soil (wet and drying) during the period. Assuming efficient thermal conductivity by the aluminium base plate (Fig. 1), the soil temperature profile at loadcell depth would be similar to that of the actual loadcell (Fig. 2). Based on this assumption, we took the soil temperature profile at loadcell depth (Fig. 2) as the loadcell temperature profile and substituted it into the temperature error effect (linear regression equation $y = -0.00185x$, $R^2 = 0.9811$, $p < 0.01$).

Data were obtained for 10 of the 12 days as the equipment was given 24 h to stabilise before collecting data. The day the microlysimeters were installed, 10 July 2008, rain was received and resulted in saturated (wet) soil at the beginning of data collection on 11 July 2008. There were no significant differences among the three microlysimeters [$F_{(2.997)} = 0.011$, $p = 0.989$] and the three outputs were consequently averaged to give one set of results for each day (e.g. Fig. 3).

Since the soil samples were isolated from belowground moisture, and we compensated for the temperature error effect on loadcell output, we could assume that the resultant mass change from the microlysimeters reflected "non rainfall" atmospheric water input and subsequent evaporation. Figure 3 shows "non rainfall" atmospheric water cycles for 12 and 13 July and clearly distinguishes two phases typical of the cycle: input (y) and evaporation (z). Classification denotes the phase dominant process, reflecting either a net gain or loss of water from the sample. The jagged graph (Fig. 3) indicates that the system is dynamic, with alternate periods of input and evaporation regardless of the phase. According to

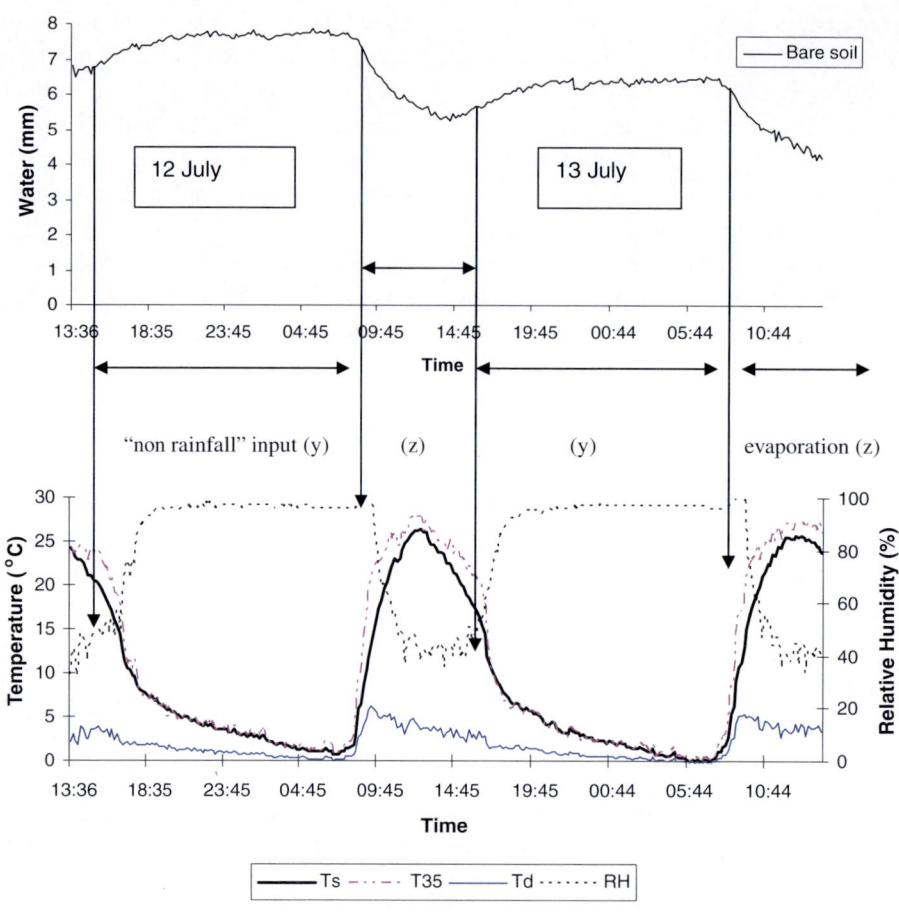

Figure 3

Typical "non rainfall" atmospheric water cycles with associated meteorological data, for 12 and 13 July 2008. T_s Soil surface temperature, T_{35} air temperature, T_d dew point temperature, RH relative humidity at soil surface

AGAM and BERLINER (2004), the evaporative phase precedes and determines "non rainfall" atmospheric water input into soil e.g. evaporation on 12 July was from the morning until mid afternoon of the same day, while "non rainfall" atmospheric water input was from mid afternoon on the 13th until the morning of the 14th July 2008.

Ambient humidity at the soil surface and air temperature 35 mm above the soil surface reveal the different phases when water input or evaporation dominate (Fig. 3). An increase in air and soil temperature decrease ambient humidity, generating a vapour gradient from soil to atmosphere—evaporative phase—resulting in a net loss in sample (soil) water content (Fig. 3). A decrease in air and soil temperatures is accompanied by an increase in ambient humidity, reversing the vapour gradient from

soil to atmosphere—input phase—resulting in a net gain in soil water content (Fig. 3). We expected that soil surface temperature would be higher than that 35 mm above the soil surface, but the opposite was the case (Fig. 3). This may have been because of the high soil water content in the sample.

"Non rainfall" atmospheric water input into the sample was calculated from the moment input exceeded evaporation, e.g. on the 13th, this was from 1535 h until the next morning when evaporation exceeded input at 0837 h (Fig. 3). Commencement of the input phase was due to vapour adsorption because dew point temperature had not been attained by the soil surface on both days (Fig. 3). Fog was absent and therefore not included. Although conditions were more favourable for vapour adsorption than dew formation, dew nonetheless did form for a few minutes in the early morning when the soil surface temperature attained dew point temperature (Fig. 3).

"Non rainfall" atmospheric water input was calculated in two ways: net and total input. The net input method is the principle behind the manual micro-lysimeter method that was used by DANALATOS et al. (1995): net input = max (y) − min (z), where y = soil water content at the end of the input phase and z = soil water at inception of input phase.

Total "non rainfall" atmospheric water input on the other hand is a summation of all input during the input phase excluding any evaporation that may have taken place during this period. Table 1 shows net and total input for each of the ten "non rainfall" atmospheric water cycles. There were significant differences between net (0.94 mm, SD 0.11, SE 0.03) and total (3.16 mm, SD 0.77, SE 0.24) "non rainfall" atmospheric water input [ANOVA: $F_{(4, 414)} = 80.04$, $p < 0.01$] and total input was 3.4 times net input. This difference could be explained by the difference in calculation in that total input does not take evaporation during this phase into account while net input includes it, therefore net input will always be lower than total input (Table 1).

Although we made some changes to the design of the microlysimeter, the design was based on that of HEUSINKVELD et al., (2006) and was an improvement on the model by BROWN et al., (2008). Both authors performed manual calibrations that were compared to the automatic microlysimeter and results showed a high level of agreement between the manual and automatic microlysimeters. We, therefore, felt it was unnecessary to make any field calibrations and instead opted to compare our results to those in the literature; in particular, we referenced the work of DANALATOS et al. (1995), which was conducted in a similar environment using manual microlysimeters. DANALATOS et al. (1995) reported net "non rainfall" atmospheric water input into soil ranging from 0.90 to 2.60 mm per night during winter which is comparable to our results which ranged from 0.76 to 1.10 mm per night in winter. We were therefore confident that our equipment performed relatively well and that the data generated was comparable to similar studies under similar conditions.

HEUSINKVELD et al. (2006) noted the need for automated instrumentation for "non rainfall" atmospheric water studies due to some handicaps of the traditional methods (JACOBS et al., 2002). A major example of these handicaps is that the duration of "non rainfall" atmospheric water input is always predetermined by the researcher e.g. 1900–0700 h (LI, 2002) but BROWN et al. (2008) demonstrated that the input phase occurs prior to sunset and can last beyond the 0700 h mark. In the current study Fig. 3, the input phase on the 12th began around 1506 h and lasted until 0755 h the following morning, while on the 13th it began around 1535 h and lasted until 0837 h the following morning. This demonstrates that predetermined input durations are not necessarily true. It is, however, important to acknowledge that the duration of the "non rainfall" atmospheric water input is dependent on site, season and soil properties.

Increasing ground level humidity increased the cumulative "non rainfall" atmospheric water input directly into the soil (Fig. 4). As humidity at the soil

Table 1

Input into and evaporation from the soil sample in mm for the ten "non rainfall" atmospheric water cycles, July 2008

Date	Net input	Total input	Vapour adsorption	Dew	Net evaporation	Total evaporation
11	1.04	4.00	2.96	1.04	1.88	3.52
12	1.09	3.11	2.56	0.56	2.29	4.20
13	1.03	3.03	2.61	0.43	2.34	3.04
14	0.88	2.40	1.86	0.54	2.14	3.60
16	0.82	4.52	3.20	1.32	1.54	3.59
17	0.99	2.67	1.89	0.77	2.71	4.35
18	0.76	1.98	1.68	0.29	2.25	4.31
19	0.90	3.72	1.98	1.74	2.26	4.02
20	0.85	2.72	1.93	0.79	2.11	3.90
21	0.98	3.40	3.01	0.38	1.39	2.72

surface increased beyond 30%, there was an increase in soil water due to "non rainfall" atmospheric water input—vapour adsorption. Soil surface humidity above 30% could have been higher than the humidity in the soil pores resulting in a vapour pressure difference between the air and the soil. This vapour pressure difference would have resulted in vapour movement from the air into the soil—input phase. This input would, in theory, continue until humidity in the soil pores was at equilibrium with ground level humidity or saturated. At ground level humidity above 95%, cumulative "non rainfall" atmospheric water input experienced a secondary increase leading to a possible peak in soil water content under these conditions—dew formation (Fig. 3).

There has been little or no attempt to differentiate "non rainfall" atmospheric water input from the different vectors due to lack of appropriate instruments. Instead, research has tended to silo vectors together (JACOBS et al., 2002; SHARAN et al., 2007; BROWN et al., 2008). This study attempted vector differentiation in the field using data derived from the total "non rainfall" atmospheric water input calculation and definitions of the vectors given in Sect. 1.2. This differentiation was successful in that it revealed the composition of this input and allowed determination of the significance of each vector to soil microhydrology. Vapour adsorption was the dominant input vector during this period and supplied on average 2.369 mm of water (75% of total input) which was about three times more input compared to dew formation (0.785 mm of water or 25% of total input). This may have been because environmental conditions are more favourable to vapour adsorption than dew formation (AGAM and BERLINER, 2004,

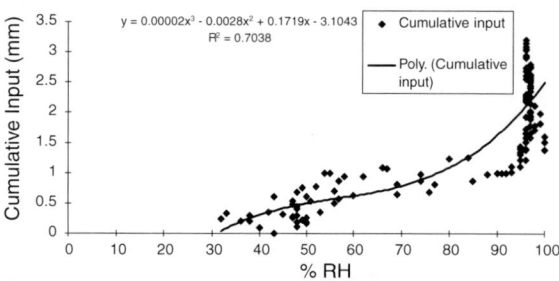

Figure 4
The effect of ground level humidity on cumulative "non rainfall" atmospheric water input directly into bare soil

2006), and this is also demonstrated in Fig. 3, where the soil surface temperature only briefly attained dew point temperature. Low dew input and duration (Fig. 3) confirm the findings of NINARI and BERLINER (2002) who concluded that dew formation is a rare occurrence on soils.

A drawback of the vector differentiation method applied to this study is that it is dependent on the sampling interval used. A relatively long sampling interval (10 min) reduced accuracy. If, for example, 9 min input were vapour adsorption and the final minute was dew, then all input in that sampling interval was classified as dew input. However, we acknowledge that a shorter sampling interval would increase accuracy of vector differentiation.

Despite the drawback mentioned above, an advantage of the automated microlysimeter method is that it enables observations of the reverse process—evaporation—and has been used in this manner to quantify small influxes (AGAM and BERLINER, 2004). Evaporation from a bare soil surface is the evaporation of water surrounding the soil particles as thin films and filling the pore spaces between them (AYDIN et al., 2005). Evaporation from bare soils is an important process although generally, soil evaporation models describe vegetated areas rather than bare soils (AYDIN et al., 2005). The automated microlysimeter method is not a model but a direct method that can be used to measure evaporation from a bare soil surface.

Evaporation was calculated from the moment evaporation exceeded input 0755 h on 13 July 2008 until the inception of the input phase 1535 h on the same day (Fig. 3). This experimental set up allowed two methods of calculation for evaporation from the soil surface—net and total evaporation similar to "non rainfall" atmospheric water input. Net evaporation is the difference between soil water content at the inception of the evaporative phase and at the end of evaporative phase; net evaporation $= a - b$, where $a =$ soil water content at the inception of the evaporative phase and $b =$ soil water content at the end of the evaporative phase.

Total evaporation, on the other hand, is the summation of all evaporation from the evaporative phase excluding any input that may have occurred during this phase. Table 1 shows evaporation from the sample for each of the ten "non rainfall" atmospheric water cycles. There were significant differences [ANOVA: $F_{(4, 414)} = 59.98$, $p < 0.01$] between total (3.73 mm, SD 0.54, SE 0.17) and net evaporation (2.09 mm, SD 0.39, SE 0.12) with the former experiencing 1.8 times more input compared to the latter (Table 1). This could have been because total evaporation is a summation of all evaporation during the evaporative phase and does not take into account any input during the phase unlike net evaporation. The net evaporation range for this study 1.39–2.71 mm day^{-1} was similar to that reported by DANALATOS et al. (1995) 0.90–2.20 mm day^{-1} under similar conditions and this demonstrates the high performance of the equipment and appropriateness of the methods used in this study.

Net and total "non rainfall" atmospheric water input into the sample was less than net and total evaporation from the same sample. These results mean there was a net loss in soil water (net method = 1.15 mm and total method = 0.57 mm) from the samples and this is justified by the fact that the study commenced with saturated (wet) soil after rains experienced on 10 July 2008. Therefore, even though "non rainfall" atmospheric water input was added into the soil, it was less than the amount of water that would be lost via evaporation from the soil due to its high water content. From 11 to 15 July there was a gradual decrease in soil water content in the sample due to high evaporation caused by the high soil water content which exceeded "non rainfall" atmospheric water input. At the same time, daily evaporation progressively decreased as the sample dried. This could have been because as the sample dried from the surface, the resulting dry layer could have acted as a buffer reducing the rate of evaporation from underlying soils.

The sudden increase in wetness from 15 to 16 July (Fig. 5) was due to a rainfall event that saturated the sample once more and almost flooded the microlysimeter as water levels almost rose to the level of the connecting tube (Fig. 1). Had this water reached this level, it would have spilled over and found its way into the loadcell compartment where it would have damaged the loadcell. After this rainfall event, evaporation once again exceeded "non rainfall" atmospheric water input leading to the observed

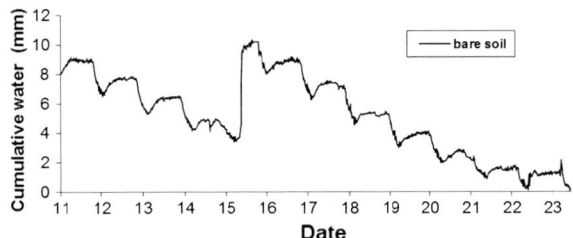

Figure 5
Diurnal changes in soil water content measured from soil in microlysimeters, Stellenbosch, July 2008

decrease in soil water content (Fig. 5). We decided to terminate the study earlier than anticipated due to weather forecasts that predicted heavy bouts of rainfall after 23 July 2008.

4. Conclusions

Caution is to be exercised in microlysimeter construction, e.g. selection of materials to limit material influence on the energy balance of the samples and temperature effects on loadcell output (STORLIE and ECK, 1996). The experimental results obtained from the automated microlysimeter developed for this study were comparable and consistent with those obtained using the manual microlysimeter method in the literature (DANALATOS et al., 1995). On average the input phase of the "non rainfall" atmospheric water cycle began late in the afternoon (1610 h) and ended after sunrise (0837 h) similar to that reported by BROWN et al. (2008). This implies that manual methods underestimate "non rainfall" atmospheric water input because the start and end times are predetermined and do not necessarily capture the full period of input.

This experimental setup enabled the determination of the composition of "non rainfall" atmospheric water directly into soil. Under the prevailing meteorological conditions at Stellenbosch during July 2008, vapour adsorption was the dominant input vector supplying three times more input compared to dew. This confirms previous observations that dew formation is a rare occurrence on soils (NINARI and BERLINER, 2002).

Despite the automated microlysimeter being the most promising in situ method for "non rainfall" atmospheric studies (AGAM and BERLINER, 2006), the instrument has some drawbacks. Firstly, because of the limited loading dish depth and capacity the instrument is not suited for heavy bouts of rainfall which could lead to flooding and damage of the instrument. Secondly, a relatively long sampling interval (10 min) reduced accuracy. If, for example, 9 min were vapour adsorption and the final minute was dew then all input in that sampling interval was classified as dew input. Despite these drawbacks, the instrument and the methods developed provide inroads into determining the significance of different water input vectors in ecological processes in arid zones.

Acknowledgments

This study was financially supported by the National Research Foundation of South Africa (NRF). The authors would also like to thank Dr. J. Irish and Prof M. Fey for their input and Glen Newins and Brian Mulder for construction and development of the equipment.

REFERENCES

AGAM, A., and BERLINER, P.R. (2004), *Diurnal water content changes in the bare soil of a coastal desert*, J. Hydrometeorol. *5*, 922-933.

AGAM, A., and BERLINER, P. R. (2006), *Dew formation and water vapour adsorption in semi arid environments – A review*, J. Arid Environ. *65*, 572-590.

ASHBEL, D. (1949), *Frequency and Distribution of Dew in Palestine*, Geographical Review, *39*, 291-297.

AWANOU, C.N. and HAZOUME, R.P. (1997), *Study of natural condensation of atmospheric humidity*, Renewable Energy, *10*, 19-34.

AYDIN, M., YANG, S., KURT, N., and YANO, T. (2005), *Test of a simple model for estimating evaporation from bare soils in different environments*, Ecol. Modelling, *182*, 91-105.

BERRY, F. A., Handbook of meteorology, (McGraw-Hill Book Company, 1945).

BEYSENS, D. (1995), *The formation of dew*, Atmos. Res. *39*, 215-237.

BERKOWICZ, S.M., HEUSINKVELD, B.G., and JACOBS, A.F.J. (2001), Dew in arid ecosystem: ecological aspects and problems in dew measurement, Proceedings, 2nd International Conference on Fog and Fog Collection, 301-304.

BOAST, C.W., and ROBERTSON, T.M. (1982), *A "microlysimeter" method for determining evaporation from bare-soil: description and laboratory evaluation*, Soil Sci. Soc. Am. J. *46*, 689-696.

BROWN, R., MILLS, A.J., and JACK, C. (2008), *Non-rainfall moisture inputs in the Knersvlakte: Methodology and preliminary findings*, Water SA. *34*, 275-278.

DAAMEN, C.C., SIMMONDS, L.P., WALLACE, J.S., LARYEAK, B., and SIVAKUMAR, M.V. (1993). *Use of microlysimeters to measure evaporation from sandy soil*, Agric. and Forest Meteorol. *65*, 159-173.

DANALATOS, N.G., KOSMAS.,C.S., MOUSTAKAS, N.C., and YASSOGLOU. N. (1995), *Rock fragments II. Their impact on soil physical properties and biomass production under Mediterranean conditions*, Soil Use Mgt. *11*, 121-126.

EVETT, S.R., WARRICK, A.W., and MATTHIAS, A.D. (1995), *Wall material and capping effects on microlysimeter temperatures and evaporation*, Soil Sci. Soc. Am. J. *59*, 329-336.

FRANCIS, M., FEY, M., PRINSLOO, H., ELLIS, F., MILLS, A., and MEDINSKI, T. (2007), *Soils of Namaqualand: Compensations of aridity*, J. Arid Environ. *70*, 588-603.

HEUSINKVELD, B.G., BERKOWICZ, S.M., JACOBS, A.F.G., HOLSTAG, A.M. and HILLEN, W.C. (2006), *An automated microlysimeter to study dew formation and evaporation in arid and semi arid regions*, J. Hydrometeol. *7*, 825-832.

HILLEL, D., Environmental Soil Physics (Academic Press, San Diego, USA, 1998).

JACOBS, A.F.G., HEUSINKVELD, B.G., and BERKOWICZ, S.M. (1999), *Dew deposition and drying in a desert system: a simple simulation model*, J. Arid Environ. *42*, 211-222.

JACOBS, A.F.G., HEUSINKVELD, B.G., and BERKOWICZ, S.M. (2002). *A simple model for potential dewfall in an arid region*, Atmos. Res. *64*, 285-295.

KIDRON, G.J. (2000). *Analysis of dew precipitation in three habitats within a small arid drainage basin, Negev Highlands, Israel*, Atmos. Res., *55*, 257-270.

LI, X.Y. (2002), *Effects of gravel and sand mulches on dew deposition in the semiarid region of China*, J. Hydrol. *260*, 151-160.

MALEK, E., McCURDY, G., and GILES, G. (1999), *Dew contribution to the to the annual water balances in semi-arid desert valleys*, J. Arid Environ. *42*, 71-80.

MONTEITH, J., UNSWORTH, M., Principles of environmental physics (Routledge New York, USA, 1990)

NINARI, N., and BERLINER, P.R. (2002), *The role of dew in the water and heat balance of bare loess soil in the Negev desert: quantifying the actual dew deposition on the soil*, Atmos. Res. *64*, 323-334.

NOFFSINGER, T.L. Survey techniques of measuring dew. In Humidity and Moisture, 2 (ed. Wexler. A..) (Reinhold, New York, 1965).

SHARAN, G., BEYSENS, D., and MILIMOUK-MELNYTCHOUK, I. (2007), *A study of dew water yields on galvanised iron roofs in Kothara (north-west India)*, J. Arid Environ, *69*, 259-269.

SCHEMENAUER, R.S., and CERECEDA, P. (1994), *Fog collection's role in water planning for developing countries*, Natural Resources Forum, *18*, 91-100.

SCOTT, D. (1962), *An instrument measuring dew deposition*, Ecology, *43*, 341-342.

STONE, E.C. (1963), *The ecological importance of dew*, The Quarterly Review of Biology, *38*, 328-341.

STORLIE, C.A., and ECK, P. (1996), *Lysimeter-based crop efficient for young highbush blueberries*, Horticultural Sci. *31*, 819-822.

WALKER, G.K. (1983). *Measurement of evaporation from soil beneath crop canopies*, Can. J. Soil Sci. *63*, 137-141.

WHITFORD, W.G., Ecology of desert systems. (Elsevier Science Ltd., 2002)

ZANGVIL, A. (1996), *Six years of dew observation in the Negev Desert, Israel*, J. Arid Environ, *32*, 361-372.

(Received November 16, 2010, revised April 14, 2011, accepted April 18, 2011, Published online May 19, 2011)

Pure Appl. Geophys. 169 (2012), 859–871
© 2011 Springer Basel AG
DOI 10.1007/s00024-011-0329-8

Chemical Composition of Dew Resulting from Radiative Cooling at a Semi-arid Site in Agra, India

ANITA LAKHANI,[1] RAVINDRA SINGH PARMAR,[1] and SATYA PRAKASH[1]

Abstract—Dew samples were collected between October 2007 and February 2008 from a suburban site in Agra. pH, conductivity, major inorganic ions (F$^-$, Cl$^-$, NO$_3^-$, SO$_4^{2-}$, Na$^+$, K$^+$, Ca^{2+}, Mg^{2+}, and NH$_4^+$), and some trace metals (Cr, Sn, Zn, Pb, Cd, Ni, Mn, Fe, Si, Al, V, and Cu) were determined to study the chemistry of dew water. The mean pH was 7.3, and the samples exhibited high ionic concentrations. Dew chemistry suggested both natural and anthropogenic influences, with acidity being neutralized by atmospheric ammonia and soil constituents. Ion deposition flux varied from 0.25 to 3.0 neq m^{-2} s^{-1}, with maximum values for Ca^{2+} followed by NH$_4^+$, Mg^{2+}, SO$_4^{2-}$, Cl$^-$, NO$_3^-$, Na$^+$, K$^+$, and F$^-$. Concentrations of trace metals varied from 0.13 to 48 μg l^{-1} with maximum concentrations of Si and minimum concentration of Cd. Correlation analysis suggested their contributions from both crustal and anthropogenic sources.

Key words: Dew, inorganic ions, trace metals, deposition fluxes, deposition velocities.

1. Introduction

Dew is formed by condensation of atmospheric water vapor on a substrate as a result of radiative cooling. The conditions causing dew are well known from both a meteorological (MONTEITH and UNSWORTH, 1990) and physical point of view (BEYSENS, 1995). Deposition of several atmospheric chemical species has been found to be enhanced when surfaces are wetted by dew (SMITH and FRIEDMAN, 1982). Dew forms at night when emission of infrared radiation from surfaces causes them to cool below the dew point. Atmospheric water vapor in contact with this cooled surface condenses and forms a water film. This condensation generally occurs on clear, atmospherically stable nights when wind speeds are

minimal. Under these conditions, atmospheric gases and particles may be transferred to the dew, where they are more readily absorbed or retained than under dry conditions, resulting in enhanced deposition. Dew is a local phenomenon, significantly influenced by microclimatic ambiance, land profile, and favorable meteorological conditions. The chemical properties of dew are driven by dissolution of surrounding gas and atmospheric particles that fall on the substrate. Studies have revealed that concentrations of chemical species in dew samples are much higher than in rain samples collected in the same areas (FOSTER *et al.*, 1990; WANGER *et al.*, 1992). Dew composition can be a good indicator of the level of atmospheric pollution in a geographical region of interest, because of the types and quantities of chemicals and materials transported by dew and the range of its interactions. Dew water dissolving water-soluble pollutants from the atmosphere may have harmful influences on the contacted materials. It has been reported that dew formed on plants and leaves enhances the dry deposition velocity of acid gases, such as HNO$_3$ and SO$_2$, and may have an important role in acid deposition to vegetation (WISNIESWSKI, 1982; WESLEY *et al.*, 1990).

Dew chemical properties have been far less investigated than its meteorological and physical properties. Early studies on dew chemistry emphasized only some inorganic ions (YAALON and GANOR, 1968; BRIMBLECOMBE and TODD, 1977). BRIMBLECOMBE and TODD (1977) found that the pH of individual dew droplets lay between 5 and 7, and that roughly equal concentrations of potassium and sodium were present in dew water. The mean pH of bulk dew water formed on the surfaces of chemically inert collectors has been reported to be 4.0 by PIERSON *et al.* (1986), who also suggested that dew water was acidified mainly by SO$_2$, although some fraction of SO$_2$

[1] Department of Chemistry, Dayalbagh Educational Institute, Dayalbagh, Agra, India. E-mail: anitasaran2003@yahoo.co.in

dissolved in dew remained as S(IV). Studies by PIERSON et al. (1986), who measured deposition of trace elements to dew, found a dependence of deposition velocity on aerosol size fraction. Elements predominantly associated with fine particles (Pb, Se, Br, and V) had the lowest deposition velocities (<0.1 cm s^{-1}), while elements with more mass on coarse particles, such as Ba, Mg, and Ca, had the highest deposition velocities (0.1–1.0 cm s^{-1}). CHAMEIDES (1987) studied generation of acid dew from dry deposition of HNO$_3$ as well as SO$_2$, using a model involving dynamical resistance, surface resistance, and reactions in dew droplets. No large variation in concentration of ions of various dew samples was reported by FOSTER et al. (1990) over a 13-month period in Indiana, USA where the mean pH of dew samples was found to be 6.82. WANGER et al. (1992) reported mean pH of dew samples of 6.37 and found that the concentrations of various ions in dew were several times those in rain, attributing this to the evaporation effect. KATAGIRI et al. (1995) pointed out that some fraction of S(IV) in dew water was present as hydroxyalkanesulfonate (HASA), which is an adduct of S(IV) with aldehydes. There are several reports about dew chemistry in urban areas such as in Chile (RUBIO et al., 2002), USA (MULAWA et al., 1967), Japan (CHIWA et al., 2003), and Jordan (JIRIES, 2001), where dew characteristics are rather different. Dew water was found to be corrosive with high ionic concentrations in Chile, very acidic in Japan, with high concentrations of sulfates and nitrates at urban and mountain sites of Mt. Gokurakuji, and slightly alkaline and weakly mineralized in Jordan. On an island (Corsica, France), dew characteristics were found to be comparable to those in Jordan, except that dew was less alkaline; water was potable with respect to the main ions and constituents investigated (MUSELLI et al., 2002). The importance of dew in the mercury cycle was investigated during three sampling periods in the Great Lakes region and one in the Florida Everglades in North America (MALCOLM and KEELER, 2002).

Recently, BEYSENS et al. (2006) investigated both the chemical and bacteriological properties of dew water and compared them with rainwater in urban area of Bordeaux, France, finding that dew exhibited ion characteristics close to low mineralized commercial spring water found in Europe. Most recently, KLIMASZEWSKA et al. (2009) applied linear discriminant analysis to study the effect of local meteorological characteristics on dew chemistry in samples collected from various sites in Poland, and POLKOWSKA et al. (2008) reported the chemical characteristics of dew water collected at several locations in Poland.

Typically, most measurements of dew chemistry have focused on the effect of dew on acid deposition. Major ions were found in dew water in equal or higher concentrations than in precipitation from the same location. Consequently, in a watershed where dew is a frequent phenomenon or has a major hydrological input, dew may also be a major source of nutrients or pollutants. There is limited research devoted to the composition and role of dew in the cycling of elements. The chemistry of wet deposition in the form of rain, dry deposition and aerosol composition has been studied extensively at the site of current study (SAXENA et al., 1991, 1996, 1997; KUMAR et al., 1993; KHARE et al., 1997; KHARE et al., 2000; SATSANGI et al., 1998; SINGH et al., 2001; PARMAR et al., 2001; LAKHANI, 2005; SINGH and KHARE, 2006; LAKHANI et al., 2007, 2008), but dew chemistry has been less explored. In the present study, dew samples collected at this semi-arid site in India were analyzed to reveal the nature of deposition. A preliminary attempt is made to quantify the relationship between aerosol concentrations and concentrations in dew water.

2. Methodology

2.1. Sampling Site

The sampling site for this study is located on our institute campus in Dayalbagh (population approximately 15,000), a small suburb lying north of Agra City (27°10'N, 78°05'E), which lies in a semi-arid zone, adjacent to the Thar Desert of Rajasthan. A map of the sampling site is shown in Fig. 1. The sampling site lies by the side of a road that carries mixed vehicular traffic, moderate (of the order of 10^5 vehicles per day) during the day and minimal (of the order of 10^2 vehicles per night) at night. The campus

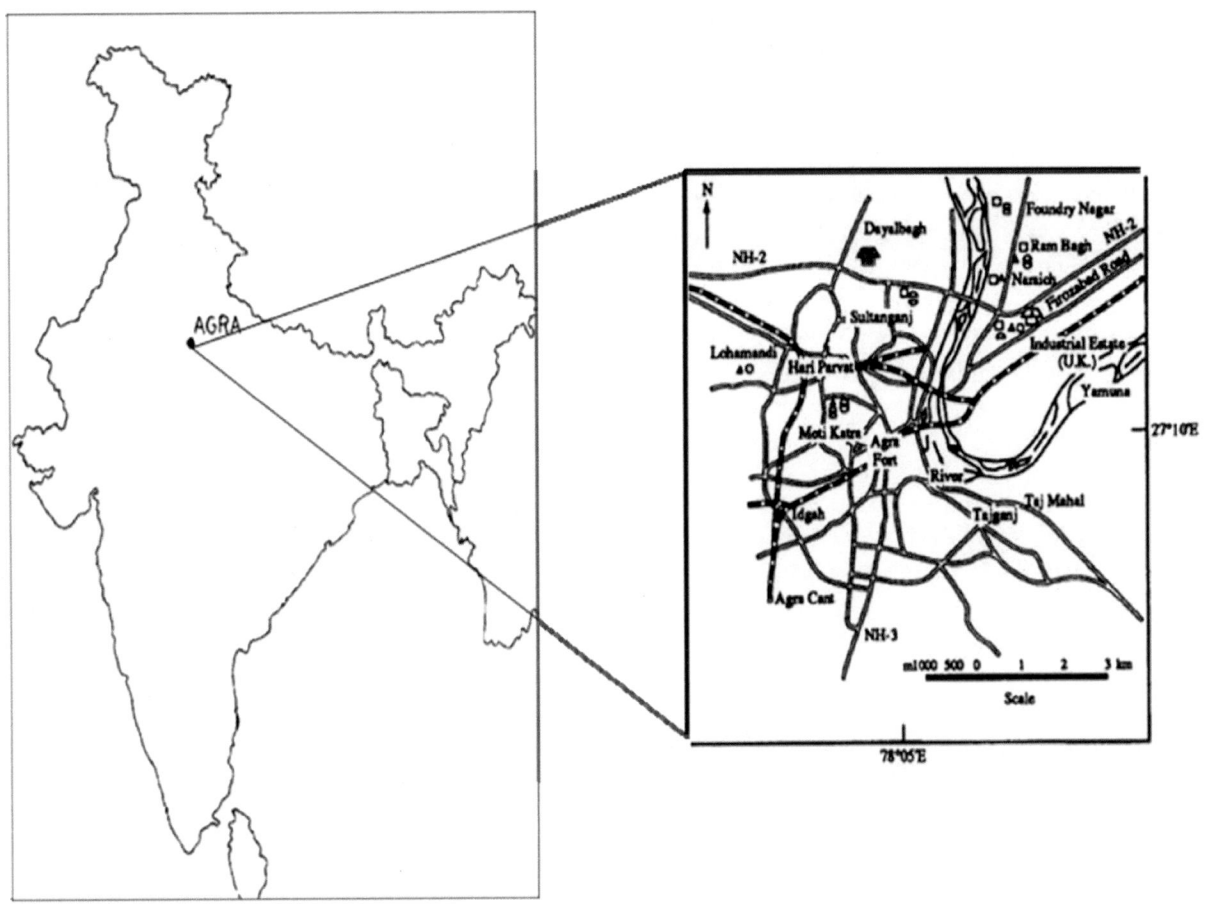

Figure 1
Map of sampling site (Agra)

lies about 2 km north of a national highway which has dense vehicular traffic (10^8 vehicles) throughout the day and night. The site is surrounded by a variety of deciduous trees and agricultural fields. The soil is sandy and calcareous by nature. There is no industry in this suburb. During the study period, temperatures ranged from 2–20°C and the average relative humidity (RH) was 70%.

2.2. Dew Collection and Measurement

Dew occurs mostly between October to March, with maximum accumulation in December and January. Maximum dewfall occurs at height of 100 cm above ground and decreases exponentially downwards. Since one effect expected of dew is to dissolve previously deposited material and to react

with the surface on which it forms, dew was collected on two polytetrafluoroethylene (PTFE) mounted trays of area 1,740 cm^2. The collectors were placed on a 1-m-high iron stand above the ground at the onset of dew formation. Before employing, the collectors were washed with detergent solution and thoroughly washed with deionized water until the conductivity of the rinsed water reduced to around 1 μS cm^{-1}. To eliminate any dilution of the samples from the washing water, the collectors were dried by blowing hot air. To exclude dry deposition of gases and particles prior to the onset of dew formation, collectors were deployed at 6.00 p.m. (local time) in the evening when dew formation commenced and the overnight dew samples were collected about 12 h later at 6.00 a.m. the next morning before sunrise by scrapping off with a clean Teflon scraper and transferred to a clean polyethylene bottle. The dew collected in this manner contains not only gas-phase substances and particulate matter deposited during dew formation but also dry deposition. Dew collection took place only on rainless nights to eliminate any influence of rain droplets on collected dew samples. The dew samples were weighed to determine the volume, considering the density to be equal to that of water. A total of 50 dew samples were collected during October 2007–March 2008. Maximum dew was observed in the month of January (our sampling period), between 9 p.m. to 6 a.m. Hence, for all calculation purposes, dew occurrence time was taken as 9 h.

Electrical conductivity and pH were determined immediately using a conductivity meter and a pH meter, respectively, in an aliquot of the sample. pH was measured with an Elico digital pH meter (model Li-122) and a glass and a reference electrode assembly using the two-buffer technique. Standard deviation for pH measurement was 0.05 units. For conductivity measurements, a Systronics conductivity meter (model 304) was used. Sample conductance was measured after calibration of the instrument with 0.1 N KCl at the prevailing temperature. The conductivity cell had accuracy of 0.5%. The remaining sample was then filtered with 0.45 μm pore-size membrane filter. A fraction of the sample was transferred into a clean PTFE bottle and spiked with $CHCl_3$ at ratio of 1:10 v/v to inhibit microbial growth and refrigerated at 4°C in a refrigerator. This part was used for analysis of the major anions and NH_4^+ while the other fraction of the sample was acidified to low pH (pH 2) by adding concentrated HNO_3 and stored in an acid-leached bottle for use in analysis of major cations.

Root-mean-square reproducibility in collection and analysis of physical parameters such as pH and conductivity evaluated from dew collectors deployed concurrently was 3% for amount of water deposited, 5% for conductivity, and 6% for pH. The discrepancies in the concentrations for all measured species were low (3–12%), as listed in Table 1. Conceivably the discrepancies are genuine in the sense that something might have fallen onto one collector and not the other during the night. Field blanks were also collected to detect any contamination of the samples which might have resulted due to improper washing

Table 1

Experimental uncertainties

Component	Collection variability (%)	Field blanks (μeq l^{-1})
F^-	4	0.4
Cl^-	3	0.3
NO_3^-	11	0.5
SO_4^{2-}	12	0.3
NH_4^+	7	0.4
Ca^{2+}	6	1.1
Mg^{2+}	2	0.5
Na^+	3	0.5
K^+	5	0.4

of the collectors and any impurity in the deionized water being used for preparation of solutions and sample handling. To collect field blanks, trays were exposed to the atmosphere for 10 min and then rinsed thoroughly with deionized water and stored in polyethylene bottles after their volume had been raised to 100 ml. These samples were treated and analyzed in a similar manner to the dew samples. The values of different ionic components in the field blanks are also listed in Table 1.

2.3. Sampling of Aerosols

Aerosols were also collected during this period to study the relation of dew with aerosol. Aerosol sampling was also conducted on the roof (10 m) of the faculty building on the institute campus, which was close to the dew sampling site. Sampling was performed using a four-stage cascade particle separator (CPS-105; Kimoto, Japan). The CPS has 50% efficient cutoff in aerodynamic diameter classes of 0.7–1.6, 1.6–5.4, 5.4–10, and >10 μm for the four stages. The average flow rate of the CPS was 800 ± 0.5 l min^{-1}. Flow rate was indicated by a rotameter attached to an automatic flow controller. Despite this, the flow rate was checked after every 3 h. The deviation in flow rate was very small and therefore was considered negligible. Aerosols were collected on 20×25 cm^2 Whatman 41 filter papers. Filter papers were carefully equilibrated in desiccators before and after sampling to eliminate the effect of humidity. Sampling was performed over 24 h to obtain sufficient mass of aerosol for analysis. The mass of aerosol particles collected on each stage was determined by the difference in weight before and after sampling.

2.4. Analysis

To determine the concentration of major cations and anions along with trace metals, filters were cut into two equal parts of known area. One part of the filter was extracted by ultrasonic agitation in 50 ml deionized water for 1 h. The extracts were then filtered through 0.45 μm pore-size nylon membrane filters, and the water-soluble cations and anions were determined in these filters in a similar manner to that described for the dew samples. For analysis of trace metals Cr, Sn, Zn, Pb, Cd, Ni, Mn, Fe, Si, Al, V, and Cu, the other half of the filter paper was digested by acid treatment. Each filter was folded and placed in a beaker. Then 5 ml nitric acid (12 N) was added, and the beaker was covered with a watch glass and heated slowly. Before complete evaporation, 5 ml acid was again added, and the beaker was covered with a watch glass and heated to obtain refluxing action until clear solution was obtained. The solution was filtered using 0.45 μm pore-size nylon membrane filters. The filtrate was transferred to 50-ml volumetric flask and made up with deionized water. All the trace metals were analyzed using inductively coupled plasma atomic emission spectroscopy (ICP-AES, Jobin Yuon Panorama 46P). To test for complete extraction, the residue of the filters was extracted again and analyzed by the same procedure. The levels of trace metals were below detection limit in these test samples.

Major anions (Cl^-, F^-, NO_3^-, and SO_4^{2-}) were analyzed by ion chromatograph (Dionex DX-500). Separation was accomplished using a separator column (AS4A-SC) with self-regenerating suppressor which ensured the lowest possible background noise level and detection limit. The column was protected upstream by a guard column (AG4A). A sample of 10 μL was injected. The eluent was a mixture of sodium carbonate (1.8 mM) and sodium bicarbonate (1.7 mM), passed at flow rate of 1 ml min^{-1}. All samples were first allowed to come to room temperature before analysis and were injected into the chromatograph without dilution. All concentrations were calculated based on chromatogram areas of

standards prepared daily from 100 ppm stock solution. The stock solutions were prepared weekly, spiked with $CHCl_3$, and stored at 4°C. No loss was found in 1 week. The major cations (Na^+, K^+, Ca^{2+}, and Mg^{2+}) and trace metals (Sn, Cr, Zn, Pb, Cd, Ni, Mn, Fe, Si, Al, V, and Cu) were analyzed using ICP-AES (Jobin–Yvon Panorama 46P). Typical argon flow rates were 1 l min^{-1} for the carrier, 0–1 l min^{-1} for the auxiliary plasma, and 15 l min^{-1} for the coolant plasma. NH_4^+ was analyzed spectrophotometrically by the indophenol blue method. Among the collected samples, 45 had sufficient volume for complete chemical analysis.

For quality control of the analytical results, measured and calculated electric conductivities were compared. The mean ratio of calculated to measured conductivity was observed to be 0.92 ± 0.10.

Dewfall ranged from 0.1 to 0.5 l m^{-2} during the sampling period. To account for the effect of amount of dew water on ion concentrations, the volume-weighted mean concentration (VWM) and the volume-weighted standard deviation (VWSD) were calculated for each ion (DAYAN et al., 1985). The VWM concentration was calculated using the formula $X = \sum_{i=1}^{N} X_i P_i / \sum_{i=1}^{N} P_i$,

$$\text{VWSD} = \frac{\sqrt{\left[N \sum_{i=1}^{N} P_i^2 [X_i]^2 - (\sum_i^N P_i [X_i])^2 \right]}}{(\sum_i^N P_i)^2 (N-1)},$$

where P_i is the dew amount corresponding to the ith sample, X_i is the concentration of the ith species, and N is the number of samples.

From the chemical composition of the dew samples and amount of water collected, deposition fluxes and deposition velocities for the major ions were calculated. The accumulated amounts of various species deposited per unit area in the dew, referred to as deposition accumulation, were obtained by multiplying the concentration in each sample (C_i) by the respective amount of water per unit area (V_i) and summing the products. The fluxes were obtained by dividing the accumulation by the sum of collection times (t).

$$\text{Deposition accumulation} = \sum C_i V_i,$$

$$\text{Deposition flux} = \frac{\sum C_i V_i}{\sum t}.$$

Deposition velocity (V_d) is defined as the ratio between deposition flux (F) and atmospheric concentration (C). Deposition velocity is dependent on the chemical characteristics of the measured species, the size of the particle, the nature of the surface upon which the deposition occurs, and the prevailing atmospheric conditions, such as winds, turbulence, temperature, and humidity. Dry deposition velocities may vary by orders of magnitude depending on the above factors, thus introducing large uncertainties in estimation of dry deposition rates from airborne concentrations.

3. Results and Discussion

3.1. Chemical Composition of Dew Water

The mean pH during the sampling period was calculated from the VWM concentration of H^+, i.e.,

$pH = -\log [H^+]$. Figure 2a shows a statistical summary of the ionic concentration, the VWM concentrations are shown in Fig. 2b, while their percentage contributions are shown in Fig. 2c. The Event Variation of pH is shown in Fig. 3. The horizontal line at pH 5.6 represents the reference level. The samples had pH varying from 6.3 to 7.8. The volume-weighted mean pH was 7.3, indicating alkaline nature. In the present study four anions (F^-, Cl^-, NO_3^-, and SO_4^{2-}) and five cations (Na^+, K^+, Ca^{2+}, Mg^{2+}, and NH_4^+) were quantitatively measured. To ensure the reliability of these ion data and to assess the possibility of any other ions with notable concentration that were ignored, the balance of total anions ($\sum-$) versus total cations ($\sum+$) was checked carefully. Average ionic balance on an equivalent basis at this site was ($\sum-/\sum+$) was 0.8. This indicates contribution of some unmeasured anions to

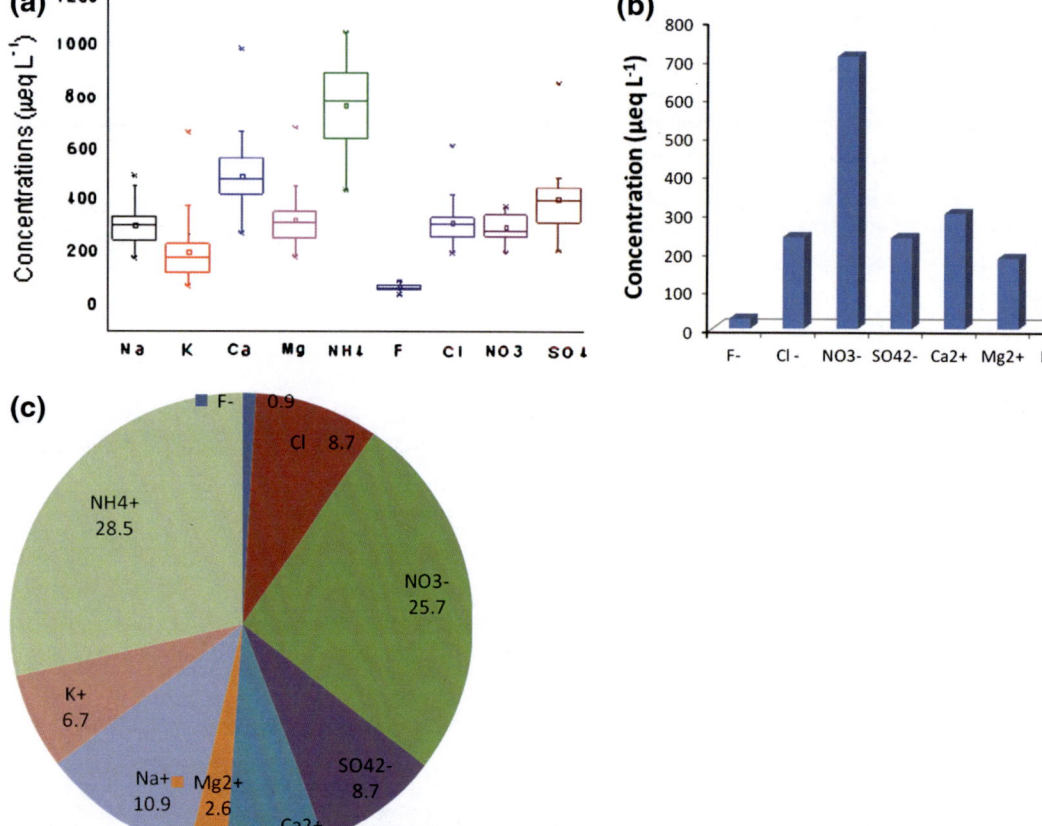

Figure 2
a Box plots of ionic constituents. **b** Volume-weighted means of ionic constituents. **c** Percentage contribution of ionic constituents

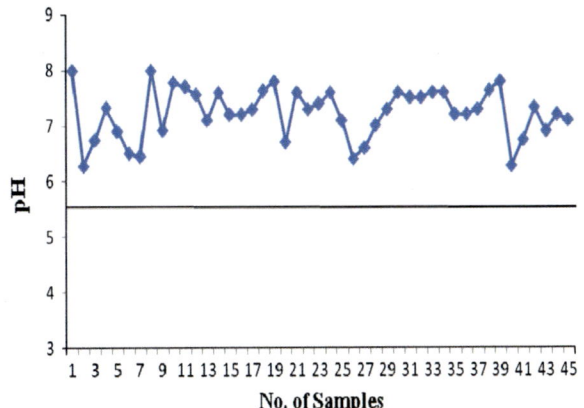

Figure 3
Event Variation of pH

Mg^{2+}) contributed 27.2% and NH_4^+ 28.5%, whereas the contribution of the anions (F^-, Cl^-, NO_3^-, and SO_4^{2-}) was relatively small, accounting for 44.1% (Fig. 2c).

3.2. Equivalent Ratios between Concentrations of Analytes

Since the site is an inland site, influence of sea salt on ionic composition is expected to be negligible. Equivalent sea salt ratios are $Cl^-/Na^+ = 1.17$, $SO_4^{2-}/Na^+ = 0.125$, $K^+/Na^+ = 0.22$, $Ca^{2+}/Na^+ = 0.44$, and $Mg^{2+}/Na^+ = 0.25$. These ratios were calculated for dew water concentrations, being in general higher than the seawater ratios (1.06, 1.39, 0.69, 1.69, and 1.1, respectively). Deviations above the seawater ratios have also been observed in precipitation studies conducted previously at this site (SAXENA et al., 1996; SATSANGI et al., 1998; SINGH et al., 2001). Cl^- in dew water may result from absorption of HCl gas by airborne liquid droplets followed by deposition and soil-derived chloride salts. Hydrogen chloride is possibly deposited via smoke from coal-fired domestic stoves, where fuel of different quality is burnt, frequently with addition of household refuse containing plastic, printed paper, wood sticks, crop residues, and cowdung cakes. The average SO_4^{2-}/Na^+ ratio was also higher than the seawater ratio, which can be attributed to emission from fuel combustion. The average K^+/Na^+ ratio was also higher; the potassium excess could be attributed to fertilizers, windblown soil and dust, and biomass burning. The higher Ca^{2+}/Na^+ ratio indicates that the major source of calcium is airborne soil, and agricultural and constructional activities carried out in the vicinity of the site. The NO_3^-/SO_4^{2-} ratio ranged over a wide interval, but in general the concentration of SO_4^{2-} was greater than that of NO_3^-.

3.3. Relationships between Chemical Species

To investigate the possible sources of ions in dew water, correlations between ionic concentrations were determined. The correlation matrix for the ion pairs is presented in Table 2. Significant correlation among the soil-derived ions Ca^{2+}, Mg^{2+} and Na^+ was observed (Ca^{2+} and $Mg^{2+} = 0.86$, Ca^{2+} and

the dew water composition. Anion deficit could be attributed to some unmeasured anions such as bicarbonate that could arise from dissolution of carbonate salts and organic anions formate and acetate, which in a previous study at this site have shown concentrations varying between 0.9–60.0 and 2.1–18.9 $\mu eq\ l^{-1}$, respectively, accounting for approximately 3.7% of total ionic concentration (KHARE et al., 2000).

The total ionic strength of dew samples calculated from the measured ionic concentrations ranged between a minimum of 4,846 $\mu eq\ l^{-1}$ and a maximum of 43,688 $\mu eq\ l^{-1}$. These values are in the same range of concentration as those derived from measurements performed on dew and fog water concentration in Delhi, India (KHEMANI et al., 1987). These values indicate high pollutant concentration in the region during the winter period. Figure 2b shows that the concentrations of major ions were in the order: $NH_4^+ > NO_3^- > Ca^{2+} > Cl^- > SO_4^{2-} > Mg^{2+} > K^+ > F^-$. The average concentrations of NH_4^+, Ca^{2+}, Mg^{2+}, and Na^+ were 780, 300, 184, and 190 $\mu eq\ l^{-1}$, respectively. NH_4^+ was the major cation and contributed most (28.5%) to the total ionic concentration; it could be derived from human and animal excrements or agricultural activities. Ca^{2+} accounted for 11% of the ionic concentrations and could be derived from dissolution of minerals $CaCO_3$, $CaCO_3 \cdot MgCO_3$, and $CaSO_4 \cdot 2H_2O$. NO_3^- was the largest anion, mainly deriving from vehicle emissions, while SO_4^{2-} could be derived from coal combustion. The cations (Na^+, K^+, Ca^{2+}, and

$Na^+ = 0.88$, Mg^{2+} and $Na^+ = 0.84$). SO_4^{2-} and NO_3^- were moderately correlated ($r = 0.68$), probably because of co-emission of their precursors (SO_2 and NO_x), and they were strongly correlated with Ca^{2+}, Mg^{2+} and Na^+ (SO_4^{2-} and $Ca^{2+} = 0.81$, SO_4^{2-} and $Mg^{2+} = 0.72$, SO_4^{2-} and $Na^+ = 0.79$, Ca^{2+} and $NO_3^- = 0.46$, Mg^{2+} and $NO_3^- = 0.55$, Na and $NO_3^- = 0.49$), suggesting similarity of their behavior in dew deposition and also indicating both natural and anthropogenic influences on dew water. These correlations also indicate that soil dust might also be a significant source of these ions apart from being formed in the atmosphere from their anthropogenic precursors. These react with atmospheric NH_3 in the gas phase to form ammonium salts, and their acidic effects are neutralized. The concentrations of NH_4^+ correlated closely with SO_4^{2-} and NO_3^- ($r = 0.91$ and 0.46, respectively). It is likely that the increase in dew pH caused by absorption of NH_3 as well as $CaCO_3$ enhanced dissolution of S(IV) and N(III), which might be derived from atmospheric SO_2 and HNO_2, respectively, as well as increasing dissolution of the weak acids formic and acetic acid that existed in the vapor phase. The $[NH_4^+]$ to $[SO_4^{2-}] + [NO_3^-]$ ratio varied from 0.19 to 3.89, and $[NH_4^+]$ was also positively correlated ($r = 0.51$) with the sum of $[NO_3^-]$ and $[SO_4^{2-}]$.

The correlation between pH and the log concentration of ions that should influence it was small and insignificant. The partial correlations of pH with SO_4^{2-} and NO_3^- controlled for Ca^{2+}, Mg^{2+}, and NH_4^+ were positive but not significant ($r_{pH;SO_4^{2-},Ca^{2+}} = 0.14$, $r_{pH;SO_4^{2-},NH_4^+} = 0.04$, $r_{pH;NO_3^-,Ca^{2+}} = 0.17$, $r_{pH;NO_3^-,Mg^{2+}}$

$= 0.20$, $r_{pH;NO_3^-,NH_4^+} = 0.11$). The NH_4^+/SO_4^{2-} ratio was 2.74 ± 0.36. The relationship between the NH_4^+/SO_4^{2-} ratio and pH was negligible, indicating that an influence of soil-derived SO_4^{2-} and direct scavenging of locally emitted NH_3 are important. pH and the NH_4^+/SO_4^{2-} ratio varied independently (1.13–5.54); consequently, the correlation coefficient was insignificant. The NH_4^+/NO_3^- and $NH_4^+/(SO_4^{2-} + NO_3^-)$ ratio also followed a similar relationship with pH ($r_{NH_4^+/NO_3^-:pH} = 0.30$ and $r_{NH_4^+/NO_3^-+SO_4^{2-}:pH} = 0.10$). Also, no relationship was evident between pH and the Ca^{2+}/SO_4^{2-} ratio. Soil is considered to be the major source of Ca^{2+} and industrial pollution the main source of SO_4^{2-} and NO_3^- particles in the atmosphere. However, dominance of these components in the atmosphere depends upon the area and their sources. The mass ratio of Ca^{2+}/SO_4^{2-} can serve as an indicator for the pH level in precipitation samples. This ratio in the dew samples collected at this site was found to be above unity (1.3).

3.4. Dew Acidification/Neutralization Process

HARA et al. (1995) suggested the use of a quantitative index pAi while discussing the acid–base relationship and the chemistry of different kinds of atmospheric water. pAi is the hypothetical pH of atmospheric water if no neutralization takes place for both sulfuric and nitric acid. This index focuses only on the acidic component, whereas the actual pH is determined by the balance between acidic and neutralizing components.

$$pAi = -\log\left[nssSO_4^{2-} + NO_3^-\right],$$

Table 2

Correlation coefficients of major cations and anions in dew (N = 45)

Component	Ca^{2+}	Cl^-	F^-	K^+	Mg^{2+}	Na^+	NH_4^+	NO_3^-	SO_4^{2-}
Ca^{2+}	1								
Cl^-	0.78**	1							
F^-	0.14	−0.03	1						
K^+	−0.20	−0.25	0.36*	1					
Mg^{2+}	0.86**	0.79**	−0.10	−0.23	1				
Na^+	0.88**	0.96**	−0.03	−0.12	0.84**	1			
NH_4^+	0.80**	0.74**	0.26	−0.32	0.67*	0.66*	1		
NO_3^-	0.46*	0.53*	0.16	−0.19	0.55*	0.49*	0.46*	1	
SO_4^{2-}	0.81**	0.65*	0.42*	−0.23	0.72**	0.79**	0.91**	0.68*	1

One tailed significance: $P = 0.01*$, $0.001**$

where $nssSO_4^{2-}$ is non-sea-salt sulfate

Since at this site the sea salt contribution is negligible, non-sea-salt sulfate has been considered as total sulfate. For all of the samples, pAi values appeared in a highly limited range (when compared with pH values) from 3.89 to 4.64, with mean value of 4.29. The difference between pH and pAi values was large, which suggests that dew water was neutralized with NH_3 or $CaCO_3$, as shown in Fig. 4.

The neutralization/acidification process of dew water can also be discussed by using the relationship between acidifying potential (AP) and neutralization potential (NP) (POLKOWSKA et al., 2008). $AP = [nssSO_4^{2-} + NO_3^-]$, $NP = NH_4^+ + nssCa^{2+}$]. The theoretical curve, linking experimental data points, can be defined as a linear equation, whose general form is $y = x$ (AP = NP). For the dew samples collected during this period, the regression equation $y = 0.470x + 0.936$ was estimated, and NP > AP applied for all samples, as shown in Fig. 5.

3.5. Deposition Fluxes and Deposition Velocities

The water condensation rate of dew varied between 2,666 and 11,333 $\mu gm^{-2}\ s^{-1}$, while the deposition flux of major cations and anions varied between 0.25 and 3.0 $\mu gm^{-2}\ s^{-1}$. The deposition fluxes of major cations and anions are listed in Table 3. The maximum deposition flux was obtained for Ca^{2+} (3.0 $\mu gm^{-2}\ s^{-1}$), while the minimum deposition flux was obtained for K^+ (0.25 $\mu gm^{-2}\ s^{-1}$).

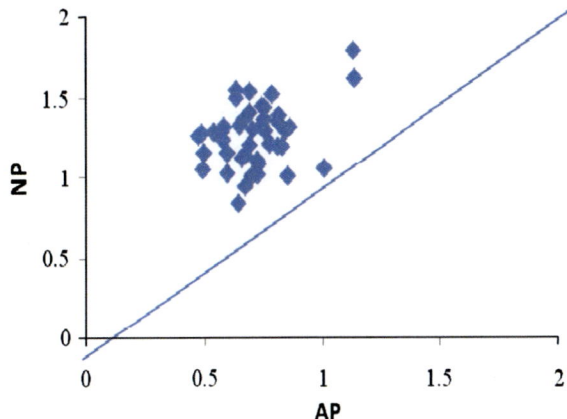

Figure 5
Relationship between AP and NP in different samples

In general, deposition fluxes decreased in the order $Ca^{2+} > NH_4^+ > Mg^{2+} > SO_4^{2-} > Cl^- > F^- > NO_3^- > Na^+ > K^+$. Deposition fluxes and velocity depend on particle size and density, being enhanced for coarse particles which are accompanied by a greater mass median diameter (MMD) (HICKS, 1986). The high deposition fluxes of Ca^{2+} and Mg^{2+} suggest their production from soil. SO_4^{2-} and NO_3^- particles, which are believed to be inputs from anthropogenic activities, showed deposition fluxes higher than and similar to those of soil-derived elements. This suggests the possibility that they are also soil derived or somehow associated with soil elements. An obvious natural SO_4^{2-} source for consideration is loading from the vast open areas of Agra and its neighborhood. Soil in this region has high concentration of SO_4^{2-}; $CaSO_4$ is added to the saline soils rich in Na^+ to make them fertile. The mechanism of SO_2 to SO_4^{2-} conversion on soil particles may also contribute to the higher deposition fluxes (WINCHESTER et al., 1986; ASHU RANI et al., 1992, SAXENA et al., 1996). NH_4^+ also shows greater deposition flux. Greater flux of ammonia is probably associated with cattle population near the sampling site, where cattle are used for plowing fields as well as in dairy activities. The deposition flux of Cl^- also resembles the deposition fluxes of the soil components. Its contribution from the sea at this inland location seems to be negligible, and since there are no major anthropogenic sources of Cl^- here, it may be anticipated that Cl^- also originates from soil.

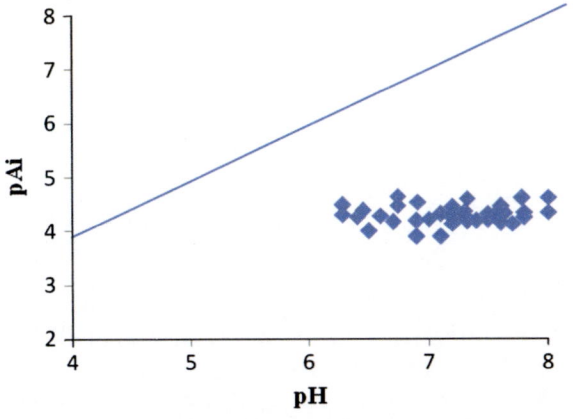

Figure 4
pAi against pH for dew samples

Table 3

Deposition fluxes ($\mu g\ m^{-2}\ s^{-1}$) of major cations and anions in dew

Variable	Minimum	Maximum	Mean ± SD
Na^+	0.18	1.5	0.43 ± 0.23
K^+	0.22	0.97	0.77 ± 0.21
Ca^{2+}	1.3	11.7	3 ± 0.28
Mg^{2+}	0.5	6.1	1.6 ± 0.16
NH_4^+	1.8	3.8	2.8 ± 1.1
F^-	0.009	1	0.5 ± 0.13
Cl^-	0.16	2.9	0.84 ± 0.53
NO_3^-	0.13	1.8	0.48 ± 0.15
SO_4^{2-}	0.23	2.6	1.04 ± 0.25

The deposition velocities of various ions are listed in Table 4. Average deposition velocities of ions to dew water ranged between from 0.3 to 4.1 cm s^{-1} with maximum deposition velocity for NH_4^+. From Table 3 it is evident that the deposition velocities were highly variable for all species. Deposition velocities, in general, decreased in the order $NH_4^+ > Ca^{2+} > Mg^{2+} > NO_3^- > SO_4^{2-} > Cl^- > K^+ > Na^+ > F^-$. In this particular region, dew deposition is generally accompanied by fog. Thus, deposition fluxes of different ionic components as well as their deposition velocities are expected to be influenced by the occurrence of fog. Fogs can influence the ambient aerosol concentrations both by accelerating the removal of particles and by providing favorable conditions for production of additional aerosol material. The mass scavenging efficiencies of radiation fogs in polluted areas depend strongly on both the size and chemical composition of the aerosol particles. The ambient aerosol in polluted areas is usually an external mixture of two distinct particle populations, one hydrophilic and the other hydrophobic. The hydrophobic particles are not scavenged by the fog and remain as interstitial aerosol. Similarly, soot particles and elemental carbon are often found to be activated to a lower extent than other particles during fog formation and are found to be most enriched in the interstitial air compared with the aqueous phase. On the other hand, elements such as sulfur, iron, manganese, and copper are found to be enriched in the aqueous phase. In the presence of fogs, aerosols with diameters larger than approximately 0.5 pm grow to sizes larger than 20 pm, and their removal is accelerated by several orders of magnitude. Hence, species expected to exist in water-insoluble form are also scavenged to a significant degree by fog droplets. Furthermore, areas characterized by an overabundance of fine particles may show small scavenging efficiencies and hence lower deposition fluxes and velocities due to competition for a small amount of condensable water. Dew can further influence the amount of deposition by changing the nature of the surface from a dry one to a wet one, as the presence of dew enhances both retention of dry-deposited particles and absorption of water-soluble gases. Hence quantification of the size dependence of the aerosol mass as well as its chemical composition and the scavenging efficiency of fogs is necessary for estimation of the deposition flux and deposition velocity of particles.

3.6. Trace Metals in Dew Water

The volume-weighted mean concentration ($\mu g\ l^{-1}$) of trace metals in dew water samples followed the order (Table 4): Si (48.3) > Al (12.7) > V(4.2) > Sn(2.4) > Ni (1.4) > Cu (1) > Zn (0.96) > Mn (0.61) > Fe (0.37) > Pb (0.33) > Cr (0.24) > Cd (0.13). The deposition fluxes of the trace metals varied from 6 to 88 (ng m^{-2} s^{-1}). The maximum deposition flux was obtained for Si and the lowest for Zn (Table 5). The deposition velocity of trace metals ranged from 0.3 to 0.09 mm s^{-1}, with maximum deposition velocity for Pb followed by Cu, Si, Mn, Cr, Ni, Al, V, Cd, Sn, Fe, and Zn (Table 5). Logarithmic correlation coefficients between trace-metal concentrations were calculated and are presented in Table 6. Al, Cd, Cr, Cu, Fe, Ni, Pb, Si, Sn, and V showed significant correlations with one

Table 4

Deposition velocities (cm s^{-1}) of major cations and anions in dew

Component	Minimum	Maximum	Mean ± SD
Na^+	0.54	1.9	0.5 ± 0.08
K^+	0.54	1.3	0.6 ± 0.12
Ca^{2+}	1.1	5.8	3.8 ± 1.5
Mg^{2+}	1.4	7.6	2.2 ± 0.83
NH_4^+	0.86	8.2	4.1 ± 1.4
F^-	0.25	1.4	0.3 ± 0.04
Cl^-	0.31	0.62	0.6 ± 0.13
NO_3^-	0.3	1.2	1.1 ± 0.3
SO_4^{2-}	0.46	2.8	0.7 ± 0.1

Table 5

Volume-weighed mean concentration, deposition flux, and velocities of trace metals

Component	Mean ± SD ($\mu g\ l^{-1}$)	Deposition flux ($ng\ m^{-2}\ s^{-1}$)	Deposition velocity ($mm\ s^{-1}$)
Cd	0.13 ± 0.02	1.1 ± 0.2	1.7 ± 0.4
Cr	0.24 ± 0.03	1.0 ± 0.1	2.2 ± 0.9
Pb	0.33 ± 0.04	49 ± 3.9	8.0 ± 2.3
Fe	0.37 ± 0.06	19 ± 1.2	0.9 ± 0.1
Mn	0.61 ± 0.02	24 ± 1.8	4 ± 1.3
Zn	0.96 ± 0.02	6 ± 0.2	0.3 ± 0.1
Cu	1.04 ± 0.3	48 ± 3.8	5.3 ± 1.2
Ni	1.4 ± 0.04	58 ± 2.9	2.0 ± 0.2
Sn	2.4 ± 0.14	21 ± 1.1	1.3 ± 0.5
V	4.2 ± 0.18	15 ± 1.8	1.8 ± 0.6
Al	12.7 ± 0.13	25 ± 2.4	1.9 ± 0.3
Si	48.3 ± 1.6	88 ± 6.2	4.7 ± 1.7

another, indicating common occurrence. The correlations among Al, Si, Mn, and Fe indicate crustal origin, and these elements appear to have been released into the local air by increased human activity as the site is in an agricultural area. The elements Pb, Cd, V, Cr, Ni, Sn, Zn, and Cu are considered to arise from anthropogenic sources such as ferrous and nonferrous foundries, tanneries, and vehicular traffic.

predominant anion was NO_3^-, followed by SO_4^{2-} and Cl^-. Dew water was found to be alkaline due to neutralization of acids by ammonia and soil constituents incorporated in dew. However, correlations between ionic species indicated both natural and anthropogenic influences on dew water. The maximum deposition flux was obtained for Ca^{2+}, followed by NH_4^+, Mg^{2+}, SO_4^{2-}, Cl^-, F^-, NO_3^-, Na^+, and K^+. The average deposition velocities of ions ranged from 0.3 to 4.1 cm s^{-1} with maximum deposition velocity for NH_4^+. This shows that dew can influence the amount of deposition by changing the nature of the surface from a dry one to wet one. The volume-weighed mean concentration ($\mu g\ l^{-1}$) of trace metals in dew water samples followed the order: Si (48.3) > Al (12.7) > V (4.2) > Sn (2.4) > Ni (1.4) > Cu (1) > Zn (0.96) > Mn (0.61) > Fe (0.37) > Pb (0.33) > Cr (0.24) > Cd (0.13). Al, Cd, Cr, Cu, Fe, Ni, Pb, Sn, and V were significantly correlated, indicating common occurrence. Correlations among Al, Si, Mn, and Fe indicated their crustal origin, while the elements Pb, Cd, V, Cr, Ni, Sn, Zn, and Cu were considered to arise from anthropogenic sources.

4. Conclusions

Dew water was found to be alkaline with mean pH of 7.3. The principal cation in dew water was NH_4^+, followed by Ca^{2+}, Na^+, and Mg^{2+}, while the

Acknowledgments

The authors are grateful to the Director, Dayalbagh Educational Institute Agra and Head, Department of Chemistry and The Department of Science and

Table 6

Correlation coefficients for trace metals in dew water

	Al	Cr	Cu	Fe	Mn	Ni	Pb	Si	Sn	V	Zn	Cd
Al	1											
Cr	0.57*	1										
Cu	0.46*	0.59*	1									
Fe	0.15	0.23	0.19	1								
Mn	−0.07	−0.03	0.06	0.64*	1							
Ni	0.72**	0.46*	0.56*	0.26	0.01	1						
Pb	0.52*	0.32	0.34*	0.14	0.01	0.79**	1					
Si	−0.02	0.11	0.20	0.65*	0.89**	0.05	0.09	1				
Sn	0.53*	0.31	0.22	0.20	0.03	0.22	0.10	0.13	1			
V	0.53*	0.31	0.33*	0.03	−0.08	0.80**	0.95**	−0.02	0.11	1		
Zn	−0.01	−0.01	0.02	0.36*	0.34**	−0.24	−0.33	0.31*	0.20	−0.42	1	
Cd	0.52	0.28	0.30	0.11	−0.03	0.78**	0.89**	0.03	0.24	0.94**	0.36	1

One tailed significance: $P = 0.01^*$, 0.001^{**}

Technology, DST project no. SR/S4/AS:207/02, New Delhi, for financial assistance.

REFERENCES

ASHU RANI., PRASAD, D.S.N., MADNAWAT, P.S.N. and GUPTA, S.K., (1992), *The role of free fall dust in catalyzing auto-oxidation of aqueous sulphur dioxide*, Atmos. Environ., *26*, 667-673.

BEYSENS, D. (1995), *The formation of dew*. Atmos. Res. *39*, 215-237.

BEYSENS, D., OHAYON, C., MUSELLI, M., CLUS O., (2006), *Chemical and biological characteristics of dew and rain water in an urban coastal area (Bordeaux, France)*, Atmos. Environ., *40*, 3710-3723.

BRIMBLECOMBE, P. and TODD, I.J., (1977), *Sodium and potassium in dew*, Atmos. Environ., *11* (7), 649-650.

CHAMEIDES,W.l., (1987), *Acid dew and the role of chemistry in the dry deposition of reactive gases to wetted surfaces*. J. Geophys. Res., *92* 11895-11908.

CHIWA, M., OSHIRO, N., MIYAKE, T., NAKATANI, N., KIMURA, N., YUHARA, T., HASHIMOTO, N., SAKUGAWA, H., (2003), *Dry deposition washoff and dew on the surfaces of pine foliage on the urban-and mountain-facing sides of Mt. Gokurakuji, Western Japan*, Atmos. Environ., *37*, 327-337.

DAYAN, U., MILLER, J. M., KEENE, W. C. and GALLOWAY, J. N. (1985) *An analysis of precipitation chemistry data from Alaska*. Atmos. Environ., *19*, 651-657.

FOSTER, J.R., PRIBUSH, R.A., CARTER, B.H., (1990) *The chemistry of dews and frosts in Indianapolis, Indiana*, Atmos. Environ., *24*(A), 229-2236.

HARA, H., KITAMURA, M., MORI, A., NOGUCHI, I., OHIZUMI, T., SETO, S., TAKEUCHI, T., DEGUCHI, T.(1995), *Precipitation chemistry in Japan 1989-1993.*, Water Air Soil Pollut., *85*, 2307-2312.

HICKS, B.B., (1986), *Measuring dry deposition: A Re-assessment of the state of the art*, Water, Air and Soil Pollution, *30* (1-2), 75-90.

JIRIES, A. (2001), *Chemical composition of dew in Amman, Japan*, Atmos. Res., *57*, 261-268.

KATAGIRI, Y., SAWAKI, N., ARAI, Y., OKOCHI, H. and IGAWA, M., (1995), Chem. Lett., *197*.

KHARE, P., SATSANGI, G.S., KUMAR, N., KUMARI, K.M. and S.S. SRIVASTAVA, (1997), *HCHO, HCOOH and CH₃COOH in air and rain water at a rural tropical site in north central India*, Atmos. Environ., *31* (23), 3867-3875.

KHARE, P., SINGH S.P., KUMARI, K.M., KUMAR A. and S.S. SRIVASTAVA, (2000), *Characterization of organic acids in dew collected on surrogate surfaces*. Journal of Atmospheric Chemistry *37*, 231-244.

KHEMANI, L.T., MOMIN, G.A., PRAKASA RAO, P.S.P., SAFAI, P.D. AND PRAKASA, P. (1987), *Influence of alkaline particulates on the chemistry of fog water at Delhi, North India*, Water, Air and Soil Pollution, *34*, 183-189.

KLIMASZEWSKA, K., SÂRBU,C., POLKOWSKA, Z., LECH D., PASLAW-SKI,p., MALEK, S., NAMIESNIK J., (2009), *Application of linear discriminant analysis to the study of dew chemistry on the basis of samples collected in Poland*, Cent. Eur. J. Chem., *7*(1), 20-30. doi:10.2478/s11532-008-0082-8.

KUMAR, N., KULSHRESTHA, U.C., SAXENA, A., KUMARI, K.M. and SRIVASATAV, S.S. (1993), *Effect of anthropogenic activity on formate and acetate levels in precipitation at four sites in Agra, India*, Atmos. Environ., *278* (1), 87-91.

LAKHANI, A., SATSANGI G.S., PARMAR R.S. and PRAKASH S. (2005), *Chemistry of sulphur and nitrogen species and other major cations/anions in fog water*. Ind. J. Radio and Space Physics, *34*, 42-49.

LAKHANI, A., PARMAR, R.S., SATSANGI, G.S. and PRAKASH S. (2007), *Chemistry of fogs at Agra, India: Influence of soil particulates and atmospheric gases*. Environ. Monit. Assess., *133*, 435-445.

LAKHANI, A., PARMAR, R.S., SATSANGI, G.S. and PRAKASH S. (2008), *Size distribution of trace metals in ambient air of Agra*. Ind. J. Radio and Space Physics, *37*, 434-442.

MALCOLM, E. and KEELER, G. (2002), *Measurements of mercury in dew: Atmospheric removal of mercury species to a wetted surface*, Environ. Sci. Technol., *36*, 2815-2821.

MONTEITH, J.L., UNSWORTH, M.H. (1990), Principles of Environmental Physics, Second Ed., Routledge, Chapman& hall, Inc., New York.

MULAWA, P.A., CADLE, S.H., LIPARI, F., ANG, C.C. and VANDER-VENNET, R.T. (1967), *Urban dew: Its composition and influence on dry deposition rates*, Atmos Environ., *20*(7), 1389-1396.

MUSELLI, M., BEYSENS, D., MARCILLAT, J., MILIMOUK, I., NILSSON, T., LOUCHE, A (2002), *Dew water collector for potable water in Ajaccio*. Atmos. Res., *64*, 297-312.

PARMAR, R.S., SATSANGI, G.S., KUMARI, K.M., LAKHANI, A., SRI-VASTAVA, S.S. AND PRAKASH, S. (2001), *Study of size distribution of atmospheric aerosol at Agra*, Atmos. Environ., *35*, 693-702.

PIERSON, W.R. BRACHACZEK, W.W., GORSE, R.A. JR. JAPAR, S.M. and NORBECK, J.M. (1986), *On the acidity of dew*. Journal of Geophys. Res. *91*, 4083-4096.

POLKOWSKA, Z., MAREK, B., KLIMASZEWSKA K., SOBIK, M., MALEK, S. and NAMIESNIK J., (2008), *Chemical characterization of dew water collected in different geographic regions of Poland*, Sensors, *8*, 4006-4032.

RUBIO, M.A., LISSI, E., VILLENA, G., (2002), *Nitrite in rain and dew in Santiago city, Chile: Its possible impact on the early morning start of the photochemical smog*, Atmos. Environ. *36*, 293-297.

SATSANGI, G.S., LAKHANI, A., KHARE, P. SINGH S.P., KUMARI, K.M. AND SRIVASTAVA, S.S. (1998). *Composition of rainwater at a semi-arid rural site in India*, Atmos. Environ. *32*(21), 3783-3793.

SINGH, S.P., KHARE, P., MAHARAJ KUMARI K. and SRIVASTAVA S.S.,(2006), *Chemical characterization of dew at a regional representative site of North-Central India*, Atmos. Res. *80*, 239 (2006).

SINGH, S.P., KHARE, P., SATSANGI, G.S., LAKHANI, A., MAHARAJ KUMARI K. AND SRIVASTAVA S.S. (2001), *Rainwater composition at a regional representative site of a semi-arid region of India*. Water, Air and Soil Pollution, *127*, 93-108.

SAXENA, A., SHARMA, S., KULSHRESTHA, U.C., and SRIVASTAVA S.S, (1991), *Factors affecting alkaline nature of rain water in Agra (India)*, Environ. Pollut., *74*, 129-138.

SAXENA, A., KULSHRESTHA, U.C., KUMAR, N., KUMARI K.M. and SRIVASTAVA S.S, (1996). *Characterization of precipitation at Agra*, Atmos. Environ., *30*, 3405-3412.

SAXENA, A, KULSHRESTHA, U.C., KUMAR, N., KUMARI, K.M., SATYA PRAKASH and SRIVASTAVA, S. S. (1997), *Dry deposition of sulphate and nitrate to polypropylene surfaces in a semi arid area of India*. Atmos. Environ., *31* (15), 2361-2366.

SMITH B.E. and FRIEDMAN E.J. (1982) Mitre corporation working paper WP82W00141, MITRE corporation, Metrek Division, 1820 Dolley Madison Boulevard, Mclean, VA 22102.

WANGER, G., STEELE, K., PEDEN, M. (1992), *Dew and frost chemistry at a mid-continental site, United States*, J. Geophys. Res., *97*, 20591-20597.

WESLEY, M.L., SISTERSON, D.L. and JASTROW, J.D., (1990), *Observations of the chemical properties of dew on vegetation that affect the dry deposition of SO₂*, J. Geophys. Res. *95*, 7501-7514.

WINCHESTER, J.W., LI, S., AND GILLETTE, D.A., (1986), Potential for airborne dust scavenging and dry deposition of SO₂. Proceedings of NAPAP Workshop on Dry Deposition, Harpers Ferry, VA, 25-27.

WISNIESWSKI, J., (1982), *The potential acidity associated with dews, frosts and fogs*, Water, Air and Soil Pollution, *17*, 361-177.

YAALON and GANOR E. (1968), *Chemical composition of dew and dry fallout in Jerusalem, Israel*. Nature, *217*, 1139-1140.

(Received November 15, 2010, revised March 23, 2011, accepted April 17, 2011, Published online May 25, 2011)

Pure Appl. Geophys. 169 (2012), 873–880
© 2011 Springer Basel AG
DOI 10.1007/s00024-011-0367-2

The Effects of Desert Pavements (Gravel Mulch) on Soil Micro-Hydrology

K. F. Kaseke,[1,3,6] A. J. Mills,[2] J. Henschel,[3] M. K. Seely,[4] K. Esler,[1] and R. Brown[5]

Abstract—The effect of desert pavements (gravel mulch) on near surface soil micro-hydrology has been inadequately studied. Micro-hydrology in arid ecosystems occurs due to a daily non rainfall atmospheric water cycle, consisting of an input phase (dew, fog, vapour adsorption) and an evaporation phase. A winter comparative study between a bare soil (control) and gravel mulch using the automated microlysimeter approach was conducted in Stellenbosch, South Africa in 2008. Results showed that dew deposition and direct water vapour adsorption were significantly higher into bare soil compared to gravel mulch. In contrast, however, soil moisture from rain persists for a longer time under gravel mulch compared to bare soil. This result suggests that the greatest impact of gravel mulch on soil micro-hydrology is towards conserving moisture and could explain why the treatment is used in dry-land agriculture in Mediterranean regions.

Key words: Gravel mulch, non rainfall atmospheric water, dew, vapour adsorption, evaporation.

1. Introduction

Desert pavements (gravel mulch) are a prominent layer of pebbles embedded in and/or scattered on the soil surface and are a characteristic feature of deserts and other arid environments (Schmiedel and Jurgens, 2004). They protect underlying soil from erosion

(Epstein *et al.*, 1966), reduce runoff, increase infiltration (Katra *et al.*, 2008) and retard evaporative losses from the soil (Kemper *et al.*, 1994; Li, 2002). Given these beneficial aspects of gravel mulch and the role they play in micro-hydrology of arid and semiarid environments, it is no surprise that traditional farming practices using gravel mulch have developed around the world, e.g. Lanzorate in the Canary Islands (Graf *et al.*, 2008) and Shatian in China (Li *et al.*, 2000).

These traditional farming practices using gravel mulch are, however, based on the belief that gravel mulch supplies additional water to the soil via enhanced nocturnal cooling (Matznetter, 1958; Acosta Baladon, 1973, 1996). Enhanced nocturnal cooling is a process where the mulch surface cools more than the surrounding environment facilitating formation of dew on its surface. According to Malek *et al.* (1999), dew formation occurs under specific conditions, is dependent on meteorological conditions and the substrate surface temperature. Therefore, dew is the condensation of water vapour into liquid droplets on a sufficiently cooled substrate surface (Stone, 1963; Beysens *et al.*, 1995). In theory, enhanced nocturnal cooling on gravel mulch could increase dew formation and stone flow would direct this water into the soil. However, to date there is insufficient experimental evidence to support this hypothesis (Graf *et al.*, 2008).

In arid environments, water scarcity is a limiting factor to arid land productivity (Louw and Seely, 1982; Whitford, 2002); therefore, any additional source (input) may have a positive impact on the ecosystem (Kidron, 2000). Non -rainfall water (dew, fog, vapour adsorption) can exceed annual rainfall and can be the sole source of water for plants in arid environments (Agam and Berliner, 2006). In the absence of rain, non rainfall water input directly into

---
[1] Department of Conservation Ecology, Stellenbosch University, P. Bag X1, Matieland 7602, South Africa. E-mail: kje@sun.ac.za
[2] Department of Soil Science, Stellenbosch University, P. Bag X1, Matieland 7602, South Africa. E-mail: mills@sun.ac.za
[3] Gobabeb Research Centre, P. O. Box 953, Walvis Bay, Namibia. E-mail: joh.henschel@gobabeb.org
[4] Desert Research Foundation of Namibia, 7 Rossini Street, Windhoek, Namibia. E-mail: Mary.seely@drfn.org
[5] Climate Systems Analysis Group, University of Cape Town, P. Bag, Rondebosch 7701, South Africa. E-mail: brown.roger@gmail.com
[6] 26 Gobvu Road, Zengeza 3, Chitungwiza, Zimbabwe. E-mail: faraikaseke@gmail.com

the soil system is, therefore, an important link in the micro-hydrology and ecology of semiarid and arid regions (DANALATOS et al., 1995; KOSMAS et al., 1998; AGAM and BERLINER, 2004).

Although small in quantity, this additional water may play a critical role in arid soil ecology (MALEK et al., 1999), providing a temporary source of water for brief periods of biological decomposition, nutrient release and temporary relief from extreme heat. Experimental evidence (DANALATOS et al., 1995; LI, 2002; GRAF et al., 2008), however, contradicts this, instead showing that gravel mulch decreases non rainfall atmospheric water input into soil. SCHMIEDEL and JURGENS (2004) reported that quartz desert pavements lowered maximum soil surface temperature compared to bare soil, but at night the reverse was true. This suggests that the development of farming methods based on desert pavements could have been because of its role in soil moisture conservation rather than water supply via enhanced nocturnal cooling.

Although not significant, GRAF et al. (2008) reported slight enhanced nocturnal condensation on lapilli mulch but the condensation gains were less than that experienced on bare soil. They attributed this to hygroscopicity of bare soil which resulted in condensation gains beyond the strict definition of dew (BEYSENS, 1995). This hygroscopicity could be due to vapour adsorption, a reversible interfacial physical process resulting from differential forces of attraction and repulsion between vapour molecules and soil particles (AGAM and BERLINER, 2006). Vapour adsorption could support rain fed vegetation during the dry summers of semi-arid Mediterranean regions (DANALATOS et al., 1995; KOSMAS et al., 2001).

Given the importance of desert pavements and non rainfall water to arid and semi-arid ecosystems it is however, still unclear what effect desert pavements have on soil micro-hydrology. Secondly, the effect of desert pavements on the composition of these water vectors is unclear and yet this might influence the organisms that will thrive in the soil. Therefore, study can contribute to our understanding of dry-land ecology and provide a basis for the management of such areas. The objectives of this paper were thus to investigate the effect of desert pavements on non rainfall water input directly into soil and the subsequent evaporation by the microlysimeter method.

2. Materials and Methods

2.1. Vector Definitions and Differentiation

Non rainfall water in arid ecosystems is supplied via three vectors, namely: fog, dew and vapour adsorption (AGAM and BERLINER, 2006). However, there is no international standard for measuring this water (ZANGVIL, 1996; BROWN et al., 2008) and there are disagreements on the very definition of the vectors (NOFFSINGER, 1965; ZANGVIL, 1996). It is, therefore, necessary to define the vectors and the method of differentiation as applied to this study, although these definitions and method have been described in greater detail elsewhere (KASEKE et al., 2011). During this study fog was not observed, and thus is not included in these definitions.

2.1.1 Dew

It is the natural condensation of water vapour into liquid droplets on a sufficiently cooled substrate surface (STONE, 1963; BEYSENS, 1995; AWANOU and HAZOUME, 1997; MALEK et al., 1999). Dew formation is dependent on the receiving substrate characteristics (BEYSENS, 1995), occurring when the substrate surface equals or falls below ambient dew point temperature but is above freezing point, otherwise frost forms (ASHBEL, 1949; BEYSENS, 1995; AGAM and BERLINER, 2006).

2.1.2 Vapour Adsorption

Vapour adsorption is a reversible interfacial physical process resulting from differential forces of attraction and repulsion between vapour molecules and soil particles (AGAM and BERLINER, 2006). It occurs as a result of vapour movement from the atmosphere into the soil due to the establishment of a vapour gradient between the two.

2.2. Theoretical Differentiation of Vectors

In theory, conditions conducive for one input vector preclude the others from occurring concurrently (BROWN et al., 2008). Dew formation occurs when the receiving substrate surface temperature equals or falls below ambient dew point (BEYSENS,

1995), and vapour adsorption occurs when a vapour gradient is established between the atmosphere and the soil, independent of dew point temperature (BROWN *et al.*, 2008). Therefore, the receiving substrate (soil) surface temperature can be used to distinguish between dew and vapour adsorption input (AGAM and BERLINER, 2006). According to our study, input that occurred when soil surface temperature was below ambient dew temperature was classified as dew and if temperature was above ambient dew point this was classified as vapour adsorption.

2.3. Site

Research was conducted in the Western Cape Province of South Africa, at the Stellenbosch University farm (S33 56.577 E 18 51.152). The site is under a Mediterranean climate and is located 50 km east of Cape Town, South Africa, 119 m a.s.l. Summers are dry and sunny averaging 30°C, while winters are warm averaging 18°C (midday) and 6°C at night; rainfall ranges between 600 and 800 mm year^{-1}.

The design specifications and principles of the microlysimeter method used in this study are described in detail (KASEKE *et al.*, 2011). However, a general description of the microlysimeter specifications is presented in this paper. The height of the microlysimeter was 280 mm and consisted of two compartments (loadcell compartment and the compartment that housed the sample dish). We used a Tedea-Huntleigh 1004 (1,500 g) single point aluminium loadcell and 2 mm polypropylene for the sample dish (diameter 140 mm and depth 35 mm). Each microlysimeter unit consisted of three microlysimeters, three Dallas Semiconductor DS18820+ temperature sensors and a Maxi Control temperature humidity combo sensor connected to a common data logger.

Two microlysimeter units were positioned into the soil at the site and loaded with river sand that had been passed through a 1.981 mm Tyler Standard Screen sieve for uniformity until the sample dishes were flush with the surface. One unit was left with bare soil (control) and was compared to the other which had a mixture of angular and rounded quartz pebbles placed on the soil surface (gravel mulch), see

Fig. 1. The Dallas Semiconductor DS 18820+ temperature sensors measured temperature on the bare soil surface (control), on the gravel mulch surface, air temperature 35 mm above the soil surface and the loadcell temperature of both microlysimeter units (to compensate for the temperature error effect, see KASEKE *et al.*, 2011). The temperature humidity combo sensor monitored relative humidity at the soil surface and the sampling interval for all instruments was set at 10 min intervals.

The equipment was setup in the field on the 10 July 2008 but we received heavy rains the same day such that the recharging system could not operate for the full night. Hence, we discarded data from that day and in the process allowed the equipment 24 h to stabilise before taking our readings the next day. At this point, our soil samples were saturated and we began our study using these saturated samples and we acknowledge that this moisture may have had a bearing on our results. Table 1 shows selected soil properties of the control soil sample.

3. Results and Discussion

KASEKE *et al.* (2011) showed two types of calculating non rainfall input directly into soil: net and total. Net input is the difference between soil water content at the end of the input phase and water content at the beginning of the same phase while taking evaporation during the phase into account (DANALATOS *et al.*, 1995). On the other hand, total input is a

Figure 1
Simulated desert pavement treatment in microlysimeter (magnification ×0.89), Stellenbosch July 2008

Table 1

Selected soil properties of the control sample

Texture	Particle density (g cm^{-3})	Bulk density (g cm^{-3})
Sand	2.60	1.31

summation of all input during the input phase excluding any evaporation that may have occurred during the phase (KASEKE et al., 2011). As a result, total non rainfall water input will always be higher than net input into the same treatment.

Figure 2 shows that the surface temperature of gravel mulch (white quartz) attained higher maxima compared to bare soil and that the former sample surface temperature profile lagged behind the trend in the latter sample (Fig. 2). In theory, materials with high albedo (reflectance) are generally cooler than materials with low albedo, which holds true for Fig. 2 until mid-afternoon. However, from late afternoon, surface temperature on gravel mulch is higher than that on bare soil. This is because gravel mulch acts as an insulator during the day and retains heat during the night (LI, 2002). Repeated measures ANOVA without replication confirms that daytime surface temperatures on gravel mulch were significantly higher than that on bare soil ($F_{(3.84)} = 56.17$, $p < 0.01$).

Figure 2 confirms that stone cover acts as a heat sink, retaining heat in the sample (JURY and BEL-LANTUONI, 1976) and possibly slowing the reversal of the vapour gradient from soil to air (evaporation to input phase). The input phase was thus expected to

commence earlier in bare soil compared to gravel mulch, facilitating an earlier occurrence of vapour adsorption into the former sample (1610 hours) compared to the latter (1710 hours). This would also translate into higher input into bare soil compared to that into gravel mulch because of the longer input duration. Repeated measures ANOVA without replication confirmed this showing that non rainfall water input into bare soil (control) was significantly more than into gravel mulch in both net and total input ($F_{(5.12)} = 290.59$, $p < 0.01$; $F_{(5.12)} = 88.96$, $p < 0.01$ respectively). Total non rainfall water input into gravel mulch (0.56 mm, SD 0.20, SE 0.06) was on average 2.60 mm (82%) lower than input into bare soil (3.16 mm, SD 0.77, SE 0.20).This result is consistent with reports by LI (2002) and GRAF et al. (2008), who concluded that desert pavements decrease non rainfall water input into soil.

It is unlikely that a 1.22°C difference in sample surface temperature (Fig. 2) could cause an 82% (2.60 mm) decrease in total input into gravel mulch compared to bare soil (3.16 mm). At best, it can only partially explain the results; however, to fully explain these results it is necessary to differentiate input from the different vectors. Table 1 shows the volumetric composition of non rainfall atmospheric water input into each of the treatments. This data was calculated from total non rainfall atmospheric water input into each sample. Environmental conditions are more conducive to vapour adsorption than dew formation (AGAM and BERLINER, 2006), and this explains why vapour adsorption into bare soil was 3.4 times more

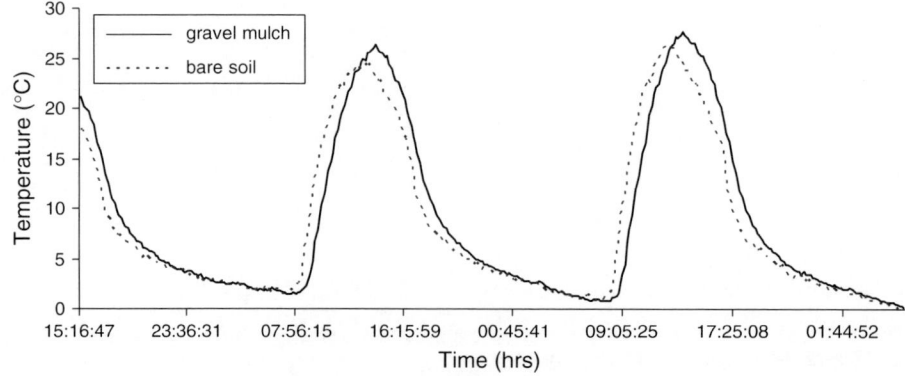

Figure 2
Sample surface temperature profiles, Stellenbosch University farm, 11–13 July 2008

than dew input. However, the contribution of dew to total input in gravel mulch was almost equal to that of vapour adsorption in the same treatment.

Repeated measures ANOVA without replication showed that vapour adsorption into desert pavement was significantly lower than that into bare soil ($F_{(5.12)} = 124.50$, $p < 0.01$). Vapour adsorption into bare soil was 7.4 times that into gravel mulch representing a 2.05 mm (87%) decrease in vapour adsorption input into the latter sample (Table 2). This could be attributed to a lower soil surface area in direct contact with the atmosphere which could have resulted in blocking of vapour adsorption into the gravel mulch treatment (DANALATOS et al., 1995). Desert pavement could have also caused compression of the soil, thereby reducing soil volume and blocking vapour transport down the profile. Stones also have a negligible adsorption capacity (DANALATOS et al., 1995). These factors combined could be responsible for the 87% decrease in vapour adsorption into the gravel mulch treatment. This decrease accounts for 78.9% of the 2.6 mm (82%) reduction in total non rainfall atmospheric water input into the gravel mulch compared to bare soil (Table 2).

The critical factor governing dew formation is nocturnal substrate (soil) surface temperature (BEYSENS, 1995). Repeated measures ANOVA without replication did not reveal any significant differences between nocturnal substrate surface temperature between the two treatments ($F_{(3.85)} = 0.08$, $p = 0.78$), see Fig. 2, and therefore, did not support the enhanced cooling hypothesis. Dew input into the two treatments was thus expected to be similar; however, repeated measures ANOVA without replication showed that dew input into bare soil was significantly higher (three times) than that into gravel mulch ($F_{(5.12)} = 12.03$, $p < 0.01$) (Table 2). A plausible explanation for this could be the theoretical basis for separation of non rainfall atmospheric water

vectors used in this study (Sect. 2.2). This is based on the assumption that conditions favourable for one vector preclude the others from occurring concurrently (BROWN et al., 2008) and that soil surface temperature could be used to differentiate between vapour adsorption and dew formation (AGAM and BERLINER, 2006). These results, however, suggest that dew input into bare soil was not exclusive of direct vapour adsorption. The input classified as dew was thus possibly a combination of the two vectors occurring concurrently, resulting in the observed higher dew input into bare soil compared to gravel mulch (Table 2). GRAF et al. (2008) reported similar results while comparing input into bare soil and stone mulch and concluded that input into bare soil was beyond the strict definition of dew as defined by BEYSENS (1995), instead they termed this hygroscopic water uptake (vapour adsorption). Gravel mulch therefore probably restricted vapour adsorption into the sample during dew formation resulting in the low input classified as dew compared to that into bare soil.

According to BEYSENS (1995), heterogeneous nucleation lowers the energy barrier required for the formation of the liquid–vapour interface which is increased by surface roughness and chemicals (organic and inorganic). The soil surface has more discontinuity and chemicals compared to gravel mulch; therefore, theoretically, it would have had a higher rate of dew formation. The droplets formed on the soil surface would be absorbed by the soil fabric (matrix), freeing the surface for further nucleation. The gravel mulch on the other hand, has a limited absorbing capacity and the droplets coalesce first to form larger droplets capable of flowing under gravity from the stone surface (stone flow) into the soil fabric to free the stone surface for further nucleation. This process is slowed by weather conditions which alternate between condensation and evaporation (KASEKE et al., 2011). The nucleation phase occurs at a faster rate than droplet growth because the receiving substrate lowers the energy barrier (BEYSENS, 1995). Therefore, bare soil would in theory experience higher input as the surface is cleared for further nucleation faster than gravel mulch.

Droplet growth is increased by calm conditions in the boundary layer (BEYSENS, 1995) and according to

Table 2

Vector input direct into bare soil and gravel mulch treatments ± standard deviation (SD), Stellenbosch July 2008

Treatment	Vapour adsorption (mm)	Dew (mm)
Bare soil	2.37 ± 0.56	0.79 ± 0.46
Gravel mulch	0.32 ± 0.12	0.25 ± 0.14

GOOSSENS (1994), stony surfaces increase turbulence in the boundary layer. The low dew input recorded in the gravel mulch could therefore also be due to gravel surface induced micro-turbulence that could have disrupted the growth of dew droplets on the mulch surface. Theoretically, any micro-turbulence that may have been experienced on the bare soil surface would be less than that experienced in the gravel mulch treatment, and hence the larger dew input into bare soil (Table 2).

In gravel mulch, vapour adsorption and dew input contributed approximately equal amounts (56 and 44% respectively) to total input while in bare soil, the former contributed significantly more than the latter (75 and 25% respectively) (Table 2). Gravel mulch could have restricted vapour adsorption into the treatment. The volumetric data on composition of non rainfall atmospheric water (Table 2) could also be explained in terms of the duration of each vector in each treatment (Table 3). The input phase in bare soil commenced on average from 1610 to 0837 hours while that on gravel mulch began an hour later (1710–0925 hours) and was 40 min shorter. There were however, periods when neither evaporation nor input was detected probably because the samples were at equilibrium with ambient humidity (GRAF et al., 2008). This lowered the active (actual) hours of input (Table 3). Based on the duration of actual input, vapour adsorption into bare soil contributed 78.43% of the total input time while dew accounted for the remaining 21.57% in the absence of fog. In gravel mulch, vapour adsorption contributed 60% of the total input time and dew accounted for the remaining 40%. These values are comparable to the volumetric percentages (Table 2), although they slightly underestimate dew input by about 4%. This is because the

method assumes equal rates of input by dew and vapour adsorption which is not necessarily true.

KASEKE et al. (2011) showed two methods of calculating evaporation from the microlysimeter method: net and total evaporation. Net evaporation is the difference between soil water content at the commencement of the evaporative phase and that at the end of the phase including input. Total evaporation is a summation of all evaporation during the evaporative phase and excludes any input that may have occurred. Repeated measures ANOVA without replication showed that both net and total evaporation from gravel mulch was significantly lower than that from bare soil ($F_{(5.12)} = 57.24$, $p < 0.01$; $F_{(5.12)} = 238$, $p < 0.01$). Net evaporation from bare soil was 2.03 times more than that from gravel mulch (2.09 mm, SD 0.39, SE 0.12 and 1.03 mm, SD 0.12, SE 0.04 respectively), while total evaporation from bare soil was 3.42 times more than that from gravel mulch (3.73 mm, SD 0.54, SE 0.17 and 1.09 mm, SD 0.15, SE, 0.05 mm respectively). KEMPER et al. (1994) reported an 85% drop in annual evaporation due to a 5 cm gravel mulch on the soil surface. The present study confirms this trend showing that gravel mulch decreased net evaporation by 51% and total evaporation by 71% compared to bare soil. The lower evaporative loss was possibly because gravel mulch reduced the soil surface in direct contact with the atmosphere, thereby shielding underlying soil from excessive wind and solar radiation, thus conserving soil moisture (DANALATOS et al., 1995; LI, 2002; KATRA et al., 2008).

The 11–15 July 2008 period shows a decrease in soil water content reflecting a net loss in water content (Fig. 3). This was because the study commenced with wet (saturated) soils due to rains experienced on 10 July when the equipment was installed. A sudden spike indicating the only rainfall event during the test period is observed on the night of the 15–16 July 2008 (Fig. 3). Bare soil had a higher water content peak than gravel mulch after the rainfall event. The difference in peaks could be attributed to a volume reduction in storage capacity caused by the stone cover (DANALATOS et al., 1995). From the 16 July until the end of the experiment, both treatments experienced a high evaporative demand resulting in net loss of soil water. It is evident that gravel mulch

Table 3

Calculated input duration of each vector into bare soil and gravel mulch

Treatment	Input	Active input	Adsorption	Dew
Bare Soil	16 h 50 min	9 h 55 min	7 h 49 min	2 h 6 min
Gravel mulch	16 h 10 min	11 h 48 min	7 h 10 min	4 h 43 min

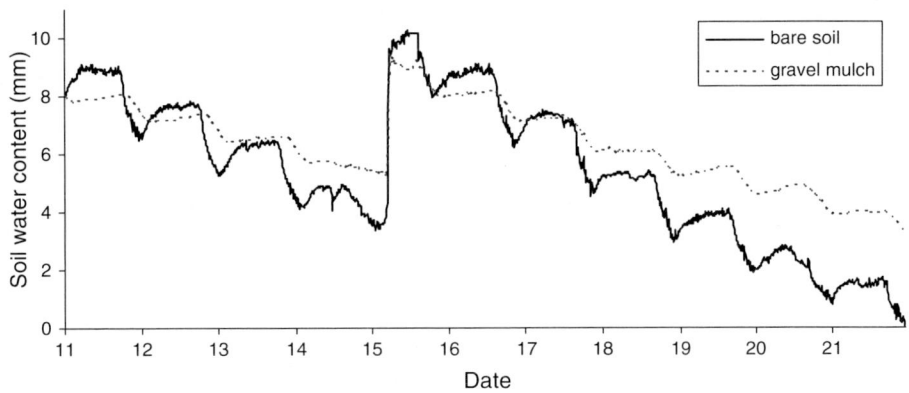

Figure 3

Diurnal changes in soil water content measured from bare soil and gravel mulch treatments using the microlysimeter method, Stellenbosch July 2008

experienced lower fluctuations in soil water content and was better at moisture conservation compared to bare soil (Fig. 3). This trend was reported by KEMPER *et al.* (1994), LI (2002) and KATRA *et al.* (2008). There was, however, no visible trend or drift in observations in both input and evaporation to the end of the research period.

4. Conclusions

The study did not yield any evidence to support the enhanced nocturnal cooling hypothesis (MATZNETTER, 1958; ACOSTA BALADON, 1973, 1996). Instead the data presented showed a significant decrease in non rainfall atmospheric water input into gravel mulch compared to bare soil. This is consistent with other findings (DANALATOS *et al.*, 1995; LI, 2002) and is possibly due to a decrease in the soil surface area in direct contact with the atmosphere and because the mulch acted as a heat sink which slowed inception of the input phase, decreased input duration and ultimately non rainfall water input into the treatment compared to bare soil.

KAPPEN *et al.* (1979) estimated that the theoretical threshold for dew availability for microbes was 0.03 mm which was exceeded nightly at Stellenbosch during winter. If expanded to dryer regions or the Mediterranean summer, this water may play a significant role in biological decomposition and nutrient release. Some authors have even suggested that it

could support the existence of rain fed vegetation during the dry summers of the Mediterranean (KOSMAS *et al.*, 1998).

Although gravel mulch did not increase non rainfall water input into soil via enhanced nocturnal cooling, it is nonetheless important for reducing evaporation. The data demonstrated that gravel mulch significantly retarded evaporation from the underlying soil, similar to reports by KEMPER *et al.* (1994), DANALATOS *et al.* (1995) and KATRA (2008). This implies that soil moisture from rain persists for a longer time under gravel mulch compared to bare soil creating benign microhabitats conducive for soil ecology especially in arid or semiarid environments. These results therefore suggest that the most important function of gravel mulch in Mediterranean areas is soil moisture conservation rather than supplying non rainfall atmospheric water. This could be important for plant growth and critical for combating desertification in arid areas (DANALATOS *et al.*, 1995).

The theoretical basis for non rainfall atmospheric vector separation has some pitfalls but nonetheless is important because it enables the determination of a proxy composition of non rainfall atmospheric water input direct into soil. This method when expanded to arid areas could enable the determination of the significance of each vector to the micro-hydrology and ecology of the soil system. Under the prevailing environmental conditions at Stellenbosch during the test period, water vapour adsorption provided the bulk of non rainfall atmospheric water input into soil.

In conclusion, the development of farming techniques that make use of gravel mulch in Lanzorate (GRAF et al., 2008) and Shatian (LI et al., 2000) were, therefore, likely based on the soil moisture conservation properties of gravel mulch rather than additional water supply to the soil.

Acknowledgments

This study was financially supported by the National Research Foundation of South Africa (NRF). The authors would also like to thank Dr. J. Irish and Prof. M. Fey for the input.

REFERENCES

ACOSTA BALADON, A. N., 1973. Cultivos Enaredos. In: GRAF, A., KUTTLER, W., WERNER, J., 2008. Mulching as a means of exploiting dew for arid agriculture? Atmos. Res. 87, 369–376.

ACOSTA BALADON, A. N., 1996. Las precepitaciones ocultas y sus aplicaciones a la agricultura. In: GRAF, A., KUTTLER, W., WERNER, J., 2008. Mulching as a means of exploiting dew for arid agriculture? Atmos. Res. 87, 369–376.

AGAM, N., BERLINER, P.R., 2004. Diurnal water content changes in the bare soil of a coastal desert. J. Hydrometeo. 5, 922–933.

AGAM, N., BERLINER, P.R., 2006. Dew formation and water vapour adsorption in semi arid environments—A review. J. Arid Environ. 65, 572–590.

BEYSENS, D., 1995. The formation of dew. Atmos. Res. 39, 215–237.

BROWN, R., MILLS, A. J., JACK, C., 2008. Non-rainfall moisture inputs in the Knersvlakte: Methodology and preliminary findings. Water SA. 34, 275–278.

DANALATOS, N. G., KOSMAS, C. S., MOUSTAKAS, N. C., YASSOGLOU, 1995. Rock fragments II. Their impact on soil physical properties and biomass production under Mediterranean conditions. Soil Use Mgmt. 11, 121–126.

EPSTEIN, E., GRANT, W. J., STRUCHTMEYER, R. A., 1966. Soil losses and crust formation as related to some soil physical properties. In: WEBB, R. H., WILSHIRE, G. H., 1983. Environmental effects of off-road vehicles. Springer-Verlag New York Inc.

GOOSSENS, D., 1994. Effect of rock fragments on eolian deposition of atmospheric dust. Catena, 23, 167–189

GRAF, A., KUTTLER, W., WERNER, J., 2008. Mulching as a means of exploiting dew for arid agriculture? Atmos. Res. 87, 369–376.

HILLEL, D., 1982. Negev: land, water and life in a desert environment. Praeger Publication, New York, USA.

JURY, W. A., BELLANTUONI, B., 1976. Heat and water movement under surface in a field soil: I. Thermal effects. Soil Sci. Soc. Am. J. 40, 505–509.

KAPPEN, L., LANGE, O. L., SCHULZE, E.D., EVENARI, M., BUSCHBOM, V., 1979. Ecophysiological investigations on lichens of the Negev Desert: IV. Annual course of the photosynthetic production of Ramalina maciformis (Del.) Bory. Flora, 168, 85–105.

KASEKE, K.F., ESLER, K.J., MILS, A.J., BROWN, R., HENSCHEL, J., SEELY, M.K., 2010. A method for the direct assessment of the "non rainfall" atmospheric water cycle: input and evaporation from the soil. Journal of Pure and Applied Geophysics, doi: 10.1007/s00024-011-0328-9

KATRA, I., LAVEE, H., SARAH, P., 2008. The effect of rock fragment size and position on topsoil moisture on arid and semi-arid hillslopes. Catena, 72, 49–55.

KEMPER, W. D., NICKS, A. D., COREY, A. T., 1994. Accumulation of water in soils under gravel and sand mulches. Soil Sci. Soc. Am. J. 58, 56–63.

KIDRON, G.J., 2000. Analysis of dew precipitation in three habitats within a small arid drainage basin, Negev Highlands, Israel. Atmos. Res. 55, 257–270.

KOSMAS, C., DANALATOS, N.G., POESEN, J., VAN WESEMAEL, B., 1998. The effect of vapour adsorption on soil moisture content under Mediterranean climatic conditions. Agri. Water Mgmt. 36, 157–168.

KOSMAS, C., MARATHIANOU, GERONTIDIS, ST., DETSIS, V., TSARA, M., POESEN, J. 2001. Parameters affecting water vapour adsorption by the soil under semiarid climatic conditions. Agri. Water Mgmt. 48, 61–78

LI, X. Y., GONG, J. D., GAO, Q. Z., WEI, X.H., 2000. Rainfall interception loss by pebble mulch in the semi arid region of China. J. Hydro. 228, 165–173.

LI, X. Y., 2002. Effects of gravel and sand mulches on dew deposition in the semiarid region of China. J. Hydro. 260, 151–160.

LOUW, G.N., SEELY, M.K., 1982. Ecology of Desert Organisms. Longman House, Essex, United Kingdom

MALEK, E., MCCURDY, G., GILES, G., 1999. Dew contribution to the annual water balances in semi-arid desert valleys. J. Arid Environ. 42, 71–80.

MATZNETTER, J., 1958. Die Kanarischen Inseln. In: GRAF, A., KUTTLER, W., WERNER, J., 2008. Mulching as a means of exploiting dew for arid agriculture? Atmos. Res. 87, 369–376.

WHITFORD, W. G., 2002. Ecology of desert systems. Elsevier Science Ltd.

SCHMIEDEL, U., JURGENS, N., 2004. Habitat ecology of southern African quartz fields: studies on the thermal properties near the ground. Plant Ecology, 153, 153–166.

STONE, E.C., 1963. The ecological importance of dew. The Quartely Rev. Bio. 38, 328–341.

(Received November 15, 2010, revised June 8, 2011, accepted June 16, 2011, Published online July 19, 2011)

Pure Appl. Geophys. 169 (2012), 881–893
© 2011 Springer Basel AG
DOI 10.1007/s00024-011-0357-4

Integration of Local Observations into the One Dimensional Fog Model PAFOG

CHRISTINA THOMA,[1] WERNER SCHNEIDER,[1] MATTHIEU MASBOU,[1] and ANDREAS BOTT[1]

Abstract—The numerical prediction of fog requires a very high vertical resolution of the atmosphere. Owing to a prohibitive computational effort of high resolution three dimensional models, operational fog forecast is usually done by means of one dimensional fog models. An important condition for a successful fog forecast with one dimensional models consists of the proper integration of observational data into the numerical simulations. The goal of the present study is to introduce new methods for the consideration of these data in the one dimensional radiation fog model PAFOG. First, it will be shown how PAFOG may be initialized with observed visibilities. Second, a nudging scheme will be presented for the inclusion of measured temperature and humidity profiles in the PAFOG simulations. The new features of PAFOG have been tested by comparing the model results with observations of the German Meteorological Service. A case study will be presented that reveals the importance of including local observations in the model calculations. Numerical results obtained with the modified PAFOG model show a distinct improvement of fog forecasts regarding the times of fog formation, dissipation as well as the vertical extent of the investigated fog events. However, model results also reveal that a further improvement of PAFOG might be possible if several empirical model parameters are optimized. This tuning can only be realized by comprehensive comparisons of model simulations with corresponding fog observations.

Key words: Fog modeling, nudging, single-column model, visibility.

1. Introduction

Fog and low clouds often yield a strong reduction of visibility, thus affecting various types of transport systems, such as car, air and ship traffic. Accurate forecasts of atmospheric conditions resulting in low visibility have, therefore, become an important issue.

Low visibility during fog events is a result of complex radiative, turbulent and microphysical processes as well as interactions between the planetary boundary layer (PBL) and the underlying surface. Therefore, the numerical forecast of fog remains a difficult task.

Intrinsically, fog is a highly three dimensional atmospheric phenomenon being characterized by horizontal heterogeneities and, in particular, strong vertical gradients of various thermodynamic quantities. In order to account for these heterogeneities in a numerical model, in principle it would be necessary to simulate the PBL by means of three dimensional atmospheric models having very high horizontal and vertical grid resolutions so that the atmospheric fine structure characterizing a particular fog event is adequately resolved. In recent years, several three dimensional fog models have been developed, e.g. MÜLLER *et al.* (2005), CAPON *et al.* (2007), MASBOU (2008) and YANG *et al.* (2009). Even though these models are capable of simulating three dimensional fog structures reasonably well, for operational fog forecasts they are presently not applied. The main reason for this is that, owing to their extreme high spatial resolutions, these models are computationally too expensive. Radiation fog usually occurs during calm atmospheric situations with rather weak horizontal forcing. Thus, it seems justified to forecast this kind of fog by means of one dimensional models without abdicating the required high vertical grid resolution.

In recent decades, numerous one dimensional fog models have been developed. The COBEL-ISBA model (BERGOT and GUEDALIA 1994; GUEDALIA and BERGOT 1994) is a one dimensional approach for the simulation of the nocturnal boundary layer coupled with the detailed soil model ISBA. A single column version of the Met. Office Unified Model has been

---
[1] Meteorological Institute, University of Bonn, Auf dem Hügel 20, 53121 Bonn, Germany. E-mail: thoma@uni-bonn.de; werner.schneider@uni-bonn.de; mmasbou@uni-bonn.de; a.bott@uni-bonn.de

developed for the prediction of radiation fog events (CLARK and HOPWOOD 2001). Based on the three dimensional limited area model HIRLAM (UNDEN et al. 2002). TERRADELLAS and CANO (2007) presented a one dimensional model for fog forecasts at central Spanish airports. BOTT and TRAUTMANN (2002) developed the one dimensional fog model PAFOG with parameterized cloud microphysics. Basically, PAFOG is a modified version of the radiation fog model MIFOG (BOTT et al. 1990) and the stratus model MISTRA (BOTT 1997), both of them including a detailed spectral treatment of cloud microphysical processes. Thus, PAFOG may be used for computationally efficient numerical simulations of radiation fog events and low level stratiform clouds. (GULTEPE et al. 2007) presented a comprehensive overview of the current state of numerical fog modeling.

For the verification of numerical fog forecasts, it seems inevitable to compare the results with corresponding field measurements. Fortunately, there exists a large number of observational studies of fog events in terms of the analysis of climatological data (TARDIF and RASMUSSEN 2007), by remote sensing techniques (CERMAK and BENDIX 2007; WESTCOTT, 2007) and additionally by in situ observations (e.g. GULTEPE and MILBRANDT 2007; GULTEPE et al. 2009). These observations serve as important tools for the tuning of different empirical parameters of a fog model. For the standard model version of PAFOG, several model parameters have been tuned by means of comprehensive model sensitivity studies where fog observations at the Lindenberg Observatory of the German Meteorological Service (DWD) have been compared with corresponding model results (see e.g. BOTT and TRAUTMANN 2002).

Although radiation fog usually forms during calm atmospheric conditions with weak horizontal heterogeneities, the corresponding horizontal forcing terms might still be of particular importance for the fog evolution. In order to integrate horizontal forcing effects as well as actual measurements in the numerical simulations, data assimilation procedures are frequently included in one dimensional models. For instance, BERGOT et al. (2005) developed a detailed procedure for the assimilation of local observations at the Paris Charles de Gaulle airport. Moreover, they implemented a 1D-VAR scheme for the assimilation

of atmospheric profiles of thermodynamic variables being modified in cloudy environments.

In this paper, an alternative data assimilation procedure for the numerical simulation of fog and low clouds will be presented. The approach consists of extending PAFOG by the nudging scheme as described in KALNAY (2003). The method concentrates on the assimilation of atmospheric profiles of temperature and specific humidity. Moreover, a procedure for the initialization of PAFOG with observed visibility data will be introduced. The goal of these model modifications is to obtain a better representation of observed data in the fog forecast. A typical application of the modified PAFOG model will be to start the fog simulation some hours before the actual time, thereby utilizing the observed data. The numerical simulations will then be extended to some hours in the future, thus yielding a short time fog forecast with consideration of measured data. The new assimilation techniques have been developed within the iPort-VIS project, part of an aviation research program funded by the German Federal Ministry of Economy and Technology in which a prognostic fog forecast system for the Munich airport site is developed. In Sect. 2 the main characteristics of the PAFOG model are shortly presented. The methods for visibility initialization and the nudging scheme for the integration of local observations are thoroughly described in Sect. 3. Section 4 deals with the numerical results of a case study. Finally, conclusions, ongoing research and future work are summarized.

2. PAFOG

The one dimensional fog model PAFOG has been developed for the numerical simulation of atmospheric processes in the PBL (BOTT and TRAUTMANN 2002). The main goal of PAFOG consists of the forecast of radiation fog and low level stratiform clouds. Owing to the one dimensional treatment, the model can deal with a relatively high vertical resolution, thus providing a detailed description of physical processes in the PBL despite the lack of simulating horizontal inhomogeneities. PAFOG can run in two different modes. For simulations of radiation fog events the lowest 200 meters of the atmosphere are subdivided into 50 equidistant layers while between 200 and 2,500 m a vertically stretching grid of 20 logarithmically equidistant layers is chosen. In the stratus mode the equidistant model region above the Earth's surface is increased from 200 to 1,500 m. The remaining parts of the grid structure are the same as in the fog mode. Cloud microphysical processes are calculated only in the equidistant model region above the Earth's surface (0–200 m for fog and 0–1,500 m for stratus). Thus, in a particular model simulation the top of the fog or stratus cloud must not exceed this region. However, in order to fulfill this requirement, the grid configuration could be freely modified. For calculations of temperature and humidity profiles within the soil, the region 0–50 cm below the Earth's surface is subdivided into 20 logarithmically equidistant layers, the first layer having a thickness of 1 mm. For radiation calculations, between 3 and 50 km a constant model atmosphere being typical for the midlatitudes is applied.

In the following, a brief summary of the most important model features of PAFOG will be given. For a detailed model description the readers are referred to the original papers by BOTT et al. (1990), BOTT (1997) and BOTT and TRAUTMANN (2002). The dynamical part of PAFOG consists of a set of prognostic equations for the horizontal wind field u and v, the potential temperature Θ and the specific humidity q. The turbulent processes in the PBL are treated by means of the 2.5 level scheme of MELLOR and YAMADA (1974, 1982) in which a prognostic equation for the turbulent kinetic energy is solved. In the stratus mode of PAFOG, a constant large scale subsidence is also taken into account. For the numerical solution of the prognostic equation set, the integration time step varies automatically between 10 s in turbulent and 60 s in stable atmospheric situations.

The radiative transfer equation is solved according to the δ-two-stream approximation PIFM2 (LOUGHLIN et al. 1997) which is a modified version of the practical improved flux method (PIFM) of ZDUNKOWSKI et al. (1982). In the solar and infrared spectral regions the extinction by different gases, aerosol particles and cloud droplets N_{act} is treated whereby both regions are subdivided into six and 12

subintervals, respectively. The transmission functions are calculated by means of the correlated k-distribution method according to FU (1991) and FU and LIOU (1992). Radiation calculations are performed every 5 min.

The soil is treated as a porous medium of dry air, water vapor, water and the soil matrix. Twelve different soil types can be chosen after PIELKE (1984). Two prognostic equations are solved for the soil temperature T_s and the volumetric moisture content η. The interaction of land-surface processes with the overlying atmosphere is modeled following SIEBERT et al. (1992a, b). In this model, the canopy is treated as a single layer including the canopy air and the foliage of the plants.

Cloud microphysical processes are treated by solving two prognostic equations for the liquid water content (LWC) and the droplet number concentration N_d. The droplet size distribution is taken as a log-normal function. The number of activated cloud droplets N_{act} is parameterized according to Twomey's relation (TWOMEY 1959)

$$N_{act} = N_a S^k \tag{1}$$

Here, N_a is the total number concentration of aerosol particles with radii exceeding 0.05 μm, S is the supersaturation, and k depends on the choice of the aerosol type being used in a particular model simulation. Four different aerosol types may be chosen: rural ($N_a = 10{,}000$ cm^{-3}, $k = 0.9$), urban ($N_a = 50{,}000$ cm^{-3}, $k = 0.9$), maritime ($N_a = 200$ cm^{-3}, $k = 0.7$) or tropospheric ($N_a = 100$ cm^{-3}, $k = 0.7$). The different N_a-values are based on JÄNICKE (1993). They may, however, be replaced by other number concentrations, e.g. if measurements are available. The calculation of the supersaturation S follows the work of SAKAKIBARA (1979), see also CHAUMERLIAC et al. (1987). Gravitational settling of cloud droplets is also considered in PAFOG.

In the present model version of PAFOG, visibility (VIS) is parameterized according to KOSCHMIEDER (1924) as

$$VIS = -\frac{\ln(\varepsilon_0)}{\beta_{sca,R} + \beta_{ext,a} + \beta_{ext,d}} \tag{2}$$

where ε_0 is a contrast threshold that is assumed to be 0.2. The scattering coefficient $\beta_{sca,R}$ treats Rayleigh scattering by air molecules with $\beta_{sca,R} = 1.227 \times 10^{-5}$ m^{-1} while $\beta_{ext,a}$ describes extinction by aerosol particles. The latter depends on the relative humidity (RH) and is calculated as function of the extinction cross section $\sigma_{ext,a}$ of the aerosol particles (SHETTLE and FENN 1979)

$$\beta_{ext,a}(RH) = N_a \sigma_{ext,a}(RH) \qquad (3)$$

The values of $\sigma_{ext,a}$ have been precalculated as function of RH and are available in PAFOG as a look-up table. During foggy situations the reduced visibility is expressed by the extinction coefficient of cloud droplets $\beta_{ext,d}$ occurring in (2). For the given log-normal size distribution of the cloud droplets this term is given by BOTT and TRAUTMANN (2002)

$$\beta_{ext,d} = \overline{Q}_{ext} N_d \frac{\pi}{4} \left(\frac{6LWC}{\pi \rho_w N_d} \right)^{2/3} \exp\left(-\sigma_c^2\right) \qquad (4)$$

Here, ρ_w is the density of water, $\overline{Q}_{ext} \approx 2$ is the mean extinction efficiency of water droplets and σ_c is the standard deviation of the droplet size distribution. In the model calculations $\sigma_c = 0.2$ has been utilized. From (2) and (4) it may be easily seen that the visibility within fog is calculated as function of the prognostic variables LWC and N_d.

3. Methods

In this section, the method will be explained how observations of visibility and the nudging scheme are included in the original PAFOG version of BOTT and TRAUTMANN (2002).

3.1. Initialization with Observed Visibility

Knowledge of the actual visibility is essential for the initialization of a fog model. Particularly, when the numerical simulations are started during a fog event, the model needs information about the initial visibility as well as the lower and upper fog boundaries. These data, which are often available by routine observations, may be used by the fog model in order to obtain a rough estimate of the initial spatial distribution and density of the fog. The visibility at a height of 2 m is denoted as VIS(2) below.

Unfortunately, PAFOG has no option to deal with missing data (e.g. missing observations of upper and lower fog boundaries) so that in these cases the model has to be started by ignoring eventually existing initial foggy conditions. To avoid this model deficiency of the original PAFOG model, a more flexible treatment for the consideration of initial visibility values being available will now be introduced. This is accomplished by distinguishing two basically different applications: First, if the model starts during foggy conditions, the initial values of LWC and N_d are calculated as function of VIS(2). Second, in fog-free conditions, the observed visibility will be used to calculate the initial aerosol concentration. The complete initialization procedure of the fog model is illustrated in the decision tree shown in Fig. 1 and the major steps will now be outlined in detail.

3.1.1 Case 1a: Ground Fog is Observed

In this situation the lower fog boundary z_b is set equal to 0 m and the top of the fog z_t as well as the visibility VIS(2) are taken from observations. If VIS(2) is not available, its value is set equal to 100 m being typical for a rather dense fog. This low visibility value has been chosen based on the experience that the model can deal better with an underestimated value than with an overestimated initial visibility. In case the top of the fog is unknown, the level where the relative humidity (RH) for the first time decreases below 95% will be used. Within the foggy layers, the initial LWC is assumed to decrease linearly from its value at 2 m height, i.e. LWC(2), to LWC = 0 at the top of the fog. With this assumption the gravitational settling of larger fog droplets is roughly taken into account. The determination of LWC(2) will be given below. The visibility is assumed to increase linearly with height. VIS at the top of the fog is calculated from (2) and (3) by utilizing the standard values of N_a (see Sect. 2) in (3). For the calculation of the initial N_d, (2) and (4) are used. In this case, the contributions of aerosol extinction and Rayleigh scattering are neglected in (2) based on the assumption that within fog VIS is mainly reduced due to the occurrence of fog droplets. This results in

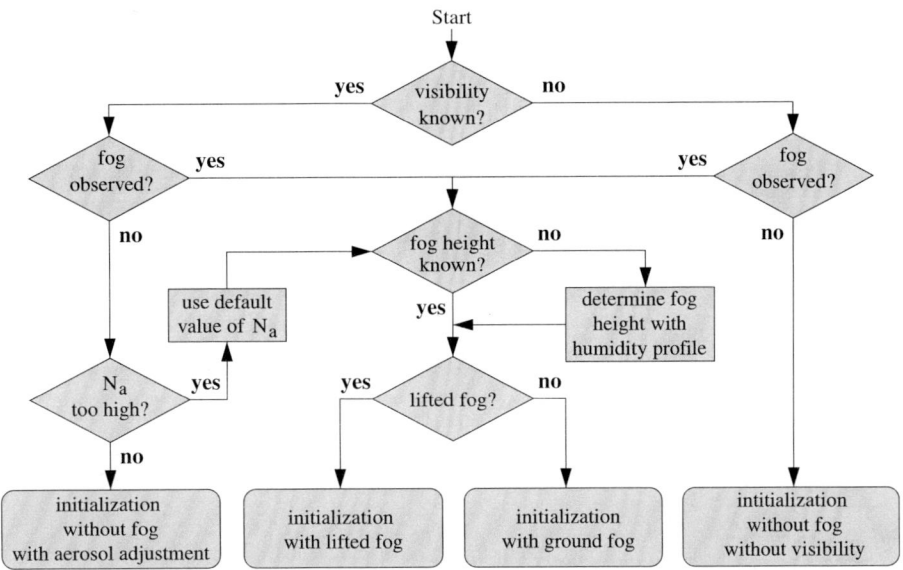

Figure 1
Decision tree for the initialization of PAFOG with observed or assumed values of visibility

$$N_d(z) = \left(-\frac{\ln(\varepsilon_0)}{\text{VIS}(z)} \frac{2\exp(\sigma_c^2)}{\pi\left(6\text{LWC}(z)\pi^{-1}\rho_w^{-1}\right)^{2/3}} \right)^3,$$

$$z_b \leq z \leq z_t \tag{5}$$

3.1.2 Case 1b: Lifted Fog is Observed

Now VIS(2) is used to calculate N_a between the Earth's surface and the lowest foggy layer from (2) and (3) as

$$N_a = -\frac{1}{\sigma_{\text{ext,a}}} \left(\frac{\ln(\varepsilon_0)}{\text{VIS}(2)} + \beta_{\text{sca,R}} \right) \tag{6}$$

If the resulting N_a exceeds the maximum value of the corresponding aerosol type [maximum values are 40,000 cm^{-3} for rural, 500,000 cm^{-3} for urban and 2,000 cm^{-3} for maritime and tropospheric aerosol particles (JÄNICKE, 1993)], then the model is assumed to be in ground fog conditions, that is VIS(2) is utilized to compute LWC (again linearly decreasing with height), and for N_a the standard values are taken. In case VIS(2) is not available, then also the standard value for N_a will be taken. Within the foggy region VIS = 100 m is again applied while LWC is linearly decreasing with height. If necessary, z_b and z_t are, analogously to the ground fog case above, determined as function of RH.

3.1.3 Case 2: No Fog is Observed

In fog-free conditions N_a is again determined by means of (6). The same treatment as in case 1b is applied if N_a exceeds the maximal possible value of the chosen aerosol type.

Finally, the initial value of LWC(2) needs to be determined. Utilizing (2) and (4), a function of the form $f(\text{LWC, VIS}, N_d) = 0$ may easily be derived. Since among the three variables only VIS is known, an additional relationship has to be provided for the determination of the remaining two variables. This relationship has been empirically derived by means of numerous sensitivity studies with PAFOG. In these model runs, a large number of data pairs of (VIS, LWC) has been obtained as function of different values of N_a. In order to reduce the effects of gravitational settling of fog droplets, the data were obtained during the formation phase of fog in a particular model layer.

As an example, Fig. 2 shows the results in the form of a scatter plot as obtained from five radiation fog events between September 2005 and October 2006. In these simulations, the rural aerosol type with standard value of $N_a = 10,000$ cm^{-3} was applied. From the data points in the figure, the solid line has been derived as the best fit for the determination of

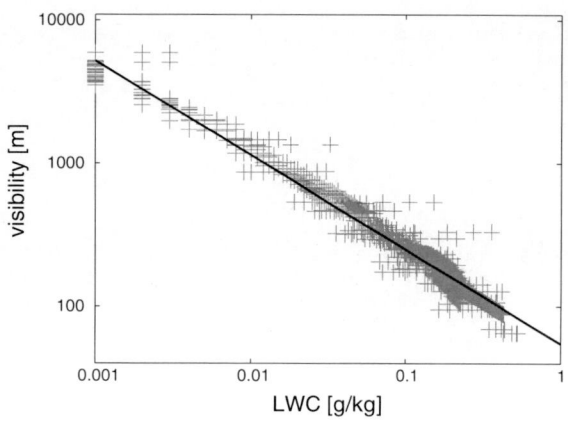

Figure 2

Scatter plot depicting the relationship between LWC and VIS for the rural aerosol type with $N_a = 10,000$ cm^{-3}

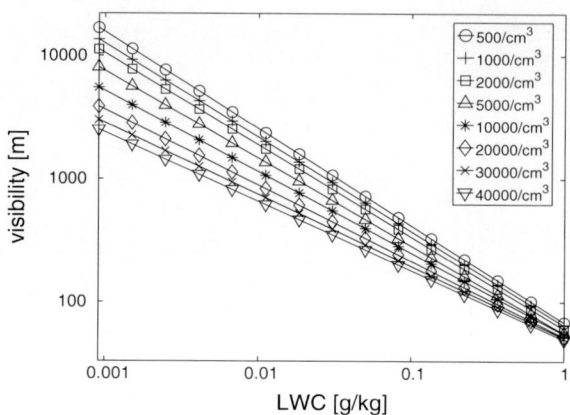

Figure 3

Best fits for the rural aerosol type for eight different values of N_a

LWC as function of VIS. Similar numerical experiments have been repeated with other values of N_a. The corresponding best fits are depicted in Fig. 3. Utilizing these data the following empirical relationship could be derived

$$\text{VIS} = a\,\text{LWC}^b \tag{7}$$

Similar studies have been repeated for the remaining aerosol types used in PAFOG (maritime, urban, tropospheric). Table 1 depicts the resulting values of the empirical coefficients. As was to be expected, a and b depend on the particular aerosol type and on the total aerosol number concentration. Inserting the values of this table into (7), with the help of linear interpolation the initial value of LWC(2) as function of VIS(2) may be obtained.

3.2. Nudging

A general problem of one dimensional fog models consists of the assumption of horizontal homogeneity of all model variables. Thus, the model is not able to treat horizontal forcing terms such as advection processes. To partially account for these effects, a nudging scheme has been implemented in PAFOG. Nudging is a very efficient method of considering measured data in numerical models. The approach is well established and widely being used in atmospheric modeling (see e.g. ANTHES 1974; STAUFFER and SEAMAN 1990; KALNAY 2003). It consists of a continuous inclusion of observational data into the

numerical simulations by adding forcing terms to the governing model equations. For the nudging of an arbitrary model variable ψ, the corresponding prognostic equation is modified according to

$$\frac{\partial \psi}{\partial t} = F + \frac{\psi_{\text{obs}} - \psi}{\tau} \tag{8}$$

where F summarizes the forcing terms of the original prognostic equation, ψ_{obs} is the observed value of ψ and τ is a relaxation time regulating the intensity of the nudging approach. Obviously, the nudging term pushes the model variables in the direction of the observations, whereas with increasing value of τ the nudging term becomes less important.

In PAFOG nudging of the vertical profiles of temperature and specific humidity is applied. In principle, for the nudging procedure it would be necessary to provide observed profiles of temperature and specific humidity for the entire vertical profile of the prognostic model part, i.e. presently the lowest 2,500 m within the atmosphere. The observed profiles are often obtained from a meteorological tower, e.g. the 100 m tower at the observatory of the DWD in Lindenberg, Germany. In this case the nudging approach can only be applied in the lowest 100 m of the atmosphere. As a consequence, it might happen that due to the nudging approach, directly above the highest nudging layer discontinuities evolve in the vertical profiles of ψ. To avoid this problem, a weighting function has been implemented in the PAFOG nudging scheme that creates artificial ψ'_{obs}-data above those layers where real observations are

Table 1

Coefficients a and b for different aerosol types and number concentrations

Aerosol type	Rural		Urban		Maritime		Tropospheric	
N_a (cm^{-3})	a	b	a	b	a	b	a	b
50					106.37	−0.73	107.22	−0.73
100					85.67	−0.75	81.60	−0.77
200					75.17	−0.75	72.29	−0.76
300					70.39	−0.74	66.51	−0.77
500	68.37	−0.79			65.31	−0.73	60.57	−0.77
1,000	64.75	−0.76			59.31	−0.71	53.93	−0.76
1,500					57.32	−0.69	51.28	−0.75
2,000	60.54	−0.75			54.49	−0.68	49.37	−0.75
5,000	55.68	−0.71	60.32	−0.67				
10,000	55.02	−0.66	58.52	−0.61				
20,000	52.11	−0.62	59.44	−0.53				
30,000	51.02	−0.58	55.61	−0.49				
40,000	49.42	−0.56						
50,000			47.40	−0.46				
100,000			51.36	−0.35				
200,000			42.97	−0.27				
500,000			40.45	−0.10				

available. With increasing height, these values converge toward the ψ-values as calculated by PAFOG. If z_m denotes the maximum height where observations are available, then ψ'_{obs} is determined according to

$$\psi'_{obs}(z) - \psi(z) = [\psi_{obs}(z_m) - \psi(z_m)]$$
$$\exp\left(\frac{z_m - z}{z' - z_m}\right), z \geq z_m \qquad (9)$$

Here, z' is the height where the difference between $\psi'_{obs}(z)$ and $\psi(z)$ has decreased to $1/e$ of its value at z_m with $z' \geq z_m$ In the present model version, $z' = z_m + 100$ m is used. This value may be adjusted if other observation profiles with different values of z_m are available for the nudging approach. As an example, Fig. 4 shows a temperature profile which above 100 m has been constructed with (9).

4. Results

In order to test the behavior of the visibility initialization and the nudging scheme in PAFOG, three different numerical experiments have been performed. In the first experiment, the original version of PAFOG was used. In the second experiment, PAFOG was extended by the visibility initialization while in the

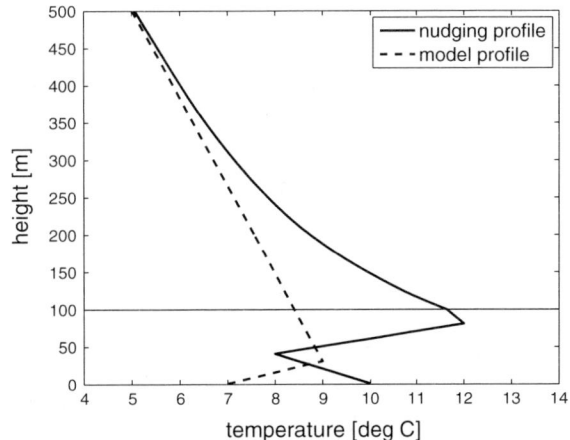

Figure 4

Example for the construction of a nudging profile, composed of observed profile data and the weighted model data after (9)

third experiment the nudging scheme was also applied (Fig. 4). For the three experiments a case study from September 14, 2005 of the Lindenberg area has been chosen. The synoptic situation was dominated by high pressure over central Europe. In the night from 13th to 14th of September, radiation fog formed over northeastern Germany between 2300 and 0000 UTC. The fog dissipated at sunrise between 0400 and 0500 UTC. The situation which may be classified as a typical radiation fog event yielded a minimum visibility of

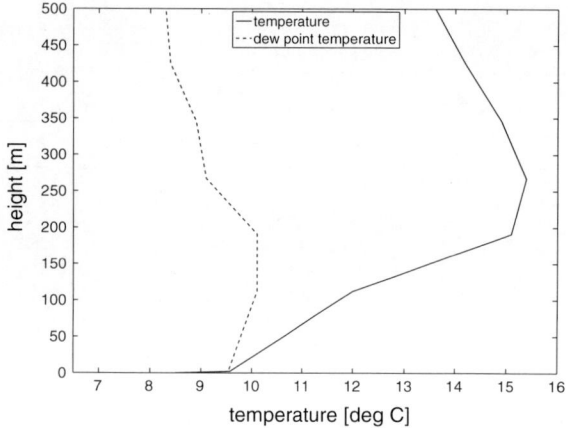

Figure 5

PAFOG input profiles for September 14, 2005 at 0000 UTC in Lindenberg from rawinsonde data and SYNOP observation. *Solid line* temperature, *dashed line* dew point temperature

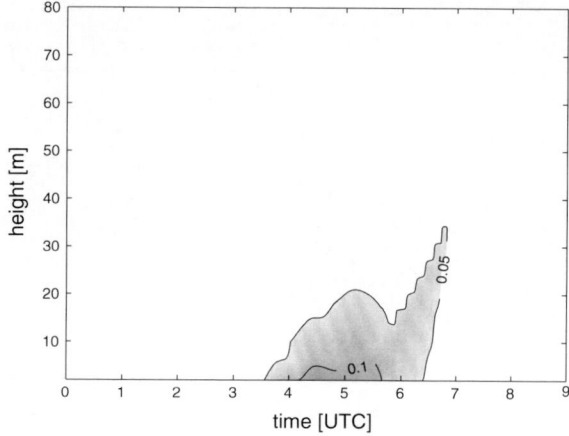

Figure 6

Contour plot of LWC (g kg^{-1}) of experiment 1

about 200 m. In all three experiments, the model was initialized at 0000 UTC using vertical profiles obtained from rawinsonde measurements performed at that time, SYNOP data were applied for the 2 m and surface values of temperature and specific humidity. Vertical profiles for experiment 3 were taken from the 100 m meteorological tower. All observation data were provided by the Lindenberg Meteorological Observatory. Figure 5 shows the vertical profiles of temperature and dew point temperature at 0000 UTC. As can be clearly seen, a relatively strong temperature inversion exists between 0 and 250 m height. Directly above the ground, the atmosphere is saturated indicating the occurrence of ground fog at model initialization time there.

4.1. Experiment 1

Figure 6 depicting the time evolution of the liquid water content as function of height reflects the typical behavior of a radiation fog event. Fog formation starts close to the Earth's surface. In the following hours the fog grows vertically up to some decameters height. After sunrise, turbulent mixing yields a relatively fast increase of the top of the fog which is caused by enhanced turbulent fluxes of latent heat and by the turbulent vertical transport of liquid water. In this sensitivity study the top of the fog increases to a maximum height of nearly 40 m at 0700 UTC.

At that time, the fog is also lifted from the ground and starts to dissipate. As no observation data about fog height and LWC are available, a verification of the model results with corresponding measurements is, unfortunately, not possible. By comparing this dissipation time with the observations, it is about 2 h too late. Although ground fog was already observed at model initialization time, in the numerical simulations fog was forming more than 3 h too late. The reason for this is that the original PAFOG version needs a relatively long spin-up time during which the vertical profiles of all model variables are adapted to one another.

4.2. Experiment 2

In this sensitivity study PAFOG was extended by the visibility initialization approach. At 0000 UTC VIS(2) = 189 m was observed in 2 m height. In the model simulations, the initial fog height was determined from the vertical profile of the relative humidity at 0000 UTC (see Sect. 3.1) resulting in the top of the fog at 32 m. The initial LWC(2) was calculated from the observed VIS(2) value yielding 0.15 g kg^{-1} (see Fig. 7). The difference of the LWC time evolution as compared to the first experiment (Fig. 6) is obvious. Already at the beginning of the numerical simulations the fog is well-developed. Moreover, the time evolution of the fog is rather different than that in Fig. 6, particularly during the fog dissipation where the top of the fog reaches up to

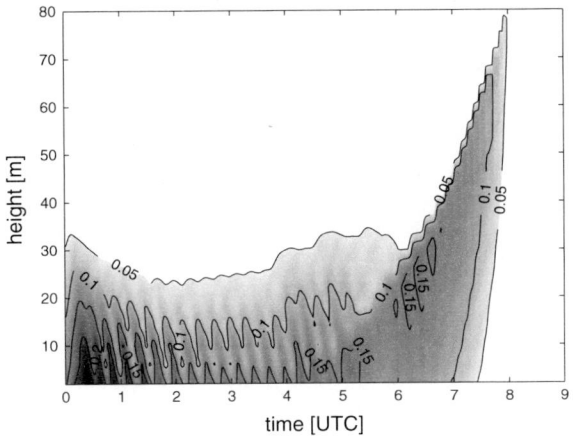

Figure 7
Contour plot of LWC (g kg⁻¹) of experiment 2

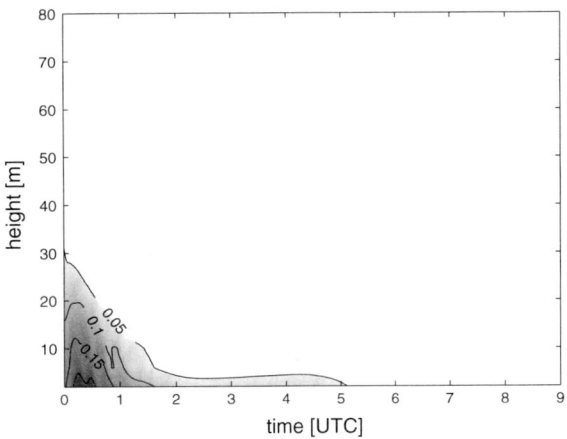

Figure 8
Contour plot of LWC (g kg⁻¹) of experiment 3

80 m height compared to 40 m in the first experiment. Finally, the fog dissipates about one hour later than in experiment 1. PAFOG produces strong fluctuations of LWC with periods of about 15 min. This effect is due to the interaction of the radiatively induced growth of fog droplets and their subsequent gravitational settling (BOTT *et al.* 1990). These fluctuations do not occur in Fig. 6 due to the much smaller LWC amounts in experiment 1.

4.3. *Experiment 3*

Experiments 1 and 2 reveal that the inclusion of the visibility initialization in PAFOG yields a distinct improvement of the formation phase of the fog event. However, by comparing the model results with observations, it also becomes evident that with this approach the dissipation time of the fog was slightly deteriorated rather than improved. A much better agreement between simulated and observed dissipation times was obtained with experiment 3 in which, in addition to the visibility initialization, the nudging scheme was also applied in PAFOG. Figure 8 depicting for this experiment the time evolution of LWC as function of height, shows a completely different behavior of LWC than in the first two experiments. Shortly after model initialization, the top of the fog decreases from 30 to less than 10 m and remains nearly constant until the fog dissipates. However, most remarkable is the time of fog dissipation at 0515 UTC which is now distinctly

earlier than without considering the nudging scheme and much more realistic.

As already mentioned above, the nudging procedure is largely controlled by the relaxation time τ. In Fig. 8, τ was chosen as 5 h. In order to obtain an impression of the model behavior as function of different values of τ, numerous PAFOG sensitivity studies have been performed, thereby varying τ between 0.5 and 60 h. As was to be expected these studies turned out that the model results are strongly affected by small values of τ. For τ-values larger than 10 h no distinct differences could be found. As an example, Figs. 9 and 10 show the time evolution of temperature and specific humidity in 2 m height and

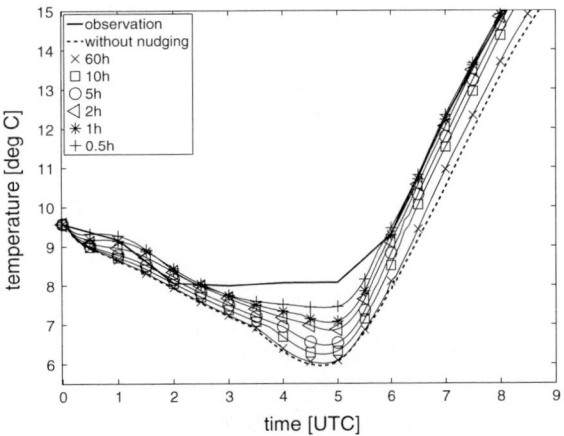

Figure 9
Time series of temperature with different values for τ; τ has been varied between 0.5 and 60 h

125

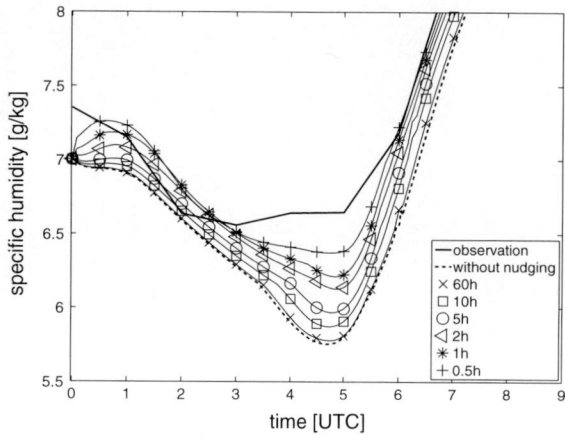

Figure 10
Time series of specific humidity with different values for τ; τ has been varied between 0.5 and 60 h

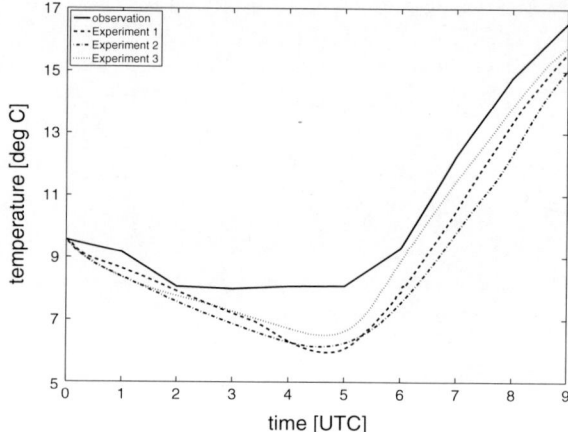

Figure 11
Time series of the observed and simulated 2 m temperature. *Solid line* observation, *dashed line* experiment 1, *dashed-dotted line* experiment 2, *dotted line* experiment 3

for the first nine simulation hours of the September 14, 2005 case study. The curves are compared with corresponding measurements at the meteorological tower of Lindenberg. It is clearly seen that for $\tau = 10$ h and $\tau = 60$ h PAFOG yields nearly identical results. In contrast to this, $\tau = 0.5$ h produces a rather strong push of the simulated temperature and humidity curves toward the observations. By recalling the motivation of the nudging approach, i.e. obtaining a slight push of the numerical simulations toward the observations without disturbing the model calculations too strongly, it has been decided that a relaxation time of five hours is appropriate. The PAFOG user may, however, modify this value if in a particular situation this seems to be appropriate.

4.4. Time Evolution of 2 m Variables

Figures 11 and 12 depict for the three experiments time evolutions of the temperature and visibility in 2 m height. Again the curves are compared with corresponding observations. Figure 11 shows that in all experiments during the early morning hours the simulated temperature is clearly too low in comparison to the observations. However, it is also seen that experiment 3 yields a distinct improvement of this model behavior. This holds in particular for the temperature values after 0600 UTC.

In Fig. 12, the visibility is plotted in the range of 0–1,000 m. It is clearly seen that in experiment 1 the

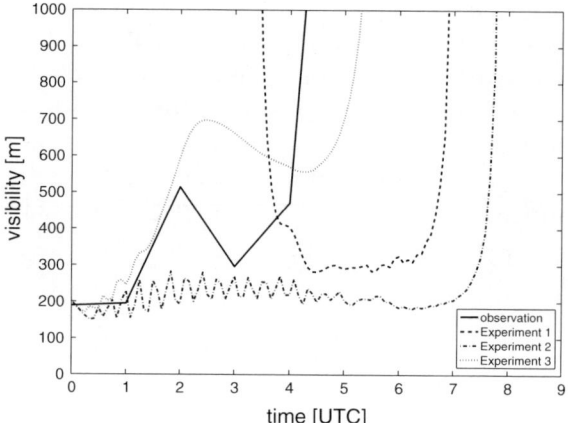

Figure 12
Time series of the observed and simulated 2 m visibility. *Solid line* observation, *dashed line* experiment 1, *dashed-dotted line* experiment 2, *dotted line* experiment 3

visibility curve does not resemble the observations. Moreover, as already shown in Fig. 6, the fog onset is not accurately represented because in this case PAFOG starts in a clear atmosphere. In experiment 2, already at the beginning of the model simulations a much better agreement of the visibility with the observations is achieved. However, with a value of about 200 m during the entire fog event, the simulated visibility is distinctly too low. Only by including the nudging scheme in PAFOG the model seems to be able to reproduce the observed visibility reasonably well. This behavior is quite remarkable since the variable VIS is not directly affected by the nudging approach.

As already mentioned, in the present version of PAFOG the visibility is calculated as function of the liquid water content and the total droplet number concentration, see (2) and (4). This treatment seems quite logical since for a given value of LWC different values of VIS might result depending on the number of particles carrying the fog water, i.e. a continental fog consisting of many rather small fog droplets yields lower visibilities than a marine fog with relatively few but large fog droplets. GULTEPE et al. (2006) confirmed this behavior by stating that visibility parameterizations based only on the liquid water content as proposed by KUNKEL (1984) yield unsatisfactory results. Instead, they recommended the inclusion of N_d as a second variable for the calculation of VIS. Based on numerous in situ measurements obtained during the Radiation and Aerosol Cloud Experiment, they developed an empirical relationship for VIS as function of LWC and N_d. By comparing their results with other existing methods, GULTEPE et al. (2006) clearly demonstrated the superiority of their visibility parameterization.

In order to test the Gultepe visibility parameterization, the method has been implemented into PAFOG. First preliminary results indicate that the visibilities as obtained by the original PAFOG and by the Gultepe method are fairly similar, the Gultepe parameterization yielding slightly higher values of VIS than the PAFOG approach. However, it is to be noted that in the PAFOG approach the visibility depends on several empirical model parameters, such as the aerosol number concentration, the standard deviation of the droplet size distribution and others. Thus, it is to be expected that a further improvement of the visibility parameterization might still be possible. To achieve this goal, it will be necessary to perform comprehensive model sensitivity studies with PAFOG utilizing the above and eventually other visibility parameterizations and to compare the model results with in situ observations. Future research with PAFOG will be directed towards this issue.

5. Conclusions

The one dimensional fog model PAFOG has been extended by two procedures for the integration of local observations into the numerical simulations. The first extension consists of a method to initialize the model with observed visibility data. This results in the possibility either to start PAFOG during foggy conditions or to use the observed visibility for the calculation of the initial aerosol number concentration. The second extension of PAFOG is given by a nudging scheme that is able to continuously integrate observed vertical profiles of temperature and specific humidity into the model simulations. These model modifications might be used to start the numerical fog simulations some hours in the past, thereby considering the observed data. The simulations will then be extended to some hours in the future, thus obtaining a short time forecast of a fog event.

Model results with PAFOG have been presented showing that the inclusion of the new methods yields distinct improvements of the fog forecast with PAFOG. When the model is started during a fog event, knowledge of the initial visibility is crucial for an accurate fog forecast. Moreover, by including the nudging scheme in PAFOG, the model yields better results regarding the time of fog dissipation. Finally, the combined inclusion of the visibility initialization and the nudging scheme resulted in a better simulation of the nocturnal cooling of the lowest atmospheric layers.

Although an improvement of PAFOG could be achieved by the integration of observational data, the model results showed clear evidence for the necessity of further model improvements. This holds in particular for the calculation of the visibility within the fog, see Fig. 12. Of particular importance will be to perform sensitivity studies with PAFOG aiming at a critical inspection of several empirical model parameters, such as the total aerosol number concentration, the form of the fog droplet spectrum in terms of the standard deviation. In these studies the model results should be compared with corresponding in situ measurements. A good example for such a field campaign is the Fog Remote Sensing And Modeling (FRAM) project which took place in Canada between 2005 and 2007 (GULTEPE and MILBRANDT, 2007; GULTEPE et al. 2009).

For the numerical simulation of fog and low visibility conditions, an accurate description of the atmospheric conditions is indispensable, both for the initial conditions of the model and for the data assimilation during the simulation time. As part of the iPort-VIS project, a new detailed observation system is currently being installed at the Munich airport site in Germany. The goal of this system is to provide PAFOG with observations as precise as possible. Presently, the nudging scheme of PAFOG is driven by direct measurements as obtained at a meteorological tower. This scheme turns out to be a suitable possibility for the integration of local observations in the model simulations. However, if the nudging scheme is to be used on an operational basis, it has to be driven by vertical profiles produced by numerical weather prediction models, such as the COSMO-DE model of DWD (Baldauf et al. 2006). For an even better representation of horizontal heterogeneities in PAFOG, additional horizontal forcing terms might be added to the governing model equations. Finally, it may be an interesting task to generate ensemble model runs with PAFOG which may be used for the production of probability forecasts. These research topics will be the focus of future model developments of PAFOG.

Acknowledgments

The iPort project is funded by the German Ministry for Economy and Technology. The authors wish to thank Michael Rohn, Wolfgang Raatz, Lucas Wenke, Björn-Rüdiger Beckmann and Peer Röhner from DWD for their cooperation in the iPort-VIS project. Wolfgang Adam from the Lindenberg observatory of the DWD is gratefully acknowledged for providing the data set of the investigated case study.

REFERENCES

Anthes, R.: 1974, *Data assimilation and initialization of hurricane prediction models.* Journal of the Atmospheric Sciences *31*, 702–719.

Baldauf, M., K. Stephan, S. Klink, C. Schraff, A. Seifert, J. Förstner, T. Reinhardt, and C.-J. Alenz : 2006, The new very short range forecast model LMK for the convection-resolving scale. In: Second THORPEX International Science Symposium. Volume of extended abstracts, Part B. pp. 148–149.

Bergot, T., J. N. Carrer, and P. Bougeault: 2005, *Improved Site-Specific Numerical Prediction of Fog and Low Clouds: A Feasibility Study.* Weather and Forecasting **20**, 627–646.

Bergot, T. and D. Guedalia: 1994, *Numerical forecasting of radiation fog. Part I: Numerical model and sensitivity tests.* Monthly Weather Review **122**, 1218–1230.

Bott, A.: 1997, *A numerical model of the cloud-topped planetary boundary-layer: Impact of aerosol particles on radiative forcing of stratiform clouds.* Quarterly Journal of the Royal Meteorological Society **123**, 631–656.

Bott, A., U. Sievers, and W. Zdunkowski 1990, *A radiation fog model with a detailed treatment of the interaction between radiative transfer and fog microphysics.* Journal of Atmospheric Sciences **47**, 2153–2166.

Bott, A. and T. Trautmann: 2002, *PAFOG - a new efficient forecast model of radiation fog and low-level stratiform clouds.* Atmospheric Research **64**, 191–203.

Capon, R., Y. Tang, and P. Clark: 2007, A 3D high resolution model for local fog prediction. In: NetFAM/COST 722 workshop on cloudy boundary layer, Toulouse, France.

Cermak, J. and J. Bendix: 2007, 'Dynamical nighttime fog/low stratus detection based on Meteosat SEVIRI data - A feasibility study. Journal of Applied Meteorology **164**, 1179–1192.

Chaumerliac, N., E. Richard, and J.-P. Pinty: 1987, *Sulfur scavenging in a mesoscale model with quasi-spectral microphysics: two-dimensional results for continental and maritime clouds* Journal of Geophysical Research **92**, 3114–3126.

Clark, P. and W. Hopwood: 2001, *One-dimensional site-specific forecasting of radiation fog. Part I: Model formulation and idealised sensitivity studies.* Meteorological Applications **8**, 279–286.

Fu, Q.: 1991, Parameterisation of radiative processes in vertically non-homogeneous multiple scattering atmospheres. Ph.D. thesis, Department of Meteorology, The University of Utah.

Fu, Q. and K. Liou: 1992, *On the correlated k-distribution method for radiative transfer in non-homogeneous atmospheres.* Journal of Atmospheric Sciences **49**, 2139–2156.

Guedalia, D. and T. Bergot: 1994, *Numerical Forecasting of Radiation Fog. Part II: A Comparison of Model Simulation with Several Observed Fog Events.* Monthly Weather Review **122**, 1231–1246.

Gultepe, I. and J. Milbrandt: 2007, *Microphysical observations and mesoscale model simulation of a warm fog case during FRAM project.* Pure and Applied Geophysics, Special Issue on fog **164**, 1161–1178.

Gultepe, I., M. Müller, and Z. Boybeyi: 2006, *A new visibility parameterization for warm fog applications in numerical weather prediction models.* Journal of Applied Meteorology **45**, 1469–1480.

Gultepe, I., G. Pearson, J. Milbrandt, B. Hansen, S. Platnick, P. Taylor, M. Gordon, J. Oakley, and S. Cober: 2009, *The fog remote sensing and modeling (FRAM) field project.* Bulletin of the American Meteorological Society **90**, 341–359.

Gultepe, I., R. Tardif, S. Michaelides, J. Cermak, A. Bott, J. Bendix, M. Müller, M. Pagowski, B. Hansen, G. Ellrod, W. Jacobs, G. Toth, and S. Cober: 2007, *Fog research: A review of past achievements and future perspectives.* Pure and Applied Geophysics **164**, 1121–1159.

Jänicke, R.: 1993, Aerosol-Cloud-Climate Interactions, Chapt. Tropospheric Aerosols, pp. 1–31. Academic Press, San Diego.

Kalnay, E.: 2003, Atmospheric Modeling, Data Assimilation and Predictability. Cambridge University Press.

Koschmieder, H.: 1924, *Theorie der horizontalen Sichtweite.* Beitraege zur Physik der freien Atmosphaere **12**, 33–35.

KUNKEL, B.: 1984, *Parameterization of droplet terminal velocity and extinction coefficient in fog models.* Journal of Climate and Applied Meteorology **23**, 34–41.

LOUGHLIN, P., T. TRAUTMANN, A. BOTT, W. PANHANS, and W. ZDUNKOWSKI: 1997, *The effects of different radiation parametrizations on cloud evolution.* Quarterly Journal of the Royal Meteorological Society **123**, 1985–2007.

MASBOU, M.: 2008, LM-PAFOG—a new three-dimensional fog forecast model with parametrised microphysics. Ph.D. thesis, Meteorological Institute, University of Bonn.

MELLOR, G. and T. YAMADA: 1974, *A hierarchy of turbulence closure models for planetary boundary layers.* Journal of Atmospheric Sciences **31**, 1791–1806.

MELLOR, G. and T. YAMADA: 1982, *Development of a turbulence closure model for geophysical fluid problems.* Reviews of Geophysics **20**, 851–875.

MÜLLER, M., M. MASBOU, A. BOTT, and Z. JANJIC: 2005, Fog prediction in a 3D model with parametrized microphysics. In: Proc. WWRP Int. Symp. on Nowcasting and Very Short-Range Forecasting, Vol. 6. pp. 6–26.

PIELKE, R. A.: 1984, Mesoscale Meteorological Modeling. Academic Press, Orlando 612 pp.

SAKAKIBARA, H.: 1979, *A scheme for stable numerical computation of the condensation process with large time steps.* Journal of the Meteorological Society of Japan **57**, 349–353.

SHETTLE, E. and R. FENN: 1979, *Models for the aerosols of the lower atmosphere and the effects of humidity variations on their optical properties.* Environmental Research Paper.

SIEBERT, J., A. BOTT, and W. ZDUNKOWSKI: 1992a, *Influence of a vegetation-soil model on the simulation of radiation fog.* Beitraege zur Physik der Atmosphaere **65**, 93–106.

SIEBERT, J., U. SIEVERS, and W. ZDUNKOWSKI: 1992b, *A one-dimensional simulation of the interaction between land surface processes and the atmosphere.* Boundary layer Meteorology **59**, 1–34.

STAUFFER, D. and N. SEAMAN: 1990, *Use of 4-D data assimilation in a limited area mesoscale model. Part I: Experiments with synoptic scale data.* Monthly Weather Review **118**, 1250–1277.

TARDIF, R. and R. RASMUSSEN: 2007, *Event-based climatology and typology of fog in the New York City region.* Journal of Applied Meteorology **46**, 1141–1168.

TERRADELLAS, E. and D. CANO: 2007, *Implementation of a Single-Column Model for Fog and Low Cloud Forecasting at Central-Spanish Airports.* Pure and Applied Geophysics **164**, 1327–1345.

TWOMEY, S.: 1959, *The nuclei of natural cloud formation Part II: The supersaturation in natural clouds and the variation of cloud droplet concentration.* Geofisica pura e applicata **43**, 243–249.

UNDEN, P., L. RONTU, H. JAERVINEN, P. LYNCH, J. CALVO, G. CATS, J. CUXART, K. EEROLA, C. FORTELIUS, J. GARCIA-MOYA, C. JONES, G. LENDERLINK, A. McDONALD, R. McGRATH, B. NAVASCUEZ, N. WOETMAN NIELSEN, V. OEDEGAARD, E. RODRIGUEZ, M. RUMMUKAINEN, R. ROOM, K. SATTLER, B. HANSEN SASS, H. SAVIJAERVI, B. WICHERS SCHREUR, R. SIGG, H. THE, and A. TIJM: 2002, 'HIRLAM-5 Scientific Documentation'. Technical report, Swedish Meteorological and Hydrological Institute (SMHI).

WESTCOTT, N. E.: 2007, *Some Aspects of Dense Fog in the Midwestern United States.* Weather and Forecasting **22**, 457–465.

YANG, D., H. RITCHIE, S. DESJARDINS, G. PEARSON, A. MACAFEE, and I. GULTEPE: 2009, *High Resolution GEM-LAM application in marine fog prediction: Evaluation and diagnosis.* Weather and Forecasting **25**, 727–748.

ZDUNKOWSKI, W., W.-G. PANHANS, R. WELCH, and G. KORB: 1982, *A radiation scheme for circulation and climate models.* Beitraege zur Physik der Atmosphaere **55**, 215–238.

(Received November 14, 2010, revised May 23, 2011, accepted May 25, 2011, Published online June 26, 2011)

Reprinted from the journal

Pure Appl. Geophys. 169 (2012), 895–909
© 2011 Springer Basel AG
DOI 10.1007/s00024-011-0327-x

Forecast of Low Visibility and Fog from NCEP: Current Status and Efforts

BINBIN ZHOU,[1] JUN DU,[1] ISMAIL GULTEPE,[2] and GEOFF DIMEGO[1]

Abstract—Based on the visibility analysis data during November 2009 through April 2010 over North America from the Aviation Digital Database Service (ADDS), the performance of low visibility/fog predictions from the current operational 12 km-NAM, 13 km-RUC and 32 km-WRF-NMM models at the National Centers for Environmental Prediction (NCEP) was evaluated. The evaluation shows that the performance of the low visibility/fog forecasts from these models is still poor in comparison to those of precipitation forecasts from the same models. In order to improve the skill of the low visibility/fog prediction, three efforts have been made at NCEP, including application of a rule-based fog detection scheme, extension of the NCEP Short Range Ensemble Forecast System (SREF) to fog ensemble probabilistic forecasts, and a combination of these two applications. How to apply these techniques in fog prediction is described and evaluated with the same visibility analysis data over the same period of time. The evaluation results demonstrate that using the multi-rule-based fog detection scheme significantly improves the fog forecast skill for all three models relative to visibility-diagnosed fog prediction, and with a combination of both rule-based fog detection and the ensemble technique, the performance skill of fog forecasting can be further raised.

1. Introduction

Fog is infrequent but it may be a very hazardous weather condition related to all forms of traffic and on health. Central guidance from the National Centers for Environmental Prediction (NCEP) on fog thresholds is being considered and particularly emphasized by the National Weather Service (NWS) of the National Oceanic and Atmospheric Administration (NOAA), and in NextGen (SOUDERS, 2010), a future Air Traffic Management System of the Federal Aviation Administration (FAA), in the United States.

However, fog is still not a part of the documents of NCEP central guidance, due to its complexity and limitation of computational resources. Instead, it is only diagnosed locally by forecasters either through subjective visibility forecasts or through other variables from model output such as MOS (Model Output Statistics). Nevertheless, effort to add it to NCEP's central guidance is considered to be important. As a step forward in response to the request from NWS and NextGen, low visibility/fog forecast has been experimentally implemented, tested and validated using NCEP operational models.

Currently, the visibility-liquid water content (LWC) relationship of (STOELINGA and WARNER, 1999) is used in horizontal visibility computations in all the NCEP models. However, studies have shown that this visibility computation has large errors, particularly in the situation of fog when droplet number concentration (N_d) is not considered (GULTEPE and ISAAC, 2004; GULTEPE et al., 2006). Besides the error from the visibility computation, a bias in the model LWC near the surface is another source of errors. The visibility computation error can be reduced by applying Gultepe's visibility versus LWC and N_d parameterization (GULTEPE et al., 2006, 2009), whereas the reduction of model LWC error is extremely difficult due to a lack of fog physics for all fog types, model bias and low resolution of the operational models. To overcome these drawbacks, we have recently developed a rule-based fog detection scheme (ZHOU and DU, 2010). The rule-based fog detection scheme is a combination of rules related to surface LWC, relative humidity with respect to water (RH), wind speed, and fog top (Zt) and base (Zb) heights for various fog types.

The second improvement effort is extending the NCEP Short Range Ensemble Forecast System (SREF, DU et al., 2006) to fog forecasting. Because

---
[1] IMSG at EMC/NCEP/NWS/NOAA, 5200 Auth Road, Camp Springs, MD 20746, USA. E-mail: Binbin.Zhou@noaa.gov
[2] Cloud Physics and SWRS, MRD, Environment Canada, Toronto, ON, Canada.

of the chaotic and highly nonlinear nature of the atmospheric system, initially small differences in either initial conditions (ICs) or the model itself can amplify over forecast time and become large after a certain period of time. Since intrinsic uncertainties always exist in both ICs and model physics, a forecast by a single model run always has uncertainties too. Such forecast uncertainties vary from time to time, from location to location, and from case to case. A dynamical way to quantify such flow-dependent forecast uncertainties is to use an ensemble forecasting (LEITH, 1974). Instead of using a single model simulation, multiple model integrations are performed that were initiated with slightly different ICs and/or based on different model configurations. Given the intrinsic uncertainties in model forecasts, and the fact that fog forecasting is believed to be extremely sensitive to the initial conditions and the physics schemes used in a prediction system (BERGOT and GUEDALIA, 1994; BERGOT, 2005), it is strongly desirable to have fog prediction as a part of the NCEP's ensemble framework.

Both the rule-based fog detection scheme and the ensemble application have been tested and evaluated in the World Meteorological Organization (WMO)'s 2008 Beijing Olympic Game Research and Demonstration Project (B08RDP) over China with 7 months of data from 13 Chinese cities (ZHOU and DU, 2010). The evaluations have shown that the rule-based fog detection scheme could improve the fog forecasting score by a factor of two while its combination with the ensemble technique could add extra value to fog predictions. After B08RDP was finished, these two techniques, individually and combined, were further tested and evaluated over North America for NCEP's regional models and the ensemble forecast system.

The objective of this paper is to evaluate (1) performance of fog prediction using the visibility-LWC relationship from the current NCEP's regional models, (2) rule-based fog detection scheme, and (3) ensemble forecast technique for fog detection over the North American domain. This paper is organized as follows: section 2 is for the configurations of the models and the ensemble forecast system involved, section 3 is for the evaluation method for the results, and section 4 is for the results and discussion, followed by the conclusion section.

2. Configurations and Methods

2.1. Configuration of Regional Models

The three regional operational models over North American domain (see Fig. 1) used for low visibility (Vis) and fog prediction verifications are (1) 13-km resolution Rapid Update Cycle model, or RUC-13 (BENJAMIN, 2003), (2) 12-km resolution North American Mesoscale model, or NAM-12 (ROGERS et al., 2005), and (3) 32-km resolution Non-hydrostatistic Mesoscale Model, or NMM-32 (JANJIĆ, 2001). RUC-13 runs hourly, specifically for aviation weather forecasts. NAM-12 runs four times (00, 06, 12 and 18Z) per day to provide central guidance to regular weather for all local forecasters in the United States. NMM-32 is NCEP's WRF (Weather Research and Forecasting) model, which is also one of the base models in the SREF system running four times (03, 09, 15 and 21Z) per day to generate both regular and aviation weather guidance. In fact, NAM-12 is also a WRF-NMM based model but runs in different horizontal resolution from NMM-32. The parameterization schemes employed in both NAM-12 and RUC-13 are listed in Table 1.

2.2. Visibility Computation Method

Currently, there is no direct fog prediction algorithm used in either the NCEP's regional models or SREF system. Instead, the visibility computation in these models is based on the algorithm of (STOELINGA and WARNER, 1999) that uses rain, snow, and cloud water or ice amount. In the case of fog, the KUNKEL (1984) Vis-LWC parameterization is employed, where the LWC is the value at the lowest model level.

2.3. Rule-Based Fog Detection Scheme

Although one hopes that LWC at the lowest model level can be explicitly used for fog calculation, experience tells us that the visibility-LWC approach doesn't work well in the operational models mainly for two reasons: one is the too coarse model spatial resolution and the other is the lack of sophisticated fog physics. As a result, LWC from the models is

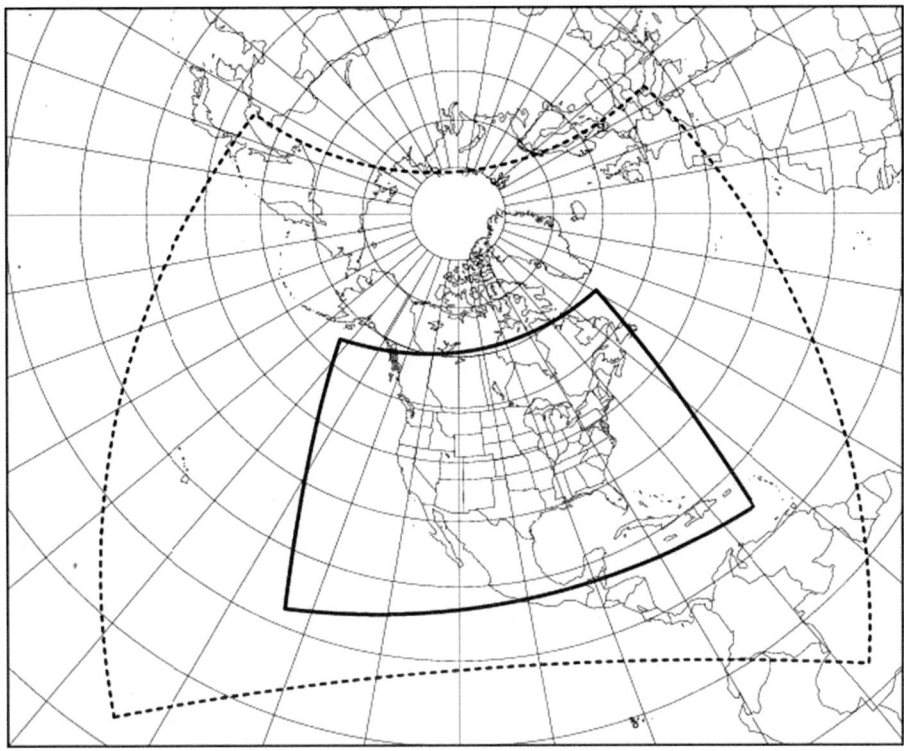

Figure 1

Model domain for NCEP regional models: *dash line* is the running domain and *solid line* is the output domain, where the low visibility and fog forecasts were evaluated

usually not reliable enough to represent fog; therefore, models tend to seriously miss the fog forecast and Vis in many cases (GULTEPE *et al.*, 2006; GULTEPE AND MILBRANDT, 2007). To better detect fog, other variables besides LWC should be considered to enhance the hit rate of fog forecasting.

For this reason, the rule-based approach used in the fog prediction is suggested for the post-processor of the NCEP models (1 for yes and 0 for no fog occurrence) given as

$$\text{LWC at model lowest level} \geq 0.015 \text{ g kg}^{-1}, \; OR \tag{1a}$$

$$\text{Cloud top height (Zt)} \leq 400 \text{ m}$$
$$\text{AND Cloud base height (Zb)} \leq 50 \text{ m, OR} \tag{1b}$$

$$\text{10m - Wind Speed}(U_{10m})$$
$$\leq 1 \text{ ms}^{-1} \text{AND 2 m-RH (RH}_{2m}) \geq 95 \, \% \tag{1c}$$

This diagnosis is similar to the conceptual scheme suggested by CROFT (1997). The LWC rule in (1a)

comes from the definition of fog visibility range. With Kunkel's equation (KUNKEL, 1984), LWC ≥ 0.015 g kg^{-1} is equivalent to visibility less than 1,000 m, that is defined as a threshold for fog definition by the World Meteorology Organization (WMO). The Zt threshold in (1b) follows the general features of fog. Observations indicate that the depth of most fogs on land is about 100–200 m (radiation fog). Some marine fogs or advection fogs are deeper, but rarely exceed 400 m. The Zb threshold in Eq. 1b reflects the lowest level height of our models. To deal with ground fog, the RH-wind rule (1c) is included. The selection of thresholds for surface wind and *RH* over large domains in a model is more difficult than those of LWC and cloud heights because fog usually occurs locally and different models can have different *RH* and wind biases. In many cases, fog was reported while the model RH was less than 100%. Thus, weak turbulence is usually a necessary condition for radiation fog formation. With appropriate thresholds for RH and turbulence intensity (e.g. those suggested

Table 1

Configuration of NAM, RUC and the SREF system

Models (members: ctr, p, n control, positive and negative perturbs)	Convection scheme	Micro-physics	Res (km)/levels	PBL	IC/BC	Long wave	Short wave
NAM	BMJ	Ferrier	12/60	MYJ	GDAS/GDAS	GFDL	GFDL
RUC	GD	Thompson	13/50	MYJ	GDAS/GDAS	RRTM	Dudhia
Eta (3: ctr, p1, n1)	BMJ	Ferrier	32/60	MYJ	NDAS/GENS	GFDL	GFDL
Eta (3: ctr, p1, n1)	KF	Ferrier	32/60	MYJ	NDAS/GENS	GFDL	GFDL
NMM (5: ctr, p1, n1, p2, n2)	BMJ	Ferrier	32/52	MYJ	GDAS/GENS	GFDL	GFDL
ARW (5: ctr, p1,n1, p2, n2)	KF	Ferrier	35/36	YSU	GDAS/GENS	RRTM	Dudhia
RSM (3: ctr, p1, n1)	SAS	ZC	32/28	MRF	GDAS/GENS	RRTM	NASA
RSM (2: p2, n2)	RAS	ZC	32/28	MRF	GDAS/GENS	RRTM	NASA

KF stands for the Kain-Fritsch scheme (KAIN and FRITSCH, 1990), SAS for the simplified Arakawa-Schubert convection scheme (KANAMITSU *et al.*, 2002), RAS for the relaxed Arakawa-Schubert (RAS) convective scheme (KANAMITSU, *et al.*, 2002), Ferrier for the Ferrier mircophysical scheme (FERRIER, 2002), ZC for the Zhao and Carr micro-physics scheme (ZHAO and CARR, 1997), YSU for the Yonsei University PBL scheme (HONG and DUDHIA, 2003), GDAS and NDAS for the Global Data Assimilation System, and NAM Data Assimilation System, MRF for Medium Range Forecast system (belong to NCEP global forecast system GFS) and GENS for the NCEP's Global Ensemble System (TOTH and KALNAY, 1993). GD stands for Grell-Devenyi convective scheme (GRELL and DEVENYI, 2002), Thompson for the Thompson micro-physics scheme (THOMPSON *et al.*, 2004), BMJ for the Betts-Miller-Janjić convective scheme (BMJ, JANJIĆ, 1996), MYJ for the Mellor-Yamada-Janjić scheme (JANJIĆ, 1996), GFDL for the Geophysical Fluid Dynamics Lab schemes (long wave: SCHWARZKOPF and FELS, 1991, short wave: LACIS and HANSEN, 1974), RRTM for the Rapid Radiative Transfer model (MLAWER *et al.*, 1997), Dudhia for the Dudhia short wave scheme (DUDHIA, 1989)

for radiation fog by ZHOU and FERRIER, 2008), fog in a model grid area can be diagnosed more efficiently. Unfortunately, the turbulence intensity was not an output from these models. An alternative approach is using a parameter related to a combination of surface RH and wind speed but no quantitative relationship between wind speed and turbulence intensity has been developed for fog formation. The forecasters usually use RH at 2 m ($\geq$90–100%, GULTEPE *et al.*, 2007; some use dew point temperature) and 10 m wind speed ($\leq$1–2 ms^{-1}) to check for fog occurrence, and these thresholds vary depending on location and model type. For centralized fog forecasting, the optimized *RH* and wind speed thresholds as 95% and 1 ms^{-1}, respectively, were usually used by local forecasters (Justin Arnott, NWS Binghamton, NY, personal communication). This application is slightly different than that of B08RDP due to a wet bias in current NCEP regional models.

2.4. *Configuration of the NCEP SREF System*

The NCEP SREF system has four base models, including 32 km Eta (BLACK, 1994), 32 km WRF-NMM (or NCEP-WRF, JANJIĆ, 2001), 32 km WRF-ARW (or NCAR-WRF, SKAMAROCK, 2005) and

45 km RSM, the regional spectral model (JUANG, 1997). Each base model is expanded with perturbed ICs to generate more ensemble members by using one control IC plus one or two pairs of positive and negative perturbed ICs (called "breeding", TOTH and KALNAY, 1993) for each of the base models. These bred ensemble members are further combined with various physical parameterizations, PBL, surface layer and radiative schemes to construct a total of 21 members in the current SREF system. The configuration of different models with different parameterizations and schemes in the SREF is also listed in Table 1, which shows that the four base models are perturbed into six, five, five, and five members, respectively. All models in Table 1 use the same Noah Land Surface Model (NOAH, EK *et al.*, 2003).

Such a combination of various models with different physical schemes and parameterization, and using a variety of ICs, the simulation errors for fog forecasting can appropriately be addressed. However, for an ensemble forecast system, the total number of members, or ensemble size, is often a concern for a particular weather's ensemble probabilistic prediction. The question is, "are the 21 members in the SREF large enough to increase the forecast skill in fog prediction?" The answer can be yes because of saturation of ensemble members. According to the theory of RICHARDSON (2001), the optimum ensemble size of an ensemble forecast system will eventually saturate. That is, an initial increase in the ensemble size has a bigger effect on the prediction skill enhancement, but if there is a further increase in the model populations, results will reach to a saturation level. At that point, there will be no more improvement in the prediction. The theory also indicates that the most effective ensemble size for a rare-occurrence weather range is within 10 members, whereas the most effective ensemble size for frequent weather events can be reduced to as low as within five members. Therefore, ensemble size of 21 for the SREF is good enough to satisfy our objectives in a fog probabilistic forecast.

2.5. Probabilistic Distribution Function Computation

In this study, fog types are not considered and the ensemble member models used are not tested for specific fog type. Thus, at the present stage, equal capability of the each member used in ensemble runs to capture fog is assumed. This means that each ensemble member in the SREF system has an equal-member weight. For such an equal member weight ensemble system, the ensemble probabilistic prediction, or probability distribution function (PDF), for a Vis range smaller than a threshold value (Vis$_t$) for each grid number (i, j) at a forecast time is given as :

$$P_{i,j}(t,vt) = \frac{1}{N} \sum_{m=1}^{N} K_{i,j}^{m}(t,vt) \text{ for Vis} \leq \text{Vis}_t$$

(or fog is predicted) (2)

where $K_{i,j}^{m}(t,vt) = 1$ if a member (m) predicts Vis $\leq$ Vis$_t$ in a grid at forecast time t. In simulations, the Vis$_t$ is set up as 500, 1,000, 2,000, 4,000, 8,000 m. The N is the ensemble size and taken as 21 in the SREF. The $P_{i,j}$ (t, Vis_t) is the ensemble probability at t for Vis $\leq$ Vis$_t$ for grid $i = 1,2, ...$nx, and $j = 1,2, ...,$ ny, where nx and ny are max size of the model area. For example, in a grid, if there are 10 members that predict fog, then the ensemble fog prediction probability is 10/21 (47.6%). Thus, for each forecast run, the ensemble probability distribution function for various visibility thresholds can be computed over the entire domain based on Eq. (2) and validated grid to grid against observations at all grids within the domain.

3. Validation Data and Evaluation Method

3.1. Validation Data

An evaluation of fog prediction over a large domain like North America is generally difficult due to a lack of direct fog observations and the fact that model-based fog value represents a grid area that cannot be interpolated to the location of the observational sites. Thus, using high resolution visibility analysis (also gridded data) from the Aviation Digital Database Service (ADDS) of the Aviation Weather Center, NCEP, as validation truth for our objective verification is appropriate. The grid space for the ADDS data is about 5 km which is routinely analyzed from more than 5,000 surface station observations over the US and Canada through a data assimilation system. The 5 km grid space is much smaller than that of the regional models. To objectively compare the grid-scaled visibility values or fog events from the regional models against the ADDS data at the same locations, the visibility/fog forecast from each model was first downscaled to match the ADDS grid values using copyg (the NCEP's grid converter; ZHOU et al., 2011) with the nearest neighbor option (no interpolation is performed because fog is considered a non-continuous feature in the horizontal direction).

3.2. Evaluation Method

The observational data period covers 6 months from Nov 1, 2009 to Apr 30, 2010. This time period is chosen because of an observed high occurrence of fog events in this period. If the observed/forecast visibility is ≤1 km in a grid, the ADDS/model grid is considered as foggy. The model forecast visibility is compared to the observed visibility in a grid as follows: if visibility is ≤1 km in both observation and model grids, this is assigned as a "hit"; if forecast visibility is ≤1 km but observed is >1 km, this is assigned as a "false alarm"; and if forecast visibility is >1 km but observed is ≤1 km, the result is assigned as a "missed alarm". Using these statistical classifications, forecast scores such as bias, probability of detection (POD), and equitable threat score (ETS) can be derived. The bias here is defined by the ratio of total forecast events divided by total observed events. If the bias is larger/smaller than 1, it means the model is over/under predicting. An over-prediction system means higher false alarms but not necessarily higher hits. In comparison to the usual threat score (TS or critical success index, CSI), the ETS has an advantage that removes the random hit contribution from the score. These traditional scores can generally be used to evaluate both a single model (deterministic) forecast and an ensemble probabilistic forecast in deterministic aspect. Since ETS is an overall score that considers combined effects (POD, false alarm rate, and missing rate, etc.), the performance ranking of evaluated models will be based on the values of ETS in the latter sections.

The traditional (deterministic) scores are usually not enough to evaluate a probabilistic forecast from an ensemble forecast system. Some other probabilistic measures, such as Brier skill score, resolution and reliability, are also required as we did during an evaluation of the ensemble fog forecast in B08RDP (ZHOU and DU, 2010). Since our purpose is to compare the fog predictions from a single and an ensemble system (not to evaluate the ensemble forecast system itself), only deterministic verification scores are evaluated. To compare a single model forecast and an ensemble probabilistic forecast in a deterministic aspect, the probabilistic visibility/fog forecast should be converted to a deterministic visibility/fog forecast with a certain probability threshold percentage. For a given percentage threshold (such as 50%), a probabilistic forecast can be viewed as a deterministic forecast in the way that an event (e.g. visibility <1,000 m) is expected to occur when the forecast probability is greater than or equal to the selected threshold. That is, if more than 10 out of 21 members in the SREF predict visibility ≤1,000 m in same grid, fog is expected in this grid by the ensemble forecast. To evaluate an ensemble forecast for a fog event over the entire PDF space, several probability thresholds such as 10, 20, 30, …

90, and 100% were selected to evaluate which ensemble probability thresholds will yield the best prediction performance.

4. Results and Discussions

4.1. Performance of Current Regional Models

In this section, first, results are presented for low visibility forecast from each regional model. The evaluation scores from NAM-12, RUC-13, and NMM-32 are illustrated in Fig. 2, from which the performance for fog range (visibility ≤1 km) can be estimated. Figure 2 shows that the general performances degrade as the visibility threshold decreases. For the visibility threshold of fog, the POD is about 25% for RUC-13, 10% for NAM-12 and only 5% for NMM-32. Since NAM is also a NMM-based regional model, it can be expected that the coarse resolution model NMM-32 has a lower hit rate (POD) or is more prone to miss the forecast than that of higher resolution (12 km) of the same model in fog prediction.

Another feature shown in Fig. 2b is that the POD for dense fog (visibility ≤0.5 km) is lower than that of shallow fog intensity (visibility >0.5 km but ≤1 km). In other words, dense fog events are more difficult to detect by these operational models in fog prediction. Figure 2a shows significant high biases for fog predictions by all three models (where bias ∼1 means no bias). A positive bias implies an over-prediction or a false alarm of fog forecast. For shallow fog, the highest bias is 3 (or 300%) for RUC-13. The bias for dense fog prediction is even larger. Such high positive biases for all models indicate that very low visibility or fog from all NCEP regional models is highly overpredicted. The low POD with high bias leads to poor general performances as indicated by ETS (Fig. 2c), where the ETS values for all three models are around 5%. These scores are similar to the single model evaluation in B08RDP. To compare the ETS values for fog prediction to those for precipitation prediction, the average precipitation forecast ETS (∼35%) from the same NCEP regional models is also marked in Fig. 2c, meaning that the ETS for fog prediction is much lower than that for

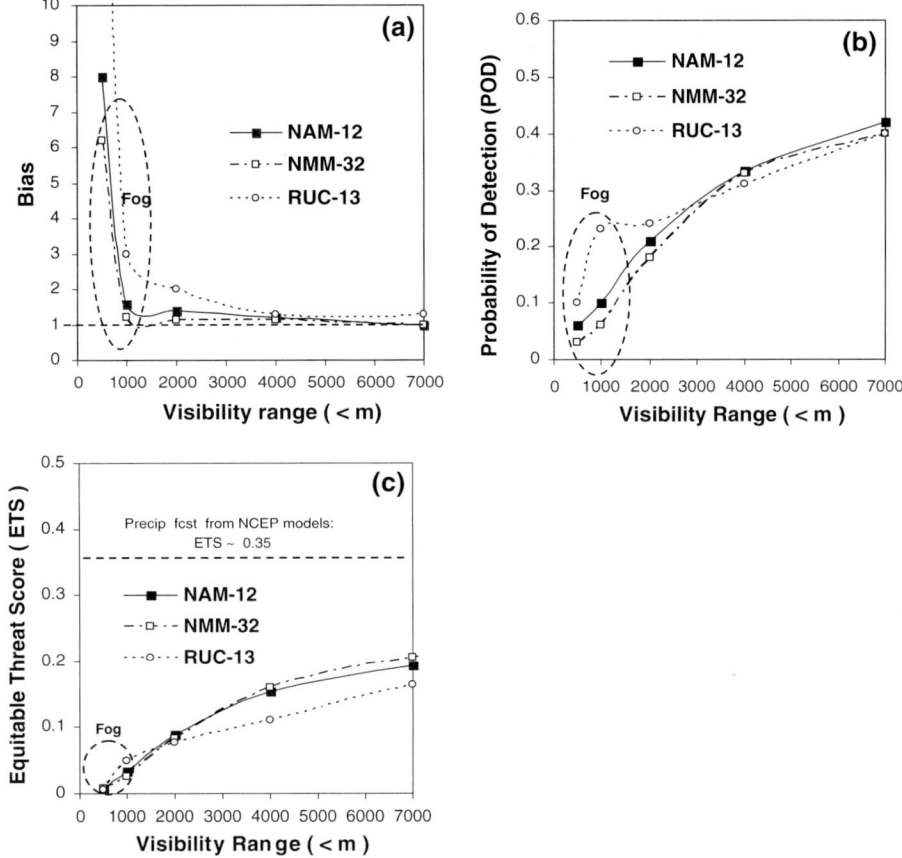

Figure 2
Tests for visibility over different thresholds (x-axis): bias (**a**), POD (**b**) and ETS (**c**) for each of the three regional models

precipitation prediction. Therefore, in order to catch up the performance of precipitation forecast at NCEP, tremendous efforts should be dedicated to improving our fog forecast. Low POD (Fig. 2b) and high bias (Fig. 2a) implies that the current models overpredict low visibility or fog occurrence in some areas but miss most of the real fog events. In fact, the visibility-diagnosed fog method is based on the LWC-rule. This is the reason why current visibility-diagnosed fog prediction, without input from other variables, has very low performance (GULTEPE *et al.*, 2006).

To examine this feature, let us further look at an east coast regional fog event that occurred on Nov 16, 2009 (Fig. 3a). This particular regional fog event covered several east coast states, including northern Florida, almost all of South Carolina and North Carolina, most of Virginia, Maryland and Delaware, extending to some regions of Pennsylvania, New

York and some of Ontario and Quebec of Canada. The visibility computations in the three models are obtained from fog LWC. The green colors (dark green, green and light green) indicate the fog intensities expressed by visibility levels and locations. Comparing the observed fog location and its intensity (Fig. 3a) with the fog visibility at 12 h forecast by NAM (Fig. 3b), one can easily notice that the NAM forecast missed most of the fog events in Virginia and North Carolina although it captured some of the fog locations in Maryland and Delaware. However, it issued false alarms for half of Pennsylvania and New York states, and most of the other northeast states as well as some regions of Canada. At hour 9, the forecast by NMM-32 (Fig. 3c) almost missed the entire fog event over the east coast. This case for Vis, again, shows a worst performance of a lower resolution model than that of higher resolution model. The RUC's 12 h forecast for this case can be seen by comparing Fig. 3a and d. The RUC forecast also missed most of the fog in Maryland, Virginia, North Carolina and South Carolina, and over-predicted the fog in Pennsylvania and New York states as well as over most of the other northeast regions, similar to the NAM forecast. This case clearly illustrates the "large false alarm" feature of low visibility and fog forecast from current models and reminds us that incorrectly predicted location and amount of grid-scaled fog LWC at the surface makes it difficult to precisely compute the visibility in the case of fog.

4.2. Suggested Improvements

Three approaches have been directed at improving the performance of the low visibility and fog forecasting in NCEP. The first is applying the rule-based fog diagnostic scheme in the three regional models, the second is conducting an ensemble fog prediction system in SREF, and the third is a combination of the rule-based and ensemble technique for fog prediction. These are explained as below.

4.2.1 The Rule Based Technique

The rule-based fog diagnostic scheme has been extensively evaluated in B08RDP in China (ZHOU and DU 2010). Because fog is extremely sensitive to surface variables, particularly to RH and wind speeds, selections of different threshold values in the rule will have significant impacts on the performance of the rule-based fog forecast. The sensitivity test of the RH-wind rule in B08RDP has shown that if the RH threshold is too large ($\sim 100\%$) or the wind threshold too small, the performance will hit a limit after which the RH-wind rule no longer has an effect on fog forecast and only the cloud rule (1b) and LWC rule (1a) play roles under such circumstances. On the other hand, if the threshold for RH is too low or the wind is too strong, the overall performance score will be even lower than that of the visibility-diagnosed method due to too many false alarms. In other words, inappropriate RH-wind thresholds may cause a negative contribution to the forecast score. Therefore,

RH and wind thresholds are critical but their appropriate thresholds are more important to a successful fog forecast. The evaluation in B08RDP also revealed that with a rule-based fog detection scheme, the prediction ETS was tripled in comparison to that with the visibility-diagnosed method, in which the RH-wind rule has most of the contribution (as large as 50%) to the skill improvement. This implies that radiation fog is the most frequent fog type since RH and calm air are two critical conditions for radiation fog. One can expect that without the RH-wind rule the models would miss at least 50% of the fog events. In this study, the rule-based fog detection scheme was further tested in each of the three regional models over North America and evaluated with the same ADDS visibility analysis data. The evaluated scores for various models are listed in Table 2, in which NAM-12 shows better POD than the other two models. Despite its higher bias, NAM-12 has best overall performance indicated by the ETS. Comparing NAM-12 and NMM-32, it is demonstrated again that the forecast skill of a higher resolution model is better than that of a lower resolution peer model for fog prediction with the rule-based detection.

4.2.2 Ensemble Fog Forecasting Technique

This technique involves the computation of the low visibility ($\leq 1,000$ and ≤ 500 m, respectively) based on ensemble predictions from the SREF system. Computing the ensemble probability for low visibility in a grid from the SREF is relatively simple: the first step is counting how many ensemble members predict low visibility in this grid, and then dividing the count by the ensemble size, 21, to obtain the probability of low visibility in this grid with Eq. 2. To use the traditional measures in evaluation of an ensemble forecast, the SREF low visibility probabilistic forecast was first converted to a deterministic forecast with a certain probability threshold. To evaluate with which ensemble forecast probability threshold the SREF has the best low visibility prediction performance, multiple forecast probability thresholds, generally from 10 to 100%, with every 10% as an interval, were selected and evaluated respectively for both visibility $\leq 1,000$ and 500 m forecasts (see Fig. 4). The results reveal that (1) the

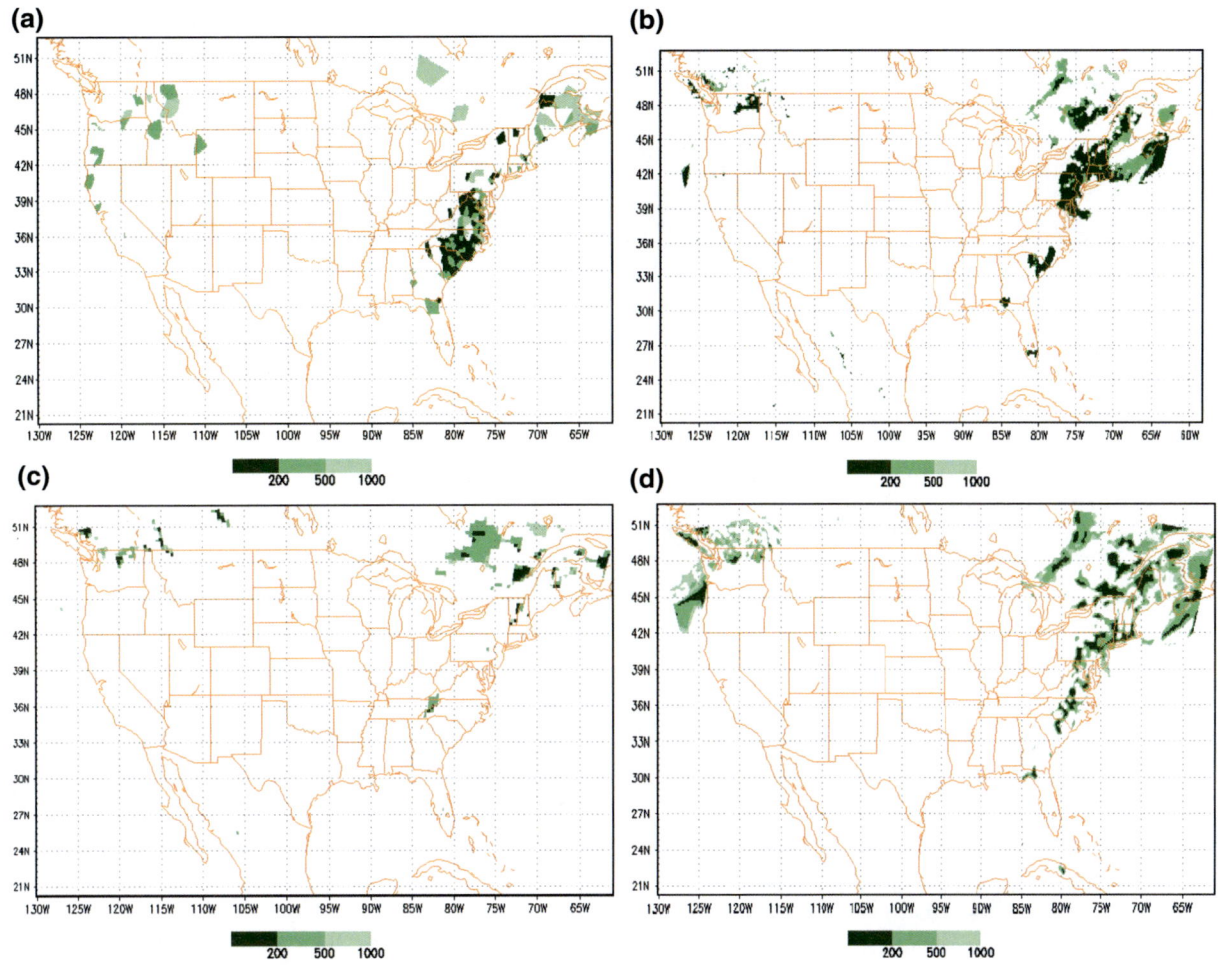

Figure 3
Nov 16, 2009 fog visibility observations from ADDS at 1200 UTC in east coast (**a**) and their 12 h forecasts from NMM-12 (**b**), 9 h forecast from NMM-32 (**c**) and 12 h forecast from RUC-13 (**d**). *Dark green* is for visibility <0.2 km, green for <0.5 km and *light green* for <1.0 km to represent different fog intensities

Table 2

Scores for rule-based fog detection method used in single models

	NAM-12	RUC-13	NMM-32
Bias	2.40	2.25	1.60
POD	0.290	0.240	0.185
ETS	0.071	0.065	0.050

performance for visibility ≤500 m forecast is consistently lower than that for visibility ≤1,000 m forecast over all of ensemble forecast probability thresholds, which means that dense fog is also more difficult to predict with an ensemble forecast system as with a single model; and (2) for different ensemble

forecast probability thresholds the SREF for both low visibility ranges (1,000 and 500 m) have different forecast performances. For a smaller forecast probability threshold, the ensemble gives a higher POD (Fig. 4b) but with a large bias as a penalty (Fig. 4a). To decrease the bias, a larger forecast probability threshold should be chosen. In this case, the forecast POD decreases accordingly. Therefore, how to choose an appropriate forecast probability threshold in fog prediction means a trade-off bias and POD. Different users may select different forecast probability thresholds based on their own unique requirements, objectives, economic values (cost-loss analysis), and decision making procedures. For

139

example, if the cost of protection is not so high in comparison to the loss, users may prefer a higher POD and may not worry about a false alarm, while others may be the opposite. If one is more concerned about POD, select a smaller forecast probability threshold; otherwise, select a larger forecast probability threshold to reduce false alarms and bias. One of the advantages of an ensemble forecast system is that it provides different users with different choices and decision making procedures based on their own needs but a single model forecast can not. Such a distribution of evaluation scores over different probability thresholds from an ensemble forecast system provides users with a decision making reference. If there is no preference, a medium range of forecast probability threshold can be selected around 40–50%, where the ensemble forecast usually has a best performance as shown in Fig. 4c. It should be noted that such a 40–50% probability range is a common feature for all of probabilistic forecast systems (WILKS, 2006).

4.2.3 Integrated Technique

This technique is a combination of the rule-based fog detection into the SREF system. The method is as follows: first apply the rule-based fog detection in each of the ensemble members from the SREF to determine whether this member predicts fog in a grid, and then use this to compute the ensemble PDF for fog occurrence with Eq. 2, based on how many ensemble members have fog occurrence in the same grid. To determine if one issues a fog forecast in a grid depends on what probability threshold is chosen. To evaluate which probability threshold in the third effort (i.e. combination of rule-based diagnosis and ensemble) has the best fog prediction performance, different probability thresholds were tested and shown in Table 3. One can see that comparing to the low visibility ensemble prediction, the fog ensemble prediction combined with the rule-based fog detection has a similar distribution of score over different forecast probability thresholds (comparing Table 3 and Fig. 4): both POD and bias (Table 3, row 2 and row 3) consistently decrease as the forecast probability threshold increases. Particularly, the ETS score (Table 3, row 4) has its best value near 40%. If

choosing a smaller probability threshold, the bias will be very high although it can raise the POD. To reduce the bias or false alarms, a larger probability threshold should be used. To see how this works in an actual ensemble fog forecast, let us see the SREF fog prediction for the same case in Fig. 3. Figure 5 shows the 9 h forecast of fog ensemble PDF from the SREF over North America valid at 12Z, Nov 16, 2009. The regions where fog most likely occurred are marked with cyan-orange-red colors. Comparing the observation in Fig. 3a and the PDF forecast in Fig. 5, it can be seen that fog events on the east coast are covered by yellow–red colors (ensemble probability larger than 70–100%), in North Carolina by cyan-yellow colors (larger than 50–70%) and in South Carolina by cyan color (larger than 40–50%), significantly improving the fog predictability in comparison to the single models as shown in Fig. 3b, c, and d. Having a closer look at Fig. 5, It can be noticed that many regions are colored with low PDF (10–20%). If selecting a higher probability threshold value, e.g. 40% (cyan color in Fig. 5), the false alarm regions with small PDF (10–20%) can be filtered out, leading to better agreement with observations (Fig. 4a) and improving the ensemble forecast performance.

4.3. Comparison of the Three Techniques

The comparisons of bias, POD and ETS among the three techniques are summarized in Fig. 6. As a reference, the scores of visibility method with single mode and ensemble are also indicated. For the first technique with rule-based fog detection scheme applied in the each model, although a small bias is added in comparison to the visibility method (black bars compared to grey bars for NAM-12, RUC-13 and NMM-32 in Fig. 6a), more POD and much bigger ETS are rewarded (black bars compared to grey bars for NAM-12, RUC-13 and NMM-32 in Fig. 6b, c), increased by almost 100% of ETS scores for NAM-12 and NMM-32, 30% for RUC-13. The reason for better performance with the rule-based fog detection is that fog has various types and each type of fog has its particular formation and development mechanism. The visibility-diagnosed forecast from current regional operational models at NCEP is based on the LWC rule, which may not efficiently capture

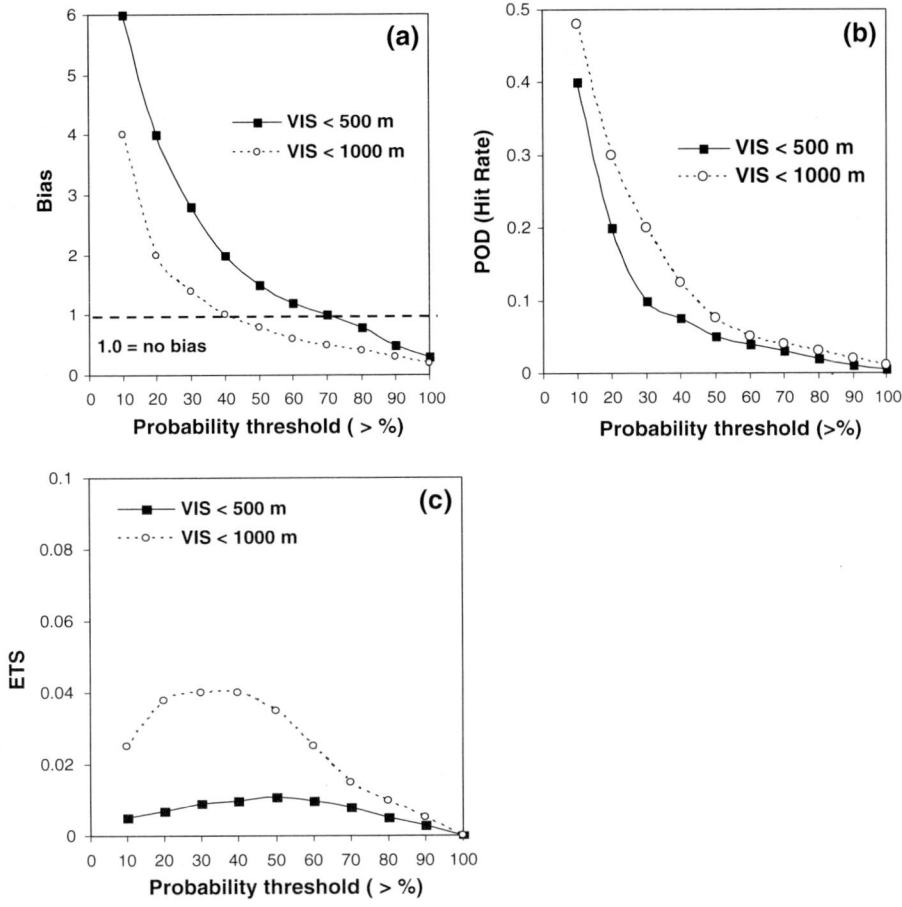

Figure 4
Scores for low visibility ensemble probabilistic prediction from the SREF: Bias (**a**), POD (**b**) and ETS (**c**) under different forecast probability thresholds as in *x*-axis

Table 3

Scores for fog probabilistic prediction for different probability thresholds from the SREF combined with the rule based fog detection method

	10%	20%	30%	40%	50%	60%	70%	80%	90%	100%
Bias	12.0	4.5	2.3	1.3	1.1	0.9	0.7	0.5	0.3	0.2
POD	0.62	0.49	0.40	0.25	0.22	0.15	0.12	0.10	0.05	0.02
ETS	0.03	0.04	0.05	0.06	0.05	0.04	0.04	0.03	0.02	0.01

all types of fog. For local fog or radiation fog, it more locally forms and develops, and in most situations, is grid-scaled weather which may not be adequately represented by the cloud schemes employed in the operational models. On the other hand, any operational model presents certain degrees of model bias, particularly, in the surface humidity, temperature and wind speed forecasts. Such biases lead to miss or false prediction of grid-scaled fog in the models in many situations and reduce the forecast POD and overall performance as a result.

Since NMM-32 is one of the base models in the SREF system, it is possible to compare the performances of the visibility-diagnosed fog detection ($\leq$1,000 m) between the SREF and the single model NMM-32 forecasts at same resolution (32 km) for the

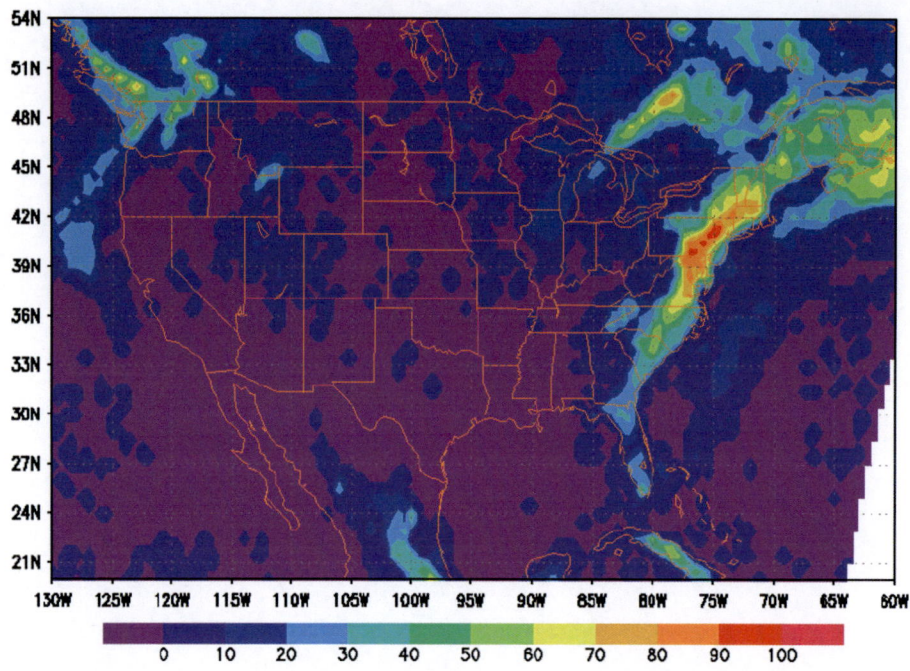

Figure 5
9 h fog ensemble probability forecast from the SREF issued at 03Z, Nov 16, 2009, valid at 12Z on the same day. The color bar is the ensemble probability indicator

second effort. Although bias reduction is not so significant for the SREF from its base model NMM-32 (compare light grey bars for NMM-32 and SREF-32 in Fig. 6a), the increases in POD (compare light grey bars for NMM-32 and SREF-32 in Fig. 6b) and ETS (compare light grey bars for NMM-32 and SREF-32 in Fig. 6c) in the SREF forecast for low visibility are obvious in comparison to those of NMM-32.

With further combination of rule-based fog detection into the ensemble in the third effort, extra POD and ETS scores were added (black bars compared to grey bars for SREF-32 in Fig. 6b and c), although a bias is expected (black bar compared to grey bar for SREF-32 in Fig. 6a). This demonstrated the better performance of ensemble over single model fog prediction over North America. It is of interest to observe that the overall score ETS of the SREF prediction, particularly with rule-based fog detection, still can not beat the same scores for NAM-12 and RUC-13 (black bar for SREF-32 compared to blacks for NAM-12 and RUC-13 in Fig. 6c). The results (ETS) of the ensemble (10 members with 15 km resolution) in B08RDP are found to be better than

what is obtained from the SREF presented in this paper. It should be kept in mind that the resolution of the current SREF is 32 km, which is much lower than that of NAM-12, RUC-13 and the ensemble in B08RDP. This implies that an increase in the resolution of the ensemble system for fog prediction is an effective way to further raise its performance after the ensemble size has reached a saturation size. The horizontal resolution of the current SREF is still not high enough to skillfully predict local grid-scaled fog events even with a better fog detection scheme. This once again prompts us to increase the horizontal resolution to get a better performance for fog ensemble prediction from the SREF in the near future.

To demonstrate meteorologically why an ensemble forecast works better than a single forecast (in same model resolution), two aspects need to be explained. First, fog is a threshold weather event that is extremely sensitive to model ICs, which in general can have some errors. Small errors in the ICs will lead to totally different fog forecasts. After the ICs are perturbed around their control values in an ensemble system, the forecast can effectively

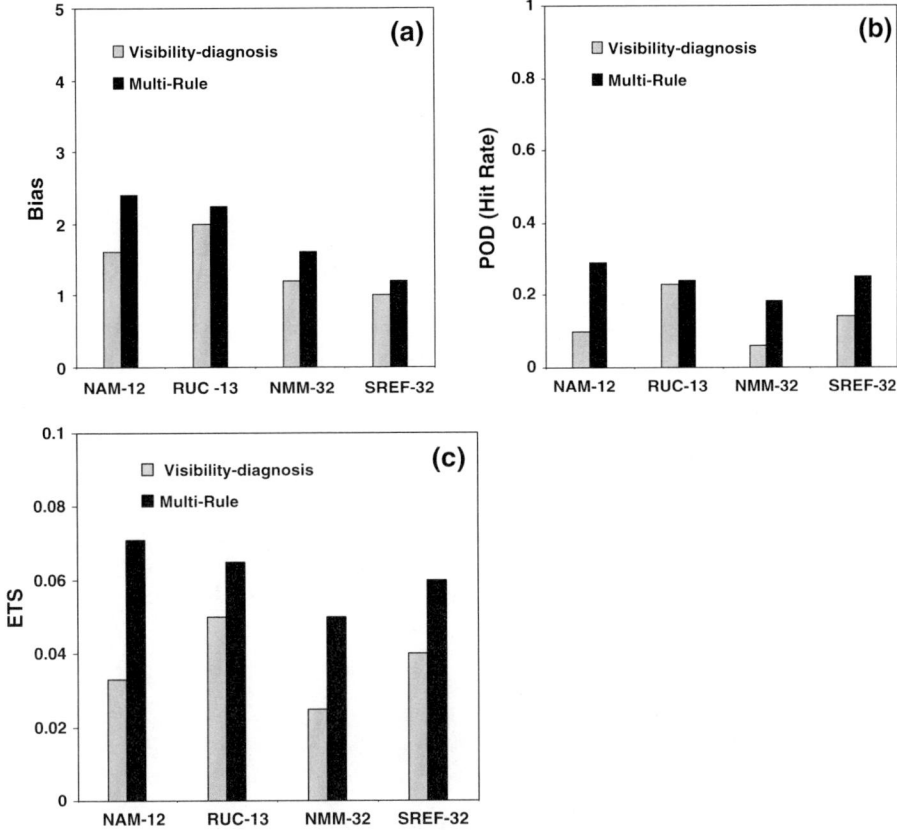

Figure 6
Bias (**a**), POD (**b**) and ETS (**c**) of fog prediction from NMM-12, RUC-13, NMM-32 and SREF with multi-rule fog detection and visibility-diagnosis schemes

encompass all possible IC values that fog may meet in the forecast. Thus, the chance of correctly forecasting fog can be significantly increased. Second, fog has various types but one model or one scheme employed may not deal with all fog types. In many cases, a model performs well for a specific fog type (SHI *et al.,* 2010) but it may not work well at all the times and over all the locations. Therefore, it is suggested an ensemble forecasting can do a better job for fog forecasting compared to the use of single model based predictions.

5. Conclusion

The operational forecasts from NCEP's three regional models, NAM-12, RUC-13 and WRF-NMM-32, over North America were evaluated against the ADDS data (observations) from November 2009 to April 2010, and their performances of low visibility and fog prediction were estimated.

The results show that the performances of the fog prediction from current models still need significant improvements. The reason may be that these models are unable to predict correct locations and intensities of fog events due probably too-coarse model resolutions, missing appropriate fog physics in the models (GULTEPE and MILBRANDT, 2007), and model numerical bias.

In order to improve the low visibility and fog prediction to meet the new request of NextGen of the FAA, three efforts have been made at NCEP; (a) develop an application of a rule- based fog detection scheme, (b) develop an application of multi-model and multi-physics SREF system, and (c) integrate these two applications. The rule-based fog detection includes LWC, cloud and RH-wind parameters to enhance fog detection. The ensemble application is used to address the errors and uncertainties in initial conditions, model systems, and physical schemes, and it is believed that fog is extremely sensitive to these conditions.

The validations suggested the following conclusions

- The rule-based fog detection scheme applied in the regional models doubled their forecast skill scores in comparison to visibility-diagnosed forecast from the same models.
- Ensemble fog prediction from the SREF also enhanced the prediction performance even if only with visibility-diagnosed fog detection in the SREF models. The reason is that the ensemble system can effectively encompass the perturbed initial conditions and capture various fog types with multi-models and multi-physics schemes.
- Combining rule-based fog detection into the ensemble prediction from the SREF, extra score was added to the forecast. The evaluation also indicated that if the ensemble size has been large enough, an increase in its resolution is one of critical and effective way to further raise the performance of ensemble fog prediction.

In the future, observations collected during an ice fog project (GULTEPE, *et al.,* 2008) will be tested for model performances in the cold climates. Although rule-based scheme improves the performance of fog prediction, it only predict occurrence of fog, no fog intensity can be diagnosed. In overcome this drawback, some new technique based on ZHOU and FERRIER (2008) has been suggested (ZHOU, 2011). The next step is testing and evaluation of the new scheme with both single model and the ensemble system at NCEP.

Acknowledgments

This research is, in part, in response to requirements and funding by the Federal Aviation Administration (FAA). The views expressed are those of the authors and do not necessarily represent the official policy or position of the FAA. Our special appreciation is given to the Aviation Weather Center for providing the Aviation Digital Data Service (ADDS) data to support this study.

REFERENCES

BENJAMIN, S.G., SMIRNOVA, T.G., BRUNDAGE, K.J., WEYGANDT, S.S., DEVENYI, D., SCHWARTZ, B.E. SMITH, T.L. (2003), *Application of the rapid update cycle at 10-13 km—initial testing.* Preprints, 16th Conference on Numerical and Weather Prediction, Amer. Meteor. Soc Jan. Seattle. WA.

BERGOT, T. and GUEDALIA, D. (1994), *Numerical forecasting of radiation fog. Part I: Numerical model and sensitivity tests,* Mon. Wea. Rev. *122,* 1218–1230.

BERGOT, T, CARRER, D., NOILHAN, J. and BOUGEAULT, P. (2005), *Improved site-specific numerical prediction of fog and low clouds. A feasibility study,* Wea. Forecasting *20,* 627–646.

BLACK, T.L., (1994), *The new NMC mesoscale Eta Model: description and forecast examples,* Wea. Forecasting *9,* 265–278.

CROFT, P.J., PFOST, R.L., MEDLIN, J. M., and JOHNSON, G.A. (1997), *Fog forecasting for the southern region: a conceptual model approach.* Wea. Forecasting, *12,* 545–556.

DU, J., McQUEEN, J., DIMEGO, G., TOTH, Z., JOVIC, D., ZHOU, B. and CHUANG, H. (2006), *New dimension of NCEP SREF system: inclusion of WRF members.* Report to WMO Export Team Meeting on Ensemble Prediction System, Exeter, UK, Feb. 6–10, 2006.

DUDHIA, J., (1989), *Numerical study of convection observed during the winter monsoon experiment using a mesoscale two-dimensional model,* J. Atmos. Sci. *46,* 3077–3107.

EK, M.B., MITCHELL, K.E., LIN, Y., ROGERS, E., GRUNMANN, P., KOREN, V., GAYNO, G., and TARPLEY, J.D. (2003), *Implementation of Noah land surface model advances in the National Centers for Environmental Prediction operational mesoscale Eta model,* J. Geophys. Res. 108 (D22), 8851.

FERRIER B.S., (2002), *A new grid-scale cloud and precipitation scheme in the NCEP Eta model.* Technical report, Spring Colloquium on the Physics of Weather and Climate: Regional weather prediction modeling and predictability.

GRELL, G.A., and DEVENYI, D. (2002), *A generalized approach to parameterizing convection combining ensemble and data assimilation techniques,* Geophys. Res. Lett. 29 (14), Article 1693. doi:10.1029/2002GL015311.

GULTEPE, I., PEARSON, G., MILBRANDT, J.A., HANSEN, B., PLATNICK, S., TAYLOR, P., GORDON, M., OAKLEY, J.P. and COBER, S.G. (2009), *The fog remote sensing and modeling (FRAM) field project,* Bull. of Amer. Meteor. Soc. *90,* 341–359.

GULTEPE, I., and MILBRAND, J. (2007), *Microphysical observations and mesoscale model simulation of a warm fog case during FRAM project,* J. of Pure and Applied Geophy. Special issue on fog, edited by Gultepe I. *164,* 1161–1178.

GULTEPE, I., MÜLLER, M.D. and BOYBEYI, Z. (2006), *A new visibility parameterization for warm fog applications in numerical weather prediction models,* J. Appl. Meteor. Clim. *45,* 1469–1480.

GULTEPE, I., and ISAAC, G.A. (2004), *An analysis of cloud droplet number concentration (Nd) for climate studies: emphasis on constant Nd.* Q. J. Royal Met. Soc. *130,* Part A, 602, 2377–2390.

GULTEPE, I., PAWGOSKI. M., and REID, J. (2007), *Using surface data to validate a satellite based fog detection scheme,* J. of Weather and Forecasting *22,* 444–456.

GULTEPE, I., MINNIS, P., MILBRANDT, J., COBER, S.G., NGUYEN, L., FLYNN, C., and HANSEN, B. (2008), *The Fog Remote Sensing and Modeling (FRAM) field project: visibility analysis and remote sensing of fog n Remote Sensing Applications for Aviation Weather Hazard Detection and Decision Support*. Preprints, Edited by Wayne F. Feltz; John J. Murray, ISBN: 9780819473080, Proceedings of SPIE Vol. *7088* (SPIE, San Diego, CA), 204 pp.

HONG, S-Y., and DUDHIA, J. (2003), *Testing of a new non-local boundary layer vertical diffusion applications*. Paper# 17.3, 20th Conference on Weather Analysis and Forecasting/16th Conference on Numerical Weather Prediction, Amer. Meteor. Soc. Jan. Seattle, WA.

JANJIĆ, Z.I, GERRITY J.P., Jr., and NICKOVIC, S. (2001), *An alternative approach to nonhydrostatic modeling*, Mon. Wea. Rev. *129*, 1164–1178.

JANJIĆ, Z.I. (1996), *The surface layer in the NCEP Eta model*, Reprints, *11th Conference on Numerical Weather Prediction*, Amer. Meteor. Soc. 354–355, Norfolk, VA.

JUANG, H.-M.H., HONG, S.-Y. and KANAMITSU, M. (1997), *The NCEP regional spectral model: an update*, Bulletin Amer. Metero. Soc. *78*, 2125–2143.

KAIN, J.S., and FRITSCH, J.M. (1990), *A one-dimensional entraining/ detraining plume model and its application in convective parameterization*, J. Atmos. Sci. *47*, 2784–2802.

KANAMITSU, M., and Coauthors, (2002) *NCEP Dynamical* Seasonal Forecast Sy*stem 2000*. Bull. Amer. Meteor. Soc. *83*, 1019–1037.

KUNKEL, B.A., (1984), *Parameterization of droplet terminal velocity and extinction coefficient in fog models*, J. Climate Appl. Meteor. *23*, 34–41.

LACIS, A.A., and HANSEN, J.E. (1974) *A parameterization for the absorption of solar radiation in the earth's atmosphere*, J. Atmos. Sci. *31*, 118–133.

LEITH, C.E., (1974) *Theoretical skill of Monte Carlo forecasts*, Mon. Wea. Rev. *102*, 409–418.

MLAWER, E.J., TAUBMAN, S.J., BROWN, P.D., IACONO, M.J. and CLOUGH, S. A. (1997) *Radiative transfer for inhomogeneous atmospheres: RRTM, a validated correlated-k model for the longwave*, J. Geophys. Res. *102*, (D14), 16,663–16,682.

RICHARDSON, D. S., (2001), *Measures of skill and value of ensemble prediction systems, the interrelationship and the effect of ensemble size*, Quart. J. Royal Meteor. Soc. *12*, 2473–2489.

ROGERS, E., Ek, M., FERRIER, B.S., GAYNO, G., LIN, Y., MITCHELL, K., PONDECA, M., PYLE, M., WONG, V.C.K., and WU, W.-S., (2005), *The NCEP North American Mesoscale Modeling System:*

final Eta model/analysis changes and preliminary experiments using the WRF-NMM. Paper# 4B.5, *17th Conference on Numerical Weather Prediction*. Amer. Meteor. Soc., Washington D.C.

SCHWARZKOPF, M.D., and FELS S.B., (1991), *The simplified exchange method revisited: An accurate, rapid method for computation of infrared cooling rats and fluxes*. J. Geophys. Res. *96*, 9075–9096.

SHI, C., WANG, L., ZHANG, H. and DENG, X., (2010), *Experiments on fog prediction based on multi-model*, this issue.

SKAMAROCK, W.C., KLEMP, J.B., DUDHIA, J., GILL, D.O., BARKER, D.M., WANG, W. and POWERS, J.G. (2005), *A description of the Advanced Research WRF*, Version 2, NCAR Technical Note.

SOUDERS, C.G. and coauthors, (2010), *NextGen weather requirements: an update*. Preprint, *14th Conf. on Aviation, Range, and Aerospace Meteorology*, Atlanta, GA, Amer. Meteor. Soc.

STOELINGA, T.G. and WARNER, T.T. (1999), *Nonhtdrostattic, mesobeta-scale model simulations of cloud ceiling and visibility for an east coast winter precipitation event*, J. Apply. Meteor. *38*, 385–404.

THOMPSON, G., RASMUSSEN, R.M. and MANNING, K. (2004), *Explicit forecasts of winter pre-cipitation using an improved bulk microphysics scheme. Part 1: Description and sensitivity analysis*, Mon. Wea. Rev. *132*, 519–542.

TOTH, Z. and KALNAY, E. (1993), *Ensemble forecasting in NMC: The generation of perturbations*, Bull. Amer. Meteor. Soc. *74*, 2317–2330.

WILKS, D.S., (2006), *Statistical methods in atmospheric sciences*, 2nd edn, International Geophysics Series, Academic Press, *59*, 627.

ZHAO, Q. and CARR, F.H., (1997), *A prognostic cloud scheme for operational NWP models*, Mon. Wea. Rev. *125*, 1931–1953.

ZHOU, B. and B. FERRIER, S. (2008), *Asymptotic analysis of equilibrium in radiation fog*. J. Appl. Meteor. and Clim. *47*, 1704–1722.

ZHOU, B and DU J. (2010), *Fog prediction from a multimodel mesoscale ensemble prediction system*, Wea. Forecasting *25*, 303–322.

ZHOU, B, DU J., LIU S. and DiMEGO, G. (2011), *Verifications of simulated radar reflectivity and echo-top Forecasts at NCEP*, Paper# P.90, 24th Conf. on Weather and Forecasting, Amer. Meteor. Soc, 23–27, Jan 2011, Seattle, WA.

ZHOU, B., (2011), *Introduction to a new fog diagnostic scheme*, NCEP Office Note 466, US Department of Commerce, NOAA, NWS, NCEP, 33.

(Received November 8, 2010, revised March 17, 2011, accepted April 9, 2011, Published online May 19, 2011)

Reprinted from the journal

Pure Appl. Geophys. 169 (2012), 911–926
© 2011 Springer Basel AG
DOI 10.1007/s00024-011-0365-4

Deep Radiation Fog in a Wide Closed Valley: Study by Numerical Modeling and Remote Sensing

J. Cuxart[1] and M. A. Jiménez[1]

Abstract—The Ebro river basin, in the northeastern part of the Iberian Peninsula in Europe, very often experiences radiation fog episodes in winter that can last for several days. The impact on human activities is high, especially on road and air transportation. The installation in July 2009 of a WindRASS in the area, which is able to work in the presence of fog, now allows inspecting the vertical structure of the temperature and wind profiles across the roughly 300-m-thick fog layer. We present a case study of a long-lasting (60 h) deep radiation fog that took place in December 2009 to obtain a deeper understanding of the dynamic processes governing such persistent fog. Field observations of vertical profiles of temperature, wind and turbulent kinetic energy are compared with a high-resolution mesoscale simulation, satellite imagery of fog distribution and observations taken in the area to understand why the fog is so persistent and how it dissipates only for a short period in the afternoon despite intermittent turbulence within the fog deck. The confinement of the fog inside a practically closed basin allows us to study the relevant physical processes in the establishment and subsequent evolution of the fog episode using a limited-area mesoscale model. The contribution of the WindRASS measurements allowed us to validate the numerical simulations, particularly inspecting the role of turbulence that can link the bottom and top of the fog through moderate episodic mixing. The fog layer has very weak winds inside, but is well mixed and experiences intermittent top-bottom turbulence generated in its upper part by convection due to radiative cooling and by wind shear due to the topographically generated flows that blow just above the top of the fog.

Key words: Ebro Basin, mesoscale modeling, radiation fog, satellite imagery, turbulence, WindRASS.

1. Introduction

Fog, namely radiation fog, is a frequent phenomenon in the Ebro Basin during the winter, which negatively affects many aspects of daily life by causing delays and hazards in road and air traffic. Such fog events, confined by the topography, tend to be rather persistent, and the question arises, what are the mechanisms that sustain the fog for a long time and what is the interaction with the air outside the fog layer, in particular what is the effect of the wind shear at the top of the atmospheric boundary layer (ABL), here defined by the top of the fog? The installation of a combined acoustic and electromagnetic vertical profiling system (WindRASS) made possible, for the first time, an investigation of the turbulent dynamics during persistent fog events in the Ebro Basin and should allow for a deeper understanding of the meteorological process making this fog so persistent.

The behavior of the ABL in the presence of fog, contrarily to a situation with clear skies, is conditioned by the fact that the surface interacts radiatively with the cloud and—as long as the fog is present—the typical diurnal cycle of ABL height is strongly dampened, such that the inversion height, which normally forms the upper limit of the fog deck, becomes almost constant. The resulting evolution of the fog is mainly driven by the phase changes of the fog water and its microphysics, radiation budget, advection, turbulence and the effects that the complex terrain may have on the fog layer. An extensive review of the current knowledge on fog and the techniques for its study is given in Gultepe *et al.* (2007), where a description of the different types of fog and the main relevant mechanisms can be found. Some of the challenges identified there related to the processes are a proper description of the microphysics, a further understanding of the role of turbulence and the interaction with the surface.

In this article we aim to address this challenge and attempt to provide a contribution to a better understanding of the different physical processes that

---
[1] Grup de Meteorologia, Dpt. Física, Univ. de les Illes Balears, Cra. Valldemossa km 7.5, 07122 Palma de Mallorca, Spain. E-mail: joan.cuxart@uib.cat

should lead to an improvement in our skills to model such fog events in the Ebro basin and elsewhere, where similar conditions lead to persistent radiation fog periods. We provide a detailed analysis of a case study that compares a mesoscale numerical simulation with available observations, which allows us to test the usefulness of the newly installed WindRASS system for increasing our understanding of fog evolution and dynamics in the Ebro valley. Our high-resolution mesoscale simulation has fine vertical resolution and explicitly represents the water deposition, as indicated by BERGOT et al. (2007).

In this study we address one particular type of fog, a wide basin confined radiation fog, which recurrently occurs in winter in the Ebro Basin, at the northeastern part of the Iberian Peninsula. This basin has a triangular shape closed by the mountain ranges of the Pyrenees at the north (with peaks above 3,000 m), the Iberic System following the NW-SE axis (with summits above 2,000 m, parallel to the Ebro river) and the Catalan Coastal Range following the NE-SW axis, with mountains above 1,000 m (Fig. 1a). The bottom of the basin is a wide plain with a height of about 300 m above sea level (asl), and we concentrated on the Eastern region around Lleida in this work. The river reaches the Mediterranean sea through narrow gorges and makes the basin a practically closed structure. This makes the basin structure different from the cases of the Po (WOBROCK et al., 1992) and the Seine (HAEFFELIN et al., 2010) basins, which are open to the sea, with frequent radiation-advection fog. The case of VAN DER VELDE et al. (2010) is a pure radiation fog over flat terrain (The Netherlands).

The climatology of fog occurrence (Fig. 1b) shows that most of the fog days occur between November and February (78% of the total amount), with fog in December and January usually lasting for 1 or more days without dissipation. This radiation fog, usually very deep and often with a strong inversion topping it linked to a high-pressure system, is sometimes called high-inversion fog (HOLETS and SWANSON, 1981). MARTÍNEZ et al. (2008) and CUXART et al. (2011) indicated that 52% of the nights in the area for the 1996–2007 period correspond to boundary-layer-dominated regimes, with 14% of the total being radiation fog situations. CERMAK et al. (2009)

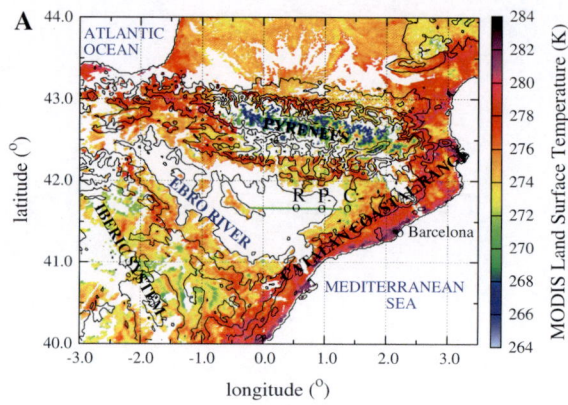

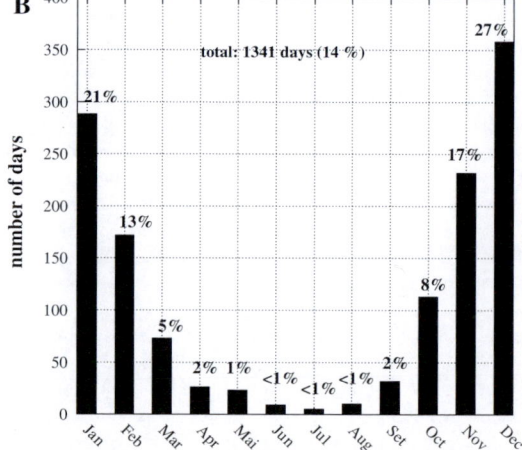

Figure 1

a Topography of the Ebro Valley (surface elevation contours are plotted every 400 m) and Land Surface Temperature derived from MODIS at 2140 UTC of 11 December 2009 indicating (in *white*) the area covered by fog. The *points* indicate the locations of the stations discussed in the text (*R* Raimat, *P* El Poal, *C* Cervera). The *green line* indicates the vertical cross section in Fig. 6. **b** Statistics of fog occurence in the period 1983–2007 for the station of Lleida

showed in their statistical study with Meteosat imagery and ground data that this area has a local maximum of fog occurrence in the northwestern Mediterranean area in winter.

Usually shallow fog precedes the formation of deeper or boundary-layer fog (FITZJARRALD and LALA, 1989). For the first type, the necessary humidity for fog formation can be supplied by evaporation from plants in the evening or from dew later at night, although quantitative estimations are still needed. The development of deeper boundary-layer fog implies that it becomes optically thick, and most of its dynamics become dominated by the processes at the top of the layer (DUYNKERKE, 1999). The relative role of each of these processes is a subject of research, usually through numerical modeling. For the transition from shallow to moderately deep fog (depths around 100 m), NAKANISHI (2000) showed in a Large-Eddy Simulation (LES) study that heating by condensation is about half of the radiative cooling at the fog top, resulting in moderately sustained top-bottom convection. ZHOU AND FERRIER (2008) theoretically described how a 100-m-deep fog in a persistent stage results from the equilibrium of cooling, gravitational settling and turbulence. They excluded deeper fog layers—like the one studied here—whose dynamics compare more to those of a stratocumulus cloud, dominated by radiative cooling at the top and having linked top-bottom convection.

The mature Ebro Valley fog is even deeper than the ones just mentioned. Once set, it is driven by processes at the top of the deck. It is confined by the slopes of the surrounding mountain ranges, and the fog layer does not move in any prevailing direction. Instead, the slopes that are above the fog are still subject to the diurnal cycle and generate within-basin motions that interact with the inversion at the top of the fog. This interaction may partly explain the evolution of the fog at its mature stage. The high-resolution simulation allows explicitly representing the different physical processes and inspecting their relative importance in different life phases of the fog. The novel in-fog observations of the WindRASS provide information that is analyzed first, and the model results are put in context related to the observations. This work intends to explain which relevant processes are involved in complex terrain fog and the relative importance of each.

2. Data and Methods

A 60-h-long episode of fog in December 2009 is studied here. It was the first multi-day fog episode in which the new Scintec WindRASS was fully operational in the Raimat station (41°41′N, 0°34′E) on the outskirts of the town of Lleida. This was a unique persistent fog event during the winter season

2009–2010 when the frequency of fog was considerably lower than average. The study combines observational (surface stations and remote sensing data) and modeling information (outputs of a high-resolution mesoscale simulation), and provides a good opportunity to inspect the evolution of a fog layer with special focus on the role of the different physical processes. Observational data gathered locally were provided by the Meteorological Service of Catalonia (SMC), which has an extensive network of Automatic Weather Stations (AWS) in the Eastern Ebro Basin with an average distance of 10–20 km between stations; 23 of them are located in our area of interest. Some local observers also reported about meteors (including fog and visibility). Finally, the Spanish Meteorological Agency has an official station in Lleida with human observers. The use of imagery provided by the satellites Aqua and Terra complements the available observational information. A summary of the technical data is given in Table 1.

2.1. WindRASS

WindRASS™ (Scintec AG, Rottenburg, Germany) is a newly developed instrument that extends the capabilities of a conventional Radio-Acoustic Sounding System (RASS), allowing the measurement of the vertical profile of wind speed, wind direction and virtual temperature across the ABL. Sound waves emitted by a Sodar (1,650–2,750 Hz) generate density heterogeneities that are sampled by electromagnetic (EM) waves emitted in the vertical and four other tilted directions at 1,290 GHz. The Doppler shift of the backscattered EM waves is analyzed and, together with the time travel of the pulse, allows computing the wind and virtual temperature profiles from the profile of the speed of sound. This is different from the conventional RASS, for which radioacoustic sounding is used for temperature measurements only.

The characteristics of the WindRASS (Scintec™) are summarized in Table 1. The device has been compared to direct measurement systems, and the quantitative comparison gives good confidence in the WindRASS measurements (CUXART et al., 2011). These comparisons have shown that it is possible to

Table 1

Relevant parameters of the data and methods used

Windrass

Technical characteristics: Sodar at 1,650–2,750 Hz, RASS at 1.290 GHz

Primary variables: wind (accuracy 1 m s^{-1}, 15°), virtual temperature (acc 1K), σ_w (m s^{-1})

Operation mode: $\Delta z = 10$ m, vertical range: 40 to 360 m a.g.l., profiles: 10 min averages

Meso-NH model

Discretization: $\Delta x = \Delta y = 2$ km; Δz (at 3 m) $= 1.5$ m, Δz (at 100 m) $= 7$ m, $\Delta t = 1$ s

Domain: central point: 42.2 N/0.51 W; Nx $= 360$; Ny $= 240$, Nz $= 85$; Top of the domain: 9,000 m

Physical schemes: TKE turbulence scheme, 3 layer soil-vegetation scheme, mixed-phase microphysics, radiation in bands

Integration: 60-h run, single domain, ECMWF lateral and surface boundary conditions, 60 computing hours in the ECMWF computer

MODIS

Sensor: MODerate resolution imaging spectroradiometer

Satellites: Aqua and Terra

Nominal resolution at nadir: 1 km^2

Number of images used during the episode, covering the whole area of interest: 18

Field used: land surface temperature provided by USGS

Nominal accuracy: 1 K

Automatic weather stations

Wind speed and direction

Temperature and humidity

Pressure and global solar irradiance

Rainfall

30 min averages available

Human reports

Lleida station of the Spanish Met. Agency: 4 to 20 UTC control of meteors, including meteors and visibility

Reporters to the Catalan Met. service: one daily report at 7 UTC, summarizing the previous 24 h

reconstruct profiles of the wind vector and the virtual temperature even if natural reflectors such as temperature inversions or turbulence are absent. The above-mentioned comparison with other devices allows considering the WindRASS a trustful provider of quantitative information of the lower atmosphere.

In Raimat the WindRASS is set to provide profiles routinely every 10 min, with a vertical resolution of 10 m and a vertical range between 40 and 360 m above ground level (a.g.l.). Data at the upper levels are usually of worse quality than closer to the ground because of a smaller number of returning echoes, advected away by the wind. It permits inspecting the

thermal stratification of the first 360 m a.g.l. and the profiles of the wind speed and direction, even within a dense fog layer extending well above the range of sampling. Its capability of providing estimations of the standard deviation of the vertical velocity provides a likely intensity and time evolution of the turbulence, and also detects strong wind shear across the top of the fog.

2.2. Satellite Images

The land surface temperatures (LSTs) estimated from satellites are a useful validation tool for models, especially when the meteorological situation is dominated by the local circulations (JIMENEZ *et al.*, 2008). Eighteen fields from a MODerate resolution Imaging Spectroradiometer (MODIS, SALOMONSON *et al.*, 1989)—on board of the polar Aqua and Terra satellites—are used (version 5 data level 2), covering the area of interest at an approximate nominal horizontal resolution of 1 km^2 (at nadir).

The MODIS LST product that we use is calculated using the Generalized Split-Window method. It performs corrections for atmospheric effects based on the differential absorption in adjacent infrared bands and requires land surface emissivity as an input (WAN and DOZIER, 1996). The retrieval of the land surface emissivity is based on the Vegetation Cover Method (CASELLES *et al.*, 1997), using land cover types, the atmospheric column of water vapor and the lower air temperature, and have an estimated accuracy of 1 K (COLL *et al.*, 1995).

An estimation of the horizontal extension of fog is provided by counting the blank pixels in the LST field. These are indicated in Fig. 2e. The confinement within the basin allows us to expect that most of those blank pixels correspond to fog, but blank pixels due to other reasons, such as passage of high clouds or pixels with wrong information, can also be included, lowering the level of confidence in this quantity.

2.3. Observer Fog Reports

Human reports of fog are given at the official meteorological observatory of the Spanish Agency of Meteorology in Lleida town, at a distance of 8 km from the Raimat site. They also indicate when water

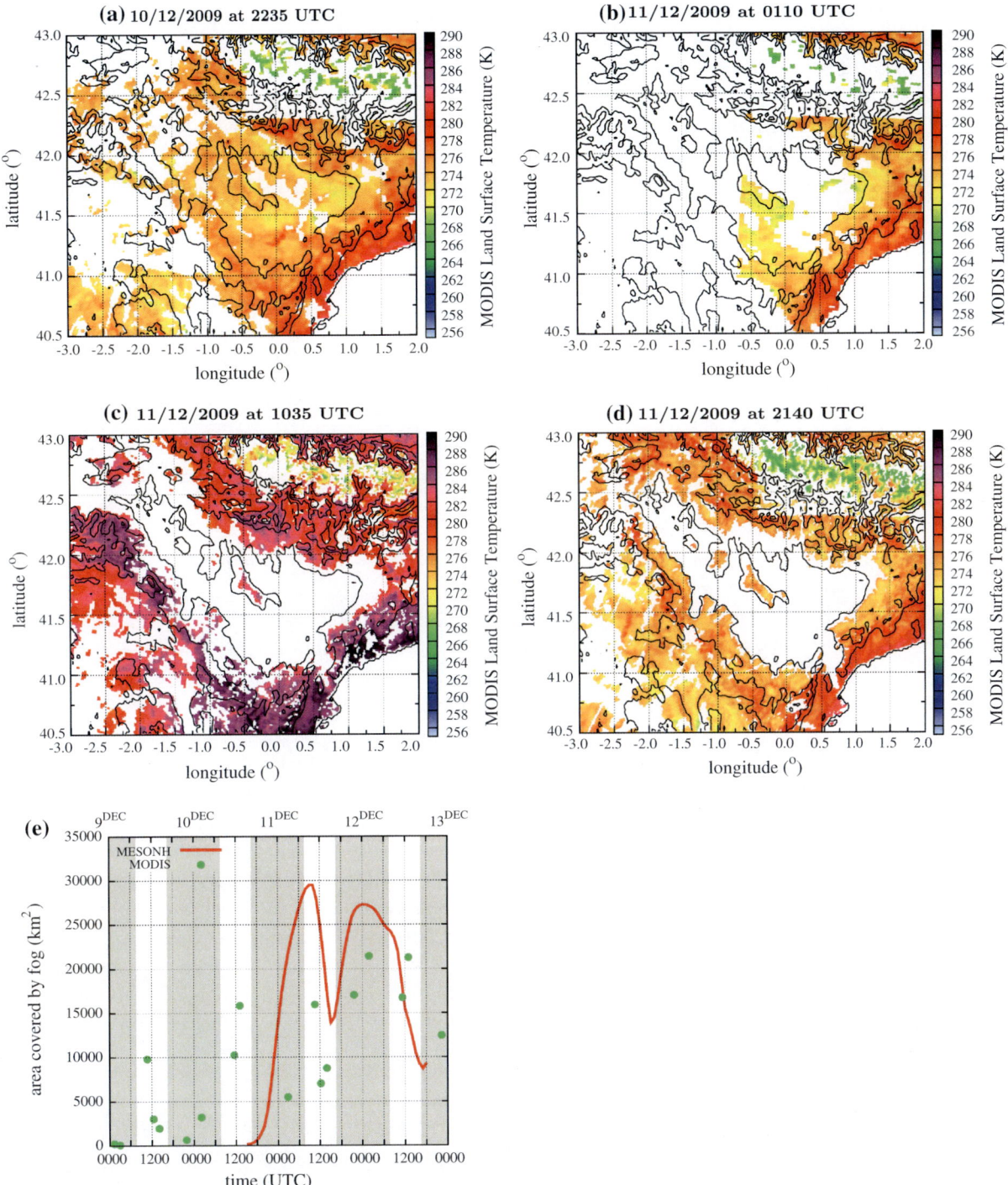

Figure 2

Land Surface Temperature derived from MODIS: **a** 10/12/2009 at 2235 UTC; **b** 11/12/2009 at 0110 UTC (data only in the eastern area); **c** 11/12/2009 at 1035 UTC and **d** 11/12/2009 at 2140 UTC. **e** Evolution of the area covered by fog as seen by the MesoNH model and MODIS. *Shaded areas* indicate nighttime. Evolution of the amount of surface covered by fog as seen by MNH, MSG and MODIS deposition from fog is collected in the rain gauge. Furthermore, the Catalan Meteorological Service has a network of collaborators that provide visual reports of meteors in the area, including fog. Both sources provide valuable information confirming the existence of fog during the selected period.

2.4. Network of Automatic Weather Stations

The Meteorological Service of Catalonia (SMC) has a network of automatic weather stations (AWS) with sensors for temperature, humidity, wind speed and direction, pressure and rainfall, and some for solar irradiation. In the area surrounding our site (the geomorphological unit called "Plain of Lleida") there are 23 AWSs with an average distance between them of 10 km. No data on visibility are available; however, the curves of humidity and solar radiation fit well with the description given by the official human reports. Fog precipitation was collected in most of the rain gauges of the area during the study period, with values up to $0.5 \, 1 \, m^{-2}$ per day of dense permanent fog, and the recorded winds were extremely weak.

2.5. Numerical Model

Meso-NH is a non-hydrostatic, fully compressible model (LAFORE et al., 1998) intended for the study of meteorological mesoscale and microscale phenomena. It uses a turbulence kinetic energy (TKE) scheme (CUXART et al., 2000), with a turbulent Prandtl number that depends on the vertical stability and a buoyancy-based mixing length (BOUGEAULT and LACARRERE, 1989). The physics package also comprises a three-layered soil-vegetation scheme (NOILHAN and PLANTON, 1989), cloud microphysics (PINTY and JABOUILLE, 1998) and a radiation scheme in bands (MLAWER et al., 1997). This package is now an option for some operational European models.

The model has been extensively used for LES studies (JIMÉNEZ and CUXART, 2005 or CUXART and JIMÉNEZ, 2007) and high-resolution mesoscale studies (JIMÉNEZ et al., 2008; MARTÍNEZ et al., 2010). For mesoscale modeling of high-pressure situations, the usual setup that is used here comprises 2 km of horizontal resolution for large basins, 3 m vertical resolution close to the ground and timesteps of the order of the second to avoid numerical instabilities that can develop over mountain slopes. Since the observed fog has some periods below 0°C, the chosen microphysics were a mixed scheme allowing for generation of rain, ice, snow and graupel. Sedimentation of droplets and snow is allowed (BERGOT et al., 2007).

The simulation ran between 10 December at 1200 UTC and 12 December at 1800 UTC. It used $360 \times 240 \times 85$ grid points, a time step of 1 s and 60 computing time hours in the European Center for Medium-range Weather Forecasts (ECMWF) computer, where 4 nodes in 64 processors were used. It was one single domain with boundary conditions from ECMWF analyses. A summary of its characteristics for this case is given in Table 1.

3. Observed Evolution of the Fog Layer

This section documents the observed evolution of the fog in the Ebro basin using satellite imagery to describe the surface covered by the fog over the whole basin, together with the local information at Raimat and surroundings, taking advantage of the WindRASS and the local dense network of AWS.

The episode fell in the time frame from 9 to 13 December 2009, which is the lifespan of a high-pressure system in the area, setting during the evening of the 8th when westerly winds weakened and ending during the 13th, when a synoptic northwesterly flow arrived in the basin. The pressure reduced to the sea level oscillated between 1,024 and 1,030 hPa until it weakened to 1,021 hPa during the night of the 11th to 12th and to 1,014 hPa the night of the 12th to the 13th.

3.1. Fog Extension as Seen by Satellite Imagery

Figures 2a–d show five MODIS images of the LST, corresponding to some instants of the evolution of the fog. White areas are missing LST data that we may identify as fog or clouds. These images are selected because they do not seem to have high clouds over the basin—at least the elongated kind of cirrus clouds across the area—and we consider them to be a good illustration of the area under fog or low stratus resulting from fog elevation. In Fig. 2e the green dots give an estimation of the time evolution of the area covered by fog or low clouds, indicating more than 20,000 km^2 at times of larger coverage. As mentioned in the previous section, high clouds or bad pixels during the fog episode make these values only a first approximation.

The first image corresponds to 10 December, 2235 UTC (in this area UTC and local are the same). At that time the fog areas (in white) were restricted to the lower parts of the basin. Shortly after, on 11 December, 0110 UTC, the fog occupied the bottom part of the Eastern basin, progressing upslope. At 1035 UTC it had practically filled up the whole bottom of the basin, also extending eastwards (the western part of this figure was outside of the satellite image). The fog (or low stratus) stayed in place the whole day (not shown), but it contracted in the slopes during the late afternoon. There was no good image for the late afternoon, but the image at 2140 UTC indicated less white area than in the former one.

3.2. Observed Evolution by the Automated Weather Stations

The evolution of the temperature, humidity, wind and solar radiation for three selected AWSs of the SMC network are shown in Fig. 3 (positions indicated in Fig. 1a). Station Cervera (C) is in the eastern limit area of the fog, station Poal (P), at the lowest point of the line, where the fog can last very long, and Raimat (R) was our sampling site. They are located approximately on an east-west line, following the main longest slope of the area.

The first night (8–9) showed that at the lower parts of the basin (El Poal) fog or mist may have been set already (high humidity, very low wind speed, small temperature drop and reports of observers), whereas at the surrounding places there may have been only shallow fog near the sunrise. The solar radiation was able to warm the surface with air temperatures in the afternoon of the 9th near 12°C and humidities below 70%. All the stations registered solar irradiation above 350 W m^{-2}. The winds had significant speeds and well-defined directions (not

shown) following the known local patterns (CUXART et al., 2011).

The second night (9–10) allowed the establishment of fog after midnight on the plain, but not on the surrounding slopes (Cervera), where the drainage winds may allow the surface air to stay warmer and well mixed with the air above. In fact, no rainfall was recorded in Cervera, contrarily to the other two stations. However, the fog was not dense enough to prevent the surface temperature from falling sustainedly during the night. The fog had dissipated the next morning except close to the rivers.

The third night (10–11) allowed a strong surface cooling only until midnight when the fog was set over the whole plain. Cooling ceased—there was even warming at some stations—and the temperature stayed near 2 to 4°C almost everywhere. At the same time, the wind speed collapsed to very weak values with erratic directions (not shown).

The subsequent evolution (until midnight of the 13th) was that of a well-set fog layer, with practically no diurnal cycle of humidity and temperature, with calm or very weak wind. There was surface lifting of the fog in the afternoon (human reports) for a short time as the radiation was more vertically incident on the fog layer, allowing for values not larger than 100 W m^{-2}, not enough to destroy the fog, which reached the ground again as the sun descended towards the horizon. The only exceptions were the stations near the limits of the deck, where the fog was shallow, and formed and dissipated diurnally. As described previously, these dynamics stayed in place for the number of days that the high-pressure synoptic conditions last.

3.3. Observer Fog Reports

Reports of fog are given at the official meteorological observatory of the Spanish Agency of Meteorology in Lleida town, at a distance of 8 km of the Raimat site. Dew and mist were indicated on the morning of 9 December, and fog with a visibility below 500 m was reported in the last night hours and first morning hours of the 10th. The fog set again on the evening of the 10th and stayed continuously over Lleida from 8 p.m. of the 10th to 12 a.m. on the 13th, with more than 60 h of continuous fog reported. The

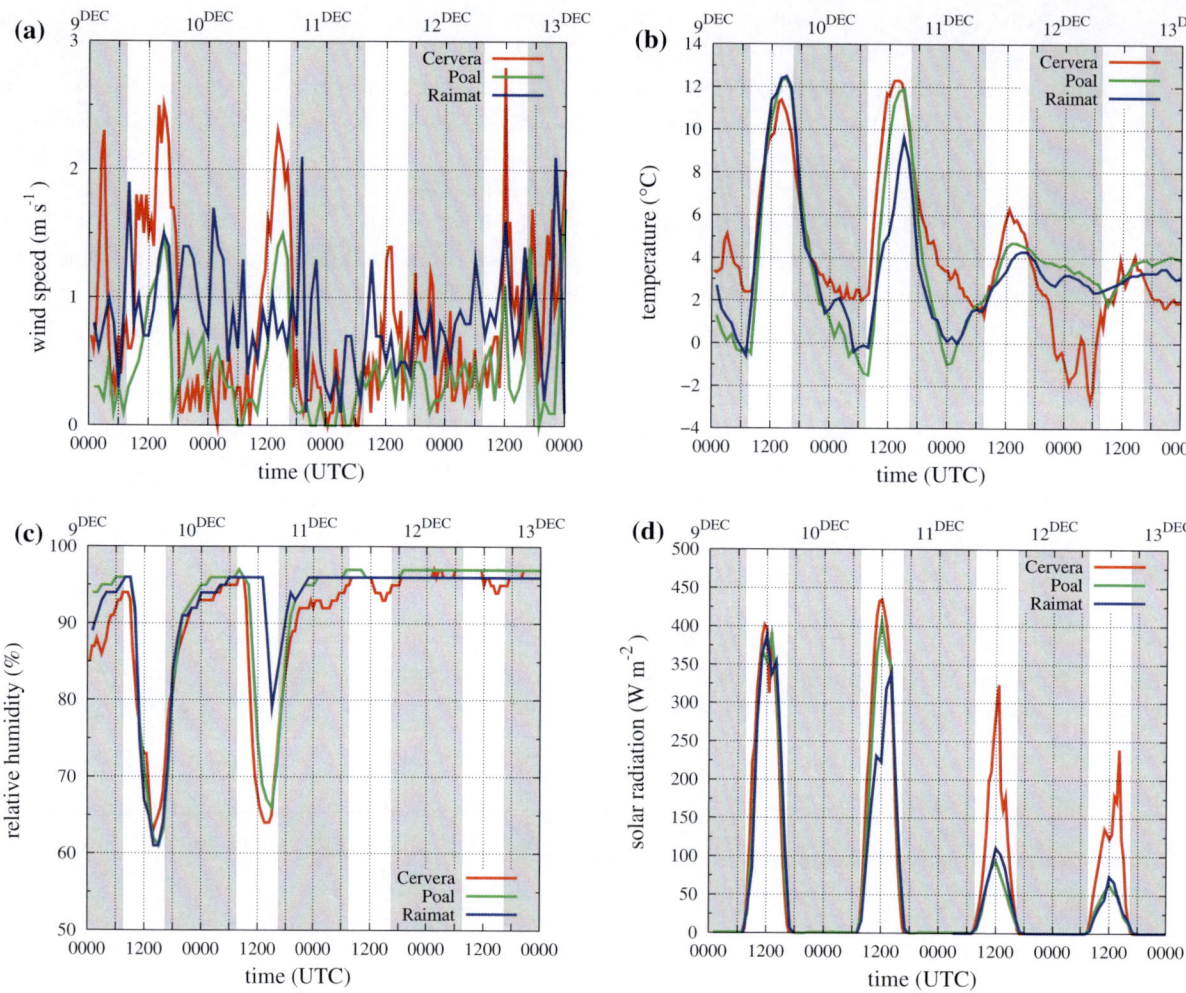

Figure 3
Evolution of three selected AWSs between 9 and 13 December: Raimat (at the WindRASS site), El Poal (at a low part of the basin) and Cervera (at the slopes, in the limit of the fog deck). **a** Wind speed, **b** temperature, **c** relative humidity, **d** solar radiation. *Shaded areas* indicate nighttime. See locations in Fig. 1a

first part of this event had visibilities between 40 and 200 m a.g.l., whereas the second part reported some elevation of the fog deck and visibilities at the ground level increased to 500 m to even 1.5 km in the late afternoon. Precipitation collected in the official rain gauge at Lleida amounted to half a liter during the episode.

The collaborators of the Catalan Meteorological Service from several stations in the area indicated general morning fog with no precipitation associated with the morning of the 9th and permanent dense fog (visibility below 500 m) from 10 to 12 December (explicitly described as "precipitating fog" by many observers). There were some reports of fog lifting during the afternoon with visibilities increasing

slightly above 1 km. Stations at the limit of the fog area showed a sharp transition between those collecting fog water and those that did not. One observer reported the limit of the fog at the outskirts of the town of Juneda on the slopes, extending from there continuously downslope in the valley direction. Inspection of recorded precipitation in rain gauges of the area indicated values of several tenths of liters per square meter in most of the stations affected by fog.

3.4. WindRASS Column Evolution over Raimat

The establishment of the fog deck as seen by the WindRASS is displayed in the left column of Fig. 4

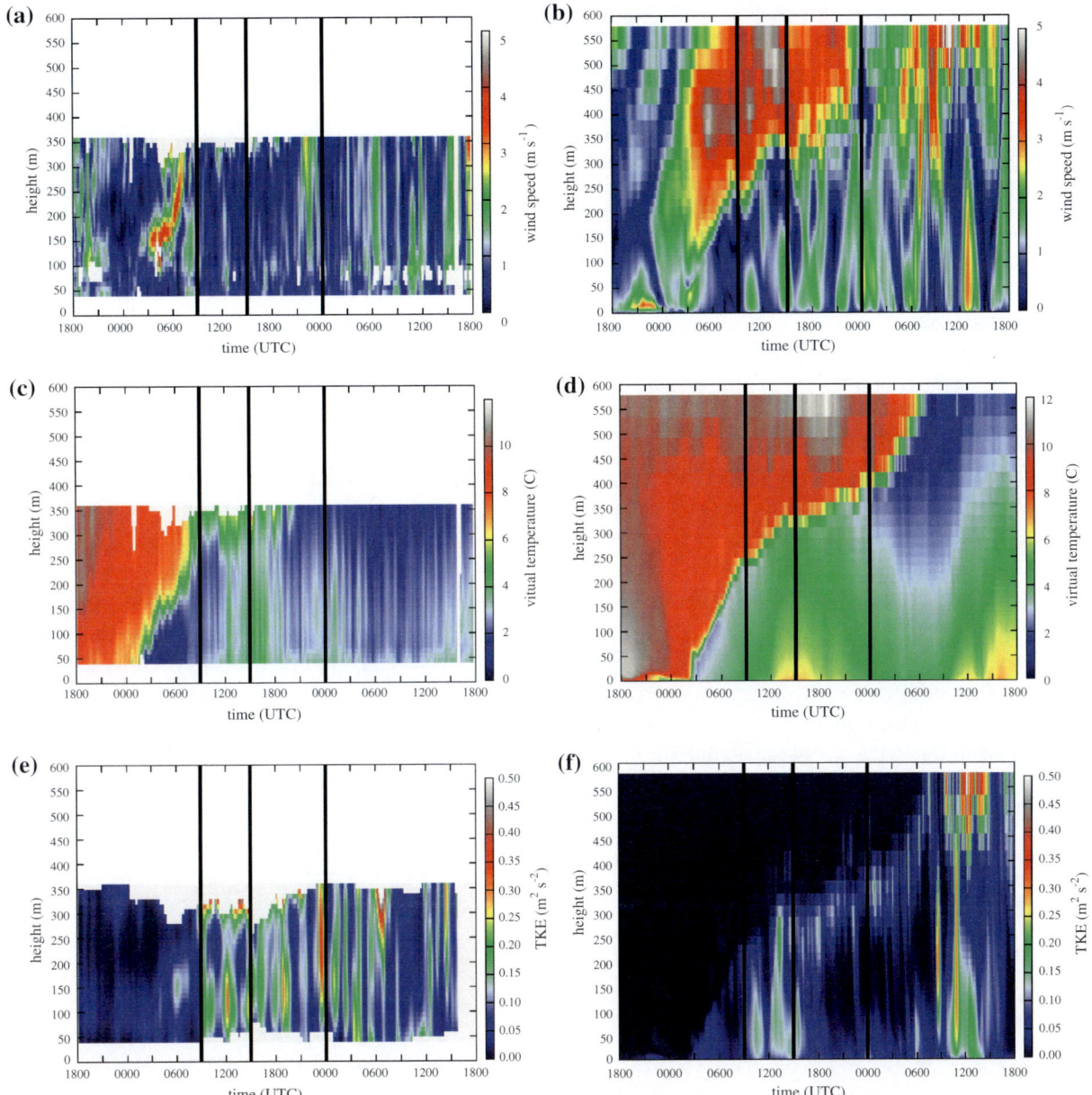

Figure 4

Comparison of the temporal evolution for 1800 UTC of the 11th to 1800 UTC of the 12th December 2009 of the profiles of the WindRASS (*left*) and the MesoNH model (*right*) for Raimat (location in Fig. 1a). **a** and **b** Wind speed; **c** and **d** virtual temperature; and **e** and **f** turbulence kinetic energy. The *vertical lines* indicate the times of the budgets in Fig. 5

for the period 1800 UTC of the 10th (no fog on the site) to 1800 UTC of the 12th (permanent fog). The features previously indicated by the satellite images and the AWS network were confirmed, and some insight into the vertical structure was gained. Before the establishment of the fog after midnight of the

11th, the wind in Raimat in the first 200 m a.g.l. was westerly (not shown) with speeds below 2 m s^{-1}, and the virtual temperature showed a diurnal cycle delayed by the fog of the previous night, with small values for the TKE, indicating that convection was very weak.

Fog developed from 0 to more than 360 m a.g.l. in a few hours during the second part of the night as is clearly seen in the virtual temperature plot, with an indication of significant wind shear at the top of the cloud and weak turbulence associated inside the fog. Once the fog was set (the top was above 360 m a.g.l.), the wind was practically calm as indicated by the erratic behavior of the wind direction (not shown). The whole column was practically isothermal and warmed after noon when radiation was able to penetrate the cloud.

A striking feature of the WindRASS observation is that, once the fog layer had developed above the upper range of the WindRASS, there was intermittent turbulence over the whole column during day and night indistinctly (at a rate of about one intense episode per hour), with estimated values up to 0.5 m^2 s^{-2}, larger than normal nighttime values and of the same order that can be found in a winter convective boundary layer. The TKE had the largest values at the top of the sampled deck, and the most likely hypothesis is that it was generated at the top of the fog deck by radiative and evaporative cooling, similarly to a stratocumulus cloud. These features will be looked for in the mesoscale simulation in the next section.

4. Physical Processes as Seen by a High-Resolution Simulation

4.1. Vertical Structure of the Fog Deck over Raimat

The model captures the development of the fog after midnight of the 11th and provides an evolution of the different variables very similar to the observed one. In Fig. 4, the WindRASS and the modeled evolutions of the vertical profiles of wind speed, virtual temperature and TKE are compared. Model results are plotted until higher elevation to provide complementary insight into this evolution.

Both the WindRASS and the model saw a similar setup of the fog during the second half of the first modeled night. However, the fog developed higher than the WindRASS range, and the model indicated that the top of the fog during the day after could be located between 350 and 400 m a.g.l., while during the beginning of the second night it developed even

higher. Unfortunately, there was no supplementary observational evidence of the height of the fog. The fog layer had weak winds in all its depth, and the model generated significant speeds over the deck, from the northwest, with values of about 4 m s^{-1}.

The vertical virtual temperature gradients were very weak, with the cooler layers near the top of the cloud, and the solar radiation was able to reach the ground for some hours after noon, when the model was generating turbulence near the ground. The model overestimated the values of virtual temperature, a behavior that is usually linked to a different response of the soil in the model and reality, but that also could be attributed to an excess of entrainment of warm air from above across the top of the fog deck.

In fact, turbulence in the model was mostly restricted to the upper part of the deck with some episodic intrusions downward, whereas the WindRASS observed intermittent but very frequent turbulence events, not only at the top part, which were more intense and well correlated with the vertical velocities (not shown). Therefore, the model only partially captures the dynamics of the turbulence, which it locates mostly at the top part, like in the majority of the studies cited in the introduction.

The temperature and TKE budgets from the model are shown in Fig. 5 for three selected times on 11 December, indicated in the lines in Fig. 4. The budgets at 0900 UTC corresponded to the final time of the initial fast development of the fog during the first simulated night. The cloud grew until 240 m a.g.l., and the temperature budget indicated that the main factors were at the upper part of the cloud, with radiative cooling being compensated by condensation (microphysics) and explicit motions inside the cloud top (horizontal and vertical advection). The corresponding TKE budget indicated that turbulence, a minor factor at this time, was basically produced by buoyancy in the interior of the cloud (convective circulations originating at the cloud top) and by turbulence transport and shear production at the upper layer.

At 1500 UTC the model allowed the radiation to reach the ground. The cloud was now 300 m thick and lifted 20 m above the ground. The temperature budget indicated that radiative cooling at the top was still the main factor, now compensated by warming

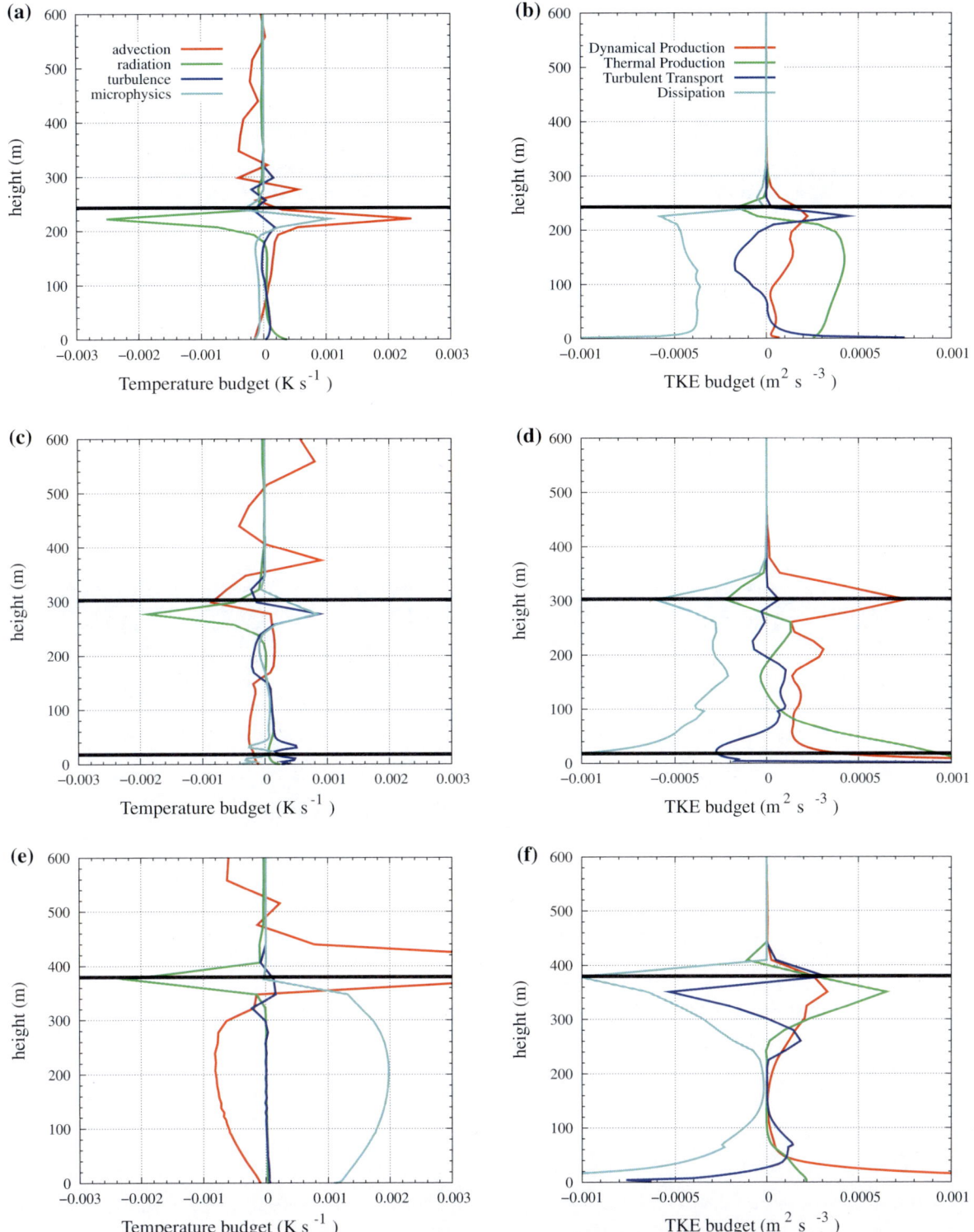

Figure 5
Temperature (*left*) and TKE (*right*) budgets in Raimat (see location in Fig. 1a) for 11 December at **a** and **b** 0900 UTC; **c** and **d** 1500 UTC, and **e** and **f** 2400 UTC. The *horizontal black lines* indicate the thickness of the fog

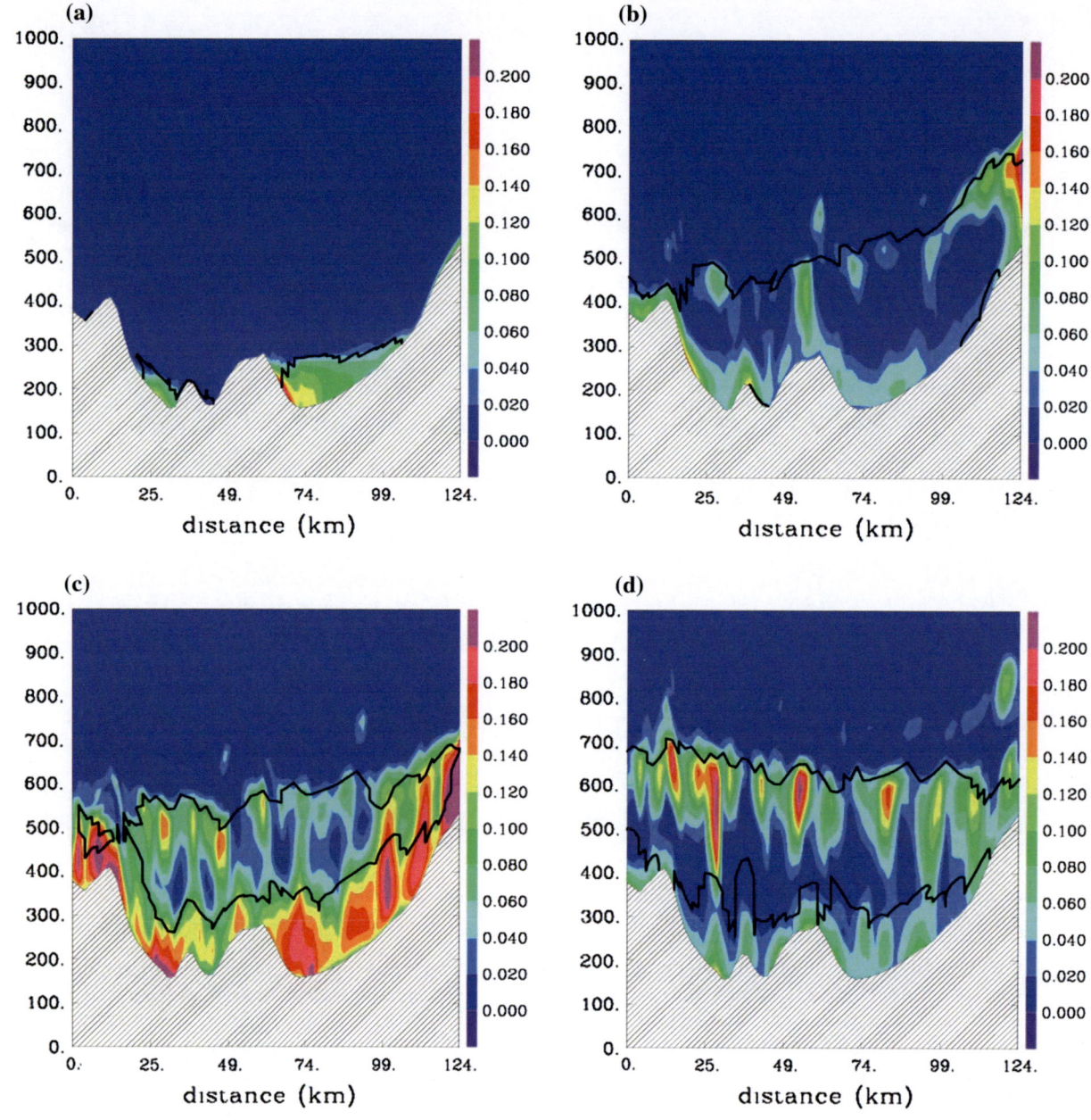

(a) **(b)**

(c) **(d)**

distance (km)

Figure 6
Vertical cross sections following the *green line* in Fig. 1a of the fog limits (in *lines*) and the TKE (m^2 s^{-2}, in *colors*) for 11 December at **a** 0000 UTC, **b** 0900 UTC, **c** 1500 UTC and **d** 2400 UTC. The Y axis represents the height above sea level (in m)

due to turbulence transport and condensation. The TKE budget indicated that most of the turbulence transport at the top layer was generated by wind shear across the cloud top, which was still growing. Near the ground, surface heating and wind shear explained

the local maxima of turbulence, which was the main factor below the cloud.

At 0000 UTC of the second modeled night, the cloud had risen up to 380 m a.g.l. and was in contact with the ground again. It was in a fast growing phase, entraining dry air from above similarly to a strato-cumulus cloud with a low Richardson number across the inversion. The cloud interior was dominated by condensational warming and the top by radiative cooling. The only compensation factors were the explicit motions that allowed the cloud to evolve and be turbulent. The TKE budget indicated that both thermal and dynamic production were important at the top of the fog, whereas dynamic production dominated in the surface layer.

4.2. *Evolution of the Horizontal Structure of the Fog Layer*

The simulation produced a three-dimensional evolution of fog that could only be partially validated with the available observational information. Figure 2e compares the surface covered by fog as seen by Aqua and Terra satellites and the model. While the evolution of the model area of fog was less irregular than the satellite, the qualitative evolution was similar, with the model slightly overestimating the amount of surface covered by fog. Let us remind that the simulation started at 1200 UTC the 2nd day, so that the first part of Fig. 2e does not have model data.

Four vertical cross sections at different times over the line marked in Fig. 1a are shown in Fig. 6. They allowed inspecting the time and space evolution of the fog deck. Here the bold black line indicates the limits of the cloud water contents (values different to zero) in the model. During the first modeled night the fog was shallow, in contact to the ground and confined by the topographical features of the bottom of the basin (Fig. 6a). There was turbulence inside the fog deck and generation by wind shear related to the flows over the slopes.

The following morning (Fig. 6b), the fog had increased its depth well above the minor features of the basin bottom, and the top of the fog deck approximately followed the main topographical slope at the eastern part. TKE was generated at some slopes and at some spots at the cloud top, but there were areas of the cloud without any significant turbulence.

After noon (Fig. 6c) the fog had slightly increased its height, while the top of the cloud started losing the connexion with the topographical profile below. The solar radiation had eroded the fog near the ground.

There was turbulence under the cloud and at some spots at the top of the cloud, sometimes connecting.

The following midnight (Fig. 6d), the fog top became flatter and disconnected from the ground, and there was downward growth to the ground at some elevated points. The turbulence was dominated by the production at the top responding to radiative cooling and wind shear, and it was mainly decoupled from the ground, although some connections could take place, generating mixing along all the layer in some spots of the basin.

5. Discussion

The long-lasting fog event studied here provided the opportunity to deepen our understanding of the relevant physical processes in fog, taking advantage of the available state-of-the-art tools, such as the Meso-NH high-resolution simulation with explicit physics and the WindRASS wind and virtual temperature profiles, which can now also be obtained during foggy conditions with this new instrument. Our study differs from that by van der Velde et al. (2010) in the complexity of the terrain and modeling approach. Their study was carried out over flat terrain, where radiation fog was not restricted by the valley walls that can be taller than the ABL, and they focused more strongly on the sensitivity of the fog modeling to the characteristics of the parameterization schemes in the models used. Contrastingly, here we put our efforts into understanding the different physical processes in a complex terrain, a less studied phenomenon.

Even if the typical behavior of the fog in the Ebro valley is known by local forecasters thanks to long-term experience by the inhabitants of the area, and some climatological studies are available (Vicente-Serrano et al., 2010), studies on the physical processes governing the phenomenon are yet to be done. In this discussion we aim to comment on the time and space evolution of the fog layer and to inspect what the relevant physical processes may be, hoping that the findings can be extended to similar events in other closed basins.

The availability of numerical model simulations, WindRASS profile measurements and surface observations allowed us to combine these different sources of information, permitting us to derive a more consistent picture of dynamic processes in fog. WindRASS profiling of fog is new in the basin and permits validating the model results in a more rigorous way than has been possible so far. We followed here a similar rationale as FITZJARRALD and LALA (1989): they defined the first three stages of a radiation fog event as the study of the fog onset, development and final thickness. They did not treat internal fog dynamics, microphysics and fog dissipation, and we partially addressed the internal fog dynamics. Their complex terrain problem was not confined as in the case treated in this work.

The time evolution of the fog for this event matches with the common knowledge in the area. After the installation over the area of a high pressure system, fog appears at the end of the first night, it dissipates the day after and forms again during the second night to last until a change in meteorological conditions occurs. The evolution in the daytime is such that, in the central hours of the day, the dissipation of radiative energy at the top of the fog bank may erode the fog layer and even allow direct radiation to reach the ground, causing a short-lived lifting of the cloud during the afternoon. Both the WindRASS observations and the model results reproduced this sequence of events, documented as well by the available observations shown here.

The depth of the fog is less known, and the present study indicates that the cloud can be very thick. The first day represents the development to a moderate height between 200 and 300 m a.g.l. Then the top of the cloud evolves as entrainment at the top may occur by shear and convection, which brings drier air into the cloud, which saturates and makes the cloud grow. The entrainment of air may come from slope flows over the surrounding slopes and is favored by the quiescent fog layer below, which enhances the wind shear across the fog top. The depth rises in the model up to 500 m a.g.l. at the end of the run.

The horizontal extension seems limited to the lower parts of the basin in the beginning but, as the time advances, the fog extends to almost all the valley with an upper limit near 700 m asl, and the top of the fog seems to lose connection with the minor topographic features of the valley bottom.

Observations in the area indicate that at areas over the slope where the fog is shallow, the fog follows a diurnal cycle with dissipation in the central hours of the day even if the fog stays downslope.

The fog layer has no mean motion, most likely because of its topographic confinement, and the recorded wind speeds are of the order of 1 m s^{-1} from very variable directions, which may also indicate some limitations of the WindRASS under weak wind conditions. The fog layer is well mixed by turbulence and is topped by a well-defined inversion of ≈ 5 K across a 50-m-deep layer. This moderate inversion can be eroded by turbulence-generating processes, as will be described in more detail below. In the fog layer energy and matter are only coupled with air from the free atmosphere above through the mixing across the upper inversion (entrainment flux).

The inspection of the temperature and TKE budgets showed that the role of the explicit motions at the top of the cloud and the wind shear above the inversion are major players in the vertical evolution of the fog, since advection of temperature and dynamic production of turbulence are key parameters at the top of the fog. Convection is mainly caused by radiative cooling just below the cloud top, whereas condensation tends to compensate this cooling at the same level, helped by the turbulence that redistributes heat vertically.

The time picture of the event indicated that in the development phase it is very similar to what Nagasaki described in his LES study, an interplay between radiation cooling and warming by condensation and advection, with turbulence redistributing the imbalances vertically. In the almost steady part between noon and midnight, an equilibrium seems to be established between radiation cooling and condensation and turbulence warming. In the 300-m-deep fog, the turbulence budget indicates that it is fully coupled from top to the ground. In the periods when the cloud lifts some 20 m above the ground, turbulence heating dominates near the ground in response to the radiation that is able to reach the surface. However, the cloud is still controlled by the processes at its top.

The main controllers of the growing phase, as seen by the simulation, are the advections explicitly generated in the model just above the fog layer, most likely linked to the topography. These within-basin mesoscale motions may be ruling the amount of entrainment across the fog top, even changing the strength of the thermal inversion. This process should be studied through suitable experimental campaigns, a big challenge because of the extreme difficulties of measuring inside the fog layer.

6. Conclusions

A 60-h lasting fog event has been studied through high-resolution numerical modeling with the support of a novel remote sensing device (the WindRASS by Scintec) and the use of the available conventional in situ instrumentation. The radiation fog is confined by the slopes of a practically closed mesoscale basin and does not experience any advection. Above the fog layer, mountain slopes under clear-sky conditions experience the diurnal cycle of temperature and energy fluxes, and generate motions above the fog that interact with the thermal inversion at its top.

The fog layer is practically isothermal and with frequent turbulence inside. It is dominated by the processes at its top, namely radiational cooling that generates downwards convection (or vertical advection) and explicit motions that create wind shear and transport heat. Condensation and turbulence transport also act in the interior of the cloud. The relative importance of each of these factors depends on the life stage of the fog, in the growing phases dynamics overpassing radiation and in the quasi-steady phases, radiation cooling being compensated basically by turbulence transport and condensation.

This study of a confined fog in a basin has allowed us to inspect what the likely relevant processes are in a pure radiation fog under the effect of upper flows generated topographically. The Wind-RASS only allowed seeing above the fog layer in the initial phase of the development. The subsequent deductions are based on the model outputs and their agreement with the available WindRASS and surface observations. To confirm or discard the importance of mesoscale basin flows in the evolution of the basin radiation fog, future research should focus on specific field experiments, since numerical models may be very sensitive to the defaults of their parameterizations when run under such conditions.

Acknowledgements

ECMWF and AEMET are thanked for the access to computing time. We also thank Jordi Cunillera and Antonio Gázquez (SMC) for access to data from the SMC Network and from the special devices installed in Raimat, Armand Alvarez (AEMET) for his support and knowledge of the local meteorology, Felipe Molinos for the initial treatment of the WindRASS and fog data, Daniel Martinez for insightful comments on the manuscript, and the Meso-NH team in Meteo France and Laboratoire d'Aérologie. This work was partially funded by SMC and through project CGL2009-12797-C03-01 of the Spanish Government also supplied with European FEDER funds. Data from MODIS are distributed by the Land Processes Distributed Active Archive Center (LPDAAC) located at the US Geological Survey (USGS) Earth Resources Observation and Science (EROS) Center (http://www.lpdaac.usgs.gov).

REFERENCES

BERGOT T., TERRADELLAS E., CUXART J., MIRA A., LIECHTI O., MUELLER M., and WOETMANN NIELSEN N. (2007), Intercomparison of single-column numerical models for the prediction of radiation fog. J. Appl. Meteorol. Clim., 46, 504-521.

CASELLES V., VALOR E., COLL C., and RUBIO, E. (1997), Thermal band selection for the PRISM instrument 1.Analysis of emissivity-temperature separation algorithms, J. Geophs. Res., 102, 11145-11164.

CERMAK J., EASTMAN R. M., BENDIX J., and WARREN S. G. (2009), European climatology of fog and low stratus based on geostationary satellite observations, Q. J. R. Meteorol. Soc., 135, 2125-2130.

COLL C., CASELLES V., GALVE J.M., VALOR E., NICLOS R., SANCHEZ J.M., RIVAS R. (2005), Ground measurements for the validation of land surface temperatures derive from AATSR and MODIS data, Remote Sens. Environ., 97, 288-300.

CUXART J., CUNILLERA J., JIMÉNEZ M.A., MARTÍNEZ D., MOLINOS F., PALAU J.L. (2011), Study of mesobeta basin flows by remote sensing, Bound.-Layer Meteorol. (in press)

CUXART J., JIMÉNEZ M.A. (2007), Mixing Processes in a Nocturnal Low-Level Jet: An LES Study, J. Atmos. Sci., 64, 1666-1679.

CUXART J., BOUGEAULT P., REDELSPERGER J.-L. (2000), A turbulence scheme allowing for mesoscale and large-eddy simulations, Q. J. Roy. Meteorol. Soc., 126, 1-30.

DUYNKERKE P.G. (1999), Turbulence, Radiation and fog in Dutch Stable Boundary Layers, Boundary-Layer Meteorology, 90, 447-477.

FITZJARRALD D.R., and LALA G.G. (1989), Hudson Valley Fog Environments, J. Appl. Meteorol., 28, 1303-1328.

GULTEPE I., TARDIF R., MICHAELIDES S. C., CERMAK J., BOTT A., BENDIX J., MÜLLER M. D., PAGOWSKI M., HANSEN B., ELLROD G.,

JACOBS W., TOTH G., and COBER S. G. (2007), Fog Research: A Review of Past Achievements and Future Perspectives, Pure Appl. Geophys., 164, 1121-1159

HAEFFELIN M., BERGOT T., ELIAS T., TARDIF R., CARRER D., CHAZETTE P., COLOMB M., DROBINSKI P., DUPONT E., DUPONT J.-C., GOMES L., MUSSON-GENON L., PIETRAS C., PLANA-FATTORI A., PROTAT A., RANGOGNIO J., RAUT J.-C., RÉMY S., RICHARD D., SCIARE J., and ZHANG X. (2010), PARISFOG: Shedding New Light on Fog Physical Processes, Bull. Amer. Meterol. Soc, 91, 767-783.

JIMÉNEZ M.A., MIRA A., CUXART J., LUQUE A., ALONSO S., GUIJARRO J.A. (2008), Verification of a clear-sky mesoscale simulation using satellite-derived surface temperatures, Mon. Wea. Rev., 136, 5148-5161.

JIMÉNEZ M. A., and CUXART J. (2005), Large-Eddy simulations of the stable boundary layer using the standard Kolmogorov theory: Range of applicability, Bound.-Layer. Meteor., 115, 241-261.

LAFORE J.P., STEIN J., ASENCIO N., BOUGEAULT P., DUCROCQ V., DURON J., Fisher C., Héreil P., Mascart P., Pinty J.-P., Redelsperger J.-L., Richard E., Vilá-Guerau de Arellano J. (1998), The Meso-NH atmospheric simulation system. Part I: Adiabatic formulation and control simulation, Ann. Geophys., 16 90-109.

MARTÍNEZ D., JIMÉNEZ M.A., CUXART J., MAHRT L. (2010), Heterogeneous Nocturnal Cooling in a Large Basin Under Very Stable Conditions, Bound.-Layer Meteorol., 137, 97-113.

MARTÍNEZ D., CUXART J., and CUNILLERA J. (2008), Conditioned climatology for stably stratified nights in the Lleida area. Journal of Weather and Climate of the Western Mediterranean, Tethys, 5, 13-24.

MLAWER, E.J., TAUBMAN S.J., BROWN P.D., IACONO M.J., and CLOUGH S.A. (1997), Radiative transfer for inhomogeneous atmospheres: RRTM, a validated correlated-k model for the longwave, J. Geophys. Res., 102, 16663-16682.

NAKANISHI M. (2000), Large-eddy simulation of radiation fog, Bound.-Layer Meteor., 94, 461-493.

NOILHAN J., PLANTON S. (1989), A simple parameterization of land surface processes for meteorological models, Mon. Wea. Rev., 117, 536-549.

PINTY J.-P., and JABOUILLE P. (1998), A mixed-phase cloud parameterization for use in mesoscale non-hydrostatic model: simulations of a squall line and of orographic precipitations. Proc. Conf. of Cloud Physics, Everett, WA, USA, Amer. Meteor. Soc., August 1999, 217-220.

SALOMONSON V.V., BAMES W.L., MAYMON W.P., MONTGOMERY H., and OSTROW H. (1989), MODIS: advanced facility instrument for studies of the Earth as a system. IEEE Trans. Geosci. Remote Sens., 27 145-153.

VAN DER VELDE I.R., STEENEVELD, G.J., WICHERS SCHREUR, B.G.J. and HOLTSLAG, A.A.M. (2010), Modeling and forecasting the onset and duration of severe radiation fog under frost conditions, Mon. Wea. Rev., 138, 4237-4253.

VICENTE-SERRANO, S.M., LOPEZ-MORENO, J.I., VEGA-RODRIGUEZ, M.I., BEGERIA, S., CUADRAT, J.M. (2010), Comparison of regression techniques for mapping fog frequency: application to the Aragon region (northeast Spain), International Journal of Climatology, 30: 935-945.

WAN Z., and DOZIER, J. (1996), A generalised split-window algorithm for retrieving land-surface temperature from space, IEEE Trans. Geosci. Remote Sens., 34, 892-905.

WAN Z., and LI Z.-L. (1997), A physics-based algorithm for retrieving land-surface emissivity and temperature from EOS/MODIS data, IEEE Trans. Geosci. Remote Sens., 35, 980-996.

WOBROCK W., SCHELL D., MASER R., KESSEL M., JAESCHKE W., FUZZI S., FACCHINI M.C., ORSI G., MARZORATI A., WINKLER P., G. ARENDS B.G., and BENDIX J. (1992), Meteorological characteristics of the Po Valley fog, Tellus B, 55, 469-488.

ZHOU B., and FERRIER B.S. (2008), Asymptotic Analysis of Equilibrium in Radiation Fog, J. Appl. Meteor. Climatol., 47, 1704-1722.

(Received October 28, 2010, revised May 15, 2011, accepted June 16, 2011, Published online July 24, 2011)

Pure Appl. Geophys. 169 (2012), 927–939
© 2011 Springer Basel AG
DOI 10.1007/s00024-011-0356-5

❙ Pure and Applied Geophysics

Urbanization Effects on Fog in China: Field Research and Modeling

Zi-hua Li,[1] Jun Yang,[1] Chun-e Shi,[2] and Mei-juan Pu[3]

Abstract—Since the policy of "Reform and Open to the Outside World" was implemented from 1978, urbanization has been rapid in China, leading to the expansion of urban areas and population synchronous with swift advances in economy. With urban development underway, the urban heat island (UHI) and air pollution are being enhanced, together with vegetation coverage and relative humidity on the decrease. These changes lead to: (1) decline of annual fog days in cities (e.g. In Chongqing, so-called city of fog in China, the annual fog days have reduced from 100–145 in the 1950s to about 20–30 in the 2000s); (2) decrease in fog water content (FWC) and fog droplet size, but increase in fog droplets number concentration [e.g. Jinghong, a city in Yunnan province, the average FWC (the droplet diameter) during an extremely dense fog episode with drizzle was 0.74 g/m^3 (28.6 µm) during the 1968/69 winter and 0.08 g/m^3 (6.8 µm) in another extremely dense fog episode during the 1986/87 winter, correspondingly, the fog droplets number density had increased from 34.9 to 153 cm^{-3}]; (3) decrease in fog water deposition (FWD) (e.g. the annual mean FWD measured in Jinghong had dropped from 17.3 mm in the 1950s to 4.4 mm in the 1970s and less than 1 mm in the 1980s, and no measurable FWD now.); (4) decrease in visibility in large cities (e.g. in Chongqing, the annual average visibility had decreased from 8.2–11.8 km in the 1960s to 4.9–6.5 km in the 1980s, and around 5 km in recent years); and (5) increase in the ion concentrations and acidity in fog water in urban areas [e.g. the average total ion concentration (TIC) in the center of Chongqing was 5.5 × 10^4 µmol/L, with mean pH value of 4.0, while the corresponding values are 9.7 × 10^3 µmol/L and over 5.5 in its rural area]. These changes endanger all kinds of transportation and human health. This paper summarized the authors' related studies, including observations and numerical simulations to confirm the above conclusions.

Key words: Urbanization, urban heat island, air pollution, fog physics, fog chemistry.

1. Introduction

Climate change is the result of discharging more and more anthropogenic greenhouse gases into the air and has been one of the greatest challenges in the world. Since the starting of the "Reform and Open to the Outside World" policy in China, cities have experienced explosive growth. Changes in urban eco-environment also caused changes in local climate, e.g. the change in the fog features in and around cities, including not only the shortening of visible distance in fog, which poses a bigger threat to all kinds of transportation, but also abundant pollutants dissolved into fog water, which are extremely harmful to human health. Therefore, observational studies of fog events are receiving more and more attention of meteorologists.

The study on fog in Europe began in 1917 (Taylor, 1917). In the early 1970s, Roach et al. (1976) and Brown and Roach (1976) made field observations in the Cardington of England; Pilie et al. (1975) made field observations in Chemung River Valley near Elmira, N.Y. They also did modeling research on fog. From the end of the 1970s an in situ radiative fog observational project was implemented for several years in Albany, N.Y.(Jiusto and Lala, 1983). These efforts made clear the basic physics of fog genesis and dissipation and its main influencing factors. Later, extensive observational studies were performed on fog microphysical structure and its evolution in the Po Valley in Italy in 1989 and 1994 (Noone et al., 1992; Wendisch et al., 1998). More recently, fog observational studies, together with numerical simulations, were conducted in various areas of the world (Ma et al., 2003; Fahey et al., 2005; Klemm et al., 2005; Eugster et al., 2006; Gultepe and Milbrandt, 2007, Gultepe et al., 2009). However, little is reported about comparative studies

---
[1]   School of Atmospheric Physics, Nanjing University of Information Science and Technology, Nanjing 210044, China. E-mail: lizihua1936@sina.com
[2]   Anhui Institute of Meteorological Sciences, Hefei 230031, China.
[3]   Jiangsu Meteorological Bureau, Nanjing 210008, China.

of the physicochemical features of fog in different decades.

Since 1958 Chinese scientists have devoted themselves to experiments on artificial fog dispersal and to studies of fog physics (LI, 2001). Since the 1980s the important projects performed in China have included the studies of fog at the Shuangliu Airport, Chengdu (GUO et al., 1989), sea fog around the Zhoushan Islands (YANG et al., 1989), radiation fog in the Xishuangbanna region (HUANG et al., 1992, LI et al., 1999a), urban fog in Shanghai (GUO et al., 1990), fog hazards in the urban Chongqing (LI et al., 1993), fog along the Shanghai-Nanjing Highway (LI et al., 1999b), fog in Nanjing (LI et al., 1999c; LI U et al., 2010) and fog in the Nanling mountains, Guangdong (WU et al., 2004). Most of these projects were carried out by means of advanced equipment, including multiple items, revealing the physico-chemical properties of urban fog in China and basically clarifying the genesis/dissipation physics and principal affecting factors. Furthermore, field observational experiments had been conducted on such events in different decades for comparative purposes over Shanghai, Chongqing, Nanjing and Xishuangbana.

The authors participated in several of the above-mentioned Chinese projects and did comparative research into the measurements in different historical periods, discovering that with urban development the fog structure and genesis/lysis processes varied greatly, and the change in the ecological environment has led to change in fog features. And the latter, in turn, affects the former. This work introduces briefly the changes in Chinese urban areas after 1978, with focus on the fog effects of the changes, which have been demonstrated via numerical simulation. The next section will introduce the consequences including urban heat island (UHI), air pollution, urban dry island, and urban wind field from the urban expansion. Fog impacts of urban development will be given in Sect. 3, including the decrease in the annual fog days in urban area, increase in the number density of droplets and decrease in FWC, and the increase of ion concentrations and acidity in fog water. Section 4 will present the confirmation of the above fog impacts by numerical simulations, including urban expansion and

its UHI, aerosol particles, vegetation effects. Summary and conclusions are given in Sect. 5.

2. Consequences from the Urban Expansion

Since 1978 great changes have happened to cities in China, e.g. the tremendous growth of urban sprawl, population, vehicles and economy—the growth tends to be a few times or even tens of times the original. Take some cities as examples: for Chongqing, an important industrial city in southwest China, its urban area was 30 km^2 with a population of 300,000 in 1949; the urban area expanded to 41.5 km^2 and its population reached 420,000 in 1978; since 1978 the city has gone through rapid development, and at present the urban area has reached 1,435 km^2 with 6,710,000 inhabitants plus 4,420,000 persons having temporary residence permit. For Hefei, the capital of Anhui province, it had an urban area solely of 5 km^2 with a population of ~60,000 in 1949; the area (population) increased to 56 km^2 (770,000) in 1980; the urban area reached 225 km^2, with its population increased to 1,300,000 in 2005. As for Chengdu, the capital of Sichuan province, it entered a stage of rapid development, starting from 1978, and particularly after 1990 its urban area experienced explosive growth, reaching 382.5 km^2 in 2003 from 74.4 km^2 in 1990. Not only have the provincial capital cities and cities directly under the central government, but also county-level cities and towns experienced rapid development, the example being the county-level city of Bozhou in Anhui. Its urban area was 5.5 km^2 and the population was 62,000 in 1978, but in 2004 the area exceeded 30 km^2 with a population of approximately 300,000. Another striking sample is the town of Jinghong in the Xishuangbana region, Yunnan. It was a countryside town with a population below 1,000 in 1952, with only six good-looking storey houses made of tiles and bricks, surrounded by virgin forest. Since 1978 the town has undergone great change, having a population of 477,000, with the urban area expanding to 83.6 km^2 and many buildings in 2007. It is now a medium-scale city.

Some factors in relation to fog features have experienced great changes as a result of the expansion of urban area, including:

(1) The effect of the urban heat island (UHI) gets enhanced, causing air temperature to rise appreciably, especially at night. Take the city of Hefei for example. Since the decade of 1976–1985 to 1996–2005, the average lowest temperature has risen from 11.96 to 13.06°C (See Table 1 in SHI *et al.*, 2008), with increasing difference in minimum temperature between cities and its neighboring towns in Anhui, showing enhanced UHI at night (See Fig. 8 in SHI *et al.*, 2008). Similarly, comparison of air temperatures in Chengdu and its neighboring town (Pengzhou) yields the yearly difference of 0.2–0.5°C during 1960–1979, and 0.8–1.1°C during 1980–2003 (HAO *et al.*, 2007). The phenomenon of increasing UHI is very prevalent in China, which is believed to inhibit fog formation.

(2) Air pollution becomes increasingly serious. Take the Chongqing for example; since 1978, the city has grown explosively, leading to ever-increasing air contamination and multiple air pollution indexes (API) exceeding the third-class national standard. Sulfur dioxides (SO_2) concentration and the number density of aerosol particles are steadily decreasing outward from the urban center, with the values in the urban center being more than two times those of the urban fringe (LI *et al.*, 1996). The increase in the number concentration of aerosols has its consequences on fog as follows: (1) the particles absorb surface long-wave radiation, thereby reducing radiative cooling at night promoting conditions which are less favorable for fog production (LI, 2001; LIU *et al.*, 2005) and (2) as condensation nuclei, aerosols impact fog microphysical structure (OKE, 1987) and the ion concentrations of fog water (LI *et al.*, 1996; PU *et al.*, 2008), especially those aerosols consisting dominantly of sulfate with diameter less than 3 μm. These aerosols are also responsible for an increase in acidity of fog water. The development of factories, *inter alia* plants, mines and busy traffic causes the increase of gaseous pollutants of SO_2 and NO_x. This, when absorbed by fog water, results in the growth of electric conductivity and ion density.

(3) Urban dry island effect becomes prominent. This is the synthesized result of decreasing surface evaporation and increasing UHI. With the growth of urban areas and increasing population, the high-density houses and cement roads increased appreciably, while the green grass area inside the city decreased. The decrease of grass land reduces evaporation, and thereby diminishes relative humidity. Again, we take Chengdu for example. Starting from the end of 1980, the annual mean relative humidity was decreasing from year to year and the value dropped from 84% in 1990 to 75% in 2003, indicating that the dry island effect is on the increase (HAO *et al.*, 2007). Similarly, in Hefei, the average RH at 08LST decreased from 84.1% in 1976–1985 to 82.7% in 1996–2005 (SHI *et al.*, 2008).

3. Fog Impacts of Urban Development

After development of a city, changes in the above elements exert great effects on fog features. The development results directly in (1) diminution of annual fog days, lowering of fog water content (FWC), decrease in droplet size and increase in the number density of fog droplets, and thus reducing visibility in fog, and (2) an increase in the ion concentration and acidity of fog water. Evidently, changes in urban eco-environment lead to variation of fog features, and the latter, in turn, influences the former.

3.1. Decrease in the Annual Fog Days in the Urban Area

Chongqing was famous for abundant fog and thus was called a city of fog. Statistics show that the

Table 1

Interdecadal variation of average annual fog water deposition (mm) at Jinghong

Decade	1954–1959	1960–1969	1970–1979	1980–1989	1990–1995
Deposition	17.3	10.4	4.4	0.9	0.2

annual fog days have reduced in Chongqing during the past 60 years. The annual fog days were 100–145 in the 1940s and 1950s, began gradually declining from the late 1950s, became almost constant in the 1970s and began to decline again in the mid-1980s. Entering the new century, the number of annual fog days was decreasing sharply to about 24 days (Fig. 1). However, for its suburb station, Beipei, and the neighboring city, Fuling, for example, the number was on a steady increase in the 1950s and increased rather appreciably after the 1970s. This indicates that the decrease in the number of fog days in Chongqing can not be attributed to the change of a large-scale climate pattern but to the development of urban areas and population, together with the increase of mines and factories that discharge large amounts of air pollutants and reinforce UHI.

Also, Fig. 1 shows that there is a sharp drop in the annual average visibility, starting from the early 1960s, with the values being 8.2–11.8 (4.9–6.5) km in the 1960s (1980s) and shortened by 1.48 km in 1990 compared to the figure of 1980. And in recent years the annual mean was about 5 km. The interannual decline trend is associated with the explosive growth of industries in the city leading to ever-increasing air pollutants.

The annual fog days, decreasing with socioeconomic prosperity, occurred not only in metropolises but also in medium- and small-scale towns. Shi et al. (2008) studied the changes of annual fog days in two classes of cities in Anhui Province. One class is for cities established earlier, e.g., Hefei, Wuhu and Huangshan, and the other class is for cites set up relatively later, e.g., Bozhou, Chizhou and Chaohu. To remove the impacts of large-scale climate change, time series of differences in climate elements between urban sites and their neighboring town sites were used. Figure 2 gives the evolutions of the differences in the annual fog days between the two-class cities and their neighboring towns, indicating that for the first class the general trend was on the decrease after 1980, similar to the case reported by (SACHWEH and KOEPKE, 1995) from their study in Germany. This illustrates that the drop in annual fog days was caused by the expansion of urban area; however, for the second class the annual fog days were on the increase from the 1970s or 1980s, but after 1995 the number was decreasing markedly. The second class of cities were at town-level and became medium-level cities only after the 1990s; their economy developed slowly before that and rapidly thereafter leading to the expansion of urban area and explosive growth of population. It follows that the number was gradually increasing when the economy was under slow development of the second class but the number was inevitably on the decrease when the urban area has expanded greatly.

The decrease of annual number of fog days occurs not just in a few cities. In their work, WANG et al. (2005) constructed the time-varying trend of the number by means of the first- and second-order principal value functions fitted using the 1961–2000 data from 604 stations nationwide. Results show that for the number on a 40-year scale, it is marked by the extremely distinct second-order trend, significant at the 0.001 level, i.e., the number is on the increase (decrease) before (after) 1981.

To sum up, the decline of annual fog days was related to the expansion of urban areas. With urban development underway the urban heat island (UHI) and air pollution are being enhanced, especially the increase in the number concentration of aerosols, which absorb surface long-wave radiation to diminish radiative cooling at the surface. In addition, with urban development, the urban dry island became prominent, which was a result of the decrease of vegetation coverage and the increase of temperature at night.

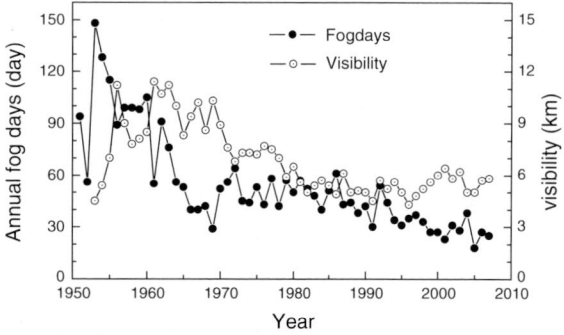

Figure 1
The interannual variation in visibility and annual fog days between 1950 and 2010 for the city of Chongqing

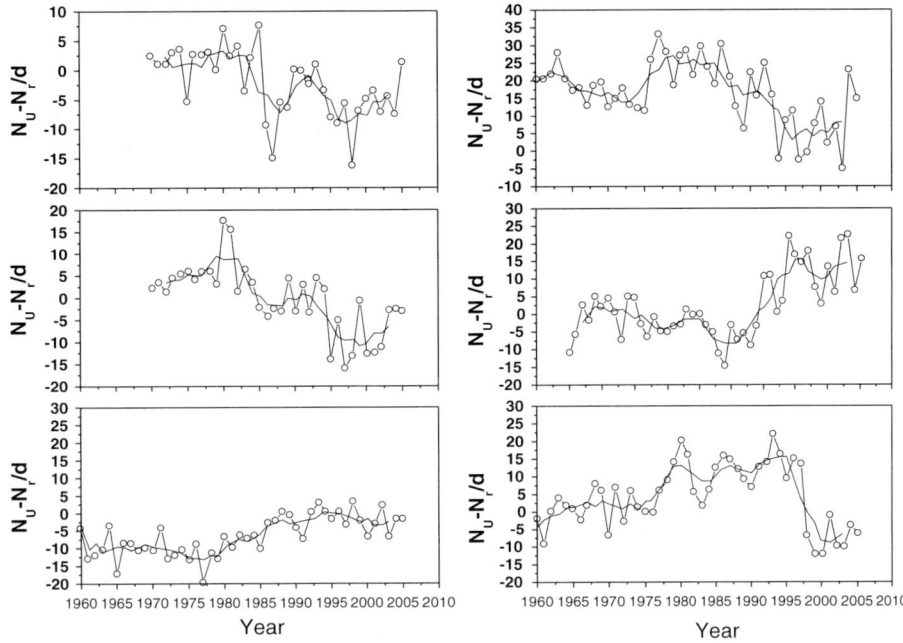

Figure 2

Difference in annual number of fog days between urban site and its neighboring town sites *Solid line* designates a 5-year moving mean and *dashed line* is interannual difference in the number of fog days, with **a** for Hefei with Feidong and Feixi; **b** Huangshan with Shexian and Xiuning; **c** Wuhu with Dangtu and Wuhu county; **d** Chizhou with Qingyang and Zongyang; **e** Chaohu with Wuwei and Hexian; **f** Bozhou and Guoyang

3.2. Increase in the Number Density of Droplets and Decrease in FWC

The field observations of the fog microstructure in Chongqing in the winter of 1989–1990 (Li and Wu, 1995; Li, 2001) show that from the urban fringe to center the number density of fog droplets was increasingly bigger as opposed to the fog droplet size and FWC that became progressively smaller. In the urban center, the average fog droplet number concentrations exceeded 500 cm^{-3}, while the average diameter was only 4.4 μm, with the FWC below 0.1 g/m^3. The FWC and fog droplet size was apparently lower than those measured in Canada (GULTEPE and MILBRANDT, 2007, GULTEPE *et al.*, 2009), but the fog number concentration was nearly two times of that in Canada. The principal factors affecting the microphysical structure of fog in Chongqing are the serious air pollution and strong UHI. Because of heavy contamination there are more condensation nuclei in the air, leading to a larger number density of droplets to compete for vapor. Furthermore, the high-density aerosols above the city

reduced the radiative cooling by weakening the long-wave radiation at the surface and enhancing the inverse radiation at night. Owing to intensified UHI the night temperature is higher by at least 2°C in the center than in the urban fringes. Therefore, the weak cooling leads to lower super-saturation of air, at the same time, more small droplets compete for vapor, preventing droplets from growth via condensation, so that droplets are fine and FWC is small.

Analysis of observations made in different decades in the same region reveals the effects of urban development upon fog microstructure. Since the 1960s, three field observations of fog had been made at Jinghong of the Xishuangbana, Yunnan province, using the same instruments. During the 1968/69 winter, a dense fog, accompanied by drizzle, was observed. The average FWC during the extremely dense fog episode was 0.74 g/m^3, with the droplet diameter at 28.6 μm and the fog droplet number density of 34.9 cm^{-3}, on average; however, the corresponding figures being 0.08 g/m^3, 6.8 μm and 153 cm^{-3} on average, for a dense fog observed in the 1986/87 winter, indicate that in less than 20 years the FWC dropped by about one order of magnitude and the number concentration rose almost fivefold. The average FWC of an extremely dense fog episode during the 1968/69 winter was much higher than the high occurrence frequency FWC measured during continental fog in eastern Canada by GULTEPE et al. (2009) (Fig. 17a in GULTEPE et al., 2009); however, it was comparable to those maxima in a fog case in GULTEPE et al. (2009) (Fig. 7c in GULTEPE et al., 2009). The average FWC during an extremely dense fog episode in the 1986/87 winter was comparable to those high occurrence frequency FWC in eastern Cadana (Fig. 17a in GULTEPE et al., 2009). The third project of fog droplet size distribution measurement was implemented on 23–30 November 1997. Four fog events occurred during that period; however, the FWC was too small to measure by the droplet collector. So the liquid spot absorber was utilized to measure the FWC instead. Evidence suggests that the maximal (minimal) FWC was 0.061 (0.008) g/m^3— much lower than those procured in the 1986/87 winter (LI et al., 2008). As shown in the interannual variation of fog water deposition (including minute quantity of dew), which was recorded by a rainfall recorder at Jinghong, the FWC was on the decrease. Table 1 shows that the annual mean fog water deposition was as high as 17.3 mm in the 1950s, dropped to 4.4 mm in the 1970s and less than 1 mm in the 1980s. Figure 3 depicts the interannual variation in fog water deposition for January and November, indicating its sharp diminution to nearly

zero after 1978. Now, the fog water deposition is too small to measure by routine instruments.

3.3. Increase of Ion Concentrations and Acidity in Fog Water

Urban development would lead not only to the drop of the number of fog days and FWC but also to changes in its chemical composition.

Based on the chemical constituents in fog water of Chongqing, we conducted a comparison of the ion concentrations between urban and rural areas. Results show that the concentrations are in a tendency of increase from far suburb to near suburb to urban fringe and to the urban center (Table 2), with pH value decreasing from the far suburb to the urban area (Fig. 4a). The total ion concentration is maximal in the urban center, reaching 5.5×10^4 μmol/L, on average, with the maximum about 1% of fog water weight, and the acidity (in the urban area) is the highest, with the mean pH value reaching 4.0. From Fig. 4b, we can see that the pattern of SO_2 concentration is extremely analogous to that of acidity, decreasing outward from the urban center. As indicated earlier, the number density of aerosols decreases outward from the urban center. This similarity shows SO_2 and aerosols to be the background for acid fog occurrence. Also, it indicates the dependence of high-density ion concentrations upon serious air pollution in the urban center.

The relationship between air pollutants and ion concentrations in fog can be seen more directly from the evolution of ion concentrations and pollutant

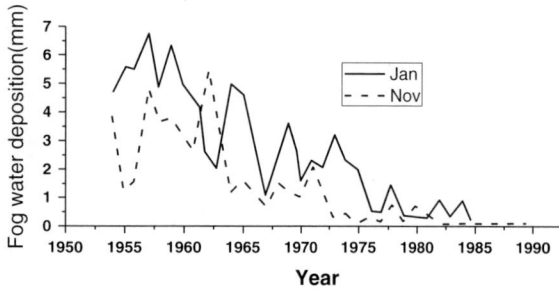

Figure 3
Interannual variation in fog water deposition (mm) in January (*solid*) and November (*dashed line*) for the Jinghong city, Xishuangbana, Yunnan Province in 1954–1996 (from GONG and LING, 1996)

Table 2

Major ion concentrations (μmol/L) and their total in fog water of Chongqing

Ions	SO_4^{2-}	NO_3^-	NH_4^+	Ca^{2+}	$\sum (+, -)$
Stations					
City center (stadium)	11444.8	1382.3	10750.0	5612.5	55364.7
City fringe (Shapingba)	9198.4	734.5	4684.2	4226.2	45766.2
Suburb (Guangyangba)	7606.8	633.9	5180.6	1343.8	29080.6
Far suburb (Simianshan)	2640.9	116.9	540.0	363.2	9710.1

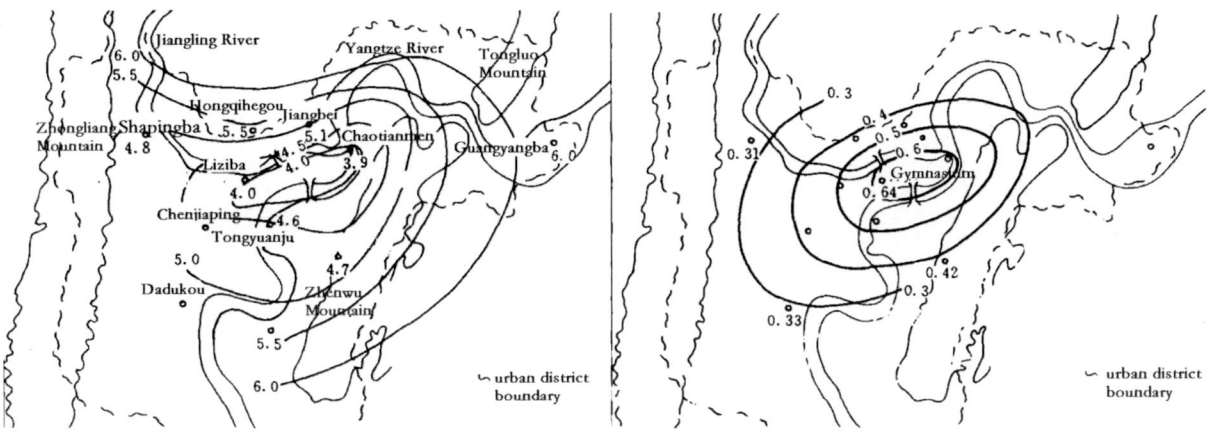

Figure 4
The pattern of fog water pH values in (**a**) and SO_2 concentrations (mg/m^3) in (**b**) for Chongqing and its vicinity. (Li *et al.*, 1996)

densities in a fog event. On 25–27 December 2006 an advection—radiation fog event occurred in Nanjing. During this fog event, 13 fog water samples were collected on a continuous basis, together with measurements of the concentrations of SO_2, NO_2 and PM_{10}. Figure 5 presents the time-dependent ions and pollutants concentrations in the air (Pu *et al.*, 2008). It can be seen that the concentrations decrease from the late evening, reaching their minima from midnight to dawn, and increase after sunrise, maximizing after noon hours. This pattern relates to anthropogenic emissions. For example, during the daytime, human activities result in an increase in anthropogenic emissions, especially in vehicle exhausts and industrial emissions, which accumulate under the stable stratification of fog weather; during the night, human activities decrease and so do the anthropogenic emissions. Comparing the evolution of the ions concentrations of fog water with the air pollutants concentrations (PM_{10}, SO_2 and NO_2), we can find that the two trends are in broad agreement, for example, the peak concentrations of SO_2 and NO_2 between 1300 and 1900 BST, December 25 corresponds to the peak ions concentrations of SO_4 and NO_3 between 1430 and 2300 BST, the same day. Then, all decline and begin to rise in the morning of day 26. This indicates that the ions in fog water are mainly from the air pollutants in the near-surface layer. Again, detailed examination of Fig. 5 shows that the peaks of air pollutants occur a little ahead of those of ions concentrations due to the fact that it

takes some time for the pollutants to be dissolved in fog water.

In conclusion, the city has grown explosively, along with the development of industry, transportation, and house construction, leading to serious air pollution, e.g. increase of gaseous pollutants of SO_2 and NO_x and aerosol particles. When absorbed by fog water droplets, those pollutants result in an increase of electric conductivity and ion density of fog water. This would have adverse effects on human health.

4. Numerical Simulations of Fog Effects of Different Impacting Factors

To demonstrate the fog effects of urban development, a series of sensitive numerical experiments have been conducted with a 3D fog model (Shi *et al.*, 1996a), with particular focus on the impacts of UHI effect, air contamination and changes in vegetation features upon fog genesis and development after urbanization (Li *et al.*, 1997; Shi *et al.*, 2001; Huang *et al.*, 2000).

The 3D fog model consists of a set of equations of horizontal wind field (u, v), the potential temperature (θ), the specific humidity (q), the mixing ratio of liquid water (C), continuity, static equilibrium, heat transfer of soil and energy budget on the Earth's surface. The long-wave radiation calculations are done with the five-wave-range model put forward by Roach and Slingo (1979). The short-wave radiation

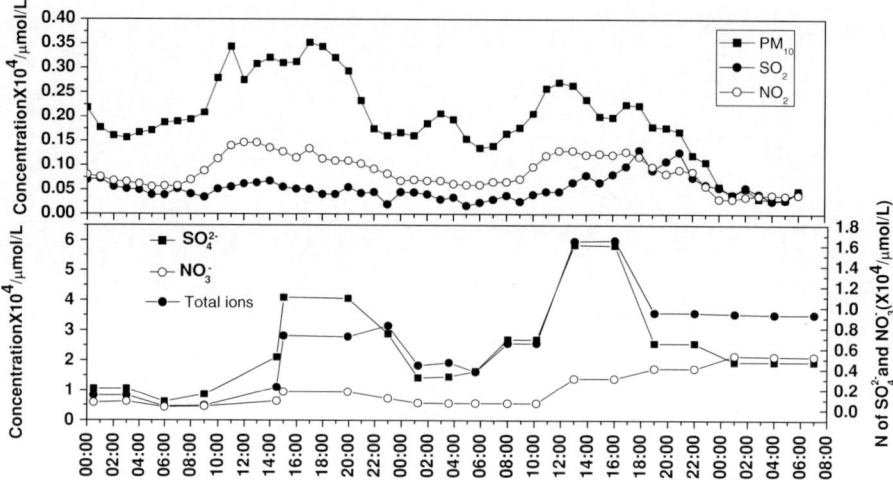

Figure 5
Evolution of air pollutants (SO$_2$, NO$_2$, PM$_{10}$) in (**a**) versus that of ion concentrations (total ion concentration, SO$_4^{2-}$, NO$_3^-$), in (**b**) in the fog event at Nanjing on December 25–27, 2006. (Pu *et al.*, 2008)

is calculated by the empirical formula of MAHRER and PIELKE (1977) and Oliver *et al.* (1978). To consider the effect of complex terrain, both the terrain following coordinate system (TFCS) and the local Cartesian coordinate system (LCCS) are used. The model domain consisted of 22 × 17 grids spanning the entire urban Chongqing in 1990s, with a horizontal resolution of 1 km. In the vertical, the PBL was divided into 19 levels in the TFCS, with the heights being from surface to 1,500 m. The underlying surface is classified into four kinds: urban, farmland, forest, water, which are described by different parameters, e.g. roughness, humidity of soil, heat conductive coefficient, artificial heat sources, etc. (SHI *et al.*, 1996a). Most parameters in the model were adopted from Pielke (1984).

A field experiment for fog was conducted between 15 Dec 1989 and 15 Jan 1990 in Chongqing. Thirteen meteorological observation stations and three fog microphysics observation sites were set up for ground-level observations. Four sounding sites for atmospheric boundary layer measurement were set up using the Automatic Data Acquisition System (ADAS, AIR company, America). Based on the observational data, local circulation and fog evolution in Chongqing was analyzed and simulated (LI *et al.*, 1994, 1997; SHI *et al.*, 1996b). The simulated results were basically consistent with the observations,

demonstrating that the model was capable of local fog simulation.

4.1. Simulation of the Fog Effects of Urban Expansion and Its UHI

In the default run, the simulated FWC could reach 0.3–0.4 g/kg in urban district at the mature stage. In the test run, the urban sprawl was expanded to four times the original size but the terrain was kept unchanged, the simulated FWC dropped greatly in the urban area to 0.1 g/kg in most segments, with the fog height descending ∼50 m (LI *et al.*, 1997). Evidently, after the expansion of the urban area, not only does the UHI intensify, but also the vegetative area decreases, which leads to evaporation decline, and hence drop in moisture in the air, so that the formation and development of fog are prevented.

To study the effect of UHI on fog, the UHI was excluded by assuming the anthropogenic heat sources in the urban center identical to those at the suburbs. The results show that the urban liquid water mixing ratio would be doubled, while the fog top is almost unchanged, suggesting that the Chongqing fog would be even heavier without UHI and updraft would be weakened over the urban area, the Yangtze and Jialing Rivers. These indicate that the fog impacts of UHI are twofold: (1) the UHI circulation can enhance

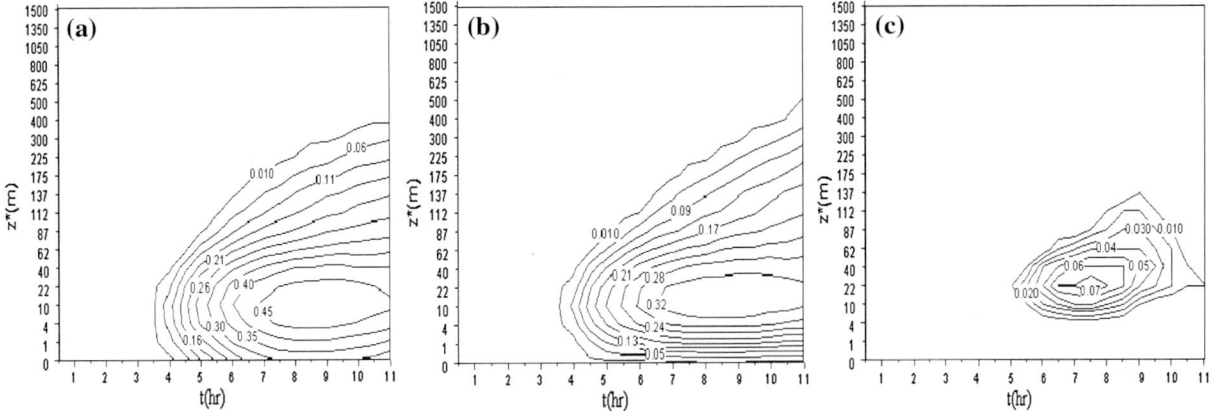

Figure 6

The time-height cross-section of FWC (g/kg) in the urban center of Chongqing from the experiments on **a** no aerosols, **b** the size spectrum n (D, z) and **c** the spectrum n (D, z) × 5. (Fig. 2 in Shi *et al.*, 2001)

the air convergence over the urban center and the Yangtze River to carry vapor and fog droplets to the higher level from low levels in favor of fog development in the vertical and (2) higher temperature in the urban area prevents vapor condensation, decreasing FWC at surface.

4.2. Sensitivity Study of Fog Effects of Aerosol Particles

Recently, aerosols are appreciably augmented due to urban expansion and economic prosperity. As mentioned in Sect. 3, aerosols impact fog in three aspects: the microphysical effect, the chemical effect and the radiative effect. In this sub-section, focus is on simulating the fog impacts of radiation effect produced by aerosols.

To investigate the impacts of the radiation effect of aerosol on radiation fog, the radiative effect of aerosols was added to the 3D fog model aforementioned (SHI *et al.*, 2001) by using the wide range model of RODGERS and WALSHAW (1966). The size spectrum is the one derived in the field experiment in Chongqing and takes the form:

$$n(D, z) = 2.866 D^{-2.95} \exp(-z/H_p) \qquad (1)$$

where, H_p is the atmospheric scale height and is set to 1,200 m and D is the diameter of the particle. The aerosols are of radius in the range of 0.01–10.0 μm. According to chemical constituents of the Chongqing aerosols, the particles are separated into two types in

the model processing, with 60% (40%) for hygroscopic (dry dust) particles. Three experiments are conducted by changing the number density of aerosol, with results given in Fig. 6.

Figure 6 shows the time-height cross-section of FWC at the urban center of Chongqing under three conditions. It can be seen that when the aerosol-related radiative effect is considered, the FWC reduces noticeably from the spectrum n (D, z), with the maxima dropping by ∼1/3 and the time of surface fog occurrence delayed approximately 1 h, when the number concentration is increased fivefold, no surface fog is observed except a low cloud that contains little water and when the number density of aerosols further increases to tenfold, neither fog nor low cloud appears. This demonstrates that radiative effect of aerosol did prevent fog formation and development.

4.3. Simulative Research into Vegetation Effects upon Fog

In recent years the vegetation area was on the decrease as the city was under development, which has a prominent impact on fog genesis and development. For instance, Xishuangbana is a fog-famous region of tropical forests, and its capital, Jinghong, had an annual mean of 166 fog days in the 1950s, with the number decreasing progressively from the end of the 1970s when the urban area started expanding, leading to the diminution of rain-fed forests. The annual mean of fog days was only 58 days in 1990–1995, together with great drop of FWC. Changes in the ecological environment and the sharp decrease in tropical forests in particular are likely to be the non-negligible causes of the continually decreasing fog days at Jinghong.

Forested areas play an important role in fog development. Plants take in water through roots and lose it via evapotranspiration, with evaporation being as strong as ~ 20 times as that from bare surface, equivalent to the evaporation from an equal-size reservoir (ZHOU and SU, 1994). Tree evapotranspiration humidifies air and reduces temperature. Because of the small thermal capacity of vegetation, lower temperature can be observed in the tree canopy than on the bare ground in the night. The increase in moisture and decrease in temperature are favorable conditions of fog genesis.

To reproduce effect on fog of vegetation a vegetation parameterization scheme is added to the 3D fog model (HUANG et al., 2000). In the scheme, with vegetation available, the leaf area is assumed to be connected to the surface directly, and we are allowed to use the leaf-surface temperature in the canopy as the underlying surface temperature.

Plant influencing fog features are associated mainly with vegetation coverage (σ_f), plant height (h) or surface roughness (Z_0), leaf area index (LAI) and leaf reflectivity (α_f). These vegetation parameters used in the model are given in Table 3.

The simulated region includes the Jinghong city and tropical rain-fed forests around.

In the simulation, σ_f is taken to be 0.95 and 0.20, the former corresponding to the vegetation state of Jinghong in the 1950s. The result shows that the bigger vegetation cover corresponds to earlier (later) fog genesis (dispersal), larger liquid water content and higher fog top.

To further explore the fog impacts of vegetation, we calculated the radiative cooling rate at the vegetation top for different vegetation coverage σ_f. The top radiative cooling rate reaches $-1.7°C/h$ for $\sigma_f = 0.95$, while it is only $-1.0°C/h$ for $\sigma_f = 0.20$. The minimal temperature on the top leaf surface is lower by 3°C at $\sigma_f = 0.95$ than at $\sigma_f = 0.20$. This demonstrates that the strong radiative cooling at the top of trees favors fog genesis, causes fog to form earlier and last longer.

The simulations agree with reality. The Jinghong city has made great progress economically since 1978, with greatly reduced vegetation cover. The Mengyang town, 13 km from Jinghong, keeps a practically unchanged tropical forest area as before. Jinghong has conspicuous rise in annual mean temperature from 1978, with 0.7°C increase between 1978 and 1995, compared to no distinct change in Mengyang. The winter-half-yearly mean humidity at 2000 BST, which is associated directly with fog genesis, dropped at Jinghong from 85% in the 1950s to 73% in the 1990s, as compared to almost no variation at Mengyang, where the 10-year mean humidity was 81% in the 1960s, decreasing to 80% in the 1990s. Therefore, the number of annual fog days at Jinghong is on the decrease from year to year but that of Mengyang, although fluctuating, experiences no big variation. As shown in Fig. 7, the numbers are comparable for both areas in the early 1960s, but in the 1990s the number is ~ 60 days at Jinghong in comparison to >100 days for Mengyang. No doubt, the Jinghong fog trend bears a relation to the urban development and vegetation diminution.

Except for the annual number of fog days differing in the two areas, the genesis and dispersal times also show discrepancy, the fog genesis (dispersal) is 3 h later (1 h earlier) at Jinghong than at Mengyang, thus the Jinghong fog duration being about half of that at Mengyang, with FWC

Table 3

Vegetation parameters

Land use type	Coverage σ_f	Reflectivity	Emissivity	LAI	Plant height (m)
Paddy field	0.50	0.20	0.95	4	0.5
Tropical forest	0.95	0.16	0.97	7	10
Rubber tree	0.95	0.10	0.95	5	12

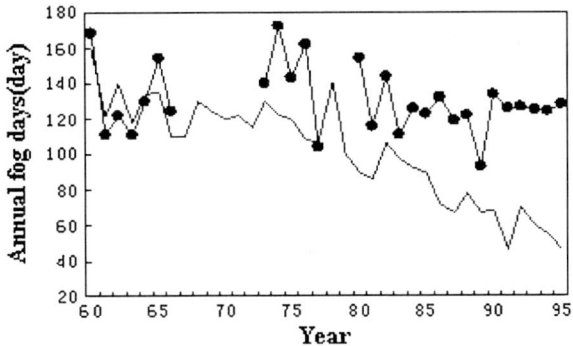

Figure 7
Interannual variation in annual fog days for Jinghong (*solid*) and Mengyang (*solid dot chain*) from 1960 to 1995 (HUANG *et al.*, 2000)

significantly lower at Jinghong. Evidently, decrease in vegetation cover has effect on fog genesis and dispersal.

5. Conclusions and Discussions

Associated with explosive urban development, are intensified thermal island effect and lessened vegetation cover, which leads to evaporation decline, and hence drop in moisture in the air, so that the formation and development of fog are prevented. In recent years, the vegetation area was on the decrease as most cities were under development. Tree evapotranspiration humidifies air and decreases temperature, which has prominent impacts on fog genesis and development.

The rapid urban development also results in pronounced increase in air pollutants (increase of SO_2, NO_2, aerosol particles, and other pollutants). In recent years, aerosols generally increased obviously due to urban development, which can be reflected by the decrease of visibility (SHI *et al.*, 2008). The fog impacts of aerosols include: (1) the microphysical effect. Aerosols can serve as condensation nuclei, so the increasing aerosols will lead to larger fog droplet number density, thereby altering the microphysical structure of fog; (2) radiative effect. The increasing aerosols will definitely change the environmental temperature, thus affecting the genesis/lysis physics of fog.

Numerous observations show that these changes in urban area exert great influence on fog features, including:

(1) Decline of annual fog days in cities, especially for Chongqing and Jinghong in Yunnan province: in the city of Chongqing, named city of fog in China, for example, the annual fog days were 100–145 in the 1950s but only 20–30 after 2000; In Jinghong, the annual fog days was more than 100 before 1978 but just about 60 in recent years.

(2) Decrease in fog water content (FWC) and fog droplet size as well as increase in fog droplet number concentration: take Jinghong in Yunnan province for example, during the 1968/69 winter the FWC during an extremely dense fog episode, accompanied by drizzle was 0.74 g/m^3, with the droplet diameter at 28.6 μm and the number density of 34.9 cm^{-3}, on average, with corresponding figures being 0.08 gm^{-3}, 6.8 μm and 153 cm^{-3}, for a dense fog observed in the 1986/87 winter.

(3) The annual mean fog water deposition measured by rainfall gauge in Jinghong observation station was as high as 17.3 mm in the 1950s, dropped to 4.4 mm in the 1970s and was less than 1 mm in the 1980s, but unfortunately, it was too small to be recorded in recent years.

(4) The visibility dropped sharply in large cities, e.g. Chongqing, with the values being of 8.2–11.8 (4.9–6.5) km averaged over the 1960s (1980s) and the annual mean was about 5 km in recent years.

(5) Meanwhile, the ion concentrations and acidity in urban areas were in great increase: the average total ion concentration in the center area of Chongqing is 5.5 × 10^4 μmol/L, and mean pH value is 4.0, but 9.7 × 10^3 μmol/L and greater than 5.5 in far suburb.

Utilizing a 3D fog model over complex terrain, sensitive simulations were conducted on the issues resulted by the city development:

(1) Firstly, the model was run with a bigger urban area. The results show that the fog formation is lighter, with lower FWC at the ground level than the default run.

(2) Secondly, numerical study was conducted for effects of aerosols on fog. The simulation focus on simulating fog impacts of radiation effect produced by aerosols. The results suggested that the FWC declined with the increase of aerosols. When the number concentration of aerosols was increased fivefold, no surface fog was observed. This suggests

that the radiation effect of increasing aerosol is one of the important reasons which led to the decrease of urban fog. However, we still do not know how the increase of aerosols influence the fog microphysics. To study the microphysics effects of aerosol on fog, detailed microphysics needs to be added into the model. We may do this in the future.

(3) Thirdly, numerical experiments were done with the fog effects of vegetation. The results show that the bigger vegetation cover corresponds to earlier (later) fog genesis (dispersal), and larger liquid water content. The simulations agree with reality of Jinghong city. Jinghong city has made great progress economically since 1978, with greatly reduced vegetation cover. With the urban development, Jinghong has a conspicuous rise in annual mean temperature, decrease in RH, and evident decrease in the number of annual fog days.

To sum up, we summarized some climate changes of fog characters in China and conducted some sensitive simulations to test the effects of urbanization on fog. However, there are still some related issues need to make clear. For example, it was shown by observations that the gases pollutants (e.g., SO_2, NO_2) have an acid effect and Ca^{2+} has a neutralizing effect, while we still do not know how these pollutants enter into the fog water, what the chemical process is, and how they influence the acidity of fog water. To make clear these questions, it is necessary to integrate the fog microphysics model and chemical model to do simulation on these processes.

Acknowledgments

This work is supported jointly by the National Natural Science Foundation of China (40975085, 40775010), the Scientific Project for Public Welfare specific to Meteorologists (GYHY200906012).

REFERENCES

BROWN, R. and ROACH, W. T. (1976), *The physics of radiation fog. II: A numerical study*, Quart. J. Roy. Meteor. Soc *102*(432), 335–354.

EUGSTER, W., BURKARD, R., HOLWERDA, F., SCATENA, F.N., and BRUIJNZEEL L.A. (2006), *Characteristics of fog and fog water fluxes in a Puerto Rican elfin cloud forest*, Agricultural and Forest Meteorology *139*, 288–306.

FAHEY, K. M., PANDIS, S.N., COLLETT, J.L., and HERCKES, P. (2005), T*he influence of size-dependent droplet composition on pollutant processing by fogs*, Atmospheric Environment 39, 4561–4574.

GONG, S.X. and LING, S.H. (1996), *Fog decreasing in Xishuangbana region*, Meteorology 22(11), 10–14 (in Chinese).

GULTEPE, I., and MILBRANDT, J. A. (2007), *Microphysical Observations and Mesoscale Model Simulation of a Warm Fog Case during FRAM Project*, Pure and Applied Geophysics *164*(6), 1161–1178.

GULTEPE, I., PEARSON, G., MILBRANDT, J. A., HANSEN, B., PLATNICK, S., TAYLOR, P., GORDON, M., OAKLEY, J. P., and COBER, S. (2009), *the Fog Remote Sensing and Modeling Field Project*, Bulletin of the American Meteorological Society 90,341–359.

GUO, E.M., YU, X.R., and LI, Y.H. (1989), *The macKroscopic structure of fog in the Shuangliu Airport, in the Symposium of National Cloud Physics and Weather Modification*. (Beijing: China Meteoro. Press, 1989) pp. 35–38 (in Chinese).

GUO, E.M., LIU, Y.G., and SU, J.X. (1990), *On the macroscopic structure of fog over the Huangpu River*. J. Beijing Inst. Meteor., *1*, 46–49 (in Chinese).

HAO, L.P., FANG, Z.F., LI, Z.L., LIU, Z.Q., and HE, J.H. (2007), *The inter-annual climate change and heat island effect of Chengdu during the recent fifty years*. Meteorological Science, *27*(6), 648–654 (in Chinese).

HUANG, J. P., LI, Z. H., HUANG, Y. R., HUANG, Y.S. (2000), *A three-dimensional model study of complex terrain fog*, Chinese J. Atmos. Sci. *24*(3), 219–233.(in Chinese).

HUANG, Y.R., XU, W.R., LI, Z.H., FAN, L., and HUANG, W.J. (1992), *An observation and analysis on the radiation fog in Xishuangbana*, Acta Meteor. Sinica *50*(1), 112–117 (in Chinese).

JIUSTO, J.E. and LALA, G.G. (1983), Radiation fog field programs: Recent studies. ASRC-SUNY Pub., No. 869.

KLEMM, O., WRZESINSKY, T., and SCHEER, C. (2005), *Fog water flux at a canopy top: Rirect measurement versus one-dimensional model*, Atmospheric Environment 39, 5375–5386.

LI, Z.H. (2001), *Studies of fog in China over the past 40 years*, Acta Meteor. Sinica 59(5), 616–624

LI, Z.H., ZHANG, L.M., and LOU, X.F. (1993), *The macro- and micro-structure of winter fog in the Chongqing urban district and the physical cause fog its formation*, Journal of Nanjing Institute of Meteorology *16*(1), 48–54.

LI, Z.H., SHI, C.E., CAO, B.M. (1994), *Study of winter fine-day local circulation structure in the urban area of Chongqing*, J. Nanjing Inst. Meteor. *17*(2), 232–237.

LI, Z.H., and WU, J. (1995), *Winter fog droplet spectrum features in urban area of Chongqing*, J. Nanjing Inst. Meteor. *18*(1), 46–51 (in Chinese).

LI, Z.H., DONG, S.N., and PENG, Z.G.(1996), *Spatial/temporal variabilities of chemical constituents fog water sampled in Chongqing*, J. Nanjing Inst. Meteor., *19*(1), 63–68 (in Chinese).

LI, Z.H., SHI, C.E., and LU, T.S. (1997), *3D model study on fog over complex terrain. Part II: Numerical experiment*. Acta Meteor. Sinica, *11*(1), 88–94.

LI, Z.H., HUANG, J.P., HUANG, Y.S., YANG, Z.Y., and WANG, Q. (1999a), *Study of the physical process of winter valley fog in the Xishuangbana region*. Acta Meteor. Sinica, *13*(4), 494–508.

LI, Z.H., HUANG, J.P., SUN, B.Y., and PENG, H. (1999b), *Burst characteristics during the development of radiation fog*, Chinese J. Atmos. Sci. 23(5), 623–631 (in Chinese).

LI, Z.H., HUANG, J.P., ZHOU, Y.Q., and ZHU, S.W. (1999c), *Physical structures of the five-day sustained fog around Nanjing in 1996,* Acta Meteor. Sinica *57*(5), 622–631 (in Chinese).

LI, Z.H. (2001), *Studies of fog in China over the past 40 years,* Acta Meteor. Sinica *59*(5), 616–624 (in Chinese).

LI, Z.H., YANG, J., SHI, C.E., and PU, M.J. (2008), Physics of Regional Heavy Fog Events. (Beijing: China Meteor. Press) (in Chinese).

LIU, D.Y, PU, M.J., YANG, J., ZHANG, G.Z., YAN, W.L., and LI, Z.H. (2010), *Microphysical Structure and Evolution of a Four-Day Persistent Fog Event in Nanjing in December 2006,* ACTA METEOROLOGICA SINICA *24*(1): 104–115.

LIU, X., ZHANG, H., LI, Q., ZHU Y. (2005), *Preliminary research on the climatic characteristics and change of fog in China.* Journal of Applied Meteorological Science (in Chinese), *16*(2),220–271.

MA, C. J., KASAHARA, M., and TOHNO, S. (2003), *Application of polymeric water absorbent film to the study of drop-size-resolved fog samples,* Atmospheric Environment, *37,* 3749–3756.

MAHRER, Y. and PIELKE, R.A. (1977), *The effect of topography on sea and land breezes in a two-dimension numerical model.* Mon. Wea. Rev. *105,* 1151–1162.

NOONE, K. J., OGREN, J. A., HALLBERG, A., HEINTZENBERG, J., and STROM, J. (1992), *Changes in aerosol-size and phase distribution due to physical and chemical processes in fog.* Tellus, *44*(5B), 489–504.

OKE, T.R., Boundary Layer Climates, second edn. (London: Methuen, 1987).

OLIVER. D.A. et al. (1978), *The interaction between turbulent and radiative transport in the development fog and low-level stratus,* J. Atmos. Sci. *35,* 301–316.

PIELKE, R.A. (1984), Mesoscale Meteorological Modeling, (New York: Academic Press, 1984).

PILIE, R.J., MACK, E.J., KOCMOND, W.C., ROGERS, C.W. and EADIE, W.J. (1975), *The life cycle of valley fog. Part I: Micrometeorological characteristics.* J. Appl. Meteor. *14*(3), 347–363.

PU, M.J., ZHANG,G.Z.,YAN, W.L., and LI, Z.H. (2008*), Features of a rare advection-radiation fog event,* Science in China Series D: Earth Sciences *51*(7): 1044–1052.

ROACH, W.T., BROWN, R., CAUGHEY, S.J., GARLAND, J.A., READINGS, C.J. (1976), *The physics of radiation fog: I—a field study,* Quart. J. Roy. Meteor. Soc. *102*(432), 313–333.

ROACH, W.T. and SLINGO, A. (1979), *A high resolution infrared radiative transfer scheme to study the interaction of radiation with cloud,* Quart. J. Roy. Met. Sci., *105,* 603–614.

RODGERS, C.D., and WALSHAW, C.D., (1966). *The computation of infrared cooling rate in planetary atmospheres.* Q. J.R. Met. Soc., *93,* 67–92.

SACHWEH, M. and KOEPKE, P. (1995), *Radiation fog and urban climate.* Geophys. Res. Lett., *22*(9), 1073–1076.

SHI, C.E., SUN, X.J., YANG, J., and LI, Z.H. (1996a), *3D model study ON fog over complex terrain. Part I: Numerical study.* Acta Meteor. Sinica, *10*(4), 493–506.

SHI, C.E., CAO, B.M., LI, Z.H., and LU, T.S. (1996b), *Numerical simulaiton of 3D local circulation over a complicated terrain.* Journal of Nanjing Institute of Meteorology, *19*(3), 320–328.

SHI, C.E., YAO, K.Y., and MA, L. (2001), *Numerical studies of effects of aerosons on urban fog.* Res. Climate and Environment, *6*(4), 485–492 (in Chinese).

SHI, C.E., BOTH, M., ZHANG, H., LI, Z.H. CHUNE, MATTHIAS ROTH, HAO ZHANG and ZIHUA LI (2008), *Impacts of urbanization on long-term variation of fog in Anhui Province, China,* Atmospheric Environment, *42,* 8484–8492, doi:10.1016/j.atmosenv.2008.08.002.

TAYLOR, G.I. (1917), *The formation of fog and mist.* Quart. J. Roy. Meteor. Soc., *43*(186), 2416–2468.

WANG, L.P., CHEN, S.Y., and DONG, A.X. (2005), *The distribution and seasonal variations of fog in China.* ACTA Geographia SINICA (in Chinese) *60*(4), 689–697.

WENDISCH, M., MERTES,S., HEINTZENBERG, J., WIEDDENSOHLER, A., SCHELL, D., WOBROCK, W., FRANK, G, MARTINSSON, B.G., FUZZI, S., ORSI, G., KOS, G., and BERNER, A. (1998), *Drop size distribution and LWC in Po Valley fog.* Contributions to Atmospheric Physics, *71*(1), 87–100.

WU, D., DENG, X.J., YE, Y.X., and MAO, W.K. (2004), *The study on fog-water chemical composition in Dayaoshang of Nanling mountain,* Acta Meteor. Sinica, *62*(4), 476–485 (in Chinese).

YANG, Z.Q., XU, S.Z., and GENG, P. (1989), *Springtime marine fog formation and its microphysical structure over the Zhoushan region.* J. Oceanography, *11*(4), 431–438 (in Chinese).

ZHOU, S.Z., and SU, Q., Urban Climatology. (Beijing: China Meteor. Press, 1994) (in Chinese).

(Received September 16, 2010, revised May 12, 2011, accepted May 14, 2011, Published online July 17, 2011)

Reprinted from the journal

Pure Appl. Geophys. 169 (2012), 941–960
© 2011 Springer Basel AG
DOI 10.1007/s00024-011-0340-0

Fog Simulations Based on Multi-Model System: A Feasibility Study

Chune Shi,[1] Lei Wang,[1] Hao Zhang,[1] Su Zhang,[1] Xueliang Deng,[1] Yaosun Li,[1] and Mingyan Qiu[1]

Abstract—Accurate forecasts of fog and visibility are very important to air and high way traffic, and are still a big challenge. A 1D fog model (PAFOG) is coupled to MM5 by obtaining the initial and boundary conditions (IC/BC) and some other necessary input parameters from MM5. Thus, PAFOG can be run for any area of interest. On the other hand, MM5 itself can be used to simulate fog events over a large domain. This paper presents evaluations of the fog predictability of these two systems for December of 2006 and December of 2007, with nine regional fog events observed in a field experiment, as well as over a large domain in eastern China. Among the simulations of the nine fog events by the two systems, two cases were investigated in detail. Daily results of ground level meteorology were validated against the routine observations at the CMA observational network. Daily fog occurrences for the two study periods was validated in Nanjing. General performance of the two models for the nine fog cases are presented by comparing with routine and field observational data. The results of MM5 and PA-FOG for two typical fog cases are verified in detail against field observations. The verifications demonstrated that all methods tended to overestimate fog occurrence, especially for near-fog cases. In terms of TS/ETS, the LWC-only threshold with MM5 showed the best performance, while PAFOG showed the worst. MM5 performed better for advection–radiation fog than for radiation fog, and PAFOG could be an alternative tool for forecasting radiation fogs. PAFOG did show advantages over MM5 on the fog dissipation time. The performance of PAFOG highly depended on the quality of MM5 output. The sensitive runs of PAFOG with different IC/BC showed the capability of using MM5 output to run the 1D model and the high sensitivity of PAFOG on cloud cover. Future works should intensify the study of how to improve the quality of input data (e.g. cloud cover, advection, large scale subsidence) for the 1D model, particularly how to eliminate near-fog case in fog forecasting.

Key words: Fog prediction, multi-model, MM5, PAFOG, verification, Nanjing, eastern China.

1. Introduction

Fog is a boundary-layer weather phenomenon with abundant water droplets or crystals that reduces visibility to <1 km (WMO, 1992). Low visibility on fog days can often endanger all kinds of transportation and can cause huge economic losses (Gultepe et al., 2007). Therefore, it has attracted increasing attention from meteorologists around the world. For example, the European Cooperation in Science and Technology (COST), one of the longest-running European institutions supporting cooperation among scientists and researchers across Europe, operated a project of "Short range forecasting methods of fog, visibility and low clouds" (COST-722) (Jacobs et al., 2005), which involves 14 European members and Canada. Recently, Canadian scientists operated a large project, fog remote sensing and modeling (FRAM), which utilized multi-method analysis to conduct fog observations and simulations (Gultepe et al., 2009). A research group at National Centers for Environmental Prediction (NCEP) of NOAA, the United States, developed a multi-model ensemble prediction system for fog forecasting based on multi-variable diagnosis (Zhou and Du, 2010). Some Australian scientists developed an objective Fuzzy-logic NWP fog guidance method to generate fog forecasts over a period of several years (Miao et al., 2010). During recent years, researchers in China have operated several nationwide projects for fog in different areas in China and have obtained some valuable and interesting results (Wu et al., 2004, 2007; Fu et al., 2006; Gao et al., 2007; Pu et al., 2008; Yang et al., 2009, 2010a, b; Shi et al., 2008, 2010). Recently, Niu et al. (2010) summarized the fog research activities in China over the past decades and urged that more effort be made in the future to investigate fog processes.

[1] Anhui Institute of Meteorological Sciences, Key Laboratory of Atmospheric Sciences & Satellite Remote Sensing of Anhui Province, Hefei 230031, Anhui, People's Republic of China. E-mail: chun.e.shi@gmail.com; shichune@sina.com

With the rapid development of computer technology, numerical weather prediction (NWP) models are widely used in studying and forecasting almost all kinds of high-impact weather patterns, and considerable progress has been made during the past decades. However, due to large computing resource requirements as well as the complicated formulation mechanisms, including the microphysics of droplets, aerosol chemistry, radiation, turbulence, dynamic processes of different scales and surface conditions (GULTEPE et al., 2007), the progress in numerical forecasting of fog is still slow. At present, fog has no direct operational guidance by any current NWP center and its prediction lags far behind forecasting of precipitation (ZHOU et al., 2011). Therefore, challenges remain for quantitative fog forecasting.

GULTEPE et al. (2007) summarized the achievements in fog research of the past several decades and pointed out that the future direction of studying and forecasting fog would utilize both 3D regional models and 1D fog models. It is encouraging that there are several groups in Europe who have already attempted to couple a 1D model to some 3D NWP models to forecast fog, obtaining promising results during the past several years. BERGOT et al. (2005) performed experiments on fog and cloud forecasting using a 1D model (COBEL), with initial state provided by an NWP model (ALADIN), together with a system (1DVAR) to assimilate local measurements, at Paris's Charles de Gaulle International Airport. Their research clearly demonstrated the importance of assimilation of local measurements to obtain promising results. Later, BERGOT et al. (2007) performed an intercomparison of six 1D fog models for predictions of two fog cases, a typical radiation fog and a so-called near-fog. The intercomparison revealed that those 1D models were able to reproduce some of the major features of the lifecycle of a fog layer, while the model results were considerably dispersed. In addition, their intercomparison demonstrated that high-resolution numerical models did not systematically improve the quality of fog forecasts. More recently, VAN DER VELDE et al. (2010) summarized their test results of fog forecasting based on two 3D NWP models (WRF and HIRLAM7.2) and the HIRLAM single-column model. Unlike BERGOT et al.

(2007), VAN DER VELDE et al. (2010) confirmed that high vertical resolution is essential for fog simulation.

This paper summarizes our exploratory study on the performance of fog simulation by a 3D mesoscale model (MM5) and a 1D fog model (PAFOG) for December of 2006 and December of 2007 in eastern China. Comparing with the works of BERGOT et al. (2005) and MULLER et al. (2007), the initialization of 1D model in our work is simpler. Considering the fact that we are in dire need of an objective and quantitative method for fog forecasting due to the rapid development of highway transportation in China, and because it is impossible to set up a sounding or tower observation like that at an airport along the highway, we need to explore some new ways to provide fog forecasts in a relative large domain. In addition, we need a method which offers better results than using a NWP model directly, because recent studies have revealed very poor performance of NWP models in fog forecasting (ZHOU et al., 2011). We noted that some other techniques have been developed using the output of an NWP model to diagnose fog or low clouds, such as rule-based fog forecasting (ZHOU and DU, 2010; BURROWS and TOTH, 2011), but these methods are for fog occurrence prediction, and do not predict fog intensity or visibility. Thus, the rule-based method does not meet the requirements of safe highway transportation. Although it has been argued that it is too costly to run a 1D fog model at each model grid for a large region, this method might be feasible for a small region, e.g. a city or a selected area along a highway which is prone to fog, and has obvious advantages over the rule-based forecasting. Therefore, the objectives of this study are: (1) to carry out experiments on fog prediction using a regional model (MM5) and a 1D fog model (PAFOG)–MM5 coupled system; (2) to investigate the possibility of using model profiles to run 1D fog models for fog prediction in eastern China; and (3) to identify any benefits provided by a coupled system over MM5 in fog prediction, and the advantages/disadvantages the coupled system has under different conditions based on objective verifications against both stationed and field observation fog data. The models were run daily for December, 2006, and December, 2007. Routine data of the Chinese Meteorology Agency (CMA) observational network and field observations on the Nanjing University of Information Science and Technology (NUIST) campus were used for model validations. NUIST is about 23 km away from the Nanjing observatory. NUIST and the Nanjing observatory are located in neighbor model grids in MM5. Special attention was paid to nine fog events which were captured on the NUIST campus. In addition, the impacts on fog prediction of many parameters, such as vertical resolution and physical parameterizations, have been studied by BERGOT et al. (2007), but the impacts of mesoscale background input data uncertainties, such as cloud, are still not clear in the regions of China. Sensitive testing is presented to explain how the coupled PAFOG depends on the initial conditions (ICs) and cloud cover. The next section will introduce the study region and data used, and then the model and coupling method will be given in Sect. 3. Section 4 will present the validations of model results with field and routine observations, including surface and PBL meteorology, general performance for fog occurrence and duration in Nanjing, fog distribution in eastern China, in particular detail for two typical fog cases, the impacts of initial and boundary conditions (BCs) for PAFOG and, finally, conclusions in Sect. 5.

2. Study Region and Observational Data

The region of interest covers several provinces in eastern China, including Anhui, Jiangsu, Shandong, Henan, Hubei, Jiangxi, Zhejiang and Shanghai (Fig. 1). Figure 1 also shows the locations of Hefei, the capital city of Anhui province and Nanjing, the capital city of Jiangsu province. Among those provinces, Anhui, Jiangsu and Zhejiang have higher fog occurrence frequencies, exceeding 20 days/year in most areas and even over 60 days/year for some areas (LIU et al., 2005). Usually, months from October to April are those with highest fog occurrence frequencies, especially December and January (LIU et al., 2005; ZHOU et al., 2007). The annual number of fog days is around 30 days in Nanjing, with highest occurrence in autumn and winter (TONG et al., 2009), and 22 days in Hefei. Due to the availability of data, we present the variation of monthly average fog days

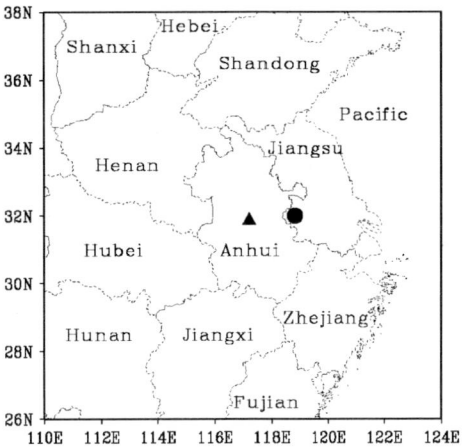

Figure 1
Domain of interest. The *solid triangle* and *circle* denote Hefei and Nanjing, respectively

in Hefei during 2005–2009 in Table 1. It can be seen that the annual average of fog days is 22.4 days and the monthly average is over 2 days from October to the next March, with the maximum in December (4.2 days). The hourly fog occurrence in Hefei during 2005–2009 is given in Fig. 2. It can be seen that the frequency is highest during the morning hours, especially around 07–09 LST (same as Beijing standard time, BST).

There are around 800 routine meteorological stations in the CMA network, providing 3-h (at 02, 05, 08, 11, 14, 17, 20, 23 LST) or 6-h (02, 08, 14, 20 LST) ground level observations, in the selected domain. In addition, field observations of fog were conducted by the NUIST fog study group on the NUIST campus (118.72°E, 32.21°N) in a northern suburb of Nanjing during the winters of 2006 and 2007 (YANG et al., 2009; LU et al., 2010). Abundant data have been accumulated, based on which a series of research results have been published, including fog physics features and mechanisms (PU et al., 2008; LIU et al., 2009; YANG et al., 2010a, b) and fog chemistry (YANG et al., 2009; LU et al., 2010). For details about the field observations, refer to YANG et al. (2009) and LU et al. (2010).

Since 08 LST is the time with the second highest fog occurrence and the most routine observational site reports, the measured visibility distributions at 08 LST were analyzed using Grads (software widely used by meteorologists) on those regional fog days which were captured by the fog research group of

Table 1

Variation of monthly fog days in Hefei during 2005–2009

Month	1	2	3	4	5	6	7	8	9	10	11	12
Fog days	4.0	2.4	2.6	0.6	0.6	0.8	0.4	1.0	0.6	2.2	3.0	4.2

3.1. MM5 Configurations

The mesoscale model employed in this study is the newest version of MM5 (v3.7.3) (GRELL *et al.*, 1994), developed jointly by Pennsylvania State University (PSU) and the National Center for Atmospheric Research (NCAR) in Boulder, Colorado. The physics schemes of MM5 selected for this study are listed in Table 2. Coarse and two-way nested grids were set up to cover a large portion of eastern Asia. The nested grid covered eastern China, including Anhui and its surrounding provinces (Fig. 1). Previous studies pointed out that model performance on fog simulation was highly sensitive to its vertical resolution (KONG, 2002; TARDIF, 2007). More recently, VAN DER VELDE *et al.* (2010) confirmed that high vertical resolution is essential for fog simulation. So, the troposphere from the ground to 100 hPa was divided into 34 sigma layers in the vertical, with 18 layers within the lowest 1,000 m to capture the turbulent processes in the boundary layer. The center of the first layer is around 4 m.

The first-guess fields and BCs for the coarse domain for every 6 h were obtained from NCAR/NCEP FNL reanalysis data ($1° \times 1°$), adjusted by 6-h (02, 08, 14, 20 LST) surface routine observational data and 12-h sounding data. Nudging or four dimensional data assimilation (FDDA) (OTTE, 2008), consisting of 6-h 3D analyses of temperature, water vapor mixing ratio and horizontal wind components for both domains, and 6-h surface analyses of horizontal wind components for the coarse domain, were used with nudging coefficients of 2.5×10^{-4} and $1.0 \times 10^{-4}\ s^{-1}$ for the winds and temperatures of the coarse and nested domains, and $1.0 \times 10^{-4}\ s^{-1}$ for humidity for both domains. The simulation period includes December of 2006 and December of 2007. The period was broken into overlapping 36 h run segments. Each period began at 08 LST and ended at 20 LST of the next day. The first 12 h of each segment was a "spin-up" period, the last 24 h were used for PAFOG.

Since MM5 output does not include visibility, we use the visibility formulation by GULTEPE *et al.* (2009) to calculate visibility based on the computed liquid water content (LWC) at the first model layer (around 4 m above the ground), that is:

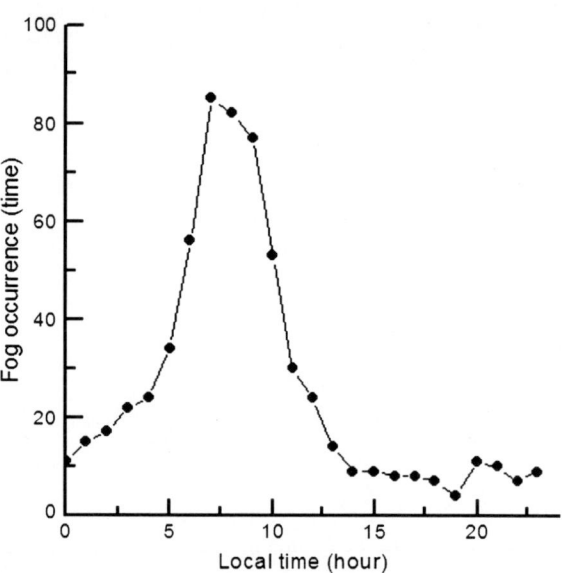

Figure 2
Hourly fog occurrences in Hefei during 2005–2009

NUIST (YANG *et al.*, 2009; LU *et al.*, 2010), based on the routine visibility observational data at stations in the domain. Figures 3 and 4 present the distributions of horizontal visibility and fog regions (red color). In Figs. 3 and 4, the threshold of observed visibility value (visibility ≤ 1 km) is used. By definition, the observed visibility is below 1 km for fog, below 500 m for dense fog and 50 m for extremely dense fog. It can be seen that the fogs observed in Nanjing were not isolated. Almost all fog cases were parts of regional fog events. This regional characteristic of fogs in eastern China has also been studied and identified by ZHOU *et al.* (2007).

3. Models and Couple Method

Two models, MM5 and a one-dimensional (1D) fog model (PAFOG), were used in this study.

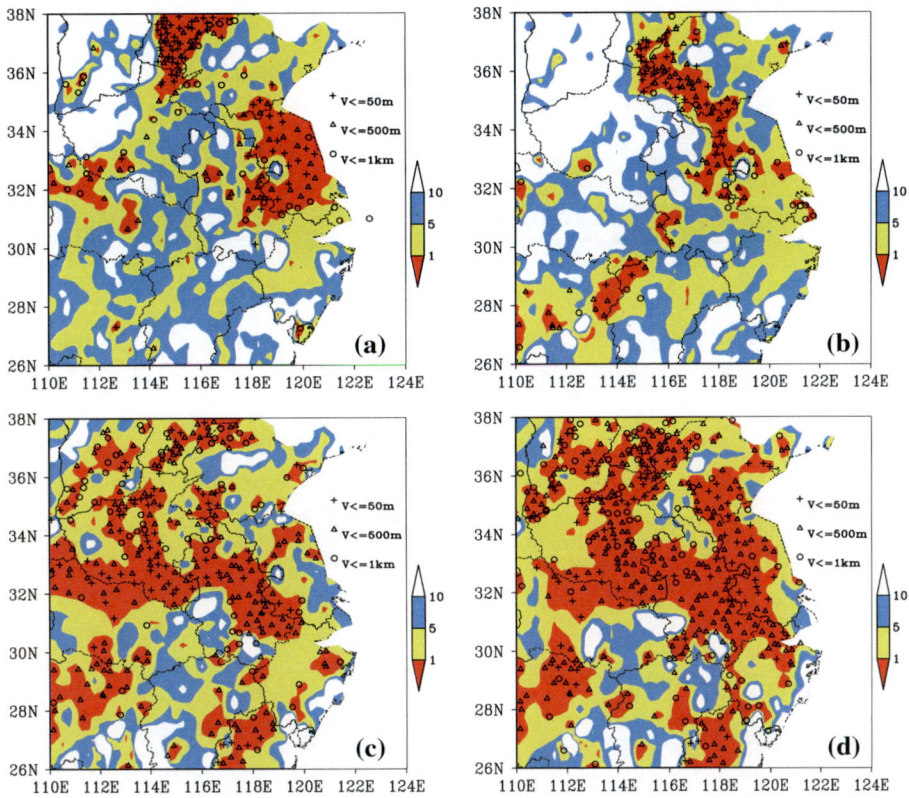

Figure 3

Surface visibilities (km) by station data at 08:00 a.m. on fog event days during December of 2006: **a** 12 December; **b** 14 December; **c** 25 December; **d** 26 December

$$V = 0.87706(\text{LWC} \times N_\text{d})^{-0.49034} \qquad (1)$$

where, V is visibility in km, LWC in g/m^3, the number density of fog droplet N_d in cm^{-3}. According to the field observations, N_d in Nanjing varied from case to case, also in different stages of the same fog case (LIU *et al.*, 2011; PU *et al.*, 2008). For example, during the fog events on 25–27 December 2006 (an advection–radiation fog event), N_d was quite large, over 300 cm^{-3} for most of the time (PU *et al.*, 2008), while it was much lower on 10–11 December 2007 (a radiation fog event) (LIU *et al.*, 2011). In this study, it was set as 300 cm^{-3} according to PU *et al.* (2008), since formula (1) was applied only for the case of 25–27 December 2006. SHI *et al.* (2010) compared this method with the traditional KUNKEL (1984) plus STOELINGA and WARNER (1999) algorithms for eastern China and found this method to be superior to the traditional method, especially for dense fog. From formula (1), it is easy to deduce that the visibility will be infinite when LWC is approaching 0, which is impossible. So, to make the visibility calculation valid, another formula of LIN (2010) for visibility based on relative humidity (RH) is used when LWC < 0.01 g/kg:

$$V = 138.061 - 29.009 \ln(\text{RH} + 15.25), \qquad (2)$$

where RH is in percentage, V in km. Formula (2) was obtained using all effective data (32,002 samples) during all the 21 fog events of November and December of 2007. In the scatter plot of RH–Vis (visibility), the visibility ranged from tens of meters to around 1 km when RH reached 100%, which suggests that the formula may have some uncertainties with very high RH ($\sim$100%).

3.2. PAFOG

The 1D model PAFOG (BOTT and TRAUTMANN, 2002) was derived from the detailed spectral

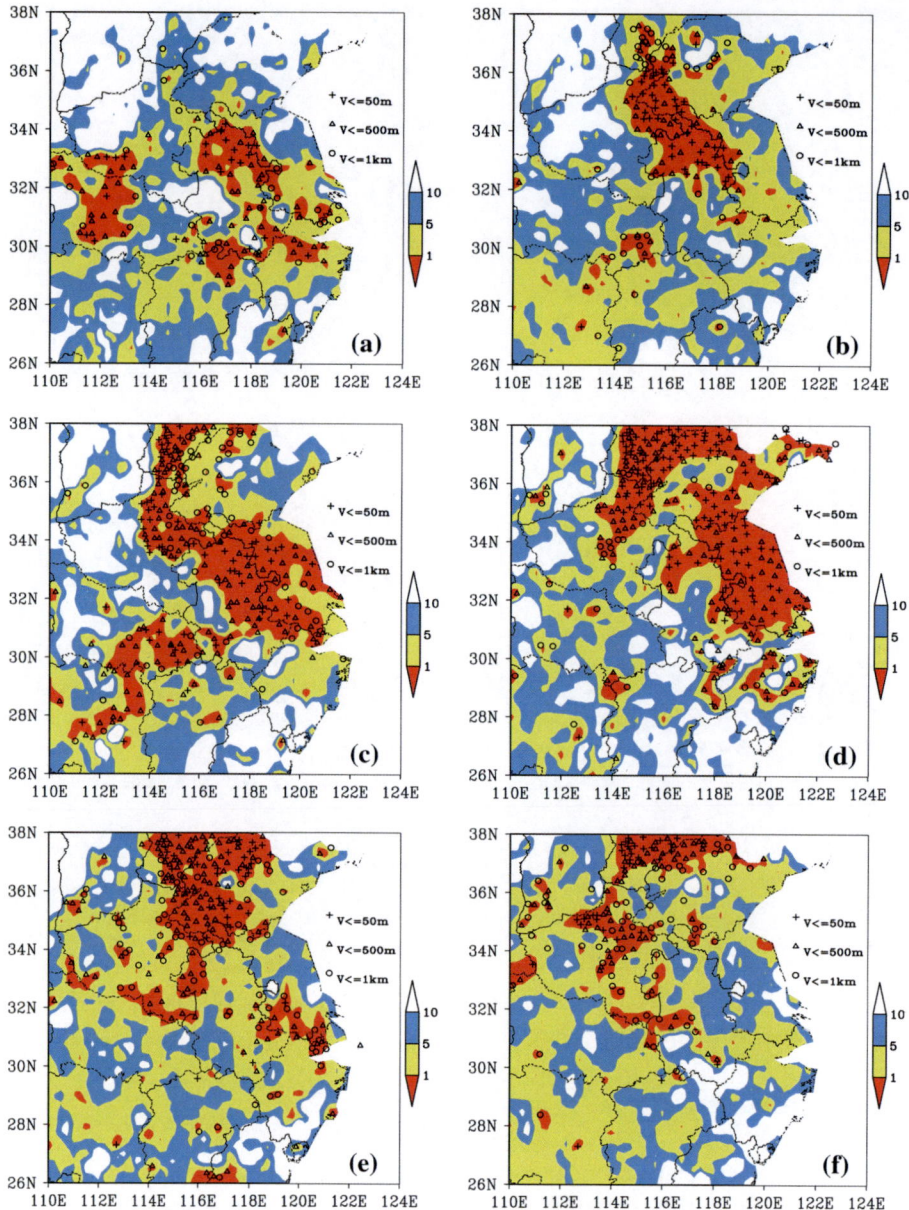

Figure 4
Surface visibilities (km) by station data at 08:00 a.m. on fog event days during December of 2007: **a** 14 December; **b** 18 December; **c** 19 December; **d** 20 December; **e** 21 December; **f** 23 December

Table 2

Configuration of MM5 inner domain

Physical process	Micro-physics	PBL	Resolution (km) and levels	Radiation	Soil	Cumulus	Micro-physics
Scheme	Simple ice	Blackadar	15 (34)	Cloud-radiation	Multi-layer	Grell	Simple ice

microphysical model MIFOG (Bott et al., 1990). It consists of four modules: the dynamic model, the microphysical model, the radiation module, and a module for low vegetation, which has already been described in the literature (Table 3). In this exploratory work, most of the parameters are the same as those in Bott and Trautmann (2002), including the major parameters for microphysics, which include $\sigma_c = 0.2$ for droplet size distribution, $k = 0.9$ for activated cloud droplets calculation ($N_{act} = CS^k$). The standard aerosol concentration (C) is adjusted to 20,000 cm^{-3} according to the observations in Nanjing (Yang et al., 2010a).

Our PAFOG model domain is divided into two sub-regions similar to those of Bott and Trautmann (2002). The lower part, reaching from the surface to z_1, is subdivided into N_1 equidistant layers, the values of N_1 and z_1 being variable. The N_1 and z_1 are set as 50 and 200 m, yielding a constant grid distance of $\Delta z = 4$ m in the lower 200 m. The second region extends from z_1 to $z_2 = 1,500$ m. Here, the grid spacing is logarithmically equidistant and the number of grid boxes is $N_2 = 20$.

Since our final goal is operational applications for fog forecasting in the future, and considering the fact that most fog events begin in the early morning (Shi et al., 2008), the simulations of PAFOG started each day 20 LST and ended at the next 20 LST, with a prediction time of 24 h in total, no matter whether there was fog or not at the start time. In fact, there was no fog at the initial time (20 LST) from the MM5 output for the fog cases in this study, even for the case on 25–27 December 2006, a long lasting advection–radiation fog. So, no initial fog was considered in this study.

The output parameters of PAFOG include not only the routine meteorological parameters (e.g. wind speed, temperature, RH, etc.), but also some special parameters for fog (e.g. fog droplets, LWC, ground-level visibility, etc.).

3.3. Coupling Method Between MM5 and PAFOG

To run PAFOG, we need to provide the ICs, including vertical profiles of temperature, specific humidity and pressure, geostrophic wind at around 850 hPa, temperature at ground surface and 30 cm below the earth's surface, volumetric moisture content of the soil, cloudiness; and the BCs, including hourly temperature, specific humidity, pressure, RH, wind speed (u and v), at the model top and geostrophic wind. Except for volumetric moisture content of the soil, which was taken from Pielke (1984), all other parameters are taken directly or indirectly from MM5 output. The low, medium and high cloud covers are diagnosed based on RH as the method used in the module of MM5toGrads in MM5 and adjusted to the corresponding heights defined in PAFOG. The geostrophic wind at 850 hPa was calculated as follows:

$$u_g = -\frac{g}{f}\frac{\partial z}{\partial y}, \quad v_g = \frac{g}{f}\frac{\partial z}{\partial x}, \tag{3}$$

where u_g, v_g are the two components of geostrophic wind, g is the gravitational acceleration, and z is the geopotential height at 850 hPa. So, PAFOG can be run for any area of interest if the method is feasible.

To further investigate the impacts of mesoscale background input data uncertainties, such as cloud, the measured profiles of temperature, humidity, wind and pressure from a radiosonde station at Nanjing observatory (118.8°E, 32.0°N) were also used to run PAFOG for two typical fog cases, and the results were validated against field observations.

Table 3

Major physics schemes of PAFOG

Physical process	Dynamic	Turbulence	Radiation	Microphysics	Vegetation
PAFOG	Bott et al. (1990)	2.5 level model of Mellor and Yamada (1974)	δ-two stream approximation of Zdunkowski et al. (1982)	See Nickerson et al. (1986) and Chaumerliac et al. (1987)	Siebert et al. (1992a, b)

4. Results and Discussions

During the simulation period (December, 2006, and December, 2007), thirteen fog events were recorded, another thirteen "near-fog" events ($V \leq 2.0$ km, RH > 90%) (BERGOT et al., 2007) were observed by Nanjing observatory and nine fog events were captured on the NUIST campus by the NUIST fog research group (YANG et al., 2009; LU et al., 2010). Utilizing MM5 and PAFOG, together with the above-mentioned coupling method, simulations were conducted daily for December of 2006 and December of 2007. In this paper, however, more attention will be paid to the nine fog events. The validation for the ground level meteorology was performed for the two study months, through calculation of some commonly-used statistical measures for temperature (T), wind speed (WS), wind direction (WD), and RH between measurements and simulation were conducted daily and monthly at all observational sites in the domain. The statistical measures include root mean square error (RMS), mean bias (MB), correlation coefficient (r), standard deviations for observation and prediction (SD_O for observation and SD_P for prediction), etc. In addition, the radiosonde data at Nanjing observatory were used to validate the MM5 simulated profiles at 20 LST in PBL, from 2 to 1,500 m, at different levels.

In the verification of fog occurrence, the thresholds of MM5 calculated LWC and RH used were:

LWC at the lowest model layer ≥ 0.01 g/kg (MULLER et al., 2007) or $\qquad$ (4)

2-m RH $\geq 98\%$ (YANG et al., 2010b)   (5)

PAFOG was applied only to NUIST and Nanjing observatory. According to the availability of field observations, the results of PAFOG were validated with the measurements from the NUIST field observations in detail for two typical cases. In addition, to fully understand the model's capability of reproducing fog events, some scores [e.g. threat score (TS), equitable threat score (ETS), false alarm rate (FAR), hit rate (HR), missing rate (MR), correct rejection rate (CRR) and bias] commonly used for verification of a binary event (yes or no) (ZHOU and DU, 2010) were calculated for Nanjing observatory for the two study months.

4.1. MM5 Performance on Surface and PBL Meteorology

For the daily statistics, MM5 performs best in ground-level temperature, with the highest correlation coefficient, most values over 0.9, most mean biases between -2.0 and $0.1°C$ (-1.0 to $0.1°C$ for those fog days). As BERGOT et al. (2007) pointed out, the high vertical resolution of NWP models may have a cold bias. There are some other factors that impact the simulated temperature at ground level, such as the moisture content of soil (ZEHNDER, 2002), which should be more important in dry areas or dry seasons. In addition, the cloud scheme employed in our model may also play a role. The cold bias shown in this study is therefore complex, and discussion of its causes in more detail is out of the scope of this study. MM5 performs worse in wind speed than in temperature. The correlation coefficients are around 0.5 with positive MBs; most MBs are within 2 m/s. The worse statistics for ground-level wind speed are often observed (e.g. SHI et al., 2010; OTTE, 2008) due to coarse horizontal resolution in comparison to complex surface or terrains for NWP models. As for RH, the correlation coefficients are around 0.7, with MBs between 3 and 25% and most around 10%. The statistics for December 2006 are given in Table 4.

Table 4

Summary of statistical measures, in comparing observations to MM5 simulations at the ground level

	MeanObs	MeanPre	RMS	R	SD_O	SD_P	MB
T	4.74	3.09	2.89	0.92	5.39	5.93	-1.65
WS	1.84	3.36	2.45	0.58	1.66	2.34	1.52
RH	67.87	78.47	19.09	0.71	21.72	19.25	10.59

Table 5

Summary of statistical measures for temperature, in comparing measurements to MM5 simulations at the different levels in PBL at 20 LST for December of 2006 and December of 2007

Height (m)	Correlation coefficient	MB (K)	RMS (K)	SD_O (K)	SD_P (K)
1,500	0.978	−0.19	0.895	3.888	3.374
1,000	0.974	−0.25	0.921	3.783	3.486
700	0.977	−0.28	0.784	3.323	3.32
500	0.972	−0.22	0.819	3.11	3.207
300	0.949	−0.2	1.023	2.917	3.083
100	0.881	−0.49	1.462	2.651	2.966
60	0.88	−0.64	1.539	2.532	2.915
20	0.896	−0.64	1.554	2.408	3.016
2	0.889	−1.01	1.846	2.641	3.331

Generally, MM5 tended to underestimate temperature and overestimate the wind speed. The simulated RH is generally higher than the observations, which might be a result of cold bias. According to statistics at 20 LST at Nanjing observatory, the MM5 model performed better with the increase of height in PBL (Table 5), which can be seen from the increasing correlation coefficient and decreasing MB, comparable SD_O and SD_P, and RMS within SD_O. Similar results are obtained for RH (not shown). Therefore, based on the criterion of Lu *et al.* (1997), the results are acceptable for temperature and RH at the surface and in PBL.

4.2. General Descriptions of Model Performances on Fog Events

Table 6 presents the descriptions of horizontal distributions of the major fog body at 08 LST in our domain of interest for each fog event, in comparison with the simulations from MM5. In the last column, bold font is used for areas consistent with the observations. As for the MM5 capability of predicting major fog, it is found that MM5 does predict regional fog or fog patches on each fog event day using the threshold of LWC. Comparing column 2 with column 3 in Table 6, however, the MM5 simulated fog regions are generally smaller than the observed fog areas. The more general situations are that the simulated and observed fog areas are partially coincident, similar to the situations discussed by Zhou *et al.*, 2011. In addition, MM5 performs

relatively well for those advection–radiation fog cases, e.g. fog events during 25–27 December 2006 and 20–21 December 2007. Although there are still some inconsistent areas in those cases, MM5 basically reproduced the major part of the fogs. For those fog cases which mainly resulted from radiation cooling, the simulated fog regions are generally smaller than the observations, or (in some cases) the simulated fog areas do not match the observations well (e.g. fog events on 13–14 December 2007). However, if we also use the threshold of RH (e.g. RH ≥ 98%), the simulated fog regions cover the observed ones for all cases and even larger areas. It is likely that if we use some additional restrictive conditions, e.g., combined with wind speed, as Zhou and Du (2010), and Burrows and Toth (2011) did, the false alarm area could be reduced. The distributions of simulated LWC and RH at 08 LST are presented for two fog cases in the following (comparing Figs. 6 and 3d, Figs. 10 and 4a).

Table 7 presents the fog durations on the NUIST campus by observation and simulations for nine fog events. In Table 7, ddThh means hour hh at date dd, and the italic bold font indicates large differences between the observations and the simulations on the NUIST campus. For MM5, there are two cases without LWC, but RH reaches 100%, very like the so-called "near-fog" state (Bergot *et al.* 2007), so we also put the time period of RH = 100% in Table 7. If we only use the threshold of LWC to decide fog production, MM5 reproduced seven out of nine Nanjing fog events successfully, although the

Table 6

General descriptions model performances for Nanjing fog events

Date	Observations	MM5
	Major fog body locations at 08 LST	Major fog body locations at 08 LST, by LWC > 0.01 g/kg
12/12, 2006	East Anhui, most Jiangsu, Hebei	Hunan, Jiangxi to **south Anhui and south Jiangsu**
12/14, 2006	Border areas around Anhui and Jiangsu, to west Shandong;	Areas around Yangtze River in **Anhui**, **central Jiangsu**
12/25–27, 2006[a]	25: Anhui, north Hubei, most Henan, Hebei, part of Shandong and Jiangsu	25: **Most Anhui** and Jiangsu, East Jiangxi, east Shandong
	26: Most of Jiangsu, Anhui, Shandong and Henan, Hebei	26: **Jiangsu**, **Anhui**, **Shandong**, east Jiangxi, east **Henan**
12/14, 2007	North Anhui, west Hubei	Central Jiangxi, north Fujian
12/18, 2007	North Anhui to west Shandong	Central Anhui to south Jiangsu
12/19, 2007	Most of Anhui and Jiangsu; North Henan to Hebei; Central Hubei	**South Anhui**, **south Jiangsu**, west Jiangxi, east **Hubei** and east Hunan
12/20, 2007	Jiangsu, Shandong, south Hebei to north Henan, north Anhui	Jiangxi, Zhejiang, south **Jiangsu**, **Anhui**, South **Henan**
12/21, 2007[a]	West Shandong to south Hebei	**West Shandong**, north Anhui, Most Jiangsu, south Zhejiang, Fujian
12/23, 2007	North Henan to south Hebei	Anhui, Jiangsu, to south Shandong, part of **Henan**

[a] Advection–radiation fog

Table 7

Comparisons of fog duration in Nanjing between observation and simulations

Date	Observations	MM5	Fog duration by PAFOG	
	Fog duration in NUIST (h)	Fog duration in NUIST (h)	V < 1 km	LWC > 0.01 g/kg
12/12, 2006	11T23–12T11	*No LWC* 11T23–12T08	11T22–12T13	11T22–12T12
12/14, 2006	13T23–14T12	14T02–09	13T22–14T11	13T22–14T11
12/25–27, 2006	24T22–27T14	24T22–25T10	24T22–25T08	24T22–25T08
		25T19–26T11	25T22–26T12	25T22–26T12 26T17–27T10
		26T19–27T10	26T17–27T10	
12/14, 2007	13T23–14T11	*No LWC* 13T23–14T08	13T22–14T11	13T22–14T10
12/18, 2007	18T03–18T11	17T21–18T09	*17T22–18T20*	*17T23–18T20*
12/19, 2007	18T16–19T12	18T23–19T08	18T22–19T12	18T22–19T12
12/20, 2007	19T17–20T16	20T00–20T10	19T21–20T11	*No LWC*
12/21, 2007	20T18–21T19	20T21–21T20	*20T22–21T07*	*21T01–21T03*
12/23, 2007	23T01–23T05	22T22–23T10	22T22–23T11	22T22–23T11

simulated fogs often dissipated earlier than the observed ones. However, by using the threshold of RH, MM5 succeeded in reproducing all fog cases in Nanjing.

For PAFOG, since it outputs both LWC and visibility, we put both in Table 7 for comparison. From Table 7, it can be seen that PAFOG captured all fog events by threshold of visibility. But there is also one case without LWC, another case in which PAFOG simulated fog lasts too long (on 18 December 2007). Further comparing the simulated fog duration with observations and MM5, one can observe that PAFOG simulated fogs usually last

longer than MM5 simulated fogs except one case (21 December 2007), that is, PAFOG simulated dissipation times are closer to observation than those of MM5.

Although both systems captured most fog cases, one may say that it is easy for a 3D model to reproduce fog because of the cold bias at the ground level. To understand to what extent MM5 and PAFOG generated "false alarm" fogs, Table 8 gives the results of objective verifications of the two models with different thresholds. In Table 8, "LWC + RH" means LWC > 0.01 g/kg or RH = 100% were used as the threshold; "LWC + Vis" means LWC > 0.01 g/kg or

Table 8

Results of objective verifications for Nanjing during the December of 2006 and December of 2007

Model (scheme)	TS	ETS	FAR	MR	HR	Bias	CRR
MM5 (LWC + RH)	0.324	0.136	0.667	0.077	0.923	2.77	0.467
MM5 (LWC)	0.370	0.217	0.583	0.231	0.769	1.846	0.696
PAFOG (LWC + Vis)	0.293	0.099	0.7	0.077	0.923	3.076	0.391

visibility < 1 km were used as the threshold. "LWC" means only LWC > 0.01 g/kg was used. It can be seen from Table 8 that MM5 (LWC) has the highest TS, ETS, MR, CRR, and the lowest FAR, HR, and BIAS. Comparing with MM5 (LWC), MM5 (LWC + RH) performs better in higher HR and lower MR, worse in other scores. PAFOG (LWC + Vis) performs the worst with the largest bias and lowest TS/ETS. So, in terms of fog occurrence in Nanjing, PAFOG does not show its advantage over MM5. MM5 (LWC + RH) does show some potential in fog simulation with the highest HR and moderate TS/ETS in this study. However, it is important to keep a balance between the two schemes (LWC, LWC + RH), for example, how to decrease its bias while keeping its high HR in MM5 (LWC + RH). In addition, it is worth noting that all schemes overestimated fog occurrence (Bias > 1 and CRR < 1) even using MM5 (LWC), which is usual in fog forecasting by NWP models (ZHOU *et al.*, 2011). Through checking the false alarm days, it was found that most false alarms occurred on those "near-fog" days. To deal with this state, one may fall back on a 1D fog model with accurate physics, like PAFOG or add some restrictive conditions to the MM5 output, like the multi-variable rule based on NWP models by ZHOU and DU (2010).

Accordingly, more attention should be paid to avoiding false alarms in fog simulation, caring not only about reproducing fog events, but also paying attention to avoiding predicting fog on "near-fog" days. Long term simulation will be helpful in this regard, instead of simulation on single fog cases.

4.3. Evaluations of Model Performance for Two Typical Fog Events

Considering the availability of data, the simulations of two fog cases were compared in detail with the field observations. The two fog cases are the advection–radiation fog occurring on 25–27 December 2006 and the radiation fog occurring on 14 December 2007.

4.3.1 Case 1: The Advection–Radiation Fog on 25–27 December 2006

An unusual dense advection–radiation fog event occurred in eastern China during 25–27 December 2006. The fog was widely spread in the horizontal dimension and deep in the vertical, covering most of Anhui province and parts of surrounding provinces at its mature stage (Fig. 3c, d), with its top maintained above 450 m at most times, maximizing at 943 m (Fig. 5; PU *et al.*, 2008). It reduced visibility to 100 m or less in much of Anhui, Henan, and Jiangsu provinces, reaching zero meters in Nanjing, the provincial capital of Jiangsu province, and lasting

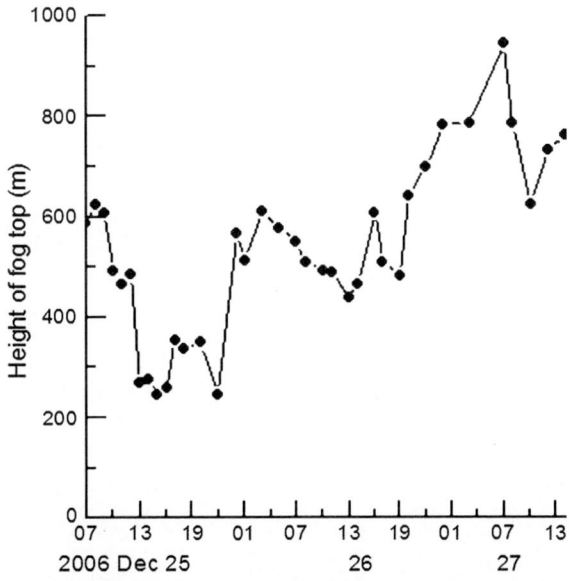

Figure 5
The measured fog top on the campus of NUIST during 25–27 December 2006

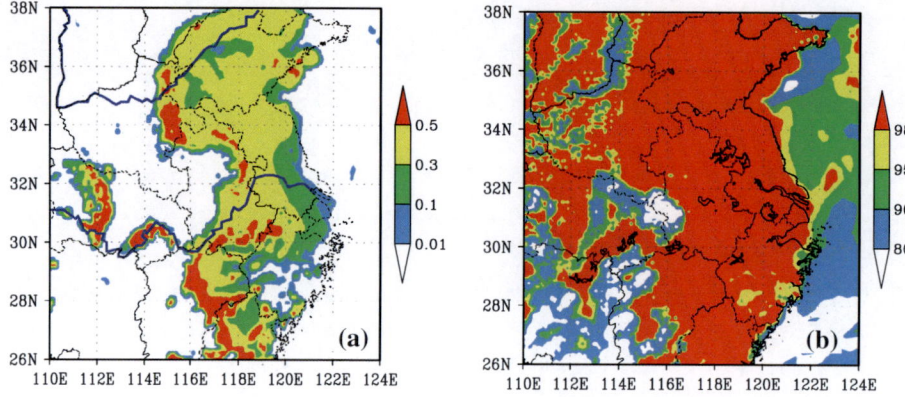

Figure 6
Distributions of MM5 simulated liquid water content (g/kg) (**a**) and relative humidity (%) (**b**) at 08:00 a.m. LST on 26 December 2006, at the lowest model layer

more than 36 h in some places (Pu *et al.*, 2008). According to the varieties of the measured microphysical parameters, Liu *et al.* (2009) divided the whole fog process into four sub-processes, which are around 24T22–25T14, 25T14–26T15, 26T15–27T00, 27T00–27T14, the numbers before/after T denote date/hour, respectively. The average LWCs in the first and second sub-process are over 0.2 g/m³, much higher than those at the last two sub-processes (0.02372, and 0.00549 g/m³). By checking the weather records, the last sub-process was actually a drizzle case, during which the instruments might not have worked properly, as shown by very low observed LWC and low observed visibility.

Figure 6 shows the distributions of simulated RH and LWC at the lowest model layer at 08 LST on 26 December. Comparing Fig. 6 with Fig. 3d, it can be seen that MM5 caught the major part of the dense fog, although it failed in the west part of the model domain. However, if RH > 98% was used, the simulated fog area covers all observed fog area with some evident false alarms.

To evaluate the capabilities of the two models in reproducing the advection–radiation fog events in the vertical dimension, the time-height cross-section of the LWC on the NUIST campus by the two models are presented in Fig. 7, which is an integration of 3-days of simulation. It can be seen that the predicted fog is disconnected. Both models reproduced three sub-fog events, and failed in the third sub-process. Comparing Fig. 5 and 7a, on 25 December, MM5

showed a 1,000 m low cloud and a 100 m fog bank, but observation showed both were combined into one 600–700 m deep fog bank. A similar situation occurred at noon on 27 December, both MM5 and observations show a tall fog top. Thus, MM5's simulation is pretty good in the view of forecasters. However, the PAFOG predicted fog heights are much lower, even with such a good background from MM5. To trace the causes, it might be related to the formation and maintainenance mechanisms of this long lasting fog event. By analysis and simulation, previous authors (Pu *et al.*, 2008; Shi *et al.*, 2010) concluded that this fog event was generated by nighttime radiative cooling, maintained and developed by large scale subsidence and the steady warm and moisture advections. However, the current PA-FOG did not include the large scale subsidence, the temperature and moisture advection. Although MM5 already simulated the low cloud at around 1,000 m on 25 December and deep fog on 27 December, the MM5 simulated LWC cannot be entered into PAFOG as input and impact PAFOG in the later time, because PAFOG gets the initial conditions at 20 LST when there is no fog in the MM5 output. Thus, the PAFOG simulated fog life cycle show characteristics of a typical radiation fog in this study. We did try to add the horizontal advection of temperature to PAFOG as Muller *et al.* (2007) did, but the results changed for the better on some days, for the worse on other days. Bergot *et al.* (2005) discussed the influence of the mesoscale flow on accurate forecasting of fog in

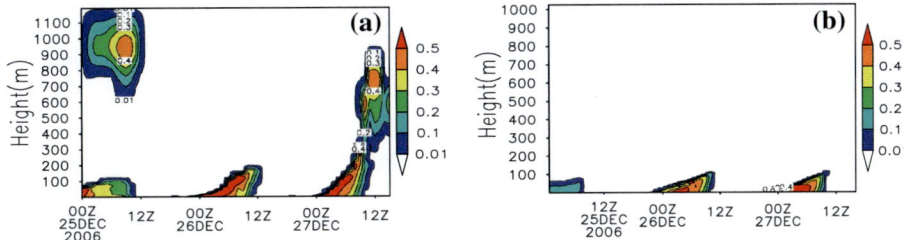

Figure 7
Time–height cross-section of LWC (g/kg) on NUIST campus by **a** MM5, **b** PAFOG. The *x*-axis is in LST (same below)

detail, and pointed out that the current NWP meso-scale model cannot give helpful information at the local scale. In conclusion, from the point of fog top height (Fig. 5), MM5 shows better performance than PAFOG, although MM5 also failed in catching the third sub-process. As for how to make full use of MM5 output in improving the PAFOG performance, we hope to address this in our future work.

Figure 8 presents the variations of simulated and measured ground-level LWC and visibility. Muller

et al. (2007) pointed out that the results of the first several hours in 1D model simulation should be thrown away due to the spin-up effect, especially under the situations in which fog already existed at the simulation beginning time. Again, Fig. 8 is an integration of the 3 days of simulations. It can be seen that both models caught the trend of surface LWC well on the first 2 days, especially for the maximum LWC time, with a correlation coefficient of 0.6 for MM5 and 0.67 for PAFOG with

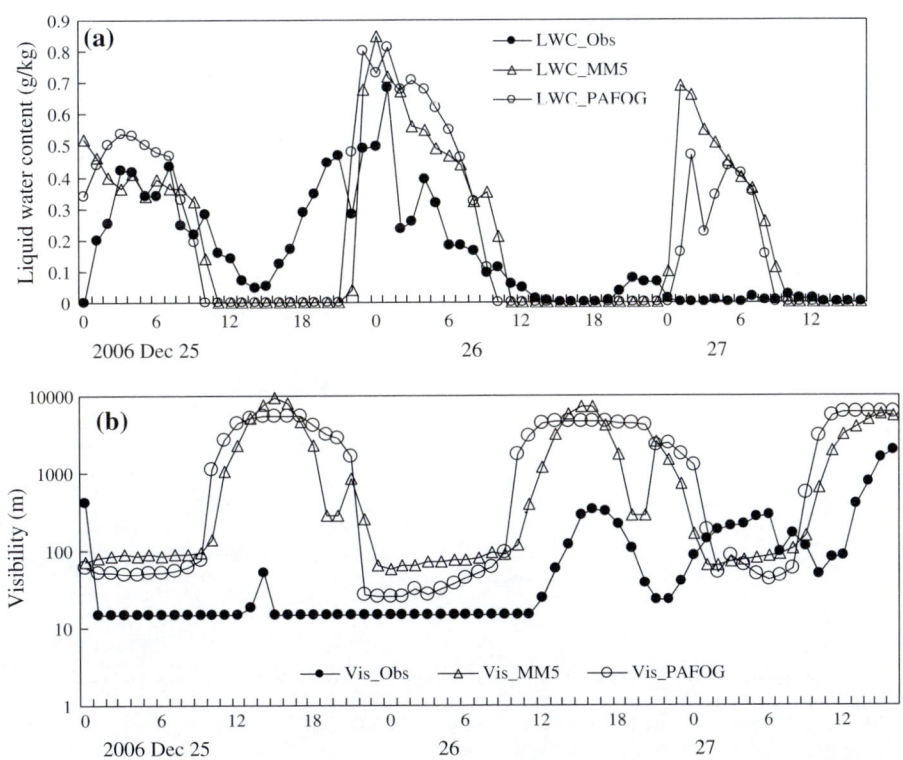

Figure 8
Simulated and measured ground-level LWC (**a**) and visibility (**b**) on the NUIST campus for case 1

measurement. In the last sub-process, the simulated ground level LWCs are much higher than the measurements. Although the PAFOG simulated fog is much thinner than the MM5 fog, its surface LWC is comparable with MM5, with a correlation coefficient of 0.88 for the 3 days. On the third day, both MM5 and PAFOG simulated LWC are much higher than the measurements. One of the reasons for overestimated LWC could be the drizzle, as we observed, during which the measurements might be affected, e.g. the instrument may not have captured the droplets in the drizzle, and drizzle also can reduce visibility to <1,000 m and is a broad-sense "fog".

The variation of visibility is consistent with that of LWC. That is, a higher LWC corresponds to a lower visibility. Although the MM5 simulated fogs were not consecutive, it did catch the major fog episodes with a little longer fog duration than PAFOG on the first day. Although the simulated visibilities by the two systems show poor correlation with observations due to the constant instrument measured visibility during the first 36 h, the MM5 and PAFOG simulated visibilities are highly correlated ($r = 0.78$). In addition, comparing Fig. 8b with 8a, it can be found that on the last day, there was no surface LWC observed (Fig. 8a), but observed visibility was still below 1,000 m, which was consistent with the drizzle recorded then.

In summary, comparing simulated ground-level LWC and visibility with measurements indicates that the performance of PAFOG highly correlated with MM5, and PAFOG showed better correlation with measurement than MM5 in surface LWC.

4.3.2 Case 2: The Radiation Fog on 14 December 2007

A strong dense fog event was recorded in Nanjing on 14 December 2007. It lasted about 14 h, with a maximum height of 600 m (Fig. 9). The fog was first formed through radiation cooling in the ground level, and then followed by a cloud layer caused by low-level cold advection (YANG et al., 2010b). At 08 LST, the regional fog event include two large fog areas, mainly covering north Anhui and west Hubei, and some small fog patches in Jiangxi, north Zhejiang, south Anhui and south Jiangsu (Fig. 4a). The fog located in north Anhui affected Nanjing. Comparing to case 1, the fog areas were smaller in the mature stage, and the fog duration was shorter in Nanjing.

Figure 10 shows the distributions of simulated RH and LWC at the lowest model layer at 08 LST on 14 December 2007. Comparing Fig. 10a with Fig. 4a, it can be seen that MM5 failed to catch the major part of the dense fog, which was located in Anhui and Hubei, neither the fog in Nanjing, magnified fog

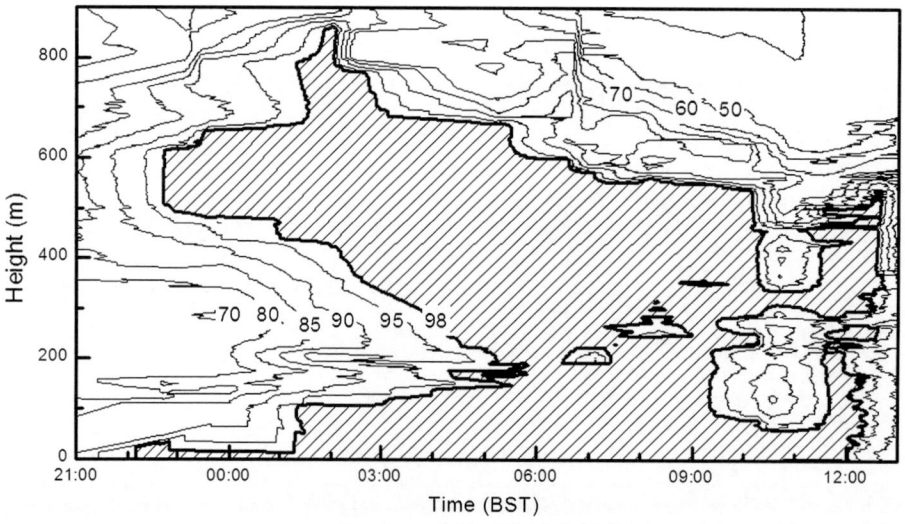

Figure 9
The measured time–height cross-section of relative humidity on the NUIST campus by sounding (the diagonal area denotes fog zone with RH > 98) (YANG et al., 2010b)

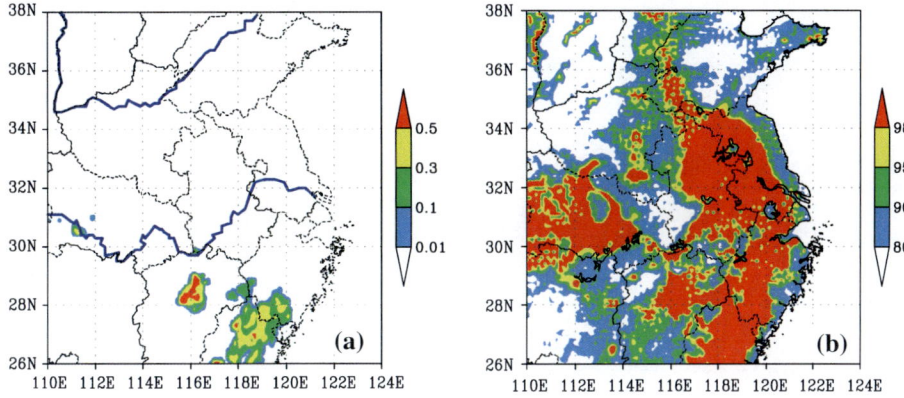

Figure 10

Distributions of MM5 simulated LWC (g/kg) and RH (%) at 08:00 a.m. LST on 14 December 2008, at the lowest model layer

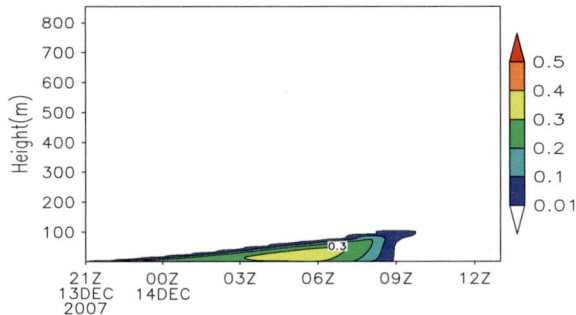

Figure 11

Time–height cross-section of LWC (g/kg) on the NUIST campus by PAFOG

patches in Fujian, if only LWC was considered. That is the situation where fog is observed in region A, while it is predicted in region B; moreover, region A and region B have almost no common area, as illustrated in ZHOU *et al.*, 2011. However, if RH > 98% was used (Fig. 10b), the simulated fog area covers all observed fog area with some false alarms.

Since MM5 failed to catch this fog event in Nanjing when the LWC-only threshold was used, we present only the time-height cross-section of the LWC on the NUIST campus by PAFOG in Fig. 11. Comparing Fig. 11 with Fig. 9 shows that PAFOG successfully reproduced the formation, development, and dissipation of the lower part of the fog, although the simulated fog dissipated earlier than actual observation and the simulated fog was thinner than observation. In addition, it was noticed in Fig. 9 that the fog did show the possibility of

dissipating a little after 09 LST, which is close to the simulated dissipation time. So, it can be deduced that the fog would dissipate earlier without the upper level fog.

Figure 12 presents the variations of PAFOG simulated and measured ground-level LWC and visibility, and MM5 simulated ground-level RH. Although MM5 did not produce LWC in Nanjing, it did forecast saturated RH. The MM5 predicted RH steady with 100% from 00 LST to 08 LST on 14 December. Although the low calculated visibilities (below 500 m) are not believable due to the uncertainty under the high RH condition of formula (2), it is reasonable to deduce that there was fog or near-fog at that time. Therefore, we can say that MM5 also has some capability in this case if using visibility or RH threshold. Again, it suggests the importance of putting forward a way to distinguish fog from near-fog based on NWP model results.

Due to the rapid radiation cooling effect of clear sky, PAFOG produced fog water soon after its startup. The PAFOG simulated LWC reaches a maximum at about 06 LST. In observations, although high humidity (RH > 98%) and low visibility ($V < 1$ km) were observed quite early (around 21 LST on 13 December) (Fig. 12b), evident liquid water was not measured until 06 LST, 14 December. Considering the fact that the concentrations of anthropogenic aerosol, especially the hygroscopic aerosols such as sulfates, are high in most of China, the hygroscopic increase in aerosol size decreases the visibility to below 1 km

191

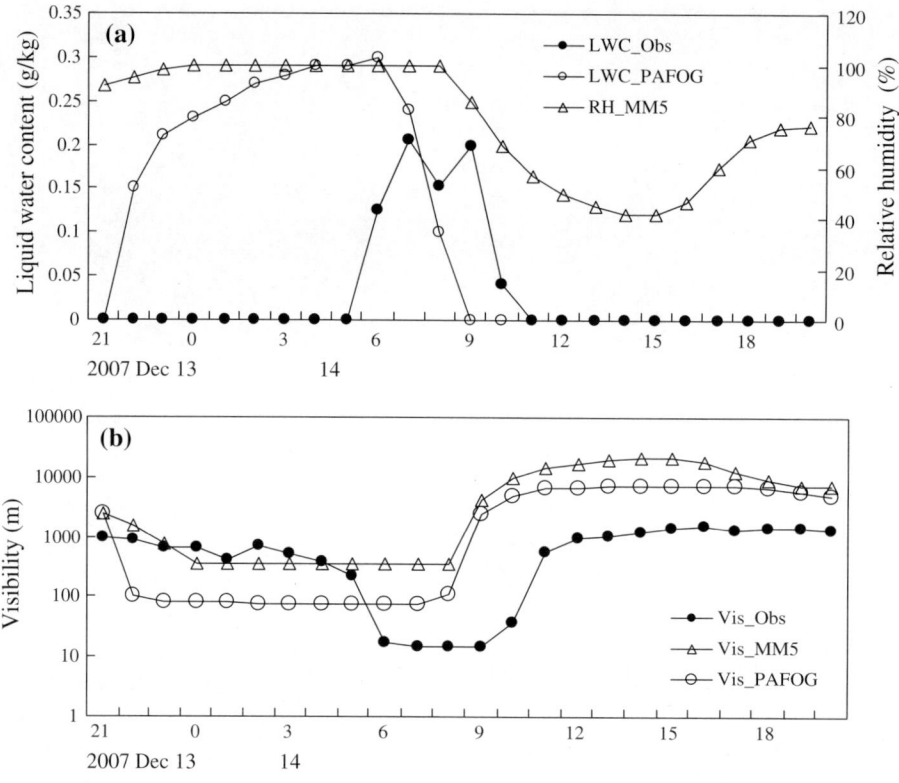

Figure 12
Simulated and measured ground-level LWC or RH (**a**) and visibility (**b**) on the NUIST campus for case 2

even without measurable LWC. For example, the observed visibility changed from fog ($V \leq 1$ km) to dense fog ($V \leq 500$ m) from 22 LST, 13 December to 05 LST, 14 December, while the measurable LWC (0.01 g/kg) appeared until 06 LST on 14 December (Fig. 12a). In Fig. 12b, due to the earlier formation of liquid water, the PAFOG simulated visibilities were lower than observations in the initial stage (before 06 LST). In the late stage (after 09 LST), the PAFOG simulated liquid water disappeared (MM5 RH decreased) quickly and simulated visibilities by the two models soon increased to over 1 km, while the observed visibilities increased slower than simulation, showing hazy weather after fog. Despite those differences, the variation trends were highly correlated with a correlation coefficient of 0.70 between PAFOG and observations, indicating that PAFOG predicted the formation and dissipation of this radiation fog well, and that it performed much better than MM5 in this case.

Through the above validations for the two cases, one can see that PAFOG performs quite well with the input from MM5, especially for the radiation fog.

4.4. Comparisons of PAFOG performance with measured and modeled IC/BC

As discussed in Sect. 4.3, it was concluded that the PAFOG performance depends highly on the MM5 output. To test the impacts of mesoscale background input data uncertainties, especially cloud, which is not a direct output of MM5 but diagnosed from MM5 output, PAFOG was also run using the observed high resolution radiosonde data at Nanjing observatory and observed cloud cover, hoping to identify the sensitivity of input data (initial state and hourly cloud) by using observed profile data, which is supposed to be very accurate.

In this run, the upper BCs are set constant, with $u_g = 2$ m/s, $v_g = 0$. The cloud cover was interpolated from the 3-h observations. For case 1, there are no cloud records at most times because of extremely dense fog in the ground level, especially on 26–27 December. So we set the low cloud cover around 10–30% on 26 December and 80% on 27 December, no medium or high clouds for those times without observations, according to the experience that too much cloud would make the fog dissipate. The assumed cloud covers are lower than MM5 simulations on 26 December and higher than MM5 on 27 December. The comparisons of the two runs with field observations were presented in this section. For convenience, the run with observed (modeled) profiles will be referred to as the test (default) run hereafter.

Figure 13 provides the ground level hourly LWC on the NUIST campus from PAFOG with two schemes (default run and test run), in comparison with measurements for case 1. It can be seen that the LWC of the test run is close to that of the default run on 25 December, but much lower/higher than that of the default run in the morning/afternoon of 26 December. The exceptionally large LWC on 26 December might be a result of the improper cloud cover, which indicates that cloud cover is a key factor affecting fog production in the 1D fog model. When compared with the observations, the LWC of the test run is closer to the observations on the first day, especially in the trends, but it is unbelievably larger than the observation on the second day. For the third day, the LWCs from both runs are much higher than observations, and the LWC from the test run is lower than that of the default run, which reflect the impacts of higher cloud cover in the test run.

Similarly, Fig. 14 presents the ground level hourly LWC on the NUIST campus from PAFOG with the two schemes (default run and test run), in comparison with measurements for case 2. From Fig. 14, we can find that: (1) the LWCs of the two runs reach maxima at almost the same time (06 LST for default run and 07 LST for test run); (2) the LWC from the test run is closer to the observation in both the maximum LWC time and fog dissipation time. So, the test run with observed ICs and cloud covers outperforms the default run with MM5 output in this case. In addition, both of the two runs fail to catch the double-layer structure of this fog event, which might be ascribed to the fact that the present 1D model does not include the effect of advection and large-scale subsidence. According to YANG et al. (2010b), this fog was first formed through radiation cooling in the ground level, and then followed by a cloud layer caused by low-level cold advection. Although YANG et al. (2010a) classified it as a radiation fog, the weak low-level cold advection did promote its development. For example, a cloud layer formed in the low level and descended downward to combine with the ground level fog. From MM5 output, although there is no low cloud formed like this in case 1 on 25 December, there did exist a thin layer of high humidity at around 600 m, which also can be seen in the PAFOG output (Fig. 15). It is reasonable to deduce that the high humid layer would form low cloud if there were appropriate cold advection, and the low cloud would descend downward if there were large scale subsidence. On the other hand, the weak cold advection should belong to a local scale

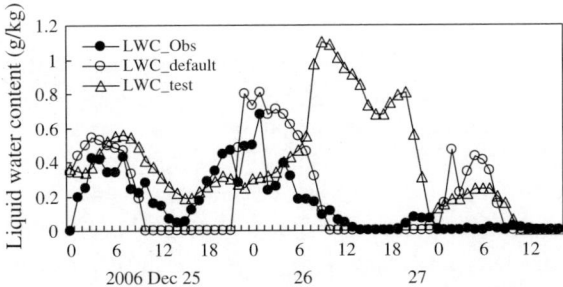

Figure 13
Ground level hourly LWC on NUIST campus from PAFOG with two schemes and measurements for case 1

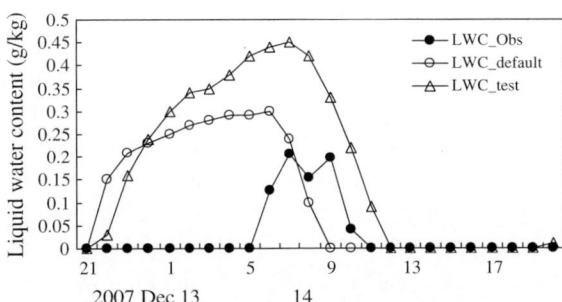

Figure 14
Ground level hourly LWC on NUIST campus from PAFOG with two schemes and measurements for case 2

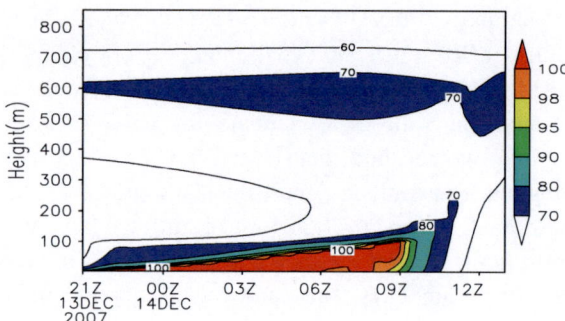

Figure 15
Time–height cross-section of RH (%) on NUIST campus by PAFOG default run

phenomenon because the fog observed at the Nanjing observatory dissipated at least 2 h earlier than at NUIST. Thus MM5 did not reproduce this local scale circulation successfully, which is a common characteristics for the current NWP models (BERGOT et al., 2005).

To conclude this section, the results of test runs demonstrated that the PAFOG results are very sensitive to the cloud amount input. For those fog cases developed or maintained by the function of advection, the background information from MM5, e.g. warm/cold advection and large scale subsidence, are critical to successful fog predictions. However, as shown by the experiment of BERGOT et al. (2005), knowing how to calculate the horizontal advection from a NWP model to improve the results of 1D model is still a challenge. The method should be site-dependent. In addition, how to prepare cloud forcing from the NWP model for the 1D model is also imperative for accurate fog forecasting.

5. Summary and Conclusions

MM5 and a MM5-PAFOG coupled systems were used to simulate regional fog occurrences in eastern China. The models were run daily for December, 2006, and December, 2007, and evaluated objectively using the fog occurrence record data at Nanjing observatory. Attention was paid to nine fog events observed in a field on the NUIST campus. Among the nine fog events, two events were selected for detailed evaluation and discussion. The main purpose is to find a best way to forecast regional fog.

Simulations of ground level meteorology were validated against the routine observations at the CMA observational network. The LWC-only method by MM5 showed the highest TS and ETS, but lowest HR; the LWC + RH method by MM5 increased HR, but decreased ETS and increased Bias/FAR as a penalty. To keep the high HR of RH + LWC method while decreasing FAR, it would be necessary to add some restrictive conditions, e.g. the rule-based method. As shown by ZHOU and DU (2010), the rule-based method with NWP models still yields high FAR. So, more attention should be paid on how to avoid false alarms in fog simulation, since all methods tend to overestimate fog occurrence, especially on near-fog days. Thus, it is imperative to find some way to remove the near-fog situation in fog forecasting.

The models reproduced almost all fog events that were observed on the NUIST campus if both thresholds of LWC and RH were used. Through validation of simulated high RH (>98%) or LWC areas against observed low visibility areas, the simulated hourly LWC and visibilities against the measurements in the field observations at NUIST, the performances of MM5 and PAFOG on fog predictability were evaluated separately for the domain of eastern China and for Nanjing. For the domain of eastern China, MM5 produced regional fog or fog patches on each fog day; the simulated fog areas cover most or part of the observed fog areas. Generally, MM5 performed better for advection–radiation fog than for radiation fog, while PAFOG showed evident advantages on radiation fog over MM5 and could be an alternative tool for forecasting fogs mainly generated by radiative cooling. From the situations in NUIST, MM5 simulated fog generally dissipated earlier than observation, while PAFOG showed evident advantages on the dissipation time, because the fog dissipation times predicted by PA-FOG were closer to observations than those of MM5.

The detailed comparisons of models results with field observations on the NUIST campus for two typical cases showed the advantages of PAFOG on the ground level LWC and visibility over MM5, especially for radiation fog. However, it is incapable of reproducing the low clouds resulted by advection, at present. The sensitive experiments for two cases ensure the quality of the MM5 modeled IC/BC for the

1D model, and demonstrate the high sensitivity of PAFOG results on the input background, such as cloud cover. Therefore, it is important to improve the quality of the input data for the 1D model, considering the impacts of large scale background, e.g. cloud cover, advection, and large scale subsidence.

Although PAFOG showed some lower performance when considering fog occurrence at Nanjing observatory, using PAFOG or some other 1D fog model is a good way to predict such a complex weather like radiation fog on land, particularly for its near-fog cases. In our future work, we will improve the coupling method, for example, by including the outside forcing (e.g. advection of temperature and vapor, large scale subsidence) in the 1D model, testing the impacts of different cloud diagnostic methods, conducting more numerical simulations for cases free of fog, especially the near-fog cases, by adopting the rule-based fog detection method by ZHOU and DU (2010), and trying the more sophisticated method developed by ZHOU and FERRIER (2008). In summary, for fog study, long term simulations should be done, instead of simulation on single fog cases, and more attention should be paid to the near-fog cases. In addition, further work must be done to improve the ground level temperature simulation.

Acknowledgments

This work was supported by funds from the National Natural Science Foundation of China (40775010), Project of New Technology Application of China Meteorology Agency (CMATG2010M16) and the Special Funds for Public Welfare of China (Grant No. GYHY200906012). We sincerely thank Professor Bott for providing the code of PAFOG and are grateful to the anonymous referees' valuable suggestions.

REFERENCES

BERGOT, T., CARRER, D., NOILHAN, J., and BOUGEAULT, P. (2005), *Improved site-specific numerical prediction of fog and low clouds: a feasibility study*, Weather and Forecasting *20*, 627-646.

BERGOT, T., TERRADELLAS, E., CUXART, J., MIRA, A., LIECHTI, O., MUELLER, M., and NIELSEN, N.W. (2007), *Intercomparison of Single-Column Numerical Models for the Prediction of Radiation Fog*, J. Appl. Meteor. Climatology *46*, 504–521.

BOTT, A., SIEVERS, U., and ZDUNKOWSKI, W. (1990), *A radiation fog model with a detailed treatment of the interaction between radiative transfer and fog microphysics*, J. Atmos. Sci. *47*(18), 2153-2166.

BOTT, A. and TRAUTMANN, T. (2002), *PAFOG-a new efficient forecast model of radiation fog and low-level stratiform clouds*, Atmos. Res. *64*,191-203.

BURROWS, W.R., and TOTH, G. (2011), Automated fog and stratus forecasts from the Canadian RDPS operational NWP model, 24th Conference on Weather Analysis and Forecasting, AMS, Seattle, Washington, USA, 23-27 January, 2011.

CHAUMERLIAC, N., RICHARD, E., and PINTY, J.-P. (1987), *Sulfur scavenging in a mesoscale model with quasi-spectral microphysics: Two-dimensional results for continental and maritime clouds*, J. Geophys. Res. *92*(D3), 3114-3126.

FU, G., GUO, J., XIE, S.P., DUAN, Y., and ZHANG, M. (2006), *Analysis and high-resolution modeling of a dense sea fog event over the Yellow Sea*, Atmos. Res. *81*(4), 293-303.

GAO, S., LIN, H., SHEN, B., and FU, G. (2007), *A heavy sea fog event over yellow sea in March 2005: Analysis and Numerical Modeling*, Advances in Atmospheric Sciences *24*(1), 65-81.

GRELL, G.A., DUDHIA, J., and STAUFFER, D.R. (1994), A description of the fifth generation Penn State/NCAR Mesoscale Model (MM5), NCAR Tech. Note NCAR/TN-398+STR, Natl. Cent for Atmos. Res., Boulder, Colo.

GULTEPE, I., TARDIF, R., MICHAELIDES, S. C., CERMAK, J., BOTT, A., BENDIX, J., MULLER, M.D., PAGOWSKI, M., HANSEN, B., ELLROD, G., JACOBS, W., TOTH, G., COBER, and S.G. (2007), *Fog Research: A Review of Past Achievements and Future Perspectives*, Pure and Applied Geophysics *164*, 1420-9136.

GULTEPE, I., PEARSON, G., MILBRANDT, J.A., HANSEN, B., PLATNICK, S., TAYLOR, P., GORDON, M., OAKLEY, J. P., and COBER S.G. (2009), *The fog remote sensing and modeling field project*, Bulletin of the American Meteorological Society *90*(3), 341-359.

JACOBS, W., MICHAELIDES, S.C., NIETOSVAARA, V., BOTT,A., BENDIX, J., and CERMAK, J. (2005). Short range forecasting methods of fog, visibility and low clouds, Second Midterm Workshop, Langen, Germany, 20 October.

KONG, F. (2002), *An experimental simulation of a coastal fog-stratus case using COAMPS (tm) model*, Atmos. Res. *64*, 205-215.

KUNKEL, B. A. (1984), *Parameterization of droplet terminal velocity and extinction coefficient in fog models*, Journal of Climate and applied meteorology *23*(1), 34-41.

LIN, Y. (2010), The parameterizations and numerical study of visibility during the fog in the winter of Nanjing and probabilistic forecast of visibility, Master Dissertation of Nanjing University of Information Science & Technology (in Chinese).

LIU, X, ZHANG, H., LI Q. and ZHU, Y. (2005), *Preliminary research on the climatic characteristics and change of fog in China*, Journal of Applied Meteorological Sciences (in Chinese) *16*(2), 220-271.

LIU, D., PU, M., YANG, J., ZHANG, G., YAN, W. and LI Z. (2009), *Microphysical structure and evolution of four-day persistent fogs around Nanjing in December 2006*, Acta Meteorologica Sinica (in Chinese), *67*(1), 147-157.

LIU, D., YANG, J., NIU, S. and LI, Z. (2011), *On the evolution and structure of a radiation fog event in Nanjing*, Advances in Atmospheric Science, *28*(1), 223-237.

LU, R., TURCO, R.P. and JACOBSON M.Z. (1997), *An integrated air pollution modeling system for urban and regional scales: 2. Simulation for SCAQS 1987*, J. Geophys Res, *102*(5), 6081-6098.

Lu, C., Niu, S., Tang, L., Lv, J., Zhao, L. and Zhu B. (2010), *Chemical composition of fog water in Nanjing area of China and its related fog microphysics*, Atmospheric Research 97(1), 47-69.

Mellor, G.L. and Yamada, T. (1974), *A hierarchy of turbulence closure models for planetary boundary layers*, J. Atmos., Sci., 31(7), 1791-1806.

Miao, Y., Potts, R., Huang, X., Elliott, G., Rivett, R. and Manickam, M. (2010), Application of Fuzzy-Logic NWP Fog Guidance to Perth Fog Forecasting Decision Support System, 5th International Conference on Fog, Fog Colloection and Dew, Munster, Germany, 25-30 July, 2010.

Muller, M.D., Schmutz, C. and Parlow, E. (2007), *A one-dimensional Ensemble Forecast and Assimilation System for Fog Prediction*, Pure and Applied Geophysics 164, 1241-1264.

Nickerson, E.C., *Richard, E., Rosset, R.* and Smith, D.R. (1986), *The numerical simulation of clouds, rain, and airflow over the Vosges and Black Forest mountains: a meso-β model with parameterized microphysics*, Monthly Weather Review 114, 398-414.

Niu, S., Lu, C., Yu, H., Zhao, L. and Lu J. (2010), *Fog Research in China: An Overview*, Advances in Atmospheric Sciences 27(3), 639-661.

Otte, T. L. (2008), *The Impact of Nudging in the Meteorological Model for Retrospective Air Quality Simulations: Part I: Evaluation against National Observation Networks*, J. Appl. Meteor. and Clim. 47, 1853-1867.

Pielk, R.A. (1984), Mesoscale Meteorological Modeling, Academic Press, Orlando 612 pp.

Pu, M., Zhang, G., Yan, W. and Li, Z. (2008), *Features of a rare advection-radiation fog event*, Science in China Series D: Earth Sciences 51(7), 1-9.

Siebert, J., Bott, A. and Zdunkowski, W. (1992a), *Influence of a vegetation-soil model on the simulation of radiation fog*, Beitr. Phys. Atmos. 65, 93-106.

Siebert, J., Sievers, U. and Zdunkowski, W. (1992b), *A one-dimensional simulation of the interaction between land surface processes and the atmosphere*, Boundary–Layer Meteorology, 59, 1-34.

Shi, C., Roth, M., Zhang, H., and Li, Z. (2008), *Impacts of urbanization on long-term fog variation in Anhui Province, China*, Atmos. Environ. 42, 8484-8492.

Shi, C., Yang J., Qiu M., Zhang, H., Zhang, S., and Li, Z. (2010), *Analysis of an extremely dense regional fog event in Eastern China using a mesoscale model*, Atmospheric Research 95(4), 428-440.

Stoelinga, M. T. and Warner, T. T. (1999), *Nonhydrostatic, Mesobeta-Scale Model Simulations of Cloud Ceiling and Visibility for an East Coast Winter Precipitation Event*, J. Appl. Meteor. 38(4), 385-404.

Tardif, R. (2007), *The impact of vertical resolution in the explicit numerical forecasting of radiation fog: A case study*, Pure and Applied Geophysics 164, 1221-1240.

Tong, Y., Yin, Y., Xu, X., Xu, M. and Xiang, Y. (2009), *Climatological characteristics of fogs in Nanjing*, Journal of Nanjing Institute of Meteorology (in Chinese) 32(1), 115-120.

Van Der Velde, I.R., Steeneveld, G.J., Wichers Schreur, B.G.J. and Holtslag, A.A.M. (2010), *Modeling and forecasting the onset and duration of severe radiation fog under frost conditions*, Mon. Wea. Rev. 138, 4237-4253.

WMO (1992), International Meteorological Vocabulary, WMO 182.

Wu, D., Deng, X., Ye, Y. and Mao, W. (2004), *The study on fog water chemical composition in Dayaoshan of Nanling mountain*, ACTA Meteorologica SINICA (in Chinese) 62(4), 476-485.

Wu, D., Deng, X., Mao, J., Mao, W., Ye, Y., Bi, Y., Tang, H., and Wan, Q. (2007), *A study on macro- and micro-structures of heavy fog and visibility at freeway in the Nanling Dayaoshan mountain*, ACTA Meteorologica SINICA 65(3), 406-415.

Yang, J., Xie, Y., Shi, C., Liu, D., Niu, S. and Li, Z. (2009), *Differences in Ion compositions of winter fog water between radiation and advection-radiation fog episodes in Nanjing*, Transactions of Atmospheric Sciences (in Chinese) 32, 776-782.

Yang, J., Niu, Z., Shi, C., Liu, D., and Li, Z. (2010a), *Microphysics of Atmospheric Aerosols During Winter Haze/Fog Events in Nanjing*, Environmental Science (in Chinese) 31(7), 1425-1431.

Yang, J., Wang, L., Liu D. and Li, Z. (2010b), *The boundary layer structure and evolution mechanisms of a deep dense fog event*, Acta Meteorologica Sinica (in Chinese) 68(6), 998-1006.

Zdunkowski, W.G., Panhans, W.-G., Welch, R.M. and Korb, G.J. (1982), *A radiation scheme for circulation and climate models*, Beitr. Phys. Atmos. 55, 215-238.

Zhou, B. and Ferrier, B.S. (2008), *Asymptotic analysis of equilibrium in radiation fog*, Journal of Applied Meteor and Clim. 47, 1704-1722.

Zhou, B. and Du, J. (2010), *Fog prediction from a multi-model mesoscale ensemble prediction system*, Wea. Forecasting 25, 303-322.

Zhou, B., Du J., Gultepe I. and Dimego G. (2011), Forecast of Low Visibility and Fog From NCEP-Current Status and Efforts, this issue.

Zhou, Z., Zhu, Y., and Ju, X. (2007), *Dense fog events in the Yangtze Delta and their climatic characteristics*, Progress in Natural Science 17(1), 66-71.

Zehnder J. A. (2002), *Simple modifications to improve Fifth-Generation Pennsylvania State University–National Canter for Atmospheric Research Mesoscale Model Performance for the Phoenix, Arizona, Metropolitan Area*, Journal of Applied Meteorology 41, 971-979.

(Received October 1, 2010, revised April 24, 2011, accepted May 6, 2011, Published online June 3, 2011)

Pure Appl. Geophys. 169 (2012), 961–981
© 2011 Springer Basel AG
DOI 10.1007/s00024-011-0393-0

A Study of Fog Characteristics Using a Coupled WRF–COBEL Model Over Thessaloniki Airport, Greece

Stavroula Stolaki,[1] Ioannis Pytharoulis,[1] and Theodore Karacostas[1]

Abstract—An attempt is made to couple the one dimensional COBEL-ISBA (Code de Brouillard à l'Échelle Locale-Interactions Soil Biosphere Atmosphere) model with the WRF (Weather Research and Forecasting)–ARW (Advanced Research WRF) numerical weather prediction model to study a fog event that formed on 20 January 2008 over Thessaloniki Airport, Greece. It is the first time that the coupling of COBEL and WRF models is achieved and applied to a fog event over an airport. At first, the performance of the integrated WRF–COBEL system is investigated, by validating it against the available surface observations. The temperature and humidity vertical profiles were used for initializing the model. The performance of WRF–COBEL is considered successful, since it realistically simulated the fog onset and dissipation better than the WRF alone. The COBEL's sensitivity to initial conditions such as temperature and specific humidity perturbations was also tested. It is found that a small increase of temperature ($\sim 1°C$) counteracts fog development and results in less fog density. On the other hand, a small decrease of temperature results in much denser fog formation. It is concluded that the integrated model approach for aviation applications can be useful to study fog impact on local traffic and aviation.

Key words: Fog, COBEL-ISBA, WRF-ARW, surface thermohydrometric conditions.

1. Introduction

Fog is a weather phenomenon that causes various problems at the airport of Thessaloniki in northern Greece. The airport is located 15 km southeast of Thessaloniki center in a fog-prone coastal region with several geomorphological complexities (valley; mountain; coastline; presence of urban and rural areas) (Fig. 1). The nearby *Anthemountas* valley and

the *Chortiatis* mountain ($\sim 1,200$ m) are believed to be of high importance, since a breeze, originating from the valley, descends from the slopes of the surrounding hills to the airport area, advecting already formed fog or providing the airport area with cold air from the valley (Stolaki *et al.* 2009).

Flight delay hours reach as much as 250 per year, whereas around 25 flights per year are cancelled due to reduced visibility conditions. Air traffic safety and operational efficiency are hindered, and the cost of such impacts imposes a great burden on the airport authorities and the airway companies. As a consequence, there is a high demand for a thorough understanding of the factors affecting the onset, the development and the dissipation stages of fog, which would lead to more accurate forecasting techniques. To successfully accomplish that, real-time measurements are needed, which, in the case of Thessaloniki's airport, are limited. Apart from SYNOPs, METARs and the routine radiosonde of the airport's meteorological station, there is no other observational data for the study of fog in the area of interest.

Under the limited availability of real-time measurements, the use of a single column model provides a valuable tool that can be used to better understand small-scale processes in the boundary layer and their related influence on fog evolution, despite the poor representation of horizontal heterogeneities. In the case of Thessaloniki Airport, the horizontal heterogeneities of its geomorphological environment need to be taken into account and this is done by coupling the 1D model with a 3D model, which provides the former with mesoscale tendencies over the surrounding region. The main advantages of such an approach are that it allows the use of an enhanced vertical resolution over the location (Tardif 2007); the assimilation of the combined available local observations with the 3D

---

[1] Department of Meteorology and Climatology, School of Geology, Aristotle University of Thessaloniki, 54124 Thessaloniki, Greece. E-mail: sstolaki@geo.auth.gr

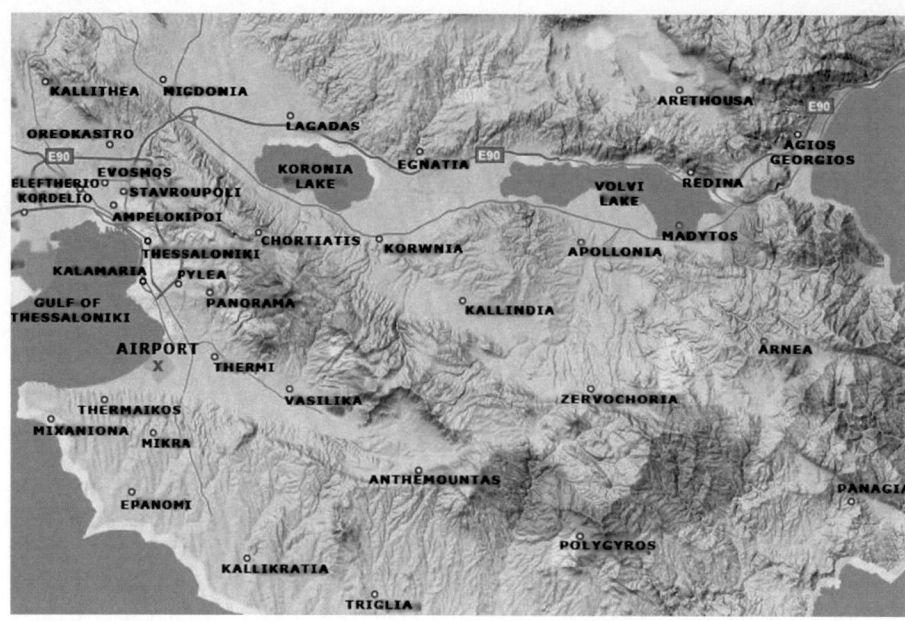

Figure 1
Relief map of the region around Thessaloniki's airport (from http://maps.google.com). The location of Thessaloniki's airport is indicated by X

model outputs, whose benefits have been outlined in several previous studies (Bergot *et al.* 2005; Bergot 2007); and the possible use of much more sophisticated parameterizations (e.g. Bott and Trautmann 2002). Bearing all these in mind, an attempt is made to couple the one dimensional COBEL-ISBA (Code de Brouillard à l'Échelle Locale-Interactions Soil Biosphere Atmosphere) model with the Weather Research and Forecasting (WRF) numerical weather prediction (NWP) model. Such an accomplishment is expected to provide enhanced capabilities for understanding and forecasting fog in the area of interest.

In the past, a few studies on fog has been carried out at or nearby Thessaloniki Airport. In this respect, Angouridakis (1973) studied the relationship between weather types and fog occurrences over Thessaloniki Airport. Angouridakis and Flocas (1983) proposed a linear equation used for forecasting radiation fog in Thessaloniki, by estimating the fog point temperature introduced by Saunders (1950) in relation to several meteorological parameters taken at 1200 and 1800 UTC on the previous day. Foris (2002) studied the climatology of fog at the airport for the period 1971–2002 and identified the most common fog types that occur in the area, that is radiation and advection fog. Houssos *et al.* (2008) examined the characteristics

of atmospheric circulation favouring the formation of fog in Thessaloniki by applying an objective analysis on 138 long duration fog events of the period 1959–2002. Last, Stolaki *et al.* (2009) have classified fog events of the period 1971–2005 into four fog types: radiation, advection, cloud base lowering and precipitation. According to the aforementioned efforts, radiation and advection fog are the most frequent fog types forming over the area.

Despite the valuable information that has been gained so far, there is no published work providing more information on the factors affecting fog formation and dissipation at this location. A high resolution state-of-the-art numerical model, such as WRF, is expected to be a useful tool for the representation of the atmospheric conditions associated with such phenomena. However, one of the most recent studies (van der Velde *et al.* 2010), that concerned the use of WRF in an effort of analyzing a case of severe radiation fog, concluded that there is still much work to be done, so that the 3D NWP models alone can represent fog formation. Taking into account the limitations that, so far, NWP models impose on fog simulation and forecasting (Teradellas and Cano 2007), the main incentive of the present work was to study a fog event that formed at

Thessaloniki Airport with the use of a one-dimensional model, where the atmospheric terms depending on the horizontal heterogeneities of the location are estimated from the outputs of an operational 3D model.

To this end, an interface between WRF and CO-BEL-ISBA models has been developed and presented for the first time. WRF is a state-of-the-art mesoscale numerical weather prediction system designed to serve both operational forecasting and atmospheric research needs (SKAMAROCK et al. 2008). On the other hand, COBEL (BERGOT et al. 2005) is a high resolution single model column, designed to study the boundary layer vertical structure at a local scale and is coupled to the land-surface scheme ISBA-DF (BOONE 1999, 2000).

COBEL-ISBA, coupled with the ALADIN model, has been in operational use since 2005 at Paris-Charles de Gaulle (Paris-CdG) Airport in France (BERGOT et al. 2005), to provide estimated times for the onset and lifting of low visibility conditions, while it is also being installed over the Paris–Orly and Lyon–Saint Exupéry airports in France (RÉMY and BERGOT 2009). Other than the coupling of CO-BEL with the ALADIN model (BERGOT et al. 2005) and MM5 (PINHEIRO et al. 2006), no other effort has been reported in the literature. Nevertheless, the coupling of 1D with 3D models is not rare. The MM5 mesoscale model has been coupled with the PAFOG (BOTT and TRAUTMANN 2002) single column model (SHI et al. 2010), and there are plans to couple the COSMO-DE of the German Meteorological Service with the PAFOG (ROHN et al. 2010).

The work presented here aims at studying the advection-radiation fog event that formed on the 20 January 2008 over the Thessaloniki Airport, by implementing a coupled WRF–COBEL model. The fog on this day occurred due to radiative and advective processes. The major goal of this work is to study this radiation-advective fog event using a combination of a high-resolution 1D fog model with a 3D NWP model output, rather than relying solely on a 3D model. The performance of the integrated WRF–COBEL system is examined and fog sensitivity to initial temperature and humidity conditions is studied in order to better understand the factors affecting fog formation and development.

2. Description of the Models and Their Coupling

2.1. The WRF-ARW Model

The non-hydrostatic Weather Research and Forecasting model with the Advanced Research dynamic solver (WRF-ARW Version 3.1.1) was utilized in the numerical experiments. It is a flexible, state-of-the-art numerical weather prediction system, designed to operate in both research and operational mode (SKAMAROCK et al. 2008; WANG et al. 2010). Nevertheless, there has been a limited use of WRF in fog modeling and forecasting.

Two 2-way interactive model domains (Fig. 2a) covering the Balkans and the major area of Central Northern Greece (Macedonia) region (Fig. 2b) at horizontal grid-spacings of 10 km × 10 km (D1) and 2 km × 2 km (D2), respectively, have been used, utilizing the staggered Arakawa C grid. Fine-resolution data (30" × 30") were used in the definition of topography, land use and soil type. Six-hourly ECMWF analyses were used as initial and lateral boundary conditions for the coarse domain, while the finer inner domain was two-way nested to the larger one. The analyses were available at 6-hourly intervals on a 0.25° lat. x 0.25° lon. regular grid. The necessary fields were retrieved at and near the surface and on the pressure levels of 1,000, 925, 850, 700, 500, 400, 300, 250, 200, 150, 100, 70 and 50 hPa. Soil temperature and moisture analyses were available on four layers down to 2.55 m. The sea-surface temperatures were also derived from the ECMWF analyses and were kept fixed to their initial values throughout the simulations. In the vertical, 39 full sigma levels (up to 50 hPa) were used by both nests. An alternative distribution of the Planetary Boundary Layer (PBL) grid has been used that consists of 9 sigma levels up to the first kilometer, instead of six originally defined by WRF. Moreover, the lowest sigma level is now located at an approximate height of about 10 m above ground, instead of around 28 m in the baseline configuration of WRF.

The microphysical processes are parameterized by the Thompson (THOMPSON 2008) microphysics scheme, while the PBL is represented by the use of the Mellor–Yamada–Janjic (Eta) TKE (JANJIC 2002)

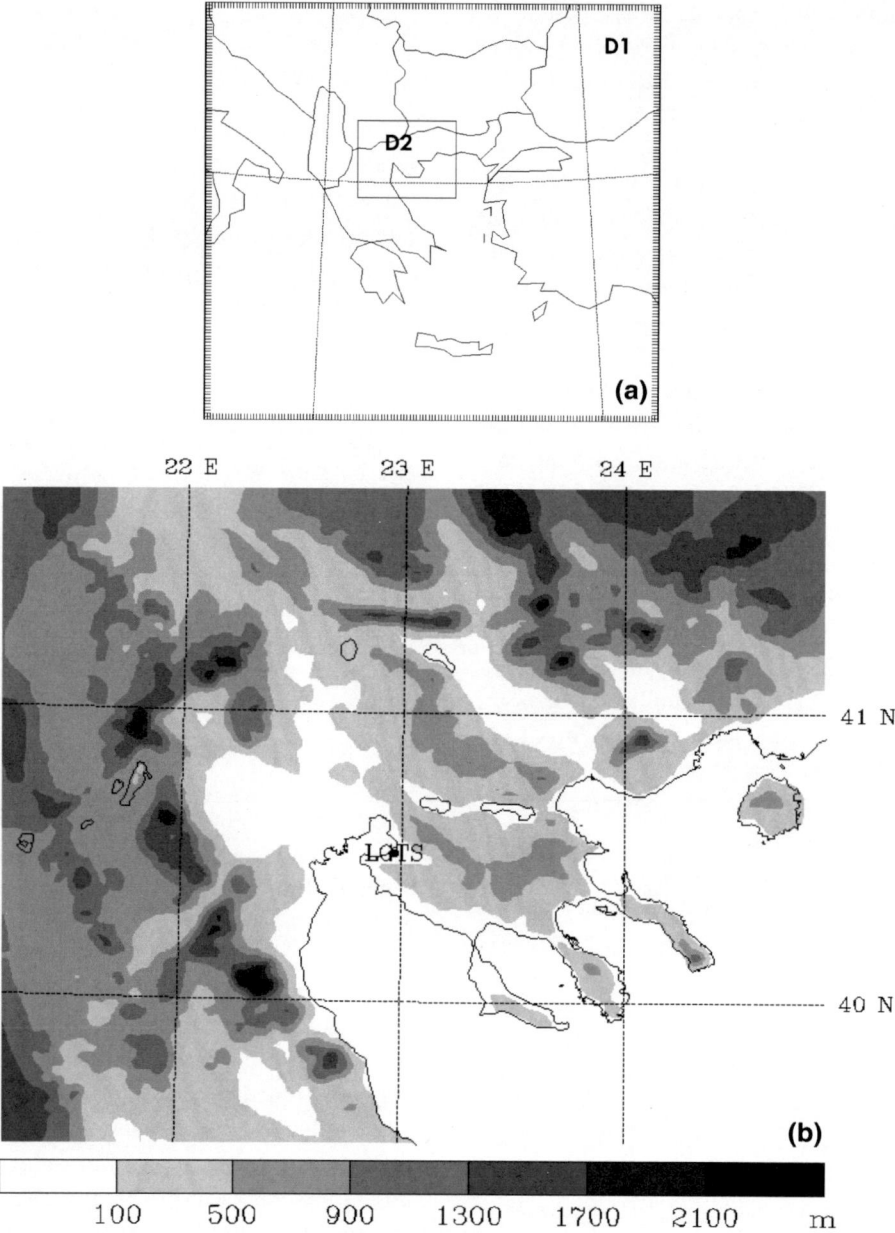

Figure 2

a The two nests used by WRF-ARW for the numerical experiments. **b** The topography of the inner domain (D2) and the location of the airport (LGTS)

PBL scheme. Cumulus convection is parameterized only in the coarse nest by the Betts–Miller–Janjic scheme (BETTS 1986; BETTS AND MILLER 1986; JANJIC 2000). The GFDL and the Monin–Obukhov (Eta) schemes were used in both nests to represent longwave/shortwave radiation and surface layer, respectively. The unified Noah land-surface model (CHEN AND DUDHIA 2001; EK *et al.* 2003), that uses four soil layers (0–10 cm, 10–40 cm, 40–100 cm, 100–200 cm), was utilized for the soil physics. It is a soil temperature and moisture model that also predicts canopy moisture, including root zone, evapotranspiration, soil drainage and runoff (SKAMAROCK *et al.* 2008).

2.2. The COBEL-ISBA Model

COBEL is a single column numerical model that was developed in collaboration among the Laboratoire d'Aérologie-Université Paul Sabatier/Centre National de la Recherche Scientifique (CNRS), Université du Québec a Montréal (U.Q.A.M., Canada) and Météo-France/Centre National de Recherches Météorologiques (CNRM). Its original aim was to simulate the evolution of the very stable atmospheric boundary layer vertical structure at the local scale.

As documented in BERGOT et al. (2005), the utilized physical package includes a detailed radiation transfer scheme, a parameterization of the boundary layer turbulent mixing under stable, neutral and unstable conditions, and a microphysical parameterization adapted to fog and low clouds. The model equations are solved on a high-resolution vertical grid which consists of 30 levels between 0.5 and 1,360 m, with 20 levels below 200 m (0.5, 1.65, 3.05, 4.74, 6.78, 9.25, 12.24, …, 200 m). It is coupled with the multilayer surface-vegetation-atmosphere transfer scheme ISBA-DF (BOONE 1999, 2000) which runs with seven soil levels, from 1 mm to 1.7 m below the surface.

COBEL has been designed to operate with a state-of-the-art 3D mesoscale NWP model (e.g. ALADIN at Météo-France) in order to take into account the effects of possible horizontal heterogeneities in the mean state of the atmosphere. The effect of these heterogeneities is treated as external forcings by the column model. These external mesoscale forcings are the pressure gradient forces (geostrophic wind), the horizontal advection of potential temperature and moisture, the vertical air velocity, the local pressure tendencies and the cloud cover and they are estimated by using the output of an NWP mesoscale model. These NWP derived input data consist of: temperature, the two horizontal wind components, potential temperature, specific humidity and pressure at 15 levels, 20, 50, 100, 250, 500, 750, 1,000, 1,250, 1,500, 2,000, 2,500, 3,000, 4,000, 5,000 and 6,000 m height above ground, as well as downward shortwave and longwave radiative fluxes and downward shortwave and longwave radiative fluxes under clear skies at the ground. The radiative fluxes under clear skies are necessary for the computation of the extinction of shortwave radiation, and the infrared radiation emitted by clouds located above the COBEL grid.

An assimilation scheme (BERGOT et al. 2005) that combines the mesoscale forcings and local observations provides the initial state of the atmospheric profiles at the initalization time of the 1D model. Although available, the radiosonde data is not used for the initialization of COBEL, since the local assimilation scheme used in the current work did not include soundings. The available local observations for Thessaloniki's airport (40.52 °N, 22.96 °E) consist of only the operational three hourly and half hourly information provided by the airport's meteorological station SYNOPs and METARs, respectively. In this study, the data for the calculation of the mesoscale forcings are provided by the WRF-ARW model.

2.3. The WRF–COBEL Coupling

An interface between WRF and COBEL has been developed for their coupling. Only the WRF output of the inner domain (D2) is used. Firstly, the WRF output on sigma levels undergoes an unstaggering from the original staggered Arakawa-C grid and followed by a bilinear horizontal interpolation to a regular lat lon grid of 0.02° (∼2 km) grid spacing. In the vertical, it is required that the 39 model sigma levels be transformed into the 15 height levels above ground (20, 50, 100, 250, 500, 750, 1,000, 1,250, 1,500, 2,000, 2,500, 3,000, 4,000, 5,000 and 6,000 m) through linear interpolation and taking into account the geopotential height of the sigma surfaces and model topography. This also applies for the WRF meteorological fields such as the horizontal wind components, the temperature, the potential temperature, the water vapor mixing ratio and pressure. To generalize the use of the interface in alternative model configurations than the one used in this study and in the case that the lowest model level is higher than the required lowest height level, the single level fields of 2 m temperature, 2 m water vapor mixing ratio, 10 m horizontal wind speed components and surface pressure are used and the vertical interpolation is performed between the single level and the lowest model sigma level. Therefore, the above mentioned parameters are postprocessed through the interface and input into COBEL and then, these parameters are interpolated into the high resolution 30 level vertical grid of the one-dimensional model.

As a next step, the WRF lack of availability of the clear sky downward longwave radiative flux at the ground, emitted by the atmosphere and the clouds, was treated. In contrast to the shortwave radiative flux under clear skies that is readily available by using the GFDL shortwave radiation scheme (SKAMAROCK et al. 2008), the clear sky longwave radiation flux is not provided by the 3D NWP model. Therefore, this parameter was estimated by selecting one among four parameterization schemes (BRUNT 1932; SWINBANK 1963; BRUTSAERT 1975; PRATA 1996). These schemes are based, either on empirical relationships derived from observed radiation fluxes, or on the radiative transfer theory (NIEMELÄ et al. 2001). In Swinbank's parameterization the downward longwave radiation under clear skies is given as a function of screen-level temperature only, whereas Brunt's and Brutsaert's formula is a function of screen-level water-vapour pressure and temperature. Prata's parameterization depends on the precipitable water content and on screen-level temperature. It is to be noted that the above parameterizations mainly provide climatological estimates of the downward longwave radiation under clear skies. In the end, Brutsaert's formula was used, since it is a formula that, in comparison with the other ones examined, has performed well in several studies under varied climatic conditions (e.g. JIMÉNEZ et al. 1987; CULF and GASH 1993; DUARTE et al. 2006; BILBAO and DE MIGUEL 2006; KRUK et al. 2009).

In this study WRF is integrated in analysis mode. This means that the 6-hourly ECMWF analyses (not the 3-hourly forecasts) are used as lateral boundary conditions of the outer domain in the WRF simulations. WRF is initialized at 0000 UTC on 20 January 2008 providing hourly output up to 21 January 2008, 0000 UTC. COBEL is also initialized at 0000 UTC on 20 January 2008, that is, before the fog event, using the hourly WRF forecasts as input data. COBEL simulations have been performed up to 24 forecast hours.

2.4. Synoptic and Mesoscale Analysis

The selected fog event formed on 20 January 2008, beginning at 0400 UTC (Greek local time = UTC + 2 h) and ending at 0920 UTC. It is a typical fog event that, according to the fog event classification methodology applied by STOLAKI et al. (2009), exhibits the characteristics of an advection-radiation fog. Fog events pertaining to the radiation category are rather dense (with minimum visibility ranging between 0 and 300 m); they last less than 5 h, while they usually occur 1–3 h before sunrise and dissipate 2–4 h after sunrise in January (STOLAKI et al. 2009). Indeed, the fog event of 20 January lasted 5.5 h and its onset and dissipation fall within the typical onset and dissipation times of radiation fog events. As illustrated in Fig. 3, the fog that extended to an area covering the whole airport, was quite dense with estimated minimum visibility reaching 100 m. At the onset of the event the visibility dropped from 3 km to 800 m and after an hour it was already 100 m. It did not change until an hour before its dissipation. Few clouds with 600 m base height had formed before the fog onset.

According to the synoptic conditions, on 19 January 2008, at the 500 hPa level, a ridge centered over Spain and northwestern Africa was affecting northern and western Greece, while a trough was extending from western Turkey to northeastern Africa. A similar synoptic pattern (Fig. 4a) prevailed at 0000 UTC on 20 January 2008 a few hours before the fog onset. The values of lower-tropospheric vertical motion (omega) extracted from the ECMWF analyses at the nearest grid point to the airport's station, reveal positive values prevailing (not shown). This is evidence for subsidence associated with the anticyclonic synoptic conditions affecting the airport before fog formation.

At the surface (Fig. 4b), these upper level conditions were associated with an anticyclone with high pressures over Greece (mslp of 1,027 hPa at Thessaloniki's airport), weak synoptic winds in the area of interest and almost clear skies. A light breeze from mainly a southeastern direction (110°–120°) was prevailing at the airport a few hours before the fog onset. This is an indication of cold air originating from the *Anthemountas* valley, being advected to the

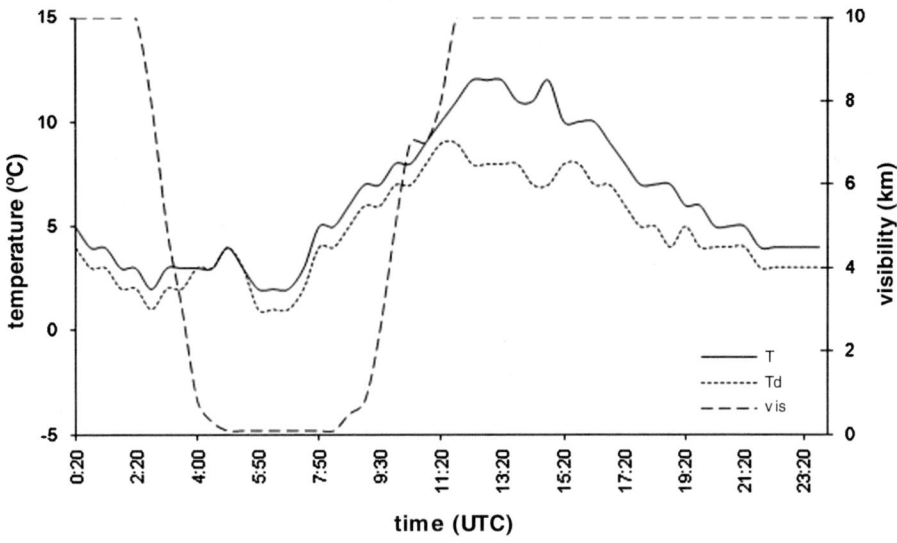

Figure 3

Temporal evolution of the 2 m temperature (T, °C), dew point temperature (T_d, °C) and visibility (vis, km) at Thessaloniki's airport on 20 January 2008 (METARs LGTS)

airport a few hours before the fog onset. The available radiosonde of 0000 UTC on 20 January 2008 (Fig. 5) indicates a thermal inversion being formed in the first 170 m above ground due to the nocturnal cooling, which is quite strong (the temperature increases with height from 4.4 to 10.2°C). Furthermore, in this column, the air is rather moist since the relative humidity reaches 93%. Over the 170 m height and up to the 850 hPa level, the air mass is close to saturation, whereas further up it is quite dry. This is in accordance with the subsidence taking place around this time due to the high pressure field over the wider area of the airport. Not surprisingly, near-surface relative humidity (RH) conditions a few hours prior to the fog onset were mostly characterized by high values (between 80 and 90%, not shown), with a gradual increase toward values at or near saturation before and during fog onset. Moreover, weak 10 m winds up to 3.1 m s^{-1} were recorded, while gradual cooling of the airmass (2 m temperature) at the rate of about 1°C h^{-1} resulted in saturation, shortly before fog formation (Fig. 3). According to the METARs, the air became saturated at 0400 UTC. The fact that, according to the World Meteorological Organization (WMO), the temperature and dew point temperature in the METAR reports are rounded to the nearest integer, should be taken into account though. Maximum temperature on

the previous day reached 12°C and dropped to about 2°C during the fog event. According to the wind components and the temperature values there is an indication that radiative cooling of the cold air mass already advected to the study area were the favorable conditions for the advection–radiation fog formation.

3. Results

3.1. WRF–COBEL Versus WRF Performance

The features of the coupled WRF–COBEL model and its performance were examined by first studying the time-height change of the liquid water mixing ratio (g kg^{-1}) (Fig. 6). The simulation presented here has been initialized at 0000 UTC before the fog onset. The threshold of 0.05 g kg^{-1} (BERGOT et al. 2007) has been used for the detection of the fog onset. It seems that the liquid water mixing ratio inside the fog layer varies between 0.05 and 0.25 g kg^{-1}, which is typical for fog (BERGOT et al. 2007). The parameter's highest values are concentrated at a height of 150 m near the surface, at the lower fog levels. It seems that the modelled fog onset and duration do not deviate significantly from the observations. WRF–COBEL simulates the fog formation about half an hour later than observed. The absence of liquid water content in the lower levels, one hour before the actual fog

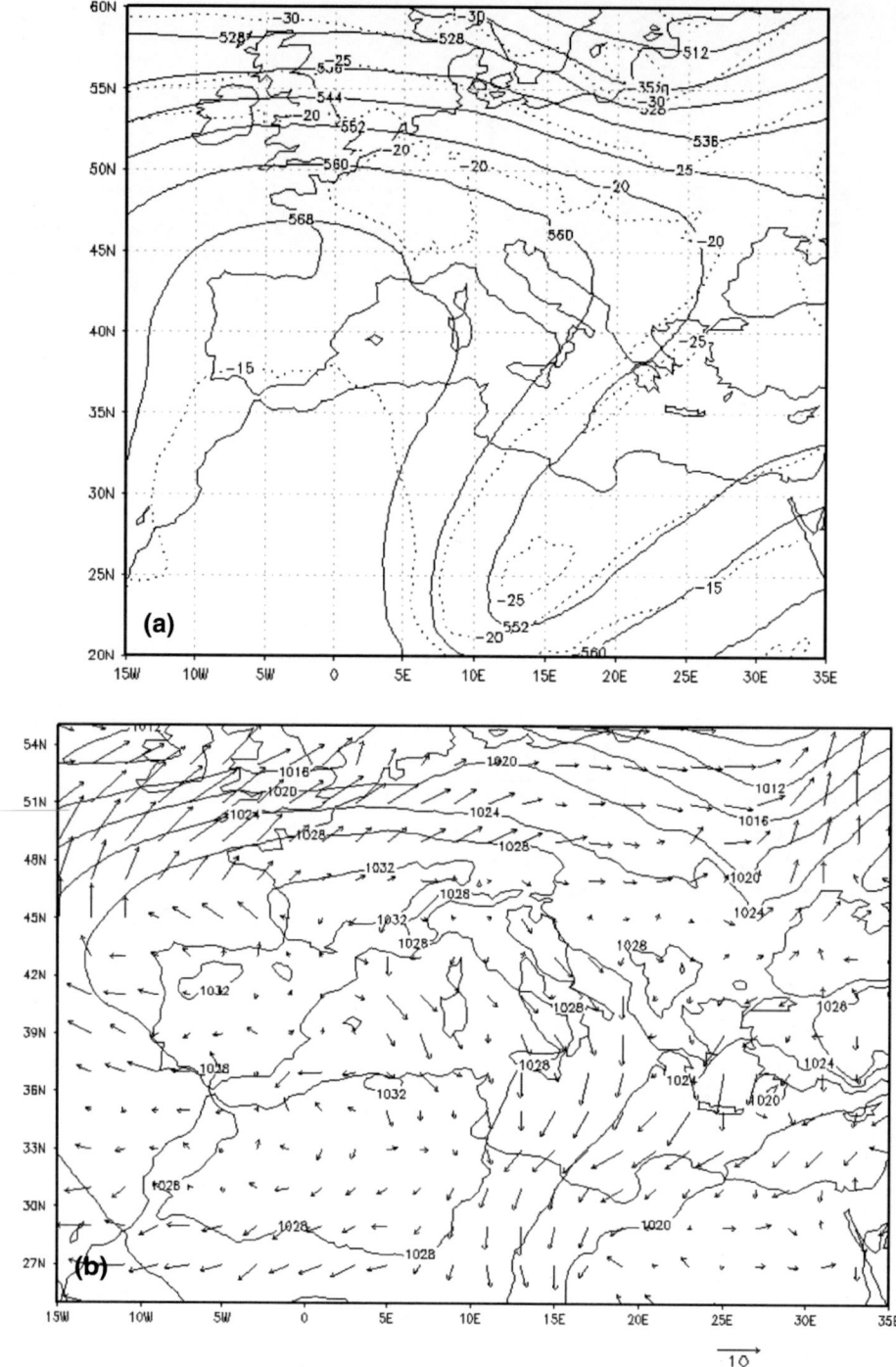

Figure 4
ECMWF analyses of **a** 500 hPa geopotential height (gpdm; *solid lines*) and temperature (°C; *dotted lines*) and **b** mean sea-level pressure (hPa; *solid lines*) and 10 m wind vectors, corresponding to 0000 UTC of the 20 January 2008. The magnitude of the vector below panel **b** corresponds to a wind speed of 10 m s^{-1}

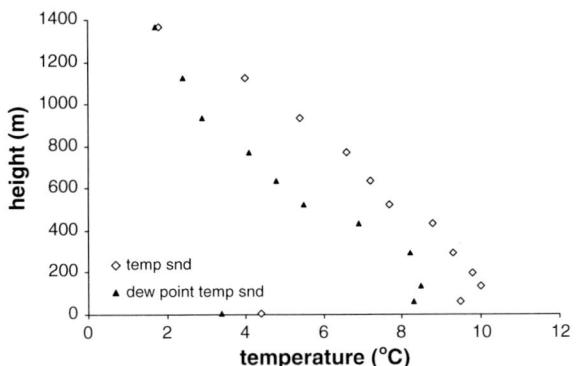

Figure 5

Temperature and dew point temperature profiles of the lower boundary layer derived from the radiosonde of 0000 UTC, 20 January 2008, at the airport of Thessaloniki

dissipation (0920 UTC), is evidence that—as will be discussed later—this might be related to the lifting of the simulated fog layer at the end of the event, although there is no information about the actual height that the fog event reached on that day. With respect to the vertical development of the condensed layer, it reached a height of 350 m.

In a next step, the coupled model was validated against actual observations. Because of the limited available actual observations though, the model validation can only be based on the comparison of the simulated 2 m temperature and relative humidity with the METARs of 20 January 2008. Figure 7 illustrates the differences between the WRF simulation, the WRF–COBEL and the observed 2 m temperature and relative humidity. In the case of the 2 m temperature, the ±0.5°C curves are also added, illustrating the uncertainty that exists in the METARs temperatures, since they are rounded to the nearest integer.

According to Fig. 7a, there is a weak cooling of about 1°C during the initial hours of the WRF simulation. The 3D model overestimates the 2 m temperature during the fog event and such a difference reaches a maximum of 6.1°C. The investigation of the WRF output did not indicate the occurrence of unrealistic warm advection during the initial hours. It is hypothesized that such an overestimation of the near surface temperature could be attributed to:

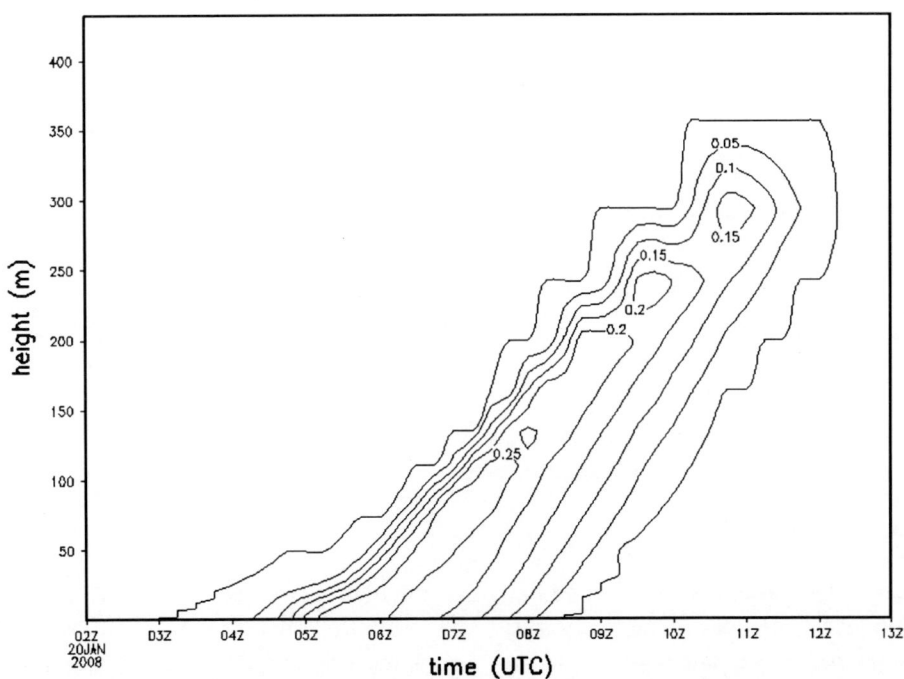

Figure 6

Time–height plot of the liquid water mixing ratio simulated by the coupled WRF–COBEL, initialized at 0000 UTC, 20 January 2008 (isolines every 0.05 g kg^{-1})

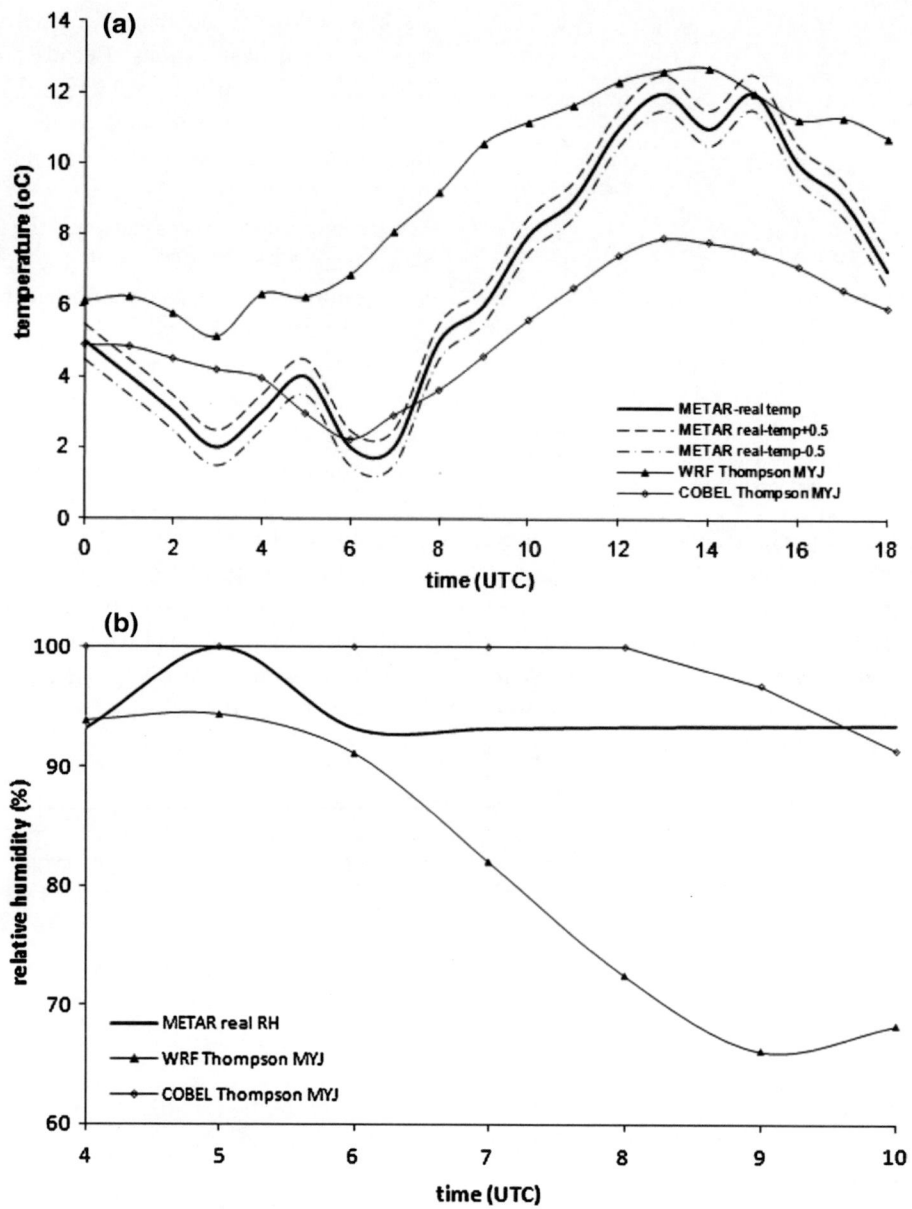

Figure 7
Temporal evolution of the observed and the WRF and COBEL: **a** simulated 2 m temperature from 0000 UTC to 1800 UTC and **b** 2 m relative humidity from 0400 UTC to 1000 UTC, of the 20 January 2008. For **a** the +0.5°C and the −0.5°C values of the observed data are plotted with *dashed* and *dashed dot lines*, respectively

(a) errors in the analysis used as initial conditions (Fig. 7a) and (b) model errors due to the land surface and boundary layer parameterizations. KALLOS *et al.* (2005) and PAPADOPOULOS and KATSAFADOS (2009) showed that the SKIRON and POSEIDON modeling systems, respectively, systematically overestimated the night-time 2 m temperature in the Mediterranean basin. Both models utilized the MYJ and OSU (which is the precursor of Noah), boundary layer and land surface schemes. Another possible reason could be the overestimation of the real existing cloud fraction during the first simulation hours. Indeed, the comparison of the WRF simulated cloud fraction with the observed one, showed that between 0000 and 0300 UTC there was an overestimation of the low (ranging between 50 and 60%) and middle (maximum of 45% at 0100 UTC) clouds fraction, since according to the observer, there were few clouds with 600 m base height prevailing during that time period. For the rest of the day and between 1200 and 1800 UTC, there is an overestimation of up to 3.7°C, although at 1500 UTC the estimated 2 m temperature is very close to the observed one. Thus, the WRF performance is much better during these hours than during the fog event.

COBEL's performance, on the other hand, is more realistic with respect to the simulation in the course of the fog event, since during that time, the absolute difference between the observed and the simulated values reaches a maximum of 2.4°C. Moreover, the model simulated well the observed minimum temperature. In a general sense, COBEL's simulated values are quite close to reality. An underestimation of 1.1–4.5°C appears between 1200 and 1800 UTC. Thus, after the fog dissipation, WRF produces more realistic results than COBEL.

At the airport of Thessaloniki the observer assesses the horizontal visibility visually, without the use of any instrument. Moreover, fog is detected when dew point temperature reaches air temperature, and thus the relative humidity equals to 100%. Therefore, the same quantitative criterion (RH ∼ 100%) is used for the detection of fog by WRF. It should be noted that the limitation of the humidity sensors near saturation should be taken into consideration in the interpretation of the results (BERGOT et al. 2005). It seems (Fig. 7b) that the actual fog event of 20 January 2008 is not represented by the 3D model. During the fog event, WRF underestimates the observed values with a mean absolute error equal to 13%. A 15% underestimation of the observed relative humidity takes place during the rest of the day. Regarding COBEL, between 0400 and 1000 UTC, the simulated relative humidity is 100%. Interpreting this result, the fact that humidity sensors do not perform very well near saturation should not be disregarded, since it imposes errors in the initialization and evaluation of the WRF and COBEL simulations.

The performance of the coupled WRF–COBEL model was further examined by comparing the radiosonde, the WRF and the WRF–COBEL temperature (Fig. 8a) and relative humidity (Fig. 8b) vertical profiles at the initialization time (0000 UTC), regarding the lowest 1.36 km. The aim of such a test was to investigate the way WRF–COBEL is initialized and works. The WRF profiles are in effect the ECMWF analyses inserted into WRF as initial conditions. In general, the structures of the two simulated temperature and humidity profiles are close enough, but they are rather different than the observed one up to around 800 m height. Such a discrepancy is probably related to errors arising in the ECMWF analyses due to the data assimilation and its model characteristics (e.g. resolution, errors in the first guess field, errors in parameterizations). It is COBEL's data assimilation scheme that contributes to the lowering of temperature near the surface, since it takes into account the real 2 m temperature value that is lower than in the ECMWF analysis (used for the WRF initialization). Therefore, the initial temperature of WRF–COBEL is lowered near the surface, but the vertical structure of the lower atmosphere (above about 800 m) is largely unchanged.

Comparing the profiles of the two models, WRF is initialized with a warmer atmosphere than WRF–COBEL and near the surface, the temperature difference between the two models reaches 2.2°C. This is in agreement with their simulated 2 m temperature differences observed and discussed earlier (Fig. 7a). The temperature inversions observed in both profiles also differ since in WRF–COBEL initial data the inversion appears from a height of 111 m, while in the WRF alone from a height of 164 m upwards. Moreover, the inversion is much more pronounced in the coupled model (~ 322 m) than in the WRF alone (~ 132 m), although within the layer, WRF exhibits a more pronounced lapse rate (0.076°C/10 m) than that of WRF-COBEL (0.047°C/10 m).

On the other hand, both models are initialized by a more humid atmosphere (Fig. 8b) than the observed one, especially below 500 m. Like in the case of the temperature, such a discrepancy seems to be driven by the fact that, in essence, the ECMWF initial temperature and humidity profiles inserted into WRF and further on into COBEL, incorporate the

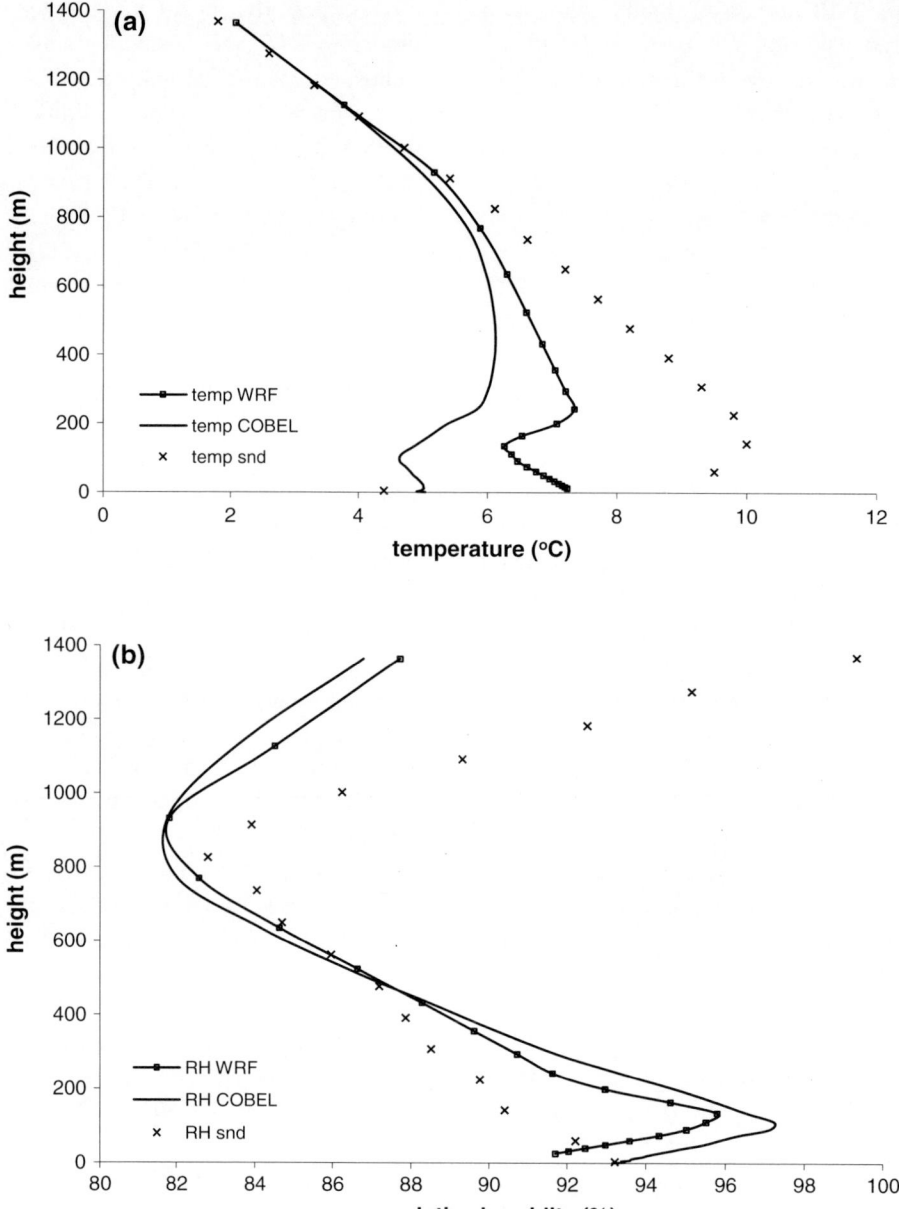

Figure 8
Observed, WRF and WRF–COBEL simulated **a** initial temperature (°C) and **b** relative humidity (%) vertical profiles at 0000 UTC of 20 January 2008

radiosonde information but preceded by the data assimilation that they undergo.

Examining the same vertical profiles for the 1200 UTC, we gain a further insight into the performance of the models. At this time, WRF is closer to reality than COBEL is (Fig. 9). This is in agreement with the comparison between the simulated and the real 2 m temperatures. On the other hand, the surface based thermal inversion up to the first 200 m that still remains in real conditions is not represented by both models. COBEL's vertical temperature profile is quite far from reality, mainly up to a height of 1,000 m, while it simulates an inversion that is formed in an upper level.

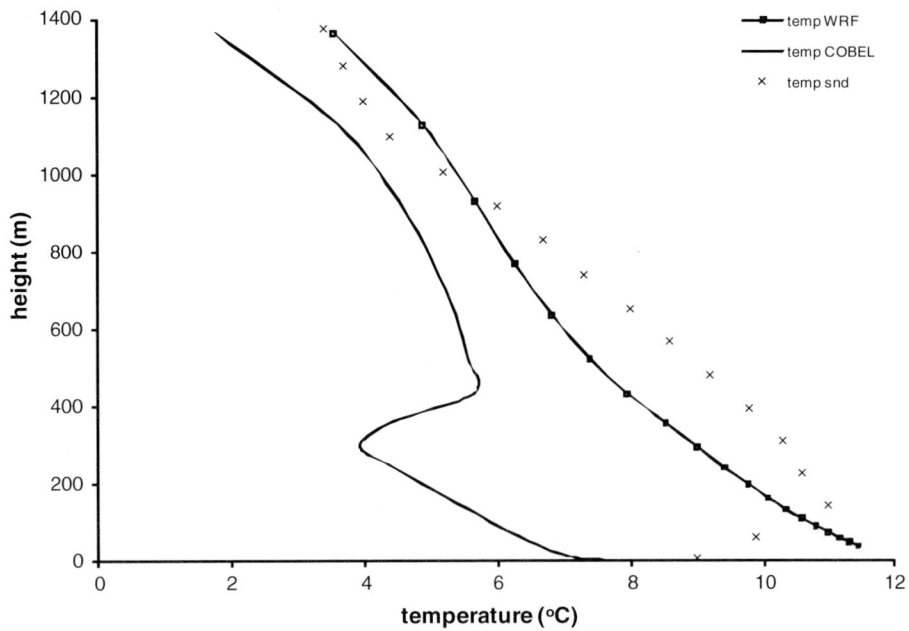

Figure 9
Observed, WRF and WRF–COBEL simulated initial temperature ($^{\circ}$C) vertical profiles at 1200 UTC of 20 January 2008

Therefore, while COBEL performed better during the fog event, at a later stage, the opposite is exhibited. The aforementioned results emphasize the importance of a correct representation of the surface atmospheric conditions—through the assimilation of local METAR observations—for a more correct fog simulation and forecasting. The initial profiles of WRF–COBEL, which were adapted to reality, favored a better fog event forecast (which is the aim of its utilization), but deteriorated at the later stages after fog dissipation. On the other hand, WRF initialized with less accurate thermo-hydrometric profiles obtained from the ECMWF analysis failed to forecast the fog event. However, it ameliorated the forecast of the atmospheric conditions in the time period of 1200–1800 UTC.

Moreover, the time and height evolution of the temperature and relative humidity as compared between WRF (Figs. 10a, 11a) and WRF–COBEL (Figs. 10b, 11b) provide more information regarding WRF's inability to capture the event. In general, WRF produces higher values of temperature and lower values of relative humidity than WRF–CO-BEL. Around the time of the fog onset ($\sim$0400 UTC), WRF produces warmer temperatures than

WRF–COBEL by 1.5°C up to almost 3°C. It is also evident that the coupled model simulates a strong inversion between 0400 and 0900 UTC, from surface up to the height of 400 m, and during the time that fog becomes denser (e.g. between 0500 and 0700 UTC), the temperature increases with a rate of 0.05°C/10 m. This inversion is probably caused by radiative cooling at the fog top. Since WRF alone did not produce a fog layer, there is also no simulated inversion aloft observed. The relative humidity values of the two simulations are closer for the first three hours of the simulation and both reach 98% up to a 200 m height, but at the fog onset (0400 UTC) and on, the atmosphere is much drier in WRF, since the relative humidity values drop to 75%. In contrast, WRF–COBEL derived values during the fog event are constantly much higher and close to saturation. The aforementioned is further evidence that WRF–COBEL exhibits a better performance than WRF alone.

3.2. COBEL Sensitivity to Initial Conditions

To assess the effect of possible temperature and specific humidity changes on the features of the fog

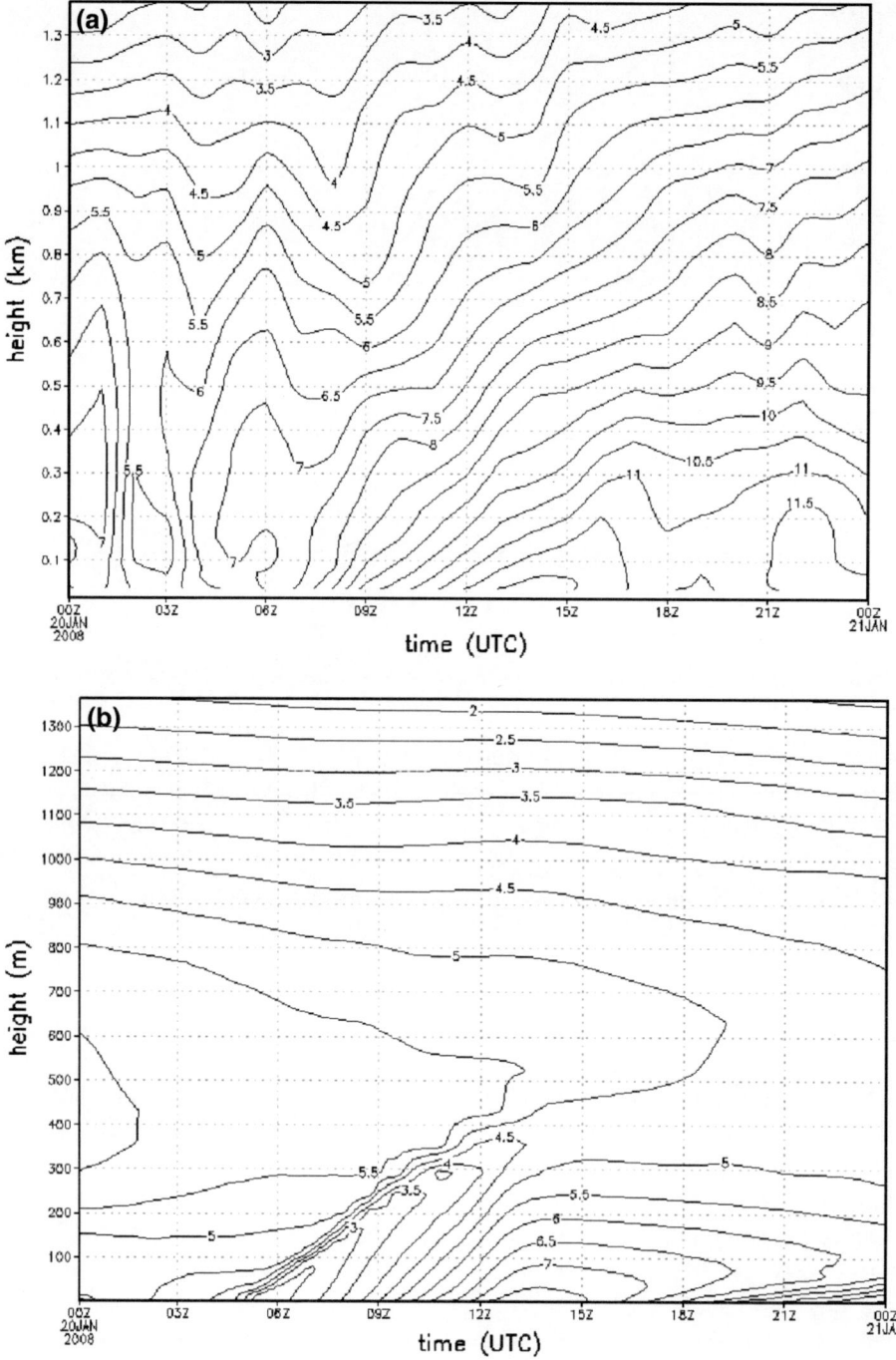

Figure 10
a WRF and **b** WRF–COBEL simulated temperature (°C) time–height plots (contour interval: 0.5°C)

studied herein, sensitivity experiments were carried out. Such an approach provides more information for a better understanding of the factors affecting fog event formation at the airport. Moreover, from what has been concluded earlier, these parameters seem to be important for all stages of fog, since small variations of them may even lead to absence of fog, as was in the case of WRF alone.

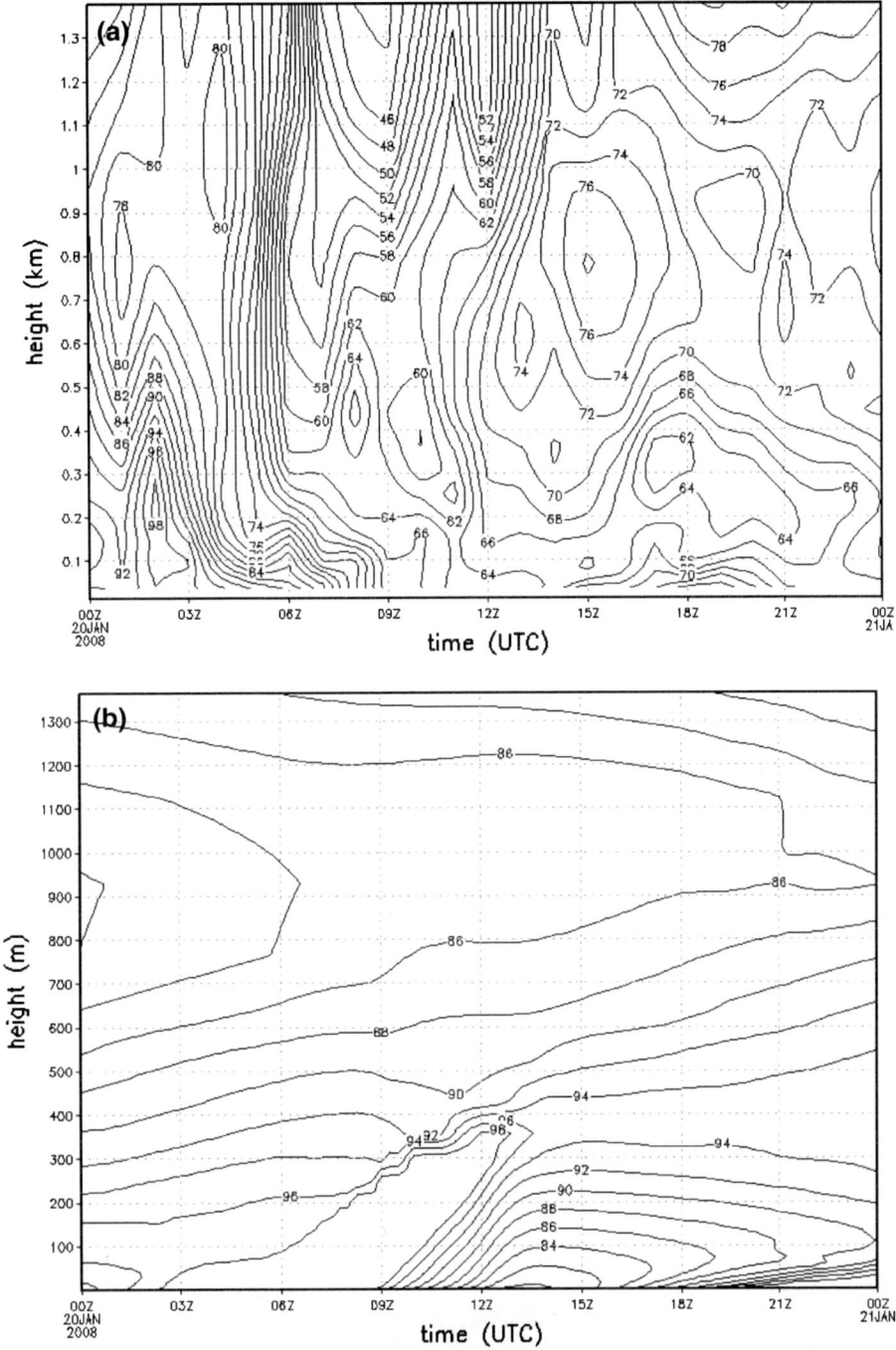

Figure 11
a WRF and **b** WRF–COBEL simulated relative humidity (%) time-height plots (contour interval: 2%)

Temperature and specific humidity profiles are two of the main meteorological parameters that are used for the initialization of COBEL-ISBA. Therefore, as a first experiment, a ±1°C temperature anomaly, in two independent runs, was imposed in the vertical column of COBEL and at the initial time 0000 UTC. Such anomalies might be considered low, but they lie within the range of day to day temperature changes, especially around the usual time of radiation fog formation. Furthermore, there was the motivation to examine whether such small temperature changes are sufficient enough to influence significantly the fog onset, development and dissipation.

Along with the vertical profiles of temperature and relative humidity, attention was paid on the liquid water mixing ratio, as a qualitative and quantitative parameter for the detection of the fog characteristics. Figure 12 illustrates the vertical profiles of the simulated liquid water mixing ratio in the course of the day, for the different experiments presented in this section. Figure 12a shows that the fog onset is delayed by almost one hour, while the vertical extent of the event is diminished by 100 m, when a 1°C temperature anomaly is introduced in the vertical, up to COBEL's top. Therefore, even if the atmosphere is moist enough for fog to form, but under a bit warmer conditions in the beginning of the second part of the night, it seems that the triggering of the mechanisms responsible for the formation of fog droplets delay one hour and fog lasts less. As it was expected, the small increase of the temperature in the atmosphere influences the fog liquid water mixing ratio inversely, since the highest value of the parameter is simulated at the first 10 m above ground, in contrast to the control run, where the maximum liquid water mixing ratio extends to a height of 150 m and persists for a longer time period.

A decrease of temperature by 1°C has opposite effects on fog characteristics. Here, the liquid water mixing ratio profile (Fig. 12b) shows that fog, under such a decrease, converts into a stratus cloud, persisting during the night of 21 January (not shown). Moreover, the cloud's liquid water mixing ratio core (0.3 g kg^{-1}) expands much more as a function of time and height (above 400 m), than in the control run. The comparison of the time and height evolution of the simulated relative humidity between the perturbed simulation and the control run indicates that the 1°C decrease forced the model to saturation. Indeed, the simulated relative humidity of the perturbed run between 0300 UTC and 0900 UTC (not shown) reached 99%, in comparison with the control run relative humidity that was at that time 96%. Moreover, the vertical extent of this 99% high

value of relative humidity was greater than in the experiment of the control run and reached a level between 400 and 700 m, which is in accordance with the liquid water mixing ratio simulated for that layer (Fig. 12b).

Similar experiments were conducted in order to examine the effect of the increase or decrease of the specific humidity upon fog formation. Experiments were conducted by imposing specific humidity anomalies as ± 0.5, ± 0.7 and ± 1 g kg^{-1} in the vertical to the COBEL's atmospheric column at initialization time (0000 UTC). Such anomalies correspond to an increase or decrease of the relative humidity in the vertical by 11, 15 and 18%, respectively. The motivation of such an investigation was to study whether changes of this magnitude have any effect on fog formation and its characteristics. Along with the examination of the effect of the aforementioned specific humidity anomalies, temperature and relative humidity changes where also considered.

The examination of the increase of specific humidity by 0.5, 0.7 and 1 g kg^{-1} in the vertical column at COBEL's initialization time had similar effects in the lower boundary layer. In all cases fog forms earlier than it actually does and at some point it converts into a stratus layer whose vertical extent varies according to the moisture perturbation. Specifically, the +0.5 g kg^{-1} increase of moisture in the vertical column (not shown) drives the formation of fog one hour earlier (0300 UTC) than in reality and until 0900 UTC it extends from the surface up to 325 m. It seems that this fog layer gradually transforms into a low cloud at 0900 UTC and during the rest of the day its depth ranges between 100 and 800 m.

The 0.7 g kg^{-1} increase in specific humidity (Fig. 13) results in fog that forms much earlier ($\sim$0200 UTC) than the actual fog onset time, while after 0500 UTC it lifts and transforms into a stratus cloud with a depth that varies between 100 and 300 m during the rest of the day of 20 January 2008. At the top of COBEL's grid the formation of a cloud layer is also observed. It seems that in the beginning of the simulation, this cloud layer does not inhibit the fog formation, but it probably affects fog's further development. The cloud layer acts as a lid,

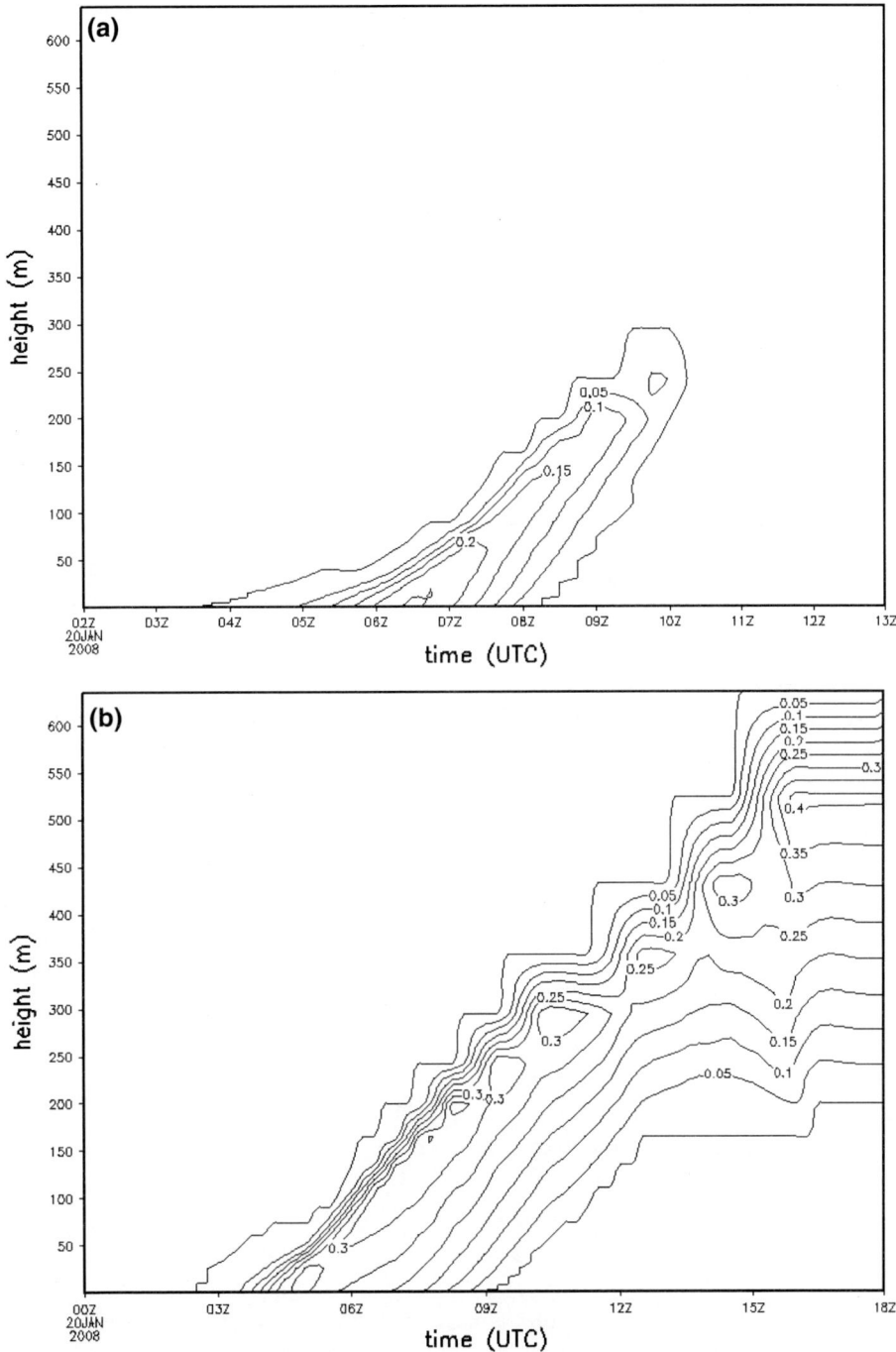

Figure 12

Time–height plot of liquid water mixing ratio (g kg^{-1}) simulated by WRF–COBEL after **a** increasing and **b** decreasing the temperature of the vertical column by 1°C at 0000 UTC initial time (isolines every 0.05 g kg^{-1})

counteracting the further cooling of the fog layer and, in combination with the increase of the near surface temperature as the day sets in, further fog development is inhibited. The +1 g kg^{-1} anomaly has similar effects on the vertical, since it results in light fog conditions near the ground between 70

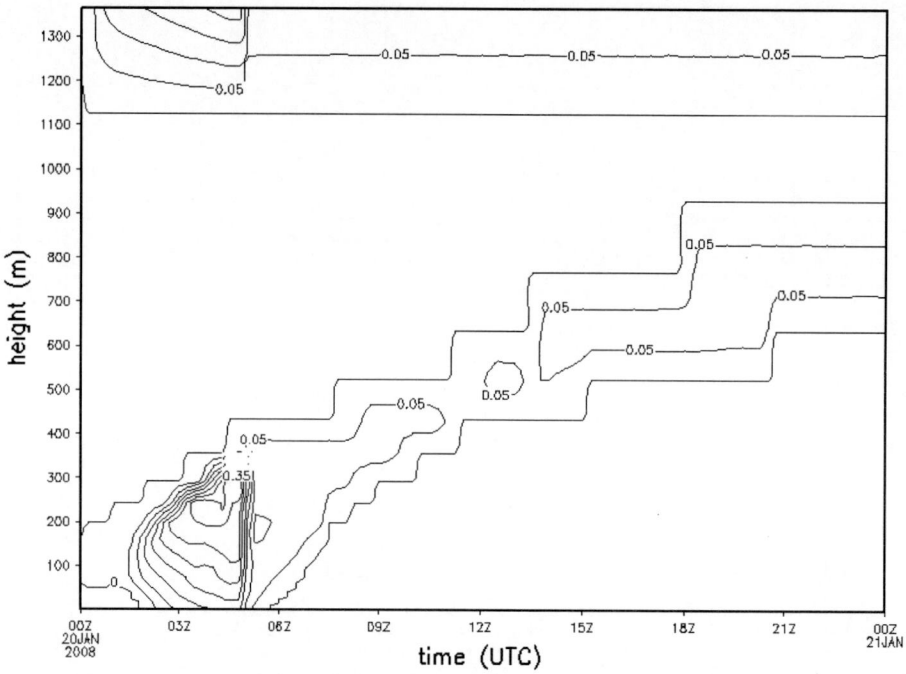

Figure 13

Time–height plot of liquid water mixing ratio (g kg^{-1}) simulated by WRF–COBEL after increasing the specific humidity of the vertical column by 0.7 g kg^{-1} at 0000 UTC initial time (isolines every 0.05 g kg^{-1})

and 225 m above ground (0.05 g kg^{-1} liquid water content). As in the previous case, these conditions prevail much earlier than the actual fog onset time. Between 0200 UTC and 0300 UTC the fog layer becomes thicker in the vertical up to 500 m and the main liquid water content core (0.35 g kg^{-1}) is identified at the level of 500 m. After 0900 UTC this fog layer is transformed into a stratus layer whose depth remains almost stable in the course of time but its base lifts gradually. Therefore, it results that considering an already high moisture content in the atmosphere at the time of COBEL's initialization, an additional moisture amount seems to drive the formation of a low cloud, adding also to the strength and depth of the fog layer that previously forms.

The gradual drying of the atmosphere in the vertical, imposed a gradual negative effect on the fog formation itself, as well as on the time of the simulated fog onset, its further vertical development and dissipation. Only the effects of the 0.5 g kg^{-1} decrease (Fig. 14) are shown here. It is obvious that such a decrease in moisture resulted in a fog layer with a limited vertical extent with respect to that of the control run. Its time of formation and dissipation

were delayed by more than 1 h regarding the actual onset and dissipation times, while it extended up to around 175 m. Therefore, a ~10% decrease of relative humidity in the atmosphere toward the end of the first part of the night seems to hinder fog formation by about 1 h. Despite this negative effect though, fog still forms.

A decrease of the specific humidity by 0.7 g kg^{-1} resulted in near-fog conditions which means fog hardly formed. In this case, the simulated relative humidity (not shown) was close to saturation (99%) only from 0500 to 0700 UTC and to a height below 100 m. Regarding the relationship between the relative humidity and the specific humidity, the decrease of the water vapor content in the atmosphere resulted in a decrease of the relative humidity in the whole vertical column and for all simulation hours, with respect to the control run. The relative humidity difference between the two runs reached 12%. In addition, there was a simulated increase of the temperature (4°C) between 0500 and 0700 UTC and within the depth that fog almost formed, in comparison with the control run temperature that was as low as 2.5°C. The above mentioned conditions

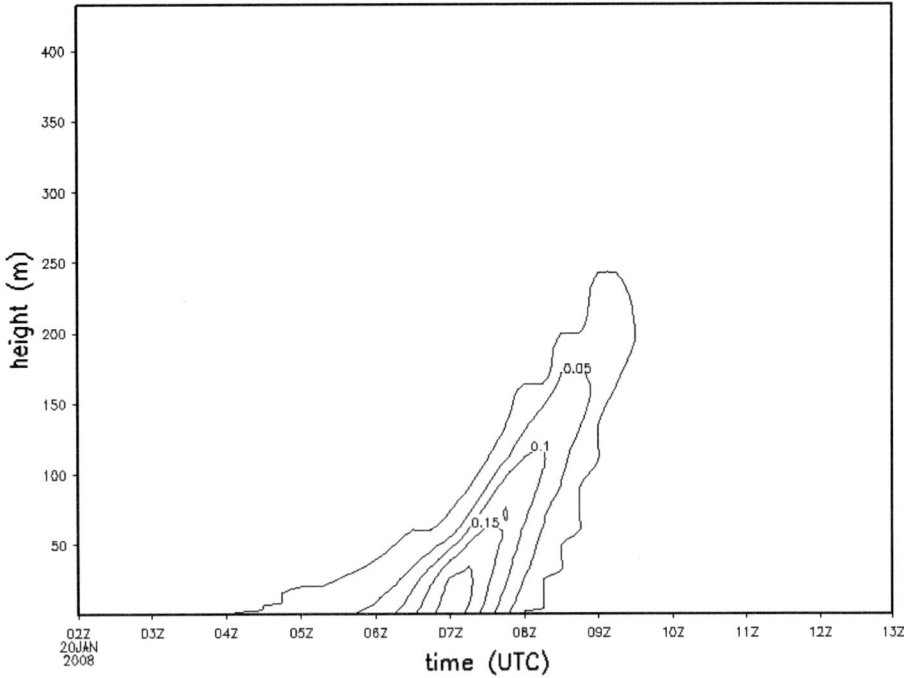

Figure 14
Time–height plot of liquid water mixing ratio (g kg^{-1}) simulated by WRF–COBEL after decreasing the specific humidity of the vertical column by 0.5 g kg^{-1} at 0000 UTC initial time (isolines every 0.05 g kg^{-1})

counteract fog formation, since, under higher temperatures, the existing fast-moving vapor molecules are not able to condensate and to form fog droplets. Drying the atmosphere by 1 g kg^{-1} resulted in the absence of fog or low clouds (not shown). Although the relative humidity, between 0500 and 0700 UTC, was quite favorable for fog formation ($\sim$99%, not shown), such a condition does not seem to be the only prerequisite for fog to form. Possibly at the same time, the increased (by 1°C) simulated temperature at the lower levels, in relation to the control run temperature, played a more important role in preventing fog formation.

4. Summary and Conclusions

In this study, a case of dense advection-radiation fog over the airport of Thessaloniki in northern Greece was presented as a benchmark for the development of the coupled WRF–COBEL numerical system. The integrated system's performance was investigated and it served as a useful tool in assessing

the fog formation sensitivity to small variations of the thermohydrometric conditions, thus adding to the knowledge on fog's characteristics at the area of study.

The main findings of the research effort presented herein can be summarized as:

1. The developed interface between WRF and CO-BEL was successful to predict fog occurrence.

2. A comparison between WRF and WRF–COBEL models provided us with strong evidence, indicating that the coupled model performs much better in capturing the fog conditions, such as the onset and dissipation times and the thermal inversion layer, than using the WRF alone. This result emphasizes the importance of integration of a high-resolution 1D model with a 3D model output comparing to solely relying on a 3D model, since this integration allows the use of a high vertical resolution along with an assimilation of local observations

3. The 1°C increase of temperature in the vertical up to 6 km height counteracts fog development, although it forms. Its formation is delayed, while

215

its vertical extent is diminished in comparison to the control run. On the other hand, a temperature decrease of only 1°C is sufficient enough for the formation of denser fog during day that it transforms into a low level cloud. Both results show how subtle are the limits between the prevalence of fog or near fog conditions and it underlines the importance of small temperature changes on influencing fog formation and development.

4. An increase in the specific humidity of about 0.5–1 g kg^{-1} results in the simulation of fog conditions prevailing much earlier than the actual fog onset time and further stimulating low cloud formation. This depends on the perturbation of the specific humidity imposed in the vertical. On the other hand, changing vapor mixing ratio by 0.5 g kg^{-1} up to 1 g kg^{-1} results in either the formation of a fog layer or its dissipation. This shows how important are the measurements to study fog characteristics.

It should be emphasized that surface boundary layer conditions are important for fog formation, development and dissipation; therefore, better observations are needed to improve the model simulations. The success of WRF–COBEL in simulating fog events in the area of interest is strongly related to a better understanding of the physical processes involved.

Future research and development on fog need detailed examination of how soil moisture, soil texture features, and vegetation affect the fog onset and development. Moreover, work toward the development of the capability to assimilate the local sounding data into COBEL would provide a useful testbed for investigating the added value of a 3D NWP model simulations. Also, performing fog experiments with modified topography, e.g., including the role of the nearby *Anthemountas* valley, will help us further understand the complex nature of the fog formation at Thessaloniki's airport.

Acknowledgments

The authors would like to acknowledge Météo-France/ Centre National de Recherches Météorologiques (CNRM) for providing the 1D COBEL-ISBA local forecast system. Many thanks are also given to Dr. T. Bergot and Dr. S. Rémy for their valuable help and support in accomplishing the adaptation and run of COBEL-ISBA at the area of Thessaloniki's airport. Furthermore, thanks are given to NCAR for the provision of WRF-ARW, as well as to the ECMWF for the gridded analyses used in this work. Last, S. Stolaki is under scholarship from the Greek State Scholarships Foundation and would like to thank it for its financial contribution.

REFERENCES

ANGOURIDAKIS, V.E., and FLOCAS, A.A. (1983), Forecasting of Radiation fog at Thessaloniki, Greece, Zeitshcrift für Meteorologie 33, 3, 176-178.

ANGOURIDAKIS, V.E. (1973), Fog and weather types in the area of Thessaloniki, Zeitshcrift für Meteorologie 7-8, 237-241.

BERGOT, T. (2007), Quality assessment of the Cobel-Isba numerical forecast system of fog and low clouds, Pure Appl. Geophys. 164, 1265-1282.

BERGOT, T., TERRADELLAS, E., CUXART, J., MIRA, A., LIECHTI, O., MUELLER, M., and NIELSEN, N.W. (2007), Intercomparison of single-column numerical models for the prediction of radiation fog, J. Appl. Meteor. Climatol. 46, 504–521.

BERGOT, T., CARRER, D., NOILHAN, J., and BOUGEAULT, P. (2005), Improved site specific numerical prediction of fog and low clouds: a feasibility study, Wea. and Forecasting 20, 627–646.

BETTS, A.K. (1986), A new convective adjustment scheme. Part I: Observational and theoretical basis, Quart. J. Roy. Meteor. Soc. 112, 677–691.

BETTS, A.K., and MILLER, M.J. (1986), A new convective adjustment scheme. Part II: Single column tests using GATE wave, BOMEX, and arctic air-mass data sets, Quart. J. Roy. Meteor. Soc. 112, 693–709.

BILBAO, J., and DE MIGUEL, A.H. (2006), Estimation of daylight downward longwave atmospheric irradiance under clear-sky and all-sky conditions, J. Applied Meteor. Climat. 46, 878-889.

BOONE, A.A. (2000), Modélisation des processus hydrologiques dans le schéma de surface ISBA: Inclusion d'un réservoir hydrologique, du gel et modélisation de la neige, Ph.D. Thesis, Université Paul Sabatier, Toulouse, France, pp. 252.

BOONE, A.A., CALVET, J.C., and NOILHAN J. (1999), The inclusion of a third soil layer in a land surface scheme using the force-restore method, J. Appl. Meteor. 38, 1611–1630.

BOTT, A., and TRAUTMANN, T. (2002), PAFOG-a new efficient forecast model of radiation fog and low-level stratiform clouds, Atmos. Res. 64, 191-203.

BRUNT, D. (1932), Notes on radiation in the atmosphere, Q. J. R. Meteorol. Soc. 58, 389–420.

BRUTSAERT, W. (1975), On a derivable formula for long-wave radiation from clear skies, Water Resour. Res. 11, 742-744.

CHEN, F. and DUDHIA, J. (2001), Coupling an advanced land-surface/ hydrology model with the Penn State/ NCAR MM5 modeling system. Part I: Model description and implementation. Mon. Wea. Rev., 129, 569–585.

CULF, A.D., and GASH, J.H.C. (1993), Longwave radiation from clear skies in Niger: a comparison of observations with simple formulas, J. Appl. Meteorol. 32, 539-547.

DUARTE, H.F., DIAS, N.L., and MAGGIOTTO, S.R. (2006), Assessing daytime downward longwave radiation schemes for clear and cloudy skies in Southern Brazil, Agric. For. Meteorol. 139 (3-4), 171-181.

EK, M.B., MITCHELL, K.E., LIN, Y., ROGERS, E., GRUNMANN, P., KOREN, V., GAYNO, G., and TARPLEY, J.D. (2003), Implementation of Noah land surface model advances in the National 28 Centers for Environmental Prediction operational mesoscale eta model, J. Geophys. Res. 108, 8851.

FORIS, D. (2002), Fog climatology for Thessaloniki's airport and feasible solutions. Proc. 4th European Conference on Applied Climatology, ECAC, Brussels, 12-15 November, CD-ROM.

HOUSSOS, E.E., LOLIS, C.J., CHARANTONIS, T., ZIAKOPOULOS, D., and BARTZOKAS, A. (2008), The main synoptic conditions favouring the formation of fog in Thessaloniki, Proceedings of the 9th Conference of Meteorology, Climatology and Atmospheric Physics, Thessaloniki, Greece, 28-31 May 2008, 561–568 (in Greek).

JANJIC, Z.I. (2002), Nonsingular implementation of the Mellor-Yamada level 2.5 scheme in the NCEP meso model, NCEP Office Note 437, National Centre for Environmental Prediction, 61 pp.

JANJIC, Z.I. (2000), Comments on "Development and Evaluation of a Convection Scheme for Use in Climate Models", J. Atmos. Sci. 57, 3686.

JIMÉNEZ, J.I., ALADOS-ARBOLEDAS, L., CASTRO-DIEZ, Y., and BALLESTER, G. (1987), On the estimation of long-wave radiation flux from clear skies, Theor. Appl. Climatol. 38, 37-42.

KALLOS G., PYTHAROULIS, I., KATSAFADOS, P., LOUKA, P., and GALANIS, G. (2005), Limited area weather forecasting for the MFSTEP activities: sensitivity and performance analysis. 4th EUROGOOS conference. 6-9 June, Brest, France.

KRUK, N.S., VENDRAME, I.F., RIBEIRO DA ROCHA, H., CHOU, S.C., and CABRAL, O. (2009), Downward longwave radiation estimates for clear and all-sky conditions in the Sertãozinho region of São Paulo, Brazil, Theor. Appl. Climatol. doi:10.1007/s00704-009-0128-7.

NIEMELÄ, S., RÄISÄNEN, P., SAVIJÄRVI, H. (2001), Comparison of surface radiative flux parameterizations: Part I. Longwave radiation, Atmos. Res 58, 1–18.

PAPADOPOULOS A., and KATSAFADOS, P. (2009), Verification of operational weather forecasts from the POSEIDON system across the Eastern Mediterranean. Nat. Hazards Earth Syst. Sci., 9, 1299-1306.

PINHEIRO, F.R., PETERSON R.G., and DE FARIAS, W.C.M. (2006), Numerical study of fog events along Rio de Janeiro coast, using

the MM5 model coupled with the unidimensional model CO-BEL, Proceedings of 8 ICSHMO, Foz do Iguaçu, Brazil, 24-28 April, 1935-1944.

PRATA, A.J. (1996), A new long-wave formula for, estimating downward clear-sky radiation at the surface, Q. J. R. Meteorol. Soc. 122, 1127–1151.

RÉMY, S., and BERGOT, T. (2009), Assessing the impact of observations on a local numerical fog prediction system, Q. J. R. Meteorol. Soc. 135, 1248-1265.

ROHN, M., VOGEL, G., BECKMANN, B.R., HNER, P.R., THOMA, Ch., SCHNEIDER, W., and BOTT, A. (2010), iPort-VIS: Site specific fog forecasting for Munich airport, Proceedings of the 5th International Conference on Fog, Fog Collection and Dew, Münster, Germany, 25-30 July, 157-159.

SAUNDERS, W.E. (1950), A method of forecasting the temperature of fog formation, Meteorol. Mag. 79, 213–219.

SHI C., W and L., ZHANG H., DENG X. (2010), Experiments on fog prediction based on multi-models, Proceedings of the 5th International Conference on Fog, Fog Collection and Dew, Münster, Germany, 25-30 July, 152-155.

SKAMAROCK, W.C., KLEMP, J.B., DUDHIA, J., GILL, D.O., BARKER, D.M., DUDA, M.G., HUANG, X-Y, WANG, W., and POWERS, J.G. (2008), A description of the Advanced Research WRF Version 3.NCAR/TN-475 + STR, pp. 113.

STOLAKI, S.N., KAZADZIS, A.S., FORIS, D.V., and KARACOSTAS, T.S (2009), Fog characteristics at the airport of Thessaloniki, Greece, Nat. Hazards Earth Syst. Sci. 9, 1541-1549.

SWINBANK, W.C. (1963), Long-wave radiation from clear skies, Q. J. R. Meteorol. Soc. 89, 339–348.

TARDIF, R. (2007), The impact if vertical resolution in the explicit numerical forecasting of radiation fog: a case study, Pure Appl. Geophys., 164, 1221-1240.

TERRADELLAS, E., and CANO, D. (2007), Implementation of a Single-Column Model for Fog and Low Cloud Forecasting at Central-Spanish Airports, Pure Appl. Geophys. 164, 1327-1345.

THOMPSON, G., FIELD, P.R., RASMUSSEN, R.M., and HALL, W. (2008), Explicit Forecasts of Winter Precipitation Using an Improved Bulk Microphysics Scheme: Part II: Implementation of a New Snow Parameterization, Mon. Wea. Rev. 136, 5095-5115.

VAN DER VELDE, I.R., STEENEVELD, G.J., WICHERS SCHREUR, B.G.J., and HOLTSLAG A.A.M. (2010), Modeling and forecasting the onset and duration of severe radiation fog under frost conditions, Mon. Wea. Rev. 138, 4237-4253.

WANG, W., BRUYÈRE, C., DUDA, M., DUDHIA, J., GILL, D., LIN, H-C, MICHALAKES, J., RIZVI, S., and ZHANG, X (2010), ARW Version 3 Modelling System's User Guide, NCAR-MMM, pp. 312.

WILKS, D.S., Statistical Methods in the Atmospheric Sciences (Academic Press, New York 1995).

(Received November 11, 2010, revised June 23, 2011, accepted June 25, 2011, Published online August 6, 2011)

Reprinted from the journal

Pure Appl. Geophys. 169 (2012), 983–1000
© 2011 Springer Basel AG
DOI 10.1007/s00024-011-0348-5

Large-Scale Environmental Influences on the Onset, Maintenance, and Dissipation of Six Sea Fog Cases over the Yellow Sea

Pengyuan Li,[1] Gang Fu,[1] and Chungu Lu[2]

Abstract—Sea fog is typically formed and developed under a set of favorable environmental conditions, which are associated with the station pressure changes, sea level pressure, winds, temperature, water vapor supply, and sea surface temperature. Understanding of these environmental factors during the evolution of a sea fog episode is crucial for forecasting the occurrence and severity of sea fogs over the ocean and adjacent coastal areas. In this study, the large-scale environment variability of six fog events over the Yellow Sea was investigated. It was realized in the present study that the northwest Pacific Ocean high (NPH) is vital to fog formation over the Yellow Sea. In our study, six fog cases can be basically divided into two types: (1) pressure-weakening type, (2) pressure-strengthening type. The former type happened in spring and the latter type in summer. Prevailing southerly winds, accompanied with the well-positioned NPH, may supply a large amount of warm water vapor for the fog formation and maintenance. The intensity of the air temperature inversion is stronger in summer cases than that in spring ones. The wind direction change from south to north and the unstable lower atmosphere may lead to fog's dissipation. This study may provide a comprehensive understanding of sea fog's onset, maintenance, and dissipation over the Yellow Sea.

Key words: Sea fog, NPH, southerly winds, air temperature inversion layer.

1. Introduction

Sea fog is a high-impact marine weather phenomenon which occurs over oceans or adjacent coastal areas, and is characterized by the atmospheric visibility below 1 km (Wang, 1985). During a sea fog event, marine or air transportations may either be delayed or cancelled. Even worse, aviation or navigation accidents often occur. At present, an accurate prediction of a sea fog event is still a challenging task to the forecasters, due to our incomplete understanding to the complexity of sea fog, as well as the environmental factors leading to its formation, maintenance, and dissipation.

One of the earliest and most notable studies on sea fog was made by Taylor (1917). His measurements were made over the Banks of Newfoundland, which is characterized by exceptionally cold water. He found that the water was colder than the air in 80% of his observed 141 cases of sea fog. However, Petterssen (1938) found that fog also formed over warm waters. Pilié *et al.* (1979) suggested that sea fog formed over the warm water was triggered by instability and mixing over warm water patches, and some radiation fog formed over land and then advected to sea by a nocturnal land breeze. Other researchers found that stratus lowering was one main mechanism for fog formed over the warm water along the California coast (Anderson, 1931; Oliver *et al.*, 1978; Pilié *et al.*, 1979; Telford and Chai, 1984). Some recent work has been conducted on the mechanism and environmental factors of sea fog formation along the California coast (Filonczuk *et al.*, 1995; Koračin *et al.*, 2001; Lewis *et al.*, 2003; Koračin *et al.*, 2005). Lewis (2004) reviewed the study history of sea fog in the UK and USA. Gultepe *et al.* (2006a) conducted a field project with respect to cloud microphysics over eastern Canada for marine fog studies. A systematic fog research review has been conducted concerning the past achievements and future perspectives (Gultepe *et al.*, 2007).

However, in the eastern Asia, sea fog received little attention before the 1980s. Wang (1985) conducted systematic research on the Asian sea fogs,

[1] Laboratory of Physical Oceanography, Department of Marine Meteorology, Ocean University of China, No. 238, Songling Road, 266100 Qingdao, People's Republic of China. E-mail: fugang@ouc.edu.cn
[2] NOAA/Earth System Research Laboratory, GSD5, 325 Broadway, Boulder, CO, USA.

especially fogs over the Yellow Sea, with a number of statistical studies and analyses and laid the solid foundation for much of what we currently know about sea fog occurrences over China coast. WANG (1985) also found that sea fog over the Yellow and East China Seas generally belongs to a type of advection fog. ZHOU and LIU (1986) confirmed that almost 80% of sea fog events over the Yellow Sea were advection fogs: more precisely, advection cooling fogs, which form as warm/moist air passes over colder water. An extensive body of work on the Yellow Sea fog begins to emerge in recent years, among which are observational studies (JING, 1980; ZHOU and LIU, 1986; DIAO, 1992; CHO et al., 2000; ZHOU et al., 2004). In the recent decade, the mechanism and more details relating to Yellow Sea fog have been explored by using state-of-the-art numerical models (FU et al., 2004, 2006; GAO et al., 2007; ZHANG et al., 2009; FU et al., 2010). FU et al. (2004, 2006) conducted a detailed study of one sea fog event over the Yellow Sea. They successfully used the high-resolution Regional Atmospheric Modeling System (RAMS, COTTON et al., 2003) to reproduce the main characteristics of sea fog over the Yellow Sea, and suggested that warmer SST hampered the formation of sea fog and the advection cooling effect played a significant role in the fog formation. GAO et al. (2007) performed a numerical study of a sea fog event over the Yellow Sea by using the MM5 (the fifth-generation Pennsylvania State University/ National Center for Atmospheric Research Mesoscale Model) and made a series of sensitivity experiments, including increasing/decreasing SST, adopting different PBL schemes, and using different horizontal resolutions. ZHANG et al. (2009) employed the Weather Research and Forecasting model version 2.2 (WRF v2.2) to investigate the mechanisms for a step-like evolution of the Yellow Sea fog season, which is characterized by an abrupt onset in April along the southern coast of the Shandong Peninsula and an abrupt, basin-wide termination in August.

Despite the previous studies, sea fog is still a difficult weather phenomenon to forecast because of multiple complex processes involved in, such as microphysical, radiation, turbulence, synoptic, and surface processes. The difficulty arises also due to the lack of observations over oceans. A comprehensive understanding of the conditions conducive to fog's onset, evolution, and dissipation from available observational data is still needed. In the present study, large-scale environmental influences, including the variation of sea level pressure, surface wind field and SLP, SST, temperature inversion layer, and relative humidity (RH) in six fog cases over the Yellow Sea are investigated. The objectives of this study are to (a) document the characteristics of six sea fog cases over the Yellow Sea using all available observations; (b) compare statistically these fog events in order to gain the common and distinct features among them; (c) lay a foundational work for the future study on sea fogs over the Yellow Sea. In Sect. 2, the data sources used in the present study will be introduced. Six sea fog events that occurred over the Yellow Sea will be examined. The overview of these cases will be given in Sect. 3. We will then analyze a set of relevant influential factors for these fog events in the following Sects. 4–8. Finally, a discussion and conclusion will be given in Sect. 9.

2. Data and Methodology

2.1. Data

The following datasets are used in the present study. (1) MODIS visible satellite imagery, which may be downloaded from http://ladsweb.nascom.nasa.gov/. (2) Daily SST dataset with the horizontal resolution of $0.5° \times 0.5°$ from National Oceanic and Atmospheric Administration (NOAA) may be downloaded from http://ftp.emc.ncep.noaa.gov/omb/pub/history/sst/. (3) Sounding data is available daily every 12-h from the University of Wyoming at the website http://weather.uwyo.edu/upperair/sounding.html. (4) Final (FNL) operational global analysis data is available at 00, 06, 12, 18 UTC every day with the horizontal resolution of $1° \times 1°$ and 26 vertical pressure levels, produced by the Global Forecast System (GFS) that is operationally run four times a day in near-real time at the National Center for Environmental Prediction (NCEP) and can be downloaded from http://dss.ucar.edu/datasets/ds083.2/. (5) NCEP Automated Data Processing (ADP) Operational Global Surface observations data, periods: 10 February 1975 to 28 February 2007, can be downloaded from http://dss.ucar.edu/datasets/ds464.0/. And (6) NCEP ADP Global surface observational weather data, during the period from 30 March 2000 to current, is used for the fog case happened in July 2008, which can be downloaded from http://dss.ucar.edu/datasets/ds461.0/.

All the datasets listed above have good quality control and are well accepted and widely used in atmospheric and oceanic science. UCAR data archive (http://dss.ucar.edu/) is a well-known and widely accepted atmospheric dataset website. It freely provides all the post-processing software to the atmosphere-related community. The Grid Analysis and Display System is used for plotting the figures in the present study.

2.2. Methodology

The features of sea fog appearance on the satellite image have been documented and clarified by some fog researchers (ELLORD, 1995; FU et al., 2006; GAO et al., 2007). ELLORD (1995) suggested that a fog event typically has a more distinct brightness and is smoother in appearance. FU et al. (2006) proposed the criteria for determining sea fog events from the satellite images. They are: (1) the cloud had no apparent cumulus structure, its cover was smooth and relatively uniform; (2) the cloud edges were distinct; (3) the movement of the cloud was not remarkable during the day, suggesting that these cloud patches were lower because high clouds typically showed significant motion during the daytime. GAO et al. (2007) also suggested that the fog patches could be identified according to the whiteness of the image with smooth texture and distinct edges. The sea fog cases in this study have been selected according to the above classifications.

3. Case Overview

The Yellow Sea (Fig. 1) is a coastal sea area with the most frequent occurrence (over 80 days a year) of sea fog in spring (March, April, May) and early summer (June, July) (ZHOU et al., 2004; ZHANG et al., 2009). It is bordered by China's coast on the west, and partially by the Korean peninsula and Japan on

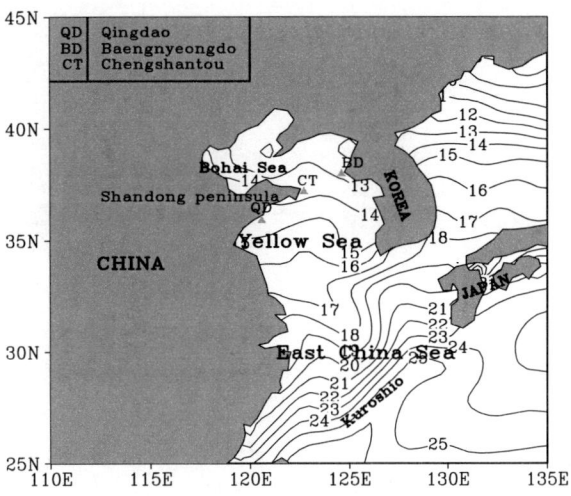

Figure 1
Horizontal distribution of SST (solid contour, 1°C interval) in the northwestern Pacific averaged from March to July during the period from 2001 to 2010, based upon NOAA daily SST dataset

the east. The Yellow Sea is located in the northwest of the Kuroshio current of the Northern Pacific. The Kuroshio current typically transports warm water from the tropical Pacific warm pool to the northern Pacific along the coasts of East Asian countries. As a result, a slantwise annually averaged SST occurs in the East China Sea (ECS) and gradually becomes a pure south-north gradient when reaching the Yellow and Bohai Seas (BHS). Along the China and Korean (west) coasts, meteorological observations from two weather stations are used in this study: Chengshangtou (CT), and Baengyeongdo (BD).

3.1. Case Selection

Six fog cases over the Yellow Sea from 2003 to 2008 were selected and verified by both satellite imagery and ground observations (Fig. 2). There are more fog events during this period. However, due to the sea fog occurring location (i.e., the Yellow Sea area), the covering area (i.e., covering at least 1/3 of the whole Yellow Sea), the duration of the fog events (i.e., lasting more than one day), and the completeness of the ground observations, only these six fog cases were selected. Two steps are taken for selecting and verifying the fog cases over the Yellow Sea.

The satellite imagery, especially the MODIS visible imagery, is used to identify the fog/stratus.

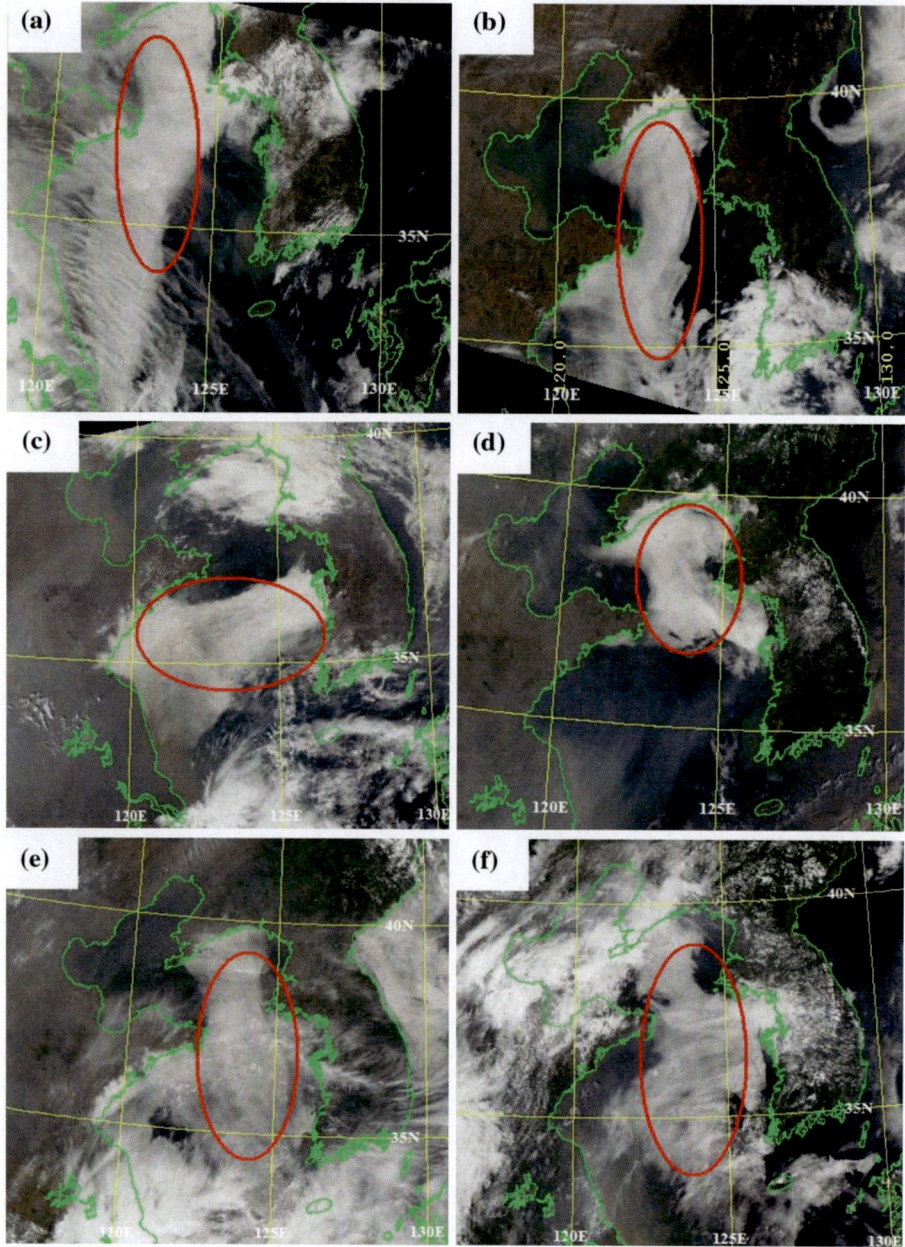

Figure 2
MODIS visible satellite imagery in which the area denoted by *red circle* is sea fog region over the Yellow Sea. **a** 0210 UTC 17 April 2003,
b 0255 UTC 11 April 2004, **c** 0455 UTC 09 March 2005, **d** 0220 UTC 23 June 2005, **e** 0225 UTC 01 June 2006, **f** 0500 UTC 07 July 2008

Figure 2 shows that the cloud edges in the elliptic circle are distinct and match the coastal line well. The cloud in six images has smooth texture and distinct brightness. The other time set of the imagery showed that the clouds did not move remarkably during the day (figures not shown). According to the classifications in Sect. 2, these cases should be fog/stratus.

Further verification is needed. In addition, these fog cases had an extension to the north in Fig. 2, except for case 2005-03-09, but disappeared in the Bohai Sea, where the SST is colder than that in the Yellow Sea (Fig. 1). The topography of the Shandong Peninsula and the local land-sea breeze circulation together played an important role in preventing warm/moist air from being transported to the Bohai Sea.

3.2. Case Verification

The horizontal atmospheric visibility observed at the ground station (CT) is shown to verify the existence of fog. Figure 3 describes the evolution of horizontal atmospheric visibility (km) of the six fog episodes, and wind field at CT station. It was observed that the atmospheric visibility displayed a continuous and very low period (near to zero) in all six fog events, with the lowest visibility observed being 0.1, 0.05, 0.05, 0.05, 0.05, 0 km, respectively, in Fig. 3a–f. The longest duration of the near-zero visibility reached 111 h (Fig. 3f), the shortest one lasted about 12 h (Fig. 3c). The mature stage can be clearly identified as the prolonged low visibility period at a station. Prior to the mature stage, the horizontal visibility decreased rapidly and the transition happened in 12 h for six fog events. It can be inferred that the transition period between the onset and mature stage at CT is very short. It is confirmed that the same situation happened during the transition from the mature to dissipation stage. We classify the whole life of fog events into three stages according to the above analysis: onset stage, mature stage, and dissipation stage. Table 1 shows the classification of six fog events. These six fog events are named by the date of the fog's mature stage, for example, in Fig. 3a, the mature stage of this case is from 01 UTC 17 to 00 UTC 18 April 2003; thus, this case is named case 2003-04-17. Based upon the same principal, the other five are named as follows: case 2004-04-11, case 2005-03-09, case 2005-06-23, case 2006-06-01, and case 2008-07-07, respectively. The wind speed at CT is under 10 ms^{-1} during the whole life of the fog event, except for case 2003-04-17 (Fig. 3a). It has been indicated by many sea fog researchers that weak wind is the most favorable condition for sea fog formation and maintenance (WANG, 1985; FU et al., 2006; GAO et al., 2007; ZHANG et al., 2009). There was no significant change for wind direction during the onset stage and for most of the mature stage for six fog events. Whereas during the transition period between the mature and dissipation stage, the wind changed from southerlies to northerlies for most of

the six fog events. The wind direction change from southerlies to northerlies is a signal of fog dissipation or lower cloud formation. The reason is that the air brought by the northerlies from the Eurasian continent pushes the moister air from the south away and this stops the water vapor source of advection cooling fog, which, confirmed by many researchers, is the most frequent occurrence of fog type that happens over the Yellow Sea. The exceptional cases are case 2005-06-23 and case 2008-07-07, in which the southerlies prevail during the whole fog life. It can be inferred from these two exceptions that only with abundant moisture supporting, fog cannot sustain. There are still more processes needed to be explored and studied for the whole fog life.

The observations of three weather stations (BD, CT, QD in Fig. 1) which provided crucial surface and upper-air sounding reports for these cases will be used for the following analysis.

4. Relative Humidity and Surface Pressure

Sea fog is a mild weather phenomenon in the marine atmospheric boundary layer (MABL). Previous studies suggested that sea fog often happens associated with the high pressure, and the subsidence associated with the high pressure system may help form and enhance the inversion layer (KORAČIN et al., 2001). Because of the lack of sounding data of CT station from the current available data sources, Baengnyeongdo station (hereafter, BD), which is close to CT station and located at the coastal area of the Yellow Sea, is used instead of CT station to study the pressure variability of the upper level layer (500, 700, 850 hPa) during the whole fog life. The surface pressure variability data uses the observations from CT station.

Figure 4 shows the time evolution of relative humidity (hereafter, RH) and surface pressure on the ground observed at CT station. It can be inferred from the evolution of surface pressure that during the onset stage, six fog events are classified into two categories: (1) the pressure-weakening type, for example, cases 2003-04-17, 2004-04-11, 2005-03-09 (Fig. 4a, b, c); (2) the pressure-strengthening type, for example, cases 2005-06-23, 2006-06-01, 2008-07-07

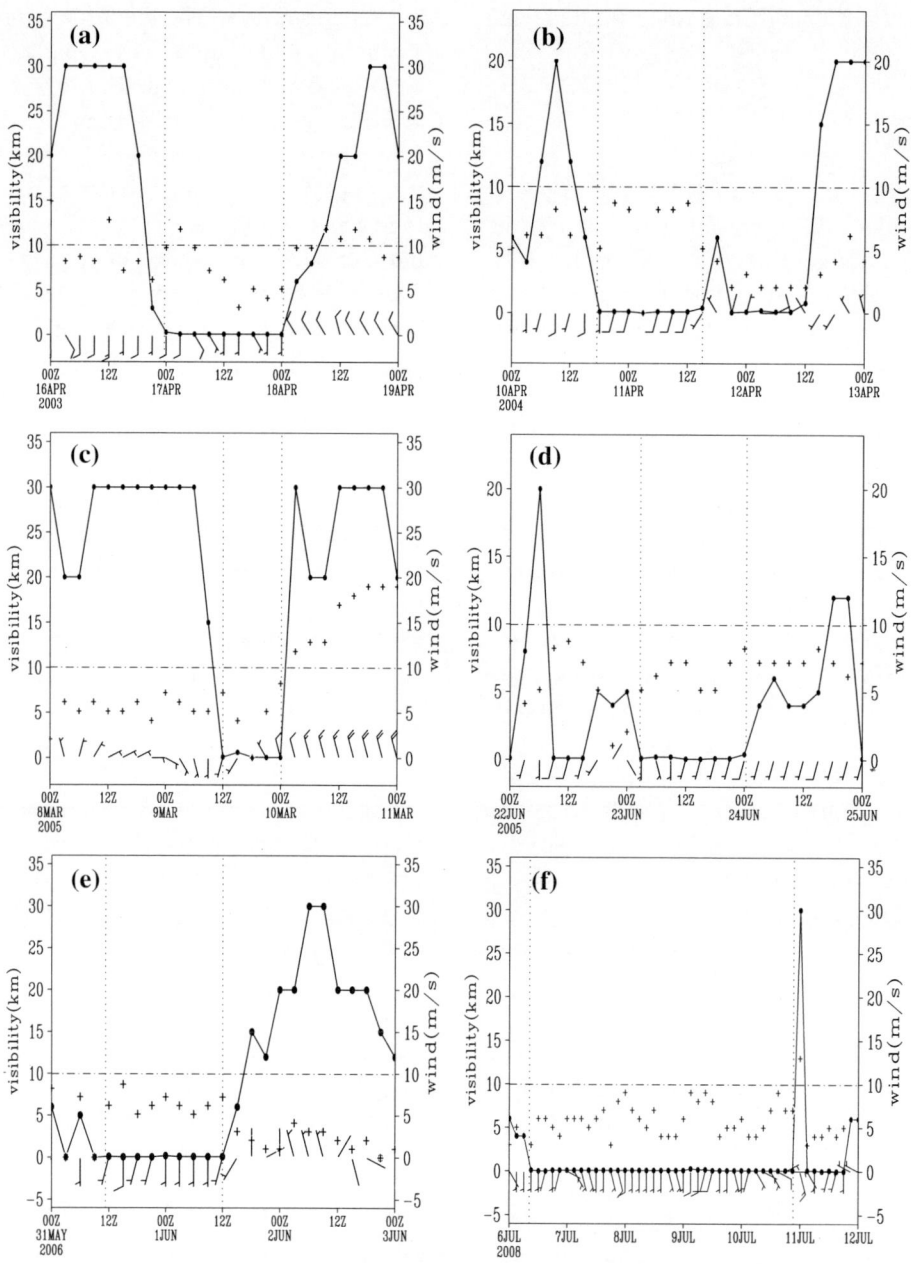

Figure 3

The evolution of atmospheric visibility at CT station, **a** case 2003-04-17, **b** case 2004-04-11, **c** case 2005-03-09, **d** case 2005-06-23, **e** case 2006-06-01, **f** case 2008-07-07, obtained from NCEP ADP datasets. *Solid line* denotes visibility (km) reported every 3 h. Plus sign denotes wind speed (ms^{-1}). *Dashed-dotted line* denotes 10 ms^{-1}. *Wind bars* denote wind speed and direction at surface. *Two vertical dotted lines* separate the fog episode into three different stages

(Fig. 4d, e, except for case 2005-03f). In the pressure-weakening type, the surface pressure reaches its minimum pressure or is gradually weakened during the mature stage, whereas in the pressure-strengthening type, the surface pressure reaches its maximum or is gradually strengthened during the mature stage. It is noteworthy that the former type happened in spring, whereas the latter one happened in summer. The range of the sea level pressure variability is 4–14 hPa during the whole fog life in all six fog

Table 1

Stage division in six fog events

Case (YYYY-MM-DD)	The developing stage/onset period	The mature stage	The dissipation stage
2003-04-17	1200 UTC 16-0000 UTC 17	0000 UTC 17-0000 UTC 18	0000 UTC 18-1800 UTC 18
2004-04-11	0900 UTC 10-1800 UTC 10	1800 UTC 10-1500 UTC 11	1500 UTC 11-1800 UTC 12
2005-03-09	1500 UTC 08-0900 UTC 09	0900 UTC 09-0300 UTC 10	0300 UTC 10-2100 UTC 10
2005-06-23	0600 UTC 22-0300 UTC 23	0300 UTC 23-0000 UTC 24	0000 UTC 24-1800 UTC 24
2006-06-01	0000 UTC 31-1200 UTC 31	1200 UTC 31-1200 UTC 01	1200 UTC 01-0600 UTC 02
2008-07-07	0000 UTC 06-0900 UTC 06	0900 UTC 06-2100 UTC 10	2100 UTC 10-0000 UTC 12

Note Beijing Local Standard Time (LST) = UTC + 8 h

events. NOONKESTER (1979) suggested that Santa Ana related fogs have been observed at Naval Ocean System Center to consist of two primary types. One type appeared as a thick (~ 200 m) layer which moved onto the coast like a front. The other type appeared as a thin (~ 50 m) layer which moved northward along the coast up from Baja, California. However, he did not suggest that the two types were associated with different seasons. One of our current cases, case 2005-03-09 (pressure-weakening type), was studied by using the Weather Research and Forecasting model version 3 (WRF V3) (LI, 2010, Ph.D. thesis). The results confirmed that the fog layer is about 200 m. It may be inferred that the turbulence stirred by the low level jet around 100–200 m may extend the fog layer to a higher level under the inversion layer. The other type, i.e., pressure-strengthening type, is associated with a high pressure system, in which subsidence may exist and suppress the fog layer vertical extension during the mature stage. This hypothesis is still needed to be verified in future study by modeling.

The situation of geopotential height (GH) variability at high levels (500, 700, 850 hPa) at BD station is identical to the sea level pressure variability (figures not shown). Namely, six fog cases also can be classified into two types, which is similar to the surface types. The GH variability of these upper levels during the whole fog stage remains the similar pattern. However, the range of the variability of GH varies among the three levels, the higher the level, the greater the range. It suggests that in higher levels, the range of pressure variability is greater, close to the ground, the range of pressure variability is smaller.

It can be seen from the evolution of RH that during the mature stage, RH is larger than 90% in six cases, and most of our six cases are larger than 95% or close to 100%. The evolution of RH has an opposite phase change with that of visibility (Fig. 3) in all six cases.

5. The Impact of Surface Processes

5.1. During the Onset Stage

The sea level pressure (SLP) and wind field averaged during the onset period is shown in Fig. 5. We can see that the northwest Pacific Ocean high pressure (NPH) centered around ($25° \sim 35°$N, $120° \sim 140°$E) dominated the lower atmospheric level of northwest Pacific Ocean during the onset period of the six fog events (Fig. 5). Owing to the existence of the NPH, the southerly winds blow over the Yellow and East China Seas. It can be inferred from the SST distribution (Fig. 1) over the Yellow and East China Seas that warm and moister air was brought by the southerlies from the south of East China Sea to the Yellow Sea. It is the favorable condition for advection cooling sea fog formation when warm air passes over cold water (WANG, 1985; DIAO, 1992; ZHANG et al., 2005). The NPH was wandering over the northwest Pacific Ocean for most of the whole fog episode of six fog cases and the intensity of the NPH changed slightly. The most important point is that the pressure gradient over the Yellow Sea remains small during the domination of the NPH. Meanwhile, the well-positioned NPH tended to prevent the weather systems from intruding

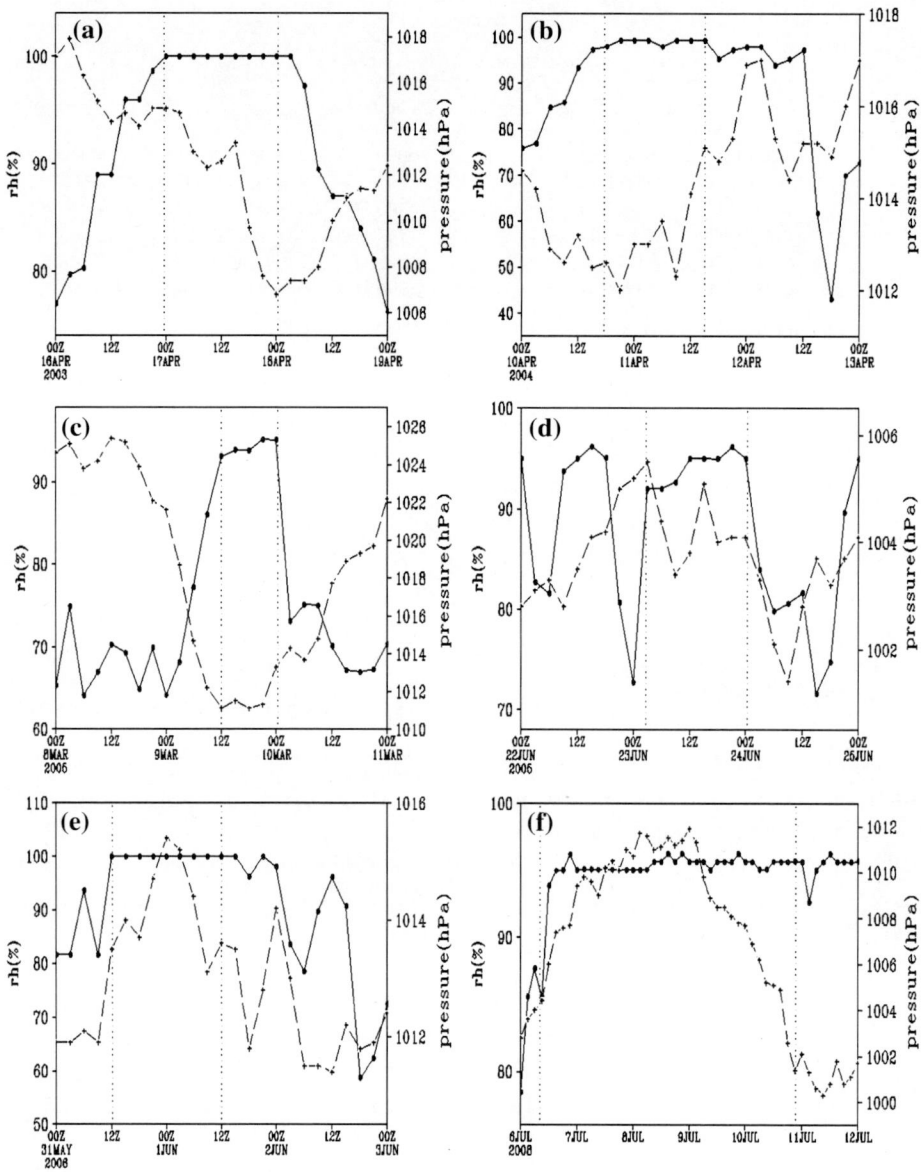

Figure 4
Relative humidity and surface pressure evolution during the whole fog episode at CT station, **a** case 2003-04-17, **b** case 2004-04-11, **c** case 2005-03-09, **d** case 2005-06-23, **e** case 2006-06-01, **f** case 2008-07-07, obtained from NCEP ADP datasets. *The solid line* denotes relative humidity. *The dash line* denotes surface pressure. *Two vertical dotted lines* separate the fog episode into three different stages

southward over the Yellow Sea. Both intensity and position of the high pressure are important for fog formation and maintenance. It is noteworthy that the pressure gradient of case 2003-04-17 over the Yellow Sea is greater than that of other cases and in turn, its wind speed is greater than that of other cases. This analysis is consistent with station observation in Sect. 3.2.

5.2. During the Mature and Dissipation Stage

During the mature stage, the surface wind field and SLP change slightly and the NPH remains an influence on the Yellow Sea area. Hence, the southerly winds prevail over the Yellow Sea area (Figures not shown) and fog becomes mature. In the end of mature stage and the whole dissipation stage,

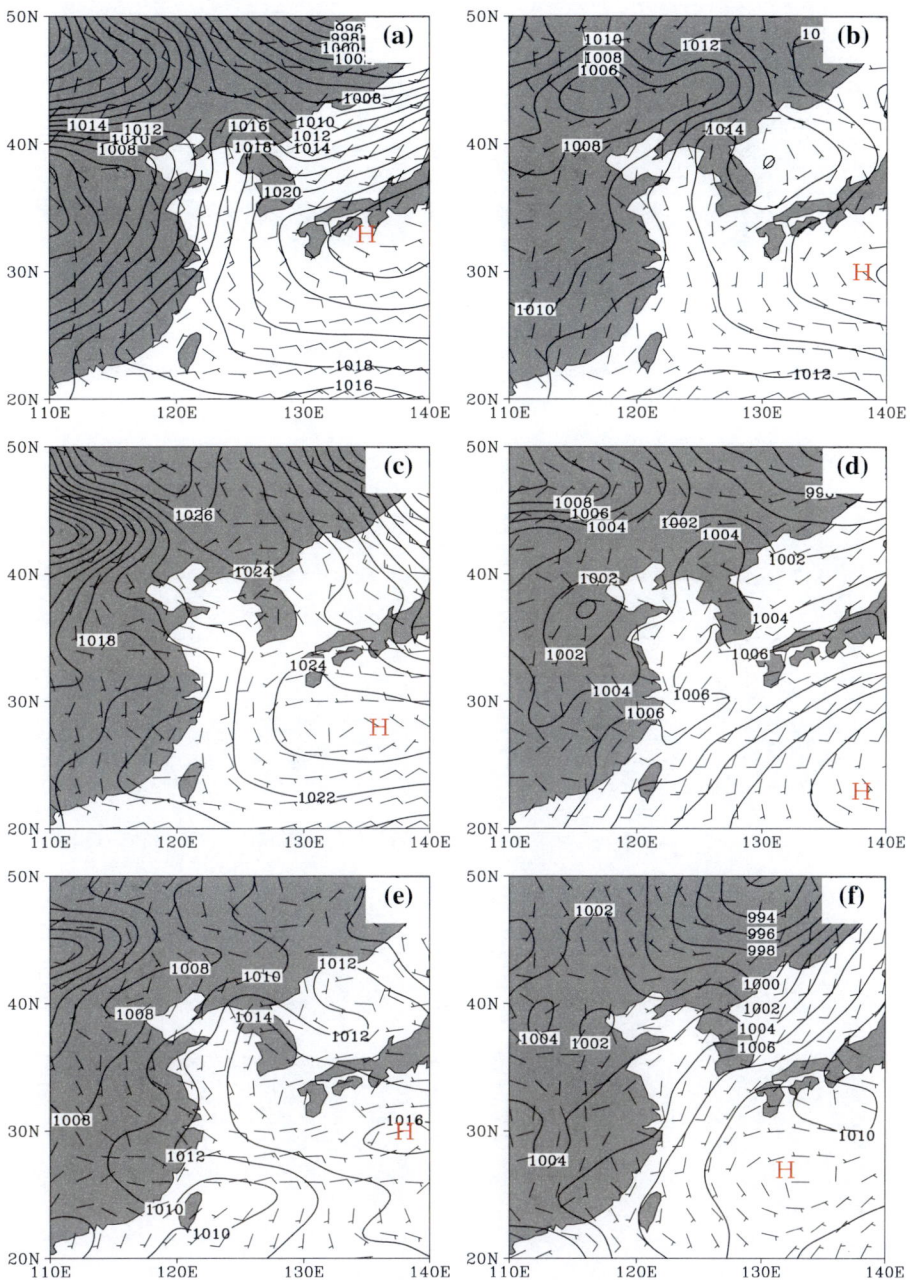

Figure 5

The sea level pressure and wind field averaged over the onset period (see Table 1 for the detailed onset period), **a** case 2003-04-17, **b** case 2004-04-11, **c** case 2005-03-09, **d** case 2005-06-23, **e** case 2006-06-01, **f** case 2008-07-07, obtained from FNL dataset. *Solid line* denotes sea level pressure, intervals 2 hPa. The *red letter* 'H' denotes high pressure. *Full and short barbs* denote 10 knots (~ 5.14 ms^{-1}) and 5 knots, respectively

the wind field and SLP distribution was shown in Fig. 6. It is observed that the NPH retreated southward or eastward in six fog cases. Because of the NPH retreat, its influence over the Yellow Sea region becomes weak. In spring, the northern weather system is stronger than that in southern tropical areas. Hence, the northern weather system intruded and dominated over the Yellow Sea region (Fig. 6a,

227

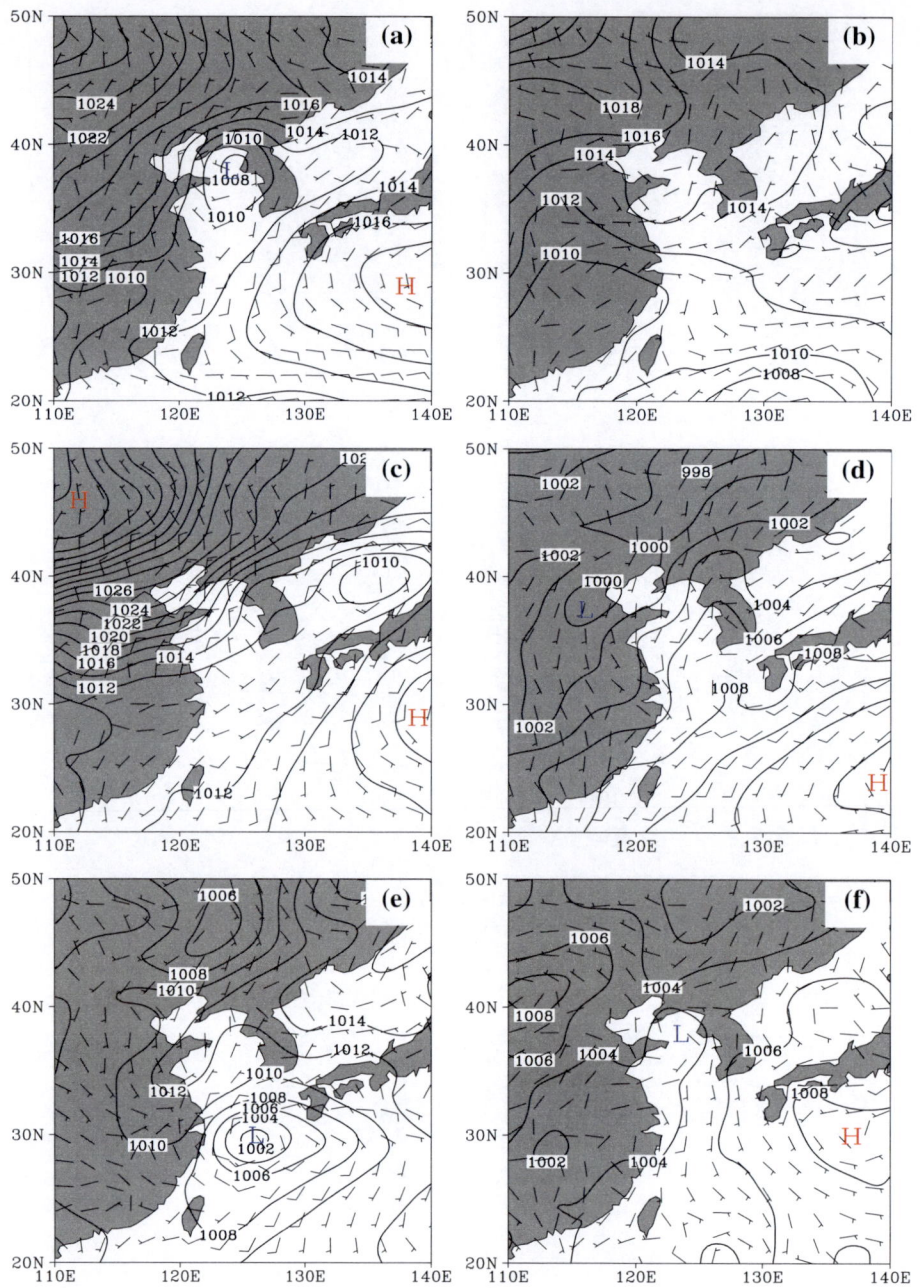

Figure 6
As in Fig. 5, but averaged over the dissipation period (see Table 1 for the detailed dissipation period), the *blue letter* 'L' denotes low pressure

b, c). In summer, weather systems in the tropical region are active and some of them may intrude and dominate over the Yellow Sea area (Fig. 6d, e, f). The intruded systems from either north or south changed the wind direction or the stability of the lower atmospheric level over the Yellow Sea region

and hence, led fog to dissipation. The reasons for fog dissipation with these above mentioned changes are that (1) the dry air brought by the northerly winds interrupts the moisture sources of sea fog and (2) the unstable lower atmosphere cannot hold the moisture and hence, fog either cannot form or becomes lower stratus. It is noteworthy that the southerly winds prevail at all times in case 2005-06-23 and case 2008-07-07. However, these two fogs finally dissipated. The reason for these two fogs' dissipation is that the stability of lower atmospheric level is destroyed by the passing low pressure system. It is confirmed by the analysis of the pressure change at CT station (Fig. 4d, f) that the low pressure passed by the Yellow Sea region during the dissipation period.

6. SST Preconditioning and Air-sea Temperature Differences

6.1. Sea Surface Temperature

The climatology of SST averaged from March to July over 10 years (2001–2010) in the northwest Pacific is shown in Fig. 1. It has been observed that the meridional temperature gradient is relatively large. The temperature difference between south of the East China Sea and north of the Yellow Sea reaches around 10°C. Meanwhile, the north-south gradient of SST over the Yellow Sea will gradually decrease from wintertime (February) to summertime (August), with the maximum difference 16°C in February and the minimum of 3°C in August in a year (figure not shown), due to the seasonal variation of solar radiation. Similar conditions have been verified by CHO et al. (2000) and observed in association with advection fog events in the northeast coastal regions of the United States (TARDIF and RASMUSSEN, 2008) and the northeastern coastal regions of Scotland (FINDLATER et al., 1989). The air brought by southerlies from the East China Sea to the Yellow Sea will be cooled and saturated quickly due to the rapid drop of SST. This is the main reason for the formation of advection cooling fog over the Yellow Sea. There is a gradual increase of fog occurrence from April to July observed at the stations around the Yellow Sea (for example, Fig. 1 in ZHANG et al., 2009). However, in the summertime (after July), there exists little difference of SST between the East China Sea and the Yellow Sea and fog events rarely occurred at that time (WANG, 1985; FU et al., 2004, 2006; ZHANG et al., 2009). Hence, the surface cooling effect is a primary mechanism for the fog formation. The local

evaporation due to the increasing SST with seasonal change may make a great contribution to the water vapor, consequently increases the occurrence of fog. However it may not be enough to offset the loss of cooling effect for fog formation. It is noteworthy in ZHANG et al. (2009) that the observations show that the fog occurrence becomes gradually increased from March to July, and the most fog occurrence happened in July. The local evaporation may make a great contribution to fog occurrence due to the above analysis of fog occurrence. However, as mentioned above, the south-north gradient of the SST gradually becomes small as the season transition from spring to summer. The greatest gradient of the SST does not correspond to the most fog occurrence.

6.2. Air-Sea Temperature Difference

To investigate the air-sea conditions during the fog onset stage, the temperature difference (TD) between air temperature (here we use 2 m temperature for air temperature) adjacent to the sea surface and SST is shown in Fig. 7. The positive value indicates that the warm air is over cold water. In this scenario, the sea surface will exert a cooling effect on warm moist air above, thus promoting condensation to occur and fog to form. The negative value indicates that cold air is over warm water. In this scenario, the sea surface heating effect makes the marine layer unstable, which is typically not favorable for advection cooling fog formation. One can see that the positive area is over the Yellow Sea for all six fog cases. The temperature difference is in the range of 1–3°C, which is in good agreement with model diagnostics in the range of 0–3°C by GAO et al. (2007). This also gives a further verification of the above analysis for surface cooling effect. In the Bohai Sea, the TD is remarkably larger than that in the Yellow Sea, and yet without much fog appearance in the Bohai Sea in these six cases (Fig. 2). It is further confirmed by observing the consecutive GOES-9 images that there is no fog appearance over the Bohai Sea. The reason for no fog occurrence in the Bohai Sea area was discussed in Sect. 3 concerning the topography of Shandong Peninsula, the land heating processes, and the local land-sea breeze circulation. The local insufficient evaporation due to the relatively low SST may be the

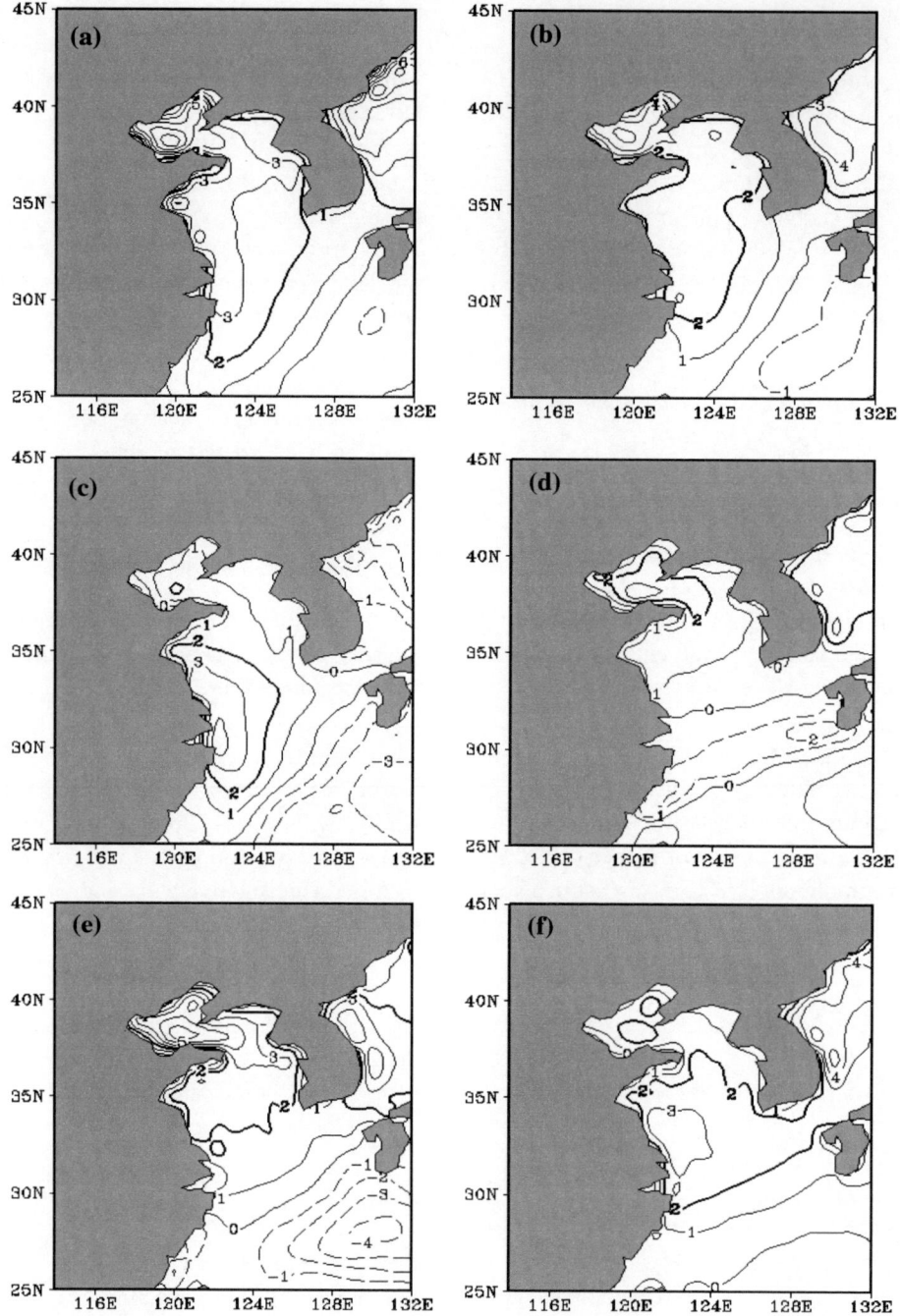

Figure 7
Horizontal distribution of air-sea temperature difference (TD) averaged over the onset period (see Table 1 for the detailed onset period), **a** case 2003-04-17, **b** case 2004-04-11, **c** case 2005-03-09, **d** case 2005-06-23, **e** case 2006-06-01, **f** case 2008-07-07, obtained from FNL dataset. *Thick line denotes* the 2°C TD

other reason. However, the TD for case 2003-04-17 (Fig. 7a) is larger than that in the other cases. GAO *et al.* (2007) suggested that TD corresponds to stronger wind in the sea fog area, according to their

model statistical result. The favorable conditions for fog formation are in the range of 0–3°C for TD and 2–8 ms^{-1} for wind speed. Combined with their statistical result, the wind speed in case 2003-04-17

(Fig. 3a) is larger than that in the other cases and the TD is the largest among all the cases. Under these conditions, fog formed. It suggested that their favorable conditions of fog formation could make a reasonable extension to the side of larger wind and TD. However, the fact is that even in case 2003-04-17, the wind speed is in the range of 2–10 ms^{-1} during most of the time period of its life, especially the mature stage, in which the wind speed is weaker than that in the onset stage.

7. Air Temperature Inversion

Figure 8 shows time evolution of the air temperature for the six fog events at BD station. One can see that there exists an air temperature inversion layer in each fog episode at BD station. The top of the inversion layer is under 850 hPa level, except for case 2005-03-09 (Fig. 8c), for which its top is near 750 hPa level. The inversion layer of case 2005-03-09 is centered over the southern area of Shandong Peninsula. We checked QD station and found that the top of the inversion layer is around 925 hPa. It is observed that at BD station, the top of the inversion layer is higher in spring fog cases (Fig. 8a, b, c) than that of summer fog cases (Fig. 8d, e, f). Because of the similar temperature difference ($\sim 6°C$) in vertical, the intensity of the inversion layer is greater in summer fog cases than that in spring fog cases. It may need more station observations or modeling work to further verify and confirm the characteristics of the inversion layer in different seasons. Combined with Sect. 1.2.1, i.e., the analysis of the surface pressure change, the temperature inversion layer intensified corresponding to the high pressure period or pressure-strengthening period, whereas the temperature inversion layer weakened or vanished during the low pressure period or pressure-weakening period. Bear in mind that BD station is located downstream of CT station (Fig. 1). Namely, the pressure change may happen later at BD station. Note that the inversion layer with an upward extension became weak in the end of case 2005-06-23 and finally vanished early on 25 June 2005. The fog in this scenario may become stratus because there is no convective limitation in the vertical direction. In other cases, case 2008-07-

07, the inversion layer vanished at about 00 UTC 11 July 2008. Hence, fog cannot hold in this scenario. From the analysis of these two cases, it is suggested that even though the moisture source is not interrupted by the wind direction change, fog cannot survive without the inversion layer.

8. Characteristics of Moisture

Figure 9 shows the horizontal distribution of relative humidity (RH) at 1,000 hPa during the onset period (see Table 1 for the detailed onset time) in six fog cases. There exists a high-value (>80%, shaded) area of RH south of the ECS and a moisture "tongue" extending from there to the north of the Yellow Sea for most of our cases. The RH gradually decreased from south to north of the Yellow Sea region. This suggests that the southern sea area contains more moisture mainly because the SST is higher in the southern sea region than that in the northern. The evaporation in the higher SST region makes great contribution to the moisture. Note that in case 2005-03-09 (Fig. 9c), the large RH area is located in the southeastern part of the Yellow Sea, which is different from that in other cases. The distribution of RH during the onset period in case 2005-03-09 may be caused by the circulation of the NPH (Fig. 5c). The southwesterly triggered by the NPH led the moisture to northeast and was finally blocked by the Korean Peninsula and hence, accumulated in the southeastern region of the Yellow Sea.

Figure 10 shows the time evolution of RH at BD station. The highest values (>90%) of RH correspond to the mature stage of these sea fogs, marked between the vertical lines on the figures. During the onset and dissipation stage, RH typically presented smaller values. The large RH area is near the ground, especially during the onset and dissipation stages. In the mature stage, the large RH area extended upward and the highest value of RH is still near the ground, except for case 2005-03-09 (Fig. 10c), in which the core is around 900 hPa level. The temperature inversion in case 2005-03-09 (Fig. 8c) is weak near the ground and its top is near 750 hPa according to the previous analysis. The weak convections in the lower level may lead moisture to the upper level due to the weak

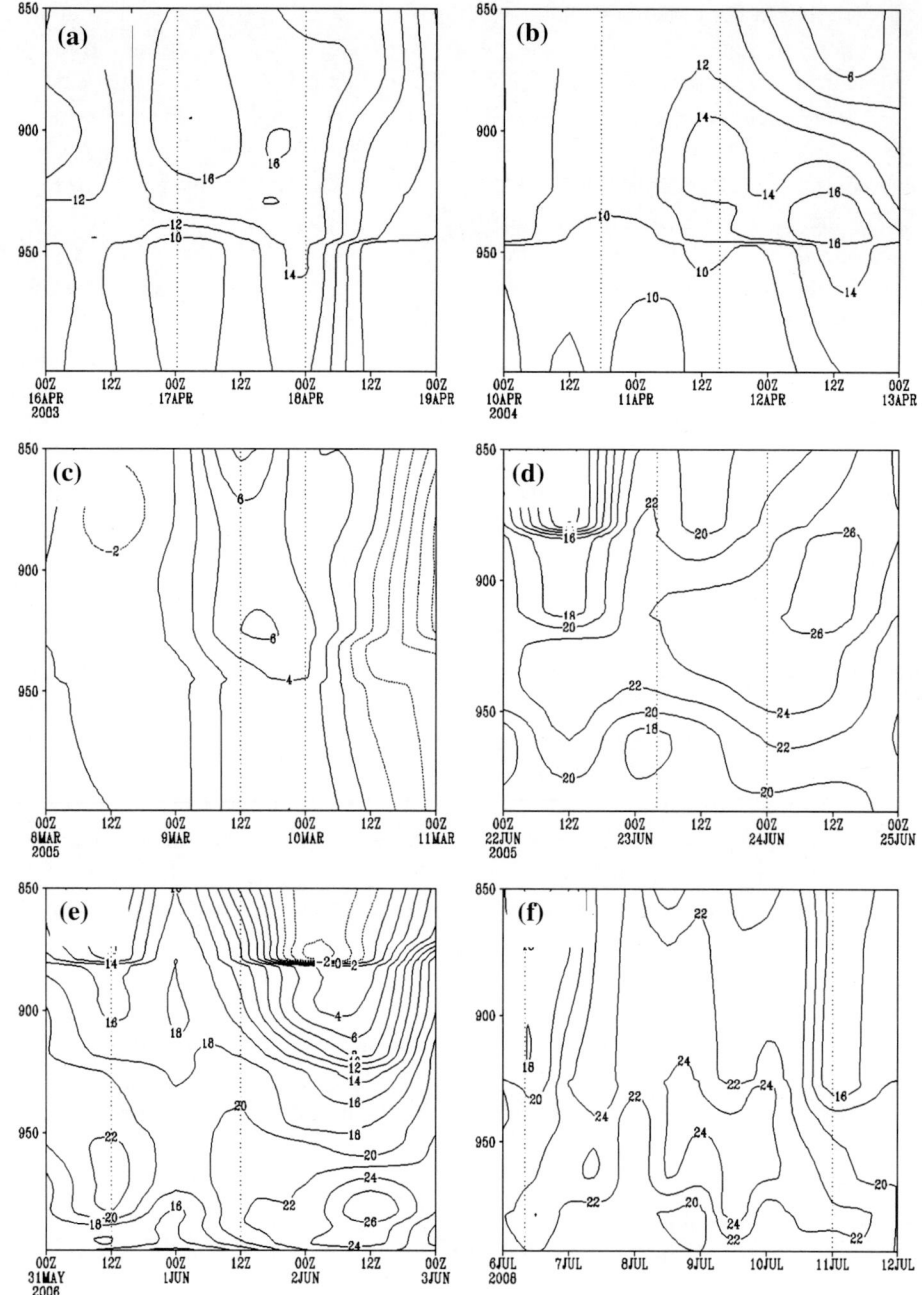

Figure 8

Time evolution of air temperature (unit, °C) for six fog events at BD station, **a** case 2003-04-17, **b** case 2004-04-11, **c** case 2005-03-09, **d** case 2005-06-23, **e** case 2006-06-01, **f** case 2008-07-07, obtained from Sounding dataset. Two *vertical dotted lines* separate the fog episode into three different stages. Temperature interval is 2°C

stability at BD station. Combined with previous analysis of the inversion layer, it is observed that the vertical extension of the large RH is associated with the intensity of the inversion layer, i.e., the stronger the inversion layer, the lower the vertical extension.

9. Discussion and Conclusion

Six sea fog episodes over the Yellow Sea were investigated in the context of the large-scale environmental influences, including the pressure change,

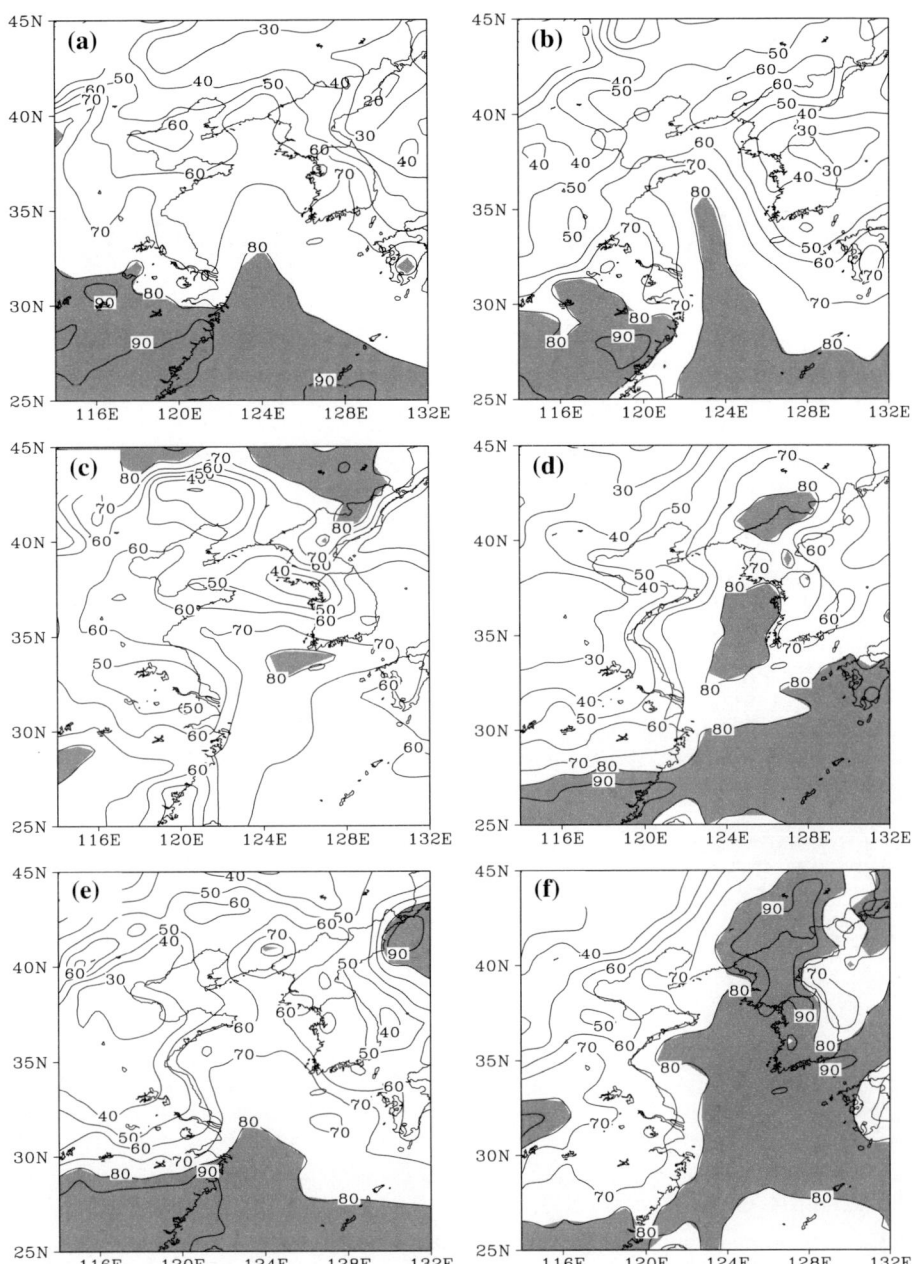

Figure 9

Horizontal distribution of relative humidity (RH) at 1,000 hPa averaged over the onset period (see Table 1 for the detailed onset period), **a** case 2003-04-17, **b** case 2004-04-11, **c** case 2005-03-09, **d** case 2005-06-23, **e** case 2006-06-01, **f** case 2008-07-07, obtained from FNL dataset. *Shaded area* denotes RH larger than 80%

the surface processes, the temperature inversion, as well as the moisture. The main findings of the present study are summarized as follows. It was found that there exist two types of sea fog in six fog cases according to the pressure change: (1) the pressure-

weakening type, (2) the pressure-strengthening type. The former happened in spring and the latter in summer. The NPH plays as the dynamic source of transporting moisture and also prevent the north or south weather systems from intruding the Yellow Sea

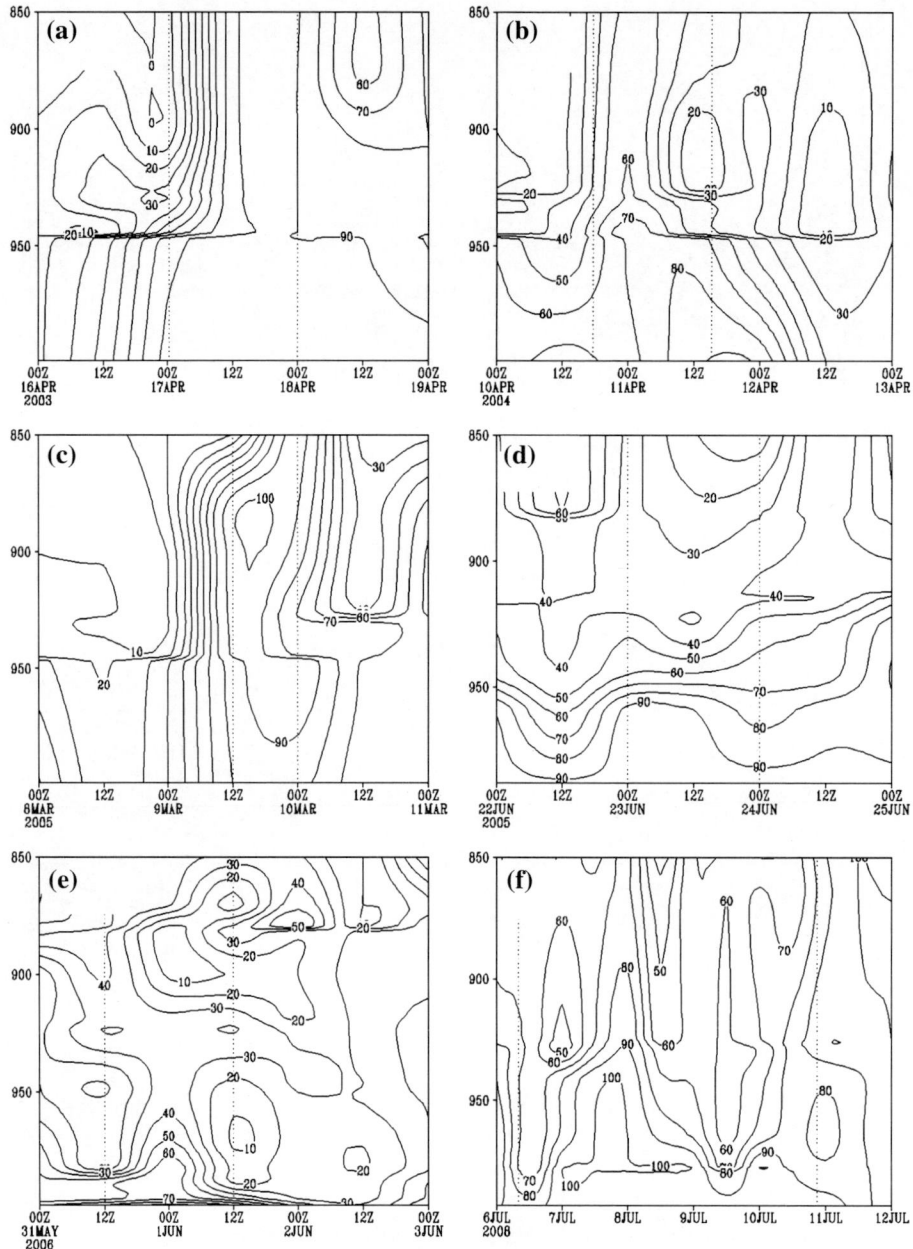

Figure 10
The evolution of RH for six fog events at BD station, **a** case 2003-04-17, **b** case 2004-04-11, **c** case 2005-03-09, **d** case 2005-06-23, **e** case 2006-06-01, **f** case 2008-07-07, obtained from sounding dataset. Two *vertical dotted lines* separate the fog episode into three different stages. RH interval is 10%

region. Also, the wind direction change takes place when the NPH retreats southward or eastward. The air temperature inversion layer is slightly stronger in summer fog cases than that in spring ones. However, the top of the inversion layer is higher in spring fog

cases than that in summer ones. The analysis of moisture supports the traditional point appreciated by many sea fog researchers, i.e., the source of the water vapor is from southern sea region of the ECS. It can be inferred that two general reasons are responsible for the fog dissipation. One is the wind direction change, which stops the transportation of the moisture. The other is the unstable lower atmosphere, which cannot hold the moisture in the lower atmospheric level, and hence, fog cannot sustain.

Some favorable conditions for fog formation are gained, based on the present study, which agreed well with previous researchers' work (WANG, 1985; FU et al, 2004, 2006; GAO et al., 2007; ZHANG et al., 2009; FU et al., 2010). They are as follows: (1) the southerly winds in the range of 2–10 ms^{-1}, (2) the well-positioned NPH, (3) the temperature inversion, (4) the south-north SST gradient. These four conditions are vital to sea fog formation over the Yellow Sea region. However, the conditions may vary differently, i.e., the intensity of the temperature inversion may vary in a large range. In the present study, we cannot give a quantitative definition of the intensity of the inversion. It remains a good topic for us to study in the near future. Many weather factors could lead to fog dissipation, such as, the low pressure system, the northerly winds. Not like other weather phenomena, fog is a shallow weather phenomenon in the MABL and a very delicate balance among all the processes. Hence, any change in the MABL could destroy the balance and lead fog to demise.

Despite the detailed analyses conducted in this study, issues related to the current analyzed environmental factors remain unsolved. For example, how does the inversion form and sustain itself? How does the modification of the inversion influence the fog? Concerning thermal dynamics: it is unclear that what roles and priorities surface cooling plays, and the radiation cooling in the whole fog episode is thus far still unknown over the Yellow Sea. There are still various other factors that influence the formation, maintenance, and dissipation of sea fogs, such as the properties of the marine layer, and the role of turbulence. We expect to use numerical models to help us explore more of these issues in the very near future.

Acknowledgments

The authors would like to express their gratitude to two anonymous reviewers for their constructive and helpful comments for the improvement of this manuscript. P. Li would like to express his sincere thanks to the China Scholarship Council for their financial support, which made his study abroad at the U.S. Federal agency NOAA possible. C. Lu's efforts towards this study were supported by NOAA. G. Fu was partly supported by the National Natural Science Foundation of China under the grant number 406750060, and the Chinese Ministry of Science and Technology under the 863 Project grant number 2006AA09Z151. This work was also partly supported by the State Oceanic Administration under the grant 908-02-03-10, and the Chinese Meteorological Administration under the grant GYHY200706031. The authors would like to thank Ann Reiser for her help in editing the paper.

REFERENCES

ANDERSON, J.B. (1931). *Observations from airplanes of cloud and fog conditions along the southern California coast*. Mon. Wea. Rev. *59*, 264–270.

CHO, Y.K., KIM, M.O., and KIM, B.C. (2000). *Sea fog around the Korean Peninsula*. J. Appl. Meteor. *39*, 2473–2479.

COTTON, W.R., PIELKE, R.A., WALKO, R.L., LISTON, G.E., TREMBACK, C.J., JIANG, H., MCANELLY, R.L., HARRINGTON, J.Y., NICHOLLS, M.E., CARRIO, G.G., and MCFADDEN, J.P. (2003). *RAMS 2001: current status and future directions*. Meterol. Atmos. Phys. *82*, 5–29.

DIAO, X.X. (1992). *Statistical analysis of sea fog near Qingdao adjacent sea*. Marine Forecasts *9*(3), 45–55 (in Chinese with English abstract).

ELLORD, G.P. (1995). *Advances in the detection and analysis of fog at night using GOES multispectral infrared imagery*. Wea. Forecast. 606–619.

FILONCZUK, M., CAYAN, D., and RIDDLE, L. (1995). *Variability of marine fog along the California coast*. Scripps Institution of Oceanography Rep. 95-2, 58 pp.

FINDLATER, J., ROACH, W., and MCHUGH, B. (1989). *The haar of north-east Scotland*. Quart. J. Roy. Meteor. Soc. *115*, 581–608.

FU, G., ZHANG, M., DUAN, Y., ZHANG, T., and WANG, J. (2004). *Characteristics of sea fog over the Yellow Sea and the East China Sea*. Kaiyo Monthly (Japan) *38*, 99–107 (Extra Issue).

FU, G., GUO, J., XIE, S., DUAN, Y., and ZHANG, M. (2006). *Analysis and high-resolution modeling of a dense sea fog event over the Yellow Sea*. Atmos. Res. *81*, 293–303.

FU, G., LI, P., CROMPTON, J., GUO, J., GAO, S., and ZHANG, S. (2010). *An observational and modeling study of a sea fog event over the Yellow Sea on 1 August 2003*. Meteorol. Atmos. Phys. *107*, 149–159.

GAO, S., LIN, H., SHEN, B., and FU, G. (2007). *A heavy sea fog event over the Yellow Sea in March 2005: analysis and numerical modeling*. Adv. Atmos. Sci. 24, 65–81.

GULTEPE, I., MULLER, M.D., and BOYBEYI, Z. (2006a). *A new warm fog parameterization scheme for numerical weather prediction models*. J. Appl. Meteor. *45*, 1469–1480.

Reprinted from the journal

GULTEPE, I., TARDIF, R., MICHAELIDES, S.C., CERMAK, J., BENDIX, J., MULLER, M.D., PAGOWSKI, M., HANSEN, B., ELLORD, G., JACOBS, W., TOTH, G., and COBER, S.G. (2007). *Fog research: a review of past achievements and future perspectives*. Pure Appl. Meteor. *164*, 1121–1159.

JING, C. (1980). *A preliminary analysis of sea fog off the Qingdao coast*. Meteorological Monthly 65, 6–8 (in Chinese with English abstract).

KORAČIN, D., LEWIS, J., THOMPSON, W., DORMAN, C., and BUSINGER, J. (2001). *Transition of stratus into fog along the California coast: observations and modeling*. J. Atmos. Sci. *58*, 1714–1731.

KORAČIN, D., BUSINGER, J., DORMAN, C., and LEWIS, J. (2005). *Formation, evolution, and dissipation of coastal sea fog*. Boundary Layer Meteorol. *117*, 447–478.

LEWIS, J., KORAČIN, D., RABIN, R., and BUSINGER, J. (2003). *Sea fog off the California coast: viewed in the context of transient weather systems*. J. Geophys. Res. *108*(D15), 4457.

LEWIS, J., KORAČIN, D., and REDMOND, K. (2004). *Sea fog research in the UK and USA: historical essay including outlook*. Bull. Am. Meteorol. Soc. *85*, 395–408.

LI, P. (2010). *Observations of sea fog over the Yellow Sea and the modeling study based upon WRF model*, Ph.D. thesis.

OLIVER, D., LEWELLEN, W., and WILLIAMSON, G. (1978). *The interaction between turbulent and radiative transport in the development of fog and low-level stratus*. J. Atmos. Sci. *35*, 301–316.

PETTERSSEN, S. (1938). *On the causes and the forecasting of the California fog*. Bull. Am. Meteor. Soc. *19*, 49–55.

PILIÉ, R.J., MACK, E.J., ROGERS, C.W., KATZ, U., and KOCMOND, W.C. (1979). *The formation of marine fog and the development of fog-stratus systems along the California coast*. J. Appl. Meteor. *18*, 1275–1286.

TARDIF, R. and RASMUSSEN, R. (2008). *Process-oriented analysis of environmental conditions associated with precipitation fog events in the New York city region*. J. Appl. Meteorol. Climatol. *47*, 1681–1703.

TAYLOR, G.I. (1917). *The formation of fog and mist*. Quart. J. Roy. Meteor. Soc. *43*, 241–268.

TELFORD, J. and CHAI, S.K. (1984). *Inversions and fog, stratus and cumulus formation in warm air over cooler water*. Bound.-Layer Meteor. *29*, 109–137.

WANG, B., *Sea fog* (China Ocean Press 1985).

ZHANG, H., ZHOU, F., and ZANG, X. (2005). *Interannual change of sea fog over the Yellow Sea in spring*. Oceanol. Limnol. Sin. *36*, 36–42 (in Chinese with English abstract).

ZHANG, S., XIE, S., LIU, Q., YANG, Y., WANG, X., and REN, Z. (2009). *Seasonal variations of Yellow Sea fog: observations and mechanism*. J. Climate. *22*, 6758–6772.

ZHOU, F. and LIU, L. (1986). *Comprehensive survey and research report on the water areas adjacent to the Changjiang River estuary and Chejudo island marine fog*. J. Shandong Coll. Oceanol. *16*, 114–131 (in Chinese with English abstract).

ZHOU, F., WANG, X., and BAO, X. (2004). *Climatic characteristics of sea fog formation of the Huanghai Sea in spring*. Acta Oceanol. Sin. *26*, 28–37 (in Chinese with English abstract).

(Received October 28, 2010, revised April 17, 2011, accepted May 3, 2011, Published online June 14, 2011)

Pure Appl. Geophys. 169 (2012), 1001–1017
© 2011 Springer Basel AG
DOI 10.1007/s00024-011-0358-3

A Comparison Study Between Spring and Summer Fogs in the Yellow Sea-Observations and Mechanisms

Suping Zhang,[1] Man Li,[1] Xiangui Meng,[1] Gang Fu,[1] Zhaopeng Ren,[1,2] and Shanhong Gao[1]

Abstract—New observations from buoys and soundings reveal the discrepancies in air–sea interface and in vertical structures between spring (April to May) and summer (July) fogs in the Yellow Sea. Spring fogs are shallow with a robust temperature inversion, dry layer and cold phase (surface air temperature or SAT is lower than sea surface temperature or SST); summer fogs are deep with weaker stability, indistinct fog top and warm phase (SAT > SST). Along with numerical simulations, conceptual models for the mechanisms of temperature inversion are suggested. The land–sea contrast is responsible for the robust temperature inversion in spring, and the deep southerlies derived from the east Asian summer monsoon and the adiabatic sinking from the western Pacific subtropical high contributes to the weaker inversion in summer. The dry layer above the sea fog top intensifies the long-wave radiative cooling effect to lead to the cold phase in spring fogs. The radiative cooling is weaker in summer fogs resulting in SAT > SST.

Key words: Yellow Sea fog, mechanism, atmospheric boundary layer, longwave radiation.

1. Introduction

Sea fogs often cause losses of shipping communities and disrupt aviation and other socioeconomic activities over the ocean and in coastal regions due to low visibility.[1] The Yellow Sea, which is surrounded by mainland China to its west and by the Korean Peninsula to the east, is a heavy fog region especially from April to July (Zhang *et al.*, 2009). Figure 1 shows the number of fog days as a function of calendar month based on 30 years of station observations. Stations on the northwest (NW) Yellow Sea coast typically record

---

[1] Fog is reported when visibility <1,000 m.

---

[1] Physical Oceanography Laboratory, and Ocean–Atmosphere Interaction and Climate Laboratory, Ocean University of China, Qingdao, China. E-mail: zsping@ouc.edu.cn
[2] Pingdu Meteorological Bureau, Qingdao, China.

more than 50 foggy days a year near Qingdao (QD) while a maximum of over 80 days is found at Chengshantou (CST) station in the northern Yellow Sea. During the fog season, 65–87% of the fog observations report visibility of less than 200 m at Qianliyan (QL) and Xiaomaidao (XM), with the average duration of sea fog events lasting about 2 days (Diao, 1992).

Most of the sea fogs in the Yellow Sea are known as advection cooling fogs (Wang, 1985), which form as the air that has been lying over a warm water surface is transported over a colder water surface, resulting in cooling of the lower layer of air below its dew point (Glossary of Meteorology, American Meteorological Society, http://amsglossary.allenpress. com/glossary/). Though being the advection cooling type, the fogs in summer (July) and in spring (April–May) are different in circulation and in vertical structures. As shown in Fig. 2, fogs are shallow in spring and deep in summer. People in QD can often find that fogs are so shallow that the tops of high buildings near shore are visible while the streets below are in such a dense fog that one cannot see more than 200 m on foggy spring days.

Previous literature has proposed various mechanisms involved in the formation of sea fogs. For instance, the turbulent transfer of heat and moisture by wind shear was found for the development of sea fogs near Newfoundland (Taylor, 1915, 1917); the unusually cold temperature of the foggy air was the result of radiative cooling at the top of the fog layer on the Scottish east coast (Douglas, 1930) and the cooling effect could lead to the surface air temperature (SAT) lower than the sea surface temperature (SST) in the haar over the North Sea (Lamb, 1943). The radiative cooling at the top of the fog or stratus layer was the main mechanism for the development

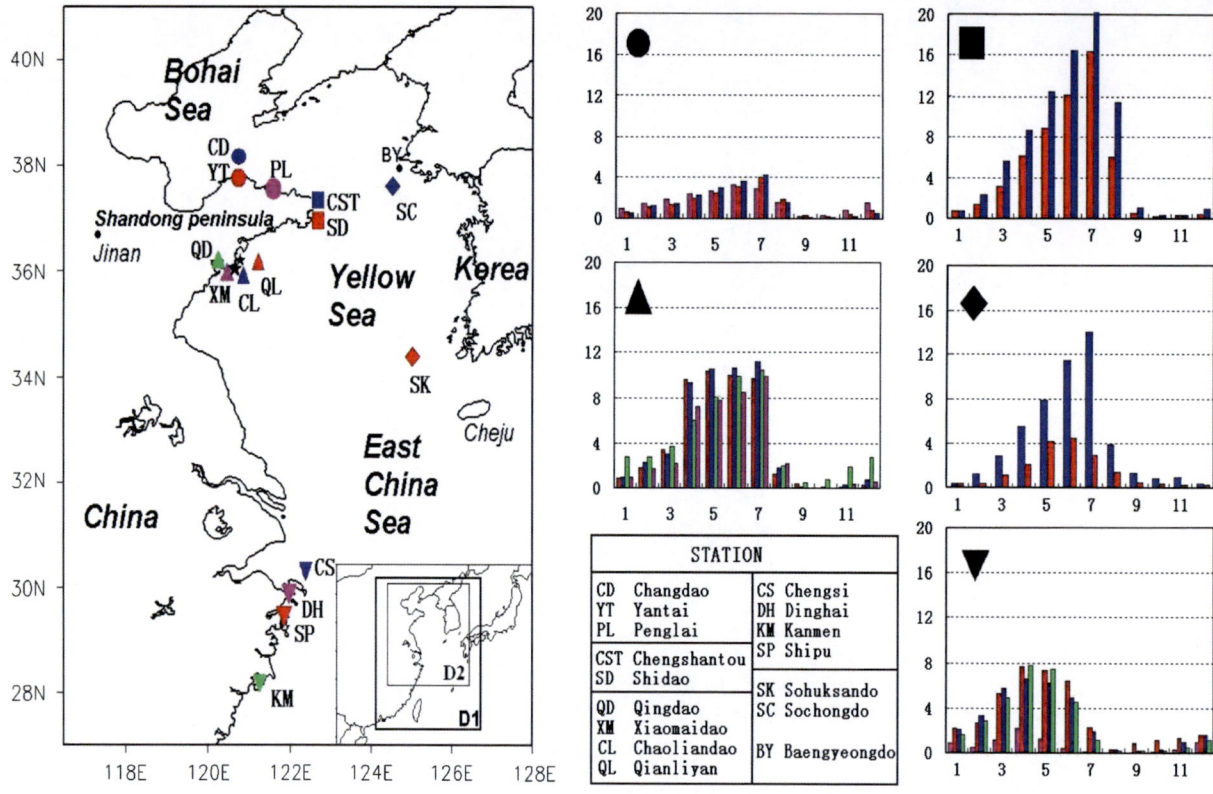

Figure 1

Climatological seasonal cycle in the frequency (day) of fog occurrence at stations adjacent to Chinese seas. Different *symbols* stand for different regions: *closed circles* Bohai Sea, *closed squares* northern Yellow Sea, *closed up triangles* NW Yellow Sea, *closed down triangles* East China Sea, *closed diamond* eastern Yellow Sea and *closed star* buoy station. *Colors* distinguish stations in each group. D1 and D2 are model outer and inner domains, respectively. (Adapted from Zhang *et al.*, 2009)

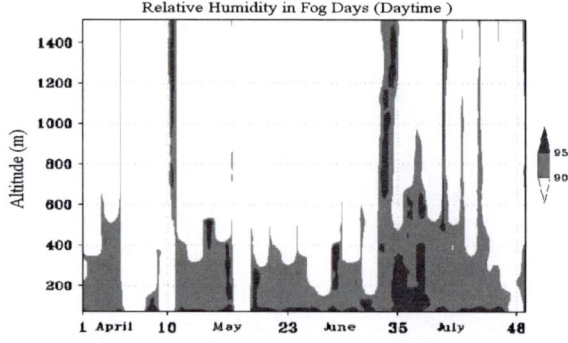

Figure 2

Seasonal variations in fog thickness (m). Data from soundings in QD including 49 fog days from 2006 to 2007

of fog-stratus systems along the coastal California (PILIÉ *et al.*, 1979; KORAČIN *et al.*, 2001; LEWIS *et al.*, 2003), and the formation of the temperature inversion was associated with sinking from the Pacific high

(LEIPPER, 1948). LEIPPER (1994), LEWIS *et al.* (2004), and GULTEPE *et al.* (2007) discussed these mechanisms in their review papers. In addition, sea fog simulations were performed to investigate the forecasting capabilities of these models. KONG (2002) suggested that the simulations were sensitive to model vertical resolutions, and a general underestimation of the cloud water content at lower levels were found in models used by YANG *et al.* (2010). Van Der Velde *et al.* (2010) showed that using usual parameterizations and configurations, current WRF-ARW might not be able to correctly predict radiation fog, yet FU *et al.* (2010) successfully simulated a sea fog event in the Yellow Sea with WRF.

It is believed that the SAT is warmer than the SST in advection cooling fogs generally (WANG, 1985). Indeed, the conditions of SAT > SST were found in many previous case studies (e.g. GAO *et al.*, 2007;

Fu *et al.*, 2010; HUANG *et al.*, 2010). The resolutions of the reanalyzed data or the SSTs retrieved from satellites used in these early studies are quite coarse compared with the spatial and temporal scales of fog patches. The situations of SAT < SST are observed more often in fog patches in spring by buoy observations recently (ZHANG and REN, 2010; GUO, 2009; QI, 2010). The buoys, available after 2008 over the Yellow Sea near Qingdao, can reveal the facts that we did not find before. The problem is: What makes the SAT > SST or SAT < SST in the Yellow Sea fogs? Why is the cold phase found usually in spring fog patches?

In the present study, we will show that the cold phase trend (SAT < SST) in spring fogs and warm phase trend (SAT > SST) in summer fogs are related to the vertical structures of the atmospheric boundary layer (ABL). In addition, the mechanisms in the formation of the temperature inversion in spring and summer fogs will be investigated. Data and the model used in the present paper are described in Sect. 2. In Sect. 3 we discuss briefly the climatology of the vertical structures in the ABL near QD and the circulations in April and July to provide readers with background information that is helpful to understand what is obtained by case studies in Sects. 4 and 5. Section 6 is the conclusions and discussions, respectively.

2. Data and Model

Four types of data are used in this study. (a) Soundings at Qingdao observatory, a station about 400 m inland from the coast, 75 m above the sea level and representative of atmospheric conditions over the northwestern (NW) Yellow Sea (ZHANG *et al.*, 2008a, b). The atmospheric pressure, temperature, humidity and wind vectors are reported at vertical resolution of 30 m from the soundings. (b) Observations from weather stations of China Meteorological Administration (CMA) from 1971 to 2000, including four island stations (CD, CL, QL, and CS) and nine coastal stations. Data for two Korean weather stations, Sochongdo (SC) and Sohuksando (SK), are from 1983 to 2002 (GAO *et al.*, 2007) (Fig. 1). (c) Buoy data including sea surface temperature (SST) at 1 m below the sea surface, surface air temperature (SAT) at 2 m above sea surface, visibility and winds. (d) Reanalyzed

products of JRA25 (Japanese 25 year Reanalysis, SAKAMOTO *et al.*, 2007) on a 1.25 grid.

We use the Weather Research and Forecasting (WRF-ARW) model version 3.1 (WRF v3.1) to support and validate the results from the observations. Two-way nested domains are used, with the outer and inner domains corresponding roughly to D1 and D2 in Fig. 1. The horizontal grid-sizes are 30 and 10 km for the outer and inner domains, respectively. The domain has 41 unevenly spaced full-sigma levels in the vertical with a high resolution in the boundary layer (21 levels below 0.8 sigma level or about 2 km). The National Centers for Environmental Prediction Final (NCEP ds083.2 FNL) Global Tropospheric Analyses on one grid every 6 h is used as the initial and lateral boundary conditions. Two control runs are operated to simulate the spring fog case from May 2 to 3 and the summer fog case from July 7 to 12, 2008, respectively. Furthermore, we perform two sensitivity experiments by switching off the shortwave and longwave radiation in WRF v3.1. The main physics options of the model are as follows:

Microphysics option:	WSM 5-class scheme (HONG *et al.*, 2004)
Longwave radiation option:	RRTM scheme (MLAWER *et al.*, 1997)
Shortwave radiation option:	Dudhia scheme (DUDHIA, 1989)
Surface-layer option:	Monin–Obukhov scheme (MONIN and OBUKHOV, 1954)
Land-surface option:	Thermal diffusion scheme
Boundary-layer option:	Yonsei University (YSU) scheme (HONG *et al.*, 2006)
Cumulus option:	Kain-Fritsch scheme (KAIN and FRITSCH, 1990, 1993)

3. The Climatology of the Circulations and the Vertical Structures in the ABL

From April to May, there exists a high pressure region over the Yellow and East China Sea surfaces, which is derived from the land-sea contrast (ZHANG et al., 2011). Moisture is advected to the NW Yellow Sea from the southern Yellow Sea and the northern East China Sea by the local southeasterlies on the west flank of the high (Fig. 3a). The high is a shallow one and disappears at about 925 hPa where westerly winds blow from warm continent to cold sea (Fig. 3b). The westerly flows, both warm and dry, overlay the cold/wet sea surface to form a robust temperature inversion and dry layer. In July, the southerly winds from the summer monsoon prevail in east Asia (Fig. 3b–d), which brings a huge amount of moisture to the Yellow Sea to produce a deep wet layer and contributes to the weaker stability in the ABL since the air masses in the lower level are from the southern ocean all the way

possessing similar thermal characteristics (ZHANG et al., 2009).

The percentage of the temperature inversion days is in excess of 50% and the boundary layer stratification is extremely stable with $\frac{\partial \theta}{\partial z} > 10\,\mathrm{K/km}$ from 100 to 400 m in April (Fig. 4a, b). Capped by a strong stable layer, fog, once formed, can hardly develop to higher levels. Whereas the stability becomes weaker in July, which is favorable for moisture moving upward, and deep fog may form with a plenty supply of moisture. Indeed, the wet layer is much deeper in July with specific humidity >12 g/kg below 600 m (Fig. 4b). With the differences in the stability and in the thickness of the wet layer, the summer fogs are generally deeper in height than the spring fogs.

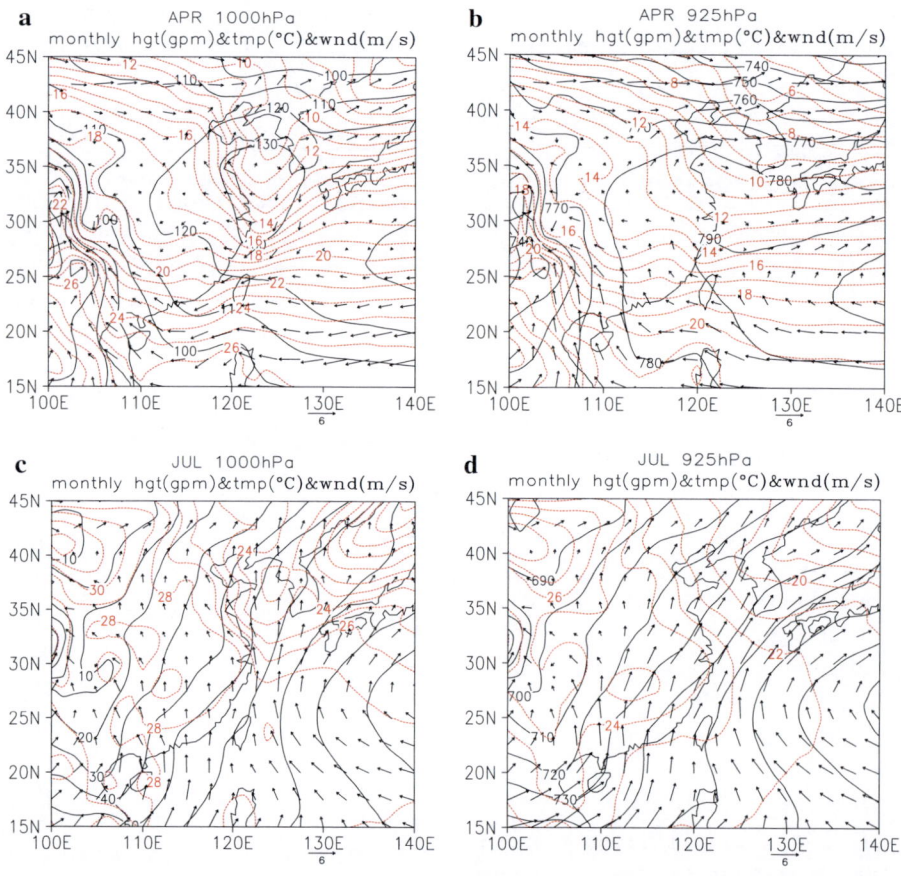

Figure 3
Climatological monthly mean winds (vectors in m/s), temperature (*red dotted contours* in °C) and geopotential height (*solid contours* in gpm). **a** 1,000 hPa in April; **b** 925 hPa in April; **c** 1,000 hPa in July and **d** 925 hPa in July. All based on the JRA-25 climatology

4. Case Studies: Observations

We investigate two fog cases occurred on 2–3 May 2008 (spring fog case hereafter) and on 7–12 July 2008 (summer fog case hereafter), which were representative of typical spring and summer advection cooling fogs, respectively. On 2 May, the southeasterlies on the west flank of a surface (1,000 hPa) anticyclone in the Yellow Sea transported moisture to the NW Yellow Sea from the southern Yellow Sea and the northern East China Sea (Fig. 5a). Pronounced horizontal temperature gradients existed along the west coast of the Yellow Sea indicating the land-sea temperature contrast. The winds at 925 hPa blew from warm continent to the cold sea (Figure not shown). On 7 July, the Yellow Sea was under the control of a high ridge stretching from the western Pacific subtropical high. The southerlies advected moist air from the southern oceans all the way to the Yellow Sea (Fig. 5b). Being associated with the east Asian summer monsoon, the southerlies produced deep wet layer in the ABL. These synoptic situations were compatible with the climatological patterns in April and in July as shown in Fig. 3a–d, favorable for the formation of sea fogs. The foggy patches detected from satellites are shown later in the left panels of Fig. 6.

4.1. Buoy Observations

At 00 UTC 2 May, the fog had not been observed yet near QD (visibility >3,000 m), the SAT was higher than the SST and the surface winds were southeasterly (upper panel in Fig. 7). The SAT fell rapidly while the visibility dropped sharply from 06 to 09 UTC. Note that the differences between the

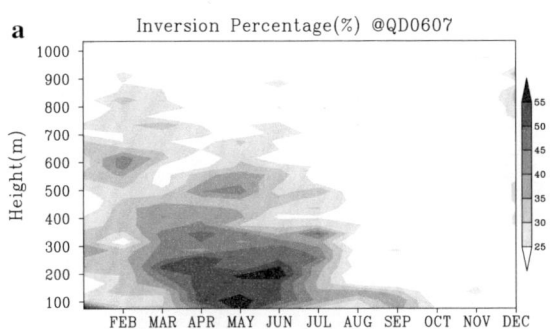

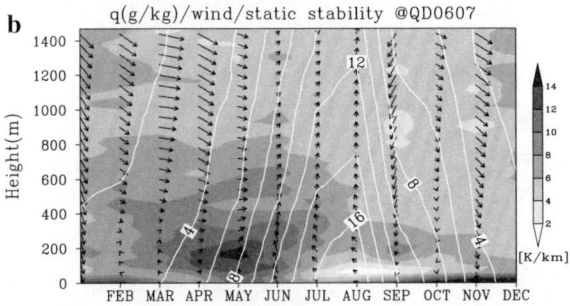

Figure 4

a Days with inversion occurrence (%). **b** Monthly-mean specific humidity (*white contours* in g/kg), winds and static stability ($\frac{\partial \theta}{\partial z}$ in shadings). All data are based on daily soundings of L-band radar at QD from 2006 to 2007

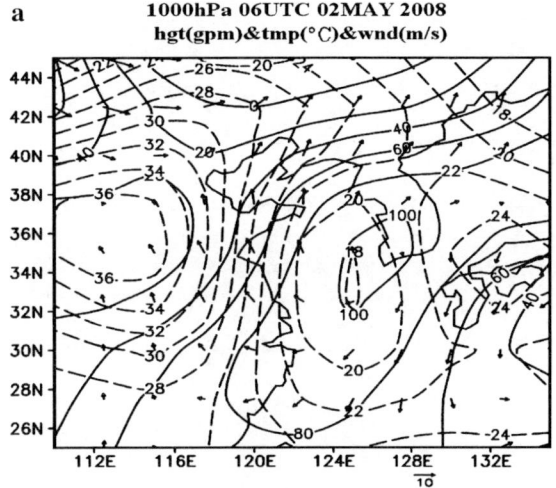

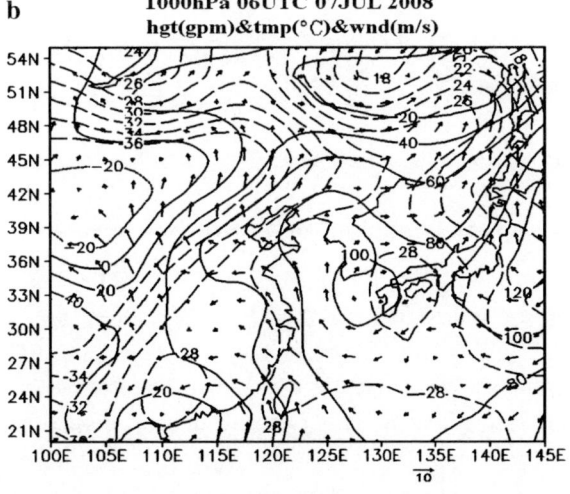

Figure 5

1,000 hPa Geopotential height (*solid line*), air temperature (*dashed line*) and winds (vector), **a** 06 UTC, 2 May 2008; **b** 06 UTC, 7 July 2008 (Data from JRA-25)

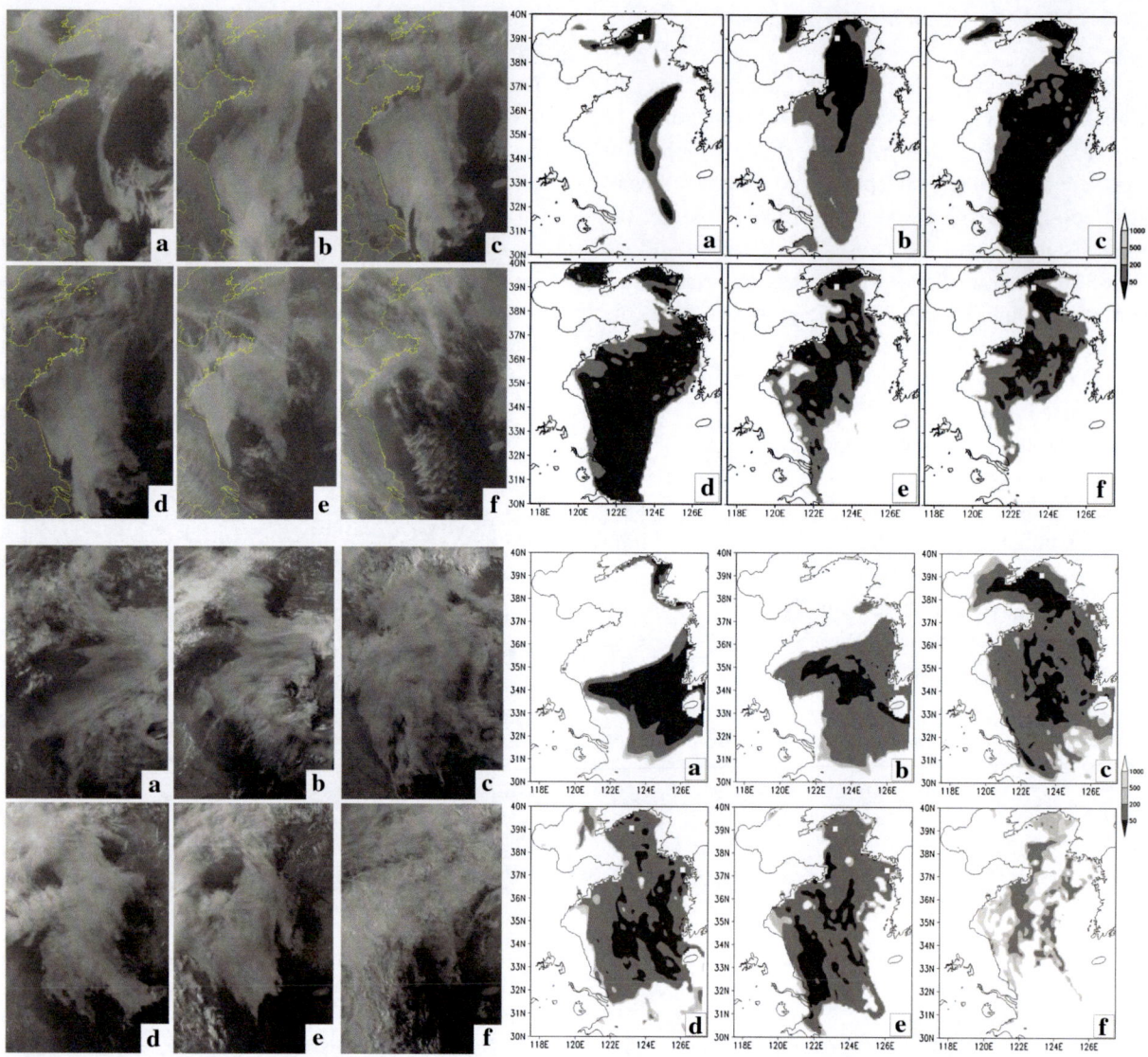

Figure 6

The *left upper panel*: FY-2D visible cloud images for the spring fog case from 1 to 3 May, 2008; the *right upper panel*: simulated visibility in fog patches (*shaded*/m). **a** 05/01 06 UTC; **b** 05/02 00 UTC; **c** 05/02 06 UTC; **d** 05/02 09 UTC; **e** 05/03 03 UTC; **f** 05/03 06 UTC. The *left lower panel*: GOES-9 visible cloud images for the summer fog case from 7 to 10 July, 2008; the *right lower panel:* simulated visibility in fog patches (*shaded*/m). **a** 07/07 00 UTC; **b** 07/07 05 UTC; **c** 07/08 00 UTC; **d** 07/08 09 UTC; **e** 07/09 09 UTC; **f** 07/10 00 UTC

SAT and the SST reduced since the SST did not change much, and the SAT became lower than the SST at 17–18 UTC when the visibility was minimum. The air temperature rose and the difference of SAT–SST became larger during the fog dissipation stage.

In the summer fog case, the SAT kept about 1.5°C warmer than the SST throughout the fog duration, though the SAT fell too at the beginning of the fog (lower panel in Fig. 7). To confirm these results, we

investigated more fog cases using buoy data, and found that the SATs were lower and close to or even less than the SSTs in spring fogs and positive differences of SAT–SST were common in summer fogs (Fig. 8).

4.2. Soundings

In the spring fog case, the temperature inversion was robust from 200 to 500 m in height and the fog (relative humidity RH >90%) was trapped under the temperature

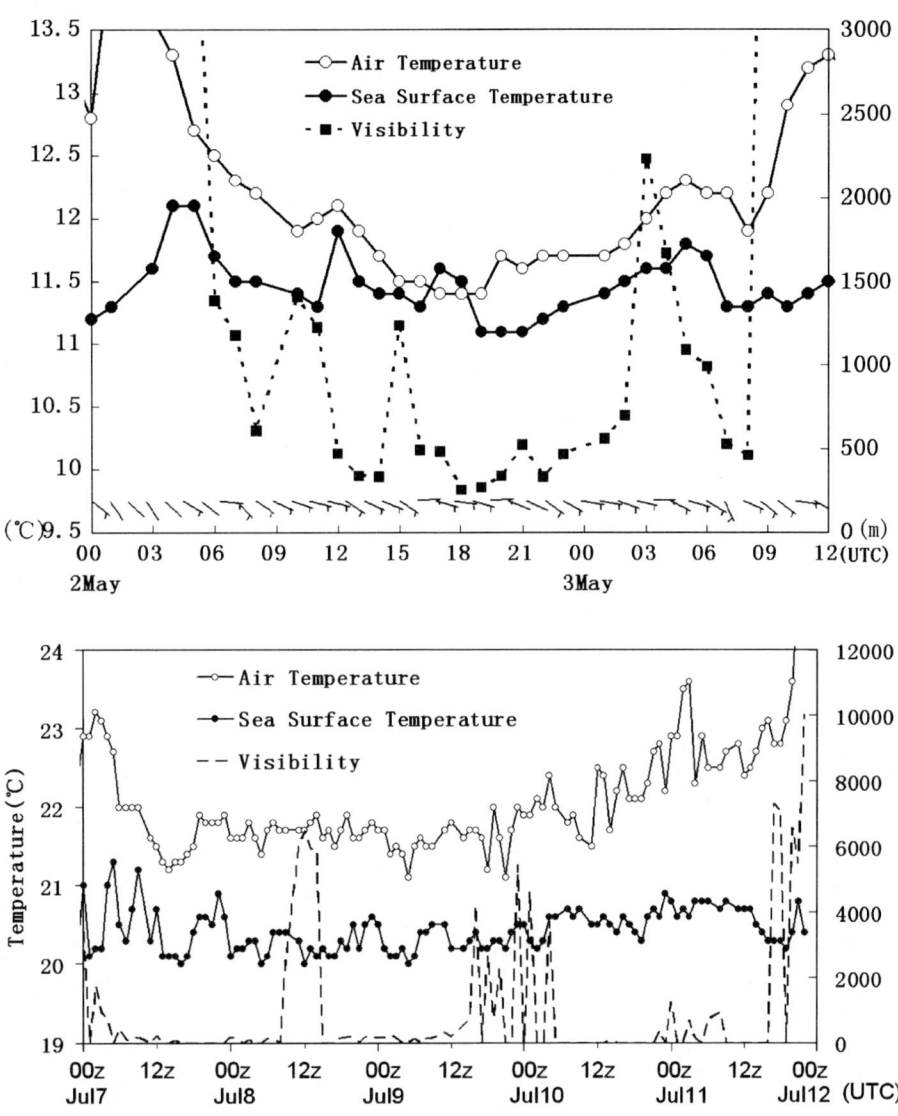

Figure 7

The observations from buoys near QD. SAT (*line with space circle*), SST (*line with solid circle*), and horizontal visibility (*dashed line*). The upper panel is for spring fog case and the *lower panel* is for summer fog case. The wind vectors in the *upper panel* are observed on 20 m observation tower at Chidao Island near QD

inversion (Fig. 9a). The RH decreased sharply upward, and was less than 50% at 500 m and less than 20% at 900 m, indicating a marked dry layer over the fog. The existence of the dry layer can intensify the outgoing longwave radiation from the fog top, possibly leading to SAT < SST, and this mechanism was largely responsible for the development of fog in the North Sea (LAMB, 1943; FINDLATER *et al.*, 1989).

On 7 July 2008, the temperature inversion was much weaker compared with its counterpart in the

spring fog. The RH decreased upward so slowly that it remained greater than 60% above 1,200 m, depicting the absence of dry layer or the indistinctive fog top (Fig. 9b). The radiative cooling at the upper level of the fog was weakened due to the plenty of moisture above the temperature inversion, and the SAT did not lower as much as in spring fog. We will discuss these respects by means of numerical simulation and experiment in the following section.

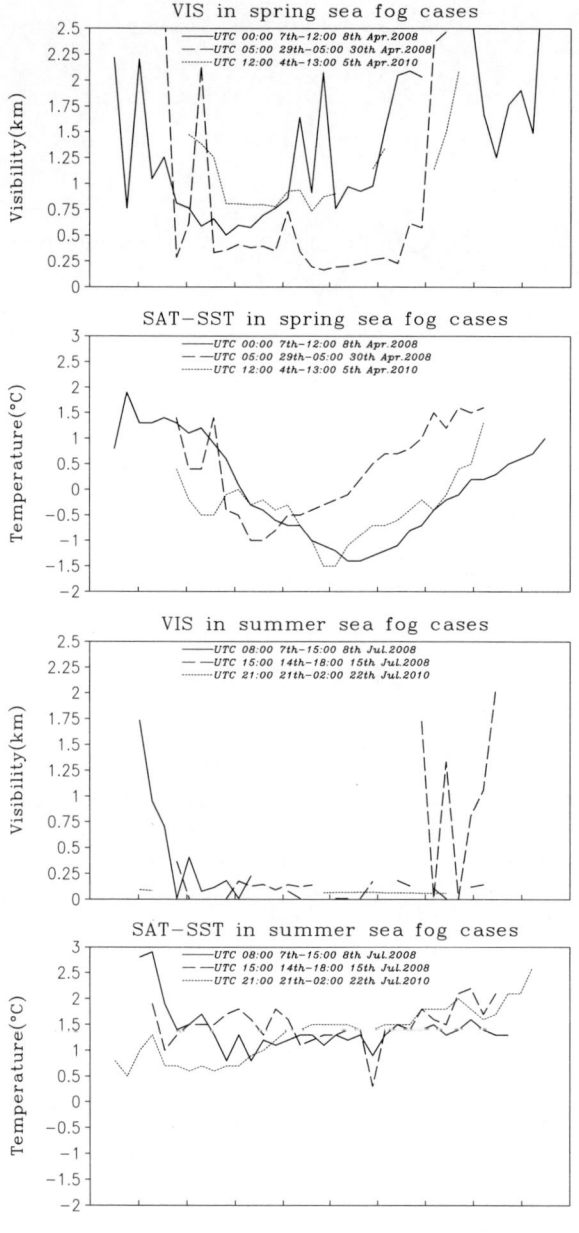

Figure 8
The visibility (km) and SAT–SST (°C) in spring sea fog cases and in summer sea fog cases (some data are default)

5. Case Study: Numerical Simulations and Experiments

5.1. Control Runs

The simulated fog patches are identified as the cloud water mixing ratio (q), and at the model's lowest level is greater than 0.01 g/kg^2. The horizontal visibility in fogs can be calculated from STOELINGA and WARNER (1999) based on Koschmieder formulation (KOSCHMIEDER, 1924):

$$X_{VIS} = -\frac{\ln(0.02)}{\beta} \qquad (1)$$

in which, X_{VIS} is the visibility (km) and β the extinction coefficient (km^{-1}) affected by cloud water (βcw), cloud ice (βci), snow (βsnow) and rain water (βrain). In spring and summer fogs, the algorithm only refers to cloud liquid water:

$$X_{VIS} = -\frac{\ln 0.02}{144.7(\rho \cdot q)^{0.88}} \qquad (2)$$

where $\rho \cdot q = $ LWC (g/m^3), q is equivalent to the cloud water mixing ratio (g/kg), ρ the density of the air (g/m^3), and LWC the liquid water content. The simulated fog patches and the corresponding visibilities of the two cases are shown in Fig. 6 (right panels). By comparing with the satellites (left panels in Fig. 6), we find that the modeling results can reflect the beginning, development and decay phases of these fogs basically, though there are some discrepancies in the distributions of fog patches. The synoptic-scale circulations from the models are also in agreement with NCEP reanalysis primarily (Figures not shown). In this present study we focus on the vertical structures in the ABL and the effect of radiative cooling based on the model output.

In the spring case, the fog is limited to below 250 m with a dense vertical temperature gradient above reflecting the strong inversion (Fig. 10a). The cold phase (SAT–SST <0) appears during the development of fog, though SAT–SST >0 before the fog formation. Compared with the spring case, the fog in the summer case is deeper at about 300–400 m with a weaker vertical temperature gradient and positive differences of SAT–SST (Fig. 10c). These simulated results are basically consistent with the soundings and buoy observations. The simulated cloud water mixing ratios are compatible with previous studies by models (TOKINAGA and XIE, 2009; KORAČIN et al., 2005; GAO et al., 2007; FU et al., 2010), but much larger than

---

[2] The threshold $q = 0.01$ k/kg corresponds to the visibility of 1,000 m according to Eq. (2).

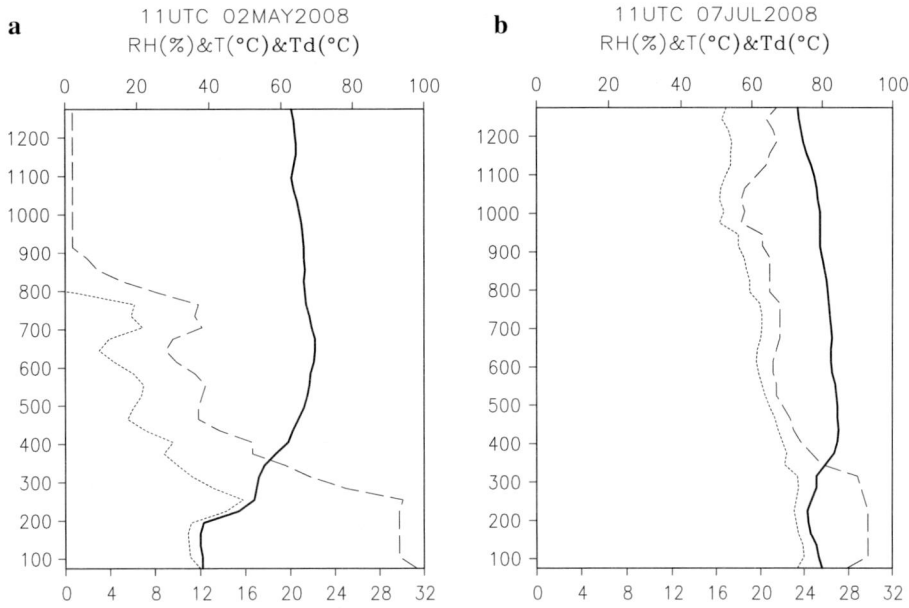

Figure 9

Soundings at QD observatory. Temperature (*solid*), dew point temperature (*dotted*), and relative humidity (*dashed*). **a** 11 UTC 2 May 2008; **b** 11 UTC 7 July 2008

those retrieved from fog measuring instrument (0.01–0.3 g/kg, GULTEPE *et al.*, 2009). We will discuss this matter later in the last section.

The backward trajectories of air masses from the modes indicate the processes of formation of temperature inversion. In the spring fog case, the air masses at 10 m moving from the northern East China Sea to the NW Yellow Sea cool and moisten gradually by the modification of the cold sea surface, while the air parcels at 500 m are dry (RH <60%) and are heated gradually by the warm continent due to taking an inland route (Fig. 11a). The warm/dry land-based air masses overlay the cold/wet sea-based air masses near the coast of QD, making pronounced temperature inversion of about 8°C/500 m at 21 UTC 2 May. The relative humidity is only about 25% at 1,000 m marking the existence of the dry layer.

In the summer fog case, the air masses below 1,000 m have a long sea-based history, and the temperature and humidity of air masses at different altitudes do not differ from each other as obviously as in spring fog (Fig. 11b). Note that the air masses at 1,000 m warm by adiabatic sinking from above, probably associated with the subtropical high ridge, the mechanism that contributes to the temperature

inversion near the California coast too (LEIPPER, 1994). The intensity of the temperature inversion is about 3°C from surface to 500 m, much weaker than that in spring fog. These simulated results are compatible with the synoptic and climatological analysis and help us understand the observational results.

5.2. The Numerical Experiments

As mentioned above, the SAT may decrease by the intensified radiative cooling in addition to the modifications by the cold sea surface in spring fogs, whereas the radiative cooling is not so strong in summer fogs that the SAT does not change much. To investigate the radiative cooling effect, we perform two numerical experiments by switching off radiation in the WRF (Spring_EXP and Summer_EXP). Without the effect of radiation, the SAT keeps about 2°C warmer than the SST in the Spring_EXP (Fig. 10b). Note that fog patches shrink heavily in the Spring_-EXP, depicting the vital importance of the radiation cooling for spring fogs (Fig. 12a). The SAT also rises in Summer_EXP but not as obviously as in Spring_-EXP, and the summer fog patches change slightly (Fig. 12b).

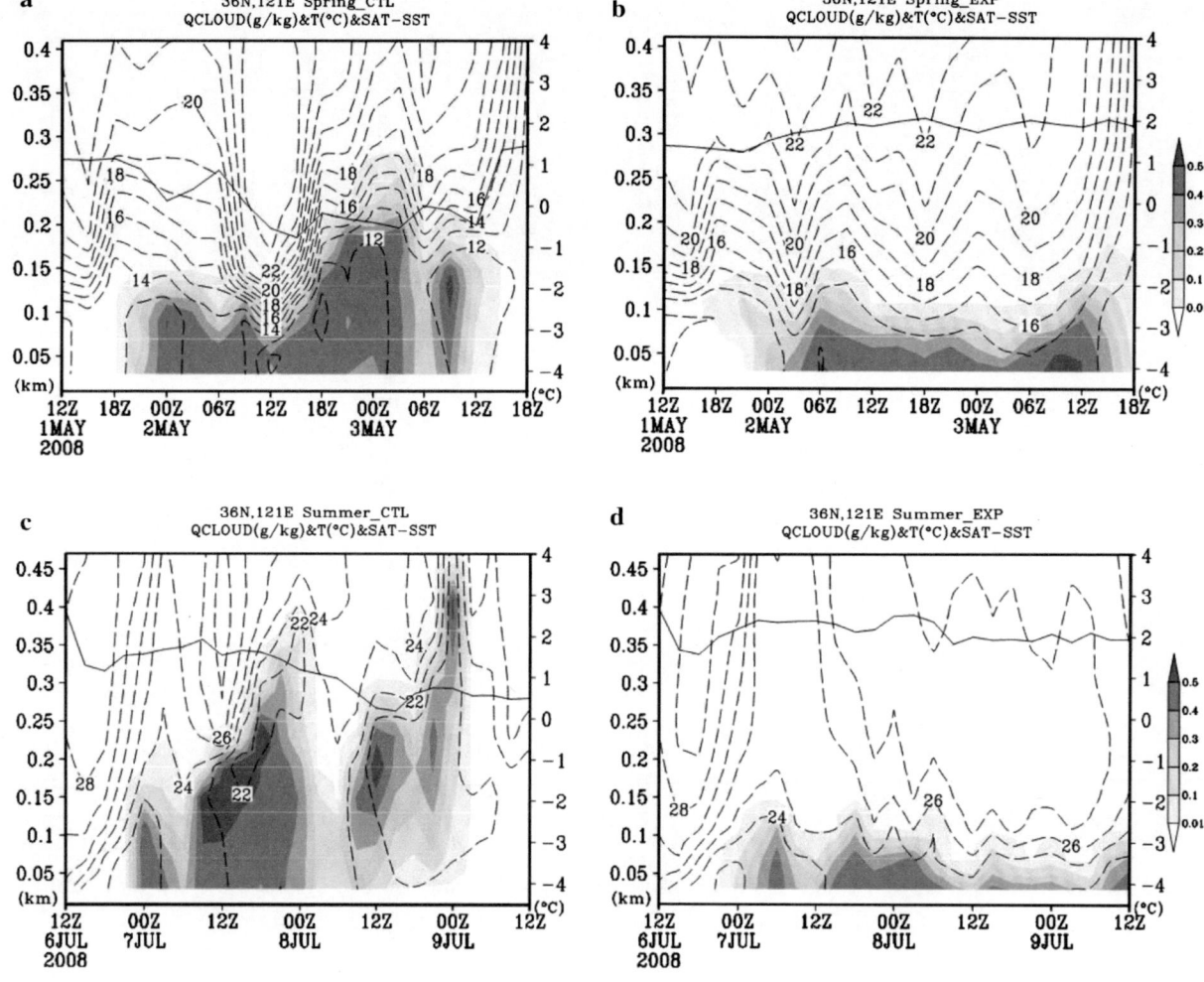

Figure 10
The time-vertical section of the cloud water mixing ration (*shaded* when >0.01 g/kg), temperature (*contours* in °C) and SAT–SST (*solid line* in °C) at 36°N, 121°E for spring fog case CTL run (**a**), EXP run (**b**); for summer fog case CTL run (**c**), EXP run (**d**)

The differences in temperature between the EXP and the CTL runs (T_EXP-T_CTL) are in Figs. 13a, c at local nighttime in consideration of longwave radiation (local time: UTC + 8 h). Generally, the temperature decreases both in spring and in summer fogs influenced by the radiative cooling effect, but the air cools more (up to 9°C) in spring fog than in summer fog (up to 3°C), reflecting the stronger net radiation emitting out of the former. In addition, the cooling effect gets weaker downward and the temperature at the lower level may keep unchanged or changed slightly by radiation if the fog is deep enough, resulting in SAT > SST. For example, the temperature decreases less than 1°C near the bottom

and about 3°C at the upper level of summer fog (Fig. 13c). The cooling effect at the fog top can influence the surface air temperature with turbulence if the fog is shallow enough. In the lower level of the spring fog, the temperature decrease about 4°C (Fig. 13a), and this value is large enough leading to SAT < SST because the difference between the two is about 2°C outside the fog. Note that the SAT > SST and SAT < SST can appear at the same time but in different fog patches due to the variations in their heights. As simulated by GUO (2009) and QI (2010) for four different spring fog cases, the SATs are lower (higher) than the SSTs in fog patches of 50–200 m in height (in fog patches

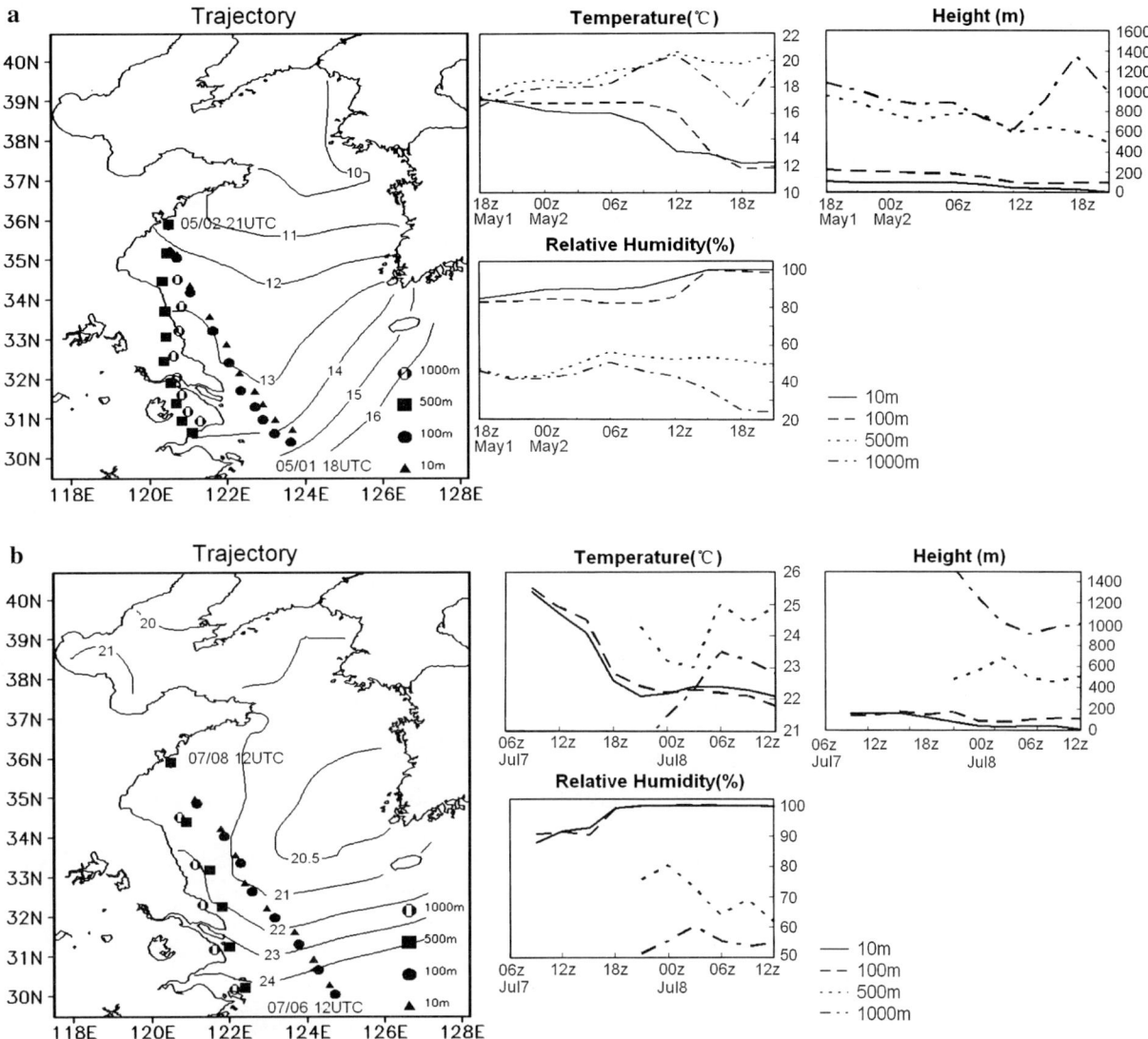

Figure 11

Trajectories of air masses (*different symbols* stand for different heights) back-tracked from QD and SST (°C) at 18 UTC 1 May 2008 (**a**) and at 12 UTC July 6 2008 (**b**), and the displacements are at 3 h intervals (*left panels*). The variations of temperature, relative humidity and the elevation along the trajectories (*right panels*). All based on control runs

great than 200 m in height) in each individual fog case.

Radiative cooling is favorable for the occurrence of condensation, thus reducing the water vapor masses and the dew-point temperature (Td). The differences in Td between the EXP and the CTL (Td_EXP-Td_CTL) are much pronounced in spring fog. The Td increases about 4–5°C at no-fog spots in the EXP run (Fig. 13b), which is associated with the warming (Fig. 13a). The Td rises only about 1–2°C in

Summer_EXP (Fig. 13d), consistent with the less obvious warming and the slightly changed fog patches. The turbulence (the Richardson Number Ri <0.25, STULL, 1988) is depressed in Spring_EXP and in Summer_EXP and consistent with the shallowed fog layer (Fig. 14a–d). The weakened turbulence is unfavorable for the development of fog, either.

HUANG *et al.* (2010) suggested that the warm temperature advection plays a part for keeping SAT > SST in a fog event near the Chinese coast

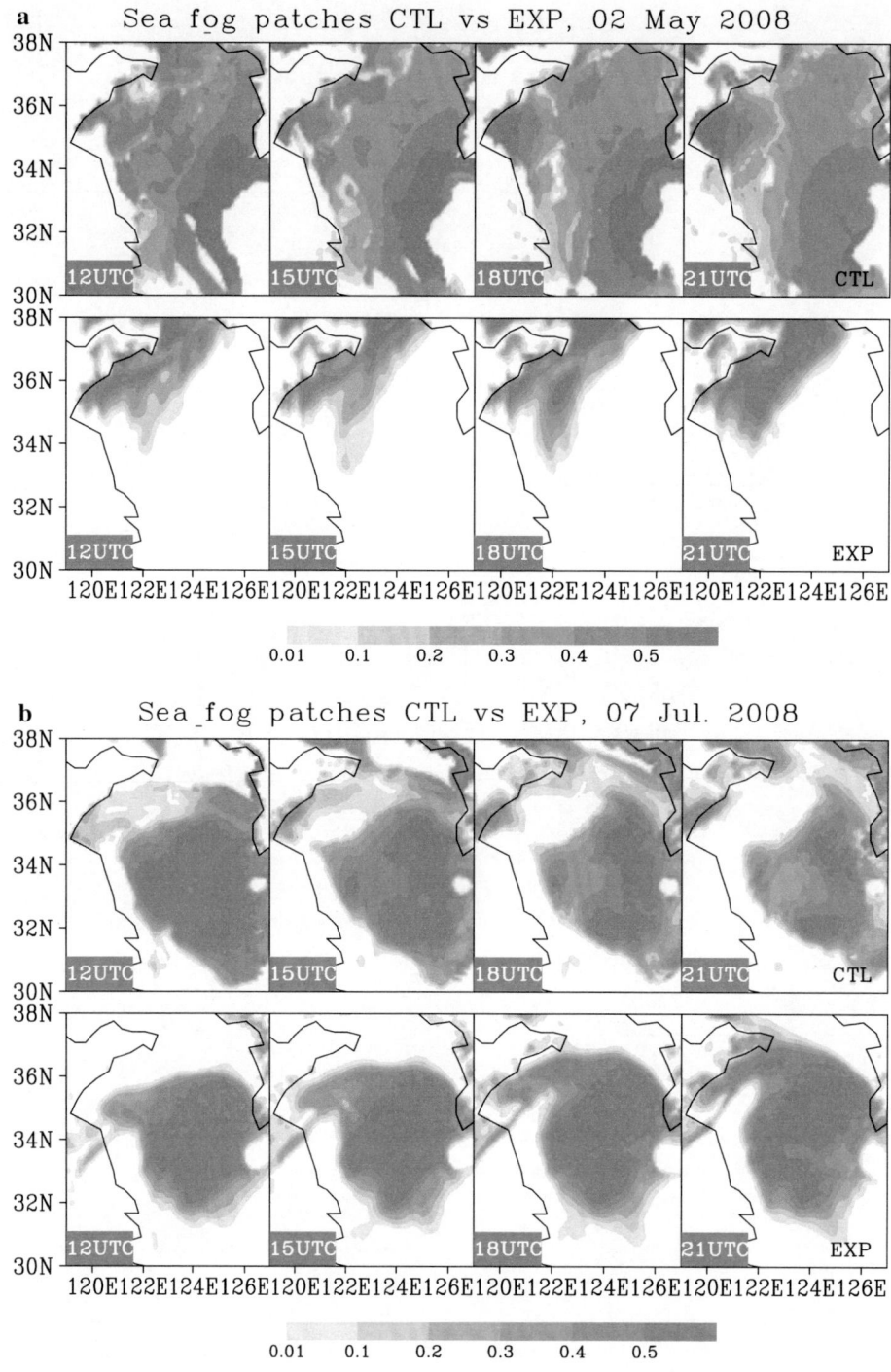

Figure 12
The fog patches for spring (*upper panel*) and summer (*lower panel*) fog cases. The first line is CTL run and the second line is EXP run in each panel. The *shaded* is cloud water mixing ratio >0.01 (g/kg)

of the South China Sea where fogs occur sometimes in winter and spring. For these two cases (the spring and summer fog cases in CTL runs), the SATs

continue to fall in fogs though they are accompanied by positive temperature advection (Fig. 15a, b). Furthermore, the warm temperature advection seems

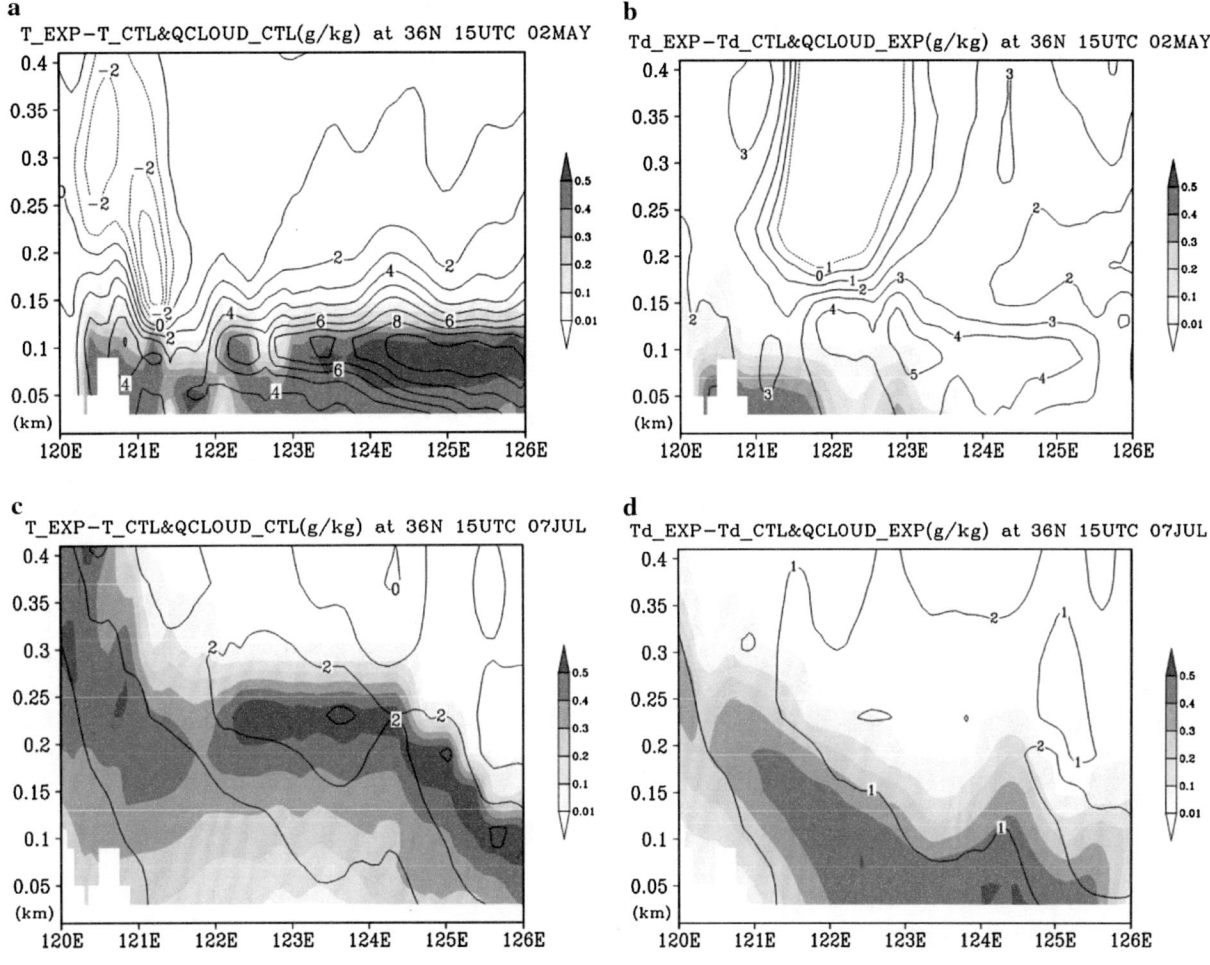

Figure 13

The difference in temperature (**a**, **c**) and in dew-point temperature (**b**, **d**) between the EXPs and CTLs (*contours* at 1°C intervals); the cloud water mixing ratio in the CTLs (**a**, **c**) and in the EXPs (**b**, **d**) (*shaded* in g/kg). (All along 36°N)

to be dominant in fog patches (Fig. 15c–d). Note that the temperature variability is larger in spring fog than in summer fog, which may be associated with the different physical processes, such as the variations produced by the differences in the intensity of radiative cooling at fog tops.

6. Conclusions and Discussions

Based on observations and model output, discrepancies are found in air–sea interface and in vertical structures between spring (April–May) and summer (July) fogs in the Yellow Sea, though they are all related to advective cooling. The major

findings from the current work can be listed as follows:

1. Climatologically, spring fogs are associated with local southeasterlies on the west flank of the high pressure over the Yellow and East China Seas and the warm offshore westerlies at about 925 hPa, while summer fogs are related to the east Asian summer monsoon that is large scale and provides deep and plenty of moisture. These discrepancies are responsible primarily for the robust temperature inversion from April to May and the weaker stability in July in the ABL, and for the formation of shallow fogs in spring and deep fogs in summer. The analyses of two fog cases on 2–3

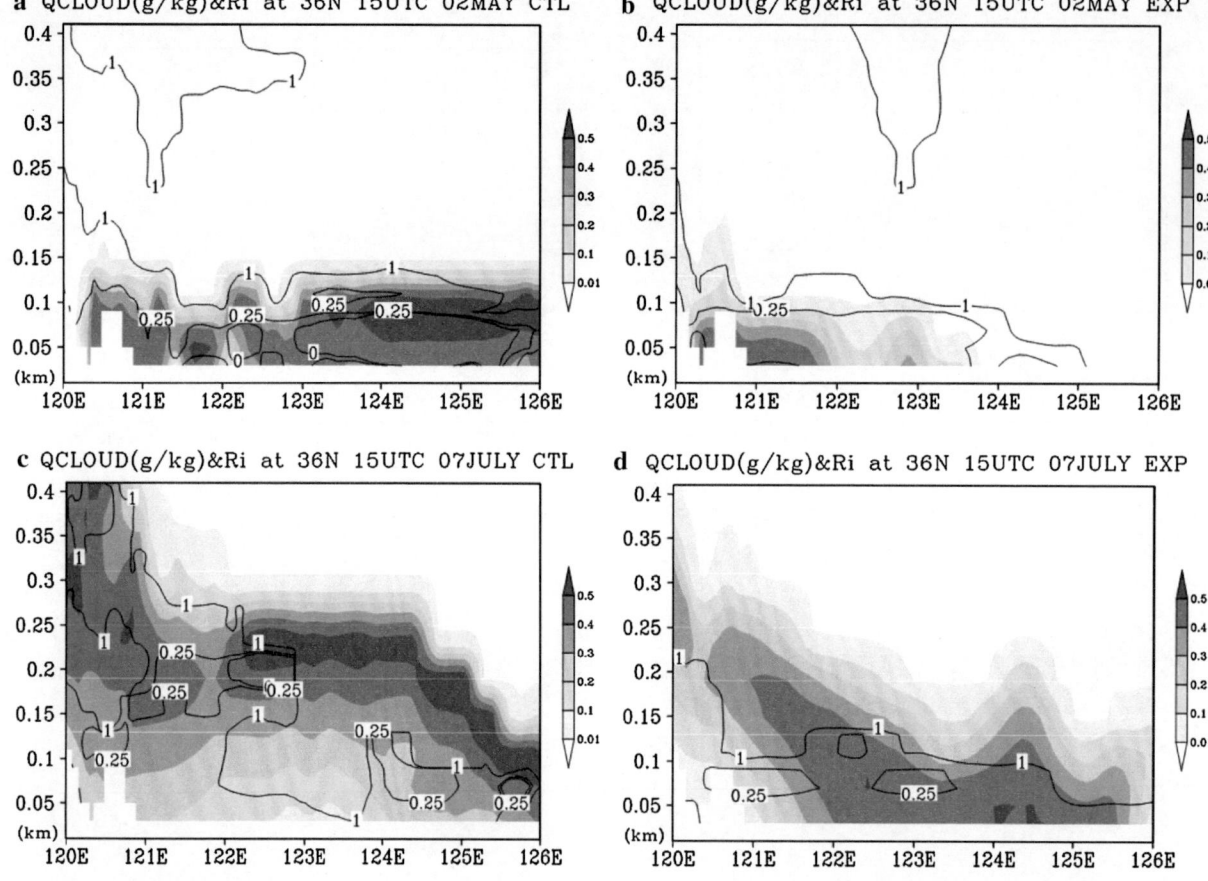

Figure 14

The Richardson number (*contours*) and the cloud water mixing ratio (*shaded* in g/kg) along 36°N in the CTLs (**a, c**) and in the EXPs (**b, d**)

May and on 7–12 July support the climatological results.

2. A conceptual model for the mechanisms of the temperature inversion are suggested. In spring, the land is warmer than the Yellow Sea. Air masses near the sea surface driven by the southerly winds cool gradually by the modification of the cold sea as they move northward, while those at about several hundred meters high get warmer since they have a long land-based history, thus forming pronounced temperature inversion and dry layer over the NW Yellow Sea. In summer, controlled by the east Asian summer monsoon, air masses in the ABL are primarily derived from the southern oceans and expose similar thermal features. Weaker temperature inversion is related to the adiabatic sinking from the subtropical high and the cooling by cold sea surface. The cold air from

northeast Asia intrudes onto the warm sea surface to form inversion sometimes (ZHANG, 2010).

3. Radiative cooling effect is robust due to the dry layer at the top of fogs in spring, which is crucial for the development of fogs. The radiative cooling at the fog top is prone to promote turbulence in the fog layer. Without the cooling, the temperature may rise and the turbulence may be reduced in fog layer (Figs. 13, 14) resulting in the heavily shrunk fog patches in the spring case, whereas the radiative cooling is weaker due to the indistinct fog top and the impact on fogs is not remarkable in summer case.

4. The cold phase (SAT < SST) often occurs in spring fog patches, while the warm phase (SAT > SST) is common in summer fog patches. The SAT in fogs is largely influenced by the radiative cooling effect and the height of fog

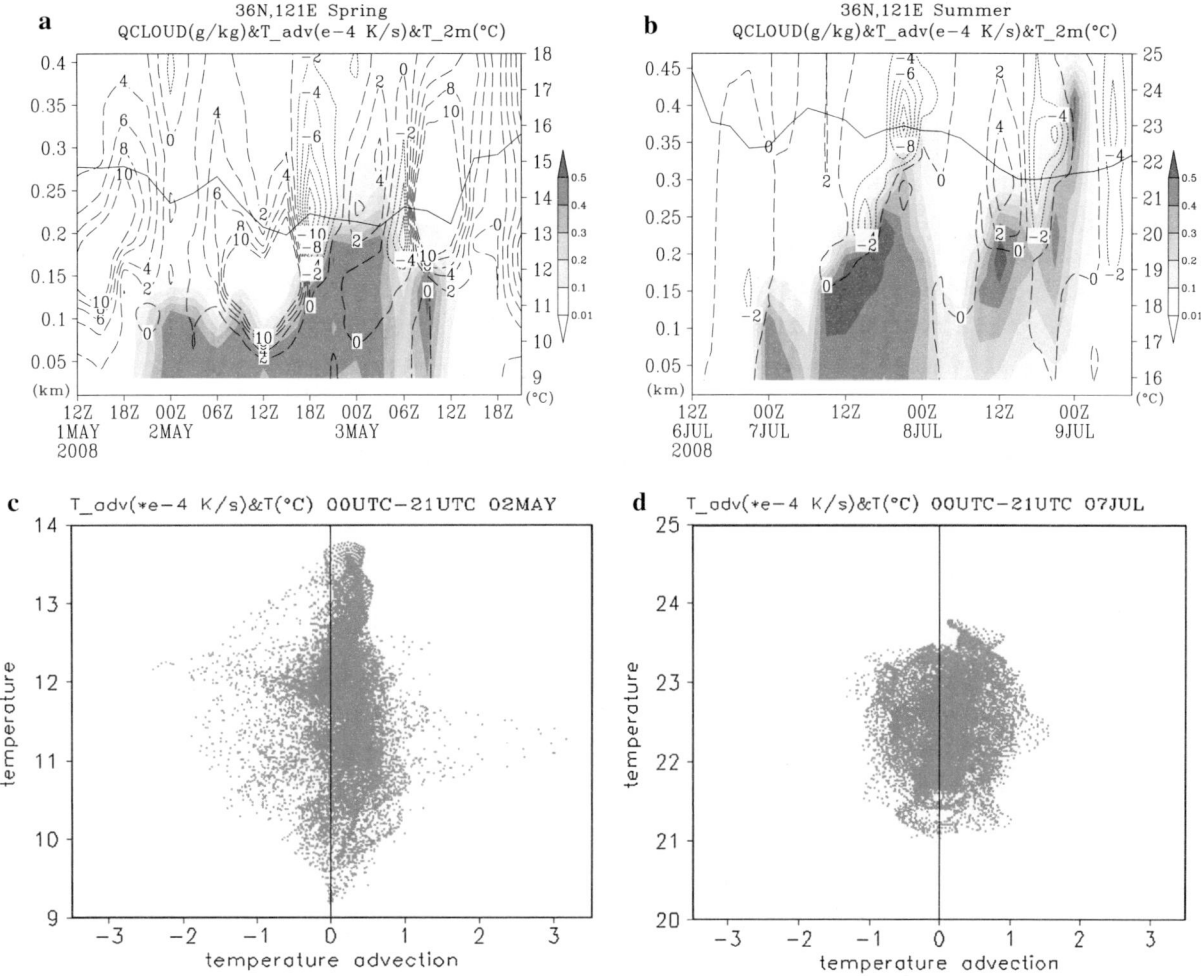

Figure 15

The temperature advection (*contours*), the air temperature at 2 m (*solid line*) and the cloud water mixing ratio (*shaded*) for the spring fog case (**a**) and the summer fog case (**b**). The scatters of temperature at 2 m with temperature advection at 10 m in fog patches for the spring fog case (**a**) and the summer fog case (**b**). All data are based on the control runs

layers. In spring, the SAT declines, due to the intensified radiative cooling, so much that it becomes very close to or less than SST in shallow fog; summer fogs are deep with weaker radiative cooling, thus keeping SAT > SST easily. Temperature advections seem to be having no evident relations to the SAT in the Yellow Sea fogs. Note that the cold phase was not found due to the coarse resolutions of the data used in the previous literatures about fogs over the Yellow Sea. The buoys can provide successive observations of SAT and SST with an interval of 10 min to trace the variations in the lowest levels of fogs. With the help of model simulations and experiments, we

investigated the physical processes that were responsible for the variations. In this sense, our study filled a gap in the lower levels of fogs over the Yellow Sea.

The Yellow Sea is a heavy fog area, but successive in situ observations are rare for a long period of time. Previous literature investigated sea fogs mainly based on observations from the coastal stations and the daily averaged SSTs from satellites. After 2008, buoys were available in the northwestern Yellow Sea. Both buoy and model data show that the cold phase (SAT < SST) can occur sometime during the development stage of fog process, though SAT > SST is dominant in the beginning and decaying stages and outside fog patches. Some authors have suggested to classify sea fogs into warm/cold types based on SAT > SST/SAT < SST (e.g. Huang *et al.*, 2010). Caution should be paid for such classification if depending on daily averaged SST retrieved from satellites or on discontinuous reports from ships.

The simulated cloud water mixing ratio values (q, up to 0.6 g/kg) were comparable with previous model studies but were much larger than the observations (Gultepe *et al.*, 2009). Figures. 13 and 14 depict that the larger values (q > 0.4 g/kg) appear mainly at the upper level of fog layers due to the radiative cooling. The ratios are less than 0.3 g/kg at the bottom of the fog layer, basically consistent with the observations. Figure 12 shows that q is larger than 0.3 g/kg in some fog patches, which is possibly related to the vertical resolution of the model because fog simulations are very sensitive to model's vertical resolution (Kong, 2002). In addition, the LWC usually peaks at the upper level of fogs and decreases downward. As suggested by Zhou and Ferrier (2008), the LWC at 32 m is greater than 0.4, about 0.2 g/kg larger than that at 2 m (2008). The lowest level of the model used in this present study is about 30 m above sea surface, which may be one of the reasons of the higher value of q in Fig. 12.

The turbulence by wind shear is also important for fog formation, especially in the initial stage of fog, but too strong or too weak turbulence are unfavorable to the sea fog (Hu *et al.*, 2006). According to the balance theory suggested by Zhou and Ferrier, (2008), if the turbulence intensity (K) inside a fog is smaller than the turbulence threshold (Kc), i.e. K < Kc, the fog persists, and otherwise the fog dissipates. Kc is a function of the depth of fog (H) and the temperature (T) in fog; small Kc corresponds to shallow fog and less cooling rate. In our Spring_EXP, the fog patches shrink heavily, which could be related to small Kc. K can exceed Kc easily leading to the shrink of fog. While in Summer_EXP, fog patches vary slightly, which could be associated with larger Kc corresponding to the deeper fog layer and less obvious cooling rate.

Gultepe *et al.* (2009) suggested a better parameterization of visibility with two variables, Vis = $f(\text{LWC}, N_d)$, in which N_d is the droplet number

concentration. In the present study, we calculate the visibility based on Eq. 2 which is a one-variable function, Vis = $f(\text{LWC})$. The parameter N_d can be obtained in the predicted data of WRF. Our next work will involve the new parameterization and compare the visibilities with the observations to improve the skill of fog forecast.

Acknowledgments

This work is supported by NSFS 40975003, MOE 2009013211008, GYHY (QX) 2007-06-31, and MOST 2006AA09Z149 Programs.

References

Diao, X. X. (1992), *Statistical analysis of sea fog near Qingdao adjacent sea*. Marine Forecasts. *9* (3), 45–55 (in Chinese).

Douglas, C. (1930), *Cold fogs over the sea*, Meteor. Mag. *65*, 133–135.

Dudhia, J. (1989), *Numerical study of convection observed during the winter monsoon experiment using a mesoscale two-dimensional mode*, J. Atmos. Sci. *46*, 3077–3107.

Findlater, J., Roach, W. T., and McHugh, B. C. (1989), *The haar of north–east Scotland,* Quart. J. Roy. Meteor. Soc. *115*, 581–608.

Fu, G., Li, P. Y., Crompton, J. G., Guo, J. T., Gao S. H., and Zhang, S. P. (2010), *An observational and modeling study of a sea fog event over the Yellow Sea on 1 August 2003*, Meteor. Atmos. Soc. *107*, 149–159.

Gao, S. H., Lin, H., Shen, B., and Fu, G. (2007), *A heavy sea fog event over the Yellow Sea in March 2005: analysis and numerical modeling*, Adv. Atmos. Sci. *24*, 65–81.

Gultepe, I., Tardif, R., Michaelides, S. C., Cermak, J., Bott, A., Bendix, J., Muller, M. D., Pagowski, M., Hansen, B., Ellrod, G., Jacobs, W., Toth, G., and Cober, S. G. (2007), *Fog research: a review of past achievements and future perspectives*, Pure Appl. Geophys. *164*, 1121–1159.

Gultepe, I., Pearson, G., Mi lbrandt, J. A., Hansen, B., Platni ck, S., Taylor, P., Gordon, M., Oakley, J. P. and Cober, S. G. (2009), *The fog remote sensing and modeling field project*, Bull. Amer. Meteor Soc., *90*, 341–359.

Guo, J. T. (2009), Study of formation and development mechanisms of sea fog: observational analysis and numerical modeling, Ph.d. Thesis, Ocean University of China (in Chinese).

Hong, S. Y., J. Dudhia, and S. H. Chen (2004), *A revised approach to ice microphysical processes for the bulk parameterization of clouds and precipitation*, Mon. Weather Rev., *132*, 103–120.

Hong, S. Y., Y. Noh, and J. Dudhia (2006), *A new vertical diffusion package with an explicit treatment of entrainment processes*, Mon.Weather Rev., *134*, 2318–2341.

Hu, R. J., Dong, K. H., and Zhou, F. X. (2006), *Numerical experiments with the advection, turbulence and radiation effects in the seafog formation process*, Advances in Marine Science *24*, 156–165(in Chinese).

HUANG, J., WANG, B., ZHOU, F. X., HUANG, F., LU, W. H., HUANG, M. G., HUANG, H. J., YANG, Y. Q., and MAO, W. K. (2010), *turbulent heat exchange in a warm sea fog event on the coast of South China*, Chinese Journal of Atmospheric Sciences *134*, 716–725(in Chinese).

KAIN, J. S., and FRITSCH, J. M. (1990), *An one-dimensional entraining/detraining plume model and its application in convective parameterization*, J. Atmos. Sci. *47*, 2784–2802.

KAIN, J. S. and FRITSCH, J. M. (1993), convective parameterization for mesoscale models: the Kain-Fritcsh scheme, the representation of cumulus convection in numerical models (eds. Emanuel K. A. and Raymond D. J.) (Am. Meteor. Soc) pp. 246.

KONG, F. Y. (2002), *An experimental simulation of a coastal fog-stratus case using COAMPS(tm) model*, Atmos. Res. *64*, 205–215.

KORAČIN, D., LEWIS, J., THOMPSON, W. T., DORMAN, C. E., and BUSINGER, J. A. (2001), *Transition of stratus into fog along the California coast: observations and modeling*, J. Atmos. Sci. *58*, 1714–1731.

KORAČIN, D, BUSINGER, J., DORMAN, C., and LEWIS, J. (2005), *Formation, evolution and dissipation of coastal sea fog*, Boundary-Layer Meteorology. *117*, 447–478.

KOSCHMIEDER, V. H. (1924), *Theorie Der Horizontalen Sichtweite*, Beitr. Phys. Atmos. *12*, 33–53.

LAMB, H. (1943), Haars or North Sea Fogs on the Coasts of Great Britain (Meteorology Office Publication M.M. 1943) pp. 24.

LEIPPER, D. F. (1948), *Fog development at San Diego, California*, J. Mar. Res. *7*, 337–346.

LEIPPER, D. F. (1994), *Fog on the U.S. West Coast: a review*, Bull. Am. Meteor. Soc. *75*, 229–240.

LEWIS, J., KORAČIN, D., RABIN, R., and BUSINGER, J. (2003), *Sea fog off the California Coast: viewed in the context of transient weather systems*, J. Geophys. Res. *108*, 4457–4473.

LEWIS, J., KORAČIN, D., and REDMOND, K. (2004), *Sea fog research in the UK and USA: historical essay including outlook*, Bull. Amer. Meteor. Soc. *85*, 395–408.

MLAWER, E. J., TAUBMAN, S. J., BROWN, P. D., IACONO, M. J., and CLOUGH, S. A. (1997), *Radiative transfer for inhomogeneous atmospheres: RRTM, a validated correlated-K model for the longwave*, J. Geophys. Res. *102*, 16,663–16,682.

MONIN, A. S., and OBUKHOV, A. M. (1954), *Basic laws of turbulent mixing in the surface layer of the atmosphere (in Russian)*, Contrib. Geophys. Inst. Acad. Sci. USSR *151*, 163–187.

PILIÉ, R. J., MACK, E. J., ROGERS, C. W., KATZ, U., and KOCHMOND, W. C. (1979), *The formation of marine fog and the development of fog-stratus systems along the California coast*, J. Appl. Meteor. *18*, 1275–1286.

QI, Y. L. (2010), Study on the Formation Mechanism of Typical Advection Fog Occurred over the Yellow Sea, Master Thesis, Ocean University of China (in Chinese).

SAKAMOTO, K., TSUTSUI, J., KOIDE, H., SAKAMOTO, M., KOBAYASHI, S., HATSUSHIKA, H., MATSUMOTO, T., YAMAZAKI, N., KAMAHORI, H., TAKAHASHI, K., KADOKURA, S., WADA, K., KATO, K., OYAMA, R., OSE, T., MANNOJI, N., and TAIRA, R. (2007), *The JRA-25 reanalysis*, J. Meteor. Soc. Japan *85*, 369–432.

STOELINGA, M. T. and WARNER, T. T. (1999), *Nonhydrostatic, mesob-eta-scale model simulations of cloud ceiling and visibility for an east coast winter precipitation event*, J. Appl. Meteor. *38*, 385–404.

STULL, R. B.(1988), An Introduction to Boundary Layer Meteorology. Atmospheric Sciences Library, Dordrecht: Kluwer.

TAYLOR, G. I. (1915), *Eddy motion in the atmosphere*, Philos. Trans. Roy. Soc. London Ser. A *215*, 1–126.

TAYLOR, G. I. (1917), *The formation of fog and mist*, Quart. J. Roy. Meteor. Soc. *43*, 241–268.

TOKINAGA, H. and XIE S. -P. (2009), *Ocean tidal cooling effect on summer sea fog over the Okhotsk Sea*, J. Geophys. Res., *114*, D14102, doi:10.1029/2008JD011477.

VAN DER VELDE. I. R., STEENEVELD, G. J., WICHERS SCHREUR, B. G. J., and HOLTSLAG, A. A. M. (2010), *Modeling and forecasting the onset and duration of severe radiation fog under frost conditions*, Mon. Wea. Rev. *138*, 4237–4253.

WANG, B. H., Sea Fog (China Ocean Press, Beijing 1985) (in Chinese).

YANG, D., RITCHIE, H., and GULTEPE, I. (2010), *High-resolution GEM-LAM application in marine fog prediction: evaluation and diagnosis*, Weather and Forecasting *25*, 727–748.

ZHANG, S. B. (2010), Study on the formulation of sea fog over the Yellow Sea related to high pressure, Master Thesis, Ocean University of China (in Chinese).

ZHANG, S. P., REN, Z. P., LIU, J. W., YANG, Y. Q., and WANG, X. G. (2008a), *Variations in the lower level of the PBL associated with the Yellow Sea fog—new observations by L-band radar*, J.Ocean Uni.China *7*, 353–361.

ZHANG, S. P., YANG, Y. Q., WANG, X. G., and WEN, J. S. (2008b), *Seasonal variations in the atmospheric stratification and relations with the Yellow Sea fog season*, Periodical of Ocean University of China *38*, 689–698 (in chinese).

ZHANG, S. P., XIE, S. P., LIU, Q. Y., YANG, Y. Q., WANG, X. G., and REN, Z. P. (2009), *Seasonal variations of Yellow Sea fog: observations and mechanisms*, J. Climate, *22*, 6758–6772.

ZHANG, S. P., and REN, Z. P. (2010), *The Influence of thermal effects of underlaying surface on the spring sea fog over the Yellow Sea—observations and numerical simulation*, Acta Meteorologica Sinica *68*, 116–125 (in Chinese).

ZHANG, S. P., LIU, J. W., and XIE, S. P. (2011), *the formation of a surface anticyclone in the spring Yellow and East China Seas*, J. Meteo. Soc. Japan *89*, 119–131.

ZHOU, B. B., and FERRIER, B. S. (2008), *Asymptotic analysis of equilibrium in radiation fog*, J. of Appl. Met. And Clim. *47*, 1704–1722.

(Received November 16, 2010, revised May 13, 2011, accepted May 24, 2011, Published online July 29, 2011)

Reprinted from the journal

Pure Appl. Geophys. 169 (2012), 1019–1036
© 2011 Springer Basel AG
DOI 10.1007/s00024-011-0341-z

Exploring Fog Water Harvesting Potential and Quality in the Asir Region, Kingdom of Saudi Arabia

P. Gandhidasan[1] and H. I. Abualhamayel[1]

Abstract—During the last decade, the exploitation of the existing water resources in the Asir region of the Kingdom of Saudi Arabia has considerably increased due to both the decrease in annual precipitation and the added population pressures from the growing tourist industry. To face the conventional water shortage, attention has been mainly focused on desalination of water. To save the region from severe water shortage, additional new water sources that are low-cost and renewable must be identified. There exists an alternative source of water such as fog water harvesting. Fog forms in the Asir Region more frequently between December and February compared to the other months of the year. This paper presents the study of the climatic conditions in the Asir region of the Kingdom to identify the most suitable location for fog water collection as well as design and testing of two large fog collectors (LFCs) of size 40 m² along with standard fog collectors (SFCs) of 1 m² in that region. During the period from 27 December 2009 to 9 March 2010, a total of 3,128.4 and 2,562.4 L of fog water were collected by the LFC at two sites in the Al-Sooda area of the Asir region, near Abha. Experimental results indicate that fog water collection can be combined with rain water harvesting systems to increase water yield during the rainy season. The quality of the collected fog water was analyzed and compared to the World Health Organization (WHO) drinking water standards and found to be potable. An economic analysis was carried out for the proposed method of obtaining fresh water from the fog. The study suggests a clear tendency that in terms of both quality and magnitude of yield, fog is a viable source of water and can be successfully used to supplement water supplies in the Asir region of the Kingdom.

Key words: Asir region, Kingdom of Saudi Arabia, large fog collectors, fog water collection, experiments, chemical analyses, economic analysis.

Abbreviations

CRF   Capital recovery factor
FAC   First annual cost, SR
M First annual cost of the system, SR
n Life of the system, years

N First annual salvage value, SR
P' Capital cost of the system, SR
r Interest rate, %
S Salvage value, SR
Y Yield, m³/year

1. Introduction

The Middle East and North Africa are among the most arid regions in the world, with 5% of the world's population, but only 1% of global annual renewable water resources. The Kingdom of Saudi Arabia is one of the hottest and driest subtropical desert countries in the world. With an average of 112 mm of rain precipitation per annum, much of the Kingdom falls within the standard definition of desert as an area with a precipitation rate of less than 250 mm per year. Saudi Arabia's state-owned Saline Water Conversion Corporation (SWCC) has estimated that through 2020, the country will need at least $50 billion on water projects. The shortage in water supplies in the Kingdom represents a constant threat to life and development.

In 2007 the UN World Tourism Organization had predicted more than 7% annual growth in tourism industry in the Middle East. The soaring mountains, green forests, cavernous valleys, waterfalls and cool climate delight the tourists to visit the Asir region of the Kingdom of Saudi Arabia. Twenty-four massive tourism projects are planned for the Asir region and these projects are expected to have a huge impact on tourism industry in the region. Flourishing tourism in the Asir region is challenged by the scarcity of water resources.

¹ Mechanical Engineering Department, King Fahd University of Petroleum and Minerals, Dhahran 31261, Saudi Arabia. E-mail: pgandhi@kfupm.edu.sa

The region of the Asir is located in the south-western part of the Kingdom of Saudi Arabia, as shown in Fig. 1, between longitudes 41–45°E and latitudes 17–21°N. The total area of the Asir region is spread over about 68,460 km^2 which represents 4.3% of the total area of the Kingdom. Asir is one of the richest regions of the Kingdom in terms of rain. The climate condition in the Asir region varies according to the geographical and topographical conditions. The region is subject to Indian Ocean monsoons, usually occurring between October and March. Agriculture is the traditional main activity of the local economy in the Asir region. From December to February it is usually extremely cold and the visibility can be reduced to almost 0% due to fog. Fog often drifts over the mountains and through the valleys.

One of the main problems in the Asir region is the high demand for water during tourism seasons. It was recognized that the Asir region in the Kingdom is the most suitable location for the fog-collection process and could supplement the existing water sources in the region (GANDHIDASAN and ABUALHAMAYEL, 2007; AL-HASSAN, 2009). Prerequisites for the successful implementation of a fog collection project are high incidence of fog, the presence of wind during the fog episode, and a suitable site. The origin of the fog, wind speed during fog events, and elevation are major determinants in the volume of water that can be collected (LOUW et al., 1998). Selection of a suitable site is vital to the success of a fog collection project.

2. Non-conventional Water Sources in the Asir Region

The use of non-conventional water resources is important to complement the usual fresh water sources in water scarce Asir region. The Asir region

Figure 1
Study area location

depends on many resources of water supply including drinking water, which is supplied by a saline water conversion workstation in Al-Shagig in Jizzan. It also depends on wells. The desalinated water, cloud seeding, rainwater harvesting, water obtained by fog capturing, etc. are included under the designation of non-conventional water.

2.1. Cloud Seeding

In order to meet the demand for water supplies in the Asir region, a cloud seeding experiment program was suggested by Aksakal (AKSAKAL, 1998). Cloud seeding produces an increase in rainfall but it is an extremely difficult task and the process itself is very expensive. Earlier studies and experiments conducted in some countries have demonstrated that cloud seeding can increase the quantity of rainfall between 5 and 20% over large areas. In the Asir region, with an annual rainfall of about 230 mm, this will do little good. This may increase runoff into existing reservoirs so that the water would be available for irrigation purposes. However, the fact is that the rain would fall on parched terrain and be totally unavailable for managed use. This water could never be piped to the distant small remote villages due to high costs. No evaluation of the work was performed by Aksakal for the Asir region, so it is difficult to know what the increased water, if any, would have cost.

2.2. Rainwater Harvesting

The rainwater can be collected from roofs by means of containers placed at locations where dripping is most intense. With an average roof size of 50 m^2 and the total annual average precipitation of about 200 mm, it should be theoretically possible to collect about 10,000 L of water per year, but the precipitation is unpredictable. The fact is that the Asir region does not get enough precipitation at any period during the year.

2.3. Fog Water Harvesting

In the last several years, fog collection projects have been successfully implemented in arid regions of many countries in the world. Fog water harvesting can be considered as potential sources of fresh water for fog-prone areas such as the Asir region. The amount of water that can be harvested depends on surface area of the fog collector, the efficiency of collection, the density of fog and wind speed. No electricity is needed to power the fog water harvesting system, which makes it eco-friendly and low-cost, and suitable for areas with no power infrastructure. The present research investigates the operation of fog water collection in the Asir region of the Kingdom of Saudi Arabia.

3. Climate Study

Fog can occur in the Asir region by advection, but such episodes have a very irregular pattern. One of the objectives of this research is to study the climatic conditions by utilizing the meteorological data relevant to the Asir region of the Kingdom obtained from the Presidency of Meteorology & Environment (PME) covering the period from 1995 to 2007 (Communication with the Presidency of Meteorology & Environment Protection, National Meteorology & Environment Center, Ministry of Defence & Aviation, Jeddah, Saudi Arabia, March 2008).

A detailed study has demonstrated that three cities in the Asir Region need serious attention, namely Abha, Al-Baha and Khamis Mushait. The mean and minimum temperatures, the maximum and mean relative humidity, the mean wind speed and the rainfall for the above three cities from 1995 to 2007 were studied and the results are shown in Figs. 2, 3, 4. Analysis of the data of Abha shows that it has the lowest mean and minimum temperatures compared with Al-Baha and Khamis Mushait. Further, Abha has the highest maximum and mean relative humidities and the highest mean wind speed. A comparison of the February data show that the mean wind speed is 23 and 10% higher in Abha than in Al-Baha and Khamis Mushait, respectively. Abha is more humid (maximum relative humidity) at 91.7 versus 77% in Al-Baha and 88.4% in Khamis Mushait for January. Abha is cooler (minimum ambient temperature) at 7.4°C than Al-Baha at 10.5°C and Khamis Mushait at 8.6°C for the month

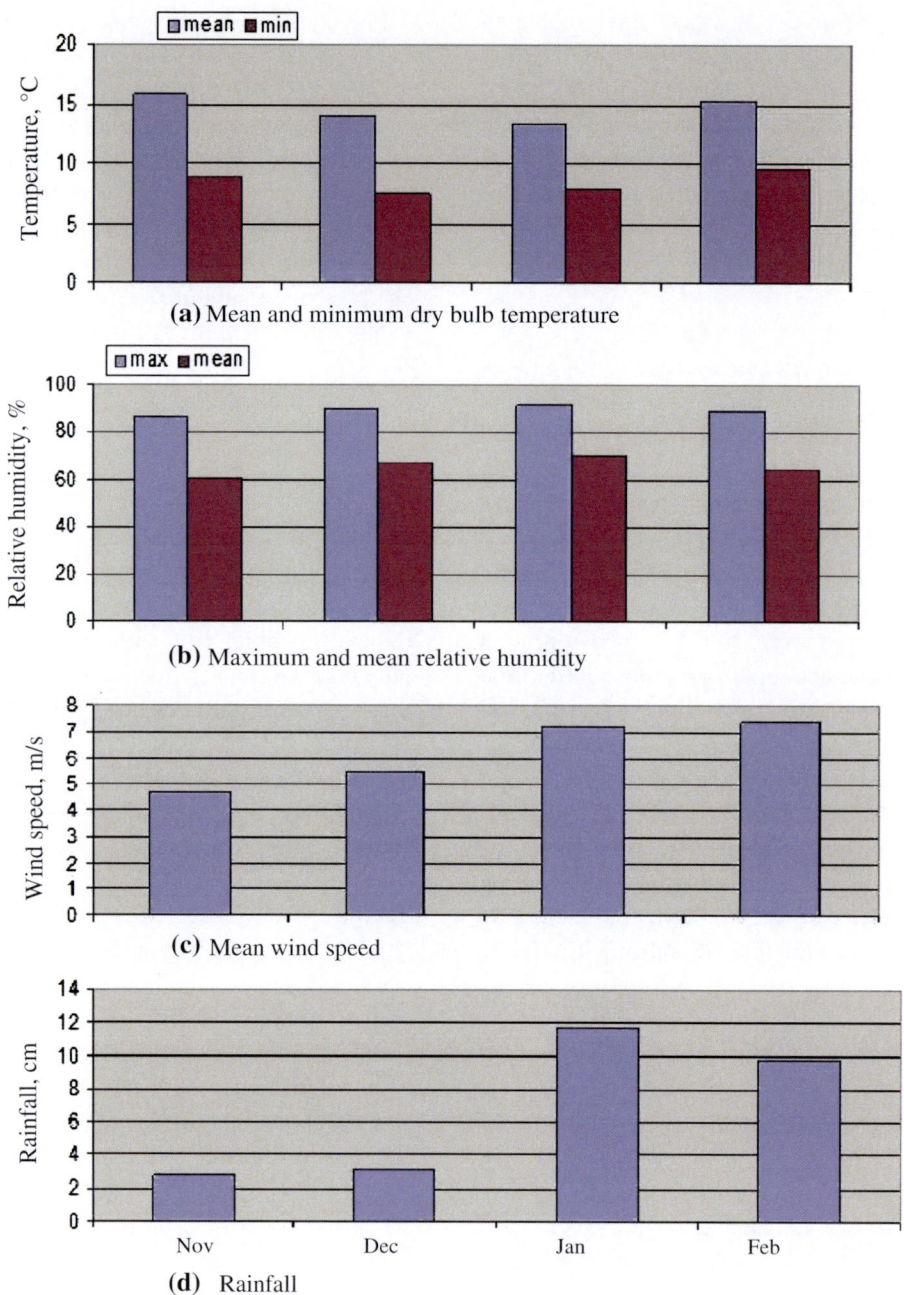

(a) Mean and minimum dry bulb temperature

(b) Maximum and mean relative humidity

(c) Mean wind speed

(d) Rainfall

Figure 2
Meteorological data for Abha

of December. According to the climatic data, high precipitation occurs in the month of January in all three cities. The highest precipitation recorded in Abha was 117 mm, while Al-Baha received 55 mm and Khamis Mushait 69 mm. The forced convection due to orographic lifting in southwestern mountainous region close to the Abha area triggers heavy rainfall. The mean value of wind speed at Abha varied from 4.7 to 7.4 m/s, the minimum ambient temperature from 7.4 to 9.6°C and the maximum

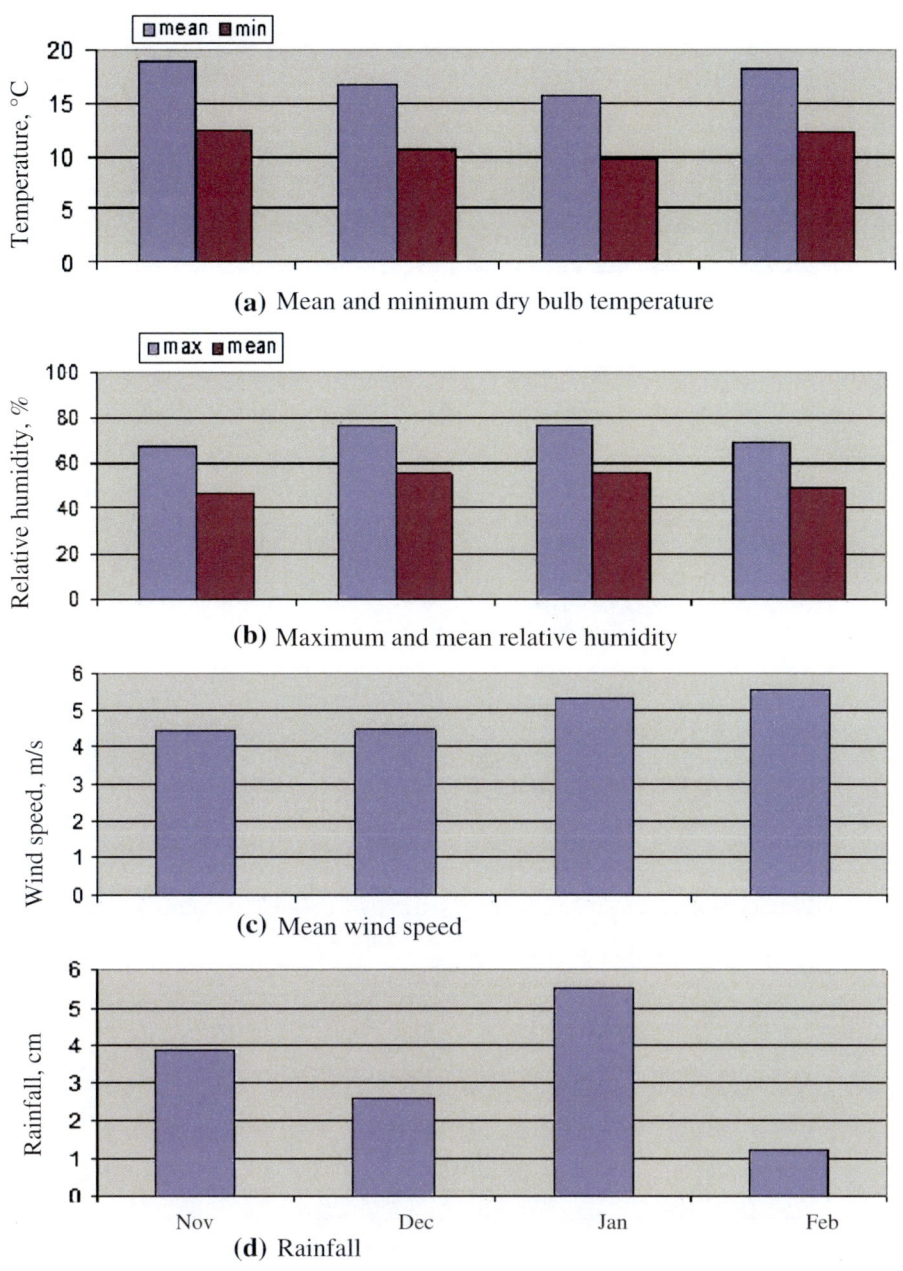

(a) Mean and minimum dry bulb temperature

(b) Maximum and mean relative humidity

(c) Mean wind speed

(d) Rainfall

Figure 3
Meteorological data for Al-Baha

relative humidity from 86.9 to 91.7%. These data provide the first comprehensive description of the meteorological conditions in Abha. The high humidity, high wind speed, and the low ambient temperature are the conditions that support the frequent fog formation in the Abha area.

4. Identification of Sites and LFC Site Selection

Conditions for fog formation in the Abha area are very good in many sites. In order to identify areas with the greatest potential for fog water collection, field research was conducted. The investigators

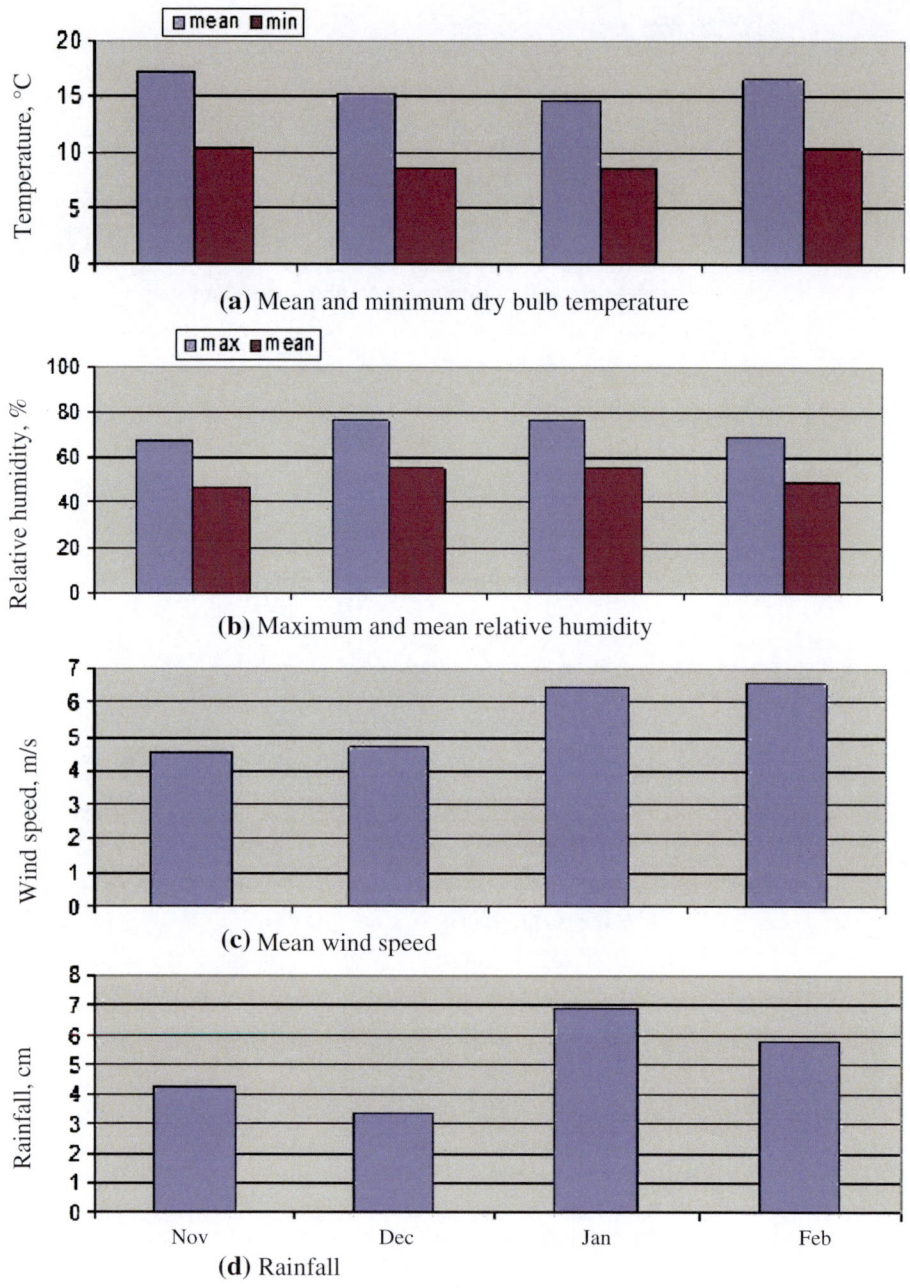

Figure 4
Meteorological data for Khamis Mushait

visited many sites in that area. The following important criteria were taken into account when selecting a suitable site for a fog water collection system (OLIVIER, 2004):

- The potential for collecting large volumes of water.
- Terrain features and accessibility.

- Security.
- Availability of qualified staff to record water collection rates.

The investigators made contact with the General Secretary of the Chamber of Commerce & Industry in Abha and the Dean of the Prince Sultan College for Tourism and Business, to identify the potential sites for fog collection. The discussion was also held with the local community. Seventeen sites, stretching from approximately 18.2°N and 42.5°E in the Asir region were initially identified. The location, roads, and available infrastructure places considerable limits on the selection of sites. Therefore, some places may have the potential for fog water collection in the Asir region but it is not feasible to collect fog water in those sites. Further, the fog collectors must be protected from monkeys living in the area of the sites.

Because the LFCs are long and require space for guy wires for the posts, at least 25 m of flat vacant land is needed for the erection of a LFC. Another important aspect is ownership of land where the fog collectors will be installed. In this respect, discussion was held with the office of the Governor of the Asir Region in Abha and their branch office in Al-Sooda area regarding the site selection for the erection of fog collectors. It is interesting to note that Al-Sooda has the highest altitude (about 3,015 m) among the other sites and is about 15 km away from Abha. Finally, five different sites were identified to erect twelve SFCs and these five sites are about 2–3 km away from each other. Polypropylene mesh nets with a 35% shade coefficient were used for SFCs as they are inexpensive, durable, non-toxic and not subject to corrosion to literally catch every drop of water in the fog. These sites were tested using SFCs with instruments especially designed to collect and measure fog water on an experimental basis. The SFC results suggest that the Glider club area and the InterContinental Hotel area have a greater potential for fog water harvesting than other sites. In November 2009, the final locations for the two LFCs were chosen.

5. Design and Site Features of the LFC

A LFC can have water collecting surface area between 40 and 50 m^2. The 40 m^2 size is stronger and recommended for use in high winds (SCHEMENAUER et al., 2005).

The general requirements considered for the design and construction of the LFC are:

- Installation and connection of panels must be quick and simple.
- The complete system must be easily built or assembled on site.
- The assembly must require little skill and not be labor intensive.
- The capital investment must be low.
- The maintenance and repairing costs must be minimal.
- No energy must be needed to operate the system or transport the water.

LFC panels must be designed to withstand the structural stresses imposed by wind, humidity, and friction between different parts. The area around the LFC must be relatively clear to allow a free flow of light to moderate winds during the presence of the fog.

The LFC design is quite complicated compared to the design of SFC (GANDHIDASAN and ABUALHAMAYEL, 2007). The size of the LFC selected for this study is 20 m wide by 2 m high with a surface area of 40 m^2 to intercept the droplets of fog and bases 2 m above the ground to maximize the exposure to the wind. Since 20 m is very long, the design is made of eight sections with 2 m $\times$ 2.5 m, as shown in Fig. 5. The LFCs are flat rectangular nets supported by a metal post at both ends and arranged perpendicular to the direction of the prevailing fog-bearing wind. The posts have hooks installed to attach the support cables. The LFC is firmly supported since the wind load on the LFC is high. The material used for the base is I-beam steel structure. The frame is made of metal for rigidity to secure the polypropylene mesh. The mesh is a roll 2 m wide. This mesh has the main lines running in a horizontal direction and the "V" shape of the weave pointing down when installed. The folded edges of the mesh are folded over the top frame and fixed tightly onto the frame. This anchors the mesh and allows it to be pulled horizontally. The double layers of mesh are pulled down as tightly as possible and fixed firmly onto the bottom frame. To fix the mesh net on the frame, aluminum strips are used on each side of the frame by bolts. LFC components such as the mesh, the cables, the posts, the turnbuckles and all other hardware used are of good quality. Thus, the whole structure is made rigid and the LFC can withstand winds up to 25 m/s.

The fog water collects on the mesh and flows under the influence of gravity into aluminum trough below the frame secured for collection. As the fog and strong winds come indiscriminately, the trough is situated exactly in the middle of the base of the frame, so as to be able to collect the fog water from either front or back side of the frame. The trough has a slight slope, so that the water drains toward the opening of the drainpipe, which is connected to a 75 mm PVC pipeline. This pipeline conducts the collected water into the trough to a closed PVC tank of 250 L.

6. Manufacturing and Erection of Two LFCs

After obtaining permission from the relevant authorities to erect the two LFCs, construction commenced in December 2009 at two sites in the Al-Sooda area to evaluate the efficiency of fog water collection. Metal posts with a wall thickness of 6 mm are selected. In order to avoid rust on the metal posts, it is painted. Iron loops are welded onto metal posts at the points where cables have to be attached. In order to evaluate the effectiveness of fog water collection at the sites, two 20 m × 2 m size of LFC with identical local collection materials were manufactured. The design of these two LFCs was the same with the same type of mesh net used with SFCs.

The mesh nets of each LFC were connected together with steel cables with anchored the system to the ground. Cables are 6 mm diameter, stranded, galvanized steel and attached to the eye-bolts. Turnbuckles are used to do the final tightening of the cables. All of the materials to construct, operate, and maintain the LFC system are available locally. One LFC was installed in the Glider club area and another LFC was placed in the Hotel area. The erection work at the Glider club site is shown in Fig. 6. The set-ups were ready for conducting the experiments for fog collection. One LFC and a SFC were installed at both sites for testing purposes. The identical collectors at both sites are named as follow:

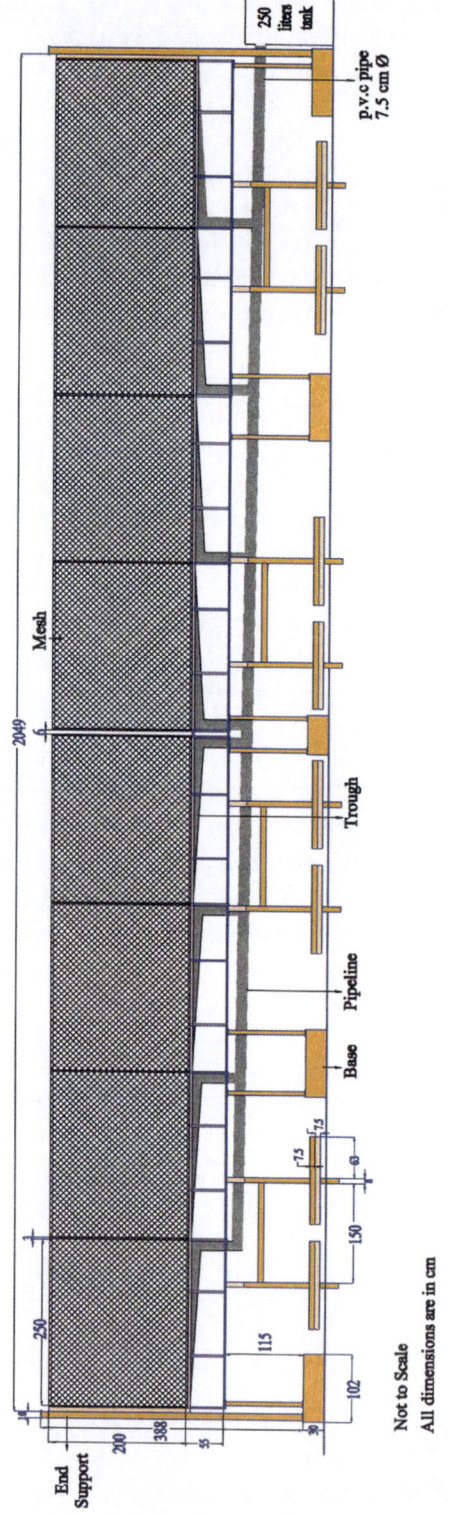

Figure 5
The design of the LFC

Figure 6
Erection of the LFC at the Glider club site

- SFC # 1 and LFC # 1—collectors tested at the Glider club area.
- SFC # 2 and LFC # 2—collectors tested at the Hotel area.

Since 27 December 2009, experiment was started and water collected by the net on 28 December 2009.

7. Results and Discussion

Experiments were conducted to record water collection rates by measuring the volume of water collected on one SFC and LFC erected at two sites. Relative humidity, temperature and the wind speed were also recorded at 07:00, 14:00 and 19:00 h. The vast majority of the fog collector data and weather observations were of good quality, although some gaps in the data occurred.

Experiments were conducted from 27 December 2009 to 23 March 2010 but data from 27 December 2009 to 9 March 2010 are used for the analysis since there were no fog events after 7 March 2010. The summary of various events taken place during the

Table 1

Summary of various events

Event	Number of days	Percentage of days
No fog	27	37.5
Fog	38	52.8
Fog and rain	7	9.7
Total	72	100.0

above experimental period of 72 days are shown in Table 1. It should be noted that both fog and rain water were collected by the fog collectors. There was no fog for 27 days whereas there was fog with rain during 7 days.

It is interesting to note that 9.7% of days during the experiments had fog with rain. Water yields, originating from a combination of both rainfall and fog collection, were also recorded on certain days. Since no data on rain are available, it is recommended not to include the fog data of days with rain. Water collected on a non-rain or dry day was attributed to fog only. This obviously under-estimates the contribution of fog to the total volume of water collected by assuming that there is no fog present during

Table 2

Comparison between yields at the Glider club and the Hotel sites

Period	LFC		SFC	
	Glider club site (L)	Hotel site (L)	Glider club site (L)	Hotel site (L)
28 Dec. 2009 to 10 Jan. 2010	692.8	645.9	41.0	41.0
11 Jan. 2010 to 26 Jan. 2010	1,329.4	1,037.7	78.5	66.9
28 Jan. 2010 to 10 Feb. 2010	633.4	529.8	60.5	46.0
11 Feb. 2010 to 22 Feb. 2010	185.0	131.0	19.3	14.0
23 Feb. 2010 to 07 March 2010	287.8	218.0	30.5	21.7
Total (L)	3,128.4	2,562.4	229.8	189.6
Average (L/day)	82.33	67.43	6.05	5.0
Average (L/m^2/day)	2.06	1.69	6.05	5.0
Ratio	1.22	1.0	1.21	1.0

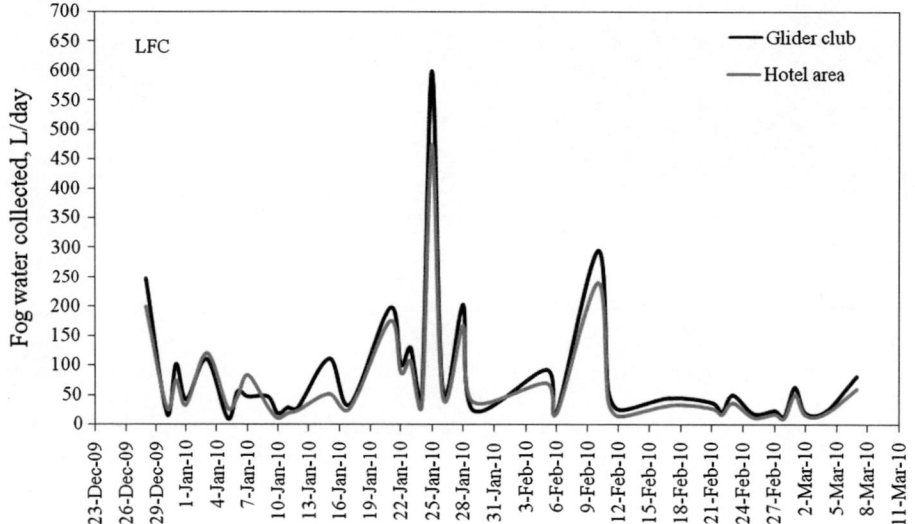

Figure 7

Comparison of the daily fog water collection rates by LFCs

a rain event. During this period, fog water was collected on 38 foggy days from LFCs and SFCs. The overall comparison between fog water yields in liters at the Glider club and the Hotel sites is given in Table 2. At the Glider club site, a total of 3,128.4 L of water was collected from the LFC # 1 during the experimental period. This gives an average yield of 2.06 L m^{-2} day^{-1}. Similarly, at the hotel site, a total of 2,562.4 L of water were collected from the LFC # 2 during the same period. The comparison between yields at the Glider club site and the Hotel site reveals that the fog water yield at the former site was considerably higher. This is probably due to the higher

wind speed and the higher water content of fog at the Glider club site. During the same period, 229.8 and 189.6 L of water were collected from SFC # 1 and SFC # 2, respectively.

The experimental results of the LFCs and SFCs are presented in Figs. 7 and 8, respectively. Although average collection rates of about 1.9 L m^{-2} day^{-1} were recorded on LFCs, this was found to be an underestimation of actual yields. The actual volume of yields collected at these two sites may be even greater than that indicated above. The 250 L storage tank regularly overflowed during nights with heavy fog and pools of water were often observed under the

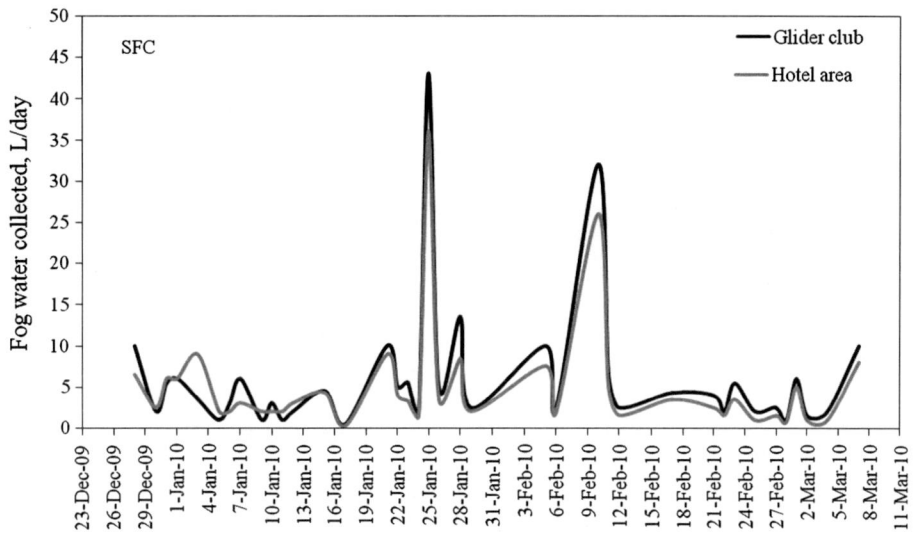

Figure 8
Comparison of the daily fog water collection rates by SFCs

collector early the next morning. Taking these inaccuracies into account, it was estimated based on the previous tests on the SFCs that the actual yields probably averaged about 4.0 L m^{-2} day^{-1} from the LFCs.

Figure 9 shows the relationships between the amount of fog water collected by the LFC and SFC at the Hotel site and the meteorological variables. When the ambient temperature is between 5 and 10°C, the water collection rate is high. Similarly, when the relative humidity is higher than about 90%, the collection rate is high and the maximum collection is produced when the relative humidity is close to 100%. It appears that high fog collection rates start with the wind speed of 2 ms^{-1} and there is no clear trend of increasing fog collection rate with increasing wind speeds greater than 2 ms^{-1}.

Frequency and intensity of water collected on fog days at the Glider club and the Hotel sites by LFCs is shown in Fig. 10. There is no doubt that the frequencies of fog on both sites are almost the same but the intensity is different. Both figures clearly show another differentiating feature between the two sites. There are more days of little water collection (of less than 25–50 L day^{-1} by the LFC) on the Hotel site, whereas the Glider club site nearly always produces more than 100–125 L day^{-1} by the LFC in spite of having the equal number of fog days.

7.1. Simultaneous Testing of SFC and LFC

Experiments were conducted with SFCs and LFCs simultaneously but the volumes of water collected in L m^{-2} day^{-1} are different and the results at the Glider club and the Hotel sites are shown in Figs. 11 and 12, respectively. The obtained results show marked differences between the SFC and LFC as well as between the two sites in the amount of water collected from the fog. 1 m^2 SFCs produced more water per unit area than the 40 m^2 LFCs. This surprising result is probably due to more favorable exposure rather than a better design of the SFC. Also, the higher evaporation and wind losses of adhered droplets are a factor in catch efficiency. LFCs tend to lose more, because droplets have to travel a much longer distance to the collection trough. A similar type of result is reported by SCHEMENAUER et al. (1988).

7.2. Fog and Rain Yields

In order to identify the role of rain in the fog water harvested, the number of days with rain was counted. During the seven rain days, there was heavy rain on 09 February 2010 and we were unable to measure since the storage tank continuously overflowed. Drizzle combined with fog occurred on 3 days and

265

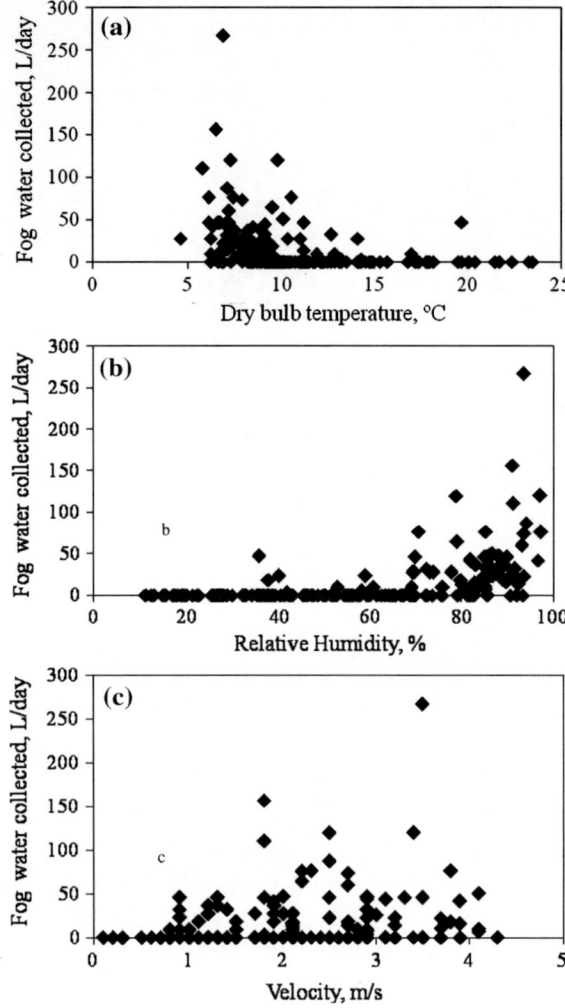

Figure 9
Daily fog water collection rates by the LFC at the Hotel site
a Plotted against temperature. **b** Plotted against relative humidity.
c Plotted against wind speed

the yields were recorded. The results are given in Table 3. It was not possible to determine the contribution of rainfall to the total amount of water collected by the LFCs. On the remaining 3 days, up to 2:00 PM there was no rain but after 4:00 PM there was rain with fog.

8. Physico-chemical Analyses of Fog Water Collected From the LFCs

Water quality is of vital importance when supplying water for drinking. A program for fog water quality monitoring is, therefore, carried out to identify any contaminants from the collector material and atmospheric deposition. Chemical analysis as well as concentration of major cations and anions of fog water collected was performed and the quality of fog water discussed is based on two sets of three water samples. The samples were analyzed for total dissolved solids (TDS), total hardness, electrical conductivity, pH, and major ions. Samples were analyzed for anions by Ion Chromotography. Total dissolved solids and total suspended solids were determined gravimetrically. Total hardness was calculated using calcium and magnesium concentrations determined by ICP-AES.

Tables 4, 5, 6 show the results of the analyses of the fog water samples collected at these two sites. The first set of three fog water samples (S-1 to S-3) were collected after two days of the experiments. S-1 was collected from the Glider club site, whereas the other two samples were collected from the Hotel site on different days. The second set of three water samples (S-4 to S-6) were collected after a month of the experiments. Sample S-4 was collected during the rainy day and samples S-5 and S-6 were collected on a foggy day at the Glider club and the Hotel sites, respectively. Water collected at these sites was pure having very low concentrations of sulphates, chlorides, dissolved calcium, magnesium, sodium, potassium, iron, manganese and nitrates. The results from the chemical analyses of the fog water collected are compared with WHO's drinking water standards (World Health Organization, 2006). In general, the quality of water collected is very good and it is interesting to note that all values are below WHO guidelines.

Although, all parameters are below WHO guidelines, some of the parameters are high in the first set of three water samples due to contamination by windblown dust and dirt deposited on the LFCs compared to the second set of three samples. The second set of three water samples are thus a true reflection of the quality of the water collected on the LFCs. The difference in water quality is marked. The TDS have decreased from almost 850 to 250 mg L^{-1}. Hence, some mechanism will have to be introduced to allow the first water collected during a wet event to be discarded.

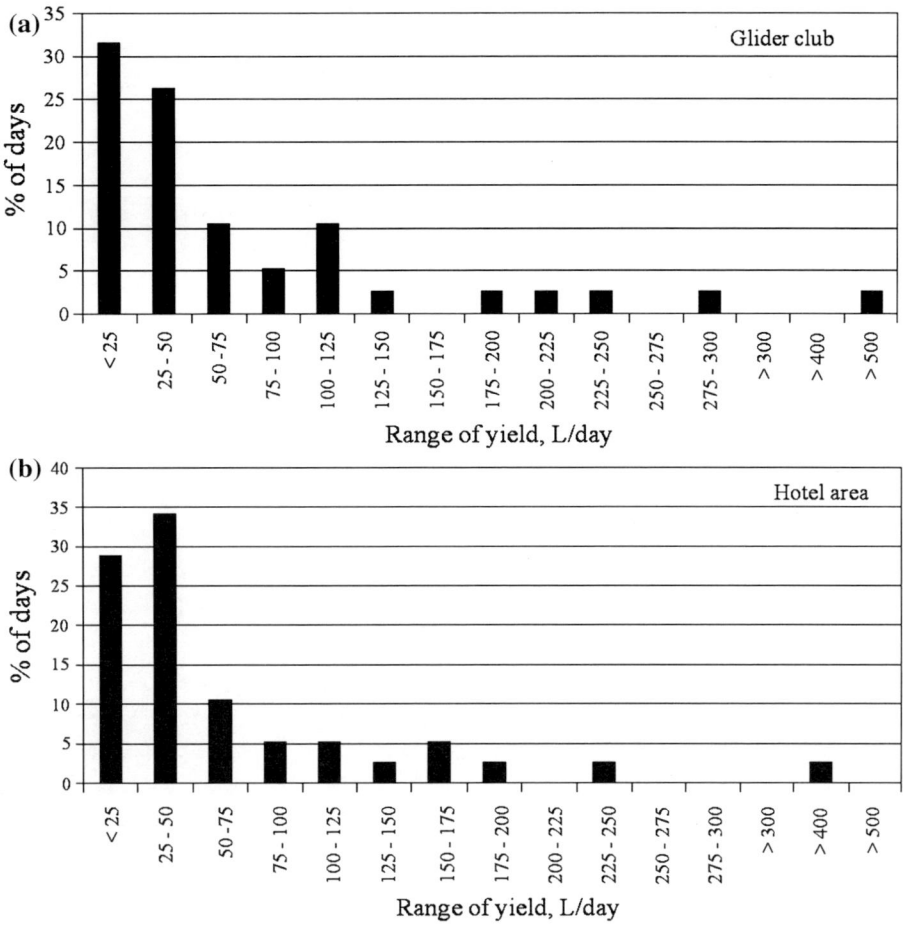

Figure 10
Frequency and intensity of water collected on fog days by LFCs **a** Glider club site. **b** Hotel site

9. Economic Analysis

The fog water collection system proposed in this study is examined from a socioeconomic standpoint. The parameter used in the analysis for evaluating the fog water supply is the cost of water consumed, which is calculated as Saudi Riyal (SR) per m^3 of water consumed. The total cost for the proposed fog water harvesting system comprises not only the structures but also the piping system. However, the cost of fence and other related costs are not included. The yield potential is extremely important, for carrying out the economic analysis of the fog water resources.

In the Kingdom, the desalinated water is priced less compared with actual production cost. Among the various viable methods for supplying water, the main water supply in the Asir region relies on water-tank trucks in most districts. The economic comparison between the tanker transport system and the fog collection system is very difficult because the government subsidizes diesel.

9.1. Factors Affecting Economic Estimates

The following factors affect the cost estimates in the analysis:

- The collection efficiency of the fog water collection system.
- The length of the main pipeline that carries the water from the collection system to the storage tank.
- Manpower, material costs, and the ease of site access.

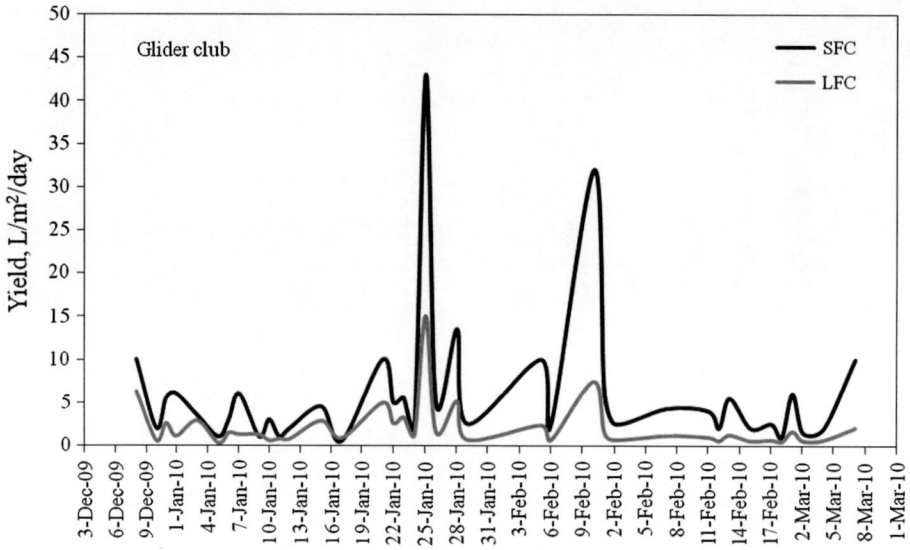

Figure 11
Comparison of daily fog water collection rates by LFC and SFC at the Glider club site

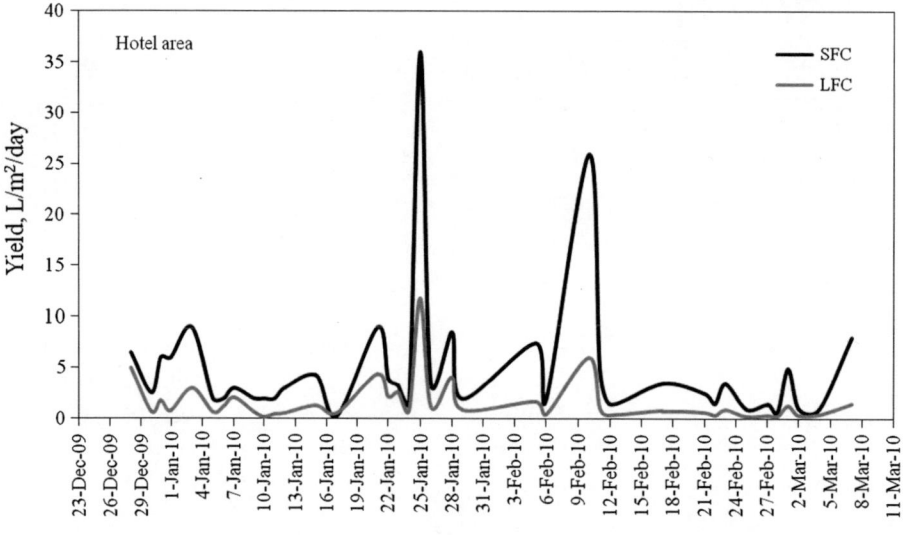

Figure 12
Comparison of daily fog water collection rates by LFC and SFC at the Hotel site

Table 3

Fog and rain yields at the Glider club and the Hotel sites

Date	Water collected on the LFC (L/day)		Water collected on the SFC (L/day)	
	Glider club site	Hotel site	Glider club site	Hotel site
29 December 2009	186.8	101.2	8.3	3.9
02 January 2010	165.6	142.6	5.5	6.0
07 February 2010	186.4	155.2	20.0	16.0

Table 4

Results of physical parameters of fog water samples from the LFCs

Sample	pH	E.cond (mS/cm)	TSS (mg/L)	TDS (mg/L)	HCO$_3$ (mg/L)	CO$_3$ (mg/L)
S-1	6.72	0.96	2.0	850	68.998	<0.020
S-2	6.68	0.97	29.0	834	55.720	<0.020
S-3	6.76	0.97	18.0	874	54.095	<0.020
S-4	7.43	0.23	3.0	158	40.347	<0.020
S-5	7.42	0.40	34.0	278	63.200	<0.020
S-6	7.24	0.39	2.0	318	71.716	<0.020
WHO (World Health Organization, 2006)	6.5–8.0	–	–	1,000	–	–

Table 5

Results of anions analyses of fog water samples from the LFCs

Sample	Cl (mg/L)	F (mg/L)	NO$_2$ (mg/L)	NO$_3$ (mg/L)	SO$_4$ (mg/L)
S-1	74.728	0.5611	3.9467	181.5989	68.998
S-2	74.204	0.4455	4.5311	178.6458	55.720
S-3	77.158	0.4271	4.5838	184.5233	54.095
S-4	13.686	0.0508	0.5540	23.0030	40.347
S-5	24.553	0.0981	1.2607	49.3663	63.200
S-6	27.538	0.0899	8.0672	44.0433	71.716
WHO (World Health Organization, 2006)	250	1.5	–	50	50

Table 6

Results of metals analyses of fog water samples from the LFCs

Sample	As (mg/L)	Ca (mg/L)	Cd (mg/L)	Cr (mg/L)	Cu (mg/L)	Fe (mg/L)	Hg (mg/L)	K (mg/L)
S-1	<0.007	120.0	<0.0009	<0.0004	<0.003	<0.0003	<0.005	9.30
S-2	<0.007	120.9	<0.0009	<0.0004	<0.003	<0.0003	<0.005	9.46
S-3	<0.007	123.8	<0.0009	<0.0004	<0.003	<0.0003	<0.005	9.62
S-4	<0.007	25.7	<0.0009	<0.0004	<0.003	<0.0003	<0.005	1.62
S-5	<0.007	49.3	<0.0009	<0.0004	<0.003	<0.0003	<0.005	2.97
S-6	<0.007	50.3	<0.0009	<0.0004	<0.003	<0.0003	<0.005	3.47
WHO (World Health Organization, 2006)	0.050	200	0.005	0.05	1.0	0.3	–	–

Sample	Mg (mg/L)	Mn (mg/L)	Na (mg/L)	Pb (mg/L)	Se (mg/L)	Si (mg/L)	Sr (mg/L)	Zn (mg/L)
S-1	6.75	0.696	47.5	<0.008	<0.011	0.203	0.454	0.045
S-2	6.88	0.692	48.7	<0.008	<0.011	0.130	0.458	0.029
S-3	7.00	0.668	49.9	<0.008	<0.011	0.136	0.461	0.017
S-4	0.966	0.003	7.05	<0.008	<0.011	0.128	0.102	0.051
S-5	1.69	<0.0001	12.7	<0.008	<0.011	0.145	0.185	0.041
S-6	1.64	0.002	12.3	<0.008	<0.011	0.245	0.199	0.019
WHO (World Health Organization, 2006)	125	0.3	200	0.05	0.01	-	-	-

9.2. Physical Characteristics of the Pilot System

Assume a population of 350 people living in a remote village and the village needs 30 L of water per day per person. That is the fog water collection system must able to provide 10.5 m^3 of water per day.

9.3. Initial Investment and Operating Cost Analysis

Assume every square meter of the fog collection system can yield a minimum of 4 L. The area of fog collecting panels that can yield 10.5 m^3 per day is 2,625 m^2. Assume that water is channeled into a regulated reinforced concrete storage tank with a

Table 7

Cost analysis for a 2 m × 20 m fog collection system

Sl. no.	Item name	Item cost, SR.	Quantity	Total cost, SR.	% Cost
1	Screen	2.50	40	100	1.32
2	Screen frame	1,500	1	1,500	19.74
3	Collector structure	2,000	1	2,000	26.32
4	Erecting mechanism	1,000	1	1,000	13.15
5	Primary PVC pipe connections	2,000	1	2,000	26.32
6	Miscellaneous	1,000	1	1,000	13.15
				7,600	100

capacity of 75 m^3, which approximately holds the equivalent of 7 days' supply and the water chlorination system.

It is assumed that an array of fog collection consist of 66 LFCs, each with a surface area of 40 m^2 plus primary pipelines, connections and primary storage tank. It is assumed that the main pipeline that carries water from the fog collection panels to the storage tank is 5 km long and 63 mm in diameter, consisting of PVC pipes, along with necessary pressure release valves. The cost of mesh is taken as SR.2.50 m^{-2} (US\$ 1 = SR.3.75) as per the present market price in the Kingdom. Table 7 shows the cost analysis for a 2 m × 20 m fog collection system. The fog collection system setup for a 40 m^2 will be SR.7,600 per screen. The most expensive items are the main pipeline that carries water from the fog collection panels to the storage tank and the fog collector structure. They represent about 52% of the total investment in the fog water harvesting system. For 66 LFCs of 40 m^2 the total investment will be SR.501,600 that will produce 10.5 m^3 of water per day.

In addition to the above initial investment costs, the cost of storage tank of SR.2,000 along with the main pipeline cost of SR.10/m to the length of 5 km is required. Since the system consists of many LFCs, individual pipelines to the main storage tank are also required. It is assumed that the required length of pipe is needed to be 500 m at a cost of SR.10/m which will add an additional cost of SR.5,000. Ten pressure release valves at the cost of SR.300 have to be considered. Hence, the total investment cost will be SR.561,600.

The maintenance cost involves that of attendants to inspect the collector drains, cable tension, cable fasteners, setup cleaning and the maintenance of the net. The cost of employing is SR.2,000 per month for each attendant for a period of 3 months during which there is possibility of obtaining fog. This results in SR.24,000 per cycle of operation assuming that four attendants are employed, as additional cost of maintenance to those involved in the initial setup cost. Hence, the total cost of fog collection system for the first year will be SR.585,600 and the cost of producing water from the fog collection system will be SR.619.68/m^3.

9.4. Cost Analysis Based on Salvage Valve Consideration

The estimated cost of material required for the fabrication, installation and maintenance in the above section is for one cycle of operation. As most of the components of the system are reusable for many years, a more appropriate cost analysis of the system will be the consideration of salvage value of the materials after an assumed life cycle of the system. The life for these systems will be generally 5 to 20 years. Hence, the cost analysis is calculated for a life cycle of 5, 10, 15, 20 years while considering interest rates of 5 and 8%.

If P' is the capital cost of the system and CRF is the capital recovery factor, then the first annual cost of the system can be determined by (HASNAIN and ALAJLAN, 1998),

$$M = P'(\text{CRF}) \tag{1}$$

where,

$$\text{CRF} = \frac{r(1 + r)^n}{(1 + r)^n - 1}. \tag{2}$$

The salvage value of the system is considered to be 50% cost of usable materials saved even after the

Table 8

Cost analysis for a 2 m × 20 m fog collection system with salvage value considered

n (years)	r (%)	CRF	SSF	S	N	M	AMC	AC	Cost/m^3
5	5	0.231	0.181	55,660	10,074	128,575	12,858	85,773	90.77
10	5	0.130	0.080	27,830	2,226	72,358	7,236	51,764	54.78
15	5	0.096	0.046	18,553	853	53,434	5,343	40,224	42.57
20	5	0.080	0.030	13,915	417	44,528	4,453	35,066	37.11
5	8	0.250	0.170	55,660	9,462	139,150	13,915	97,405	103.07
10	8	0.149	0.069	27,830	1,920	82,933	8,293	63,396	67.09
15	8	0.117	0.037	18,553	686	65,122	6,512	53,081	56.17
20	8	0.102	0.022	13,915	306	56,773	5,677	48,535	51.36

system life is over. If S is the salvage value of the system, the first annual salvage value can be determined as:

$$N = S(\text{SSF}) \qquad (3)$$

where

$$\text{SSF} = \frac{r}{(1+r)^n - 1}. \qquad (4)$$

As every system requires some maintenance, the annual maintenance cost should also be considered. Keeping this in view, the annual maintenance cost has been taken as 10% of the first annual cost (FAC). If the annual yield of the system is Y, then, the actual annual cost of the system is given by:

Annual cost = First annual cost + Annual maintenance cost − Annual salvage value
Product cost per cubic meter = Annual cost/Y

Table 8 predicts the effect of different parameters, such as life of the system, interest rate and annual maintenance cost, on the cost of water obtained from the proposed fog collection system. The total estimated cost of the fog collection system is SR.556,600 which is obtained from Table 7, without land values, as land cost is insignificant for remote areas.

If one assumes that the life of the system under consideration has a life cycle of 10 years, with 5% interest rate, the overall cost of water for the case will be SR.54.78/m^3; if the system life is further assumed to be 15 years for the same interest rate, the cost will be SR.42.57/m^3 of water produced. It is to be noted that only a 3-month period of the fog is assumed in the analysis. If the government assumes the cost of the storage tank and pipeline network to individual

houses, the unit cost to the consumer will be reduced; users would pay only operational and maintenance costs for the water.

10. Conclusions

The feasibility study of obtaining fresh water from the fog in the Asir region of the Kingdom of Saudi Arabia was successfully conducted. In order to evaluate the extraction of water from fog, the climatic conditions by utilizing the meteorological data relevant to southwestern region of the Kingdom were studied. It was identified that Al-Sooda is the suitable location for fog collection and it has the highest altitude among the other sites. The results show that there is sufficient fog water available in this region to merit the construction of LFCs. It is found that when the ambient temperature is between 5 and 10°C, the water collection rate is high. Similarly, when the relative humidity is higher than about 90%, the collection rate is high and the maximum collection is produced when the relative humidity is close to 100%. There is no clear trend of increasing yield with increasing wind speeds greater than 2 ms^{-1}. By carefully noting the times of rainfall events, fog water collection can be combined with rain water harvesting system to increase water yield during the rainy season.

The collected water meets the World Health Organization's drinking water standards for potable water and the value of pH varies from 6.68 to 7.43. An economic analysis for the proposed method of obtaining fresh water from the fog reveals that if one assumes the life of the system has a life cycle of

10 years, with 5% interest, then the overall cost of water production from the fog is found to be SR 54.78/m^3. However, it is conceivable that special conditions in terms of remoteness of settlement may make the technique an interesting alternative in the Asir region.

Acknowledgments

The authors are grateful for the financial support provided by King Abdulaziz City for Science and Technology (KACST) through Project No. AR-26-25 and the facilities provided by the King Fahd University of Petroleum and Minerals (KFUPM).

REFERENCES

GANDHIDASAN, P., and ABUALHAMAYEL, H. I. (2007), Fog collection as a source of fresh water supply in the Kingdom of Saudi Arabia, Water and Environment Journal, 21, 19-25.

AL-HASSAN, G. A. (2009), Fog water collection evaluation in Asir region—Saudi Arabia, Water Resource Management, 23, 2805-2813.

LOUW, C., van HEERDEN, J., and OLIVIER, J. (1998), The South African fog-water collection experiment: Meteorological features associated with water collection along the eastern escarpment of South Africa, Water SA, 24, 269-280.

AKSAKAL, A. (1998), Rainfall amount in Saudi Arabia and a technique to increase the rainfall by cloud seeding, The Arabian Journal for Science and Engineering, 23, 101-119.

Communication with the Presidency of Meteorology & Environment Protection, National Meteorology & Environment Center, Ministry of Defence & Aviation, Jeddah, Saudi Arabia, March 2008.

OLIVIER, J. (2004), Fog harvesting: An alternative source of water supply on the West Coast of South Africa, GeoJournal, 61, 203-214.

SCHEMENAUER, R. S., CERECEDA, P., and OSSES, P., FogQuest fog water collection manual (2005).

SCHEMENAUER, R. S., FUENZALIDA, H., and CERECEDA, P. (1988), A neglected water resource: The Camanchaca of South America, Bulletin of the American Meteorological Society, 69, 138-147.

World Health Organization: Guidelines for drinking-water quality, Incorporating first addendum, Recommendations, 2006.

HASNAIN, S. M., and ALAJLAN, S. (1998), Coupling of PV-powered R. O. brackish water desalination plant with solar stills, Renewable Energy, 14, 281-286.

(Received November 10, 2010, revised April 27, 2011, accepted May 6, 2011, Published online June 4, 2011)

Pure Appl. Geophys. 169 (2012), 1037–1052
© 2011 Springer Basel AG
DOI 10.1007/s00024-011-0342-y

Ion Composition of Fog Water and Its Relation to Air Pollutants during Winter Fog Events in Nanjing, China

Jun Yang,[1,2] Yu-Jing Xie,[1] Chun-E Shi,[3] Duan-Yang Liu,[1] Sheng-Jie Niu,[1] and Zi-Hua Li[1]

Abstract—Intensive field experiments focused on fog chemistry were carried out in the northern suburb of Nanjing during the winters of 2006 and 2007. Thirty-seven fog water samples were collected in nine fog events. Based on the chemical analysis results of those samples and the simultaneous measurements of air pollution gases and atmospheric aerosols, the chemical characteristics of fog water and their relations with air pollutants during fog evolution were investigated. The results revealed an average total inorganic ionic concentration TIC = 21.18 meq/L, and the top three ion concentrations were those of SO_4^{2-}, NH_4^+ and Ca^{2+} (average concentrations 6.99, 5.95, 3.77 meq/L, respectively). However, the average pH value of fog water was 5.85, which is attributable to neutralization by basic ions (NH_4^+ and Ca^{2+}). The average TIC value of fog water measured in advection–radiation fog was around 2.2 times that in radiation fog, and the most abundant cation was NH_4^+ in advection–radiation fog and Ca^{2+} in radiation fog. In dense fog episodes, the concentration variations of primary inorganic pollution gases showed a "V"-shaped pattern, while those of volatile organic compounds (VOCs) displayed a "Λ"-shaped pattern. The dense fog acted as both the source and sink of atmospheric aerosol particles; fog processes enhanced particle formation, leading to the phenomenon that the aerosol concentration after fog dissipation was higher than that before the fog, and at the same time, mass concentration of PM_{10} reached the lowest value in the late stage of extremely dense fog episodes because of the progressive accumulated effect of wet deposition of large fog droplets. Both air pollution gases and aerosols loading controlled the ion compositions of fog water. The Ca^{2+} in fog water originated from airborne particles, while SO_4^{2-} and NH_4^+ were from both heterogeneous production and soluble particulate species.

Key words: Fog water, ion concentration, pH value, air pollutant, atmospheric aerosol, Nanjing.

[1] Key Laboratory of Atmospheric Physics and Environment, School of Atmospheric Physics, Nanjing University of Information Science and Technology, No. 219 Ningliu Road, Nanjing 210044, China. E-mail: jyang@nuist.edu.cn
[2] Key Laboratory for Cloud Physics and Weather Modification of China Meteorological Administration, Beijing 100081, China.
[3] Anhui Institute of Meteorological Sciences, Hefei 230031, China.

1. Introduction

Fog is an aerosol system consisting of liquid or solid hydrometeor particles that are suspended and slowly deposited in the surface layer of the atmosphere. Fog usually forms in the conditions of stable stratification, weak winds, and shallow mixing layer, which are conducive to the accumulations of the primary air pollutants in the surface layer, resulting in an increase in the concentration of air pollutants. In addition, due to the high humidity, the small size, large specific surface area and long residence time of fog droplets, the fog droplets work as "micro-reactors" and are capable of transforming some primary pollutants into secondary pollutants, e.g., the aqueous-phase oxidation of SO_2 produces sulfate (SEINFELD and PANDIS, 2006). Therefore, fog processes were related frequently with some severe air pollution events, having severe adverse effects on human health, such as the infamous London smog fog in 1952 (WILKINS, 1954).

The environmental, ecological and health effects of a fog process are directly determined by chemical characteristics of fog water as well as changes in the concentrations of pollution gases and aerosols in the fog. Studies on fog water chemistry started in the 1950s, and it has drawn more and more attention since the 1980s. Numerous studies show close relations between chemical characteristics of fog water and atmospheric pollutants. CASS (1979) pointed out that the high concentrations of sulfate aerosols in the Los Angeles region correlated significantly with the occurrence of fogs. WALDMAN *et al.* (1982) analyzed the fog water sampled in Los Angeles and Bakersfield and found that the acidity of fog water and the ion concentrations of sulfate, nitrate and ammonium were very high, and the interaction of moisture and aerosol particles and the scavenging of gaseous nitrate were important processes affecting the chemical characteristics of fog water. MUNGER et al. (1983) argued the close correlation between chemical constituents of fog water and air pollutants. They also put forward the "smog–fog–smog cycle", which means that the high concentration of atmospheric aerosols is favorable to the formation of fogs at night and the next morning, while the increase in the concentration of aerosol particles after fogs may form smog and result in low visibility. JACOB et al. (1986) considered small changes in emission of SO_2 and NO_x that could lead to widespread acid fog. HIROSHI et al. (2006) conducted a 5 year observational study in Gunma of Japan, and pointed out that nitrates and sulfates in fog water were the primary substances responsible for the acidification of the fog water, and 95% of sulfate ions originates from air pollutants.

For the relation between fog duration and fog chemistry, PANDIS et al. (1990a, b) reported an increase of the sulfate levels by a factor of 2 during a 2 h typical radiation fog episode and also observed that the total sulfate concentration (droplet plus interstitial aerosol) started decreasing after the first hour of the fog life. For some coastal locations close to SO_2 sources, the sulfate concentration can increase 3–4 times the preexisting sulfate levels. PANDIS et al. (1992) argued that for the short-life fogs, the lifetime is sufficient to produce a large amount of sulfates, but not enough to remove most aerosol particles from the air through the wet scavenging of fog droplets. However, fog influence on aerosol concentration is different in those long-life fog processes; observations of JACOB et al. (1984) in San Joaquin Valley showed that the accumulation of aerosols in fogs was limited due to the strong wet scavenging of aerosols by fog droplets, resulting in the decline in aerosol concentration.

On the other hand, high aerosol concentration may affect the fog formation and maintenance directly. For example, based on analysis of fog climate in Anhui province in eastern China, SHI et al. (2008) found that fog dissipation time occurred later, average fog duration increased during the last 30 years. They partly ascribed those fog climate changes to the increasing aerosol concentration,

represented by the decreasing visibility therein. In addition, high levels of anthropogenic aerosols, especially a hygroscopic aerosol like sulfate, can change the light scattering in the air, especially under the condition of high relative humidity (CASS, 1979); thus, high levels of aerosols can decrease the visibility to below 1 km even without measurable liquid water content (LWC) in Nanjing, China (YANG et al., 2010). Accordingly, the one-dimensional fog model (PAFOG), which includes the impacts of aerosols, outperformed MM5 in forecasting the fog dissipation time (SHI et al., 2011). Therefore, from the view of improving fog forecasting by numerical model, it is imperative to study the chemical constitution of fog water and its relation with air pollutants. Although the present numerical weather prediction models are complicated and perfect in considering most physical process, and widely used in studying and forecasting almost all kinds of high-impact weather, it is still a big challenge for quantitative fog forecasting (TUDOR, 2010; VAN DER VELDE et al., 2010), which might be due to lack of fog chemistry in the models, especially the impacts of aerosols.

Since the 1980s, numerous field studies with a focus on fog chemistry were performed in many regions of China, including Zhoushan (MO et al., 1989), Lushan (DING et al., 1991), Anning (HUANG et al., 1992), Chongqing (LI and PENG, 1994; LI et al. 1996), southern Fujian (LIU et al., 1996), Shanghai (LI et al., 1999), Xishuanbanna (ZHU et al., 2000), Chengdu (YANAGISAWA et al., 2004), Nanling Mountains (WU et al., 2004), Urumqi (DILNUR and ABLIKIM, 2005). These studies revealed that the collected fog water samples have a wide range of pH values from 2.91 to 9.15 and the total ion concentration (TIC) in fog water appeared an increasing trend with development of cities. LI et al. (1994, 1996) performed field observations of winter fogs in Chongqing region and found that the acidity of fog water showed a declining trend from urban area to suburban one. The measurements of ZHANG et al. (2005) showed temperature inversion, accumulations of SO_2 and NO_2 in the low level over urban Beijing and its peripheral areas before the formation of a persistent dense fog event; condensation grew rapidly with the increase of the concentrations of SO_2 and NO_2 prior to and in the formation stage of fog. On the contrary, the concentrations of SO_2 and NO_2 declined in the dense fog episodes.

As mentioned, the chemical characteristics of fog water vary with locations, fog physics, and are largely impacted by ambient factors. The Yangtze River Delta is one of the areas with the most dense population, most concentrated city agglomerations, and mostly active economy in China. The mean annual number of fog days over this delta region is about 20 days and even more than 60 days in some locations, in which the dense fog (visibility less than 500 m) days account for 29.6%. On average, 43.3% of the total fog events occur in the months of November to January (ZHOU et al., 2007). To reveal the interactions between air pollutants (pollution gases and aerosol particles) and fog water, field observational projects were conducted in Nanjing in the winters of 2006 and 2007. Thirty-seven samples of fog water were collected in nine radiation and advection–radiation fog events, among which continuous samples were collected through the extremely dense stages in six fog events. Continuous measurements of aerosols and pollution gases were obtained synchronically with fog water sampling. In this paper, we analyzed chemical characteristics of fog water in winters in details, and focused on the variation of pollutants during fog lifetime and its relation with the ion compositions of fog water.

2. Field Experiments and Analysis Methods

During November and December of 2006 and 2007, comprehensive field experiments were conducted at a flat lawn on the campus of Nanjing University of Information Science and Technology ($32°12'09''$N, $118°42'25''$E; 22 m above sea level), and this area is unique from several perspectives. It is located in the northern suburb of Nanjing, 1.3 km west to Ningliu highway and 5–8 km southwest of heavy pollution sources such as petrochemical factories, a steel plant, a thermal power station and a nitrogenous fertilizer plant in Jiangbei Industrial Park of Nanjing. The emissions of SO_2, NO_x and NH_3 are 2.0×10^5, 7.5×10^4, and 3.0×10^4 tons/year, respectively (Lu et al., 2010). Table 1 lists the instruments used in the experiments. Fog water was

collected using an Active Fog water Collector (DE-MOZ et al., 1996); the sampling time ranged from 1.5 to 3.0 h for one sample, and the sampling fog water volume was not less than 10 ml. A FM-100 fog droplet spectrometer was used to measure the liquid water content (LWC) and droplet size. Aerosol particles were sampled using a four-stage Anderson Cascade Sampler (WU et al., 2003). The cut offs for each stage were 2.1–1.1 (Stage 1), 1.1–0.65 (Stage 2), 0.65–0.43 (Stage 3), and less than 0.43 μm (Base filter), and hourly PM_{10} mass concentration was measured by a MP101M EX analyzer (ZOLGHADRI and CAZAURANG, 2006). Gas constituents such as SO_2, NO, NO_2 and O_3 were measured using a DOAS System (KIM and KIM, 2001), and VOCs using a ppbRAE Plus Sensor (LAW et al., 2003).

The pH values and electrical conductivities of fog water samples were measured in situ, the fog water was then filtered with organic microporous membranes and stored in a refrigerator at temperature 4°C for chemical analysis. Aerosol samples were extracted from filtration membranes in deionized water using an ultrasonic bath for 30 min, and subsequently shaking for another 1 h, and then the solutions were filtrated with 0.22 μm microporous membranes for anion and cation analysis. NH_4^+, Ca^{2+}, Na^+, Mg^{2+}, K^+, F^-, Cl^-, NO_3^- and SO_4^{2-} in fog water and in the soluble constituents of aerosols were determined using capillary electrophoresis (VALSECCHI and POLESELLO, 1999). Observations, chemical analyses and data quality control were performed following the requirements of acid deposition monitoring network in East Asia (THE SECOND INTERIM SCIENTIFIC ADVISORY GROUP MEETING OF ACID DEPOSITION MONITORING NETWORK IN EAST ASIA, 2000) and the GAW Precipitation Chemistry Program of WMO (WORLD METEOROLOGICAL ORGANIZATION, 2004).

Fog water was sampled in nine fog events (covering seven radiation fogs and two advection–radiation fogs) in the winters of 2006 and 2007, totally acquiring 37 samples of fog water (23 radiation fog samples, 14 advection–radiation fog samples). Table 2 gives the sampling time and pH value of each fog water sample in various fog events. Definitions of "fog", "dense fog" and "extremely dense fog" in Chinese specifications for surface meteorological observation are used in this paper

Table 1

Instrumentation and measurements for the winter fog field observational project in Nanjing

Instrument	Model	Manufacturer	Measurement	Time resolution	Remarks
Active fog water collector	CASCC	Self-made, China	Fog water	/	
Fog droplet spectrometer	FM-100	DMT Inc., USA	Droplet Size, LWC	1 s	2–50 μm, 20 channel
Conductivity meter	DDP-210	Shanghai Kangyi Inc., China	Electrical conductivity of fog water	/	Detection limit:0–2 × 10^4 mS/m
pH Meter	PHS-P	Shanghai Kangyi Inc., China	pH value of fog water	/	
Capillary electrophoresis instrument[a]	PC 800 processing system ultraviolet sensor fused silica capillary	Waters Inc.	Fog water ions	/	Detection limit: Cl^-, SO_4^{2-}, NO_3^-, F^-: 0.4, 0.8, 0.8, 0.12 mg/L; K^+, Ca^{2+}, Na^+, Mg^{2+}: 0.2, 2.0, 1.0, 0.4 mg/L
DOAS system[a]	AR500	OPSIS AB Inc., Sweden	SO_2, NO_2, O_3, NO	1 min	Detection limit: SO_2, NO_2, O_3, NO: 1, 1, 2, 2 μg/m^3; accuracy: ±2% for SO_2, NO_2, O_3, ±3% ~ ±15% for NO
VOC sensor[a]	ppbRAE Plus PGM7240	RAE Inc., USA	VOCs	1 min	Detection limit: 1 ppb
Visibility sensor	ZQZ-DN2	Jiangsu Province Radio Science Institute, China	Visibility	1 min	≤1,000 m: ±10%
Aerosol analyzer[a]	MP101M EX	Environment-S.A. Inc., France	PM_{10} mass concentration	1 h	Detection limit: 0.5 μg/m^3
Particle impactor[a]	FA-3	Liaoyang Kangjie Instrument Institute, China	Aerosol particle sampling	12 h	4 stages

[a] Used in 2007

Table 2

Fog water samples

Fog event	Date	Sampling time	pH	Sample no.
1	December 12, 2006	0440–0800	6.49	1
	December 12, 2006	0720–0820	7.06	2
	December 12, 2006	0820–0950	6.72	3
2	December 14, 2006	0830–1100	6.16	4
3[a]	December 25, 2006	0030–0300	6.63	5
	December 25, 2006	0300–0600	6.90	6
	December 25, 2006	0600–0900	7.27	7
	December 25, 2006	0900–1430	6.73	8
	December 25, 2006	1430–1955	5.08	9
	December 25, 2006	1955–2300	5.34	10
	December 25–26, 2006	2300–0115	5.15	11
	December 26, 2006	0115–0334	5.70	12
	December 26, 2006	0335–0537	5.14	13
	December 26, 2006	0537–1009	4.72	14
	December 26, 2006	1010–1600	4.89	15
	December 26–27, 2006	1601–0059	5.39	16
	December 27, 2006	0100–1550	4.24	17
4	December 14, 2007	0600–0700	6.85	18
	December 14, 2007	0700–0800	6.65	19
	December 14, 2007	0800–0900	6.57	20
	December 14, 2007	0900–1000	6.25	21
5	December 18, 2007	0800–0930	6.81	22
	December 18, 2007	0930–1100	6.65	23
6	December 19, 2007	0200–0300	7.12	24
	December 19, 2007	0300–0500	7.28	25
	December 19, 2007	0500–0700	6.76	26
	December 19, 2007	0700–0900	5.70	27
	December 19, 2007	0900–1300	4.90	28
7	December 19–20, 2007	2300–0030	4.64	29
	December 20, 2007	0030–0300	4.62	30
	December 20, 2007	0300–0430	4.89	31
	December 20, 2007	0430–0600	4.38	32
	December 20, 2007	0600–0700	4.33	33
	December 20, 2007	0700–0800	5.52	34
	December 20, 2007	0800–1000	4.11	35
8[a]	December 21, 2007	1200–1600	6.04	36
9	December 23, 2007	0114–0530	6.88	37

[a] Advection–radiation fog process

(China Meteorological Administration, 2007). By definition, the visibility observed by Vaisala FD12 type visibility sensor has to be below 1,000 m for fog, 500 m for dense fog and 50 m for extremely dense fog.

Because Na^+ in atmospheric aerosols mainly originates from sea salt particles, it is often used as the conservative trace factor for the latter, and the non-sea salt equivalent concentrations for sulfate and calcium ions can be calculated by following expressions (Hara et al., 1995):

$$nss\ SO_4^{2-} = SO_4^{2-} - Na^+ \left(SO_4^{2-}/Na^+\right)_{seawater}$$

$$nss\ Ca^{2+} = Ca^{2+} - Na^+ \left(Ca^{2+}/Na^+\right)_{seawater}$$

where $(SO_4^{2-}/Na^+)_{seawater} = 0.12$ and $(Ca^{2+}/Na^+)_{seawater} = 0.044$ (Emerson and Hedges, 2008) denote mean equivalent concentration ratios of the two species of ions in seawater.

The pH value of fog water is the joint effects of various cations and anions; it depends not only on the effects of acidic ions, e.g., sulfate and nitrate ions, but also on the neutralization of alkaline substances, e.g., ammonium and calcium. Sulfate and nitrate ions are the major acidic ions in clouds, fogs and precipitation in the atmosphere, therefore the unneutralized acidity of fog water can be expressed by pAi, which is semi-quantitatively defined with equivalent concentrations of non-sea salt sulfate and nitrate ions (Hara et al., 1995; Polkowska et al., 2008):

$$pAi = -\log\left(\left[nss\ SO_4^{2-}\right] + \left[NO_3^-\right]\right)$$

3. Chemical Characteristics of Fog Water

3.1. pH and pAi Values and Electrical Conductivity

Table 3 presents arithmetic mean concentrations of primary ions of winter fog water and associated chemical characteristic parameters. It can be seen that the pH values ranged from 4.11 to 7.28, with a mean value of individual samples being 5.85. According to the definition of acid rain (pH < 5.6) and the fact that acid rain with pH value lower than 4.6 will badly imperil forests, crops and material, we grouped the samples based on pH values and obtained 16 samples (43.2%) having pH between 4.6 and 5.6, four samples (10.8%) with pH lower than 4.6 and four samples (10.8%) with pH higher than 7.0. Overall, the average acidity of winter fog water over the all samples was neutral, but the possibility for acidic fog was about four times that for alkaline fog. The ratio of acidic fog water samples to total numbers of fog water samples was 61.5% for the fog events of December 25–27, 2006 (Fog event 3) and 66.7% for the fog event of December 19–20, 2007 (Fog event 7). Both are obviously higher than the mean of 43.2%. Therefore, the longer a fog lasts, the higher possibility of acidic fog.

Table 3

Chemical compositions of winter fog water in Nanjing in 2006 and 2007

Category	Variable unit	pH	pAi	Conductivity (mS/m)	Cl^- (meq/L)	SO_4^{2-} (meq/L)	NO_3^- (meq/L)	F^- (meq/L)	NH_4^+ (meq/L)	K^+ (meq/L)	Ca^{2+} (meq/L)	Na^+ (meq/L)	Mg^{2+} (meq/L)	TIC (meq/L)	nss SO_4^{2-} / SO_4^{2-}	nss Ca^{2+} / Ca^{2+}	Cl^- / Na^+
Radiation fog	Mean	5.97	2.46	47.0	0.88	4.20	0.55	0.40	3.55	0.32	3.85	0.49	0.33	14.58	0.99	1.00	6.75
	Std. dev.	1.05	0.36	27.5	0.65	3.34	0.62	0.26	4.66	0.25	3.78	0.98	0.35	12.51			
	Max	7.28	3.18	140.0	2.60	12.13	2.20	0.96	16.00	0.99	15.68	3.62	1.36	40.78			
	Min	4.11	1.91	14.3	0.19	0.61	0.06	0.05	0.09	0.06	0.46	0.02	0.04	1.59			
Advection–radiation fog	Mean	5.66	2.01	93.7	1.23	11.37	1.60	0.62	9.71	0.48	3.63	2.54	0.37	31.54	0.96	0.96	2.02
	Std. dev.	0.92	0.30	52.3	1.02	8.08	1.56	0.46	7.78	0.42	3.00	2.11	0.43	20.04			
	Max	7.27	2.73	197.0	4.26	32.30	5.36	1.68	29.43	1.79	11.58	6.79	1.50	78.48			
	Min	4.24	1.46	25.1	0.19	1.84	0.24	0.09	0.46	0.10	0.87	0.14	0.03	5.52			
All samples	Mean	5.85	2.29	64.7	1.02	6.99	0.96	0.49	5.95	0.38	3.77	1.28	0.35	21.18	0.98	0.98	4.91
	Std. dev.	1.00	0.40	44.5	0.82	6.59	1.18	0.36	6.69	0.33	3.45	1.81	0.38	17.70			
	Max	7.28	3.18	197.0	4.26	32.30	5.36	1.68	29.43	1.79	15.68	6.79	1.50	78.48			
	Min	4.11	1.46	14.3	0.19	0.61	0.06	0.05	0.09	0.06	0.46	0.02	0.03	1.59			

The values of pAi, without the neutralization effect of alkaline ions, ranged from 1.46 to 3.18, with an average of 2.29, which is obviously lower than 4.0, the mean pAi of rainwater during 2005–2006 (ZHENG et al. 2007). The low pAi indicates a high concentration of acidic constituents in fog water, suggesting that the fog water would be strong acidic if there were not the neutralization of alkaline constituents. Figure 1a is the scatter plot of pH and pAi for all samples. Values of pAi were concentrated in a relatively narrow and small area, suggesting that the pH value of fog water mainly depended on the concentration of alkaline ions in the fog water. DAUM et al. (1984) proposed equivalent concentration ratio FA (fractional acidity) $= [H^+]/([nss\ SO_4^{2-}] + [NO_3^-])$ to represent the ratio of unneutralized H^+ in liquid water, as in Fig. 1b. It can be clearly observed from Fig. 1b that the high pH values resulted from the neutralization effect of alkaline compositions in fog water, while the low pH values reflected the effect of acidic compositions. On average, the FA in winter fog water is two orders smaller than that in rainwater in Nanjing (ZHENG et al., 2007), which is due to the much longer residence time of fog droplets in the surface layer of a polluted atmosphere, and enough time to interact with atmospheric pollutants; therefore, the ratio of unneutralized H^+ in fog water was far lower than that in rainwater.

The electrical conductivity (EC) of fog water ranged from 14.3 to 197.0 mS/m, 8.9 times the mean value of rainwater (ZHENG et al., 2007), thus further suggesting that the TIC in the fog water was far greater than that in rainwater, and the fog water was more polluted than rainwater. It can be seen from the relation between EC and pAi (Fig. 2) that EC decreased with increasing pAi. The high concentrations of acid ions, high value of unneutralized acidity, and high concentrations of alkali neutralization ions in the fog water jointly resulted in the increase in conductivity.

3.2. Ion Composition of Fog Water

Although the mean pH value of winter fog water appeared to be nearly neutral, it did not indicate that the fog water was clean. In fact, the fog water was highly polluted, which can be seen from the high TIC

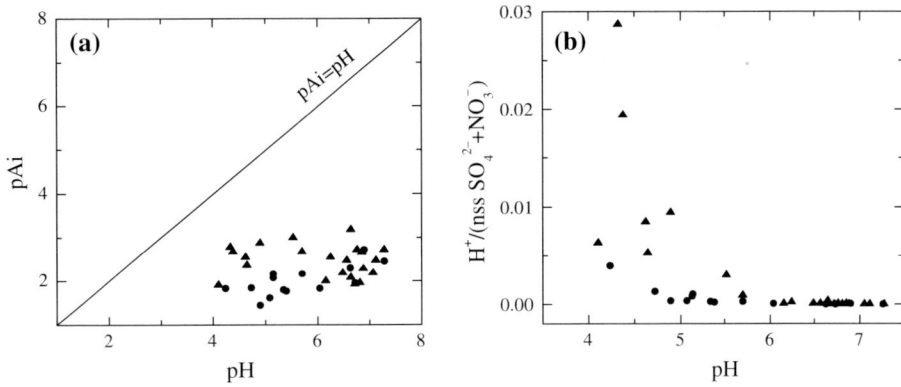

Figure 1
Relations between pH and pAi (**a**), pH and FA (**b**) in winter fog water in Nanjing for 2006 and 2007 (*closed triangles*, radiation fog; *closed circles*, advection–radiation fog)

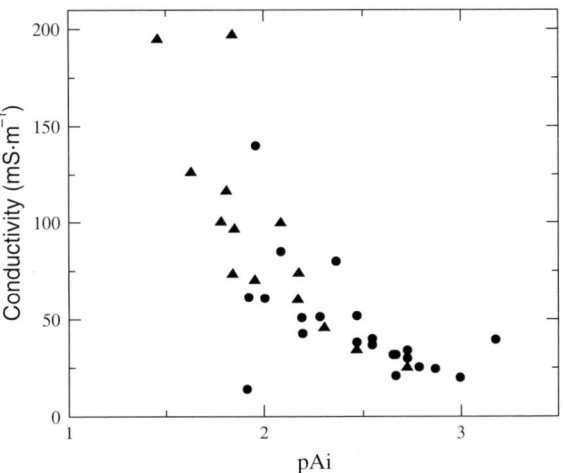

Figure 2
Relations between pAi and electrical conductivity in winter fog water (*closed triangles*, radiation fog; *closed circles,* advection–radiation fog)

values of nine ion species in Table 3. The mean value of TIC was 21.18 meq/L, 18.5 times the mean value (1.15 meq/L) of TIC in rainwater at the same site in 2005 and 2006. The anion and cation concentrations of fog water are 17.0 and 19.8 times those of rainwater, respectively, reflecting the basic characteristic that the ion concentrations in fogs are far greater than those in rain (HOLTON *et al.* 2003). The TIC level in Nanjing is higher in comparison with the ion compositions of fog water measured in most areas over the world. When comparing with the measured TIC in other areas of China, it is about

half of the TIC (43.53 meq/L) observed in the urban area of Shanghai in 1993–1995 (LI *et al.*, 1999), two-thirds of the mean TIC (33.17 meq/L) of fog water over 1994–2000 in the urban and suburban areas of Chongqing, and 5.8 times the mean TIC (3.54 meq/L) in 1999 and 2001 in the clean atmosphere in Dayaoshan of Nanling Mountains (WU *et al.*, 2004).

With respect to the contribution of different ions to the fog water chemistry, SO_4^{2-} absolutely dominated the equivalent concentration of anions, accounting for 74% of the total anion concentration and 33% of the total ion concentration, around 6.0/7.3 times those of Cl^-/NO_3^-, indicating that the acidification of winter fog water in Nanjing was principally due to sulfate. This agrees well with the fact that SO_2 is still the most predominant acidifying pollutant in industrial cities of China, such as Shanghai (LI *et al.*,1999), Chongqing (LI and PENG, 1994), and Anning (HUANG *et al.*, 1992), as well as in most areas of the world (KIM *et al.*, 2006). However, in Los Angeles and the northeast coastal region of the USA, where traffic was heavy, concentration of NO_3^- ions in fog water was more than six times that of SO_4^{2-} due to the effect of a large amount of nitrogen oxides emission (HOLTON *et al.* 2003). The nss SO_4^{2-}/SO_4^{2-} ratio was as high as 98% (the minimum value of the ratio was 90%), which indicates that the SO_4^{2-} concentration in fog water there was mainly determined by anthropogenic emission of sulfur.

The predominant cations in winter fog water in Nanjing were NH_4^+ and Ca^{2+}, agreeing well with the basic character of high concentration of $(NH_4^+ + Ca^{2+})$ in acid rain in China (DING et al., 1997). The equivalent concentration of NH_4^+ in winter fog water was 1.6 times that of Ca^{2+}; they accounted for 51 and 32% of the total cations, and 28 and 18% of the total ions, respectively. The high concentrations of NH_4^+ and Ca^{2+} jointly resulted in the high value of pH in the fog water.

Although Ca^{2+} concentration ranked second place in the list of cation concentrations, the equivalent concentration of Ca^{2+} in 47% of all fog water samples was greater than that of NH_4^+. Since nss Ca^{2+}/Ca^{2+} was about 1, the Ca^{2+} in fog water in Nanjing mainly originated from continental aerosol particles, including the construction dusts. It is noted that the Ca^{2+} concentration was several times observed to be obviously greater than that of NH_4^+ some places of China. For example, the equivalent concentration ratio of Ca^{2+} to NH_4^+ was 20.8 at the Dayaoshan site in 2001 (WU et al., 2004), and 8.1 in the urban area of Shanghai in 1993–1995 and 2.2 in Chongqing during 1984–1990 (LI et al., 1999). Although the NH_4^+ concentration of fog water were observed to be similarly greater than that of Ca^{2+} in the USA (COLLETT et al., 2002), Germany (WRZESINSKY and KLEMM, 2000), and Japan (MINAMI and ISHIZAKA, 1996), yet the content of Ca^{2+} in Nanjing winter fog water is far greater than those observed

no matter whether its absolute value or relative content.

3.3. Comparison of Chemical Characteristics between Radiation and Advection–Radiation Fogs

To clearly show the differences in the ionic compositions between radiation and advection–radiation fogs, Fig. 3 presents the arithmetic mean equivalent concentrations of various cations and anions. Except Ca^{2+}, the concentrations of the other eight ions in the fog water of advection–radiation fogs were all greater than those of radiation fogs. The average TIC was 31.54 meq/L for advection–radiation fogs and 14.58 meq/L for radiation fogs, and the former was 2.2 times the latter. Among them the concentration of Na^+, as a trace composition of sea salt, in the fog water of advection–radiation fogs was 5.2 times that of radiation fogs, thus showing a largest difference; the mean ion concentrations of SO_4^{2-} and NH_4^+, as predominant anion and cation, in advection–radiation fog water were both 2.7 times those in radiation fog water.

As for the sequence of anion concentrations, the concentrations of Cl^- and NO_3^- ranked second and third places, respectively, in radiation fog water, but changed the order in advection–radiation fog water. The average ratio of SO_4^{2-} to NO_3^- were 12.5 and 9.6 in radiation fog and advection–radiation fog,

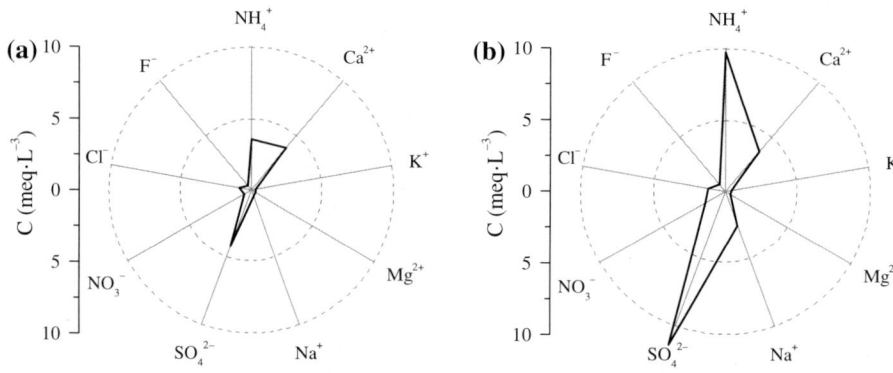

Figure 3
Average ionic equivalent concentrations (C; meq/L) in winter fog water (**a** radiation fog; **b** advection–radiation fog) in Nanjing for 2006 and 2007

respectively, indicating the effect of SO_4^{2-} on acidification was more obvious in radiation fog water than in advection–radiation fog water in this region. The cation concentration in descending sort order was successively $Ca^{2+} > NH_4^+ > Na^+ > Mg^{2+} > K^+$ in radiation fog, but $NH_4^+ > Ca^{2+} > Na^+ > K^+ > Mg^{2+}$ in advection–radiation fog. The size order of NH_4^+ and Ca^{2+} concentrations (cations of the first two largest content) was opposite in the two types of fog, and the concentration of Na^+ took the third place in the lists of cation concentrations in both radiation fog and advection–radiation fog. On the view of the influence of marine aerosols, ratios of nss SO_4^{2-}/SO_4^{2-} and nss Ca^{2+}/Ca^{2+} in advection–radiation fog water were both smaller than those in radiation fog water, indicating that the ionic composition of advection–radiation fog was a little more affected by the transport of marine aerosols than that of radiation fog. These differences led to the general lower average values of pAi and pH in advection–radiation fogs than those in radiation fogs; however, the lowest value of pH was observed in radiation fog. Because the TIC in advection–radiation fog water was obviously higher than that in radiation fog water, the EC of advection–radiation fog water was 2.0 times that of radiation fog water.

The aforementioned differences in the chemical characteristics of the two types of fog were related with the formation and development processes of the two types of fog. In comparison with advection–radiation fog, the chemical characteristics of radiation fog were more impacted by the property of local air pollutants, while those of advection–radiation fog were largely affected by the long range transport of air pollutants, besides the local pollutants. To illustrate the different transport patterns of the two types of fog, the HYSPLIT model (HYSPLIT_4, http://www.ready.noaa.gov/ready/hysplit4.html/) was used to show the 24 h back trajectory of air mass at 100 m height over the sampling site (Fig. 4). It can be seen that the back-trajectories are different between typical advection–radiation (Events 3 and 8) and radiation fog (Events 1, 4, 5 and 6), both in the direction and in the length. The 24 h back-trajectories of the advection–radiation fog events are obviously longer than those of the radiation fog events. The controlling air masses over the sampling site during

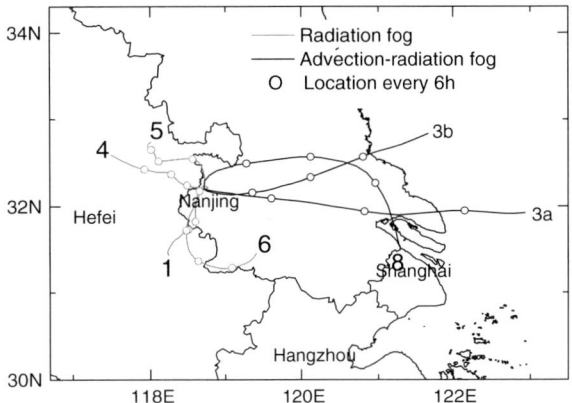

Figure 4
24 h back-trajectories ending at 0800 BST in typical fog events (Fog event numbers are shown in Table 2; *3a*: December 25, 2006, *3b*: December 26, 2006)

the two advection–radiation fog events initially originated from the east coast area, passing through the economically developed area in the low reach of the Yangtze River, carrying and continuously transporting pollutants from the upstream areas to the sampling site. Therefore, the ion concentrations in the advection–radiation fog water were higher than those in the radiation fog water at the same site. Similarly, due to the marine origin of the air masses, the mean concentration of Na^+ in the advection–radiation fog water was 5.2 times that in the radiation fog water (Table 3).

4. Relations between Air Pollution and Chemical Characteristics of Fog Water

4.1. Variations in Pollution Gases and Chemical Compositions of Fog Water during Fog Events

As the predominant anion in our observations, the production and concentration variation of SO_4^{2-} are closely related with the presence and variation of SO_2. Table 4 lists the Pearson correlation coefficients between pollution gases concentrations and visibilities in the field experiment of 2007 to reveal the variation of atmospheric pollution gases in the development and evolution processes of fogs. It can be seen that the positive correlation between SO_2 and visibility in the fourth, fifth, sixth and eighth fog events was significant passing the confidence level of

Table 4

Correlation coefficient r between visibility and gaseous concentration and its significance level α in fog events in December 2007

Fog event	SO₂		O₃		NO₂		NO		VOCs	
	r	α	r	α	r	α	r	α	r	α
4	0.66	0.00	−0.15	0.17	0.23	0.05	−0.22	0.19	−0.46	0.00
5	0.55	0.00	0.16	0.20	−0.11	0.37	−[a]	−	−0.39	0.00
6	0.59	0.00	0.45	0.00	0.44	0.00	−	−	−0.32	0.00
7	−0.03	0.74	−0.33	0.00	−0.34	0.00	−	−	−	−
8	0.53	0.00	0.52	0.00	0.57	0.00	−	−	−	−
9	−0.32	0.11	−0.12	0.56	−0.16	0.44	−	−	−	−

[a] Sample number whose concentration reached the test limit was not enough

0.01, thus suggesting that the SO_2 concentration in these fog events declined with intensifying fog and then gradually increased again with the dissipation of fog. This anti-correlation is the most evident in the fourth fog event (Fig. 5). In this case, the visibility started to decline rapidly from 1,957 m at 1534 BST December 13, then declined slowly and fluctuated with a high frequency from 1730 BST December 13 to 0548 BST December 14 when the fog developed into an extremely dense stage. The fog started to dissipate after 1000 BST December 14, meanwhile

visibility increased rapidly, subsequently increased slowly after 1130 BST, to 1,848 m at 1847 BST. During this fog event, the concentration of SO_2 declined from the initial value of 160–170 µg/m³, synchronically with the development of fog after its onset at 1830 BST. During the extremely dense fog episode (visibility <50 m) from 0548 to 1006 BST, the DOAS system failed to analyze the gas concentrations at most times due to lack of sufficient light signals under the extinction effect of droplets and aerosols; however, the concentration of SO_2 was still

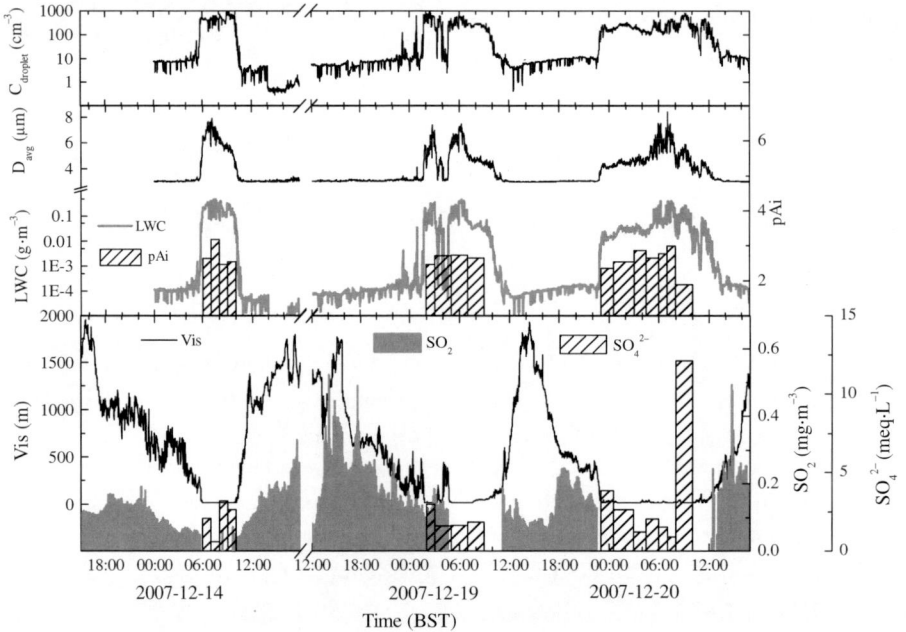

Figure 5
Temporal variations of LWC, number concentration ($C_{droplet}$), average diameter (D_{avg}), pAi, SO_2, SO_4^{2-} and visibility (Vis) in the fourth, sixth and seventh fog events

measured to decline from 50.1 μg/m³ at 0548 BST to 29.5 μg/m³ at 0932 BST. Afterwards it increased gradually with rapid increase in visibility, reaching a value higher than 270 μg/m³ after 1707 BST.

Atmospheric SO_2 mainly originates from the burning of fossil fuel. In the super dense fog events, the presence of water droplets and hydroscopic aerosols provides the condition for the conversion of SO_2 to SO_4^{2-} in aqueous-phase oxidation reaction. Figure 5 also presents variations of SO_4^{2-} concentrations in the super dense fog stages of the fourth, sixth and seventh fog events, which show a variation of "V" pattern: the SO_4^{2-} concentration usually declined with time in the early stage of super dense fog and increased in the later stage. Comparing LWC and average droplet diameter of fog in the fourth and seventh fog events (Fig. 5), in which the sample series of fog water in the super dense fog stage were sampled relatively complete, it can be seen that the reduction of SO_4^{2-} concentration in the early super dense stage corresponded to the increase process of LWC, while the lowest SO_4^{2-} concentration matched with the highest LWC. Subsequently, the reduction of LWC and average diameter, accompanied by the maintenance of a relatively high number concentration of fog droplets, corresponded to the increase in SO_4^{2-} concentration. MUNGER et al. (1983) argued that the reduction of SO_4^{2-} concentration with time in the early stage of super dense fog was caused by the dilution resulted from fog droplet growth. In this stage, all the total surface area, total liquid water volume, and total LWC increased gradually; therefore, the gravitational depositional rate of droplets also increased. Dilution by droplet growth could take place without any appreciable change in LWC if the sedimentation rate was high enough to balance the condensation rate. However, in the late stage of super dense fog, the reduction in average diameter, accompanied by high number concentrations of fog droplets, indicates the evaporation of droplets, the decline in depositional rate, and the increase in number concentration. In our study case, the predominant anion in fog water was SO_4^{2-}, its change manifested in a variation of "V" pattern of acidic ion concentration, and correspondingly the pAi exhibited a variation of weak "Λ" pattern in the super dense fog stages (Fig. 5)

NO_2 concentration showed a positive correlation with visibility (Table 4), suggesting that it had a variation pattern similar with that of SO_2 in dense fog events, i.e., the NO_2 concentration was low when the fog was dense. For the similar reason in discussing the correlation between SO_2 and visibility, the relation between NO_2 and visibility in the seventh and ninth fog processes differed to some extent from those in the other processes; however, the variation of "V" pattern of NO_2 concentration still can be identified in the super dense fog stage (Fig. 6). Under the ambient condition of dense fog, NO_2 solved in fog water to produce HNO_3. Figure 6 also presents the "V" pattern variation of NO_3^- equivalent concentration in the super dense fog stage, which is similar to that of SO_4^{2-}.

VOCs concentration was synchronically measured in the fourth, fifth and sixth fog events (Table 4), in which the measurements were relatively continuous in the fourth and fifth fog events. Table 4 indicates that the VOCs concentration was

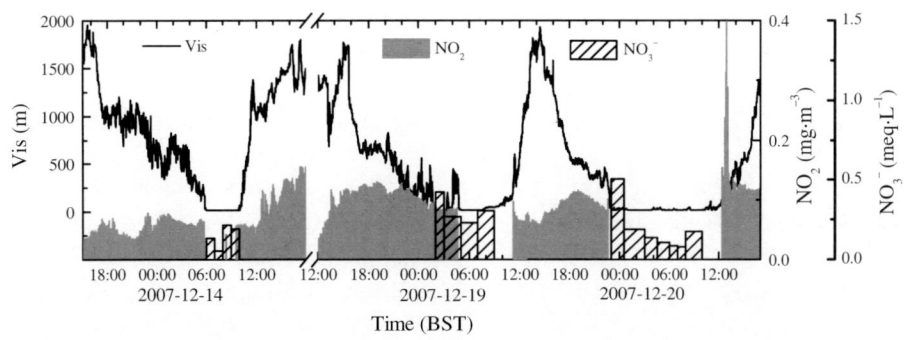

Figure 6
Temporal variations of NO_2, NO_3^- and visibility (Vis) in the fourth, sixth and seventh fog events

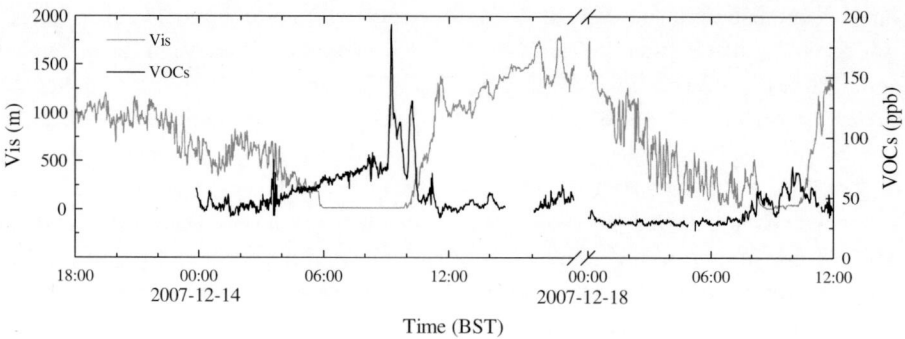

Figure 7
Temporal variations of VOCs and visibility (Vis) in the fourth and fifth fog events

anti-correlated with visibility significantly. Figure 7 presents the time-variations of VOCs and visibility during the fourth and fifth fog events. The VOCs concentration increased persistently from dense fog stage to the late super dense fog stage, and when the fog dissipated it returned to the normal level (the average VOCs concentration from December 7 to 19 was 52.5 ppb). This change pattern might be due to the strongly stable atmospheric stratification, under which the diffusion of pollutants was depressed. In addition, both the observed VOCs peaks occurred around 9 o'clock in morning in the two fog events, in accordance with the peak of human activities, indicating that the high VOCs concentration then was also related with automobile emissions. In that respect, solar radiation was weakened by dense fog; the photochemical reaction

consuming VOCs could not take place effectively. Therefore, VOCs were accumulated in the surface layer. With the dissipation of fog, the reinforcing photochemical reaction with NO_x consumed VOCs, leading to decrease in its concentration. For the above three reasons, the VOCs and visibility in the fog events showed a significantly negative correlation (Table 4).

Absolute concentration levels of NO, NO_2, O_3 and VOCs can be cross-linked with each other through gas/aqueous oxidations and photochemical reactions during fog life cycles. Figure 8 shows the variations of NO and O_3 concentrations in the fourth fog event. Under the conditions of no illumination from midnight of December 13 to sunrise (0658 BST December 14), O_3 concentration declined due to the consumption of NO and reached the minimum before the onset of super dense fog. During this period, the increase rate of NO concentration exceeded its dissipation rate, showing an increasing trend; and the increase rate of NO_2 concentration produced by the oxidation of NO was equivalent to the consuming rate by dissolving into fog water, therefore the NO_2 concentration was observed to be stable and slightly declined (Fig. 6). After the super dense fog stage, the fog dissipated rapidly, VOCs consumed NO in the photochemical oxidation reaction, thus reducing the NO concentration obviously, meanwhile increasing the NO_2 concentration, as the production of the oxidation reaction. Immediately after the stage, O_3 concentration rapidly increased due to the photolysis of NO_2 under the condition of increased illumination, and reached its maximum in the afternoon of December 14.

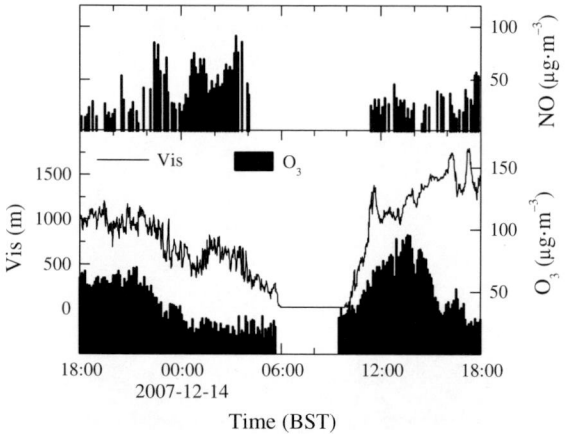

Figure 8
Temporal variations of NO, O_3 and visibility (Vis) in the fourth fog event

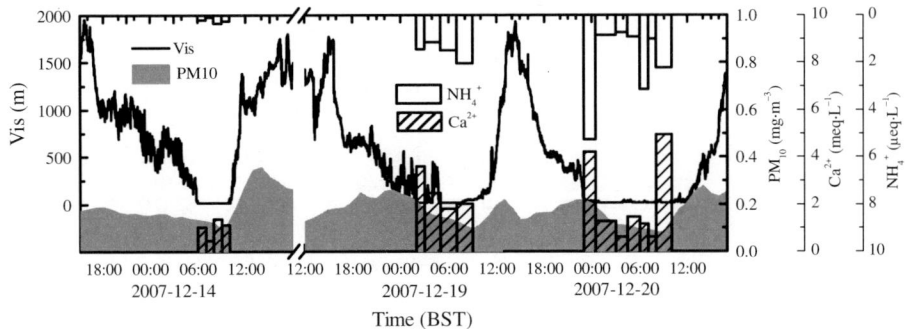

Figure 9

Temporal variations of PM_{10} and visibility (Vis) in the fourth, sixth and seventh fog events

4.2. Change in Aerosol Concentration and Its Association with Chemistry of Fog Water

Within the fog layer, the aerosol particles are reduced by nucleation growth to fog droplet and collection by fog droplets, and thus changes the chemical properties of fog water. Figure 9 displays the temporal evolution of PM_{10} mass concentration in the fourth, sixth and seventh fog events. The PM_{10} concentration declined at first and then increased with its minimum value occurring at the late stage of super dense fog, showing a "V" pattern in its variation. Figure 9 also shows increase trends in the PM_{10} mass concentration when the visibility declined from 2,000 to 200 m in the development stage of fog of the sixth and seventh fog events. In this stage, the gradually enhancing of boundary layer stability made the aerosols accumulate in the surface layer, thus the PM_{10} mass concentration exhibited an increasing trend. However, with the increases in droplet size and LWC (Fig. 5), the wet scavenging of aerosols was enhanced by the processes of nucleation, collisions, and adhesion of fog droplets. Those aerosol particles entering fog droplets were removed from the atmosphere to the ground surface by deposition of large fog droplets. It can also be seen from Fig. 9 that the beginning of the decline in PM_{10} mass concentration, and rapid increases in the mean size of droplets and fog water content were synchronic. In addition, based on the data of the six fog events of December 2007, we calculated the correlation coefficients between PM_{10} mass concentration and the number concentration of fog droplets, LWC, and average diameter of fog droplets, which are -0.42, -0.39 and -0.50

($\alpha < 5.0 \times 10^{-5}$), respectively, suggesting that the wet scavenging of aerosols by fog droplets determined the PM_{10} mass concentration in fog. The persistent wet scavenging effect led to the minimum of PM_{10} mass concentration at the late stage of super dense fog. Subsequently, with reduction in fog water content and the mean size of fog droplets, the PM_{10} mass concentration rose again.

Figure 9 also shows that the PM_{10} concentrations in the dissipation stage of fogs were higher than those in the development stages in the fourth and seventh fog events. This can be ascribed to two factors. On one hand, after dissolution of gaseous pollutants, such as NH_3 into liquid water, some aqueous-phase reactions increased the content of some constituents in fog droplets, such as $(NH_4)_2SO_4$, which can be illustrated by the increasing trend of NH_4^+ concentration in fog water in the mid-late stage of super dense fog. The Yangtze River Delta is a part of the high anthropogenic ammonia emission region (Dong et al., 2010), mainly from application of nitrogenous fertilizer and animal husbandry, for 49.3 and 44.1% of total anthropogenic ammonia emissions, respectively (Dong et al., 2009). During the period from the end of super dense fog to the dissipation stage of the fog, the evaporation of fog droplets released non-volatile constituents, such as $(NH_4)_2SO_4$, leading to the PM_{10} concentrations in the dissipation stage higher than the one before the fog (Pandis et al., 1992). On the other hand, during the dissipation stage, both the wet scavenging of aerosols by fog droplets and stability in the surface layer concurrently weakened, thus favoring to the recovery of the number concentration of atmospheric aerosol particles (PANDIS *et al.*, 1990a). Visibility increased only for a short period between the sixth and seventh fog events, and the maximum visibility was 1,900 m, lower than the visibility before the sixth fog event. The aerosol concentration in the dissipation stage of the sixth fog process did not exceed that in the development stage because the time between two fog events is too short to increase the number concentration of atmospheric aerosol to a high level.

After the dissolution of aerosols into fog water, its soluble constituents changed the chemical properties of the latter. Like SO_4^{2-} and NH_4^+, the trend of Ca^{2+} content in the fog water also exhibited a "V" pattern variation in the super dense fog stage (Fig. 9). When the PM_{10} concentrations reached minima, the ion concentrations of Ca^{2+} in fog water were larger than minima, even reached a peak. This indicated that Ca^{2+} in fog water mainly originated from atmospheric aerosol particles. Major sources of calcium in aerosols were soil dust and construction dust, especially dust particles associated with cement. On one hand, some aerosols with high calcium acted as condensation nuclei, producing the initial high concentration Ca^{2+} in fog water in the formation process of fog, and on the other hand, after fog formation the aerosols captured by fog droplets increased Ca^{2+} in fog water. However, with the progressive accumulation of wet scavenging effect in the development process of fog, the aerosol particles that originated from construction dust and soil dust persistently declined, and the relative content of Ca^{2+} in fog water was also reduced, especially in a long super dense fog stage, for example, fog event 7.

To further reveal the relations between aerosol constituents and ion compositions of fog water,

Table 5 presents soluble constituents of atmospheric aerosol particles less than 2.1 μm, ion mass concentrations of fog water in the six fog events in December 2007, and the correlation analysis results. As a whole, the relative contents of the soluble constituents of all spectral ranges were highly correlated with ion compositions of fog water, suggesting again that chemical constituents of atmospheric aerosols determined largely the ion concentrations of fog water. Among those, the effects of aerosol particles within the size section of 0.65–1.1 μm were most significant, and the effects of particles less than 0.43 μm relatively declined. This is because both the effect of aerosols as condensation nuclei and their collision effect with fog droplets reduce with decrease in particle size.

5. Conclusions

Observations of fog chemistry were carried out in Nanjing, China during the winters of 2006 and 2007, dedicated to describe the ion composition of fog water and its relations to air pollutants during fog life cycles.

Averaged pH value of winter fog water in Nanjing was 5.85 based on 37 samples and 43.2% of them are acidic (pH < 5.6). Extended the lifetimes of fog episodes tends to the formation of acidic fogs. The fog water was highly contaminated, and the averaged TIC was 21.18 meq/L. The predominant ions were SO_4^{2-}, NH_4^+ and Ca^{2+}, and the acidification of fog water was mainly due to SO_4^{2-}, especially for the radiation fog water. Sufficient neutralization of Ca^{2+} is the reason responsible for the high pH of fog water. The TICs in advection–radiation fogs were higher

Table 5

Ion mass concentrations (mg/L) of fog water, contents (μg/m³) of soluble constituents of aerosols by range of particle diameter (D; μm) and correlation coefficients (r) and confidence level (α) of the relative contents of their ion compositions (μg/m³)

D	F⁻	Cl⁻	SO_4^{2-}	NO_3^-	Ca^{2+}	K⁺	Mg^{2+}	Na⁺	r	α
<2.1	0.05	2.47	19.26	1.28	5.46	1.16	3.02	3.28	0.9626	1.27×10^{-4}
2.1–1.1	0.05	2.08	19.29	1.03	4.83	0.84	2.69	2.80	0.9629	1.24×10^{-4}
1.1–0.65	0.09	3.56	22.78	2.18	6.08	1.90	3.44	3.93	0.9653	1.01×10^{-4}
0.65–0.43	0.05	2.46	18.67	1.45	5.63	1.25	3.04	3.32	0.9635	1.18×10^{-4}
<0.43	0.03	1.77	16.30	0.46	5.30	0.64	2.92	3.06	0.9519	2.68×10^{-4}
Fog water	8.13	34.08	190.93	26.49	71.29	15.17	3.82	5.57		

than those in radiation fogs. The top one cation was NH_4^+ in advection–radiation fogs and Ca^{2+} in radiation fogs.

In dense fog processes, the principal atmospheric inorganic pollution gases showed a variation trend of "V" pattern due to aqueous-phase reaction, dilution resulted from increase in LWC and the enhancement of deposition of fog droplets. On the contrary, VOCs concentration exhibited a trend of "Λ" pattern, which was determined by the natural production, anthropogenic emission and the photochemical depletion

Dense fog on one hand was the source of aerosol particles, because the gaseous-phase and aqueous-phase chemical reactions might help to produce new aerosol particles and aerosol particles might be released to the air when the fog droplets evaporated in the dissipation stage of fog. On the other hand, dense fog was also the sink of aerosols because the wet scavenging effect of fog droplets removed aerosols out of the air.

Pollution gases and aerosol particles in the ambient atmosphere jointly determined the chemical compositions of fog water. The Ca^{2+} originated from atmospheric aerosol particles, while SO_4^{2-} and NH_4^+ came on one hand from gaseous-phase and aqueous-phase reactions, and on the other hand from atmospheric aerosol particles, especially from those whose size was within the size section of 0.65–1.1 μm.

Acknowledgments

We wish to thank Bin Zhu, Guo-Zheng Zhang, Shu-Xian Fan, Jun-Lin An, Li-Li Tang, Wei-Wei Wang, Jing Wang, and Jia-De Yan for their help in the field observations and chemical analyses of fog water and aerosol samples. This material is based on work supported jointly by the Key Project of Natural Science Foundation of Jiangsu Province under Grant No. BK2007727, the Special Fund of Industrial (Meteorology) Research for Public Welfare of China under Grant No. GYHY(QX)200906012 and GYHY(QX)200706026, the National Natural Science Foundation of China under Grant No.40775010, and the Key Basic Research Project of Natural Science of Jiangsu Provincial Colleges and Universities under Grant No. 06KJA17021.

REFERENCES

CASS, G.R. (1979), *On the relationship between sulfate air quality and visibility with examples in Los Angeles*, Atmos. Environ. *13*, 1069–1084.

CHINA METEOROLOGICAL ADMINISTRATION (2007), *Specifications for surface meteorological observation*, Part 4: Observation of weather phenomenon, QX/T48-2007, Beijing, 10 pp. (in Chinese).

COLLETT, J.L., BATOR, A., SHERMAN, D.E., MOORE, K.F. and et al., (2002), *The chemical composition of fogs and intercepted clouds in the United States*, Atmos. Res. *64*, 29–40.

DAUM, P.H., KELLY, T.J., SCHWARTZ, S.E. and NEWMAN, L. (1984), *Measurements of the chemical composition of stratiform clouds*, Atmos. Environ. *18*, 2671–2684.

DEMOZ, B.B., COLLETT, J.L. and DAUBE, B.C. (1996), *On the Caltech active strand cloudwater collectors*, Atmos. Res. *41*, 47–62.

DILNUR, T. and ABLIKIM, A. (2005), *Chemical composition of fog at Nanshan in the Urumqi and its influence on environment*, Urban Environ. & Urban Ecol. *18*, 7–9. (in Chinese).

DING, G.A., JI, X.M., FANG, X.M. and et al. (1991), *Characters of cloud-fog water in Lushan Mountain*, Acta Meteor. Sinica *49*, 190–197. (in Chinese).

DING, G.A., XU, X.B. and FANG, X.M. and et al. (1997), *Current status and future of acid rain in China*, Chinese Sci. Bull. *42*, 2076–2081.

DONG, Y.Q., CHEN, C.H., HUANG, C., WANG, H.L., LI, L., DAI, P. and JIA, J.-H. (2009), *Anthropogenic emissions and distribution of ammonia over the Yangtze River Delta*, Acta Sci. Circumst. *29*, 1611–1617. (in Chinese).

DONG, W.X., XING, J. and WANG, S.X. (2010), *Temporal and spatial distribution of anthropogenic ammonia emissions in China:1994-2006*, Environ. Sci. *31*, 1457–1463. (in Chinese).

EMERSON, S. and HEDGES, J. D. 2008: *Chemical oceanography and the marine carbon cycle*. Cambridge University Press, 13.

HARA, H., KITAMURA, M., MORI, A. and et al. (1995), *Precipitation chemistry in Japan 1989–1993*, Water, Air, & Soil Pollut. *85*, 2307–2312.

HIROSHI, T., HIROKAZU, K., KUNIHISA, K. and et al. (2006), *Long-term observation of fogwater composition at two mountainous sites in Gunma prefecture, Japan*, Water, Air, & Soil Pollut. *175*, 375–391.

HOLTON, J. R., CURRY, J.A. and PYLE, J.A. (2003), *Encyclopedia of Atmospheric Sciences*, (Academic Press, Boston).

HUANG, Y.S., GUO, H.G., LIU, F.X. and et al. (1992), *The chemical composition of radiative fog-water in industrial and non-industrial districts*, Acta Geogr. Sinica *47*, 66–73. (in Chinese).

JACOB, D.J., WALDMAN, J.M., MUNGER, J.W. and HOFFMANN, M.R. (1984), *A field investigation of physical and chemical mechanisms affecting pollutant concentrations in fog droplets*, Tellus B *36*, 272–285.

JACOB, D. J., MUNGER, J. W., WALDMAN, J. M. and HOFFMANN, M. R. (1986), *H_2SO_4–HNO_3–NH_3 system at high humidities and in fogs, Part 1, Spatial and temporal patterns in the San Joaquin Valley of California*, J. Geophys. Res. D *91*, 1073–1088.

KIM, K.H. and KIM, M.Y. (2001), *Comparison of an open path differential optical absorption spectroscopy system and a conventional in situ monitoring system on the basis of long-term measurements of SO_2, NO_2, and O_3*, Atmos. Environ. *35*, 4059–4072.

Reprinted from the journal

KIM, M.G., LEE, B.K. and KIM, H.J. (2006), *Cloud/fog water chemistry at a high elevation site in south Korea*, J. Atmos. Chem. *55*, 13–29.

LAW, T.S.C., CHAO, C., CHAN, G.Y.W. and LAW, A.K.Y. (2003), *Confined catalytic oxidation of volatile organic compounds by transition metal containing zeolites and ionizer*, Atmos. Environ. *37*, 5433–5437.

LI, Z.H. and PENG, Z.G. (1994), *Physical and chemical characteristics of the Chongqing winter fog*, Acta Meteor. Scinica *52*, 477–483. (in Chinese).

LI, Z.H., DONG, S.N. and PENG, Z.G. (1996), *Spatial/temporal variability of chemical constituents of fog water sampled in Chongqing*, J. Nanjing Inst. Meteor. *19*, 63–68. (in Chinese).

LI, D., CHEN, M.H. and SHAO, D.M. (1999), *Study on air pollution and composition of fog in Shanghai*, Shanghai Environ. Sci. *18*, 117–120. (in Chinese).

LIU, H.J., WANG, W., GAO, J.H. and et al. (1996), *Preliminary study on the characteristics of acid fog in Minnan area*, Res. Environ. Sci. *9*, 30–32. (in Chinese).

LU, C.S., NIU, S.J., TANG, L.L., LV, J.J., ZHAO, L.J. and ZHU, B. (2010), *Chemical composition of fog water in Nanjing area of China and its related fog microphysics*, Atmos. Res. *97*, 47–69.

MINAMI, Y. and ISHIZAKA, Y. (1996), *Evaluation of chemical composition in fog water near the summit of a high mountain in Japan*, Atmos. Environ. *30*, 3363–3376.

MO. T.L., XU. S.Z. and CHEN. F. (1989), *The Acidity and chemical composition of fog–water over Zhoushan region*, Shanghai Environ. Sci. *8*, 22–26. (in Chinese).

MUNGER, W.J., JACOB, D.J., WALDMAN, J.M. and HOFFMANN, M.R. (1983), *Fogwater chemistry in an urban atmosphere*, J. Geophys. Res. C *88*, 5109–5121.

PANDIS, S.N., PILINIS, C. and SEINFELD, J.H. (1990a), *The smog–fog–smog cycle and acid deposition*, J. Geophys. Res. D *95*, 8489–8500.

PANDIS, S.N., SEINFELD, J.H. and PILINIS, C. (1990b), *Chemical composition differences in fog and cloud droplets of different sizes*, Atmos. Environ. *24*, 1957–1969.

PANDIS, S.N., SEINFELD, J.H. and PILINIS, C. (1992), *Heterogeneous sulfate production in an urban fog*, Atmos. Environ. *26*, 2509–2522.

POLKOWSKA, Z., BLAS, M., KLIMASZEWSKA, K. and et al. (2008), *Chemical characterization of dew water collected in different geographic regions of Poland*, Sensors *8*, 4006–4032.

SEINFELD, J.H. and PANDIS, S.N. (2006), *Atmospheric chemistry and physics: from air pollution to climate change*, (John Wiley and Sons, Hoboken).

SHI, C.E., ROTH, M., ZHANG, H. and et al. (2008), *Impacts of urbanization on long-term fog variation in Anhui Province, China*, Atmos. Environ. *42*, 8484–8492.

SHI, C.E., WANG, L., ZHANG, H. and et al. (2011), *Fog simulations based on multi-model: a feasibility study*, Pure Appl. Geophys. *168*, this issue.

THE SECOND INTERIM SCIENTIFIC ADVISORY GROUP MEETING OF ACID DEPOSITION MONITORING NETWORK IN EAST ASIA (2000), *Technical manual for wet deposition monitoring in East Asia. Niigata-shi, Japan*: EANET.

TUDOR, M. (2010), *Impact of horizontal diffusion, radiation and cloudiness parameterization schemes on fog forecasting in valleys*, Meteor. Atmos. Phys. *108*, 57–70.

VALSECCHI, S.M. and POLESELLO, S. (1999), *Analysis of inorganic species in environmental samples by capillary electrophoresis*, J. Chromatogr. A *834*, 363–385.

VAN DER VELDE, I.R., STEENEVELD, G.J., WICHERS SCHREUR, B.G.J. and et al. (2010), *Modeling and forecasting the onset and duration of severe radiation fog under frost conditions*, Mon. Wea. Rev. *138*, 4237–4253.

WALDMAN, J.M., MUNGER, J.W., JACOB, D.J. and et al. (1982), *Chemical composition of acid fog*, Science *218*, 677–680.

WILKINS, E.T. (1954), *Air pollution aspects of the London fog of December 1952*, Quart. J. Roy. Meteor. Soc. *80*, 267–271.

WORLD METEOROLOGICAL ORGANIZATION, (2004), *Manual for the GAW precipitation chemistry programme. Guidelines, data quality objectives and standard operating procedures*, Global Atmosphere Watch, no. 160, (ALLAN, M.A. Ed.).

WRZESINSKY, T. and KLEMM, O. (2000) *Summertime fog chemistry at a mountainous site in central Europe*, Atmos. Environ. *34*, 1487–1496.

WU, S.P., CAO, J., LI, B.G. and et al. (2003), *Residues and distribution of organochlorine pesticides in airborne particles of different sizes from urban areas*, Res. Environ. Sci. *16*, 36–39. (in Chinese).

WU, D., DENG, X.J., YE, Y.X. and MAO, K.W. (2004), *A study on chemical compositions in Dayaoshan of Nanling Mountain*, Acta Meteor. Sinica *62*, 476–485.

YANAGISAWA, F., JIA, S.Y., AKATA, N. and et al. (2004), *Chemical Characteristics of Dew in Fog Collected from Jan. 2 to 4, 2002 in Chengdu*, Sichuan Environ. *23*, 62–64. (in Chinese).

YANG, J., NIU, Z.Q., SHI, C.E. and et al. (2010), *Microphysics of atmospheric aerosols during winter haze/fog events in Nanjing*, Environ. Sci. *31*, 1425–1431. (in Chinese).

ZHANG, G.Z., BIAN, L.G., WANG, J.Z. and et al. (2005), *The boundary layer characteristics in the heavy fog formation process over Beijing and its adjacent areas*, Sci. China D *35*(Suppl), 88–101.

ZHENG, Y.F., TANG, X.Y., XU, J.Q. and et al. (2007), *The analysis of precipitation acidity and chemical composition in the industrial estate located on north bank of the Yangtze River*, Nanjing, Res. Environ. Sci. *20*, 45–51. (in Chinese).

ZHOU, Z.J., ZHU. Y.J. and JU. X.H. (2007), *Dense fog events in the Yangtze Delta and their climatic characteristics*, Prog. Nat. Sci. *17*, 821–827.

ZHU, B., LI, Z.H., HUANG, J.P. and et al. (2000), *Chemical compositions of the fogs in the city and suburban of Xishuangbanna*, Acta Scientiae Circumstantiae *20*, 316–321. (in Chinese).

ZOLGHADRI, A. and CAZAURANG, F. (2006), *Adaptive nonlinear state-space modeling for the prediction of daily mean PM_{10} concentrations*, Environ. Modell. Softw. *21*, 885–894.

(Received November 11, 2010, revised April 29, 2011, accepted May 6, 2011, Published online June 7, 2011)

Pure Appl. Geophys. 169 (2012), 1053–1066
© 2011 The Author(s)
This article is published with open access at Springerlink.com
DOI 10.1007/s00024-011-0331-1

| Pure and Applied Geophysics |

Dew Formation and Chemistry Near a Motorway in Poland

GRZEGORZ GAŁEK,[1] MIECZYSŁAW SOBIK,[1] MAREK BŁAŚ,[1] ŻANETA POLKOWSKA,[2] and KATARZYNA CICHAŁA-KAMROWSKA[2]

Abstract—In this study, the influence of traffic intensity on dew formation efficiency and chemistry is presented. The measurements were conducted near the A4 motorway in SW Poland in almost flat land relief with intense agricultural activity. The dew/hoarfrost was collected by means of insulated plain passive radiative condensers at three sites: AN and AS located in the close vicinity of the motorway (30 m) on the opposite sides of the road, and AR, representing rural background conditions beyond the motorway influence. Measurements were conducted in two short campaigns in April and September 2009 with 9 and 10 collection days respectively. The average daily efficiency of dew formation was 0.179 L/m². Its value for AN, AS and AR was on average 0.170, 0.199 and 0.173 L/m², respectively. The efficiency of dew formation at measurement sites located on both sides of the road differed by up to about 200% during an individual dew episode. Maximum daily value reached 0.389 L/m². The average volume–weighted pH was acidic and ranged from 4.29 (AS) to 4.58 (AR). The electric conductivity (EC) at all measurement sites was relatively low reaching on average 55.9 μS/cm for AN, 62.2 μS/cm for AS and 35.8 μS/cm for AR. The average volume–weighted TIC parameter (total ionic content) reached the value of 0.62 meq/L (AN and AS) and 0.38 meq/L (AR). Both EC and TIC values indicated strong influence of the motorway at sites located in its close vicinity (AN and AS). Depending on airflow direction during individual dew collection events, AN or AS sites were situated alternatively on windward or leeward side of the road, which distinctly influenced dew formation and chemistry: the leeward condenser was characterized by smaller water volume, higher EC and higher TIC when compared both with its windward counterpart and the background site. The ionic structure of the collected samples was similar at all measurement sites. The largest share had NO_3^- anion (28–32%) and Ca^{2+} cation (22–25%). Thus, air pollution was relatively low in the vicinity of the A4 motorway in SW Poland not exceeding the typical values for urban background stations.

Key words: Dew formation, dew chemistry, dew sampling, traffic pollution.

[1] Department of Climatology and Atmosphere Protection, University of Wrocław, 8 A. Kosiby St, 51670 Wrocław, Poland. E-mail: grzegorz.galek@uni.wroc.pl

[2] Department of Analytical Chemistry, Gdańsk University of Technology, 11/12 G. Narutowicza St, 80233 Gdańsk, Poland.

1. Introduction

The research studies concerning efficiency of dew formation by means of insulated plain passive radiative condensers have developed for the last two decades. Construction of these devices differed in terms of many parameters. The surface of the condenser was 0.25 m² in the pilot studies conducted in 1995 in Tunisia by BEYSENS and MILIMOUK, (2000), 0.3 m² in the research studies performed in Grenoble (France) (BEYSENS *et al.*, 2003), 1 m² in the scientific surveys made in Jerusalem (BERKOWICZ *et al.* 2004), on the islands of Komiža and in Zadar (Croatia) (MILETA *et al.*, 2004), on tropical islands (Tahiti and Tuamotu Archipelago) (CLUS *et al.*, 2008), in the Netherlands (JACOBS *et al.*, 2008), in Morocco (LEKOUCH *et al.*, 2010a), and in Poland (SOBIK *et al.*, 2010), 1.4 m² in Tanzania and Sweden (NILSSON, 1996), 4 m² in Poland (NAMIEŚNIK *et al.*, 2007), 30 m² on the island of Corsica (France) (MUSELLI *et al.*, 2002). The inclination angle of the collecting surface was usually 30° (NILSSON, 1996; MUSELLI *et al.*, 2002; BERKOWICZ *et al.*, 2004; MILETA *et al.*, 2004; NAMIEŚNIK *et al.*, 2007; CLUS *et al.*, 2008; JACOBS *et al.*, 2008), which was shown to be optimal in terms of dew collection (BEYSENS *et al.*, 2003). The condensers were insulated from the influence of ground heat radiation by 2–5 cm thick layer made of styrofoam (NILSSON, 1996; MUSELLI *et al.*, 2002; BEYSENS *et al.*, 2003; MILETA *et al.*, 2004; CLUS *et al.*, 2008; JACOBS *et al.*, 2008).

The condensing surfaces were inclined in the opposite direction to the sun at the time of sunrise (NILSSON, 1996; MUSELLI *et al.*, 2002; JACOBS *et al.*, 2008) or in the opposite direction to the prevailing wind direction at night (MUSELLI *et al.*, 2002; BEYSENS *et al.*, 2003). The surface of the device was covered with polyethylene film doped with TiO_2 and $BaSO_4$

(NILSSON, 1996; MUSELLI et al., 2002; BEYSENS et al., 2003; BERKOWICZ et al., 2004; MILETA et al., 2004; CLUS et al., 2008; JACOBS et al., 2008), developed by Nilsson, Vargas et al. (VARGAS et al., 1994; NILSSON, 1996) and produced by OPUR, (2010). The efficiency of dew formation on different surfaces was compared, among others, in the works published by BERKOWICZ et al. (2007) and RUBIO et al. (2008).

The working principle of plain radiative condensers was used in the attempts to gain drinking water in the areas threatened by its deficit. Roofs of low buildings (BEYSENS et al., 2007; SHARAN et al., 2007a, b; CLUS et al., 2010; OPUR, 2010) or ground collectors (CLUS et al., 2007, 2010) were used as dew condensers.

The dew formation efficiency depends both on the surface (BRISCOE et al., 2005) on which it is formed, as well as on the series of meteorological parameters defining the state of the atmosphere. The mechanism of dew formation, from the physical point of view, was described in the works published by BEYSENS, (1995, 2006). Among the meteorological parameters that can have a positive influence on dew formation are all those favouring heat radiation from the surface, such as: cloudless weather, low pressure of water vapour in the air column, low air pollution, poor air turbulence. A second necessary condition for dew formation is the presence of water vapour in the ambient air at the contact with the cooled surface, on which water can condense. Favourable conditions in this case are: high pressure of water vapour, high relative humidity, air turbulence. Some of these factors are contradictory. For example, low pressure of water vapour in the air conducts towards radiative cooling of the surface, but is a poor source of condensation. CLUS et al. (2008) demonstrated experimentally that the humid tropical zone, which is rich in moisture, is not an optimum place for dew formation as a result of weaker radiative cooling of the surface. Similar proposals in the theoretical considerations were presented by NILSSON, (1996) using a function of air temperature. The conducted measurements indicate that the greatest efficiency of dew formation (about 0.6 L/m^2) during a single dew episode was observed in the subtropical and semi-arid tropical zones in Israel, Croatia and India (BERKOWICZ et al., 2007; MUSELLI et al., 2009; LEKOUCH et al., 2010b).

Another contradiction was noticed in the case of atmospheric turbulence. On the one hand, the lack of atmospheric turbulence favours effective cooling of the surface, but on the other hand excessive moisture depletion of the air above the surface may develop gradually along with dew. The research studies made in Ajaccio (Corsica, France) (MUSELLI et al., 2006a) indicate a positive impact of the relatively weak air turbulence (windspeed 1.0–3.0 m/s) on the efficiency of dew formation in relation to almost windless conditions (0.5–1.0 m/s). Theoretical considerations involving the assumption that relative humidity is close to 100% lead to similar conclusions (NILSSON, 1996).

The average daily efficiency of dew formation in the conducted research studies by means of plain radiative condensers was about 0.1 L/m^2 per a single night with dew. Measurements are difficult to compare, because they were conducted during various periods. Additionally, different constructions of devices were applied. The average dew efficiency values are presented in Table 1.

Dew measurements were also carried out on 0.16 m^2, thermally isolated plate of Plexiglas or Teflon (polymethylmethacrylate—PMMA or poly-tetrafluororoethylene—PTFE) used as a reference plate in Bordeaux, Grenoble, Ajaccio and Tahiti (BEYSENS et al., 2005; CLUS et al., 2008). Average dew volume from these condensers was significantly smaller in compare to bigger foil condensers.

There are not many articles concerning dew chemistry published worldwide. Such research studies were conducted, among others, in Israel (YAALON and GANOR, 1968), the USA (MULAWA et al., 1986; PIERSON et al., 1988; FOSTER et al., 1990; WAGNER et al., 1992), Jordan (JIRIES, 2001), Chile (RUBIO et al., 2002), France (MUSELLI et al., 2002; BEYSENS et al., 2006b), Japan (OKOCHI et al., 1998; CHIWA et al., 2003; TAKENAKA et al., 2003), India (SINGH et al., 2006), Germany (ACKER et al., 2008), Poland (POLKOWSKA et al., 2008), Morocco (LEKOUCH et al., 2010a), Croatia (LEKOUCH et al., 2010b). Most of the dew analysis focused on the presence of only selected compounds. Hardly any papers described, e.g. the problem of biological composition of dew (BEYSENS et al., 2006b). Dew chemistry research developed in recent decades because of the attempts to use dew as

Table 1

Dew yield from plane radiative condensers in various field campaigns

Sampling site	Samples number	Study period	Mean volume [L/m²/dew day]	Max volume [L/m²/dew day]	
Kungsbacka (Sweden)	11	14 Aug 1993–01 Sept 1993	0.145	0.21 (0.28)[a]	Nilsson, (1996)
Dodoma (Tanzania)	21	November 1993	0.057	0.082 (0.23)[a]	Nilsson, (1996)
Brive-la-Gaillarde (France)	~275	01 Jan 2000–31 Dec 2000	0.115	<0.475	Beysens et al., (2006a)
Ajaccio (France)	214	22 July 2000–11 Sept 2001	0.12	0.38	Muselli et al., (2002)
	[c]	10 Dec 2001–10 Dec 2003	~0.106	~0.332	Muselli et al., (2006a)
Osaka (Japan)	16	[c]	0.14&0.15[b]	[c]	Takenaka et al., (2003)
Bordeaux (France)	110	15 Jan 2002–14 Jan 2003	[c]	~0.22	Beysens et al., (2006b)
Jerusalem (Israel)	176	01 June 2003–31 May 2004	0.188	~0.50	Berkowicz et al., (2004)
	554	2003–2006	0.199	~0.60	Berkowicz et al., (2007)
Komiža (Croatia)	76	24 June 2003–26 April 2004	0.08	[c]	Mileta et al., (2004)
	263	07 Jan 2003–31 Oct 2006	0.108	0.592	Muselli et al., (2009)
Zadar (Croatia)	87	21 July 2003–31 May 2004	0.15	[c]	Mileta et al., (2004)
	484	07 Jan 2003–31 Oct 2006	0.138	0.406	Muselli et al., (2009)
Central Netherlands	[c]	December 2003–May 2005	0.10	[c]	Jacobs et al., (2008)
Tahiti	151	16 May 2005–14 Oct 2005	0.068	0.22	Clus et al., (2008)
Tikehau	109	21 June 2005–07 Oct 2005	0.102	0.23	Clus et al., (2008)
South–West Morocco	178	01 May 2007–30 April 2008	0.106	[c]	Lekouch et al., (2010a)
Wrocław (Poland)	421	05 Oct 2007–07March 2010	0.103	0.354	Sobik et al., (2010)
Sudetes (Poland)	55	21 June 2009–16 Jan 2010	0.190	0.452	Sobik et al., (2010)

[a] Obtained beyond cited field campaign

[b] Average dew intensity on the glass and PTFE sheet

[c] No data

a source of potable water (Beysens and Milimouk, 2000; Muselli et al., 2006b) and, on the other hand, in the context of its significant pollution compared with rainwater (Wagner et al., 1992; Rubio et al., 2002; Polkowska et al., 2008).

Because research methods differed significantly in particular campaigns, their results are difficult to compare. The surfaces used during scientific experiments on which dew condenses, do not affect the chemical composition of deposits, as evidenced by Takenaka et al. (2003). The construction of condensers and used materials can have an influence on efficiency of dew formation and indirectly on dew chemistry. For example, Beysens et al. (2006b) and Lekouch et al. (2010b) indicate an increase in electrical conductivity (EC) and pH inversely proportional to the efficiency of dew formation.

The main aim of this paper is to examine the influence of intense road traffic on dew formation efficiency and chemistry. Such research studies have not been widely reported in the literature worldwide.

2. Sampling Site

The research studies were conducted in the lowland part of the southwestern Poland near Wrocław (51°07′ N; 17°02′ E; 120 m above sea level). The analyzed area is characterized by relatively weakly varied and slightly rolling terrain. Local land elevations do not exceed a few meters. Towards E, at a distance of approximately 150 m from AN and AS sites, there is a viaduct above the motorway. The main element of the hydrological network of the analyzed area is the Bystrzyca river. The land use is dominated by agriculture on which mainly wheat, corn, rapeseed and root crops are cultivated. Forests and woodlands cover about 10% of the analyzed region (Fig. 1).

The built-up area consists mainly of small villages, which are the source of dispersed 'low emission' of pollutants. Measurement sites were located outside their direct vicinity. Kąty Wrocławskie (5,500 people, CSO, 2009) is a little town located at a distance of about 2.5 km from the

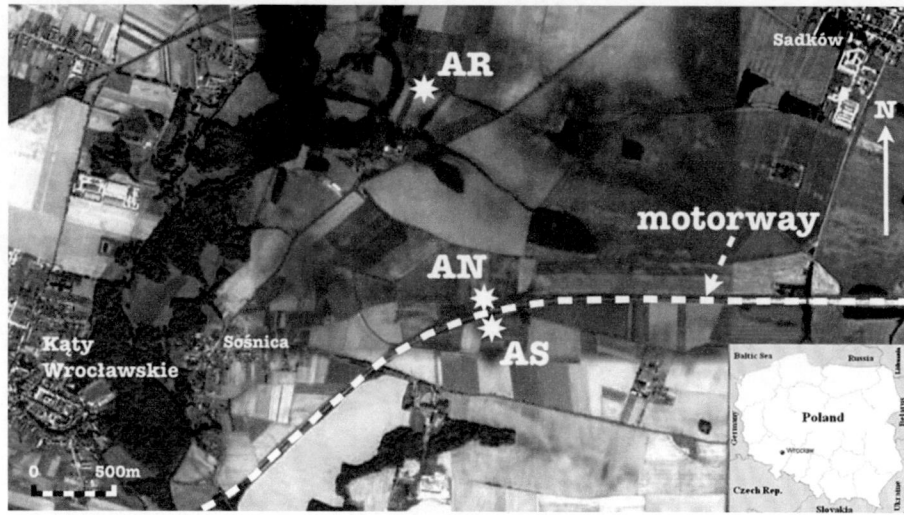

Figure 1
Location of three measurements sites. Three main types of land use are visible: agriculture areas, forests (*dark patches*) and settlements
(source: Google Earth®—modified)

measurement sites in the W to SW direction. The city of Wrocław (630,000 people, CSO, 2009) is about 10 km away from the research area (borders) with the center located in the ENE direction (18 km). Wrocław is the regional center of light manufacturing industry.

Road network in this area consists of the A4 motorway, provincial road no. 347 and two infrequently used local roads. The A4 motorway is a part of the international route E40, linking the western and eastern parts of the continent comprising a typical permanent linear source of pollution. The motorway orientation is consistent with SW–NE directions in the western part of the research area, whereas it is roughly latitudinal in the eastern part of the analyzed region. Route 347 is characterized by low traffic intensity at night, whereas local roads are very infrequently used.

3. Traffic Intensity

Traffic intensity is a factor directly influencing the amount of emitted pollutants along the communication arteries. According to the GTM 2005 (General Traffic Measurement 2005) (GDDKiA, 2010), the traffic intensity on the A4 motorway (E40) was 17,192 vehicles per day. The section of the

motorway, at which the measurement sites were located (Wrocław-Kąty Wrocławskie), was characterized by higher traffic intensity with 24,027 vehicles per day. For comparison, its average value was estimated to be 8,224 vehicles per day on national roads in Poland and 13,561 vehicles per day on international routes. Heavy motor vehicles traffic (GDDKiA, 2010) represented approximately 30% of the total traffic.

According to the report released by the General Directorate for National Roads and Motorways concerning the decade 1995–2005 (GDDKiA, 2010), an increase in traffic intensity on Polish roads by over 50% and heavy traffic by about 150% was observed. Assuming the rate of increase in the traffic intensity from the years 2003–2005 (2–3% per year) and its value of about 14% higher in the warm half of the year (April–November) compared to the annual average, traffic intensity during the 2010 growing season was 30,000–31,500 vehicles per day.

As a result of the availability of only estimated data for 2010 based on measurements made in 2005, the authors have made their own complementary observations to estimate the real traffic intensity during the field campaign. It was found that the average traffic intensity on the A4 motorway in the vicinity of measurement sites was approximately 37,500 vehicles per day. Based on the applied method, it was estimated that the average traffic intensity from dusk to dawn was about 20% lower than all day average. The share of heavy motor vehicle traffic as a whole, as in the case of GTM 2005, was about 30%.

4. Materials and Methods

Measurements were conducted in two campaigns. The first one took place between 01 April 2009 and 20 April 2009 (S1), when nine dew samples were collected at each measurement site. In the remaining days, the atmospheric deposit was not formed or research studies were not conducted. The second series of measurements lasted from 18 Sept 2009 to 28 Sept 2009 (S2). This series was not broken by any precipitation episode, so 10 sets of dew samples were collected during the period. Only on one night was dew formation not intense enough to take samples.

Days with rainfall events were excluded from the measurements in order to eliminate its impact on the chemical composition of the collected samples. Measurements were conducted by means of three identical condensers. Two of them (AN and AS) were located in the close vicinity of a motorway (around 30 m distance) on the opposite sides of the road, whereas the third one (AR) was set 1.25 km in the NNW direction (Fig. 1). AR site reflected the background level of pollution in a given area, typical for the region of Lower Silesia and close to the whole Poland average with combustion of fossil fuels and agriculture as the main sources of pollution. The deposited dew/hoarfrost was collected by means of insulated plain radiative condensers (NILSSON, 1996; MUSELLI et al., 2002; BEYSENS et al., 2003; MILETA et al., 2004) (Fig. 2). The condensing surface of 1 m^2 was covered with polyethylene film and 5 cm thick insulating layer made of styrofoam. It was placed on the height of about 1 m above the ground. The inclination angle of the surface was 15–20°. The condenser was inclined towards the west in order to minimize the warming effect of direct solar radiation after sunrise. In the lower part of the device, there was a collective pipe made of polyvinyl chloride (PVC). Application of condensing surface made of polyethylene film was connected with the method

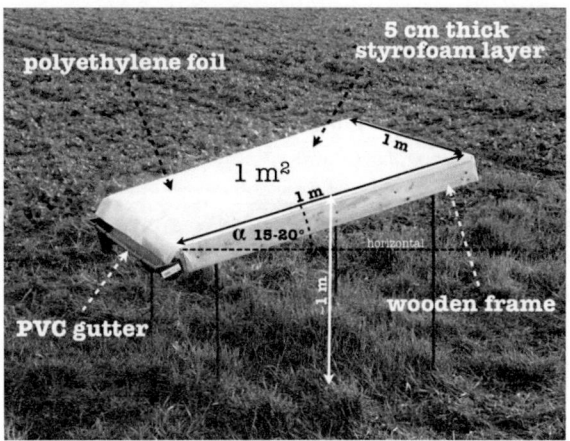

Figure 2
Construction details of dew collectors used in the experiment

used at different measurement sites in Poland (POLKOWSKA et al., 2008). Due to the short measurement series, the ageing of the film (MUSELLI et al., 2002) was not noticed.

The condenser was cleaned up with deionized water, around sunset, just before the potential start of a new dew formation episode. The dew/hoarfrost was scraped off the foil with a polyethylene scraper and collected in tightly closed polyethylene containers, not later than 30 min after sunrise. The time span between sunrise and sunset did not differ significantly in two series. The samples were stored in the dark, at about 4°C for no longer than 1 month. According to the experimental findings of SINGH et al. (2006), the above mentioned time of storage should not have a significant influence on sample degradation.

To check the potential influence of the foil used in the experiment on dew chemistry, deionised water was sprinkled onto a collector and a sample was taken. Chemical analysis of such 'blind sample' has not revealed any measurable effect on water chemistry.

Selected anions and cations were quantified against synthetic rain standard using ion suppressed chromatography (ICS 3000, Dionex Corporation, USA). This synthetic standard is Reference Material No. 409 (BCR-409, Institute for Reference Materials and Measurements, Belgium) and Analytical Reference Material Rain (National Water Research Institute, Environment Canada) (POLKOWSKA et al., 2005). The detailed analytical procedure and problems connected with it are described in NAMIEŚNIK *et al.* (2007).

Meteorological characteristics, such as: temperature, humidity, wind direction and speed, cloud coverage and other observed atmospheric phenomena were measured twice a day (during cleaning-up of the condenser and collection of deposits). Additional source of metrological data was Wrocław Strachowice—synoptic station located in a distance of about 10 km from research sites in the NE direction.

5. Meteorological Characteristics

Measurements were carried out in two research periods mainly on weather characterized by strong long-wave radiation losses. The observed various thermal and circulation conditions allowed to examine their impact on dew formation and chemistry.

In both measurement series, rainfall events did not occur. During the sampling periods, Poland was mostly in the area affected by high-pressure centers with a small horizontal pressure gradient. During the measurement series conducted in April, the prevailing wind direction measured at a synoptic station Wrocław Strachowice was east and southeast (Fig. 3a). During measurement campaign performed in September, air circulation was variable with a predominance of the inflow of air masses from the western sector (Fig. 3b).

The average air temperature measured during dew collection at heights of 100 and 5 cm above the ground was 3.5 and 0.8°C for S1 and 9.0 and 8.5°C for S2, respectively. The average relative humidity measured at the same time was 91% for S1 and 94% for S2, with a minimum of 85%. The average wind speed (averaging time—3 min) measured by means of manual anemometer in the vicinity of the condensers was higher at AN and AS sites than at AR one.

6. Results and Discussion

6.1. Dew Intensity

The volume of the deposit formed at night on the surface of condenser was measured by its collection to the polyethylene container. The average daily efficiency of dew formation for all sampling sites and two measurement campaigns was 0.179 L/m^2 (Table 2). The S2 series (0.211 L/m^2) was characterized by about 50% greater average daily efficiency of dew formation than S1 (0.141 L/m^2). Presumably such contrasting dew intensity stemmed from different meteorological conditions: significantly higher vapour pressure and slightly lower vapour deficit in the ambient air in the September series in comparison with the April one.

The highest daily efficiency of dew formation was found at the AN site (0.389 L/m^2) during S2. In both

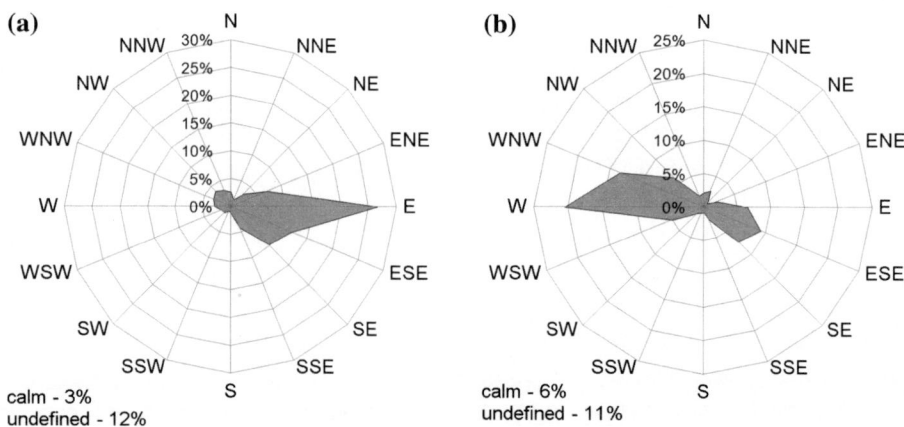

Figure 3
Wind roses at Wrocław-Strachowice synoptic station based on every 30 min measurements: **a** S1 series, **b** S2 series

Table 2

Mean values and range of main inorganic ions during S1 and S2 series

Analyzed element	Para-meter	AR			AN			AS		
Samples number	N	S1 9	S2 10	All 19	S1 9	S2 10	All 19	S1 8	S2 10	All 18
Volume [L]	Mean	0.154	0.189	0.173	0.114	0.214	0.170	0.159	0.231	0.199
	Range	0.58–0.223	0.81–0.309	0.58–0.309	0.21–0.202	0.72–0.389	0.21–0.389	0.56–0.255	0.118–0.358	0.56–0.358
pH	Mean	4.22	5.35	4.58	4.10	4.88	4.48	4.02	4.57	4.29
	Range	3.69–6.84	4.57–7.01	3.69–7.01	3.73–6.72	4.29–6.93	3.73–6.93	3.66–6.51	4.09–6.70	3.66–6.70
EC [μS/cm]	Mean	50.96	25.85	35.76	100.04	37.09	55.90	85.97	49.12	62.22
	Range	7.99–149.80	15.78–41.90	7.99–149.80	30.70–193.7	15.27–49.90	15.27–193.7	13.81–145.4	29.20–81.90	13.81–145.4
TIC [meq/L]	Mean	0.48	0.32	0.38	0.97	0.47	0.62	0.76	0.55	0.62
	Range	0.17–1.35	0.20–0.62	0.17–1.35	0.32–2.26	0.26–1.37	0.26–2.26	0.14–1.26	0.33–0.92	0.14–1.26
Cl^- [meq/L]	Mean	0.056	0.021	0.035	0.067	0.037	0.046	0.054	0.031	0.039
	Range	0.019–0.085	0.010–0.032	0.010–0.085	0.028–0.175	0.011–0.088	0.011–0.175	0.022–0.118	0.012–0.053	0.012–0.118
NO_3^- [meq/L]	Mean	0.115	0.103	0.107	0.295	0.140	0.186	0.234	0.181	0.200
	Range	0.003–0.414	0.053–0.209	0.003–0.414	0.080–0.714	0.066–0.413	0.066–0.714	0.027–0.395	0.102–0.331	0.027–0.395
SO_4^{2-} [meq/L]	Mean	0.045	0.029	0.036	0.114	0.055	0.073	0.069	0.059	0.063
	Range	0.024–0.114	0.016–0.091	0.016–0.114	0.062–0.294	0.020–0.198	0.020–0.294	0.017–0.156	0.029–0.116	0.017–0.156
H^+ [meq/L]	Mean	0.060	0.005	0.026	0.079	0.013	0.033	0.094	0.027	0.051
	Range	0.0001–0.204	0.0001–0.027	0.0001–0.204	0.0002–0.186	0.0001–0.051	0.0001–0.186	0.0003–0.219	0.0002–0.081	0.0002–0.219
Na^+ [meq/L]	Mean	0.016	0.038	0.029	0.008	0.032	0.025	0.008	0.032	0.024
	Range	0.004–0.041	0.005–0.093	0.004–0.093	0.005–0.019	0.004–0.111	0.004–0.111	0.006–0.015	0.006–0.091	0.006–0.091
K^+ [meq/L]	Mean	0.021	0.017	0.018	0.023	0.027	0.026	0.023	0.032	0.029
	Range	0.008–0.065	0.008–0.028	0.008–0.065	0.006–0.072	0.008–0.053	0.006–0.072	0.007–0.052	0.014–0.050	0.007–0.052
Ca^{2+} [meq/L]	Mean	0.108	0.068	0.084	0.274	0.108	0.157	0.208	0.123	0.153
	Range	0.025–0.372	0.023–0.147	0.023–0.372	0.108–0.849	0.047–0.361	0.047–0.849	0.040–0.354	0.026–0.256	0.026–0.354
Mg^{2+} [meq/L]	Mean	0.020	0.018	0.019	0.040	0.030	0.033	0.018	0.020	0.019
	Range	0.011–0.056	0.008–0.035	0.008–0.056	0.016–0.172	0.009–0.059	0.009–0.172	0.006–0.042	0.005–0.036	0.005–0.042
NH_4^+ [meq/L]	Mean	0.040	0.021	0.029	0.066	0.032	0.042	0.053	0.043	0.047
	Range	0.017–0.071	0.005–0.043	0.005–0.071	0.002–0.186	0.006–0.099	0.002–0.186	0.008–0.122	0.007–0.114	0.007–0.122

series of measurements, differences between average volume of dew samples collected from the sites were observed. The AS condenser reached the highest average daily efficiency during both sampling campaigns (0.159 L/m^2 during S1 and 0.231 L/m^2 during S2) (Table 2). The AN and AR condensers were characterized by the lowest average daily efficiency of dew formation (0.114 L/m^2 for AN during S1, 0.189 L/m^2 for AR during S2). Differences probably resulted from wind direction and speed. During the days with prevailing west wind direction at AR, both atmospheric calm and the lowest efficiency of dew formation in relation to other sites were noticed (Fig. 4).

Between the measurement sites located in the close vicinity of the motorway, the dependence on wind direction was noticed. The obtained sample volumes during the same dew episode differed by up to above 200% there (Fig. 4). The sampling site located on the leeward side of the motorway was characterized by the significantly lower efficiency of dew formation (compare Figs. 1, 4). It indicates the effect of air heating above the motorway with decreasing relative humidity. Indirect evidence for the existence of this effect is the observation of white dew (a meteorological phenomenon meaning 'frozen dew') on 16 April 2009 at AS site, while the normal dew appeared at AN condenser. On this day, the circulation from the eastern sector dominated.

6.2. pH Measurements

The pH of dew samples was determined in the laboratory during 4 weeks after the collection of deposits. The average volume–weighted pH was 4.48 for AN, 4.29 for AS and 4.58 for AR. The pH of individual samples ranged from 3.66 to 7.01 (Table 2). A significant difference was noticed by comparison of average pH values from both measurement series.

Taking all measurement sites into consideration, average pH value was 4.13 and 4.81 for S1 and S2, respectively. The pH value was neither depending on sample volume nor the vicinity of the motorway

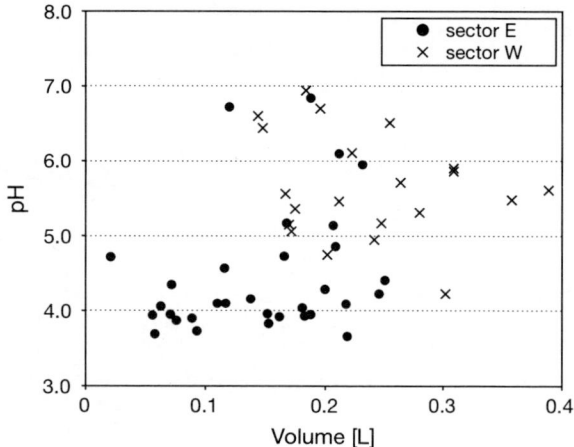

Figure 5
Dew samples pH versus water volume depending on wind direction

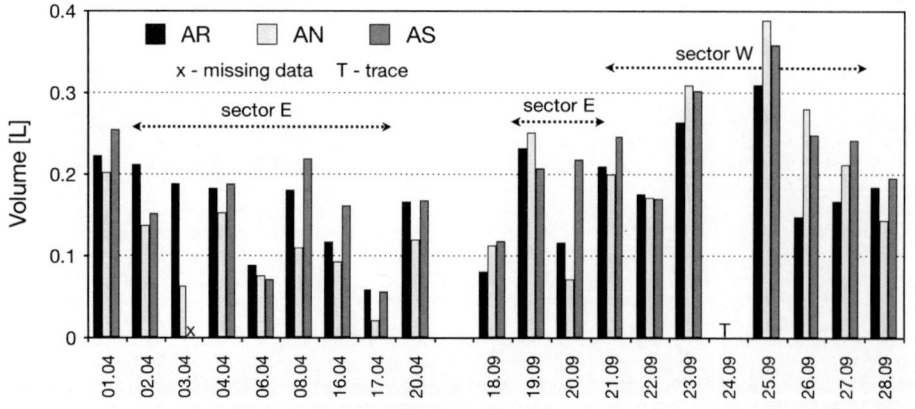

Figure 4
Samples volume on three sampling sites with general wind direction marked above the *bars*. In the remaining days there was no prevailing wind direction. All dates refer to year 2009

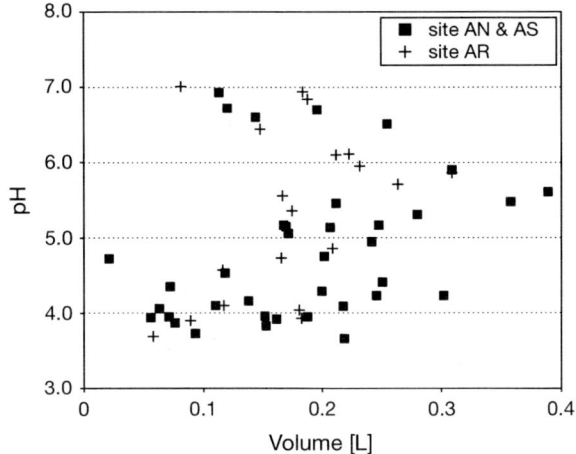

Figure 6
Dew samples pH versus water volume depending on site location

(Figs. 5, 6). Clear influence of the prevailing direction of air circulation on pH values was observed. Acidity of dew samples was lower during the advection of air from the western sector than from the eastern one (Fig. 5). It probably resulted from the fact that the city of Wrocław and the most industrialized parts of Poland with high emission rate of SO_2 are located to the east from the research area.

The average reported pH of dew samples was significantly lower when compared with literature data, where it exceeded the value of 6.0 in most cases (Foster et al., 1990; Wagner et al., 1992; Jiries, 2001; Rubio et al., 2002; Takenaka et al., 2003; Beysens et al., 2006a; Singh et al., 2006; Lekouch et al., 2010a, b). This value may result from the fact

that the majority of the above mentioned research studies were conducted in arid and semi-arid regions, where more alkaline mineral matter is present in the near-ground air layer.

The main components responsible for relatively low pH of the collected samples were NO_3^- and SO_4^{2-} ions. Both—high correlation coefficient of NO_3^-, SO_4^{2-} and Ca^{2+} as well as considerably higher ratios of SO_4^{2-}/Na^+ and Ca^{2+}/Na^+ than present in maritime aerosol—indicate an anthropogenic origin of nitrates and sulfates (Table 3). The possible source of these ions were emission from residential heating and traffic. In case of Ca^{2+} the emission took place from numerous construction sites in Wrocław and its vicinity as well as from gradual wearing out of road surfaces (Polkowska et al., 2008) rather than from wind-blown soils which are poor with minerals containing calcium. The motorway collection sites (AN and AS) were characterized by around doubled concentrations of these ions if compared with the rural background site (AR) which was independent of dew formation intensity. This last fact indicates for the additional role of the motorway nighttime traffic as a source of NO_x and SO_2 emission from vehicles and Ca from the road surface. Additional re-emission is possible from the road surface as the result of turbulence induced by fast moving vehicles.

6.3. Conductivity Measurements

The average volume–weighted electric conductivity (EC) of the collected dew samples was 55.9,

Table 3

Coefficients determining the origin of the selected ions

Analyzed parameter	Statistical parameter	AR			AN			AS		
		S1	S2	All	S1	S2	All	S1	S2	All
Cl^-/Na^+ (1.17)	Mean	5.75	1.59	3.23	7.70	4.28	5.30	6.36	3.08	4.24
	Mean/1.17	4.91	1.36	2.76	6.58	3.66	4.53	5.44	2.63	3.62
SO_4^{2-}/Na^+ (0.12)	Mean	4.47	1.77	2.84	14.24	6.22	8.62	8.51	4.31	5.80
	Mean/0.12	37.3	14.8	23.7	118.7	51.8	71.8	70.9	35.9	48.3
K^+/Na^+ (0.022)	Mean	2.04	1.20	1.54	2.59	3.12	2.96	2.84	3.03	2.96
	Mean/0.022	92.7	54.5	70.0	117.7	141.8	134.5	129.1	137.7	134.5
Ca^{2+}/Na^+ (0.045)	Mean	9.92	5.51	7.25	31.65	12.61	18.30	26.11	9.72	15.55
	Mean/0.045	220.4	122.4	161.1	703.3	280.2	406.7	580.2	216.0	345.6
Mg^{2+}/Na^+ (0.25)	Mean	2.05	1.10	1.47	4.54	4.09	4.23	2.28	1.27	1.63
	Mean/0.25	8.2	4.4	5.9	18.2	16.4	16.9	9.1	5.1	6.5

In brackets, the characteristic values for maritime aerosol were given (Polkowska et al., 2008)

62.2 and 35.8 μS/cm for AN, AS and AR sites, respectively. The determined EC values were nearly two times higher at measurement sites located in the close vicinity of the motorway, what can be treated as an evidence of its impact on the overall mineralization of dew. The EC values ranged from 8.0 to 193.7 μS/cm. The average EC value of the collected samples was significantly higher for S1, when compared to S2 (Table 2). This was partly due to the fact that inversely proportional dependence between EC values and efficiency of dew formation (Fig. 7) is generally typical (BEYSENS, *et al.*, 2006b).

In Fig. 8 significant differences between EC values determined in dew samples collected from measurement sites located in the close vicinity of the motorway are presented. These differences probably result from the stronger influence of the motorway on the condenser located on the leeward side. Heat and pollution emitted by cars had the influence on the EC values. This effect was strongly manifested when the wind direction was almost parallel to the motorway and the air stream was present on the roadway for a prolonged time (Figs. 1, 8). In case of the S1 series the AN site frequently represented the lee side of the motorway while during the S2 series such role was most often played by the AS condenser.

The average EC value determined in all samples collected from three condensers indicate rather moderate overall mineralization of the deposits. Similar results were obtained, among others, by BEYSENS *et al.*, (2006a) in Bordeaux. A few times higher EC values were reported during research studies conducted by JIRIES, (2001), MUSELLI *et al.*, (2006a), POLKOWSKA *et al.* (2008) and LEKOUCH *et al.* (2010b) in Jordan, France, Poland and Croatia, respectively. LEKOUCH *et al.* (2010a) indicated the average EC value as high as 725 μS/cm determined during the scientific survey performed in Morocco.

To check the ionic balance, the percentage ionic difference (PDI) was calculated for each collected sample. On average the PDI for individual sites was within 3–7% range, which is far below the 20% limit treated by some authors as a criterion of sufficient

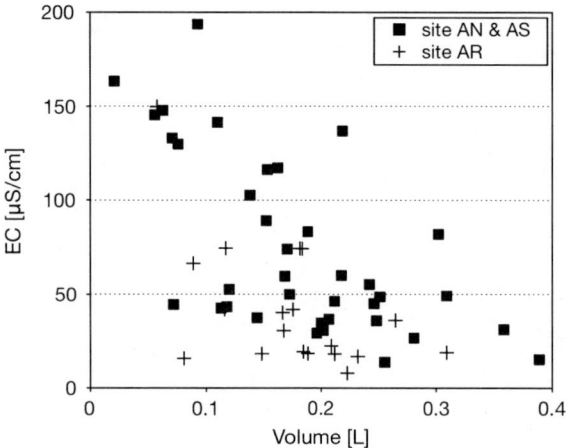

Figure 7
Dew samples electric conductivity (EC) versus water volume depending on site location

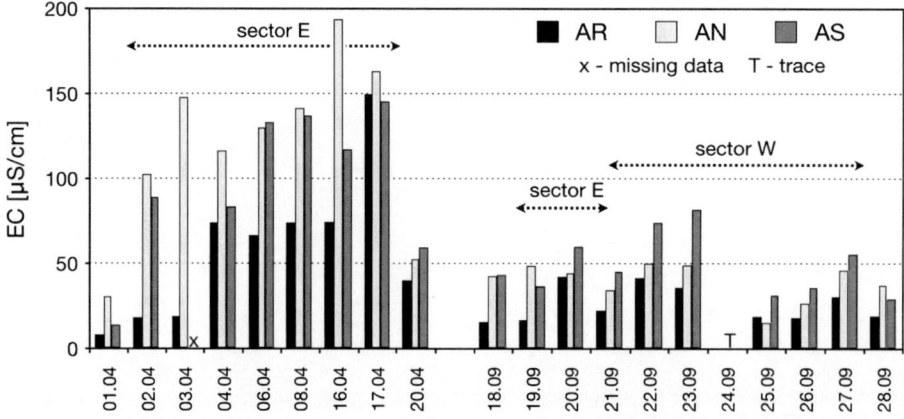

Figure 8
Dew samples electric conductivity (EC) at three sampling sites with general wind direction marked above the *bars*. In the remaining days there was no prevailing wind direction. All dates refer to year 2009

quality measurements (e.g. CINI *et al.*, 2002). Thus, presumably all major ions were analyzed allowing us to calculate properly TIC of the collected samples.

6.4. Chemical Composition

The overall contamination of dew samples was measured by TIC (total ionic content) parameter, which is defined as the sum of major inorganic ions [TIC = $\Sigma(SO_4^{2-}$, NO_3^-, Cl^-, NH_4^+, Ca^{2+}, Mg^{2+}, Na^+, K^+, H^+)]. Its average volume–weighted value was 0.62, 0.62 and 0.38 meq/L for AN, AS and AR sites, respectively. These results confirm that sampling sites located in the close vicinity of a motorway remained under its strong influence, which was manifested by almost doubling the background TIC. However, the impact of traffic intensity on the AR site was not observed. The values of TIC parameter ranged from 0.14 to 2.26 meq/L. The average pollution of the samples was significantly higher during series S1, compared to S2 (Table 2). It mainly resulted from smaller efficiency of dew formation.

Measurements results during both series indicated the rising tendency in TIC value during the following days under the conditions of unchanged circulation (Fig. 9). This effect was mainly due to the lack of rainfall, which effectively removes contaminants from the atmosphere (JIRIES, 2001). The value of TIC parameter in the samples collected from AN and

AS condensers during one dew episode differed of even about 139%. It was probably connected with the wind direction and the impact of the motorway. During the circulation from the eastern sector (except NE), the AN site was on the leeward side of the road. The opposite situation occurred during western circulation. Then the AS site was located on the leeward side of the road. The stronger impact of the motorway on TIC values at AN site resulted from the fact that air stream was present on the roadway for a longer time.

In the ionic structure meaning the relative contribution of individual ions, the largest share had anion NO_3^- and cation Ca^{2+} (Fig. 10). There were no significant differences in the ionic structure between AN, AS and AR sites, what indicated the emission of all the ions that made up the TIC parameter along the motorway.

In general, chemical composition of dew depends mainly on two mechanisms (BEYSENS *et al.*, 2006b). Firstly, chemical substances present on a collector surface are being dissolved in dew water what is favored by heterogeneous nucleation. Assuming that deposition rate of particulates (mainly mineral) during a given night is relatively stable, the observed concentrations are decreasing with increasing intensity of dew formation. Secondly, gaseous substances, e.g. SO_2 and NO_x, are absorbed by dew droplets proportionally to their concentration in the ambient

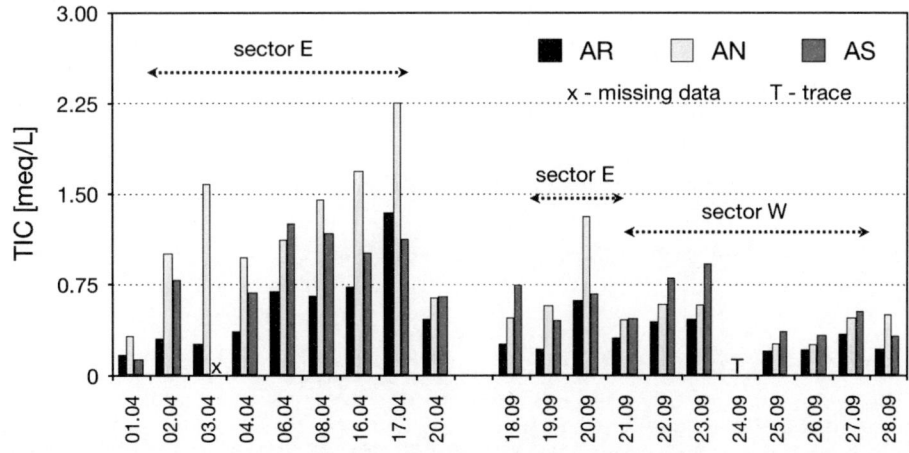

Figure 9
Dew samples TIC value at three sampling sites with general wind direction marked above the *bars*. In the remaining days there was no prevailing wind direction. All dates refer to year 2009

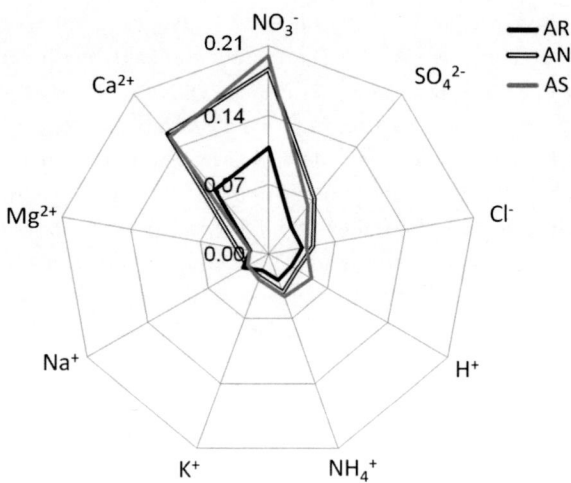

Figure 10
Dew samples main ions structure [meq/L] at three sampling sites

air. The amount of absorbed gases increases with increasing dew formation intensity and duration. If the main way leading to higher concentrations of nitrates and sulfates at AN and AS sites was the second mechanism, the observed decrease of these ions concentration with increasing sample volume would be less pronounced than in case of Ca^{2+} which is of mineral origin. However, such dependence was not observed in the gathered data. Thus, presumably the main pathway influencing dew chemical composition at sites close to the motorway is the deposition of particulate matter caused both by primary and secondary emission.

In order to determine the origin of the selected ions, their ratio to Na^+ ions was compared with the ratio characteristic for maritime aerosols (Table 3). On the basis of the obtained results, this ratio was frequently crossed. Particularly high values of these parameters were achieved for Ca^{2+}, K^+ and SO_4^{2-} ions. The higher value of average ratio of ions in the S1 series was probably connected with the prevailing circulation from the continental eastern sector.

7. Conclusions

The obtained average and maximum efficiency of dew, in comparison with the results of research studies conducted worldwide, was relatively high. Measurement results indicate that the sampling sites located in the close vicinity of the A4 motorway remained under its considerable influence. The condenser located on the windward side of the road reached up to three times higher efficiency than its leeward counterpart. Air circulation at an acute angle less than 30° to the motorway was the most favourable for such large differences.

The samples obtained from condensers were characterized by strong acidity, with an average pH value ranging from 4.29 to 4.58. Clear dependence of pH values on the direction of air circulation was observed: samples obtained during the air circulation from the eastern sector (city of Wrocław and the most industrialized parts of Poland) were characterized by particularly low pH. Any significant impact of the motorway on pH of dew was not noticed. Both EC and TIC parameters were dependant on the efficiency of dew formation and the motorway proximity. Dew samples at measurement sites located in the close vicinity of the motorway were about twice more polluted than in the background. TIC values were two times greater in the case of the condenser located on the leeward side of the road, when compared with the windward one.

The ratio of the selected ions to Na^+ indicated that they originated mainly from sources other than maritime. On the basis of EC, TIC and individual ions concentration values, one can state that air pollution was relatively low in the vicinity of the A4 motorway in SW Poland, not exceeding the typical values for urban background stations. Surprisingly, ionic structure in the motorway proximity was very similar to that at the rural background site indicating that background concentrations are mostly a result of local traffic and similar combustion processes.

Acknowledgments

This scientific work was financially supported in years 2008–2010 years by the Polish Ministry of Science and Higher Education as research project N N305 231035.

REFERENCES

ACKER, K., BEYSENS, D., MÖLLER, D. (2008), *Nitrite in dew, fog, cloud and rain water: An indicator for heterogeneous processes on surfaces*, Atmospheric Research *87*, 200–212.

BERKOWICZ, S.M., BEYSENS, D., MILIMOUK, I., HEUSINKVELD, B.G., MUSELLI, M., WAKSHAL, E., JACOBS, A.F.G. (2004), *Urban dew collection under semi-arid conditions: Jerusalem*, Proceedings of the 3rd International Conference on Fog, Fog Collection and Dew, Cape Town, South Africa, October 11–15, E4.

BERKOWICZ, S.M., BEYSENS, D., MILIMOUK-MELNYTCHOUK, I., HEUSINKVELD, B.G., MUSELLI, M., JACOBS, A.F.G., CLUS, O. (2007), *Urban dew collection in jerusalem: a three-year analysis*, Proceedings of the 4th International Conference on Fog, Fog Collection and Dew, La Serena, Chile, 23–27 July 2007, 297–300.

BEYSENS, D. (1995), *The formation of dew*, Atmospheric Research *39* (1–3), 215–237.

BEYSENS, D. (2006), *Dew nucleation and growth*, C R Physique *7*, 1082–1100.

BEYSENS, D., MILIMOUK, I. (2000), *Pour les ressources alternatives en eau*, vol. *11*, no 4, Sécheresse (translated into English by T. Fuller).

BEYSENS, D., MILIMOUK, I., NIKOLAYEV, V., MUSELLI, M., MARCILLAT, J. (2003), *Using radiative cooling to condense atmospheric vapor: a study to improve water yield*, Journal of Hydrology *276*, 1–11.

BEYSENS, D., MUSELLI, M., NIKOLAYEV, V., NARHE, R., MILIMOUK, I. (2005), *Measurement and modelling of dew in island, coastal and alpine areas*, Atmospheric Research *73*, 1–22.

BEYSENS, D., MUSELLI, M., MILIMOUK, I., OHAYONE, C., BERKOWICZ, S.M., SOYEUXG, E., MILETA, M., ORTEGA, P. (2006a), *Application of passive radiative cooling for dew condensation*, Energy 31, 1967–1979.

BEYSENS, D., OHAYON, C., MUSELLI, M., CLUS, O. (2006b), *Chemical and biological characteristics of dew and rain water in an urban coastal area (Bordeaux, France)*, Atmospheric Environment *40*, 3710–3723.

BEYSENS, D. CLUS, O., MILETA, M., MILIMOUK, I., MUSELLI, M., NIKOLAYEV, V.S. (2007), *Collecting dew as a water source on small islands: the dew equipment for water project in Biševo (Croatia)*, Energy *32*, 1032–1037.

BRISCOE, B.J., WILLIAMS, D.R., GALVIN, K.P. (2005), *Condensation on hydrosol modified polyethylene*, Colloids and Surfaces A: Physicochem Eng Aspects 264, 101–105.

CHIWA, M., OSHIRO, N., MIYAKE, T., NAKATANI, N., KIMURA, N., YUHARA, T., HASHIMOTO, N., SAKUGAWA, H. (2003), *Dry deposition washoff and dew on the surfaces of pine foliage on the urban and mountain-facing sides of Mt. Gokurakuji, western Japan*, Atmospheric Environment *37*, 327–337.

CINI R., PRODI F., SANTACHIARA G., PORCU F., BELLANDI S., STORTINI A.M., OPPO C., UDISTI R. (2002), Pantani F., *Chemical characterization of cloud episodes at a ridge in Tuscan Appenines, Italy*, Atmos Res, *61*, 311–334.

CLUS, O., SHARAN, G., SINGH, S., MUSELLI, M., BEYSENS, D. (2007), *Simulating and testing a very large dew and rain harvester in Panandhro (NW India)*, Proceedings of the 4th International Conference on Fog, Fog Collection and Dew, La Serena, Chile, 23–27 July 2007, 311–314.

CLUS, O., ORTEGA, P., MUSELLI, M., MILIMOUK, I., BEYSENS, D. (2008), *Study of dew water collection in humid tropical islands*, Journal of Hydrology, *361*, 159–171.

CLUS, O., LEKOUCH I., DURAND, M., LANFOURMI, M., MUSELLI, M., MILIMOUK-MELNYTCHOUK, I., BEYSENS, D. (2010), *Large Dew water collectors in a village of S-Morocco (Idouasskssou)*, Proceedings of the 5th International Conference on Fog, Fog Collection and Dew, Münster, Germany, 25–30 July 2010, 243–246.

CSO—Central Statistical Office (2009), Statistical Yearbook of the Republic of Poland 2009, Warsaw.

FOSTER, J.R., PRIBUSH, R.A., CARTER, B.H. (1990), *The chemistry of dews and frosts in Indianapolis*, Indiana Atmos Environ *24*(A), 2229–2236.

GDDKiA—General Directorate for National Roads and Motorways (2010), www.gddkia.gov.pl, access date 22.05.2010.

JACOBS, A.F.G., HEUSINKVELD, B.G., BERKOWICZ, S.M. (2008), *Passive dew collection in a grassland area*, The Netherlands, Atmospheric Research *87*, 377–385.

JIRIES, A. (2001), *Chemical composition of dew in Amman, Jordan*, Atmospheric Research *57*, 261–268.

LEKOUCH, I., KABBACHI, B., MILIMOUK-MELNYTCHOUK, I., MUSELLI, M., BEYSENS, D. (2010a), *Influence of temporal variations and climatic conditions on the physical and chemical characteristics of dew and rain in South-West Morocco*, Proceedings of the 5th International Conference on Fog, Fog Collection and Dew, Münster, Germany, 25–30 July 2010, 43–46.

LEKOUCH, I., MILETA, M., MUSELLI, M., MILIMOUK-MELNYTCHOUK, I., ŠOJAT, V., KABBACHI, B., BEYSENS, D. (2010b), *Comparative chemical analysis of dew and rain water*, Atmospheric Research *95*, 224–23.

MILETA, M., MUSELLI, M., BEYSENS, D., MILIMOUK, I., BERKOWICZ, S., HEUSINKVELD, B.G., JACOBS, A.F.G. (2004), *Comparison of dew yields in four Mediterranean sites: similarities and differences*, Proceedings of the 3rd International Conference on Fog, Fog Collection and Dew, Cape Town, South Africa, 11–15 October 2004, E2.

MULAWA, P.A., CADLE, S.H., LIPARI, F., ANG, C.C., VANDERVENNET, R.T. (1986), *Urban dew: composition and influence on dry deposition rates*, Atmospheric Environment *20* (7), 1389–1396.

MUSELLI, M., BEYSENS, D., MARCILLAT, J., MILIMOUK, I., NILSSON, T., LOUCHE, A. (2002), *Dew water collector for potable water in Ajaccio (Corsica Island, France)*, Atmospheric Research *64*, 297–312.

MUSELLI, M., BEYSENS, D., MILIMOUK, I. (2006a), *A comparative study of two large radiative dew water condensers*, Journal of Arid Environments *64*, 54–76.

MUSELLI, M., BEYSENS, D., SOYEUX, E. (2006b), *Is dew water potable? Chemical and biological analyses of dew water in Ajaccio (Corsica Island, France)*, J Environ Qual *35*, 1812–1817.

MUSELLI, M., BEYSENS, D., MILETA, M., MILIMOUK, I. (2009), *Dew and rain water collection in the Dalmatian Coast, Croatia*, Atmospheric Research *92*, 455–463.

NAMIEŚNIK, J., POLKOWSKA, Ż., SKARŻYŃSKA, K. (2007), *Analytics of dew and fog samples—problems and challenges*, Proceedings of the 4th Conference on Fog, Fog Collection and Dew, La Serena, Chile, 23–27 July 2007, not reviewed extended abstract, 323–326.

NILSSON, T. (1996), *Initial experiments on dew collection in Sweden and Tanzania*, Solar Energy Materials and Solar Cells 40, 23–32.

OKOCHI, H., TAKEUCHI, M., IGAWA, M. (1998), *Effect of acid deposition on urban dew chemistry in Yokohama*, Japan, 1st International Conference on Fog and Fog Water Collection, Vancouver, Canada, 19–24 July 1998, 301–304.

Reprinted from the journal

OPUR—International Organization For Dew Utilization (2010), www.opur.fr, access date 22.05.2010.

PIERSON, W.R., BRACHACZEK, W.W., JAPAR, S.M. (1988), *Dry deposition and dew chemistry in Claremont, California, during the 1985 nitrogen species methods: comparison study*, Atmos Environ *22*, 1657–1663.

POLKOWSKA, Ż., ASTEL, A., WALNA, B., MAłEK, S., MĘDRZYCKA, K., GÓRECKI, T., SIEPAK, J., NAMIEŚNIK, J. (2005), *Chemometric analysis of rainwater and throughfall at several sites in Poland*, Atmospheric Environment *39*, 837–855.

POLKOWSKA, Ż., BŁAŚ, M., KLIMASZEWSKA, K., SOBIK, M., MAłEK, S., NAMIEŚNIK, J. (2008), *Chemical Characterization of Dew Water Collected in Diffrent Geographic Regions of Poland*, Sensors *8*, 4006–4032.

RUBIO, M.A., LISSI, E., VILLENA, G. (2002), *Nitrite in rain and dew in Santiago city, Chile. Its possi- ble impact on the early morning start of the photochemical smog*, Atmospheric Environment *36*, 293–297.

RUBIO, M.A., LISSI, E., VILLENA, G. (2008), *Factors determining the concentration of nitrite in dew from Santiago, Chile*, Atmospheric Environment *42*, 7651–7656.

SHARAN, G., BEYSENS, D., MILIMOUK-MELNYTCHOUK, I. (2007a), *A study of dew water yields on galvanized iron roofs in Kothara (North-West India)*, Journal of Arid Environments *69*, 259–269.

SHARAN, G., SHAH, R., MILIMOUK-MELNYTCHOUK, I., BEYSENS, D. (2007b), *Roofs as dew collectors: I. Corrugated galvanized iron roofs in Kothara and Suthari (NW India)*, Proceedings of the 4th Conference on Fog, Fog Collection and Dew, La Serena, Chile, 23–27 July 2007, 301–304.

SINGH, S.P., KHARE, P., KUMARI, K.M., SRIVASTAVA, S.S. (2006), *Chemical characterization of dew at a regional representative site of North-Central India*, Atmospheric Research *80*, 239–249.

SOBIK, M., BŁAŚ, M., POLKOWSKA, Ż. (2010), *Climatology of dew in Poland*, 5th International Conference on Fog, Fog Collection and Dew, Münster, Germany, 25–30 July 2010, not reviewed abstract, 78.

TAKENAKA, N., SODA, H., SATO, K.,TERADA, H., SUZUE, T., BANDOW, H., MAEDA, Y. (2003), *Difference in amounts and composition of dew from different types of dew collectors*, Water, air, and soil pollution *147*, 51–60.

VARGAS, W.E., NIKLASSON, G.A., GRANQVIST, C.G., NILSSON, T. (1994), *Condensation of water by radiative cooling*, Sol Energy *5* (f), 310–317.

WAGNER, G., STEELE, K., PEDEN, M. (1992), *Dew and Frost Chemistry at a Midcontinental Site*, United States J Geophys Res *97*, 20591–20597.

YAALON, D.H., GANOR, E. (1968), *Chemical composition of dew and dry fallout in Jerusalem*, Israel Nature *217*, 1139–1140.

(Received November 17, 2010, revised March 31, 2011, accepted April 18, 2011, Published online June 1, 2011)

Pure Appl. Geophys. 169 (2012), 1067–1081
© 2011 Springer Basel AG
DOI 10.1007/s00024-011-0359-2

Water and Chemical Properties of Hydrometeors Over Central European Mountains

Marek Błaś,[1] Mieczysław Sobik,[1] Żaneta Polkowska,[2] Katarzyna Cichała-Kamrowska,[2] and Jacek Namieśnik[2]

Abstract—Atmospheric pollutants are transferred to the ground by the contribution of various types of hydrometeors. Because of the different techniques of measurement, comparative analyses between them are often neglected. Hence, the main goal is to compare water volume and chemistry of different types of hydrometeors and their role in both: water balance and pollutants deposition. The results of water input and atmospheric deposits chemistry at Szrenica Mt. during the period between 01 December 2008 and 28 February 2010 are presented. The volume-weighted total inorganic ionic content (TIC) of dew, hoarfrost and fog water (both solid and liquid) was 345, 134, 425 µeq l^{-1} respectively, while typical TIC value for precipitation was 233 µeq l^{-1}. The chemical composition of dew, hoarfrost and fog water differ significantly from each other and from precipitation, depending on background emission and atmospheric processes. Taking into account both concentration and volume of deposited water, comparison between pollutant load of hydrometeors was possible.

Key words: Fog, dew, hoarfrost, chemistry, deposition, Sudety Mts.

1. Introduction

Szrenica Mt. [1,363 m a.s.l.; $\varphi = 50°48'N$, $\lambda = 15°31'E$] is situated in the western part of the main ridge of the Karkonosze Mts., which straddle the Polish/Czech border falling steeply northward on the Polish side (1,000 m of relative height, Fig. 1). Measurements were conducted directly in the meteorological station located just below the summit at an altitude of 1,330 m a.s.l. within a small clearing with slope inclination around 10% and WSW aspect. Szrenica Mt is covered by dwarf pine 2 m high on

¹ Department of Climatology and Atmospheric Protection, University of Wrocław, 8 Kosiby Street, 51670 Wrocław, Poland. E-mail: marek.blas@uni.wroc.pl
² Department of Analytical Chemistry, Chemical Faculty, Gdansk University of Technology (GUT), 11/12 G. Narutowicza St, 80-233 Gdańsk, Poland.

average, whereas the upper tree line runs through mountainous terrain 100 m below. There is only one building (mountain hotel) placed about 100 m away from the meteorological garden. Because of the described position this site represents meteorological conditions typical for the upper parts of Karkonosze ridge.

Karkonosze Mts., as a part of the Western Sudety Mts., are located in one of the most polluted regions in Europe (Błaś and Sobik, 2003). The research area is under strong influence of the anthropogenic sources of pollutant emission to the atmosphere. SO$_2$ emission is mainly connected with combustion of brown and hard coal with high sulphur content. It is used mainly for the needs of power plants, industry and household heating. In the case of NO$_x$ emission, the road transport is the major source. As much as 97% of ammonia should be linked to the agricultural activity. Power plants and industry situated in the northern and northwestern Czech Republic, in Saxony (Germany) and the Turów Power Plant in the SW tip of Poland are decisive for air pollution in the Sudetes.

The local climate is well established on the basis of 50 years of measurements made at the meteorological observatory of the University of Wrocław situated at Szrenica Mt. (Baranowski, 1974; Migała et al., 1993, 1995). The mean annual air temperature exceeds +2.0°C, the coldest month being January (−6.3°C) and the warmest period, July and August (+10.1°C). The most frequent directions of circulation are from W–WNW sector but winds from NE–ENE form a secondary maximum (Fig. 3). The mean annual wind speed measured at the Szrenica summit is equal to 9.5 m s^{-1}, but velocities higher than 40 m s^{-1} are often recorded. The frequency of calms in all seasons is very low, ranging from 1 to 2% (Pereyma et al., 1997). Weather conditions vary

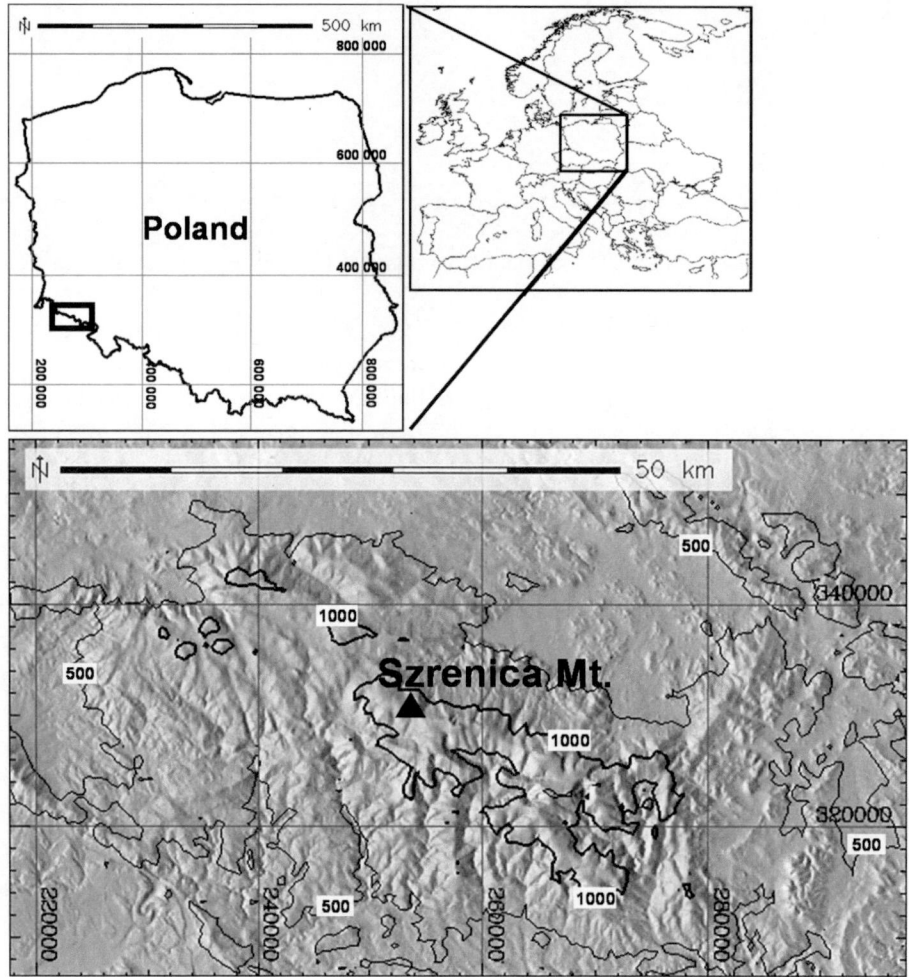

Figure 1
Location of Szrenica Mt. against larger background, WGS84 coordinates used, 500 and 1,000 m *contour lines* shown

strongly from year to year, which is characteristic for the transitional climate of Poland. The Western Sudety Mts. are under a prevailing influence of polar air masses, both of maritime (64.6%) and continental origin (29.3%; Migała *et al.*, 2002).

Atmospheric pollutants are transferred to the ground via various types of hydrometeors: atmospheric precipitation and non-precipitation atmospheric deposits, i.e. dew and hoarfrost as well as rime and liquid fog. Precipitation is the main source of water flux at Szrenica Mt. reaching 1,430 mm annually. Monthly precipitation amounts are highest in June, July and August (>150 mm monthly), and the lowest in February and March (<90 mm monthly). The seeder–feeder mechanism typically causes an increase in

precipitation at Szrenica Mt. of about 50% relative to the lowlands (Dore *et al.*, 1999; Sobik *et al.*, 2001; Błaś and Sobik, 2003). Because of this mechanism orographic cloud droplets are scavenged by upper level precipitation leading to a significant increase of both precipitation and pollutant deposition.

At Szrenica Mt. a significant part of water flux from the atmosphere is formed by direct deposition from fog. It is the most frequently observed atmospheric phenomenon, being present on an average of 45% of the time (Błaś *et al.*, 2002; Błaś and Sobik, 2003). Fog is most often of orographic origin, connected with the forced ascent of humid air on a windward slope. Thus, orographic fog is a combination of advective and adiabatic types of fog. This kind of fog is more typical for the cold season (November–April) because of the predominance of humid air masses and small diurnal temperature range (BŁAŚ et al., 2002; BŁAŚ and SOBIK, 2004).

Dew and hoarfrost are formed especially during the anticyclonic type of weather with weak wind and clear night skies (BEYSENS, 1995), rare at summit positions. The annual number of days with dew in the lowland part of Poland varies around 100–160 and decreases to 10–30 days/year in case of conspicuous mountain summits and ridges. Hoarfrost frequencies show similar spatial distribution as dew—from <30 days at well exposed mountain summits to more than 80 days annually in concave landforms and central regions of Poland (LORENC, 2004).

Research in the 1980s and 1990s further made the distinction between fog/cloud water and wet deposition (precipitation); however, the role of dew and hoarfrost was often neglected (GARLAND, 1978; HEGG and HOBBS, 1981; HICKS, 1984; DAVIDSON et al., 1985; BERG et al., 1991, WEATHERS et al., 1988; BEYSENS, 1995; HIDY, 2003).

While precipitation and fog deposition is described quite well, there exists only very limited information concerning the occurence and chemical composition of dew and hoarfrost in Poland. The aim of this work is to present information about: (1) chemical composition of the following hydrometeors, (2) the complete view about components of the pollutant deposition in the mountainous regions of the Central European mountains with Szrenica Mt. as an example, as well as (3) how chemical composition differs significantly as a result of, e.g., different circulation direction and speed, origin and the age of air masses and the depth of the atmospheric vertical mixing. On the other hand, it aims at showing the connection between the intensified deposition and the spatial pattern of mountain ecosystems destruction.

1.1. Data and Methods

The following hydrometeors: precipitation (in both liquid and solid forms), dew, hoarfrost, liquid fog and rime samples were collected daily during the period between 01 December 2008 and 28 February 2010. Fog water samples were taken daily with the use of simple passive collectors set 2 m above the

ground (Fig. 2a). Liquid fog droplets ($T > 0°C$) were deposited on 160 strings vertically oriented and forming a cone turned upside down. Strings 17.7 cm long and 0.45 mm in diameter make up a total surface deposition of 400 cm^2. A quite high angle of string arrangement and narrow drainpipe for flowing down fog water were used to effectively distinguish liquid fog deposit from precipitation. Additionally, the collector was shaped by a plastic hood to reduce rain contamination. The lack of water within the collector during rain-only conditions (without fog) is an evidence for its effectiveness.

In the case of rime ($T < 0°C$), the passive collector consists of a duralumin rod of 40 cm length and 3.2 cm in diameter, placed vertically 2 m above the ground. It was used to measure rime efficiency expressed by the weight of the ice deposit and the length of vanes of rime (Fig. 2b). In the case of both collectors, total surface deposition was estimated at 400 cm^2; however, the wind exposed (working) area constitutes half of that and corresponds to inlet surface of standard Hellmann rain gauge (200 cm^2). Finally, it allows comparing relative water flux efficiency for the following types of hydrometeors. To collect rime samples for chemical analysis, a cable 5 m long and 5 mm in diameter was stretched 150 cm above the ground. The cable was polyethylene covered to minimize potential contamination of samples (Fig. 2c).

Dew and hoarfrost samples were collected using a sampler based on the design described by MUSELLI et al. (2002). Surface of this sampler (100 cm × 100 cm) was made of rigid polyethylene foil mounted on a wooden frame, thermally isolated from the ground with 5 cm thick polystyrene foam and mounted 50 cm above the ground with the surface inclined at 15° angle in the western direction (Fig. 2d). Pollutant deposition transferred via precipitation was calculated on the basis of water volume measured by a standard Hellmann gauge and chemistry from the rain collector—a plastic container placed 1.5 m above the ground or snow cover.

Precipitation and fog (liquid and solid) water samples were collected always at the same time, 0600 UTC in the morning. In the case of dew and hoarfrost, it was changeable and took place exactly 30 min after sunrise. Because of the indispensability

Figure 2
Samplers of the following types of hydrometeors used to collect samples and to measure water flux volume: **a** liquid fog, **b** rime (water flux only), **c** rime (samples collection only), **d** dew and hoarfrost

of water volume to chemical analysis, only samples >1 ml were included in the calculations. That is the reason why nine daily samples with precipitation were excluded (daily sum <0.05 mm), 11 and 8 days with minimal efficiency of liquid and solid fog deposits, respectively (low liquid water content or/ and calm wind conditions). Furthermore, in the case of rime deposited and afterwards melted or fallen off before a given sampling term, it was also not taken into account (approximately 5 days). The collection of dew and hoarfrost samples took place only on rainless nights to eliminate the influence of precipitation. Because of that, six samples 'contaminated' by precipitation were eliminated.

To reduce the disturbing effect of ongoing dry deposition, the collecting samplers' surfaces were cleaned before exposure, rinsed with deionized water and wiped. Blank sample analysis revealed that the material of the collector's surface did not have any measurable influence on water chemistry. The volume of samples was measured (first melted of solid) just after collection, then water was transferred to polypropylene containers (50 cm^3) and stored dark in liquid state at low temperature without chemical preservatives because the analysis was performed immediately after the samples were delivered to the laboratory. Except for volume, the samples were analyzed on-site for pH, and conductivity.

Selected anions and cations were quantified against a synthetic rain standard using ion suppressed chromatography (ICS 3000, Dionex Corporation,

USA). This synthetic standard is Reference Material No. 409 (BCR-409, Institute for Reference Materials and Measurements, Belgium) and Analytical Reference Material Rain (National Water Research Institute, Environment Canada; POLKOWSKA et al., 2005). Data quality control was performed by evaluating the percentage difference of the ionic balance (PDI; CINI et al., 2002) calculated as:

$$PDI = \frac{Conc_{anions} - Conc_{Cations}}{Conc_{anions} + Conc_{Cations}}$$

The acceptability criterion was set at: PDI $\leq \pm20\%$. TIC represents the sum of liquid phase concentration of SO_4^{2-}, NO_3^-, Cl^-, H^+, NH_4^+, Ca^{2+}, Mg^{2+}, Na^+ and K^+ (MÖLLER et al., 1996).

Information regarding chemical composition of analyzed hydrometeors was interpreted with the context of meteorological data: type of air mass, back trajectories using Hysplit (Hybrid Single-Particle Lagrangian Integrated Trajectory; DRAXLER et al., 1997, 1998, 2011) as well as a calendar of atmospheric circulation types (NIEDŹWIEDŹ, 2010).

2. Results and Discussion

2.1. General Characteristics of Weather Conditions

The comparison of annual wind roses for altitude representing the Szrenica summit, prepared on the basis of data from aerological soundings for two

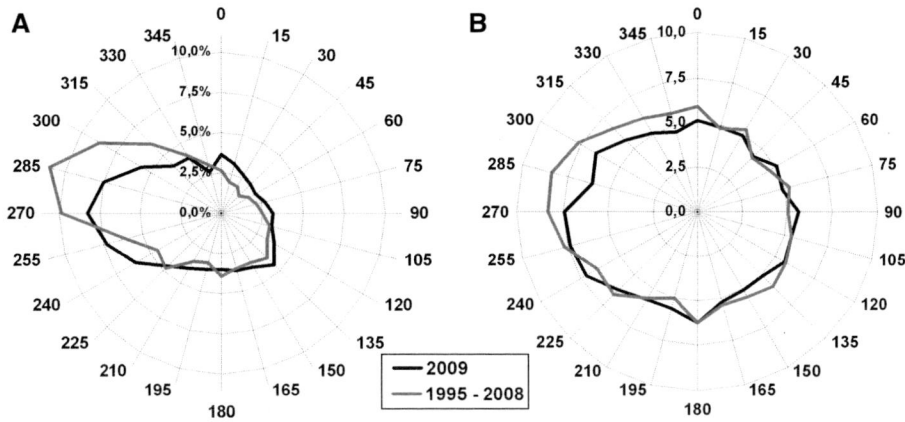

Figure 3

Geostrophic wind direction (**a**) and wind speed (m s^{-1}) (**b**) at the altitude of Szrenica Mt. based on radiosounding data with 15° angular resolution. An average conditions for two periods: 1995–2008 and 2009, based on Wrocław, Praha and Lindenberg data taken into account altogether

periods 1995–2008 and 2009 are presented in Fig. 3a. For both of them the most frequent wind directions were W and WNW with secondary maximum from SE and N–NE sectors. However, the relative contribution of the western sector in 2009 was significantly lower (about 9%) than an average with a small surplus for WSW and the whole eastern sector. It resulted from a more intense advection of continental air masses in 2009 from the east (35.7% days in 2009 and 29.2% on average in a longer period of time). Otherwise, limited advection of cool and humid polar maritime air from the western sector was observed, also connected with less vigorous circulation (57.2% days in 2009 and 64.6% on average). The decrease of annual mean wind speed for W–NW–N sector was between 1 and 2 m s^{-1} when compared with annual average (Fig. 3b). Even more atypical were relationships between precipitation and water flux from fog versus circulation types (Fig. 4). Most fog and precipitation events occurred during W–WNW–NW airflow. In the case of 2009 the sum of precipitation was close to the annual average but the dominating role of NE cyclonic circulation was unusual. During such synoptic circumstances extreme intensity of precipitation and fog water flux occur in the Karkonosze Mts. Hence, during 4 days (23–25 June and 30 May), the sum of precipitation equalled 225 mm (13% when compared with 2009 total).

2.2. Water Flux Estimates

From 01 December 2008 to 28 February 2010, appropriately 131, 238, 53 and 12 samples of precipitation, fog (altogether in liquid and solid forms), dew and hoarfrost were collected respectively on Szrenica Mt (Table 1). Total water flux of the following types of hydrometeors and recalculated into the precipitation unit are the following: 1,719 mm of precipitation, 481 and 458 mm for liquid and solid fog deposits and finally 10.2 and 4.3 mm in case of dew and hoarfrost. Actual water flux from fog, taking into account real land use category (dwarf pine), was estimated in previous measurements as 30% of precipitation (SOBIK and MIGAIA, 1993) which makes approximately 537 mm for the 15 months of the project time.

Dew was rather beyond the role of atmospheric circulation because vigorous circulation considerably limits its frequency and intensity (Fig. 4). Conditions allowing for dew (as well as hoarfrost) formation occurred in the anticyclonic conditions (72.7% dew water flux), when the Sudety Mts. were influenced by the central part of high pressure systems. A lesser probability of dew generation refers to the transition situation (when the circulation was generated in the intermediate zone between the high and low pressure systems; 23% contribution) while dew is almost neglectable during cyclonic conditions (only 4% of total dew water flux).

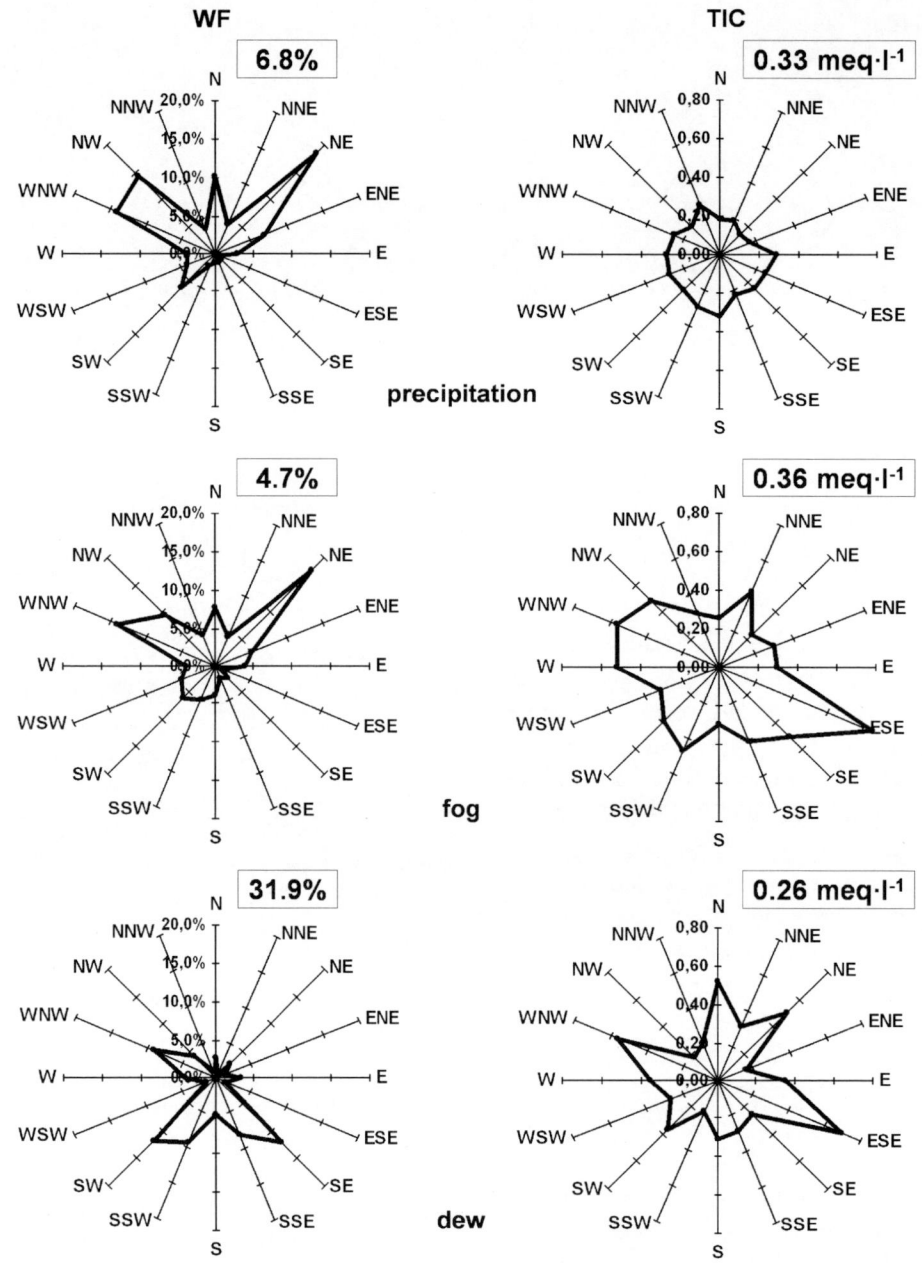

Figure 4
Relative contribution of precipitation, fog and dew in water flux (%) from the atmosphere and Total Inorganic Ionic Content (μeq l^{-1}) at Szrenica Mt. in relation to the direction of atmospheric circulation with 22.5° angular resolution. Appropriate values for undefined circulation are included separately inside *boxes*

In the case of fog, quite a large volume of water was deposited on Szrenica Mt. in the anticyclonic conditions (32%) similarly as during cyclonic and transitional type of weather (37 and 31%, respectively). This should be attributed to the important role of airflow deformation and the air ascent forced by

orography, which often takes place even during anticyclonic conditions and results from the Karkonosze Mts. range compactness and its significant relative altitude.

While data on precipitation and fog water flux to the ground in montane environment of the Sudetes is

Table 1

Conductivity, pH, total inorganic ionic content (TIC), concentrations of the major chemical components in precipitation (P), dew, hoarfrost (H) and fog water samples collected at Szrenica Mt. (rime and liquid fog deposits together due to chemical similarities) as well as total pollutant deposition

Analytes	P	Dew	H	Fog
Number of samples	131	53	12	238
Volume of water flux (mm)	1,719	10.2	4.3	938.8
Conductivity ($\mu s\ cm^{-1}$)	18.0	18.2	9.9	52.8
pH (volume weighted)	4.49	5.30	4.85	4.06
Volume weighted ($\mu eq\ l^{-1}$)				
TIC	233	345	134	425
Na^+	19.5	25.0	17.0	31.4
NH_4^+	32.5	32.3	12.4	79.5
H^+	32.5	5.0	14.0	65.8
K^+	6.9	10.1	1.7	5.9
Mg^{2+}	4.9	30.9	12.2	12.3
Ca^{2+}	36.1	67.7	19.6	52.3
Cl^-	32.3	30.6	24.5	31.9
NO_3^-	28.8	84.3	21.7	75.2
SO_4^{2-}	39.9	30.6	11.3	70.5
Total pollutant deposition ($meq\ m^{-2}$)	400.5	3.5	0.6	398.6

well established, there have not been until present any reports on dew and hoarfrost volume. According to the result of this experiment, dew and hoarfrost at Szrenica site were observed much less often than in other sites in Poland situated in the lowlands or in lower parts of the mountains. This was caused by the combined effect of the increased cloudiness, more frequent precipitation events and much more vigorous wind speed at Szrenica if compared with other sites. The average volume of water collected daily by dew and hoarfrost sampler at Szrenica was 0.193 and 0.358 mm, respectively. Surprisingly a few dew samples at Szrenica were characterized by very high water volume not observed at any other site in Poland. The largest dew volume (0.458 mm) was measured on 28 September 2009 when the Szrenica summit was cloudless and still immersed in moist air of relatively shallow atmospheric boundary layer with much drier air in the remaining part of troposphere where intense anticyclonic subsidence took place. Despite high relative humidity at measurement site, the precipitable water content (PWC) in the air column above Szrenica was very low. In such conditions the atmospheric far infra-red radiation was very low causing the radiation balance to

drop lower than during other nights with higher PWC. As the final result the extreme radiative cooling was responsible for intense dew formation.

2.3. Total Inorganic Ionic Content (TIC)

Because of substantial differences in chemical composition, dew and hoarfrost are presented separately, opposite liquid and solid deposits of fog, similar to each other. Absence of substantial differences between liquid and solid deposits of fog is characteristic for Szrenica Mt. (Sobik, 1999; Błaś et al., 2010). The question is why seasonal variations in pollutant emission seem to not influence the chemistry of liquid and solid fog deposits (typical for warm and cold half of the year, respectively). Possible explanations are as follows:

1. there are significant differences of the atmospheric boundary layer (ABL) depth depending on synoptic situation as well as seasonal and daily cycles, this depth is relatively larger during cyclonic situations, warm half of the year and daylight part of the day;
2. the ABL regional upper limit stays close to the level of the Szrenica summit being frequently below or above;
3. fog predominantly occurs at Szrenica in cyclonic synoptic situations within the all-year polluted ABL, so both rime and liquid fog deposits show similar contamination level;
4. dew and hoarfrost are formed during anticyclonic situations, so hoarfrost mainly during cool part of the year in clean air above the ABL without any air parcels entrained from below, while dew is typical for the warm part of the year when Szrenica is immersed within the ABL for the whole day or at least its significant part.

The highest values of TIC were observed in case of fog (425 $\mu eq\ l^{-1}$) and dew (345 $\mu eq\ l^{-1}$), whereas the much lower values were characteristic for hoarfrost (134 $\mu eq\ l^{-1}$) and precipitation (233 $\mu eq\ l^{-1}$, Fig. 5; Table 1). Independently of airflow direction, TIC values of precipitation were relatively stable ranging from 0.144 to 0.316 $\mu eq\ l^{-1}$ (Fig. 4). Significant maximum was observed during eastern and southern circulation, also characterized by rare precipitation events with more efficient 'wash out' processes. The

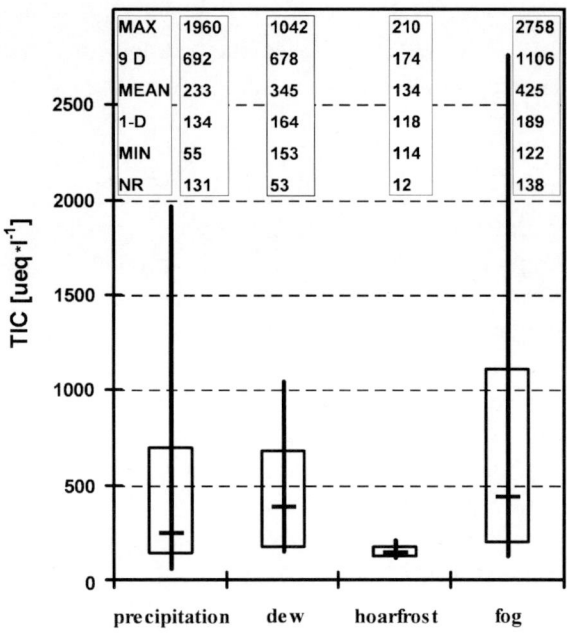

MAX	1960	1042	210	2758
9 D	692	678	174	1106
MEAN	233	345	134	425
1-D	134	164	118	189
MIN	55	153	114	122
NR	131	53	12	138

Figure 5

TIC statistical indices of the hydrometeors collected at the Szrenica Mt.: max., nine decile, volume weighted mean, first decile, min. and the number of samples (NR)

lowest TIC was evident during NE circulation with intense cyclogenesis and the heaviest precipitation. Similar relationships are evident when fog samples are taken into consideration. However, a range of TIC is wider, from 0.239 to 0.848 μeq l^{-1}.

Relatively moderate fog concentrations appear where the largest sources of pollution are found (SW–W circulation). It should be explained by frequent washing out of orographic clouds by atmospheric precipitation which leads to effective pollutant removal (JONES AND CHOULARTON, 1988; DORE et al., 1992; Sobik, 1999). Furthermore, liquid water content (LWC) of fog has also a significant influence on TIC: the smaller LWC and the higher concentration. Hence, cloud water chemistry in the context of relevant emissions and the movement of air masses does not clarify such differences satisfactorily (Kim and Aneja, 1992; Igawa et al., 2001).

During this experiment we did not measure fog liquid water content (LWC) at Szrenica but some former observations showed that LWC at this site varied typically from 0.05 to 1.0 g m^{-3} with median value around 0.20 g m^{-3} (BŁAŚ, 1997). It means that the total load of dissolved pollutants (LWC·TIC)

calculated per volume unit of fog was about 106 neq m^{-3}. Elbert et al. (2000) roughly classified available literature data from 101 fog chemistry measurements around the world. It is grounds for classifying fog pollutant loading at the Szrenica Mt. as heavily polluted, like other locations representing middle-size mountains exposed to highly polluted air from heavy industry (CLARK et al., 1990; Schemenauer et al., 1995; ACKER et al., 1998; COLLETT et al., 2002; ZIMMERMANN and ZIMMERMANN, 2002; LANGE et al., 2003; Fišák et al., 2004; AIKAWA et al., 2005; Watanabe et al., 2006).

The TIC of dew was significantly higher in comparison with precipitation and surprisingly twice as big as for hoarfrost. To explain such differences chemical and physical processes should be taken into consideration: transport, vertical redistribution, chemical transformation, removal from the atmosphere, and impact on the life-time in the case of aerosol and gases (MARINONI et al., 2004). Quite important, in the context of the level of dew and hoarfrost pollution, is the height of the mixing layer and the type of thermal stratification. The relative purity of hoarfrost can be explained by the suppression of vertical mixing during winter anticyclonic weather conditions, when the predominance of hoarfrost is evident. Stable thermal stratification in the ABL limits vertical air exchange, allowing for a high and steady pollutant concentration within the boundary layer. The Szrenica summit is located a few hundred meters above inversion and cloudiness, within subsiding air (Fig. 6a). That is the reason for the minimal concentration of pollutants with anthropogenic origin. The contribution of the major ions (SO_4^{2-} and NO_3^-) to the TIC (25%) is relatively limited, and does not confer to a strong acidity character to hoarfrost samples. The contribution of the maritime aerosol, existing in the upper parts of the atmosphere, becomes relatively more important. Such differences between aerosol composition in the free troposphere and boundary layer are reported by Raes et al. (1997), KLINE et al. (2004), MARINONI et al. (2004) and COZIC et al. (2008). Because of that, hoarfrost chemical composition was quite different from other hydrometeors (Table 1; Fig. 7). The highest contribution was typical for: Cl^-, NO_3^-, Ca^{2+} and Na^+ (18, 16, 15 and 13%, respectively). To

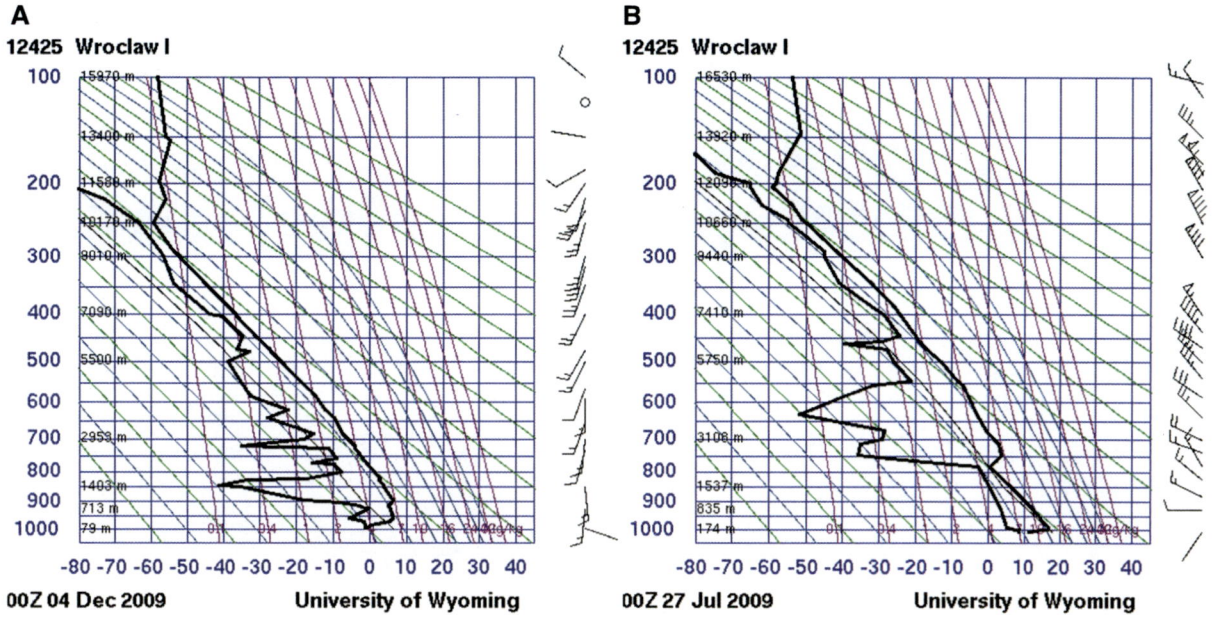

Figure 6
Aerological diagrams from Wrocław meteorological station for two selected days with: **a** hoarfrost collection—winter anticyclonic conditions (4 December 2009); **b** dew collection—summer anticyclonic conditions (27 July 2009; http://www.weather.uwyo.edu)

analyse differences in the air mass origin, backward air mass trajectories were calculated with the trajectory model HYSPLIT. Trajectories arriving at the Szrenica Mt. position at the two levels, Szrenica (1,300 m a.s.l.) and upper free troposphere (4,000 m a.s.l.), were initialized for all days with hoarfrost episodes. Back trajectories for the Szrenica level show a weak circulation (Fig. 8a). Whereas Fig. 8b confirms a significant potential influence of maritime aerosols as the air mass originated close to the surface over the Atlantic Ocean. Vigorous circulation conditions at the 4,000 m a.s.l. are responsible for horizontal transport of maritime aerosol, while air subsidence within anticyclonic system (over Szrenica) leads to its vertical redistribution. This indicates that despite the inland location of this site, maritime components are at least significant in overall deposition.

Pollutants are more efficiently transported and mixed under anticyclonic unstable conditions in the warmer half of the year, when dew is more typical ($T > 0°C$). Hence, the depth of the mixing layer is largest then, usually extending considerably beyond the altitude of Szrenica (Fig. 6b). The opposite of the winter anticyclonic situation, pollutants concentration

of anthropogenic origin becomes more important (Table 1).

To characterize some emission-deposition relationships of different forms of atmospheric water flux to the ground, the episodes with top ten concentrations separately for precipitation, fog (liquid deposit + rime) and dew were selected from the whole data set. For collection dates of these samples, 3-day back trajectories of the coming airmass were shown at the level of Szrenica observatory (Fig. 9). Each of these sub-categories presents its individual features influenced by different atmospheric processes. In the case of fog (Fig. 9b) and dew (Fig. 9c) almost all trajectories come from major source areas of pollutant emission. Fog water polluted most was collected almost entirely during the advection of humid air masses from the western sector where substantial emission takes place. As far as dew is concerned, the presented back trajectories are zonal with the western sector almost balanced by its eastern counterpart coming from densely populated industrial region of Upper Silesia as well as the remaining part of southern Poland. All ten cases of atmospheric circulation from both sectors were connected with anticyclonic weather and relatively dry air. Usually if

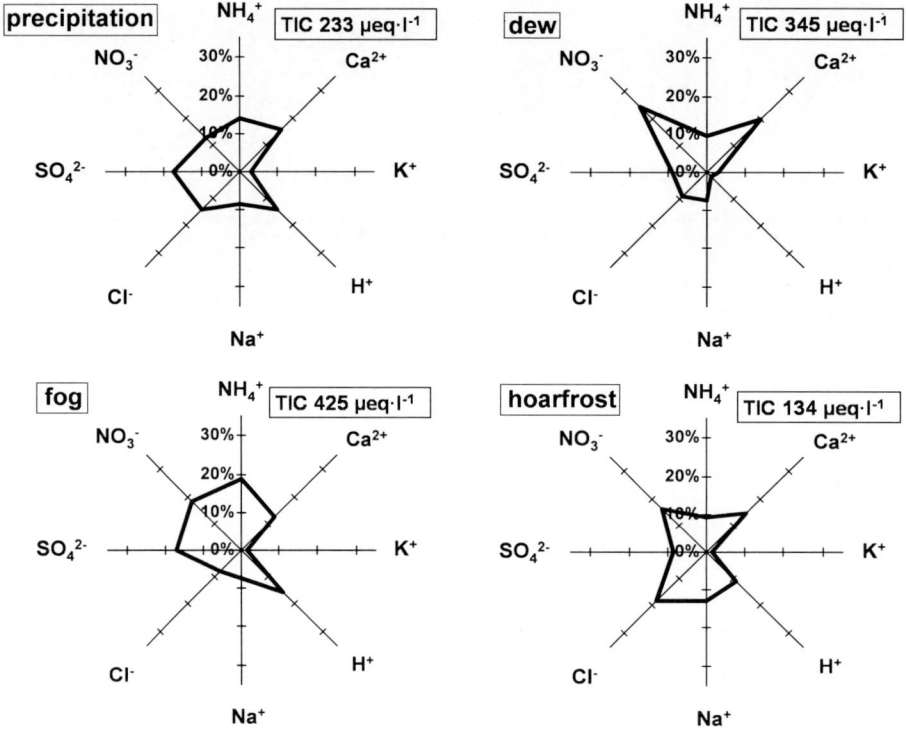

Figure 7
Relative contribution of the major chemical components in relation to TIC as a 100%

air flows from the east, it is not humid enough to produce orographic fog at Szrenica but contains sufficient amount of moisture to condense as dew at night.

The spatial pattern of back trajectories leading to the ten most polluted precipitation events is a little surprising because of its almost even distribution between all sectors (Fig. 9a). A possible explanation is the fact that atmospheric processes producing precipitation particles take place mainly in higher layers of troposphere than the level of Szrenica observatory where precipitation samples were collected. That's why the air parcels where precipitation is formed rather do not follow the presented trajectories but are transported in a different way, presumably being more concentrated in the western sector.

2.4. Ionic Composition

For fog water samples equilibrium between principal ions NH_4^+, NO_3^- and SO_4^{2-} was characteristic

(19, 18 and 17% of TIC as a 100%, respectively). The average total concentration for these three ions as well as for H^+, reflecting the dominant influence of anthropogenic sources on cloud water chemistry, reached about 69% with respect to TIC (Fig. 7). During typical W–NW wind circulation Szrenica Mt. is exposed to highly polluted air from heavy industry, densely situated at a distance of tens to hundreds of kilometres on the windward side of the mountains. Because of altitude, differences in boundary layer depth and much more vigorous circulation, especially during cyclonic weather, the chemistry of fog water collected at Szrenica Mt. may show major contribution from emission sources located at much larger upwind distances (BŁAŚ et al., 2010; MÖLLER et al., 1996). Concentration of ammonia, emitted in main part by agriculture and livestock operations are much lower than the sum of nitrite and sulphate concentration, resulting in only partial neutralization of the existing fog water acidity. Sulphates, which were the most important constituents of cloud water chemistry in the early 1990s (30%), became the third most

A

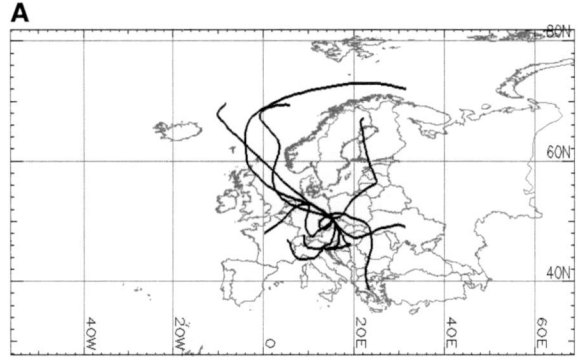

B

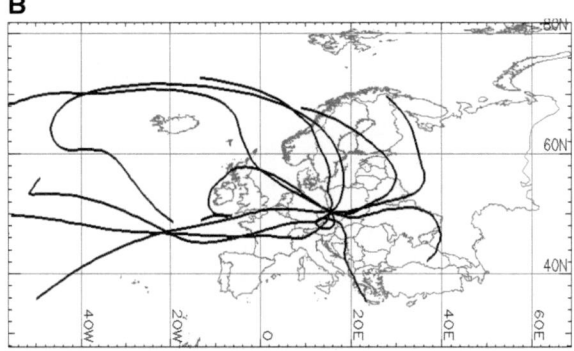

Figure 8
Compilation of 3 days back trajectories ending at Szrenica Mt., 1,300 m a.s.l. (**a**) and 4,000 m a.s.l. (**b**) at 01 UTC for all days with hoarfrost measurements

important ion, with their actual contribution being two times lower (Błaś *et al.*, 2010). Relatively high concentration of nitrates can be clarified by increasing emissions from traffic (Acker *et al.*, 1998; Mitosek *et al.*, 2004).

The following components: SO_4^{2-}, NO_3^-, Ca^{2+} and NH_4^+ were the principal ions in the chemical structure of dew and precipitation (Fig. 7). It is not surprising that sulphates are the most important constituents of precipitation (17%). Central Europe (including Poland, Germany and the Czech Republic) is one of the most heavily polluted areas on the continent—plagued for decades by choking coal dust emitted by electric power plants and district heating plants. However, a distinct reduction of SO_2 emissions, especially from power stations, is observed thanks to desulphurization methods. Emission was achieved by the use of electrostatic precipitators and circulating fluidized bed boilers, which absorb 90% of the SO_2, which would otherwise have gone into the atmosphere (Libicki, 1998). As a result, a relative

decrease (in relation to TIC) occurred in the case of sulphates in precipitation samples, from 30% in the beginning of the 1990s (Błaś *et al.*, 2002) to 17% in 2009. These changes resulted in a spectacular decrease of acidity, expressed by an increase of over half a unit in pH values (from 3.8 in 1990 to 4.49 in 2009). Besides a distinct increase is visible in the case of Ca^{2+} contribution (from 2% in the beginning of the 1990s to 15% in 2009) which might be explained by the application of $CaCO_3$ in the desulphurisation process in some electric power plants and increased construction activity in the surrounding regions. It is also evident in the case of dew samples (20%).

2.5. Pollutant Deposition

The estimation of pollutants load accumulated by a given hydrometeor is possible when the average TIC and the water flux reaching a unit area are known (Fig. 10). According to TIC value equal to 314 μeq l^{-1} and 14.5 mm of water, pollutant load accumulated by dew and hoarfrost (between 01 December 2008 and 28 February 2010) was around 4.5 meq m^{-2}. It was only 1.2% of the pollutants load delivered via atmospheric precipitation. A quite different feature was reported by Polkowska *et al.* (2008, 2009) for the European lowlands, where dew and hoarfrost are responsible at least for an additional 60% of pollutants flux into the ground when compared with precipitation. Such a discrepancy should be explain by relatively high pollutants concentration (even 6–7 times larger than precipitation) and annual water equivalent estimated by Hutorowicz (1963) as 53 mm and 14 mm, in case of dew and hoarfrost, respectively in the lowland areas. However over convex landforms in mountainous areas especially dew and hoarfrost frequency is limited due to much more windy and cloudy weather conditions.

In the case of fog, due to 274 foggy days/year accompanied by high wind speed, pollutants load is much more important but substantial differences exist depending on particular location. In this case, land use expressed by roughness is a very important factor responsible for the huge spatial differentiation in fog water flux on a microscale (distance between few and hundreds metres). At the Szrenica summit, covered

A

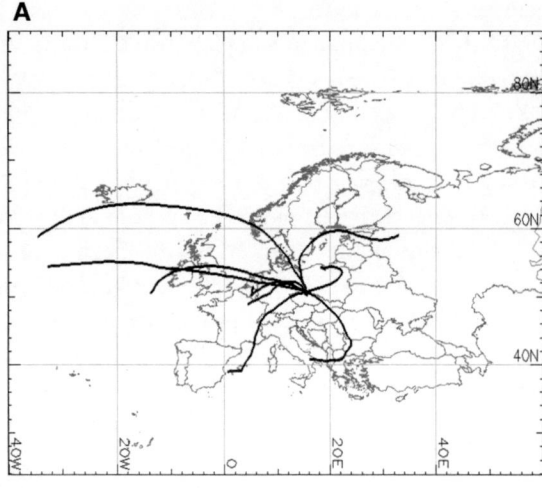

B

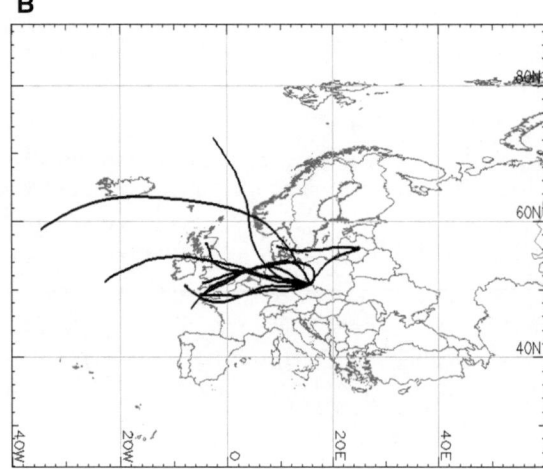

C

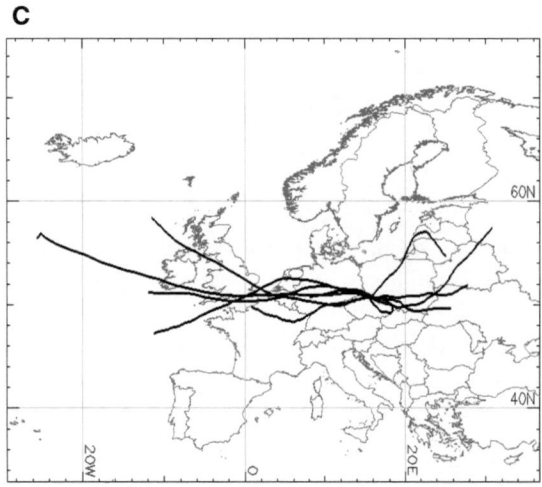

Figure 9

Compilation of 3-day back trajectories ending at Szrenica Mt., 1,300 m a.s.l., at 01 UTC for ten cases with the highest TIC values: **a** precipitation **b** fog and **c** dew

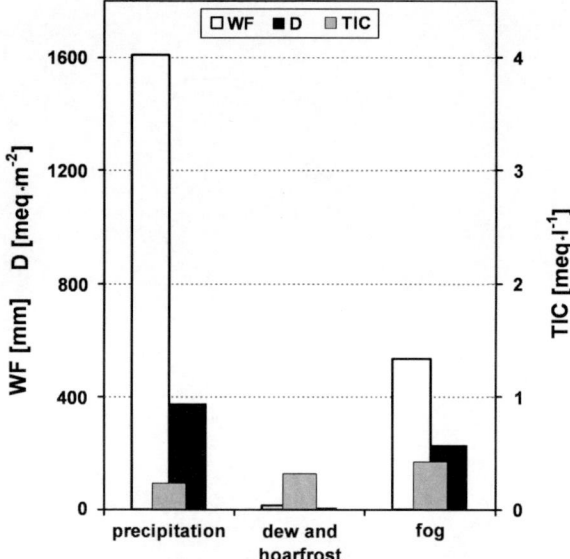

Figure 10

Water flux (*WF*), TIC, and pollutant deposition (*D*) via precipitation, fog, dew/hoarfrost sampled at the Szrenica Mt. between 01 December 2008 and 28 February 2010

by dwarf pine, water flux from fog is estimated as 30% when compared with precipitation treated as 100%. However, pollutant deposition via fog is comparable with precipitation. On well exposed ridges covered by coniferous forest, it could be expected that the role of pollutant deposition by fog may even dominate (Błaś and Sobik, 2003; Lange et al., 2003; Zimmermann and Zimmermann, 2002).

3. Summary and Conclusions

Except for dry and precipitation induced wet deposition, there also exist other pathways of atmospheric pollutant flux onto the ground. These are different kinds of atmospheric deposits in the form of hydrometeors, which do not belong to precipitation categories: dew, hoarfrost, rime and liquid fog deposit. The aim of this work was to identify how their chemical composition differs significantly as a result of, e.g., different circulation direction and speed, origin and the age of air masses and the depth of the atmospheric vertical mixing. Furthermore, their contribution in total wet deposition at Szrenica Mt.

During favourable weather conditions (stable thermal stratification, higher wind speed and lack of precipitation), the removal rate of aerosols and gaseous components from air is considerable limited and pollutants can affect distant regions, even those several hundred or thousands of kilometres away from emission sources. Higher efficiency of pollutants removal as well as transport limitations occur when precipitation is more abundant. Unexpectedly, fog and dew are better indices of long range transport of pollutants than atmospheric precipitation. Chemistry of slope fog collected at Szrenica might be used via 'back trajectories analysis' to characterize distant emission and pollutant deposition due to long range transport during the advection of humid air masses. If the coming air mass is relatively dry such role can be played by dew chemistry. Also, the relative purity of hoarfrost can be explained by the suppression of vertical mixing during winter anticyclonic weather conditions. Stable thermal stratification in boundary layer limits vertical air exchange, allowing for a high and steady pollutants concentration within boundary layer.

According to the results of this experiment, dew and hoarfrost at the Szrenica site were observed much less than in other sites in Poland situated in the lowlands. Surprisingly, isolated dew samples at Szrenica were characterized by the largest dew volume not observed at any other site in Poland. However, water flux from fog (in the form of rime and liquid deposits) tends to be an even more important pathway than precipitation. TIC of fog was much higher than precipitation with the average TIC ratio 1.8. These different concentrations are caused by the fact that fog and low cloud droplets form in the polluted environment of the ABL where condensation nuclei are abundant, while raindrops and snowflakes typically come from much cleaner higher zones of the troposphere.

In all types of atmospheric deposits, different ions play an important role: SO_4^{2-} and Ca^{2+} in precipitation, NO_3^- and Ca^{2+} in dew, Cl^- and NO_3^- in hoarfrost as well as NH_4^+, NO_3^- and SO_4^{2-} in fog. The chemical composition differs significantly as a result of, e.g., different circulation direction and speed, origin and the age of air masses and the depth of the atmospheric vertical mixing.

Taking into account both TIC and volume of deposited water, dew and hoarfrost form a neglectable path of pollutants deposition to the ground being responsible for approximately a 1.2% contribution of the deposited pollutants if compared with atmospheric precipitation. This does not apply to fog which becomes a significant or even primary component of wet deposition at mountain ridges leading to the destructive environmental results.

Acknowledgments

This scientific work was financially supported by the Polish Ministry of Science and Higher Education in years 2008–2010 as a research project (NN305 231035 and NN305 373438).

References

ACKER, K., MÖLLER, D., MARQUARDT, W., BRÜGGEMANN, E., WIEPRECHT, W., AUEL, R., KALAß, D. (1998), *Atmospheric research program for studying changing emission patterns after German unifications*, Atmospheric Environment *32*, 3435–3443

AIKAWA, M., HIRAKI, T., SROGA, M., TAMAKI, M. (2005), *Chemistry of fog water collected in the Mt. Rokko Area (Kobe city, Japan) beween April 1997 and March 2001*, Water, Air, and Soil Pollution *160*, 373–393

BARANOWSKI, S. (1974) Meteorological Bulletin, No 26 (1), Part B: Source Data, Acta Universitatis Wratislaviensis, No 206, Państwowe Wydawnictwo Naukowe, Wrocław 1974.

BERG, N., DUNN, P., FENN, M. (1991) *Spatial and temporal variability of rime ice and snow chemistry at five sites in California*, Atmospheric Environment *25A*, 915-926.

BEYSENS, D. (1995) *The formation of dew*, Atmospheric Research *39*, 215-237.

BŁAŚ, M. (1997), Experimental measurement of cloud liquid water content (LWC) in summer and winter conditions, Acta Universitatis Wratislaviensis, Prace Instytutu Geograficznego, Meteorologia i Klimatologia IV, 147-154.

BŁAŚ, M., SOBIK, M., QUIEL, F., and NETZEL, P. (2002), *Temporal and spatial variations of fog in the Western Sudety Mts., Poland*, Atmospheric Research *64*, 19-28.

BŁAŚ, M., and SOBIK, M. (2003), *Natural and human impact on pollutant deposition in mountain ecosystems with the Sudetes as an example*, Studia Geograficzne *75*, 420-438.

BŁAŚ, M., and SOBIK, M. (2004), *The distribution of fog frequency in the Carpathians*, Geographia Polonica *77*, 19-34.

BŁAŚ, M., POLKOWSKA, Ż., SOBIK, M., KLIMASZEWSKA, K., NOWIŃSKI, K., and NAMIEŚNIK, J. (2010), *Fog water chemical composition in different geographic regions of Poland*, Atmospheric Research *95*, 455-469.

CINI, R., PRODI, F., SANTACHIARA, G., PORCU, F., BELLANDI, S., STORTINI, A.M., OPPO, C., UDISTI, R., PANTANI, F. (2002),

Chemical chara-cterization of cloud episodes at a ridge site in Tuscan Appennines, Italy, Atmospheric Research *61*, 311-334.

CLARK, P.A., GERVAT, G.P., HILL, T.A., MARSH, A.R.W., CHANDLER, A.S., CHOULARTON, T.W., GAY, M.J. (1990), *A field study of the oxidation of SO₂ in cloud*. Journal of Geophysical Research *95*, 13985-13995.

COLLETT Jr, J.L., BATOR, A., SHERMAN, D.E., MOORE, K.F., HOAG, K.J., DEMOZ, B.B., RAO, X., REILLY, J.E. (2002), *The chemical composition of fogs and intercepted clouds in the United States*, Atmospheric Research *64*, 29-40.

COZIC, J., VERHEGGEN, B., WEINGARTNER, E., CROSIER, J., BOWER, K.N., FLYNN, M., COE, H., HENNING, S., STEINBACHER, M., HENNE, S., COLLAUD COEN, M., PETZOLD, A., BALTENSPERGER, U. (2008), *Chemical composition of free tropospheric aerosol for PM1 and coarse mode at the high alpine site Joungfraujoch*, Atmospheric Chemistry and Physics *8*, 407-423.

DAVIDSON, C.I., SANTHANAM, S., FORTMANN, R.C., OLSON, M.P. (1985), *Atmospheric transport and deposition of trace elements onto the Greenland ice sheet*, Atmospheric Environment *19*, 2065-2081.

DORE, A.J., CHOULARTON T.W., BROWN, R., BLACKALL, R.M. (1992), *Orographic rainfall enhancement in the mountains of the Lake District and Snowdonia*, Atmospheric Environment *26A*, 357-371.

DORE, A.J., SOBIK, M., and MIGAŁA, K. (1999), *Patterns of precipitation and pollutant deposition in the Western Sudety Mountains, Poland*, Atmospheric Environment *33*, 3301-3312.

DRAXLER, R.R., HESS, G.D. (1997), Description of the HYSPLIT_4 modeling system. NOAA Tech. Memo. ERL ARL-224, NOAA Air Resources Laboratory, Silver Spring, MD, 24 pp.

DRAXLER, R.R., HESS, G.D. (1998), *An overview of the HYSPLIT_4 modeling system of trajectories, dispersion, and deposition*, Aust. Meteor. Mag. *47*, 295-308.

DRAXLER, R.R., ROLPH, G.D. (2011), HYSPLIT (HYbrid Single-Particle Lagrangian Integrated Trajectory) Model access via NOAA ARL READY Website (http://ready.arl.noaa.gov/HYSPLIT.php), NOAA Air Resources Laboratory, Silver Spring, MD.

ELBERT, W., HOFFMANN, M.R., KRAMER, M., SCHMITT, G., ANDREAE, M.O. (2000), *Control of solute concentrations in cloud and fog water by liquid water content*, Atmospheric Environment *34*, 1109-1122.

FIŠÁK, J., TESAŘ, M., ŘEZÁČOVÁ, D., ELIAS, V., WEIGNEROVÁ, V., FOTTOVÁ, D. (2004), *Pollutant concentrations in fog and low cloudwater at selected sites of the Czech Republic*, Atmospheric Research *64*, 75-87.

GARLAND, J.A. (1978), *Dry and wet removal of sulphur from the atmosphere*, Atmospheric Environment *12*, 349-362.

HEGG, D.A., HOBBS, P.V. (1981), *Cloud water chemistry and the production of sulfates in clouds*, Atmospheric Environment *15*, 1597-1604.

HICKS, B.B., (1984), Deposition both wet and dry. [in:] HICKS, B.B., TEASEY, J.I., (Eds.), Acid. Precip. Serie 4, Butterworths, London.

HIDY, G.M., (2003), *Snowpack and precipitation chemistry at high altitudes*, Atmospheric Environment *37*, 1231-1242.

HUTOROWICZ H. (1963) Dew measurements at Olsztyn [in:] Assamblée Generale de Berkeley 1963, Gentbrugge 1964 8', UGGI Association International d'Hydrologie Scientifique 65, 352-359.

IGAWA, M., MATSUMURA, K., OKOCHI, H. (2001), *Fog water chemistry at Mt. Oyama and its dominant factors*, Water Air & Soil Pollution *130*, 607-612.

JONES, A., CHOULARTON, T.W. (1988), *A model of wet deposition to complex terrain*, Atmospheric Environment *22*(11), 2419-2430.

KIM, D.S., ANEJA, V.P. (1992), Chemical composition of clouds at Mount Mitchell, North Carolina, Tellus 44B, 41-53.

KLINE, J., HUEBERT, B., HOWELL, S., BLOMQUIST, B., ZHUANG, J., BERTRAM, T., CARRILLO, J. (2004), *Aerosol composition and size versus altitude measured from the C-130 during ACE-Asia*, Journal of Geophysical Research *109*, D19S08, doi:10.1029/2004JD004540.

LANGE, Ch.A., MATSCHULLAT, J., ZIMMERMANN, F., STERZIK, G., WIENHAUS, O. (2003), *Fog frequency and chemical composition of fog water—a relevant contribution to atmospheric deposition in the eastern Erzgebirge, Germany*, Atmospheric Environment *37*, 3731-3739.

LIBICKI, J. (1998), Brown coal in Poland today and after the 21st century, Poltegor-Project, Wrocław, Poland, pp. 1-20.

LORENC, H. (2004), Atlas klimatu Polski, IMGW Press, Warszawa 2005.

MARINONI, A., LAJ, P., SELLEGRI, K., MAILHOT, G. (2004), *Cloud chemistry at the Puy de Dôme: variability and relationships with environmental factors*, Atmospheric Chemistry and Physics *4*, 715-728.

MIGAŁA, K., PEREYMA, J., SOBIK, M., and SZCZEPANKIEWICZ-SZMYRKA, A. (1993), Climatic conditions at the Karkonosze during the warm half of the year 1992, Karkonoskie Badania Ekologiczne, I Konferencja, Wojnowice, 3-4 grudnia 1992, Oficyna Wydawnicza Instytutu Ekologii PAN, 47-70.

MIGAŁA, K., PEREYMA, J., SOBIK, M., SZCZEPANKIEWICZ-SZMYRKA, A. (1995), Współczesne warunki klimatyczne i zróżnicowanie topoklimatyczne Karkonoszy, Problemy ekologiczne wysokogórskiej części Karkonoszy (ed. Fischer, Z.), Instytut Ekologii PAN, Dziekanów Leśny, 51-78.

MIGAŁA, K., LIEBERSBACH, J., SOBIK, M. (2002), *Rime in the Gigant Mts. (The Sudetes, Poland)*, Atmospheric Research *64*, 63-73.

MITOSEK, G., DEGÓRSKA, A., IWANEK, J., PRZYBYLSKA, G., SKOTAK, K. (2004), EMEP Assessment Report—Poland, Institute of Environmental Protection, Warsaw.

MÖLLER, D., ACKER, K., and WIEPRECHT, W. (1996), *A relationship between liquid water content and chemical composition in clouds*, Atmospheric Research *41*, 321-335.

MUSELLI, M., BEYSENS, D., MARCILLAT, J., MILIMOUK, I., NILSSON, T., and LOUCHE, A. (2002), *Dew water collector for potable water in Ajaccio (Corsica Island, France)*, Atmospheric Research *64*, 297-312.

NIEDŹWIEDŹ, T. (2010), Calendar of atmospheric circulation types for the Southern Poland—Internet database, Uniwersytet Śląski, Katedra Klimatologii, Sosnowiec 2010.

PEREYMA, J., SOBIK, M., SZCZEPANKIEWICZ-SZMYRKA, A., MIGAŁA, K. (1997), Contemporary climatic conditions and topoclimatic differentiation of the Karkonosze Mts., Acta Universitatis Wratislaviensis, No 1950, Seria C. Meteorologia i Klimatologia, Vol. IV, 75-94.

POLKOWSKA, Ż., ASTEL, A., WALNA, B., MAŁEK, S., MĘDRZYCKA, K., GÓRECKI, T., SIEPAK, J., and NAMIEŚNIK, J. (2005), *Chemometric analysis of rainwater and throughfall at several sites in Poland*, Atmospheric Environment *39*, 837-855, 2005.

POLKOWSKA, Ż., BŁAŚ, M., KLIMASZEWSKA, K., SOBIK, M., MAŁEK, S., and NAMIEŚNIK, J. (2008), *Chemical characterization of dew water collected in different geographic regions of Poland*, Sensors *8*(6), 4006-4032.

POLKOWSKA, Ż., SOBIK, M., BŁAŚ, M., KLIMASZEWSKA, K., WALNA, B., and NAMIEŚNIK, J. (2009), *Hoarfrost and rime chemistry in Poland—an introductory analysis from meteorological perspective*, J. Atmos. Chem. *52*, 5-30.

RAES, F., VAN DINGENEN, R., CUEVAS, E., VAN VELTHOVEN, P.F.J., PROSPERO, J.M. (1997), Observations of aerosols in the free troposphere and marine boundary layer of the subtropical Northeast Atlantic: Discussion of processes determining their size distribution, Journal of Geophysical Research 102, NO. D17, 21315-21328.

SCHEMENAUER, R.S., BANIC, C.M., URQUIZO, N. (1995), *High elevation fog and precipitation chemistry in southern Quebeck, Canada*, Atmospheric Environment *29*, 2235-2252.

SOBIK, M., MIGAIA, K. (1993), *The role of cloudwater and fog deposits on the water budget in the Karkonosze (Giant) Mountains*, Alpex Regional Bulletin *21*, 13-15.

SOBIK, M. (1999), Meteorologiczne uwarunkowania zakwaszenia hydrometeorów w Karkonoszach, Unpublished Ph.D. Thesis, Wrocław, University of Wrocław.

SOBIK, M., NETZEL, P., and QUIEL, F. (2001), Zastosowanie modelu rastrowego do określenia pola rocznej sumy opadów atmosferycznych na Dolnym Śląsku, Rocznik Fizyczno-Geograficzny, Vol. VI, 27-34.

WATANABE, K., TAKEBE, Y., SODE, N., IGARASHI, Y., TAKAHASHI, H., DOKIYA, Y. (2006), *Fog and rain water chemistry at Mt. Fuji: A case study during the September 2002 campaign*, Atmospheric Research *82*, 652-662.

WEATHERS, K.C., LIKENS, G.E., BORMANN, F.H., BICKNELL, S.H., BORMANN, B.T., DAUBE B.C. Jr., EATON, J.S., GOLLOWAY, J.N., KEENE, W.C., KIMBALL, K.D., McDOWELL, W.H., SICCAMA, T.G., SMILEY, D., TARRANT, R.A. (1988), *Cloudwater chemistry from ten sites in Noth America*, Environ. Sci. Tech. *22*, 1018-1026.

ZIMMERMANN, L., ZIMMERMANN, F. (2002), *Fog deposition to Norway Spruce stands at high-elevation sites in the Eastern Erzgebirge (Germany)*, Journal of Hydrology *256*, 166-175.

(Received November 29, 2010, revised May 16, 2011, accepted May 25, 2011, Published online June 25, 2011)

Reprinted from the journal

Pure Appl. Geophys. 169 (2012), 1083–1091
© 2011 Springer Basel AG
DOI 10.1007/s00024-011-0345-8

Typical Insoluble Particles in Fog Water at Milešovka Observatory (Czech Republic)

Jaroslav Fišák,[1] Valeria Stoyanova,[2] Kristýna Bartůňková,[1] Miroslav Tesař,[3] and Annie Shoumkova[2]

Abstract—This study concerns insoluble chemical pollution in fog at the Milešovka Observatory in the Czech Republic. From August 2006 to July 2007, 25 fog samples at the top of Milešovka Mountain in the České Středohoří Mountain Range were collected with an active fog water collector. Water samples were filtered to obtain the insoluble particles. A range of 53–116 particles from every sample was chosen according to the quantity of the particles found in the dried filters. Altogether more than 2,000 particles were analyzed. The particles were examined with the help of a scanning electron microscope and energy dispersive X-ray spectrometer to distinguish sizes, shapes and composition. After analyzing the data, a statistical evaluation was made. The particles were categorized according to their shapes (spherical × non-spherical), sizes (coarse particles, $PM_{2.5-10}$, $PM_{1-2.5}$ and PM_1) and composition. Typical (frequently represented) particles with a content of a given element greater than 5% (element-rich particles), such as Al-, Si-, K-, Fe- or Ca-rich particles were determined. The focus was also on particles with elements rarely represented in the atmosphere, like Ni, Au, Pb, Cu, Zr and Ba. Groups of typical insoluble particles were classified according to the meteorological conditions, synoptic situations and wind directions that prevailed in the days of the fog events to find out the possible sources of this fog pollution.

Key words: Fog pollution, insoluble particles, meteorological conditions, sources of pollution.

1. Introduction

Pollution in the atmosphere significantly influences human health and ecosystems. The formation of fog and low clouds plays a significant role in washout processes in the atmosphere (Li and Shao 2009). Therefore, fog chemistry research has become an important topic of investigation in last few decades.

Particles in the atmosphere are products of both natural and/or anthropogenic sources. Natural sources of air-borne dust are sea salt, terrestrial dust, volcanoes, forest fires and bioaerosols. The dust particles usually have a size of approximately 10 μm. Anthropogenic sources are primarily products of combustion processes from vehicle traffic and power plants, cement factories, lime-kilns, quarries and mining or particles from building sites or areas devoid of vegetation. These kinds of aerosol particles are usually smaller. Another significant anthropogenic source of pollutants in the atmosphere is agriculture (Hovorka, 2010; Irz, 2010).

Determining the natural and anthropogenic sources of pollutants in the atmosphere is very difficult. Most elements found in our dataset are common components of natural ambient conditions, and most of them can also have origins as products of anthropogenic activity.

It must be noted that some particles in fog and low clouds are natural, and are necessary for fog and cloud genesis. It is important to examine the character of the material that occurs in fog water and to determine the possible sources of this pollution. The objective of this study was to distinguish natural and anthropogenic sources of insoluble aerosols collected in fog samples.

There have been numerous studies focusing on atmospheric aerosols and fog chemistry. Some of these studies have focused on the impact of air pollution to human health (Poppe and Dockery, 2006; Brook et al., 2004). These studies note that it is mainly fine particles that can have damaging effects on the respiratory and cardiovascular system. Other investigators emphasize the influence of air and fog pollution on ecosystems (Schemenauer 1986; Lin et al., 1997) or buildings (Monte and Rossi, 1997). In the Czech Republic, fog

---
[1] Institute of Atmospheric Physics, AS CR, v.v.i., 1404 Boční II, 141 31 Prague 4, Czech Republic. E-mail: fisak@ufa.cas.cz

[2] Institute of Physical Chemistry, Bulgarian Academy of Sciences (IPC BAS), Sofia, Bulgaria. E-mail: valeria@ipc.bas.bg

[3] Institute of Hydrodynamics, AS CR, v.v.i., Pod Paťankou 30/5, 166 12 Prague 6, Czech Republic.

chemistry has been studied by researchers such as FISAK *et al.*, (2001, 2002a, b, 2009).

In this study, both typical and rare insoluble particles were collected and analyzed with the help of an active fog collector, a scanning electron microscope and energy X-ray spectrometer. Typical particles are those frequently represented in the dataset (particles containing elements that are present in more than 1% of the whole dataset). Particles containing elements that are present in less than 1% are labeled as rare particles.

After statistical evaluation of the content of the elements in the particles, along with their sizes and shapes, the results were classified according to meteorological conditions—synoptic situations and wind directions that prevailed during the days of the foggy periods. The goal was to find a possible connection between meteorological conditions and potential sources of pollution in the surroundings of the Milešovka Mountain and the insoluble pollutants found in fog.

2. Site of Measurement

Fog samples were collected at the Milešovka observatory, which belongs to the Institute of Atmospheric Physics, Academy of Science, Czech Republic. This observatory lies on the top of Homonymic Mountain, which is the highest mountain of the České Stredohoří Mountain Range in the north of the Czech Republic (Fig. 1). This mountain reaches a height of 837 m above sea level. The mountains are of volcanic origin and primarily composed of igneous rocks like phonolite and basalt. There are also some fillings of claystone, coal and calcite. Milešovka Mountain has a conical shape, and it rises to 400 m above the surrounding terrain. This area is one of the windiest places in the Czech Republic. The most common wind directions are W, NW, SW, N and SE (ŠTEKL *et al.*, 2005).

Fog events occur often here. The annual number of foggy days from the period of 1961–1990 was 224.9, and they occurred mostly in the colder half of the year with a maximum in December (ŠTEKL *et al.*, 2005). The top of Milešovka Mountain has mostly advection fog that primarily belongs to the category of low clouds. Previous studies have shown that the fog and low cloud pollution in this area is higher than in other areas of the Czech Republic (FISAK *et al.*, 2002a, b). In this study pollution (ion concentration) in the fog at Milešovka Mountain was twice as high

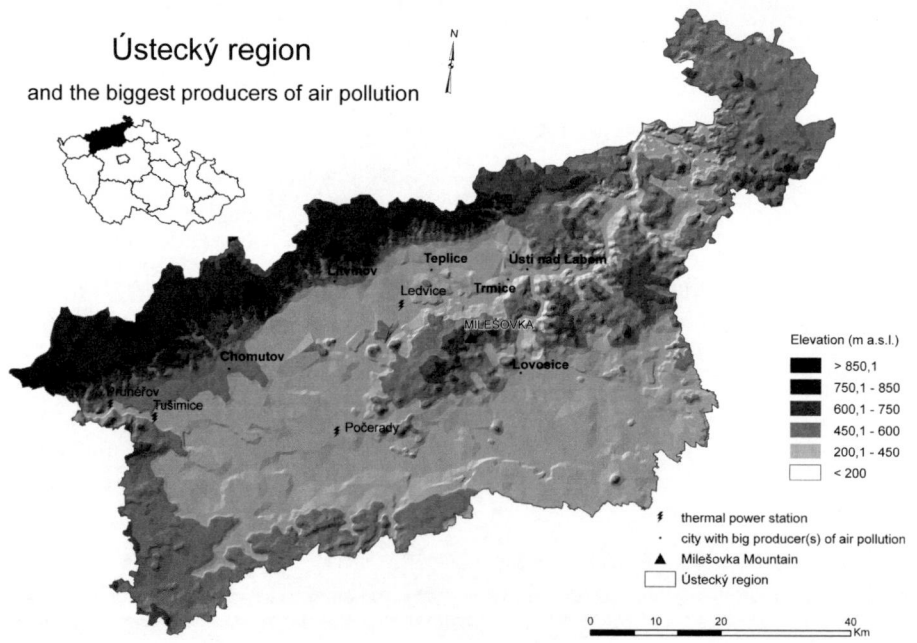

Figure 1
The map of the Milešovka mountain region

on average than that found in fog water at the Churáňov, Šumava Mountains.

The region of the České Stredohoří Mountains belongs to one of the most polluted areas in the Czech Republic. There are many thermal power stations, brown coal mines, chemical industries and large urban and industrial areas here, such as the capital city Prague, Ústí nad Labem, Most, Litvínov, Chomutov, Teplice, Litoměřice and Lovosice. Figure 1 shows a map of the Ústecký region with the biggest producers of air pollution.

3. Methods

Twenty-five samples of fog were collected in the period from August 2006 to July 2007 throughout all seasons of the year. An active fog water collector was used as a sampler (Fig. 2). The active fog water collector is described in DAUBE et al., (1987) and TESAŘ et al., (1995). At the time of collection, the following synoptic situations prevailed in most cases (according to BRÁDKA et al., 1961; CHMI, 2010): SWc2 (southwestern cyclonic synoptic situation), B (trough of depression), Wc (western cyclonic synoptic situation), NWc (northwestern cyclonic synoptic situation) and NWa (northwestern anticyclonic synoptic situation).

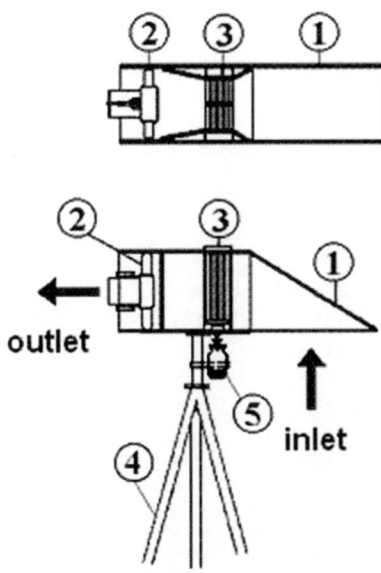

Figure 2

A scheme of the active fog water collector according to DAUBE et al., (1987) (1—sampler box; 2—fan; 3—cartridge with Teflon fibres; 4—sampler stand; 5—sample pot)

The wind directions on these foggy days were mostly NNW (north–northwest), WNW (west–northwest) and WSW (west–southwest). The wind speed ranged between 6 and 14 m s^{-1}. On 18 of the foggy days, from the total of 25, it was raining, which can instigate wash-out processes. Precipitation amounts 24 h before the fog event were up to 34.9 mm. However, the scavenging of insoluble particles by precipitation was not verified in this study.

The collected volume of fog water ranged from 50 to 400 ml. The duration of fog collection was from 3 to 47.5 h. The samples were filtered by cellulose nitrate filters with pore sizes of 0.45 μm (Sartorius AG). The filters were dried and covered with a thin carbon coating. The carbon coating was used as a conducting layer, which is necessary for analysis by SEM and EDX. Regrettably, it was not possible to cover filters by different kinds of coating (e.g., gold) in this case, which would likely be more beneficial for this kind of analysis. Therefore, no carbon could be detected, even though it is very likely that carbon is an important component of the collected aerosols. Additionally oxygen was excluded from the analysis since it did not contribute to the goals of the investigation.

A large number of particles were found on every filter, and a small fraction of these were analyzed. From each filter, 53–116 particles were analyzed. In total, more than 2,000 particles were analyzed. The particles were chosen randomly on the basis of the quantity of all the particles in the dataset. The chosen particles were analyzed for elemental composition, heavy metals, size and morphology. To determine the characteristics of the chosen particles, a scanning electron microscope (SEM) and energy dispersive X-ray spectrometer (EDX) were used. For the SEM analyses, two modes were utilized: the SEI mode helped to recognize the size and morphology of the particles, and the BEI mode (backscattering) detected heavy metals. The EDX was used to establish their elemental composition. More information about the used techniques can be found in STOYANOVA et al., (2010).

3.1. Detection Limits for the Analysis are Presented in Table 1

After collecting all the data, a statistical evaluation was made. The data were analyzed according to

Table 1

Detection limits (DL) for the analyses (ppm)

Element	DL (ppm)
Si	600
Al	300
Fe	1,300
K	300
Mg	600
Ca	300
S	600
Na	300
Ti	300
Cl	300
Zn	1,600
Mn	900
Cr	1,000
Cu	1,300
Co	800
Ag	700
Sn	600
Sb	1,300
Ni	1,800

the characteristics of the particles and the meteorological conditions. The initial analysis focused on elemental composition. The other elements in one particle together are 100% (not including carbon and oxygen). If a particle contained more than 5% of an element, it was defined as an element-rich particle.

First, the typical elements (frequently represented) in the particles were found and characterized. Then, rare elements were detected and characterized. Based on the sizes, four categories were made: PM_1, $PM_{2.5}$, PM_{10} and particles with larger than 10 μm. The shapes consisted of differentiated spherical and non-spherical particles.

Finally, the possible sources of the insoluble particles in the fog were studied. Accordingly, meteorological conditions that occurred on the fog events were used. In the first case, the particle characteristics were analyzed according to synoptic situations and, in the second case, to wind directions.

4. Results

4.1. Insoluble Particles in Fog

This section includes a characterization of the particles in the dataset. The first characteristic is the

content of the elements followed by the sizes and shapes of the analyzed particles. This information can be helpful for recognizing possible sources of insoluble particles in fog.

4.1.1 Content of Elements

The typical and rare particles were studied according to their elemental composition in the fog. Particles with a content of a certain element greater than 5% were defined as element-rich particles. Typical particles were defined as particles containing elements that occur in more than 1% of all particles. The rest were designated as rare particles. These rare particles were added to the statistics to help detect possible sources.

The most abundant particles found in the fog at the Milešovka observatory are those rich in Si, Al, Fe and K. Each of these elements appeared in more than 50% of all analyzed particles. These particles are frequently represented in natural environments. Other frequently represented elements are Mg, Ca, P, S, Na, Ti, Cl, Mn, Zn and Cr, which occur in 1–28% of analyzed particles.

The abundance of the individual elements are presented in Table 2. Most frequent elements (Si, Al, Fe and K) were often represented together in the same particles. Si-rich particles were also often rich in P, Mg, S, Na, Ti and Mn; Al-rich particles were often rich in P, Mg, S, Na and Ti; and Fe-rich particles were also rich in P, Mg, Mn and Cr. Cl was quite often found in the same particles as K and Na. Even though NaCl is highly soluble, particles of sea salt are frequent in the atmosphere and probably did not dissolve completely in the fog water. Some particles containing rare elements were found. The occurrence of these particles in the datasheet was not frequent (under 1%). Such particles were Ni, Co, Zr, Ba, Au and Pb.

4.1.2 Size and Shape

The particles were divided into categories according to their average sizes:

1. particles larger than 10 μm—coarse particles
2. particles between 2.5 and 10 μm—$PM_{2.5-10}$
3. particles between 1 and 2.5 μm—$PM_{1-2.5}$

Table 2

Table of synoptic situations and their characteristics which prevailed in days when fog was collected at Milešovka Mountain according to Brádka *et al., (1961)*

Synoptic situation	Short description
B	Trough of low atmospheric pressure keeping over the central Europe contains the atmospheric front. Individual frontal waves are approaching from Italy via Moravia over Poland. The predominant air flow is from S to SSW.
NWc	Frontal zone stretches from the Atlantic across the North Sea and Poland to east. Atmospheric fronts quickly proceed in the northwest flow and actively affect the Czech Republic territory.
NWa	Frontal zone stretches from the Atlantic across the North Sea and Poland to east. Atmospheric fronts marginally affect the Czech Republic territory only. Anticyclonic weather character is predominate in the Czech Republic
SWc2	The frontal zone is heading from the Atlantic via France towards the Baltic Sea. The fronts affect the Czech Republic territory, which is under the impact of warm maritime air from southwest.
Wc	The frontal zone stretches from the Atlantic across the British Isles to east over the Baltic. Fronts actively affect the Czech Republic territory. Warm and cold maritime air masses interchange over our territory from west.

4. particles smaller then 1 µm—PM$_1$

Coarse particles were not frequently detected. Only 4% of all analyzed particles belonged to this category. Most notable in the elemental constitution of these particles were Ca, S, Na, Cu, Ba and Au. Particles in the category PM$_{2.5-10}$ were represented by 34% in the whole dataset. Particles rich in K, Mg, Ca, S, Na, Cl, Ni and Au were frequently found to this category. Particles classified as PM$_{1-2.5}$, contained 28% of all analyzed particles were rich in Fe, Mg and Cr and contained Co, Pb and Ni. The category of the smallest particles, with an average size less than 1 µm, are supposed to have potentially anthropogenic origins (Hovorka, 2010). In this category were collected particles that contained Zr (75% of particles) and are rich in Zn (74%). Also in this category were well-represented particles rich in Si, Al, Fe, P, Ti, Mn, Cr and particles containing Ba and Pb.

Particles from the dataset were also divided into two categories according to their shape: non-spherical particles and spherical particles. Particles spherical in shape were found to be rich in zinc (more than 50%), manganese (more than 40%), titanium, iron, chrome, aluminum and silicon. For the rare elements, the greatest percentage of spherical particles contained the rare elements copper and barium, both with 20% of the spherical particles. Particles containing nickel were spherical in almost 10% of the cases. Particles containing zircon were not spherical at all, which could mean that these particles are of natural origin (zircon in small quantities occurs in most igneous rocks), although they all have sizes smaller then 2.5 µm.

4.2. Possible Pollutant Sources

The most represented elements in the dataset—Si, Al, Fe and K, and also other detected particles like Ca, Na, or Mg—belong to the most represented elements in the Earth's crust. Silicon was found in about 90% of all particles, and it is an element that is found in almost every igneous rock. Also, it is the second most represented element in the Earth's crust. Including a significant number of these elements in fog water is natural and most particles with greatest abundance of Si, Al, Fe, K, Ca, Na and Mg could originate from the Earth. On the other hand, these elements could also show up in the fog water from anthropogenic sources. A large fraction of detected particles were spherical in shape. A majority of spherical particles are considered a product of burning processes at high temperatures (about 1,200°C and higher). The fact that there are many power plants, incineration plants and other industrial works in the surrounding areas of Milešovka Mountain indicates a high probability of the presence of anthropogenic products in collected fog water.

4.2.1 Influence of Synoptic Situations to the Occurrence of Element-Rich Particles

To determine the possible sources of particles present in fog at Milešovka observatory, samples collected during the same synoptic situation were compared to samples collected during the other synoptic situation. Interest was focused on the percentage of element-rich particles occurring in single synoptic situations.

The classification used for this method was a Czech classification, the so-called Brádka classification. This classification is used by the Czech Hydrometeorological Institute (CHMI) and the view of the single synoptic situation is stated on the website of CHMI (CHMI, 2010) and in BRÁDKA et al., (BRÁDKA et al., 1961). This classification shows so-called leading fluctuations—transportation of particles in the highest levels of the atmosphere.

Synoptic situations that prevailed in the days when fog was collected on Milesovka, were mainly SWc2 (southwestern cyclonic synoptic situation— (BRÁDKA et al., 1961); B (trough of depression— (BRÁDKA et al., 1961); Wc (western cyclonic synoptic situation—(BRÁDKA et al., 1961); NWc (northwestern cyclonic synoptic situation—(BRÁDKA et al., 1961) and NWa (northwestern anticyclonic synoptic situation—(BRÁDKA et al., 1961). The characteristics of these synoptic situations are described in Table 3.

The total share of element-rich particles was divided by the share of element-rich particles in a given synoptic situation. In the following analysis, attention was given only to those situations in which two or more samples were collected. The most frequent situation was SWc2. Seven samples were collected during this situation. Three samples were collected during the synoptic situations B and Wc, and two samples during NWc and NWa. In all other situations only one sample was collected.

Figure 3 shows relationships between particles rich in given elements and the mean value of the occurrence of particles rich in given elements calculated from the whole dataset.

The following information was discovered by observed synoptic situations:

SWc2:

- particles rich in manganese and zinc exceeded the mean value more than twice
- particles rich in magnesium, chlorine, calcium and titanium did not reach the mean values of occurrence
- the occurrence of other elements was approximately around the mean value

B:

- the occurrence of particles rich in sodium, magnesium, chlorine and calcium exceeded the mean value by 1.5x
- the occurrence of particles rich in phosphorus and potassium exceeded the mean value by 1.25x

NWc:

- significantly lower than the mean value was the occurrence of particles containing magnesium, chrome, phosphorus, calcium and chloride
- no particles rich in manganese were noticed

NWa:

- no particles rich in manganese, zinc and chrome were noticed

Wc:

- only a low occurrence of particles rich in manganese, potassium and phosphorus was noticed.

Particles rich in silicon and aluminum occurred approximately in the same amount in every analyzed synoptic situation.

Table 3

Table of elements occurence in the whole dataset

Element	Particles quantity	%
Si	1,904	87
Al	1,688	77
Fe	1,321	60
K	1,091	50
Mg	618	28
Ca	523	24
P	476	22
S	449	21
Na	279	13
Ti	265	12
Cl	127	6
Mn	56	3
Zn	55	3
Cr	36	2
Ni	21	0.96
Au	20	0.91
Pb	16	0.73
Cu	10	0.46
Zr	8	0.37
Ba	5	0.23
Ag	4	0.18
Co	4	0.18
Total	2,187	

Typical particles are those with the representation of the element higher than 1% in the whole dataset and rare particles with lower representation of the element

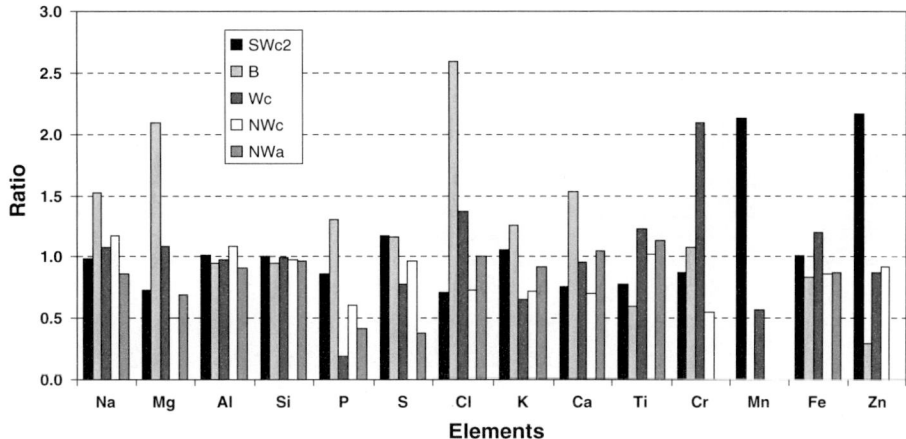

Figure 3

Occurrence of particles rich in given elements by chosen synoptic situations (ratio = total share of element-rich particles/share of element-rich particles in given synoptic situation)

4.2.2 Sectors of near ground wind flow and occurrence of element-rich particles

Three main sectors according to near-ground wind fluctuation were found: NNW (north–northwest), WNW (west–northwest) and WSW (west–southwest). Four samples were collected during the NNW fluctuation (M07-04, M07-05, M07-07 and M07-10), five samples were collected during the WNW fluctuation (M06-06, M06-07, M07-06, M07-08 and M07-13) and five samples were collected by the WSW fluctuation (M06-08, M06-10, M06-12, M07-11 and M07-12).

Attention was focused on the difference between the percentage representation of particles rich in chosen elements and spherical particles on one side and separate sectors of ground wind fluctuations on the other side. First, the percentages of the presence of element-rich particles from the whole dataset for every element were calculated. Then, the percentages of element-rich particles each of the three sectors were calculated. The percentages of element-rich particles were then divided by the whole percentages of the mean occurrence element-rich particles. Figure 4 shows the relation between the mean occurrence of element-rich particles in each sector and their mean occurrence in the whole dataset.

During the ground wind flow from sector NNW, there was a significant representation of particles rich in sodium, chlorine and chrome in the fog water

(Fig. 4). Cl-rich particles were represented 2.8× more often than the mean value. Slightly above average (1.1×–1.5×) was the presence of particles rich in phosphorus, potassium, calcium and titanium. However, the occurrence of particles rich in manganese and zinc was below the mean value (less then 0.5×), and the occurrence of particles rich in iron, magnesium and aluminum was slightly below the mean value.

In the WNW sector, the occurrence of particles rich in calcium was significantly above-average, and the occurrence of particles rich in magnesium and calcium was slightly above average. The occurrence of particles rich in sulphur, manganese, iron and zinc was slightly below average and the occurrence of particles rich in chrome was well bellow average.

In the situations when the ground wind flowed from the WSW sector, all element-rich particles were below the average value. In the WSW direction there are more agricultural than industrial zones, and therefore, the ratios in this sector are always lowest (Fig. 4).

5. Discussion and Conclusions

Twenty-five samples of fog water were collected by an active fog water collector at the observation station on the top of Milešovka Mountain in the České Středohoří Mountain Range. The samples were

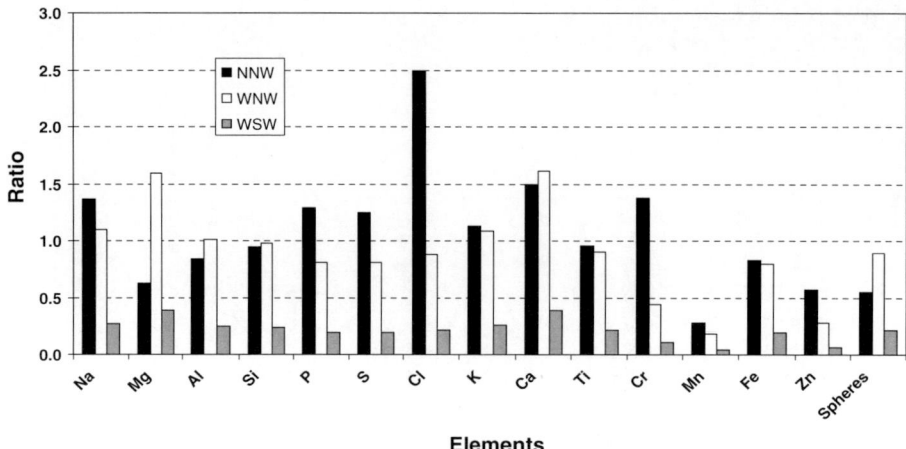

Figure 4
Occurrence of particles rich in given elements and spherical particles in wind sectors (ratio = total share of element-rich particles/share of element-rich particles in given sector of wind fluctuations)

filtered, and chosen particles (more then 2,000) were analyzed. Scanning electron microscopy and energy dispersive X-ray spectrometry were used for the analysis, and the data were compared by statistical evaluation.

Elements Ca, S and Na are represented mainly in size categories of particles with the largest sizes: $PM_{2.5-10}$ and coarse particles. Category $PM_{2.5-10}$ had approximately 40% of Ca-, S- and Na-rich particles, which is much more than in the category of coarse particles, where it is only slightly above 10%. This fact does not aid in the determination of the sources. The shape of most of the Ca-, S- and Na-rich particles is non-spherical, and only about 10% of these particles have a spherical shape, which could indicate natural sources.

Particles rich in Ca and Na were frequently collected during synoptic situation B (trough of depression) when no strong fluctuations in higher levels of the atmosphere occur. Ca and Na were also frequently represented by the WNW and NNW wind sectors. Particles rich in S were common by the sector of ground fluctuations WSW. By the same synoptic situation and in the same sector of ground fluctuation (WNW) as Ca and Na, the element K was also found, the size of which is often in category $PM_{2.5-10}$.

Iron-rich particles belong mostly to the categories of the smallest particles ($PM_{1-2.5}$ and PM_1), even if some of these particles also belong to the category $PM_{2.5-10}$. Between the iron-rich particles was a significant amount of spherical particles. An increased quantity of iron-rich particles was found in the sector of wind fluctuations WSW. The characteristics of sizes and shapes or iron-rich particles indicates that some part of these particles are of anthropogenic origin and that one of the possible sources of them could be one of the thermal power plants that lie southwest of Milešovka Mountain (Fig. 1).

Particles rich in zinc belong mostly to category PM_1, and they belong to particles that are in more than 50% spherical indicating a high probability that a great number of them are of anthropogenic origin. Zinc-rich particles are more frequent in synoptic situation SWc2. There are two great sources of zinc in the surroundings of Milešovka Mountain. The first and largest source is the power station that lies to the south of the observatory (in Středočeský kraj, which is not pictured in Fig. 1). The second is a non-ferrous metal works facility that lies to the southeast of the observatory.

The results suggest that the most abundant particles are of terrestrial origin and that they probably originate in the nearest surroundings of Milešovka Mountain. The exact identification of sources of single elements in fog based solely on the chemical composition of the particles is not possible. It is obvious that the content of the elements is influenced by the synoptic situations and wind fluctuations, but it is not always the case that concrete sources of particles can be ascertained from these indicators. Synoptic situations show air fluctuations in higher levels of the atmosphere and, therefore, indicate the transport of particles from greater distances. On the other hand, sectors of wind fluctuations show fluctuations that are affected by relief. Wind fluctuations provide more information about local sources of atmospheric particles. The content of rare elements in fog samples facilitates the identification of possible sources, but further analysis of this dataset and collection of additional fog samples would help better identify potential sources.

Acknowledgments

The results described in this paper were obtained in the frame for AS CR and BAS collaboration with support of the GACR Project No. 205/09/1918, and the IRP No. AV0Z30420517.

REFERENCES

BRÁDKA, J., DŘEVIKOVSKÝ, A., GREGOR, Z. and KOLESÁR, J. (1961). *Weather on Czech and Morivia Area in typical weather situations.* Hydrometeorological Institute, Prague (in Czech).

BROOK ROBERT, D. *et al.* (2004). *Air pollution and cardiovascular disease: a statement for healthcare professional: from the American heart association,* Circulation *109,* 2655–2671.

DAUBE B., KIMBALL K.D., LAMAR P.A., WEATHERS K.C. (1987). *Two new ground-level cloud water sampler design which reduce rain contamination.* Atmospheric Environment *21,* 893–900, doi: 10.1016/0004-6981(87)90085-0.

FISAK J., REZACOVA D., ELIAS V., TESAR M. (2001), *Comparison of pollutant concentrations in fog (low cloud) water in Northern and Southern Bohemia,* Journal of Hydrology and Hydromechanics *49*(5), 275–290.

FISAK J., TESAR M., REZACOVA D, ELIAS V., WEIGNEROVA V., FOTTOVA D. (2002a), *Pollutant concentrations in fog and low cloud water at selected sites of the Czech Republic,* Atmospheric Research *64*(1–4), 75–87.

FISAK J., REZACOVA D., WEIGNEROVA V., TESAR M. (2002b), *Pollutant concentration in fog water samples from Milesovka,* Report Series in Aerosol Science *56,* 29–34.

FISAK J., TESAR M, FOTTOVA D. (2009), *Pollutant concentrations in the rime and fog water at Milesovka Mountain,* Water, Air & Soil Pollution 194 (1–4), 273–285.

HOVORKA J., Atmospheric aerosol, *In atmosphere and climate: actual issues in meteorology, climatology and air protection (In Czech)* (Ed. Braniš, M. and Hůnová, I.) (Karolinum, Prague, 2010), pp 121–139.

LI WEIJUN and SHAO LONGYI (2009), *Characterization of mineral particles in winter fog of Beijing analyzed by TEM and SEM,* Environmental Monitoring and Assessment *161,* 565–573.

LIN Z.Q., SCHEMENAUER R.S., SCHUEPP P.H., BARTHAKUR N.N., KENNEDY G.G. (1997), *Airborne metal pollutants in high elevation forests of southern Quebec, Canada, and their likely source regions.* Agriculture and Forest Meteorology *87*(1), 41–54.

MONTE MARCO DEL and ROSSI PAOLA (1997), *Fog and gypsum crystals on building materials,* Atmospheric Environment *31*(11), 1637–1646.

POPPE III ARDEN C. and DOCKERY DOUGLAS W. (2006), *Health effects of fine particulate air pollution: lines that connect,* Air & Waste Manage Assoc. *56,* 709–742.

SCHEMENAUER ROBERT S. (1986), *Acid deposition to forests: the 1985 chemistry of high elevation fog (CHEF) project,* Atmosphere-ocean *24*(4), 303–328.

STOYANOVA V., *et al.* (2010), *SEM-EDX identification of particles from fog in an industrially polluted region of Czech Republic,* Proceedings of 10th International Multidisciplinary Scientific Geoconference "Modern Management of Mine Producing, Geology and Environmental Protection" SGEM 2010 *2,* 269–276.

ŠTEKL J. *et al.*, *Milesovka and the region of Milešovka (In Czech)* (Academia, Praha, 2005).

TESAR M., ELIAS V., SIR M. (1995). *Preliminary results of characterization of cloud and fog water in the mountains of Southern and Northern Bohemia,* Journal of Hydrology and Hydrodynamics *43,* 412–426.

IRZ—Integrated Register of Pollution (In Czech) (2010), cit. 2010-06-22, Available on:http://www.irz.cz/latky/poletavy_ prach.

CHMI—Czech hydrometeorological institute (In Czech) (2010), Typology of synoptic situation for the area of Czech Republic, cit. 2010-06-22, Available on:http://www.chmi.cz/portal/dt?menu= JSPTabContainer/P3_0_Informace_pro_Vas/P3_3_Historicka_ data/P3_3_1_Pocasi/P3_3_1_11_Typizace_situaci&last=false.

(Received November 13, 2010, revised May 4, 2011, accepted May 6, 2011, Published online June 3, 2011)

Pure Appl. Geophys. 169 (2012), 1093–1106
© 2011 Springer Basel AG
DOI 10.1007/s00024-011-0360-9

| Pure and Applied Geophysics

Environmental Role of Rime Chemistry at Selected Mountain Sites in Poland

Michał Godek,[1] Marek Błaś,[1] Mieczysław Sobik,[1] Żaneta Polkowska,[2] Katarzyna Cichała-Kamrowska,[2]
and Jacek Namieśnik[2]

Abstract—The results of field experiments on fog pollutant deposition enhanced by local mountain climate, completed by the dendrochronological analysis of the forest response, are presented in this paper. In spite of their low absolute altitude (1,000–1,600 m a.s.l), the Sudetes and the Silesian Beskid form a noticeable orographic barrier for the airflow of the humid Atlantic air masses. This results in the increase of cloudiness and fog frequency as well as both atmospheric precipitation and horizontal precipitation volume. Between January and December 2009 the daily samples of atmospheric precipitation and rime were collected on three selected mountain tops of similar height. The selected measurement sites were situated along a 300 km WNW-ESE profile parallel to the direction of the prevailing atmospheric circulation. High day-to-day variability of rime water volume, the total ionic content and chemical composition of the individual samples were typical of each measurement site and depended on the emission patterns, synoptic situation and the local climatic conditions influenced chiefly by terrain relief. Significantly larger rime efficiency and pollution deposition via fog were observed at the westernmost Szrenica Mt site rather than more to the southeast at Śnieżnik Mt and Skrzyczne Mt. This difference should be explained by more intense orographic deformation of predominant airflow from the western sector as well as the higher liquid water content of fog in the vicinity of Szrenica. Both temporal and spatial variability of fog deposition correlates closely with the health status of the drilled trees of Norway Spruce (*Picea Abies*) in the Śnieżnik Massif. The averaged annual tree rings width near the local tree line (1,350 m a.s.l.) on the summit dome of Śnieżnik decreased by 71% between 1950 and the early 1980s. This is also the area of the highest rate of atmospheric pollutant deposition due to particularly important role of fog. At an altitude of 1,200 m a.s.l. The relevant changes of ring width were different depending on slope aspect: 60% on western slopes well exposed for orographic fog formation and 42% on eastern slopes where fog deposition is less intense. The results of the dendrochronological analysis provide the evidence for the upward trend of tree rings width since 1981–1984 break through up to date, which should be attributed to the progressive reduction of pollutant emission in Central Europe.

[1] Department of Climatology and Atmosphere Protection, University of Wroclaw, Kosiby Str. 6/8, 51-670 Wroclaw, Poland. E-mail: michal.godek@uni.wroc.pl

[2] Department of Analytical Chemistry, Gdansk University of Technology, Gabriela Narutowicza Str. 11/12, 80-233 Gdansk, Poland.

Key words: Air pollution, fog deposition, rime, mountain climate, total ionic content, spruce dendrochronology.

1. Introduction

The disintegrated body of Western Europe, arrangement of the major landforms and location in a zone of prevailing westerlies result in the fact that the oceanic air masses may readily penetrate a considerable part of the continent. The greater distance from the Atlantic Ocean and the transformation of the moist air masses as they pass over the complex terrain contribute to a situation in which the climate shows more distinctly continental features in eastern and southern Poland, than in the westernmost part of the Sudetes (Wallen, 1977). Altitude is one of the predominant factors controlling the average annual fog frequency in Poland. The other factors comprise the exposure to advection of the humid Atlantic air masses and the absence of other mountains on the windward side which would be able to block or transform the existing airflow. Hence, a peripheral location (taking into account the direction of the atmospheric circulation) is particularly conducive to the presence of fog, which does not exist on the lee side where the air subsides. It should be mentioned that the middle-sized mountains in the southern part of Poland (e.g. the Sudetes, the Western Carpathians) are the European leaders of fog frequency [(296, 284, 274 and 247 days with fog per year at Śnieżka, Praded Mt., Szrenica Mt. and Lysa Hora Mt., respectively (Błaś and Sobik, 2000, 2004; Migała *et al.*, 2002)]. The highest fog frequency is typical for the colder half of the year, when the condensation level goes down because of the seasonal decrease in the solar radiation and predominant stable stratification with Stratus and Stratocumulus clouds below 2,000 m.a.s.l. (Błaś et al., 2002).

The direct deposition of fog/cloud droplets to vegetation can be an important contributor to hydrological and chemical inputs (Lovett, 1984; Pahl et al., 1994; Weathers et al., 1995; Zimmermann and Zimmermann, 2002; Blas and Sobik, 2003). Rime is one of two possible forms of the fog deposit. In contrast to the liquid fog deposit, it appears in temperatures below 0°C in the solid state, having the form of white or milky, opaque granular ice deposit. Rime is formed by the rapid freezing of supercooled water drops as they touch exposed objects (e.g. trees, bushes, grass, stones or snow cover). Apart from the fog frequency, some other factors which determine the fog deposit efficiency are: the cloud liquid water content, fog droplet size and the wind speed. Wind speed has a double role to play because it influences the number of droplets depositing on the receptor within a time unit, and secondly, the droplets' momentum increases, thus the effectiveness of the collisions with the receptor rises (Lovett, 1984). It must be emphasized that the discussed factors describe the potential share of the atmospheric deposit in the total water flux. In order to obtain the true representation, the impact of the land use on the deposit efficiency must be accounted for. The largest amounts of the real deposit are connected with the presence of receptors in the form of isolated trees and bushes situated at convex landforms, which are very effective in capturing the fog/cloud droplets impacted by the wind. The specific character of the fog deposition process (accounting for the inertial collisions and sedimentation) has been described in detail by Lovett (1984); Weathers et al. (1995); Sobik et al. (1998).

Historically, Poland belongs to the group of the European countries with the highest sulphur and nitrogen emission. This is because coal is the main fuel used in energy production, industry and in non-industrial combustion (Mitosek et al., 2004). Emission abatements from the European domestic sources caused large environmental benefits and resulted in the decrease of areas with the exceeded critical levels and loads. However, in Poland the total area where the critical loads are exceeded is still considerable.

Fog and clouds play an important role as processors of atmospheric aerosols and soluble gases. The role they play in particle removal by fog drop deposition is of particular interest. Hence, fog chemistry is a useful tool for complementary interpretation and identification of long range transport of air pollutants. In different locations, the contribution of liquid/solid (rime) deposition to total deposition (including precipitation) varies from a few percent (in the lowland part of Poland) up to 70% in the mountainous regions (Baranowski and Liebersbach, 1978; Błaś et al., 2003). Deposition of pollutants by fog is the second reason, and locally, even the most important reason, causing deforestation in the western Sudetes and the Western Carpathians.

On the other hand, very high deposition rates may influence forest growth. There is a clear relationship between the decreasing annual tree rings width in the forest stands at sites with high intensity of more polluted fog/low level of cloud deposition. Most of all, this phenomenon can be observed in Europe, NE part of North America (Huettl et al., 1993) and in the Norilsk and Lake Baikal area in Siberia (Schweingruber, 2007). The dependency between the pollutant deposition and the width of annual rings in Poland has been determined using the example of spruce in the Outer Western Carpathians and the Karkonosze Mts. in the Sudetes (Feliksik, 1995; Moravčik Černý, 1995) as well as pine (Danek, 2007; Krapiec, et al., 2001; Malik, 2009; Wiśniowski, 2001) and Douglas fir (Feliksik and Wilczyński, 2003) in a non-mountainous area of Poland. At all sites mentioned before, the fog deposition intensity is moderate or even marginal, whereas the atmospheric precipitation is the major pathway of pollutant deposition. Dendrochronological studies on subalpine spruce conducted in 1999 in the Karkonosze Mts was the first approach to determine the role of cloud deposition in the study area (Godek et al., 2009). Moreover, it should be remembered that the size of the annual tree rings is diversified year by year, also under the influence of the changeable meteorological conditions (Fritts, 1976). However, the temporary anomalies in the weather conditions are repeated at various locations within the same small area, thus the course of the dendroscales for the specified trees' population of the same species is very similar.

Whereas, the presence of the non-climatic factors of the environmental stress, e.g. atmospheric pollution, may significantly reduce the size of the rings' annual growth, to the degree dependable on the microscale diversity of the pollutant deposition level. When the stress impact ceases, the trees again increase the size of their annual rings (FELIKSIK, 1995). The observed variability of annual tree-rings width responds to different environmental stimuli. This response is fast enough to be an important indicator of the forest ecosystem degradation (VINŠ and POLLANSCHÜTZ, 1977) and is accompanied by the changes of other dendrological factors, like: height increments, needle production and needle density (FERRETTI *et al.*, 2002). Different important stress-inducing factors affecting the trees, which long-term influence should be excluded, comprise the mechanic impact of the snow cover, mass movement of the ground, feeding of herbivores, habitat competition, presence of dangerous fungi and insect pests, and more rarely, seismic

and volcanic activity, cyclic forest fires and glacier movement (SCHWEINGRUBER, 1996).

The objective of this study is to explain the meso- and microscale spatial distribution of pollutant deposition via fog/rime in complex mountain terrain. On the other hand, it aims at showing the connection between the intensified deposition and the spatial pattern of forest destruction.

2. Area of Research

2.1. Morphological and Geological Characteristics of the Area

The measurement sites were selected so as to represent the following: distinctly convex landform, similar absolute altitude accompanied by a significant relative altitude (800–900 m) as well as the zone of the upper tree line (Fig. 1; Table 1). They were located along the NW–SE profile line, i.e. parallel to

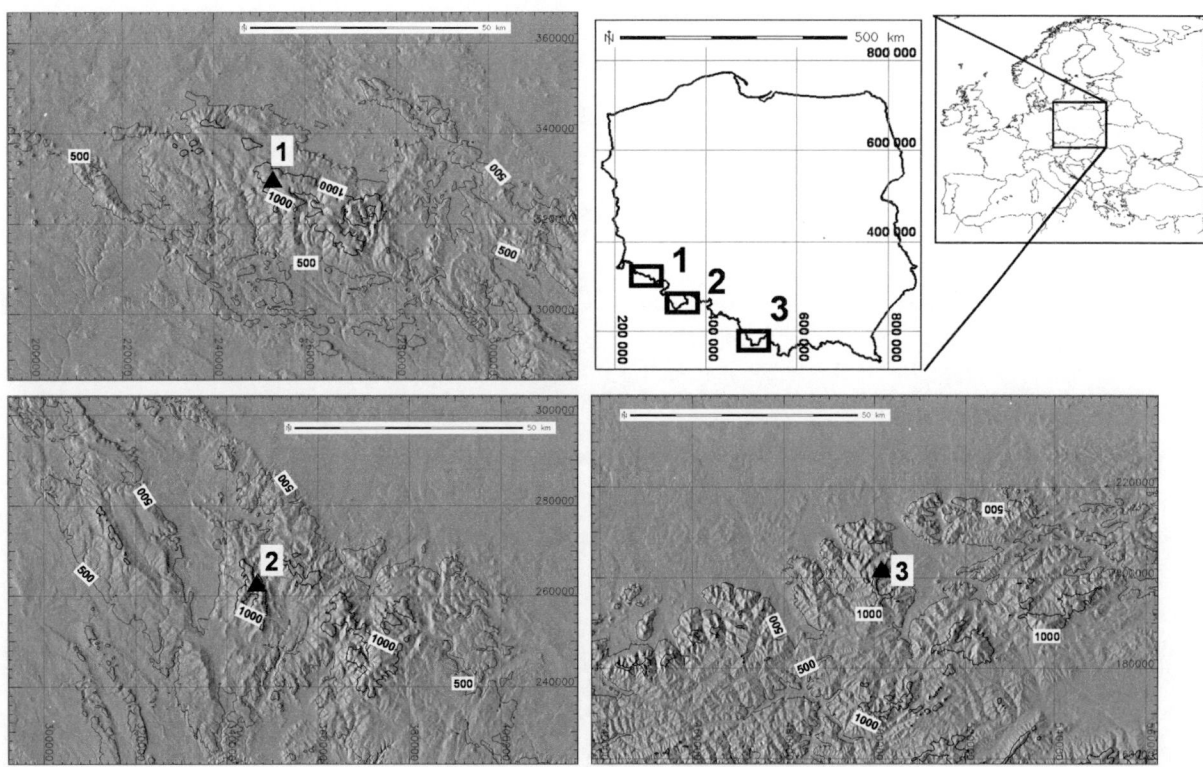

Figure 1
Location of the sampling sites (see also Table 1)

Table 1

Detailed characteristics of the sampling sites (see Fig. 1)

Site no.	Sampling site	Code	Geographic region and coordinates	Altitude [m a.s.l.]	Landform type	Land-use	Site description
1	Szrenica	SZ	Karkonosze Mts 50°48″ N 15°31″ E	1,330	Convex, mountain ridge	Around 100 m above tree line, dwarf pine patches; uninhabited area	Situated in the western part of Karkonosze, the highest range of the Sudety Mts with a mean altitude of 1,200–1,400 m a.s.l.; stretch from NW to SE for almost 30 km
2	Śnieżnik	SN	Śnieżnik Massif 50°12″ N 16°50″ E	1,230	Convex, upper part of slope	Large clearing in spruce forest, around 100 m below tree line; uninhabited area.	Situated on the western macro-slope of Śnieżnik Massif in the eastern part of the Sudety Mts., an inselberg-type massif with altitude exceeding 1,400 m a.s.l.; oval massif base with a diameter above 25 km.
3	Skrzyczne	SK	Silesian Beskid 49°41″ N 19°02″ E	1,250	Convex, mountain ridge	Large clearing in spruce forest; uninhabited area	Situated in the highest part of the Silesian Beskid, i.e. westernmost range of the Polish Western Carpathians with a maximal altitude exceeding 1,200 m a.s.l.; stretch from SW to NE for above 20 km.

the prevailing direction of the atmospheric circulation of maritime origin. The selected mountain tops belong to the Sudetes (Szrenica Mt. and Śnieżnik Mt.) and the Western Carpathians (Skrzyczne Mt.) and they are situated 100–150 km away from each other. The passive rime collectors were always installed on a slope with western aspect, a bit below a mountain summit.

Similarly to the substantial part of the Karkonosze Mts., Szrenica Mt. is built of igneous rocks of the soft granite Karkonosze-Izera block (ŻELAŹNIEWICZ, 2005). The limited annual buffering capacity of acid compounds by the bedrock—at the level of only 20–30 mMol H^+ m^{-2} $year^{-2}$ (NILSSON and GRENNFELT, 1988)—contributes to the high level of soil acidification in this region. The Śnieżnik Massif is built of the acid orthogneisses of the Orlica-Śnieżnik complex, with a significant share of schists and the local contribution of marbles in the lower parts (ŻELAŹNIEWICZ, 2005). Skrzyczne Mt. together with the central part of the Silesian Beskid is built mainly of sandstone of godul and istebnian layers (CHOWANIEC, 1991). The increased calcium carbonate content of the sandstone bedrock provides better

buffering against acid deposition than the metamorphic and igneous rocks of the Sudetes.

2.2. Climatic and Environmental Conditions

Karkonosze Mts., Śnieżnik Massif and the Silesian Beskid are under the influence of the western zonal atmospheric circulation. Because of similar altitude, the average annual air temperature in all three measurement sites is ca. 2.0°C with the lowest monthly average in January (ca. −6.0°C; PEREYMA et al., 1997). The prevalence of the advection of maritime air masses, high relative altitude and the distinctly convex landform translate into heavy and frequent atmospheric precipitation (ca. 1,400 mm) and a significant share of fog deposits in the total water flux from the atmosphere. The most frequent category of fog deposits is rime; potentially it may occur in all months of the year (only apart from August). On Szrenica Mt., out of 274 foggy days annually, rime was observed in 170–180 instances. The primary period of its appearance starts in October and finishes in May. Apart from the fog frequency and its liquid water content, the fog deposit efficiency is also influenced by the wind speed, which is considerably diversified along the analysed measurement profile. The Karkonosze Mts. which comprise Szrenica Mt. are characterized by the highest speeds out of all the mountain sites of the continental Europe. The average annual wind speed on the Szrenica Mt. exceeds 9.4 m s^{-1}. The airflow deformation in the Karkonosze Mts. takes place in the mesoscale under the influence of the compact and significantly eminent mountain ridge perpendicular to the prevailing wind direction, this results in the forced ascent of air induced by land morphology at the distant forefield of the Karkonosze Mts. (DRUKMAN et al., 1997). On the remaining more isolated massifs, the wind speed was clearly lower reaching 4.1 m s^{-1} on the Śnieżnik Mt. and 6.8 m s^{-1} on the Skrzyczne Mt. Moreover, it must be emphasized that during frequent rime formation in the cold half of the year, atmospheric circulation is more dynamic; therefore, the wind speed is clearly higher then, being respectively 11.0 m s^{-1} (Szrenica Mt.), 4.5 m s^{-1} (Śnieżnik Mt.) and 7.6 m s^{-1} (Skrzyczne Mt.).

In practice, vegetation type (in this case spruce trees) is critical in determining the pollutant deposition input via riming In the natural conditions, the spruce formed the dense stand of conifers only in the upper subalpine forest, over 950–1,050 m a.s.l. The spruce monocultures currently present in the lower altitudes, to the large extent have the genotype of the lowland origin; therefore, they establish an unstable element of the ecosystem which is poor in terms of the habitat (MATUSZKIEWICZ, 2002).

2.3. Emission Background

The research area is under the strong influence of the anthropogenic sources of pollutant emission to the atmosphere. SO$_2$ emission is mainly connected with combustion of the brown and hard coal with high sulphur content. It is used mainly for the needs of power plants, industry and household heating. In case of NO$_x$ emission, the road transport is the major source. As much as 97% of ammonia should be linked to the agricultural activity (EMEP Status Report, 2010; EMEP Data Report, 2010).

Following source areas: the Turów Power Plant in Poland, pollutants originating in the Żytawska Valley, northern and northwestern Czech Republic and eastern Germany are decisive for the condition of atmosphere in the Sudetes. Whereas, the emission sources connected with the Upper Silesian Industrial Region, Rybnik Coal Area, Ostrava-Carvina Coal Region and Cracow have the dominant influence on the Silesian Beskid (Fig. 2). The average annual concentration of SO$_x$ in the air as modelled by FRAME (Fine Resolution Atmospheric Multi-pollutant Exchange), fluctuates in the research area between ca. 2.5 μg S m^{-3} and 6.0 μg S m^{-3}. The wet deposition of S–SO$_4{}^{2-}$ exceeds 10 kg S year^{-1}, N–NO$_3{}^-$ is between 5.5 and 6.5 kg N year^{-1}, and N–NH$_4{}^+$ appropriately 8.5–10.0 kg N year^{-1} (KRYZA et al., 2009, 2010).

Between 1978 and 1984, the massive dying out of spruce stands took place in the Sudetes. As much as 13,500 hectares of the forest, over 92% of the coniferous trees were destroyed in the SW part of Poland, particularly in the Sudetes. In the 1980s and early 1990s strong acidification of soil in the Sudety Mts brought the pH values of soil and ground water below 4.0. This situation resulted in disorder of nutrients circulation, mainly in the case of nitrogen, phosphorus, calcium and magnesium (STRZYSZCZ, 1995). Spruce trees were visibly damaged: showing

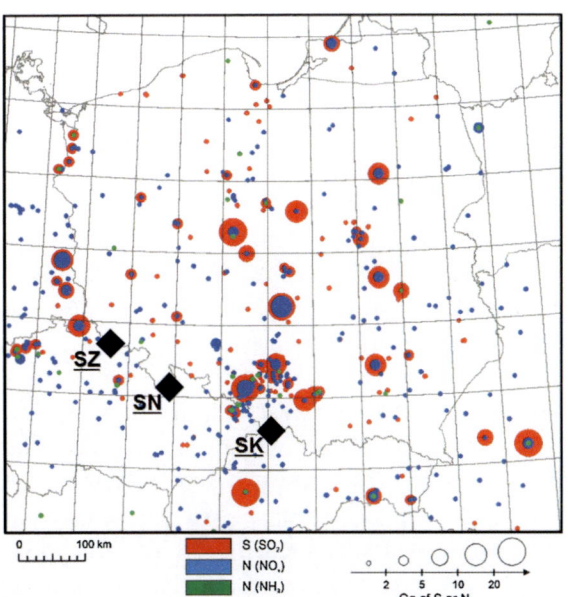

Figure 2
Location of the main sources of pollutant emission on the study area in 2002

the needles turning yellow, withering growth cone and crown thinning. Degradation of spruce ecosystems resulted in formation of exposed edges of forest stands, which even quadrupled the efficiency of wet deposition and its input on soil chemistry (JAła and BŁAŚ, 2000). At that time, the region of Poland, the Czech Republic and Germany borderland was named the "Black Triangle". In turn, the ecological disaster in the Outer Western Carpathians has grown stronger for the last two decades as an effect of later soil acidification process mainly related to larger buffer capacity of the local bedrock and location to the east in relation to the Sudety Mts. Moreover, between 2006 and 2009, mass attacks of bark beetles and other insects took place here, largely caused by the extremely dry summer in 2006. It was proved that despite the ongoing improvement of air and precipitation quality in the study area, there still occurs significant soil degradation level, limiting trees increments. Because of the abatement strategies—introduced since the turn of the 1980s and the 1990s of the last century in former eastern-block countries like Poland, Czechoslovakia and Eastern Germany—a dramatic decrease of SO2 emmission took place. The SO2 emission, which amounted in Poland to 1131 Gg in 2007, has decreased by about 72% since 1980. The scale of emission abatements was even larger in Czech Republic and Germany (87 and 86%, respectively; OLENDRZYŃSKI et al., 2009). Following such transformation of emission, substantial changes of chemical composition were observed, e.g. threefold decline of sulphate concentration in both cloud water and precipitation samples at the Szrenica Mt in the Western Sudety Mts. (BŁAŚ et al., 2008). Decreasing pollutants emission resulted in improvement of trees condition—particularly in the Sudety Mts—still seems to be too early to define, if it is sufficient factor to prevent a risk of forest degradation, like in the Western Carpathians. Therefore there is lime fertilizing of soil recommended, particularly for the subalpine spruce forest sites with shallow and often eroded soil (OLEJARSKI et al., 2004). The observed systematic improvement of groundwater quality in the Sudety Mts should be treated as a symptom of decreasing environmental pollution. It is optimistic, that groundwater pH, even

within the degraded area of the Karkonosze Mts more and more often reaches values above 6.0 (KRYZA et al., 2005).

2.4. Sampling Methods

Rime samples were collected daily from January to December 2010, at three different locations in Poland, with the application of the passive collector consisting of a 200 cm long plastic rod having 30 mm in diameter. In order to reduce the disturbing effect of the ongoing dry deposition, the samplers were cleaned before exposure, rinsed with de-ionized water and wiped dry. The blank samples analyses showed that the material of the collectors surface did not have any measurable influence on the rime samples chemistry. The measurements of the atmospheric precipitation sum were carried out with the use of the standard Hellmann gauge. To study the synoptic scale airflow in order to estimate pollutants contribution, back-trajectories and a calendar of atmospheric circulation were used (NIEDŹWIEDŹ, 2010). The samples were collected in the morning after each deposition event and then stored at low temperature without chemical preservatives because the analyses were performed either directly on-site, or immediately after the samples were delivered to the laboratory. The samples were analyzed on-site for pH, volume and conductivity. The selected anions and cations were quantified against synthetic rain standard using ion suppressed chromatography (ICS 3000, Dionex Corporation, USA). This synthetic standard is the Reference Material no. 409 (BCR-409, Institute for Reference Materials and Measurements, Belgium) and Analytical Reference Material Rain (National Water Research Institute, Environment Canada) (POLKOWSKA et al., 2005).

Cores from the trunks of the Norway Spruce (*Picea abies*) were collected 130 cm above the ground level in the highest parts of the Śnieżnik Mt. with the use of a 40 cm long increment borer. The collected population represents four research sites, on each of them eight cores from the mature trees of various levels of damage were collected. Two sites were situated on the windward and leeward slope of the Śnieżnik Mt, more than 10 m below the summit dome fold. Highly stunting, thin spruce forest dominates here with numerous visibly damaged top parts of crowns (broken or withered tops), crown thinning and needles turning yellow, flag habit of crown and single entirely dead trees. Their average height is between 10 and 15 m. The following two sites were situated ca. 100 m below the top. The high, upper subalpine spruce forest grows on the leeward slope, the trees are above 30 m high with dense crowns. Damage and distortion of the tree crowns are rather rare; no dead specimens were recorded. Highly stunted, thin spruce forest (average tree height of 10 m) is characteristic for the windward site. The majority of trees grow in biogroups with dominating flag crown habit. As in 1950, all drilled trees had over 20 shaped annual rings.

After high resolution scanning, the dendrochronological samples were subjected to analysis with the application of the PAST V4 and OSM3 software. It was assumed, as the starting point, that the average 5 year size of annual rings at the research site is 100% as in 1950 and constitutes a reference point for the subsequent years. Year 1950 seems as a reasonable reference time prior to major industrialization in the region resulting in the massive increase in atmospheric emission of sulphur. European SO_2 emission was, moreover, stable between 1900 and 1950 and increased clearly in the 1960s and 1970s (MÖLLER, 1984). It was mainly attributed to the consumption of crude oil, while the contribution of hard coal was 68% in 1950 and decreased to 46% in 1975. Furthermore, a more sudden increase of sulphur dioxide emission in Poland from 1975 to 1985 was linked with increase of coal and energy production by a factor of 6 and 2 for lignite and hard coal, respectively.

3. Results and Discussion

3.1. Rime Efficiency—Spatial Distribution

From 1 January 2009 to 31 December 2009, appropriately 110, 80 and 69 samples of rime were collected on the Szrenica Mt., Śnieżnik Mt. and Skrzyczne Mt (Table 1), respectively. The primary period of rime occurrence were the following months: January–March and November–December. Besides, some rime incidents were recorded in April, May and June. Measurements of ice deposit mass were performed only once a day at 6:00 UTC; however, in the case of rime deposited and afterwards melted or fallen off before a given sampling term, it was not taken into account. In 2009, its total highest weight deposited on the passive collector surface was recorded on the Szrenica Mt and amounted to 14.6 kg. This means rime efficiency over four times higher than at the two remaining measurement sites. This clearly increased rime frequency and efficiency results from the Szrenica Mt. location in the NW limit part of the measurement profile, where the advection of a humid polar-maritime air mass is recorded more frequently. This means that while overcoming the subsequent mountain ranges along the measurement profile, gradual dewatering of the flowing air mass must take place as well as raising of the orographic cloud base. Moreover, the location of the Karkonosze Mts. on the NW limit of the Sudetes, combined with their high relative altitude, result in strong airflow deformation of the drifting humid air masses. At all sites, rime is formed most frequently along with the circulation from the NW—23.6% cases on the Szrenica Mt. to 20.3% on the Skrzyczne Mt. and most rarely along with the directions from the eastern sector (Figs. 3, 4). The secondary maximum of rime efficiency for the S–SW sector is

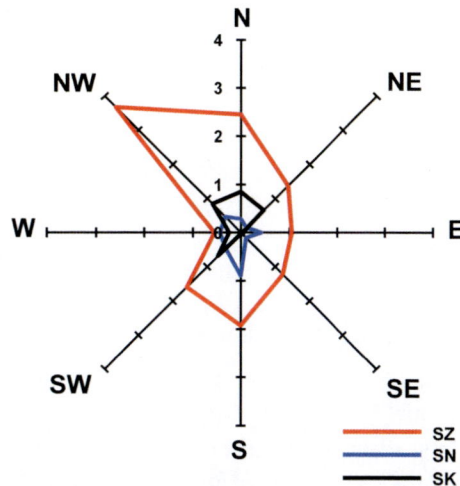

Figure 3
Relative contribution of rime water flux (total volume of water after rime melting) [l] versus air circulation direction at Szrenica Mt. (SZ), Śnieżnik Mt. (SN) and Skrzyczne Mt. (SK)

335

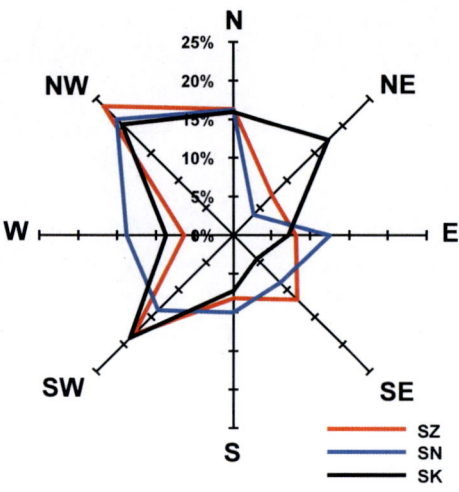

Figure 4

Rime frequency in relation to direction of air circulation [%] at Szrenica Mt. (SZ), Śnieżnik Mt. (SN) and Skrzyczne Mt. (SK)

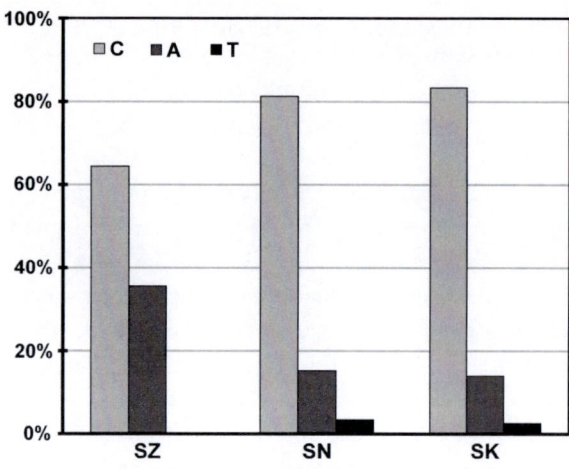

Figure 5

Contribution of accumulated rime water flux (total volume of water after rime melting) [%] versus synoptic conditions (*C* Cyclonic, *A* anticyclonic, *T* transition) at Szrenica Mt. (SZ), Śnieżnik Mt. (SN) and Skrzyczne Mt. (SK)

connected with the frequent formation of a specific orographic cloud called foehn wall.

Meanwhile, the measured sums of atmospheric precipitation differ less because the highest value for the Szrenica Mt. (1,531 mm) is only 30% larger than at Śnieżnik Mt. where precipitation was the lowest. Slighter differences in the case of atmospheric precipitation result from the fact that (particularly in the warm part of the year) it is to a lesser extent dependable upon both the macroscale aspect and orographic airflow deformation. In a more continental climate, in the SE part of the sampling profile (Skrzyczne Mt.), convective rainfall makes a more important contribution to the total annual rainfall and it does not generally occur in the presence of the orographic clouds.

A bit surprising are the results of the rime efficiency comparison while providing for the synoptic situation. Three types of synoptic situation were distinguished: cyclonic, anticyclonic and transition (when the circulation was generated in the intermediate zone between the high and low pressure systems; Fig. 5). While the share of days with the cyclonic and anticyclonic situation in 2009 is more or less equal, clear domination of rime is marked at two sites during the cyclonic days: 83% over the Skrzyczne Mt and 81% on the Śnieżnik Mt. The low pressure systems are characterized by the neutral or potentially unstable stratification which forces the air flow over the

mountain ridge. Whereas, on the Szrenica Mt. the share of the cyclonic circulation falls to 64% for the benefit of the anticyclonic days (36%). Moreover, larger rime mass was deposited on the Szrenica Mt. in the anticyclonic conditions, only then at the remaining sites during all synoptic conditions together. This should be attributed to the more important role of airflow deformation, which often takes place even during anticyclonic conditions, resulting from the Karkonosze Mts. range compactness and its significant relative altitude. This is supplemented by the higher frequency of the Stratus or Stratocumulus clouds observed in the Czech Basin at stable stratification (and even inversion) in the cold part of the year.

3.2. Rime Chemical Composition

Inorganic chemical composition, conductivity, pH, TIC (total inorganic ionic content) and calculated load of pollutants for rime and precipitation, collected at three mountain tops are reported in Table 2 (see also Fig. 6). The most acidic samples were obtained at the Szrenica Mt. (pH = 4.21) with strong variations of individual samples from 3.22 to 7.86. The highest mean pH values were observed at the Śnieżnik Mt. (pH = 5.12). Heavily influenced, by acid anthropogenic emissions, rime samples have high concentrations

Table 2

Selected physico-chemical properties of rime samples (percentages of ion concentrations calculated according to ionic charge)

Parameters	Szrenica	Śnieżnik	Skrzyczne
N	110	80	69
Total volume [dm³]	14.58	3.28	3.44
pH[a]	4.21	5.12	4.92
Conductivity[a] [μS/cm]	45.9	54.7	61.4
SO_4^{2-} [%]	11.0	15.5	18.5
NO_3^{-} [%]	17.7	14.3	10.7
Cl^- [%]	10.8	11.6	14.7
PO_4^{3-} [%]	1.1	1.3	0.0
NH_4^+ [%]	16.0	20.4	22.3
Ca^{2+} [%]	8.3	12.1	6.2
Na^+ [%]	13.0	11.1	9.3
K^+ [%]	1.7	3.4	4.3
Mg^{2+} [%]	2.8	2.5	1.7
H^+ [%]	16.9	7.6	12.3
Avg. TIC[a] [meq/dm³]	0.36	0.42	0.52
Total load [meq]	5.28	1.85	1.54

N Number of samples

% According to ionic charge

[a] Volume weighted averages

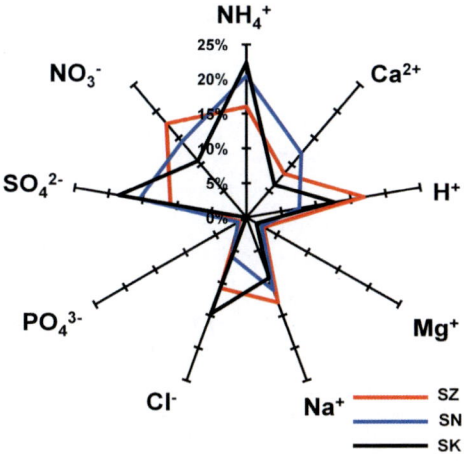

Figure 6

Relative contribution of major rime chemical components [%] at Szrenica Mt. (SZ), Śnieżnik Mt. (SN) and Skrzyczne Mt. (SK)

of nitrate, sulphate and chloride. Ammonium concentrations represent neutralizing inputs from ammonia either to precursor aerosol particles.

The highest values of TIC were observed at the Skrzyczne Mt. (0.524 meq l⁻¹) and subsequently at the Śnieżnik Mt. (0.427 meq l⁻¹). The lowest TIC is characteristic for Szrenica Mt. (0.362 meq l⁻¹). Such differences are connected with liquid water content

changes (higher at Szrenica Mt.) and proximity to emissions sources in prevailing wind conditions (more important in the case of Skrzyczne Mt.). Szrenica Mt. is the only site in Poland with long-term continuous fog chemical measurements, so TIC measured in 2009 is approximately four times lower in relation to the early nineties (SOBIK, 1999; BŁAŚ *et al.*, 2008). Decreasing trend of dissolved pollutants content expressed by TIC should be explained by structural changes in industrial sector resulting from economy transition. Hence, substantial reduction of gaseous emissions has been observed, with SO_2 being reduced more significantly (MITOSEK *et al.*, 2004). However, one should stress, that present pollutant loadings measured at three analysed stations are still classifying as heavily polluted in comparison with fog chemistry measurements around the world (ELBERT *et al.*, 2000).

The volume weighted TIC values were approximately two times higher when compared with bulk precipitation at the same time. Fog water, originating in low-level air, has higher pollutant concentration due to the more polluted nature of the boundary layer than the free atmosphere. On the other hand, precipitation particles are formed due to the processes occurring within deeper air layer, often reaching the middle or even the upper part of the troposphere, where concentration of pollutants emitted close to the ground is much lower (SOBIK, 1999). That is the reason, why TIC values in case of precipitation were more constant between analysed locations and also why cloud chemistry can be a more sensitive indicator of the composition of regional emission than precipitation (BŁAŚ *et al.*, 2010).

Independent of direction of airflow, TIC values of rime are rather similar ranging, which is typical especially for Szrenica Mt., where TIC stays within the range between 0.26 and 0.43 meq l⁻¹. At the Śnieżnik Mt. and Skrzyczne Mt. a weak maximum is observed for W–NW sector (0.66 meq l⁻¹) and NE–E atmospheric circulation (0.58 meq l⁻¹), respectively. The observed increase of rime pollutant concentration in some circulation patterns might be explained by rare precipitation events which do not lead to sufficient pollution removal (SOBIK, 1999; BŁAŚ *et al.*, 2010). Meanwhile, relatively low rime concentrations appear where the largest sources of

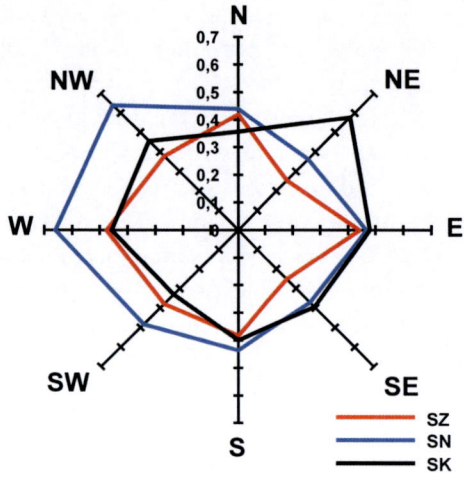

Figure 7
Volume weighted average TIC (Total ionic content) versus direction of air circulation [meq l^{-1}] at Szrenica Mt. (SZ), Śnieżnik Mt. (SN) and Skrzyczne Mt. (SK)

pollution are found: SW–W circulation at Szrenica Mt. and NW–N sector at Skrzyczne Mt (Fig. 7). It should be explained by relatively frequent washing out of orographic clouds by atmospheric precipitation which leads to effective pollutant removal (JONES and CHOULARTON, 1988; DORE et al., 1992; SOBIK, 1999). WRZESINSKY and KLEMM (2000) suggest that the liquid water content (LWC) of fog has also a significant influence on TIC concentration: the smaller the LWC, the higher the concentration. Hence, cloud water chemistry in the context of relevant emissions and the movement of air masses do not clarify such differences satisfactorily (KIM and ANEJA, 1992; IGAWA et al., 2001). The highest TIC concentrations are related to thin low-level stratiform clouds formed below an elevated thermal inversion. The vertical dispersion is limited and the existing emission is trapped in sub-inversion layer leading to high pollution concentration in cloud.

Taking into account the dominant ions in rime chemical composition, there are important differences between the Szrenica Mt. and two remaining sites. At Szrenica Mt. equilibrium between nitrate, chloride and ammonium is characteristic. Rime composition at Śnieżnik Mt. and Skrzyczne Mt. is ammonium dominated (20 and 22%, respectively) with a secondary maximum of sulphate. It is well known that sulphate, nitrate, ammonium and

hydrogen are the principal ions reflecting the dominant influence of anthropogenic sources on rime chemistry. The average aggregated concentration for these ions was similar at all sites and reached about 60% of TIC. Despite the inland location of the study area, maritime components (chloride and sodium) have at least significant contribution. Firstly, this can be explained by the high elevation of sampling sites, which corresponds to the upper part of the atmospheric boundary layer with almost free-atmosphere conditions of air transport. Secondly, during winter conditions, advection from the Atlantic Ocean becomes vigorous and dilution by vertical mixing in stable atmospheric conditions is much more limited. Hence, chemical composition of rime samples at three mountain top locations suggests solute contributions from emission sources located at much larger upwind distances.

Rime deposition of the sum of ions at Szrenica Mt. was 2.8 and 3.4 times higher when compared with Śnieżnik Mt. and Skrzyczne Mt. It was especially due to much more frequent fog, higher wind speed and liquid water content, while the TIC value was the lowest. Approximately 60% of total deposition via rime in the case of Szrenica Mt. occurs during the circulation from the NW–N sector and almost 25% is connected with S–SW–W sector. A contrasting spatial distribution of the deposited pollutants is visible at Skrzyczne Mt., resulting from outnumber rime frequency during inflow of humid polar-maritime air masses from the wide northern sector NW–N–NE (70%), 13% (forming secondary maximum) is connected with SW sector. Pollutant deposition via rime at the Śnieżnik Mt. is under the influence of the wide western sector (S–W–N) with maximum from S direction (21%) (Fig. 8).

3.3. Dendrochronological Research

The release of such heavy load of pollutants may lead to drastic drop of pH in the soil layer as well as in the surface water. This is particularly dangerous in the areas characterised by low environmental buffering capacity of acidic substances (MELACK et al., 1982), which is typical for igneous bedrock areas, e.g. built of granite or gneiss (only 20–30 mMol m^{-2} year^{-1}; AGREN, 1988; NILSSON and GRENNEFELT,

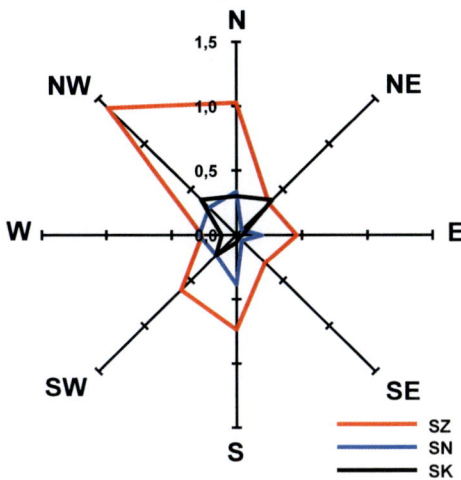

Figure 8
Total pollutant deposition via rime on the surface of passive collector versus direction of air circulation [meq·collector surface^{-1}] at Szrenica Mt. (SZ), Śnieżnik Mt. (SN) and Skrzyczne Mt. (SK)

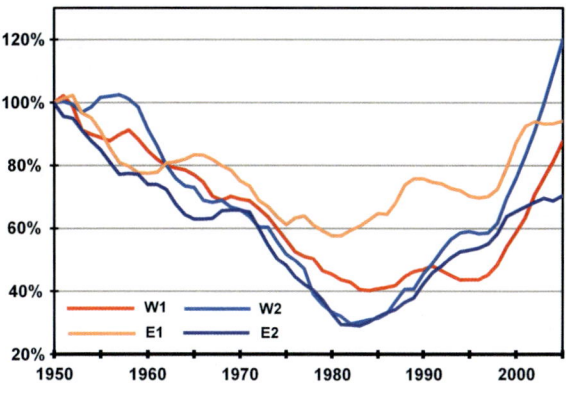

Figure 9
Moving averages (5 years) of relative thickness increase [%] of Norway spruce (*Picea abies*) from the Śnieżnik Massif. *W1* Subalpine spruce forest, 1,230 m a.s.l., western windward slope; *E1* Subalpine spruce forest, 1,230 m a.s.l., eastern leeward slope; *W2* Upper tree line, 1,350 m a.s.l., western windward slope; *E2* Upper tree line, 1,350 m a.s.l., eastern leeward slope

1988). For this reason, strong acidification along the whole vertical profile of the soil layer is observed in a major part of the study area. As a result, during the 1970s and 1980s, the soil layer pH value decreased rapidly by 0.5 unit and the forest damage was observed on a large scale (SOBIK, 1999). Previous dendrochronological investigation in the subalpine zone of the Karkonosze Mts. proved an intense decrease of tree ring width by 70% between 1950 and the mid 1980s—with the phase of systematic increase after this period (GODEK *et al.*, 2009). For the need of current study dendrochronological sampling was conducted at the Śnieżnik Massif which is situated in the central part of the study area and seriously influenced by atmospheric pollutant deposition. The achieved dendrochronological data enable reconstruction of changes in ecosystem's condition during the last six decades. Whereas chemical data from the 2009 show a close relationship between the huge spatial differences of pollutant deposition (in a meso- and microscale) and the negative changes of the trees vitality. The highest pollutant deposition rate is observed at convex landforms on their windward side where fog deposition tends to be the main component of the total flux of pollutants to the ground. Furthermore, vegetation itself exerts an important influence on the hydrological and chemical inputs, particularly in forested areas where vegetation efficiently intercepts fog droplets.

The dendrochronological research carried out showed high spatial and temporal variability of trees vitality within the Śnieżnik Massif (Fig. 9). From the beginning of the second half of the twentieth century, the falling tendency in the size of the annual rings was marked which persisted until mid 1980s (Figs. 10, 11). It overlaps with the gradual development of the power industry and constantly growing pollutant emission and deposition. Thus, the year 1950, i.e. the average value for 1948–1952, has been assumed as the reference point for further considerations concerning trees' vitality. In order to obtain reliable and comparable information, the dendrochronological samples were taken only from the mature individuals of the Norway spruce (*Picea abies*) which were over 80 years old. The obtained data was presented in the form of 5 year moving average of annual increments. While the size of the annual increments systematically decreased after 1950, the rate of the decrease closely depended on the location. The largest reduction in the annual increments was measured in the W2 and E2 sites situated at the upper tree line, where the share of fog in the total water flux and pollutant deposition is the most distinct in the entire vertical profile. At the peak of the ecological disaster, the decrease of the annual increments by around 70% as compared to the values in 1950 was

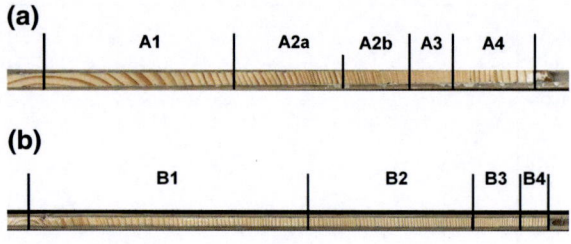

Figure 10

Dendrochronological samples from two sites at the Śnieżnik Mt.: **a** windward slope with W exposure (W1) **b** leeward slope with E exposure (E1); 1, intensive growth; 2, decreasing vitality; 3, maximal reduced increments; 4, increasing vitality

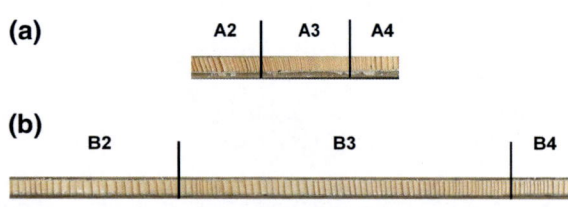

Figure 11

Comparison between maximal reduced spruce vitality period at Śnieżnik Mt.: **a** windward slope with W exposure (W1) **b** leeward slope with E exposure (E1)

recorded there. From that time on, the systematic increase of trees' vitality can be observed. In 2005, the size of the annual increment reached appropriately 120 (W2) and 71% (E2) of the 1950 values. It is connected with substantial reduction of air pollutant emission over the past two decades. The largest reductions covered SO_2 with the most remarkable improvements in the Upper Silesia and the "Black Triangle". Since 1990, a substantial decline is observed with a further downward trend continuing up to date. However, MILL et al., 2003; MILL, 2006) showed that in spite of nearly 70% reduction of sulphur emission, the threat of acidification affecting forests and other nature ecosystems is still relatively high in the research area, concerning 40% of the forested terrain.

The sequence of annual increments at the sites situated lower (W1 and E1) is a bit different. Here, the minimum recorded in 1980s is not so deep and the slope exposure to the prevailing wind direction plays a more significant role in their diversity. The decrease of trees' vitality by 60% was found on the windward slope of W exposure, whereas on the leeward slope (E exposure) only by 40%. Clear differences between

them are visible in Figs. 6, 9. On the leeward slope, in the peak stage of trees' vitality weakening, the annual increments are three times broader on average than the increments recorded on the windward side. At present, with the progressing reduction of atmospheric pollutant emission and deposition, the average widths of increments have similar values. Moreover, a short (6 week long) experiment was conducted and it aimed at checking the dependency between the rime efficiency and the atmospheric circulation direction. It showed that rime was 30% less efficient at the E1 leeward site compared to the W1 windward site. It is because fog is more frequent with higher liquid water content when a station is a windward one and is not so likely to form, when the descending airflow at leeward side is observed. In the highest parts, the differences are subtle and they become more remarkable with the decreasing height.

4. Conclusions

The efficiency of fog-induced horizontal precipitation differs from site to site depending on the altitude, aspect, relative height of a windward slope as well as the screening effect of the surroundings relief. At three mountain stations (Szrenica Mt., Śnieżnik Mt. and Skrzyczne Mt.), rime forms a considerable component of water transfer from the atmosphere to the ground and due to high concentration of chemical constituents it is an important pathway of pollutant deposition processes. The highest annual total of water flux and pollutant deposition from rime, observed at the Szrenica Mt., was due to its north-westernmost position along NW–SE profile in relation to the prevailing wind direction and, to certain extent, due to the "sheltering" of Śnieżnik Mt. and Skrzyczne Mt. by successive hill peaks on their windward side. This effect took place throughout the whole year, but it is more pronounced during winter because of a higher orographic cloud frequency and the very efficient scavenging of cloud droplets by snow crystals, which have a high surface area to volume ratio. That is also the reason why TIC is a bit higher during the winter season.

The dendrochronological record taken from the trunks of the subalpine Norway Spruce (*Picea abies*) showed that the most difficult period for the trees' growth was the first half of the 1980s, when the trees' vitality (width of annual increments) was reduced in the area of the upper tree line in the Śnieżnik Massif by as much as 70% comparing to 1950. It must be emphasized that the spatial variability of this parameter rather closely relates to the role of rime in shaping the atmospheric pollutant deposition rate. It is particularly clear while comparing sites representing alternatively the windward and the leeward side (Fig. 11). Because of a significant emission reduction, from the beginning of the 1990s systematic improvement of the forest health status is observed at all sites. However, both its earlier deterioration and the present amelioration show a significant spatial diversity. At present, due to monocultures in majority of the forest ecosystems existing within the research area, the potential gradation of the insect pests and as the further consequence the risk of forest die-back still remain the significant problem.

Acknowledgments

This scientific work was financially supported by the Polish Ministry of Science and Higher Education between 2008 and 2010 as a research project (N N305 231035).

REFERENCES

AGREN, CH. (1988), *Critical loads. Figures continuing downwards*, Acid News *3*, 1–4.

BARANOWSKI, S., and LIEBERSBACH, J. (1978), *The intensity of different kinds of rime on the upper tree line in the Sudety Mountains*, J Glaciol. *19*, 489–497.

BŁAŚ, M., and SOBIK. (2003) Natural and human impact on pollutant deposition in mountain ecosystems of the Sudetes, In Man and climate in the XX century (ed. Pyka, J.) (Acta Univ. Wratisl. 2542, Studia Geograficzne 75, Wrocław), pp. 420–438.

BŁAŚ, M., and SOBIK, M. (2000), *Fog in the Giant Mountains and selected European massifs*, Opera Corcontica *37*, 35–46.

BŁAŚ, M., and SOBIK, M. (2004), *The distribution of fog frequency in the Carpathians*, Geographia Polonica—General and Applied Climatology: Selected Aspects *77*, 19–34.

BŁAŚ, M., SOBIK, M., QUIEL, F., and NETZEL, P. (2002), *Temporal and spatial variations of fog in the Western Sudety Mts., Poland*, Atmos. Res. *64*, 19–28.

BŁAŚ, M., SOBIK, M., and TWAROWSKI, R. (2008) *Changes of cloud water chemical compositions in the Western Sudety Mountains, Poland*, Atmos. Res. *87*, 224–231.

BŁAŚ, M., POLKOWSKA, Ż., SOBIK, M., KLIMASZEWSKA, K., NOWIŃSKI, K., and NAMIEŚNIK, J. (2010), *Fog water chemical composition in different geographic regions of Poland*, Atmos. Res. *95*, 455–469.

CHOWANIEC, J. (1991) Region karpacki, In Budowa Geologiczna Polski, vol. VII Hydrogeologia (Ed. Malinowki, J.) (Wydawnictwa Geologiczne, Warszawa), pp. 204–215.

DANEK, M. (2007), *The influence of industry on scots pine stands in the south-eastern part of the Silesia-Kraków Upland (Poland) on the basis of dendrochronological analysis*, Water Air Soil Pollut. *185*, 265–277.

DORE, A.J., CHOULARTON T.W., BROWN, R., and BLACKALL, R.M. (1992), *Orographic rainfall enhancement in the mountains of the Lake District and Snowdonia*, Atmos. Environ. *26A*, 357–371.

DRUKMAN, I., MIGAŁA, K., and SOBIK, M. (1997) Selected Characteristics of wind speed structure in the West Karkonosze Mts., In Climatological Aspects of environment protection in mountain areas (ed. Migała, K.). Acta Univ. Wratisl., Prace Instytutu Geograficznego, s. C, Meteorologia i Klimatologia, IV, Wrocław) pp. 67–73.

ELBERT, W., HOFFMANN, M.R., KRAMER, M., SCHMITT, G., and ANDREAE, M.O. (2000), *Control of solute concentrations in cloud and fog water by liquid water content*, Atmos. Environ. *34*, 1109–1122.

EMEP Data Report, Acidifying and eutrophying compounds and particulate matter (Norwegian Institute for Air Research 2010).

EMEP Status Report, Transboundary Acidification, Eutrophication and Ground Level Ozone in Europe in 2008 (Norwegian Meteorological Institute 2010).

FELIKSIK, E. (1995), *Dendrochronological monitoring of the threat to the forest of Western Beskids created by industrial immisions*, Zpravodaj Beskydy *7*, 23–34.

FELIKSIK, E., and WILCZYŃSKI, S. (2003), *Tree ring as indicators of environmental change*, Electronic Journal of Polish Agricultural Universities, series Forestry 6/2.

FERRETTI, M., INNES, J.L., JALKANEN, R., SAURER, M., SCHÄFFER, J., SPIECKER, H., and VON WILPERT, K. (2002), *Air pollution and environmental chemistry—what role for tree-ring studies?* Dendrochronologia. *20*, 159–174.

FRITTS, H.C., Tree Rings and Climate (Academic Press, London 1976).

GODEK, M., MIGAŁA, K., and SOBIK M. (2009), *Air pollution and forest disaster in the Western Sudetes in the light of high elevation spruce tree ring data*, TRACE Tree Rings in Archeology, Climatology and Ecology, Procc. of the DENDROSYMPOSIUM 2008, Zakopane. *7*, 121–126.

HUETTL, R.F., and MUELLER-DOMBOIS, D. (1993) Forest decline in the Atlantic and Pacific Region (Springer Verlag, Berlin).

IGAWA, M., MATSUMURA, K., and OKOCHI, H. (2001), *Fog water chemistry at Mt. Oyama and its dominant factors*, Water Air Soil Pollut. *130*, 607–612.

JAIA, Z., and BŁAŚ, M. (2000), *Choisen chemical characters of soils against a background of a wet deposition of air pollution in the Sudetes*, Opera Corcontica *37*, 69–78.

JONES, A., CHOULARTON, T.W. (1988), *A model of wet deposition to complex terrain*, Atmos. Environ. *22* (11), 2419–2430.

KIM, D.S., and ANEJA, V.P. (1992), *Chemical composition of clouds at Mount Mitchell, North Carolina*, Tellus *44B*, 41–53.

KRĄPIEC, M., and SZYCHOWSKA-KRĄPIEC E. (2001), *Tree-ring estimation of the effect of industrial pollution on pine (Pinus*

Reprinted from the journal

sylvestris) and fir (Abies alba) in the Ojców National Park (Southern Poland), Nature Conservation *58*: 33–42.

Kryza, H., Kryza, J., and Marszaiek, H. (2005) Wody podziemne Karkonoszy, In Karkonosze, Przyroda Nieożywiona i Człowiek (ed. Mierzejewski M. P.)(Wrocław) pp. 453–486.

Kryza, M., Błaś, M., Dore, A.J., and Sobik, M. (2009) Application of a Lagrangian Model FRAME to estimate reduced nitrogen deposition and ammonia concentrations in Poland, In atmospheric ammonia: detecting emission changes and environmental impacts. Results of an expert workshop under the convention on long-range transboundary air pollution (ed. Sutton, M.A.) (Springer) pp. 357–366.

Kryza, M., Błaś, M., Dore, A.J., and Sobik, M. (2010), *Fine-resolution modeling of concentration and deposition of nitrogen and sulphur compounds for Poland—application of the FRAME model*, Arch. Environ. Prot. *36*(1), 49–62.

Lovett, G.M. (1984), *Rates and mechanism of cloud water deposition to a subalpine balsam fir forest*, Atmos. Environ. *18*, 361–371.

Malik, I., Danek, M., and Krąpiec, M. (2009) Air pollution recorded in scots pine growing near a chemical plant, preliminary results and perspective (Upper Silesia, southern Poland), In trace tree rings in archeology, climatology and ecology, vol.8, Procc. of the DENDROSYMPOSIUM 2009, Otočec, Slovenia, (ed. Levanic, T.) (GFZ Potsdam, Scientific Technical Report STR, Potsdam) pp. 41–45.

Matuszkiewicz, J.M. (2002) Zespoły leśne Polski (Wydawnictwo Naukowe PWN, Warszawa).

Migaia, K., Liebersbach, J., and Sobik, M. (2002), *Rime in the Giant Mts. (The Sudetes, Poland)*, Atmos. Res. *64*, 63–73.

Mill, W. (2006), *Temporal and spatial development of critical loads exceedance of acidity to Polish forest ecosystems in view of economic transformations and national environmental policy*, Env. Sci. Pol. *9*, 563–567.

Mill, W.A., Schlama, A., Twarowski, R., Błachuta, J., and Stasyewski, T. (2003) Modelling and mapping of critical thresholds in Europe. CCE Status Report 2003, National Focal Centre Report—Poland (Bilthoven, Netherlands).

Mitosek, G., Degórska, A., Iwanek, J., Przybylska, G., and Skotak, K. (2004) EMEP Assessment Report—Poland. (Institute of Environmental Protection, Warsaw).

Möller, D. (1984), *Estimation of the global man-made sulphur emission*, Atmos. Environ. *18*, 19–27.

Moravčik, P., and Černý, M. (1995) Forest die-back affected regions of Czech Republic, In Acid Reign'95, acidification in the black triangle (ed. Grennfeld, P.) (Proceeding from the 5th International Conference on Acid Deposition, Göeteborg).

Niedźwiedź, T. (2010) Kalendarz typów cyrkulacji atmosfery dla Polski południowej—zbiór komputerowy [Calendar of atmospheric circulation types for the Southern Poland—Internet database] (Uniwersytet Śląski, Katedra Klimatologii, Sosnowiec).

Nilsson, J., and Grennfelt, P., Critical loads of Sulphur and Nitrogen (Nordic council of Ministers, Report, Copenhagen 1988) pp. 8–57.

Olejarski, I., Oszako, T., and Hilszczańska, D. (2004), *Evaluation of health status of forest ecosystems in Sudetes after ecological*

disaster with special regard to revitalization of forest soils and its influence on regeneration, Opera Corcontica *41*, 421–433.

Olendrzyński, K., Kargulewicz, I., Skośkiewicz, J., Dębski, B., Cieślińska, J., Olecka, A., Kanafa, M., Kania, K., and Saiek, P. (2009) Poland's National Inventory Report 2009 (National Administration of the emission traiding Scheme, Institute of Environmental Protection, Warszawa,) pp. 189.

Pahl, S., Winkler, P., Schneider, T., Arends, B., and Schell, D. (1994), *Deposition of trace substances via cloud interception on a coniferous forest at Kleiner Feldberg*, J Atmos. Chem. *19*, 231–252.

Pereyma, J., Sobik, M., Szczepankiewicz-Szmyrka, A., and Migaia, K. (1997), *Contemporary climatic conditions and topoclimatic differentiation of the Karkonosze Mts.*, Acta Univ. Wratisl., Prace Instytutu Geograficznego IV, 75–94.

Polkowska, Ż., Astel, A., Walna, B., Maiek, S., Mędrzycka, K., Górecki T., Siepak, J., and Namieśnik J. (2005), *Chemometric analysis of rainwater and throughfall at several sites in Poland*, Atmos. Environ. *39*, 837–855.

Schweingruber, F.H. (1996) Tree Rings and Environment Dendroecology (Swiss Federal Institute for Forest, Snow and Landscape Research, Birmensdorf).

Schweingruber, F.H. (2007) Wood Structure and Environment (Springer Series in Wood Science, Berlin).

Sobik, M. (1999) Meteorologiczne uwarunkowania zakwaszenia hydrometeorów w Karkonoszach (Unpublished Ph.D. Thesis, Wrocław, University of Wrocław).

Sobik, M., Dore, A.J., and Migaia, K. (1998) Influence of orography on wet deposition patterns in the Western Sudetes, In Geoecological Problems of the Karkonosze Mountains, Transborder Biosphere Reserve Karkonosze/Krkonoše, pp. 97–108.

Strzyszcz, Z. (1995), Warunki glebowe a zamieranie drzewostanów w Karkonoskim Parku Narodowym, Geoekologiczne problemy Karkonoszy, Poznań, 89–94.

Vinš, B., and Pollanschütz, J. (1977), *Erkennung and Beurteilung immissionsgeschädigter Wälder anhand von Jahrringanalysen. [Identification and assessment of damaged forests on the basis of tree-ring analysis]*, Alg. Forstzeit. *6*, 146–148.

Wallen, C.C. (1977) World Survey of Climatology Volume 6—Climates of Central and Southern Europe (ed. Landsberg, H.E.) (World Meteorological Organisation, Geneva).

Weathers, K.C., Lovett, G.M., and Likens, G.E. (1995), *Cloud deposition to a spruce forest edge*, Atmos. Environ. *29*, 665–672.

Wiśniowski, Z. (2001) Dendrochronologiczno-geochemiczna analiza przemysłowej degradacji środowiska na przykładzie lasów Puszczy Wkrzańskiej (aglomeracja szczecińska) (Państwowy Instytut Geologiczny, Warszawa).

Wrzesinsky, T., and Klemm, O. (2000), *Summertime fog chemistry at a mountainous site in central Europe*, Atmos. Environ. *34*, 1487–1496.

Żelaźniewicz, A. (2005) Przeszłość geologiczna, In Przyroda Dolnego Śląska, (ed. Fabiszewski, J.) (PAN, Wrocław) pp. 61–134.

Zimmermann, L., Zimmermann, F. (2002), *Fog deposition to Norway Spruce stands at high-elevation sites in the Eastern Erzgebirge (Germany)*, J Hydrol. *256*, 166–175.

(Received December 2, 2010, revised May 18, 2011, accepted May 25, 2011, Published online June 26, 2011)

Pure Appl. Geophys. 169 (2012), 1107–1119
© 2011 Springer Basel AG
DOI 10.1007/s00024-011-0351-x

A Fuzzy Logic Fog Forecasting Model for Perth Airport

Y. Miao,[1] R. Potts,[1] X. Huang,[2] G. Elliott,[3] and R. Rivett[3]

Abstract—Perth Airport is a major airport along the southwest coast of Australia. Even though, on average, fog only occurs about twelve times a year, the lack of suitable alternate aerodromes nearby for diversion makes fog forecasts for Perth Airport very important to long-haul international flights. Fog is most likely to form in the cool season between April and October. This study developed an objective fuzzy logic fog forecasting model for Perth Airport for the cool season. The fuzzy logic fog model was based on outputs from a high-resolution operational NWP model called LAPS125 that ran twice daily at 00 and 12 UTC, but fuzzy logic was employed to deal with the inaccuracy of NWP prediction and uncertainties associated with relationships between fog predictors and fog occurrence. The outcome of the fuzzy logic fog model is in one of the four categories from low to high fog risk as FM0, FM5, FM15 or FM30, intended to map to approximate fog probability of 0, 5, 15 and 30%, respectively. The model was found useful in its 5 year performance in the cool seasons between 2004 and 2008 and required little recalibration if mist was treated as if it were also a fog event in the skill evaluation. To generate an operational fog forecast for Perth Airport, the outcome of the fuzzy logic fog model was averaged with the outcomes of two other fog forecasting methods using a simple consensus approach. Fog forecast so generated is known as the operational consensus forecast. Skill assessment using frequency distribution diagram, Hansen and Kuiper skill score, and Relative Operating Characteristic curve showed that the operational consensus forecast outperformed all three individual methods. Out of the three methods, the fuzzy logic fog model ranked second. It performed better than the other objective method called GASM but worse than the subjective method which relied on forecaster's subjective assessment. The skills of the fuzzy logic fog model can be further improved with the tuning of fuzzy functions. In addition, similar models can be customised for other airports. The study also suggested the use of the simple consensus approach to enhance forecasting skills for other stations or weather phenomena if there were two or more independent forecasting methods available.

Key words: Fog, fog forecasting, fuzzy logic, NWP, consensus, Perth Airport.

1. Introduction

Fog forecasting is an important issue at Perth Airport (31.58°S, 115.49°E, altitude 20 m), the largest international airport along the southwest coast of Australia. Although fog occurs relatively infrequently there, the consequence of an unforecast fog event could be very serious. The airport is not equipped for planes to land or take off in low visibility and the closest airport for large airplane diversion is more than 1,200 km away. As a result, airlines pay special attention to the official Perth Airport aviation fog forecasts issued at the Western Australia Regional Forecasting Centre (WARFC) located in Perth.

The meteorological conditions leading to fog events at Perth Airport are quite varied. On the large scale, WARFC identified a variety of synoptic patterns conducive to fog, ranging from cut-off lows, post-frontal and ridging in cool months to troughs and pre-trough/pre-frontal in warmer months. Clearly fog types are not restricted to radiation fog only, although the evolution of these patterns often leads to radiative cooling through reduced winds and cloud cover overnight. Perhaps not coincidently, precipitation is prevalent in many cool-month patterns, contributing to fog formation through additional moisture that it supplies to the ground and air. At local scale, detailed mesoscale numerical weather prediction (NWP) modeling by Golding (1993) for a fog event at Perth Airport demonstrated the complex interaction between the prevailing and local winds. The study suggested that the mixing and convergence between the onshore moist westerly flow and the opposing

---
[1] The Centre for Australian Weather and Climate Research, Bureau of Meteorology, GPO Box 1289, Melbourne, VIC 3001, Australia. E-mail: y.miao@bom.gov.au
[2] National Meteorological and Oceanographic Centre, Bureau of Meteorology, GPO Box 1289, Melbourne, VIC 3001, Australia.
[3] Western Australia Regional Office, Bureau of Meteorology, PO Box 1370, West Perth, WA 6872, Australia.

cool easterly drainage flow frequently determined the location and timing of fog formation. Thus, meteorological complexities at multi-scales make fog forecasting for Perth Airport difficult.

Due to the complexity and scarcity of fog events at Perth Airport, a few known statistical or rule-based fog forecasting methods, such as REGANO's (1997) analog method that relied heavily on current observations to infer fog occurrence next morning, were found not as useful when applied to Perth Airport. For a long time, therefore, synoptic pattern recognition and matching combined with physical reasoning by forecasters was used as the primary fog forecasting approach for Perth Airport. This is known as the synoptic pattern matching subjective method or simply the subjective method.

The use of NWP models in fog forecasting is an active research area, naturally so, given the high expectation of model performance with the increasingly powerful supercomputers and more sophisticated data assimilation techniques. The reality, however, is that even the most advanced operational NWP models cannot directly simulate such fog processes as droplet microphysics, aerosol chemistry and turbulence (GULTEPE et al., 2007). Some weather service operational centres such as UK Met Office and NOAA National Centers for Environmental Prediction (NCEP) tried alternative ways of producing fog and visibility forecast fields using proxy data such as relative humidity and liquid water content, but the performance was not convincing (CLARK et al., 2008; ZHOU et al., 2010). A cheaper and popular alternative to 3D modelling is the 1D single column fog NWP modelling that is forced by a 3D NWP model and aided by local observations in initialisation. The 1D models of COBEL-ISBA (BERGOT and GUEDALIA, 1994), PA-FOG (BOTT et al., 1990), and the 1D version of the UK Met Office's Unified Model (CLARK and HOPWOOD, 2001) are examples used operationally at various airports. Nonetheless, the performance of this type of modelling remains unconvincing. A comparison study of six different 1D models by BERGOT et al. (2007) revealed significant divergence in fog prediction results even for radiation fogs. GULTEPE et al. (2007) attributed the model deficiency to the (i) assumption of horizontal homogeneity and (ii) inability to parameterize turbulence in strongly stable conditions, among

other factors. Therefore, to a large extent, forecasters don't normally use fog predictions directly from NWP models. Rather, they make fog risk assessment through synergetic consideration of both NWP predicted fields and physical reasoning. Hence, often the NWP products are used on a qualitative and subjective basis.

An attempt has been made at the Australia Bureau of Meteorology (BoM) to quantitatively use direct outputs of NWP models to produce guidance of weather forecast such as probability of fog. The Generalized Analogue Statistics Model (GASM, DAHNI and STERN, 1995) was used until recently. This analogue method uses both the current analysis and NWP prognoses of future weather conditions for comparison with similar past events obtained from a climate database. Although showing some skill for Perth Airport fog forecasting, this requires human intervention and can be labour-intensive.

A new fog prediction method is developed in this study that applies fuzzy logic to the objective use of NWP model fields for fog forecasts. Fuzzy logic is adopted to cater for model inaccuracy and the inexact nature of relationships between a predictor and fog risk forecast. Use of fuzzy logic has gained popularity in recent years in weather forecasting. BARDOSSY and DUCKSTEIN (1995) used the fuzzy rule-based method in the classification of atmospheric circulation patterns. HANSEN (2007) reported a sophisticated analogue method using the fuzzy logic concept to forecast ceiling and visibility for 190 airports in Canada. Details of our fuzzy logic based NWP fog model are discussed in "Fuzzy Logic NWP Fog Forecasting Model" below.

For operational fog forecast, forecasters at WARFC used outcomes from all three methods, i.e., the fuzzy logic fog model, GASM and the subjective method. A challenge was to decide how to optimally combine the three outcomes and a simple consensus approach was adopted to produce an operational consensus forecast. The rationale behind choosing this approach is discussed in "Operational Consensus Forecast". The skill of the fuzzy logic model is evaluated and the relative skills among the three methods and the operational consensus forecast are compared in "Performance Evaluation and Discussion". A summary is provided at the end.

2. Definitions

2.1. Fog Event

The WMO meteorological definition of a fog is that the prevailing horizontal visibility is less than 1,000 m and mist is for visibility greater than 1,000 m but less than 5,000 m (WMO, 1995). However, because aviation forecasting is more concerned about the weather conditions below the Special Landing Alternate Minima (SLAM), and because of difficulty in ascertaining whether visibility reduction over a sector could lead to a widespread fog event, WARFC and this paper define a fog event as when the minimum visibility for any sector is at or below 2,000 m, i.e., SLAM for Perth Airport. Accordingly, in this paper we define a mist event as the minimum visibility being larger than 2,000 m but equal to or less than 5,000 m.

2.2. Fog Seasons

Two distinct "fog seasons" are defined at Perth Airport, a cool season from April to October inclusive and a warm season for remaining months. Based on observations at Perth Airport during the 23 year period from 1988 to 2010, there are about 11 fog events (or 1.5 per month) in the cool season and only one fog event (or 0.3 per month) in the warm season (Fig. 1). Thus, the development of the fog model in this study is restricted to the fog-prone cool season only.

2.3. Fog Forecast Products

The standard ICAO Terminal Area Forecasts (TAFs) forecast fog at Perth Airport when the expected fog probability is 30% or higher. For fog probability of less than 30%, a unique forecast product called Code Grey is issued to major airlines in Australia. In practice, the fog probability of 5, 10 or 20% is forecasted on Code Grey. The early afternoon forecasts of TAF and Code Grey valid from 0600 coordinated universal time (UTC) are critically important to international long-haul flights arriving in Perth next morning. The time zone in Perth is 8 h ahead of UTC. Hence, 0600 UTC corresponds to

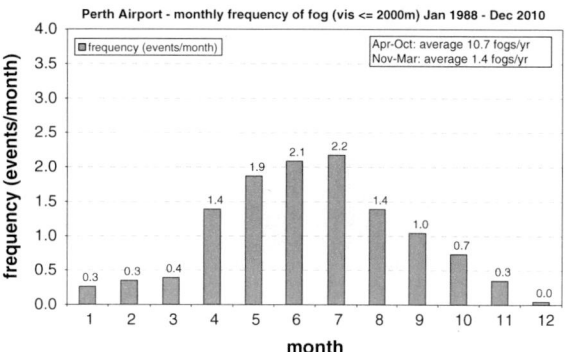

Figure 1
Perth Airport monthly distribution of the observed fog events based on the 23 year observations from Jan 1988 to Dec 2010

2 pm Perth local time—the western standard time (WST). For the 0600 UTC issue of forecasts, fog forecasting decision is one of these three—not forecasting fog at all (hereafter as NoFog), forecasting less than 30% fog probability on Code Grey (hereafter as CG), and forecasting 30% or higher fog probability on TAF (hereafter as Prob30).

3. Fuzzy Logic NWP Fog Forecasting Model

The NWP fields as input to the fuzzy logic fog forecasting model (hereafter fuzzy model) came from the LAPS125 model which was part of the LAPS modelling system developed at BoM (Puri et al., 1998). LAPS125 was chosen for its high-resolution and comprehensive diagnostic fields. It has a horizontal resolution of 0.125° (~12.5 km), 29 levels in vertical, running twice daily at 00 and 12 UTC out to 48 forecast hours.

To make the best use of the NWP model fields in any fog model, two areas of uncertainties should be understood before they can be dealt with. The first area of uncertainties relates to how well a NWP can predict a weather parameter. For example, NWP models normally predict the mean sea level pressure field better than other surface fields such as the screen level temperature and dew point. The second area of uncertainty is how well a predictor relates to a predictand. This relationship uncertainty is irrespective of how well a NWP model predicts the predictor. For example, even the observed screen level dew point depression of close to zero does not always lead to fog formation at Perth Airport. Detailed study is needed to quantify these two areas of uncertainties.

Fuzzy logic provides a means to deal with these uncertainties and still produce useful fog prediction (MURTHA, 1995). Fuzzy logic or the fuzzy set theory was first introduced by ZADEH (1965) and provides a simple way to arrive at a definite conclusion based upon vague, ambiguous, imprecise, noisy, or missing input information. Contrary to what the name may imply, fuzzy logic is a precise logic of imprecision (ZADEH, 1999). It is a logical system involving fuzzy sets which are sets whose elements have degrees of membership ranging from 0 to 1. In this study, fuzzy sets are those of fog risk categories with which a fog predictor has an associated membership function. A set of fuzzy rules such as inference and averaging are involved to produce a composite fuzzy membership function for each fog risk category before a final outcome as one of the fog risk categories is produced. This process of aggregating membership functions through a set of fuzzy rules before arriving at an output is known as defuzzification.

The following steps are involved with the development of this fuzzy model. They will be discussed in the same order.

- Determine fog predictors
- Define fog risk categories
- Assign fuzzy membership functions
- Defuzzification to a final outcome

3.1. Fog Predictors

Three criteria are used to determine whether a weather parameter can be used as a fog predictor by this fog model. The first is that the frequency of fog occurrence has to be sensitive to the value change of the parameter. This test is normally performed on observational data. The second is that the NWP model should have acceptable skills in the prediction of this parameter. Lastly the sensitivity of fog frequency to the change of the modelled value should be similar to that to the change of the observed value.

Both observed and predicted data are used in the search for fog predictors. The observational dataset included the radiosonde temperature and dew point profiles for fog events in the cool seasons from 1992 to 2001, and the observed average wind speed at 00, 12 and 18 UTC at around 300 m for fog events in the cool seasons from 1996 to 2001. LAPS125 fields provided the predicted data in the cool seasons of 2001 and 2002.

Before fog predictors are introduced, some terms that will be used in the predictor definitions and throughout the paper are explained below:

- Unless otherwise stated, all mentioned height is above the ground level, not mean sea level.
- Dew point depression (DPD) is the air temperature minus dew point temperature.
- Overnight consists of the hours from 12 to 24 UTC (or from 8 pm to 8 am WST).
- Initial model error is the modelled minus the observed dew point depression at the analysis time at around 60 m. This initial error is assumed to be the same at all levels considered (60, 200 and 600 m) and forecast hours up to 24 h. This assumption of the same error persisting at all levels and forecast hours is crude but it provides a means to account for model error given the limited information available.
- The corrected dew point depression at a forecast hour at a level has had the initial model error subtracted.
- Dew point depression trend to a forecast hour is the difference between DPD at a forecast hour and DPD at analysis. If DPD is $4°$ at $+12$ forecast hour and $2°$ at analysis, then the DPD trend to $+12$ forecast hour is $2°$, i.e. $(DPD_{+12} - DPD_{00} = 2)$. This indicates that the air is forecast to dry with time.

Extensive investigation of the observational and the modelled datasets in the cool season identified the following four fog predictors for the fuzzy model based on the 00 UTC LAPS125 model prediction (hereafter 00 UTC Fuzzy Model).

- Mean overnight wind speed at 200 m (μ_1 in Eq. 1)
- Overnight minimum of the corrected dew point depression at 200 or 600 m (μ_2)
- Corrected dew point depression at the $+24$ forecast hour at 600 m (μ_3)

- Minimum of the dew point depression trend to +12 or +24 forecast hours at 200 or 600 m (μ_4)

It is obvious that the last three of these predictors are related to each other since they all involve dew point depression. Although each is included for different contributions, their interdependency is taken into consideration as reduced weight when applying fuzzy rules in the final defuzzification stage to work out the forecast outcome.

The absence of screen level dew point depression as a predictor is because at Perth Airport even the observed dew point depression generates high false alarms. This may be more pertinent to Perth Airport than other places because many rain events correspond to low dew point depression but don't necessarily lead to fog events.

For the 12 UTC Fuzzy Model, some changes are made to the predictor list to increase the emphasis on observations due to the shorter lead time to fog onset. The list is as follows.

- Mean overnight wind speed at 200 m (μ_5)
- Overnight minimum of the corrected dew point depression at 200 m (μ_6)
- Overnight minimum of the dew point depression trend at 200 m (μ_7)
- Observed dew point depression at 600 m at the analysis time of 12 UTC (μ_8).

3.2. Fog Risk Categories

To match with WARFC's fog forecast decision categories of NoFog, CG and Prob30 as explained in "Fog Forecast Products", the output of the Fuzzy Model is defined as one of the four fog risk categories FM0, FM5, FM15 and FM30, representing negligible, very low, low and moderate fog risk, intended to map to approximate fog probabilities of 0, 5, 15 and 30% respectively. When applied in the operational forecast environment, FM0 and FM5 are assigned to NoFog, FM15 to CG and FM30 to Prob30 forecast decision. The resulting fit between the risk categories and the probabilities as well as the suitability of assignment to a forecast decision are evaluated in "Performance of the Fuzzy Model" when the performance of the Fuzzy Model is verified over a 5 year period.

3.3. Fuzzy Membership Functions

After the synergetic assessment of the fuzzy relationships between fog predictors and fog risk categories, all fuzzy membership functions of fog predictors in the fuzzy sets of fog risk categories are derived.

These fuzzy membership functions are expressed as

$$F_k(\mu_i) = \begin{cases} 1 & \mu_i <= x1_{ki} \\ y1_{ki} + \frac{y1_{ki}-y2_{ki}}{x1_{ki}-x2_{ki}}(\mu_i - x1_{ki}) & x1_{ki} < \mu_i < x2_{ki} \\ 0 & \mu_i >= x2_{ki} \end{cases} \tag{1}$$

where subscript $i = 1, 2,..., 8$ corresponds to the predictor of $\mu_1, \mu_2, ..., \mu_8$ respectively. Subscript $k = 5, 15, 30$ refers to the fog risk category of FM5, FM15 and FM30 respectively. $F_k(\mu_i)$ is the membership function of the predictor μ_i in the fuzzy set of the fog risk category k. Figure 2a–d provide graphical representation of those membership functions of μ_i in fuzzy sets of the FM5, FM15 or FM30 fog risk categories in the 00 UTC Fuzzy Model.

3.4. Defuzzification to a Final Outcome

A few steps are involved in the defuzzification process to produce a final outcome of the Fuzzy Model. For each of the FM5, FM15 and FM30 fog risk categories, the first step is to find a minimum function out of the three functions corresponding to the three predictors of μ_2, μ_3 and μ_4 that all involve dew point depression. This inference operation is then followed by a simple averaging operation between the function of μ_1 and the minimum function just acquired. A composite fuzzy function for each of the three fog risk categories FM5, FM15 and FM30 is thus achieved while the composite fuzzy function for the F0 fog risk category is effectively one minus the composite function for F5. To work out the final outcome of the Fuzzy Model, a decision tree as shown in Fig. 3 is used based on the comparison between each composite function and the corresponding activation threshold (set as 0.725) for each fog risk category. For example, for a given set of composite fuzzy function of (0.45, 1.0, 1.0, 0.33) corresponding to the set of fog risk of (FM0, FM5, FM15, FM30), the final outcome should be FM15.

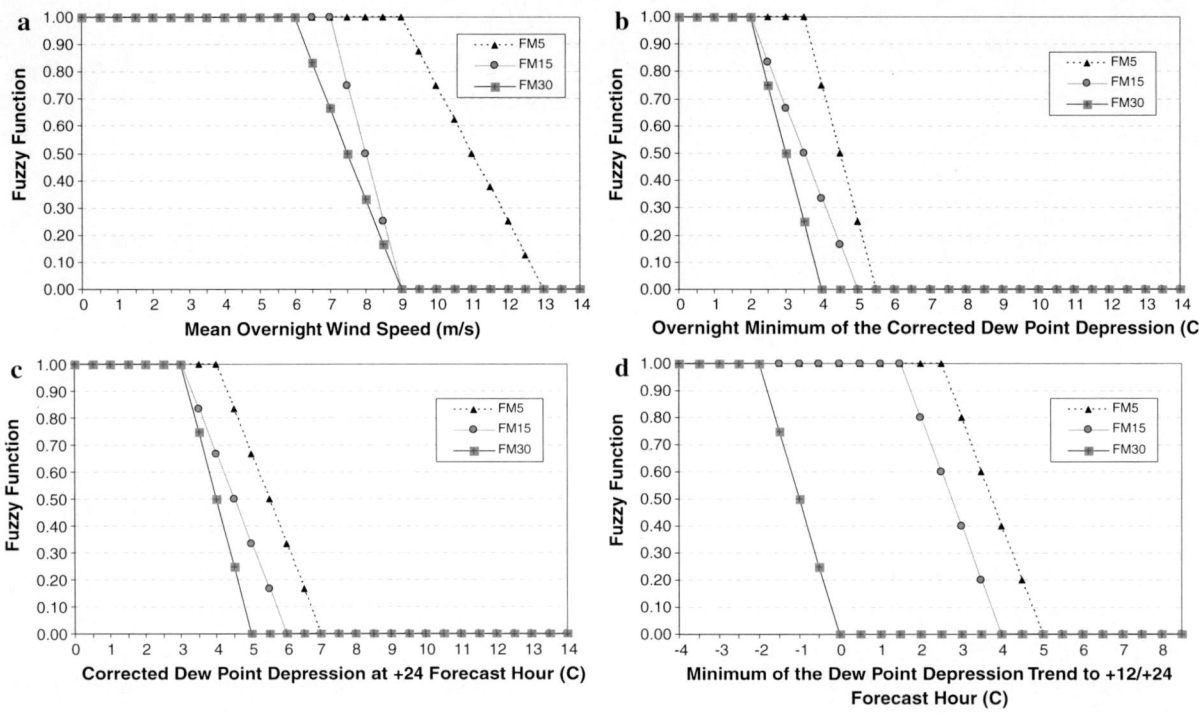

Figure 2

a The 00Z Fuzzy Model fuzzy membership functions of the mean overnight wind speed at 200 m in the fuzzy sets of FM5, FM15 and FM30 fog risk categories. **b** The 00Z Fuzzy Model fuzzy membership functions of the overnight minimum of the corrected dew point depression at 200 or 600 m in the fuzzy sets of FM5, FM15 and FM30 fog risk categories. **c** The 00Z Fuzzy Model fuzzy membership functions of the corrected dew point depression at the +24 forecast hour at 600 m in the fuzzy sets of FM5, FM15 and FM30 fog risk categories. **d** The 00Z Fuzzy Model fuzzy membership functions of the minimum of the dew point depression trend to +12 and +24 forecast hours at 200 or 600 m in the fuzzy sets of FM5, FM15 and FM30 fog risk categories. The negative value suggests moistening trend, which favours fog

4. Operational Consensus Forecast

In addition to the work on the Fuzzy Model, there has been work in recent years to develop a more structured forecast process for fog at Perth Airport using the consensus of three fog forecasting methods to give an operational forecast. The operational forecast generated with the consensus approach is known as the operational consensus forecast or simply the consensus forecast. We briefly present details below on consensus as we compare forecast performance for the Fuzzy Model and the consensus forecast as well as inter-comparison of all methods involved.

Apart from the Fuzzy Model, the consensus forecast includes the GASM analogue method and the subjective forecast method (as described in "Introduction") but with some elaboration of the two steps involved. The first step is for a forecaster to identify if

and how well the current synoptic pattern matches with any of the known fog-conducive patterns. It then allows the forecaster to make further subjective assessment of fog risk likelihood, taking into consideration the result of the pattern matching. As a general rule, a forecaster is not allowed to reduce the fog risk if there is partial or full pattern matching. However, a forecaster can elevate the fog risk based on physical reasoning and other evidence, even if the pattern matching exercise fails to identify any pattern that matches with the current one. Although this method is subjective in nature, it is not designed to overrule other forecasting methods. Rather its forecast outcome is given the same weight as the objective methods such as the Fuzzy Model.

The use of a consensus in formulating the operational fog forecasts was prompted by the many examples in the literature that demonstrate its usefulness (e.g., THOMPSON, 1977) and the success of

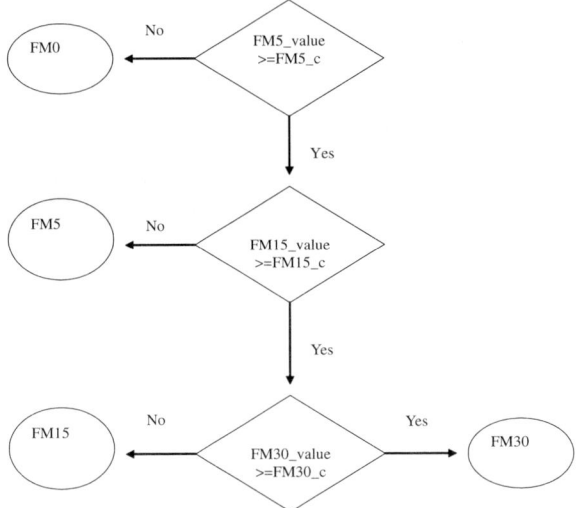

Figure 3

Decision tree to determine outcome of the 00Z Fuzzy Model. Here FM5_value, FM15_value and FM30_value are the composite fuzzy functions for a fog risk category of FM5, F15 and F30 respectively. FM5_c, FM15_c and FM30_c are the activation thresholds for the respective fog risk categories of FM5, FM15 and FM30. They are all valued at 0.725 in this study

consensus forecasts used operationally in BoM. WOODCOCK and ENGEL (2005) developed an objective consensus forecasting system (OCF) that employed routinely available bias-corrected direct model outputs (DMO) and model output statistics (MOS) and combined them using a simple weighted average algorithm. They showed that the OCF system not only outperformed the individual DMO and MOS forecasts of the screen-level temperature maxim and minima, it outperformed the official forecasts at over 70% of sites. A similar approach has now been successfully employed in the hourly OCF forecasts (ENGEL and EBERT, 2007).

The generation of the consensus forecast for Perth Airport involves the following steps. The first is to assign an outcome from each fog forecasting method to a value of 0, 1 or 2, corresponding respectively to one of the three possible forecast decisions of NoFog, CG or Prob30. Three numerical outcomes are then added together to generate a simple sum. The next procedure is effectively a simple averaging of the sum. Lastly, the average is checked against the outcome of the subjective method. It is adjusted up to be the same as the outcome of the subjective method if it was lower initially. On the other hand, the result is

unchanged if it equals to or is higher than the outcome of the subjective method.

5. Performance Evaluation and Discussion

The performance of the Fuzzy Model, the consensus forecast and the relative skill of the Fuzzy Model against other methods applied to the consensus forecast are assessed based on their operational use at WARFC in the 5 year period between 2004 and 2008. Timing is very important in fog forecasting but the accurate prediction of fog onset and clearance is extremely difficult due to the lack of effective guidance apart from fog climatology and forecaster experience. As a result, the assessment presented in this study is restricted to fog occurrence only. A positive forecast of fog indicates that fog is expected to occur sometime between 12 and 24 UTC (or 8 pm and 8am WST) and for purposes of verification, a positive observation of fog indicates that fog was observed at least once during the period.

Apart from not considering the exact timing of fog, the scope of the evaluation is further restricted to the 06 UTC issue of forecasts and the cool season. The 06 UTC forecasts of CG or TAF that rely on the 00 UTC Fuzzy Model output are very important for the planning of the long-haul flights to Perth Airport. The restriction to the cool season (April to October) is because fog is rare in the warm season and the Fuzzy Model was developed for the cool season only.

The performance of the Fuzzy Model is first assessed through a frequency distribution histogram, with focus on how well the model can differentiate fog risk on a given day. Next, this ability of the Fuzzy Model to differentiate fog risk is compared against that of the consensus forecast. The relative performance of the Fuzzy Model against other methods applied to the consensus forecast is then discussed using contingency tables and the associated skill scores.

5.1. Performance of the Fuzzy Model

One way to verify multi-category forecasts such as the fog risk forecasts of FM0, FM5, FM15 and FM30 produced by the Fuzzy Model is through the frequency distribution approach (MURPHY and WIN-KLER, 1987). If *Ni* represents the number of forecasts in the forecast category *i*, and *Oi* is the number of events observed out of these intended *Ni* forecasts, then the frequency of events for this forecast category *i* should be *Oi/Ni* × 100%. The frequency distribution is simply a histogram that plots relative frequency for each forecast category. If each forecast category is assigned a numerical probability value, then the distribution can be used to assess the validity of the assignment as well as provide evidence for recalibration. An ideal distribution is the one that the frequency for each category is the same as the assigned probability value.

Figure 4a shows a desirable trend of frequency increasing with fog risk with the frequency of around 2% at the lower-end fog risk categories of FM0 and FM5, increasing to about 10% at FM15 and 15% at FM30. This result vindicated the ability of the Fuzzy Model to provide reliable fog risk forecasts in the operational environment. However, these frequencies did not match up perfectly with the assigned numerical probability values of 0, 5, 15 and 30% for the respective fog risk categories of FM0, FM5, FM15 and FM30. Before discussing the possible action to take to improve the Fuzzy Model, one practical factor in the operational environment should be considered. This is about how to define a positive forecast.

The above frequency distribution shows a positive forecast if fog was observed overnight. In practice, however, forecasters have little skill to differentiate forecasts for fog and mist. Hence, when a fog forecast is issued, forecasters would consider it a successful forecast if mist instead of fog actually occurred. In this sense then, it may be more appropriate to use the frequency distribution of fog/mist to re-examine whether the assignment of numerical probability values is appropriate before deciding if recalibration or other actions are required.

The distribution considering a positive forecast as when either fog or mist was observed is shown in Fig. 4b. Not surprisingly, all frequencies have increased when compared with those in Fig. 4a. The standout increase occurred at FM15 from 10 to 15% and at FM30 from 15 to over 20%. By considering both fog and mist as a hit, the observed frequencies at FM5 and FM15 now align well with

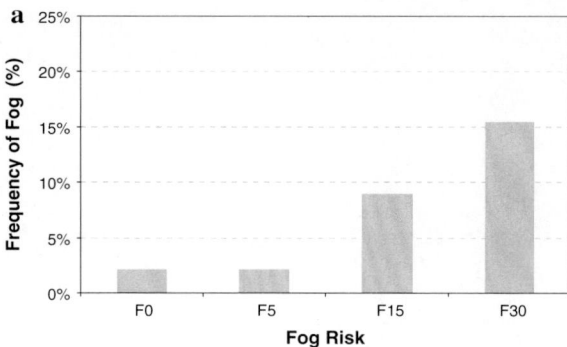

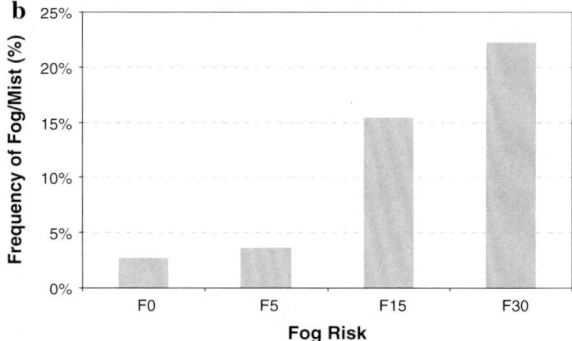

Figure 4
a Frequency distribution of fog for each fog risk forecast category produced by the Fuzzy Model for the cool seasons of the 5 year period between 2004 and 2008. **b** Frequency distribution of fog or mist for each fog risk forecast category produced by the Fuzzy Model for the cool seasons of the 5 year period between 2004 and 2008

the assigned probability. The observed frequency at FM30 is lower than the forecast probability of 30%, and thus it would be appropriate to adjust this down to around 25%. The observed frequency at FM0 of around 2–3% may seem close to the forecast probability of 0% but the large number of forecasts in this category of NoFog meant that a number of unforecast fog or mist events would fall into this category. This weakness of the Fuzzy Model can be improved through the careful tuning of the fuzzy functions so that the cross-validated performance results have none or very few fog or mist events fall into this FM0 category.

5.2. *Fuzzy Model Versus Consensus Forecast*

Having demonstrated the usefulness of the Fuzzy Model in differentiating fog risk, it would be interesting to know how the model compares with the consensus forecast in this aspect. For a direct comparison, the Fuzzy Model outcomes were first assigned to one of the three forecast decisions of NoFog, CG and Prob30. The Fuzzy Model results including the number of forecasts, the observed fog events and the frequency of fog for each fog decision are listed in Table 1 while the same results but for the consensus forecasts are provided in Table 2. Note that there is some difference in the number of records between the Fuzzy Model and the consensus forecast due to missing Fuzzy Model records on some days.

The two frequency distributions in Fig. 5 both show increased fog frequency with increasing fog forecast probability. The main difference occurred at the NoFog and Code Grey forecast decisions. At NoFog, the frequency is about 2% for the Fuzzy Model but only 0.4% for the consensus forecast. This small percentage difference at NoFog would amount to 11 fewer unforecast fog events if using the consensus forecast instead of the Fuzzy Model outcome. With regard to the difference at CG, although not small, its practical impact in the operational environment is not considered to be significant.

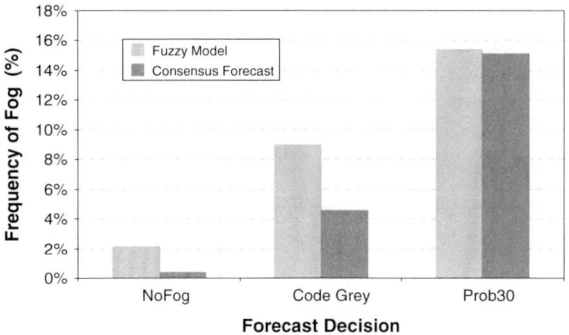

Figure 5

Comparison between the Fuzzy Model (*light shaded*) and the operational consensus forecast (*dark shaded*) of the frequency distribution of fog for each forecast decision category for the cool seasons of the 5 year period between 2004 and 2008

Table 1

Dataset for the cool seasons of the 5 year period between 2004 and 2008 of fog forecasts and observations for each equivalent forecast decision that would be made by the Fuzzy Model if the fog risk categories FM0 and FM5 were assigned to NoFog, FM15 to Code Grey and FM30 to Prob30 forecast decision

Forecast decision	Number forecasts	Number fog observed	Frequency of fog (%)
NoFog	613	13	2.1
Code Grey	156	14	9.0
Prob30	162	25	15.4

Table 2

Dataset for the cool seasons of the 5 year period between 2004 and 2008 of fog forecasts and observations for each forecast decision made by the operational consensus forecast

Forecast decision	Number forecasts	Number fog observed	Frequency of fog (%)
NoFog	537	2	0.4
Code Grey	176	8	4.5
Prob30	298	45	15.1

5.3. Multi-Method Comparison with Skills Scores and the ROC Curve

The better performance of the consensus forecast over the Fuzzy Model is perhaps not surprising, given that the consensus forecast is built on the outcomes of the Fuzzy Model as well as the other two methods, namely GASM and the subjective method. It would be interesting to compare the relative skills of the Fuzzy Model against the other two methods. To facilitate the inter-comparison, it is necessary to convert the multi-category forecasts to a series of Yes/No dichotomous fog forecasts so that the conventional and effective performance metrics can be applied. To verify dichotomous forecasts, a contingency table such as illustrated in Table 3 is needed to list all the essential elements such as hits, misses, false alarms and non-events (or correct-negatives) that will be used to produce various skill scores. From the contingency table, the following performance measures can be produced.

$$POD = \frac{A}{A + C} \qquad (2)$$

$$FAR = \frac{B}{A + B} \qquad (3)$$

$$POFD = \frac{B}{B + D} \qquad (4)$$

$$HKSS = POD - POFD \qquad (5)$$

where

POD is the probability of detection or hit rate, ranging from 0 to 1 with 1 being the perfect score.

Table 3

Contingency table for assessing Yes/No dichotomous fog forecasts

Fog forecast	Fog observed		
	YES	NO	
YES	A (hits)	B (false alarms)	A + B (forecast fog)
NO	C (misses)	D (non-events)	C + D (forecast no-fog)
	A + C (observed fog)	B + D (observed no-fog)	A + B+C + D (Total)

FAR is the false alarm ratio, ranging from 0 to 1 with 0 being the perfect score.

POFD is the probability of false detection. It ranges from 0 to 1 with 0 being the perfect score. It is often used in conjunction with POD to plot the relative operating characteristic (ROC) curve which is used widely for probabilistic forecasts. Its other use is in the calculation of HKSS.

HKSS is the Hanssen and Kuipers discriminant or Hanssen and Kuipers skill score ranging from −1 to 1 with 0 indicating no skill and 1 being a perfect score. It uses all elements in the contingency table and answers the question about how well the forecast separates the "yes" events from the "no" events. WOODCOCK (1976) found this score to be most useful for comparing different forecasting techniques.

Two types of Yes/No dichotomous fog forecasts are generated for all methods and the consensus forecast. The first type treats all forecast decisions of CG and Prob30 as Yes forecast, and NoFog forecast decision as No forecast. The second type treats the Prob30 forecast decision as the only Yes forecast, leaving CG and NoFog forecast decisions as No forecast.

Table 4 lists all elements and the performance measures corresponding to each type of Yes/No dichotomous fog forecasts for the three methods and the consensus forecast. Regardless of which type to use for the skill assessment, the rank of skills from low to high is consistently GASM, the Fuzzy Model, the subjective method and the consensus forecast. For example, for the type that treated CG and Prob30 as Yes forecast, the respective POD values were 0.66, 0.75, 0.89 and 0.96. With FAR values showing little difference among the three methods and the consensus forecasts, the overall HKSS score that takes both hits and false alarms into account confirmed the ranking with the respective values of 0.33, 0.43, 0.52 and 0.55.

Another popular measure of the skill of probabilistic forecasts is using relative operating characteristic (ROC) curve which plots POD on *y*-axis and POFD on *x*-axis (MASON, 1982). It is often used to measure how well the probabilistic forecasts discriminate between events and non-events (the resolution). Because it is independent of bias it can be viewed as a kind of *potential* skill. The area under the ROC curve is a scalar measure (symbolized as AUC)

Table 4

Verification statistics of all methods contributing to the operational consensus forecast as well as the consensus forecast itself after converting multi-category forecasts into Yes/No dichotomous fog forecasts

Method	YES forecast	NO forecast	A	B	C	D	POD	FAR	POFD	HKSS
GASM	CG + Prob30	NoFog	33	294	17	590	0.66	0.90	0.33	0.33
GASM	Prob30	CG + NoFog	17	91	33	793	0.34	0.84	0.10	0.24
Fuzzy Model	CG + Prob30	NoFog	39	279	13	600	0.75	0.88	0.32	0.43
Fuzzy Model	Prob30	CG + NoFog	25	137	27	742	0.48	0.85	0.16	0.32
Subjective	CG + Prob30	NoFog	48	330	6	554	0.89	0.87	0.37	0.52
Subjective	Prob30	CG + NoFog	31	132	23	752	0.57	0.81	0.15	0.42
Consensus	CG + Prob30	NoFog	53	421	2	535	0.96	0.89	0.44	0.52
Consensus	Prob30	CG + NoFog	45	253	10	703	0.82	0.85	0.26	0.55

A Hits, *B* false alarms, *C* misses and *D* correct negatives

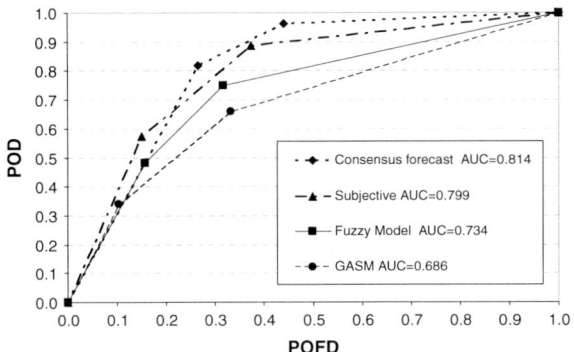

Figure 6

ROC curves of fog forecast decisions produced by GASM (*dashed line and dot symbols*), the Fuzzy Model (*solid line and square symbols*), the subjective method (*dot dashed line and triangle symbols*), and the consensus forecast (*dashed line and diamond symbols*) for the cool seasons of the 5 year period between 2004 and 2008. The area under each curve (AUC) is provided in the respective legend label

that is frequently used to summarize the resolution. The perfect value is 1.0 and the no-skill value is 0.5. Figure 6 confirms that the consensus forecast has the most skill with AUC of 0.814. It is followed closely by the subjective method (AUC of 0.799), the Fuzzy Model (AUC of 0.734) and GASM (AUC = 0.686).

One of the satisfying findings from the above performance inter-comparison is that out of the two independent forecasting methods, the Fuzzy Model has better skills than GASM in practically all performance measures. The subjective method was treated as if it was another independent method contributing to the consensus forecast, but in practice, forecasters may have already known the outcomes of the other two methods before making their subjective assessment. Therefore it is not totally surprising to see that it had better skill than the other two methods. The other pleasing part of the findings is the effectiveness of simple averaging to generate the consensus forecast. Through simple averaging, the consensus forecast outperformed all individual methods that contributed to it. This confirms the findings of THOMPSON (1977).

6. Summary

This study developed an objective and automated fog model for Perth Airport. Fuzzy logic is employed

to factor in model inaccuracy and uncertainties associated with relationships between fog predictors and fog occurrence. The 12.5 km resolution LAPS125 regional operational model provided the NWP inputs to the fuzzy logic fog model. The identified predictors include wind speed at 200 m above ground and an array of corrected dew-point depressions at 200 or 600 m above the ground. Fuzzy membership functions of each fog predictor in fuzzy sets of fog risk categories of FM5, FM15 and FM30 were derived. The final outcome is produced through the defuzzification process involving generating composite fuzzy functions for fog risks of FM5, FM15 and FM30 based on a set of fuzzy rules such as interference and averaging, deducting the composite fuzzy function for FM0 from that of FM5, conducting cross-checking of the respective composite fuzzy function with a threshold, and applying a decision tree to arrive at the final outcome.

A simple consensus approach was used to generate an operational forecast by synthesizing outcomes of the three available fog forecasting methods at WARFC, namely GASM, the Fuzzy Model, and the subjective method. To generate the operational consensus forecast, the outcomes of the three methods were simply averaged although more weight was given to the subjective method if it was found that the averaged result was lower than the outcome of the subjective method.

To demonstrate the utility of the Fuzzy Model, the performance of the 00 UTC Fuzzy Model in the cool season over a 5 year period between 2004 and 2008 was evaluated using a frequency histogram. This showed increasing fog frequency with forecast fog probability. In validating the forecast, if an observed fog is taken to include fog or mist, then the assigned fog probabilities of 5 and 15%, for FM5 and FM15, respectively, are well calibrated. For a forecast outcome of FM30 the forecast fog probability should be reduced from 30 to 25%. The frequency of around 2% at the lowest fog risk category FM0 is an undesirable finding. This FM0 category is associated with the NoFog forecast decision in the operational environment thus any observed fog events in this category meant that they would be unforecast. Due to the large number of forecasts normally associated with this category, this small percentage translates to quite a few missed fog events. Rather than calibrating this category, improvement should be made by fine-tuning the fuzzy functions in the Fuzzy Model to minimize the missed events.

The skill of the Fuzzy Model to differentiate fog risk is compared with that of the consensus forecast using the frequency distribution histogram of fog decision. It was found that the consensus forecast reduced the frequency of observed fogs in the NoFog forecast decision category to almost zero, a significant improvement over the Fuzzy Model.

The relative skill among the three forecasting methods of GASM, the Fuzzy Model and the subjective method that contribute to the consensus forecast are then assessed with conventional performance measures such as POD, FAR and HKSS after converting multi-category forecasts into two types of dichotomous Yes/No fog forecasts, one considering CG and Prob30 as Yes forecasts, the other considering Prob30 only as Yes forecast. These different performance measures consistently pointed to the better performance for the consensus forecast, followed closely by the subjective method, the Fuzzy Model and GASM. The same conclusion can be derived from the ROC curves. It is satisfying to note that the Fuzzy Model outperformed GASM. Although the subjective method performed better than the Fuzzy Model, it had the advantage of accessing results of the two independent methods. The superiority of the consensus forecast vindicated the choice of simple averaging and the robustness of the consensus approach.

Using fuzzy logic to post-process NWP model output can be a cost-effective way to develop an objective fog forecasting tool for any NWP model and location. Indeed, a similar fuzzy logic based fog model has been developed for Sydney Airport although fog predictors and fuzzy functions were quite different from those used for Perth Airport. The key to develop a successful fuzzy logic based NWP fog forecasting model lies with the detailed study of the local observations and model characteristics at that site. Future work will include the refinement of the fuzzy-functions to minimize the frequency of observed fog for the FM0 fog risk category, and the development of an ensemble of fuzzy logic fog models based on different NWP models and OCF forecasts.

Improvement of forecasting performance can be achieved with the consensus approach of simple averaging without the improvement of skills of individual forecasting tools. This study provided another example to support the statement. As long as there are at least two useful and independent forecasting methods available, this approach can be applied to improve the forecasting performance of fog or other weather phenomena such as thunderstorm. One advantage of the approach is that it can be adjusted easily if the availability of forecasting tools changes over time. As of December 2009, the GASM forecasting tool became unavailable to WARFC and the operational forecast was changed to exclude this input. A preliminary evaluation of the operational forecast performance since the change shows that there has been no adverse impact on the performance.

Acknowledgments

The authors wish to thank Frank Woodcock, Grahame Reader and Philip Riley for their constructive comments and suggestions during the internal review process at BoM. The authors also thank all WARFC forecasters for the wisdom and contribution to the operational consensus forecast. Thanks also go to an anonymous reviewer who made valuable suggestions and helped to improve the quality of the paper.

REFERENCES

BARDOSSY, A. and DUCKSTEIN, L. (1995), *Fuzzy rule-based classification of atmospheric circulation patterns*, International Journal of Climatology *15*, 1087–1097.

BERGOT, T. and GUEDALIA, D. (1994), *Numerical forecasting of radiation fog Part I: Numerical model and sensitivity tests*, Mon. Wea. Rev. *122*, 1218–1230.

BERGOT, T., TERRADELLAS, E., CUXART, J., MIRA, A., LIECHTI, O., MUELLER, M., and NIELSEN, N. W. (2007), *Intercomparison of single-column numerical models for the prediction of radiation fog*, J. Appl. Meteorol. Climatol. *46*, 504–521.

BOTT, A., SIEVERS, U. and ZDUNKOWSKI, W. (1990), *A radiation fog model with a detailed treatment of the interaction between radiative transfer and fog microphysics*, J. Atmos. Sci. *47*, 2153–2166.

CLARK, P. A. and HOPWOOD, W. P. (2001), *One-dimensional site-specific forecasting of radiation fog Part I, Model formulation and idealized sensitivity studies*, Meteorological Applications *8*, 279–286.

CLARK, P.A., HARCOURT, S. A., MACPHERSON, B., MATHISON, C. T., CUSACK, S. and NAYLOR, M. (2008), *Prediction of visibility and aerosol within the operational Met Office Unified Model Part I: Model formulation and variational assimilation*, Q. J. R. Meteorol. Soc. *134*, 1801–1816.

DAHNI, R. R. and STERN, H. (1995), *The development of a generalised UNIX version of the Victoria Office's operational analogue statistics model*, BMRC Research Report No. 47, pp 34.

ENGEL, C. and EBERT, E. (2007), *Performance of hourly operational consensus forecasts (OCFs) in the Australian region*, Weather and Forecasting *22*, 1345–1359.

GOLDING, B. W. (1993), *A study of the influence of terrain on fog development*, Mon. Wea. Rev. *121*, 2519–2541.

GULTEPE, I., TARDIF, R., MICHAELIDES, S. C., CERMAK, J., BOTT, A., BENDIX, J., MULLER, M. D., PAGOWSKI, M., HANSEN, B., ELLROD, G., JACOBS, W., TOTH, G. and COBER, S. G. (2007), *Fog research: a review of past achievements and future perspectives*, J. Pure Appl. Geophys. *64*, 1121–1159.

HANSEN, B. (2007), *A fuzzy logic-based analog forecasting system for ceiling and visibility*, Weather and Forecasting *22*, 1319–1330.

MASON, I. (1982), *A model for assessment of weather forecasts*, Aust. Met. Mag. *30*, 291–303.

MURTHA, J. (1995), *Application of fuzzy logic in operational meteorology*, Scientific Services and Professional Development Newsletter, Canadian Forces Weather Service, 42–54. http://chebucto.ca/Science/AIMET/archive/murtha.pdf.

MURPHY, A.H. and WINKLER, R.L. (1987), *A general framework for forecast verification*, Mon. Wea. Rev. *115*, 1330–1338.

PURI, K., DIETACHMAYER, G., MILLS, G.A., DAVIDSON, N.E., BOWEN, R.A., and LOGAN, L.W. (1998), *The new BMRC Limited Area Prediction System, LAPS*, Australian Meteorological Magazine Vol *47*, No 3, 203–223.

REGANO, L. (1997), A fully automated forecasting aid for fog in Australia, Meteorological note 212, Bureau of Meteorology, Australia.

THOMPSON, P. D. (1977), *How to improve accuracy by combining independent forecasts*, Mon. Wea. Rev. *105*, 228–229.

WMO (1995), WMO No. 306, Manual on Codes, International Codes, Volume I.1, Part A, Alphanumeric Codes.

WOODCOCK, F. (1976), *The evaluation of yes/no forecasts for scientific and administrative purposes*, Mon. Wea. Rev., *104*, 1209–1214.

WOODCOCK, F. and ENGEL, C. (2005), *Operational consensus forecasts*, Weather and Forecasting, *20*, 101–111.

ZADEH, L. A. (1965), *Fuzzy sets*, Information and Control *8*, 338–353.

ZADEH, L. A. (1999), *From computing with numbers to computing with words—from manipulation of measurements to manipulation of perceptions*, IEEE Transactions On Circuits and Systems—I: Fundamental Theory and Applications, *45*, 105–119.

ZHOU, B., DIMEGO, G., and GULTEPE, I. (2010), *Forecast of low visibility and fog from NCEP—current status and efforts*, proceeding of the 5th International Conference on Fog, Fog Collection and Dew, 25–30 July 2010, pp 105–108, available at the web link of http://meetingorganizer.copernicus.org/FOGDEW2010/FOGDEW2010-57-8.pdf.

(Received November 15, 2010, revised March 21, 2011, accepted April 25, 2011, Published online June 12, 2011)

Pure Appl. Geophys. 169 (2012), 1121–1135
© 2011 Springer Basel AG
DOI 10.1007/s00024-011-0325-z

Marine Boundary Layer Structure for the Sea Fog Formation off the West Coast of the Korean Peninsula

CHANG KI KIM[1] and SEONG SOO YUM[1]

Abstract—Marine boundary layer (MBL) structure for the formation of sea fogs off the west coast of the Korean Peninsula are examined for the investigation period from January 2002 to August 2006, using the meteorological data measured at a buoy and the vertical sounding data measured at an island in this region. There is the total of 3,294 vertical soundings during the investigation period. Based on these vertical soundings, the MBL structure is classified as convective boundary layer (CBL; when inversion exists aloft but at altitudes lower than 3 km, 1,618 soundings), stable boundary layer (SBL; when inversion base is at the surface, 655 soundings) or near-neutral boundary layer (NNBL; when there is no inversion or inversion base is higher than 3 km altitude, 1,021 soundings). Under the CBL condition, the most frequently formed lower level cloud is stratocumulus but fogs do form in spring and summer months mostly as warm sea fogs [TSST (=T–SST) < 0]. Under the SBL condition, stratus and cold sea fogs (TSST > 0) are the most frequently found lower level clouds. The effects of turbulence, advection and radiation on sea fog formation vary with turbulence strength, represented by bulk Richardson number, R_b. For cold sea fog cases, in the highly turbulent regime ($R_b < 0.03$), strong turbulent cooling and drying are canceled out by equally strong or even stronger warm and moist advection, and thus the additional radiative cooling turns out to be critical in the successful formation of fog. In the weak turbulent and non-turbulent ($R_b > 0.30$) regimes, the effects of turbulence decrease dramatically and so do the advection effects but radiative cooling is still strong, again making it the crucial reason for the successful formation of cold sea fogs. On the other hand, the turbulent moisture supply from the warmer sea surface is the crucial factor for the formation of warm sea fogs while turbulent warming and radiative cooling largely cancel each other out and the advection effects are negligibly small.

Key words: Sea fogs, west coast of Korea, marine boundary layer, turbulence, advection, radiation.

1. Introduction

Sea fog forms due to air-sea interactions in the marine boundary layer that lead to water vapor saturation. For example, over the eastern subtropical Pacific off the California coast, heat and moisture are supplied from the ocean surface and are confined in the marine boundary layer due to the strong subsidence inversion formed in the downward branch of a subtropical high, often leading to the formation of low level clouds or sea fogs (ALBRECHT *et al.* 1985, 1988; STULL 1988; KLEIN and HARTMANN 1993; KAGAN 1995; KLEIN *et al.* 1995; NORRIS 1998a, b). Over the western Pacific, sea fog and stratus occurrences increase in summer, when the monsoonal circulation that brings warm air over the relatively cooler ocean surface facilitates the formation of a thermally stable layer (KLEIN AND HARTMANN 1993; NORRIS and LEOVY 1994). Several studies suggested that turbulent mixing is vital in cooling and moistening of low level air to form sea fogs. For instance, PILIE *et al.* (1979) stressed the importance of turbulent mixing in the formation of the southern California sea fogs. LEIPPER (1995) also suggested that heat and moisture exchange occurring via turbulence over the cold sea surface might be a crucial factor for the formation of sea fogs over the San Francisco Bay area. KORACIN *et al.* (2005a) showed, through back-trajectory analysis, that an air mass that is nearly saturated by turbulent cooling over a cold sea surface could move over a warm sea surface and achieve complete saturation due to turbulent fluxes of heat and moisture. FINDLATER *et al.* (1989) also showed, through observation and numerical modeling, that *Haar*, the sea fog in Scotland, is formed by turbulent cooling of warm air over a relatively colder sea surface.

[1] Department of Atmospheric Sciences, Yonsei University, 262 Seongsanno, Seodaemun-gu, Seoul 120-749, Korea. E-mail: ssyum@yonsei.ac.kr

In addition to turbulence effect, several numerical studies suggested that radiative cooling of air could stabilize the boundary layer and ultimately dominate the thermal structure under light wind conditions over land (e.g., GARRATT and BROST 1981; ANDRE and MAHRT, 1982; GOPALAKRISHNAN et al. 1998; KRISHNA et al. KRISHNA 2003; SAVIJARVI 2006). Observations by ESTOURNEL et al. (1986) showed that radiative cooling under weak wind conditions might contribute to cooling of air within a stable boundary layer. Furthermore, DUYNKERKE (1999) concluded, based on numerical simulation, that radiative cooling had a positive effect on the formation of radiation fog within a stable boundary layer at Cabauw in the Netherlands. COANITC and SEGUIN (1971) also emphasized the importance of radiative cooling in the surface layer over water surface with numerical simulation. Furthermore radiative cooling was found to play an important role in forming sea fogs within a thermally stable marine boundary layer in a similar manner to radiation fog over the land (KORACIN et al. 2005b).

Over the western Yellow Sea off the Chinese coast, the occurrence of sea fog increases abruptly in April when a stable boundary layer is formed by strong warm advection from the Chinese continent (ZHANG et al. 2009). Recently, KIM and YUM (2010) examined statistical characteristics of the sea fogs that form over the other side of the Yellow Sea off the Korean west coast. They classified sea fogs as cold and warm sea fogs, based on the differences between air temperature (T) and sea surface temperature (SST) (TSST = T–SST) at the onset time of sea fog formation (i.e., cold sea fog if TSST > 0°C and warm sea fog if TSST < 0°C) and associated the occurrence statistics with relevant meteorological variables. However, an elaborate and comprehensive effort to understand the formation mechanism of these fogs was not attempted.

This study is an extension of the work by KIM and YUM (2010). Here we examine the physical processes that are relevant to sea fog formation over the eastern Yellow Sea off the Korean coast. In particular, we focus on the influence of turbulent mixing of heat and moisture, temperature and moisture advections, and radiative cooling. Section 2 explains the data used in this study. Marine boundary layer characteristics are given in Sect. 3. The influences of turbulence, advection and radiation on cold and warm sea fogs are discussed in Sect. 4. Finally, summary and conclusions are given in Sect. 5.

2. Data

The surface meteorological variables analyzed in this study are wind direction (WD), wind speed (WS), air temperature (T), dew point temperature (T_d), and SST measured every hour at the buoy near Dukjeok Island (Fig. 1) during the investigation period from January 2002 to August 2006. Vertical soundings measured at Baeknyeong Island at 09 and 21 LST are used to examine the structure of the marine boundary layer. The vertical resolution varied for each sounding but the average for all soundings was 156 m below 800 hPa altitude. As illustrated in Fig. 1, however, the location of the rawinsonde site is about 200 km away from the buoy location. In order to confirm that the air mass is similar between the buoy and rawinsonde locations, we examined the back-trajectories at both locations on cold sea fog days, produced by HYSPLIT model (DRAXLER and ROLPH 2011); they were indeed similar, in that most of the air masses originated from the open sea and the T variations along the back-trajectories were very close to each other (Fig. 2). The cloud classification is made hourly at the weather station that conducts the rawinsonde measurement but it does not necessarily indicate that the cloud classes at buoy are the same as these. Synoptic variables used to characterize the marine boundary layer are the temperature advection at 1,000 hPa ($AT_{1,000}$, °C h^{-1}), 975 hPa (AT_{975}, °C h^{-1}), 950 hPa (AT_{950}, °C h^{-1}), and 925 hPa (AT_{925}, °C h^{-1}); the surface divergence (D_S, s^{-1}); and the omega velocity at 700 hPa (ω_{700}, mb day^{-1}). All of these values are derived from the 6-h Global Data Assimilation System (GDAS) output with 1° horizontal resolution produced by the National Center for Environmental Prediction (NCEP). During the investigation period, there were 27 instances of cold sea fog and eight of warm sea fog, all observed in the 7 months from February to August and none in the other months of the year (KIM and YUM 2010).

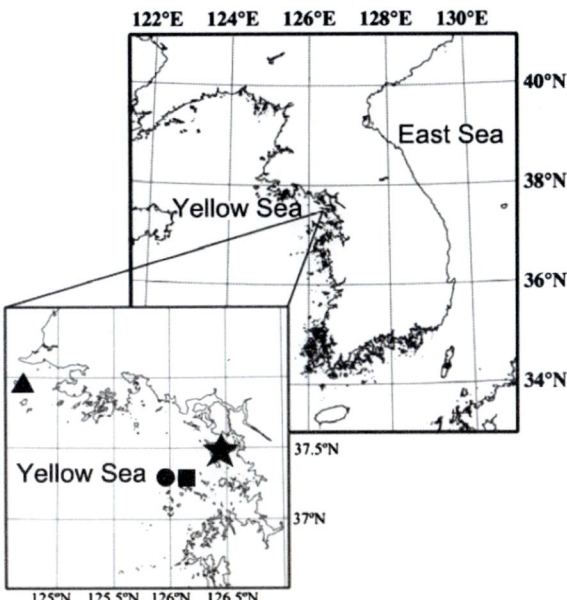

Figure 1
The locations of Incheon International Airport (IIA) and the other measurement sites. The *asterisk*, the *closed circle*, the *triangle*, and the *rectangle* indicate IIA, the buoy location near Dukjeok Island, the rawinsonde site at Baeknyeong Island, and the lighthouse at Seonmi Island, respectively

3. Marine Boundary Layer

3.1. Definition of Marine Boundary Layer

In general, marine boundary layers can be classified as convective, stable, or near-neutral boundary layers (GARRATT 1994). A convective boundary layer (CBL) is defined as the layer below a capping inversion and has a nearly dry adiabatic lapse rate, whereas a stable boundary layer (SBL) is the shallow layer between the cold surface and the relatively warmer air. In this study, all vertical soundings are classified as either inversion cases, when there exists a layer that shows the vertical potential temperature gradient greater than 0.5 K $(100 \text{ m})^{-1}$ (JOHNSON and O'BRIEN 1973), or no-inversion cases, when inversion does not exist or exists but at an altitude higher than 3,000 m. All inversion cases are sub-classified as SBL, when the inversion base is at the surface, or as CBL otherwise. No-inversion cases are referred to as near-neutral boundary layers (NNBL), based on the assumption that buoyancy is nearly zero in no-inversion cases. The top height of a CBL is equivalent to the inversion base height: On the other hand,

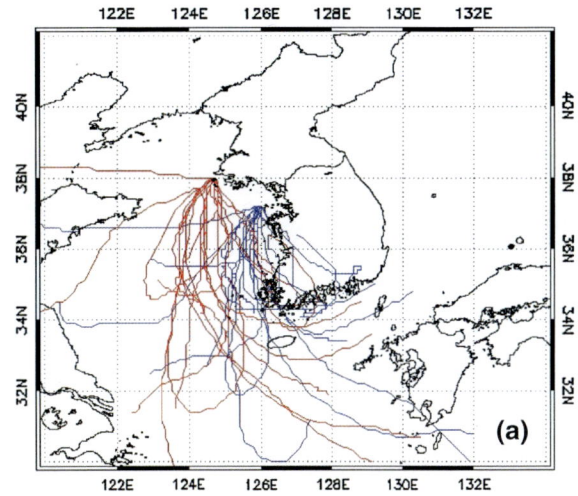

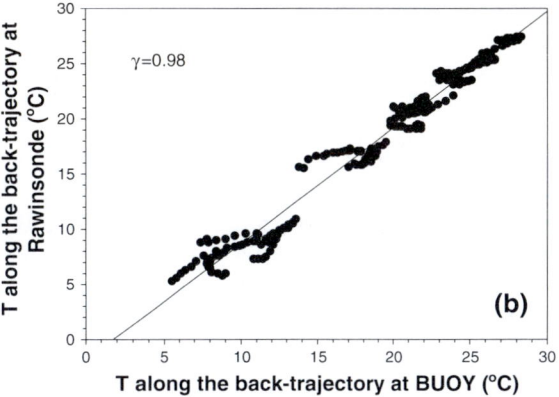

Figure 2
The 72 h air mass back-trajectories produced by the HYSPLIT model (DRAXLER and ROLPH 2011) at the buoy and the rawinsonde site for the 27 cold sea fog cases (**a**), and scatter plot of Ts along the back-trajectories at the two sites (**b**). All 27 cold sea fog cases are shown together in (**b**)

the depth of the SBL is generally defined as the height from the surface to the top of the surface inversion. Note that in our analysis the term 'surface inversion' does not necessarily mean that the inversion base is at the sea surface, since the rawinsonde site is located 158 m above mean sea level.

Cloud classes can be used to characterize the synoptic characteristics of the marine boundary layer. Specifically, clouds observed at 09 and 21 LST (rawinsonde measurement times) are classified into six classes: no-cloud, overcast, upper level clouds, stratocumulus, stratus, and fog. Upper level clouds include cirrus, cirrostratus, cirrocumulus, altostratus, altocumulus, cumulus, and cumulonimbus. The designation 'overcast' is used when the cloud amount is 10 and the observer cannot clearly identify the cloud morphologically. Note that 'overcast' is distinguished from 'stratus' and 'stratocumulus'; these two classes are designated only when cloud classes can be clearly identified as such.

3.2. Statistical Distribution

A total of 3,294 vertical soundings are observed during the investigation period. According to the above classification criteria, the numbers of CBL, SBL, and NNBL are 1,618, 655, and 1,021, respectively. Figure 3a shows the monthly average relative frequencies of each of the three boundary layer types for each month, the monthly mean TSST, and the depth of CBL and SBL. As expected, the occurrence of SBL increases with increasing TSST (in the summer season), and CBL appears most frequently when TSST is negatively large. The relative frequency of NNBL increases when the magnitude of TSST is small, which is consistent with SMEDMAN et al.'s (1994) observation which showed that a near-neutral marine atmospheric boundary layer is formed over the Baltic Sea usually when buoyancy inducing sensible heat flux is small. The depth of SBL is much shallower than that of CBL, and its monthly variation is also relatively small. On the other hand, the depth of CBL varies with TSST, such that the CBL depth is thicker for negatively large TSST and vice versa (Fig. 3b). However, the inversion strength is almost uniform at 1.5 K $(100$ m$)^{-1}$ regardless of the sign of TSST.

Table 1 summarizes the boundary layer classification for each cloud class. About 81% of the vertical soundings observed during sea fogs (103 out of 127) can be classified as CBL or SBL. Cold sea fogs form to about 50% under SBL and to about 50% under NNBL/CBL conditions. On the other hand, warm sea fogs form predominantly under CBL. However, in any boundary layer type sea fog occurrence is least likely (see the relative frequency), which suggests that sea fogs form under special conditions that is not frequently observed. Another thing to note is that the most frequently found cloud class under the CBL condition is stratocumulus. The CBL topped by stratocumulus is often called stratocumulus-topped boundary layer

(STBL) (STULL 1988). STBL is of utmost importance because of its climatic implication (e.g., ACKERMAN et al. 1995; STEVENS et al. 2003): aerosol induced perturbation of cloud microphysical and radiative properties of these clouds significantly affect the earth radiation budget.

3.3. Convective Boundary Layer

In Fig. 4, the relative frequency of CBL in each month (RF_C) is correlated with the monthly averages of several meteorological variables, such as TSST, $AT_{1,000}$, AT_{975}, AT_{950}, AT_{925}, D_S, and ω_{700}. RF_C is large in cold months (see month numbers 1, 2, 10, 11 and 12) when TSST is negatively large (Fig. 4a) and cold advection is strong at near surface levels (i.e., 1,000 and 925 hPa) (Fig. 4b). Moreover, subsidence is strong at 700 hPa in these months (Fig. 4c). Due to these clear seasonal trends, RF_C shows strong correlations with TSST, $AT_{1,000}$, AT_{925} and ω_{700} but a weak positive correlation with D_S (Fig. 4c). These relationships suggest that cold advection in low altitudes and/or downward motion at 700 hPa may provide good conditions for the formation of CBL overcoming the effect of heat supply from the warmer sea surface. The California coast is well known for these conditions. NEIBURGER (1960) showed that subsidence at 700 hPa and thermal advection are responsible for the high frequency of subsidence inversions that form a CBL.

Monthly distribution of each cloud class only during the CBL condition is shown along with monthly mean TSST in Fig. 5. Relatively frequent occurrences of the 'no-cloud' class in spring and fall (Fig. 5a) are consistent with the climatological analysis by LEE (1995), who reported that the weather in this region was generally clear in spring and fall due to migratory anticyclones that often appear in these seasons. In summer, heavy rain comes frequently in Korea due to the monsoonal frontal system called *Jangma*. The peak value for the 'overcast' class in June and July appears to be the result of these heavy precipitating clouds, assuming that the 'overcast' observations mostly consist of stratiform clouds associated with the frontal system. 'Upper level cloud' is observed consistently throughout the year.

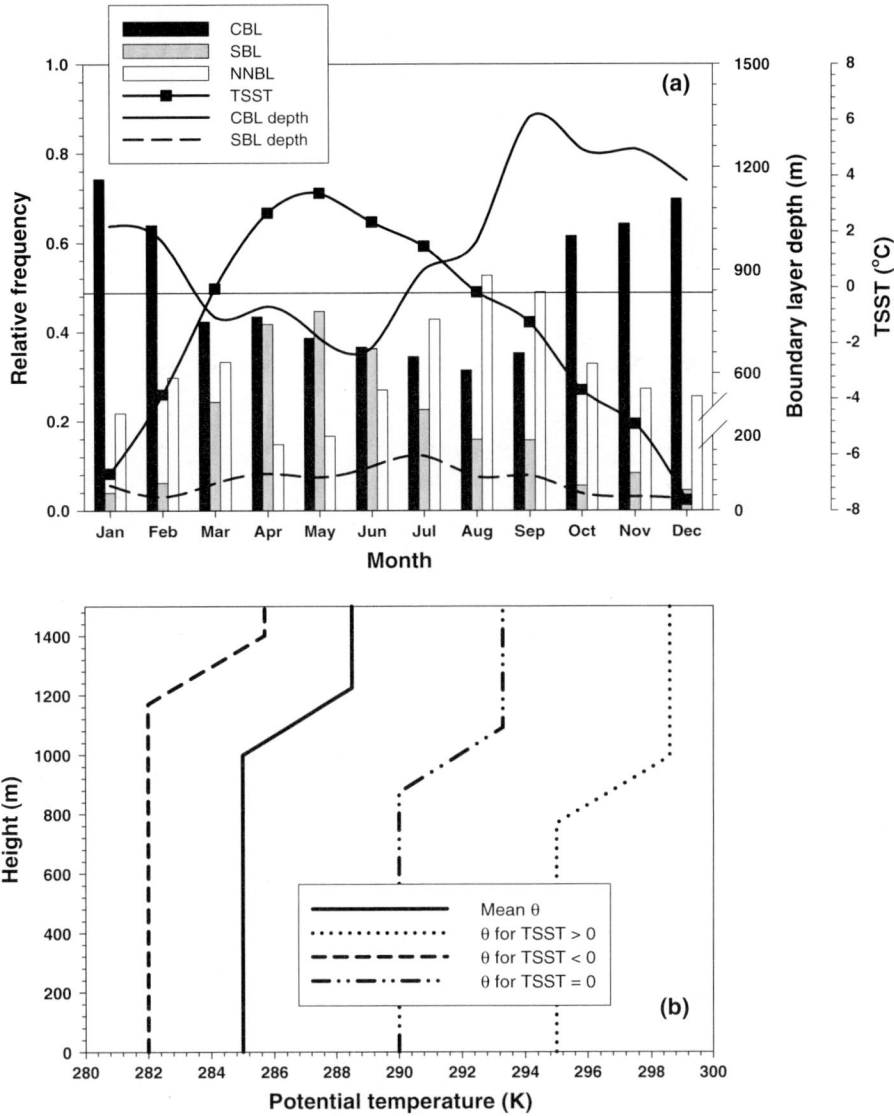

Figure 3

Monthly averages of relative frequencies of the three boundary layer types, TSST calculated with T at the buoy, and convective and stable boundary layer depths (**a**), and the mean vertical soundings of potential temperature only for the convective boundary layer (CBL) (**b**). In (**b**), the *solid line* indicates the mean potential temperature profile for all CBL cases. The *dotted*, *dashed* and *dash-dotted lines* indicate the mean profiles for positive, negative, and nearly zero TSST, respectively

On the other hand, 'lower level cloud' shows a clear seasonal trend that matches with RF_C (Fig. 3a). In detail, 'lower level cloud' is dominated by stratocumulus, especially in colder months and fogs are observed only in spring and summer when the frequency of 'lower level cloud' is relatively low (Fig. 3b). Therefore, it can be said that fogs are scarcely formed under CBL condition.

3.4. Stable Boundary Layer

Contrary to CBL, the relative frequency of SBL in each month (RF_S) increases with TSST (Fig. 6a). Moreover, strong warm advection near the sea surface increases the chance for SBL formation, as suggested by rather high $\gamma[RF_S-AT_{1,000}]$ and $\gamma[RF_S-AT_{975}]$ of 0.74 and 0.66, respectively (Fig. 6b). RF_S

Table 1

Summary of boundary layer type classification for each cloud class

	Convective boundary layer	Stable boundary layer	Near-neutral boundary layer	Total
No-cloud	264 (16.3/48.0) 876	123 (18.7/22.4) 78	163 (16.0/29.6)	550 (100)
Overcast	149 (9.2/45.4) 612	86 (13.1/26.2) 132	93 (9.1/28.4)	328 (100)
Upper level cloud	678 (42.0/47.4) 923	318 (48.3/22.2) 120	434 (42.5/30.3)	1,430 (100)
Stratocumulus	395 (24.3/60.4) 1,261	35 (5.8/5.9) 113	219 (21.4/33.7)	649 (100)
Stratus	81 (5.0/38.6) 670	41 (6.2/19.5) 92	88 (8.6/41.9)	210 (100)
Cold sea fog	25 (1.5/26.3) 412	49 (7.4/51.6) 106	21 (2.1/22.1)	95 (100)
Warm sea fog	26 (1.6/81.2) 981	3 (0.5/9.4) 87	3 (0.3/9.4)	32 (100)
Total soundings	1,618 (100)	655 (100)	1,021 (100)	3,294 (100)

The two numbers in parentheses indicate the relative frequency of the corresponding cloud class for each boundary layer type and the relative frequency of the corresponding boundary layer type for each cloud class, respectively. The number in the second row of each cloud class indicates the average boundary layer depth in m

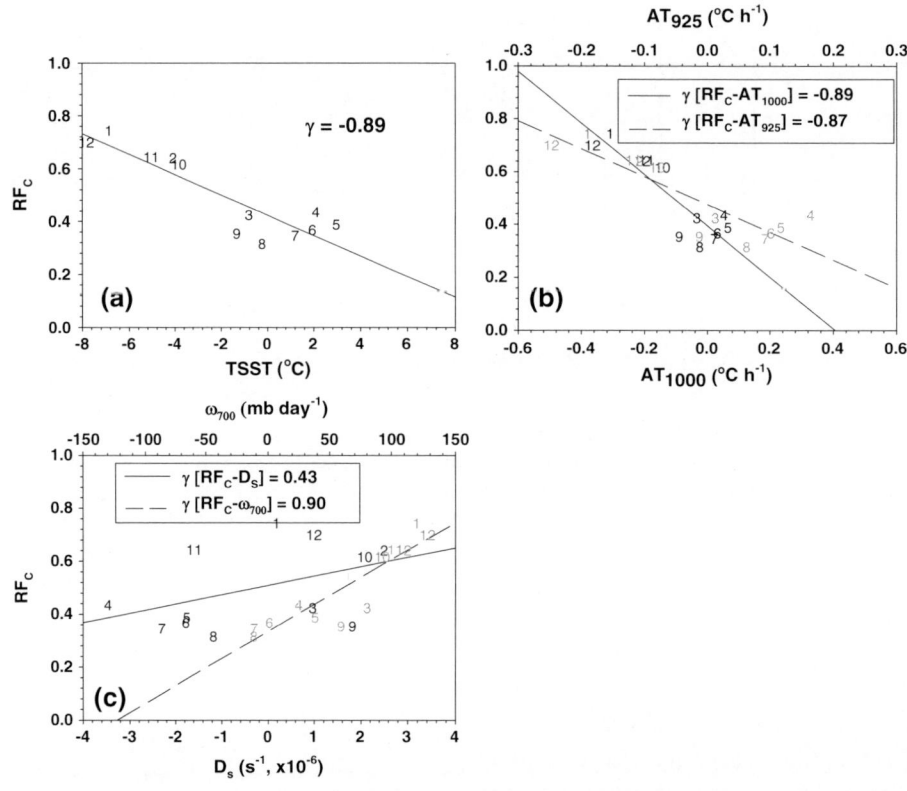

Figure 4

RF_C (relative frequency of CBL) for each month as a function of the monthly mean value of various meteorological variables: TSST (**a**); $AT_{1,000}$ and AT_{925} (**b**); D_S and ω_{700} (**c**). The *symbols* indicate the month number

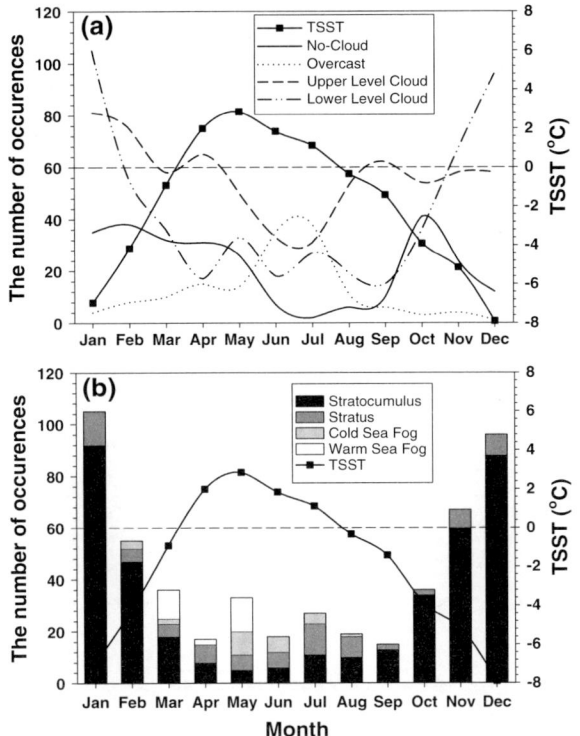

Figure 5
Monthly averages of TSST and monthly occurrence of different cloud classes (No-cloud, Overcast, Upper Level Cloud and Lower Level Cloud) (**a**); and detailed distribution of cloud types (Stratocumulus, Stratus, Cold sea fog and Warm sea fog) that belong to Lower Level Clouds (**b**) for the convective boundary layer (CBL)

seems to be only weakly correlated with the other variables, D_s, ω_{700}, AT_{950}, and AT_{925} (Fig. 6c, d). These results may imply that warm advection near the colder sea surface is important in SBL formation while synoptic forcings are not. The effect of the warm advection has been of interest in studies of summertime mid-latitude marine boundary layers. KLEIN and HARTMANN (1993) and NORRIS (1998b) found that warm advection over increasingly cold SST often induced stratus or fog in the western part of mid-latitude north Pacific.

As with CBL, during the SBL condition, the occurrence frequency of 'no-cloud' is the highest in April, while 'overcast' shows a peak in June (Fig. 7a). The relative frequency of lower level clouds is generally consistent with TSST ($\gamma = 0.84$, not shown), and lower level clouds are primarily of stratus and fog classes (Fig. 7b). Especially in

summer months more than half of lower level clouds are cold sea fogs. This is similar to the results from KLEIN and HARTMANN (1993), and NORRIS and LEOVY (1994), who found that stratus and fog are observed more frequently than stratocumulus in the western part of mid-latitude north Pacific, owing to the strong stability produced by warm air masses advected from subtropical regions.

3.5. Marine Boundary Layer During Sea Fogs

Now we examine how the structures of MBL during sea fogs differ from the mean state, and what causes these differences. Figure 8a shows the monthly mean MBL depth and its anomalies, i.e., the deviation during sea fogs from the monthly mean. In the calculation of the monthly mean MBL depth, both the CBL and SBL depths are considered together. The monthly mean MBL depth generally follows that of CBL, i.e., thicker in winter months than in summer, since SBL itself is much shallower than CBL and its seasonal variation is small. Interestingly, during cold sea fogs in February and March, the MBL depth is much shallower than the mean (Fig. 8a) and TSST is positive, which is opposite to the monthly mean (add the anomaly to the mean in Fig. 8b). This may be due to the anomalous warm advection near the surface (Fig. 8c; see that the addition of mean and anomaly are positive in these 2 months). Meanwhile, the anomaly of subsidence at 700 hPa, ω_{700}, is strongly negative (Fig. 8d), indicating ascending motion, maybe due to strong warm advection. The anomalies during cold sea fogs in summer are relatively small but mostly in the same direction as in February and March. Sea fogs do not occur from September to January (KIM and YUM 2010). ZHANG et al. (2009) showed, using observations and numerical simulations, that sea fogs off the east coast of China were formed below the inversion that is induced by warm advection from the Chinese continent. The stabilized air below the inversion can achieve vapor saturation readily, which is a crucial factor in the formation of cold sea fogs.

In contrast to cold sea fogs, the MBL during warm sea fogs becomes much thicker than the mean (Fig. 8a). The average TSST during these months is positive or zero, but the negatively large anomaly of TSST actually indicates that TSST is negative during

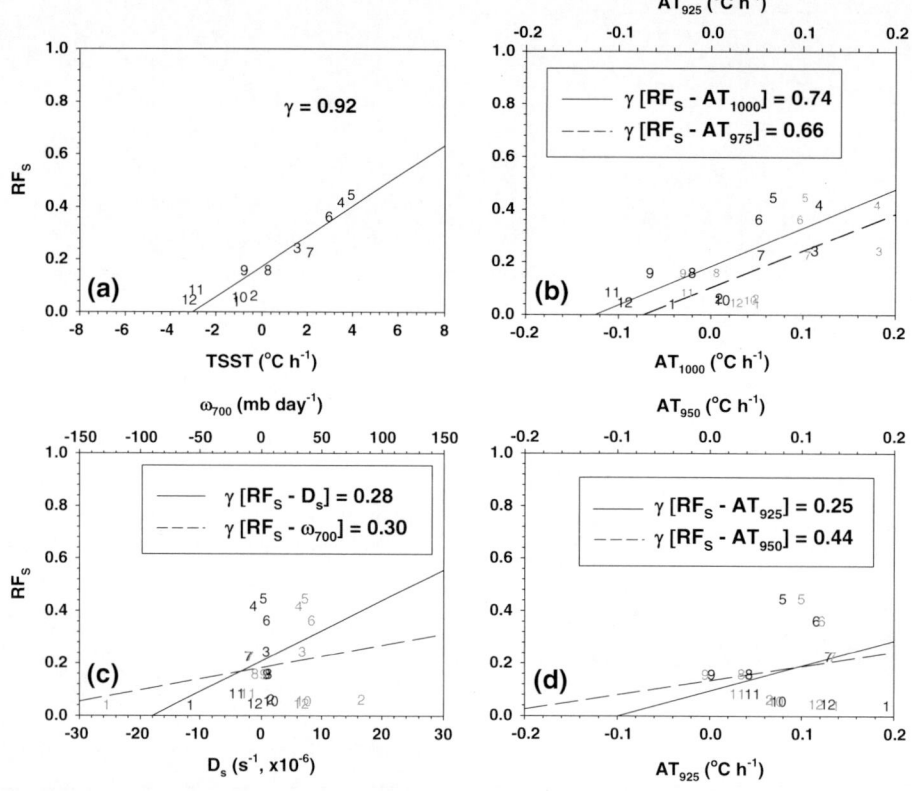

Figure 6
Same as Fig. 4 except for stable boundary layer (SBL). Added are the scatter plot of RF_S versus AT_{925} and RF_S versus AT_{950} (**d**)

warm sea fogs (Fig. 8b). This may be due to the anomalous cold advection near the surface (Fig. 8c; see that the addition of mean and anomaly are negative). In addition, the anomaly of ω_{700} during warm sea fogs is comparable to or larger than the monthly mean (Fig. 8d). This enhanced subsidence might play an important role in producing a capping inversion, below which a well-mixed layer would be maintained for a long time, eventually forming warm sea fogs within this mixed layer; moreover, the positive buoyancy flux generated by the cold advection over the warm sea surface might facilitate turbulent mixing.

4. The Effect of Turbulence, Radiation, and Advection on Sea Fog Formation

Now we examine how the low-level air above the sea surface attains water vapor saturation to form sea fogs. Warm advection over a cold sea surface may

produce SBL; in a sense, this is similar to surface inversion over land that is formed by radiative cooling and turbulence in the nocturnal boundary layer (STULL 1988). LALA et al. (1975) and DUYNKERKE (1999) emphasized the effects of radiation and turbulence on fog formation in the nocturnal boundary layer. Temperature and moisture advections will obviously affect the tendencies of T and T_d (GARRATT 1994). PAGOWSKI and MOORE (2001) insisted that horizontal advection be of the same order of magnitude as turbulent mixing for PBL study. Here we attempt to identify the role of turbulent mixing, radiative cooling, and advections of heat and moisture in forming sea fogs.

The cooling rate (CR) within the MBL can be expressed as the heat balance of all the forcings (STULL 1988):

$$\frac{\partial T}{\partial t} = CR = CR_T + CR_A + CR_R + CR_{RE}, \quad (1)$$

where CR_T, CR_A and CR_R are the cooling rates by turbulence, advection and radiation, respectively. The

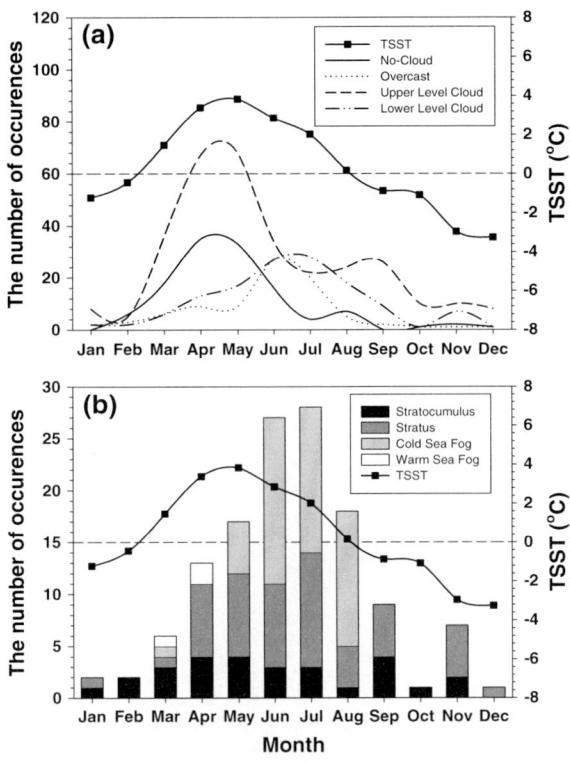

Figure 7
Same as Fig. 5 except for stable boundary layer (SBL)

last term on the right side of Eq. 1, CR_{RE}, represents the residual that is imposed to close the thermodynamic equation. The CR for each sea fog is calculated using the difference between the T at the onset of fog formation and the T 3 h before. The moistening rate (MR) is defined similarly to CR:

$$\frac{\partial T_d}{\partial t} = MR = MR_T + MR_A + MR_{RE}, \quad (2)$$

where MR_T and MR_A are the moistening rates by turbulence and advection, respectively. MR is calculated from the changing rate of T_d.

The turbulence cooling rate (CR_T) and turbulence moistening rate (MR_T) are given by:

$$CR_T = \frac{\partial(\overline{w'T'})}{\partial z} = \frac{\overline{w'T'}_{MBL} - \overline{w'T'}_S}{z_{MBL}} = \frac{-\overline{w'T'}_S}{z_{MBL}}, \quad (3)$$

and

$$MR_T = \frac{L_v}{C_p} \cdot \frac{\partial(\overline{w'q'})}{\partial z} = \frac{L_v}{C_p} \cdot \frac{\overline{w'q'}_{MBL} - \overline{w'q'}_S}{z_{MBL}}$$
$$= -\frac{L_v}{C_p} \cdot \frac{\overline{w'q'}_S}{z_{MBL}}, \quad (4)$$

where $\overline{w'T'}$ and $\overline{w'q'}$ are the vertical turbulent heat and moisture fluxes, respectively, with w being the vertical wind component and q being the water vapor mixing ratio; L_v is the latent heat of vaporization, C_p is the heat capacity at constant pressure, and the subscript S and MBL indicate the surface and the MBL depth during sea fogs, respectively. The CR_T and MR_T are equivalent to the vertical gradients of kinematic turbulent heat flux and moisture flux, respectively. The difference in turbulent flux between the MBL top and the sea surface (i.e., $\overline{w'T'}_{MBL} - \overline{w'T'}_S$ and $\overline{w'q'}_{MBL} - \overline{w'q'}_S$) reduces simply to the surface flux under the given condition of no turbulence at the MBL top before the formation of cloud or fog (STULL 1988).

Bulk aerodynamic equations are employed to calculate the kinematic sensible heat (SH) flux and latent heat (LH) flux every hour, using T, T_d, SST, and wind measured at the buoy:

$$\overline{w'T'}_S = C_H(SST - T)U, \quad (5)$$

and

$$\overline{w'q'}_S = C_E(q_S - q)U, \quad (6)$$

where U is the horizontal wind speed at the reference level (5 m above mean sea level, i.e., the height of the buoy). Similarly T and q at the reference level are used. The q_s is assumed to be the saturation vapor mixing ratio at the temperature of SST. Eddy exchange coefficients for heat (C_H) and moisture (C_E) are from the study by SMITH (1988).

The advection terms CR_A and MR_A in Eqs. 1 and 2 are, respectively, equivalent to $AT_{1,000}$ and $AT_{d1,000}$, which are the 6-h GDAS output variables obtained at the time closest to the onset time of fog formation. The radiative cooling rate (CR_R) is calculated hourly by the radiative transfer model (RTM) (CHOU et al. 2001), since no relevant radiation observation data are available. The vertical sounding measured at the rawinsonde site is used as an initial condition for running the RTM; some adjustments are required, however, since no measurement data are available between the surface altitude of the rawinsonde site (158 m) and the sea surface, as stated in Sect. 3. The T and T_d measured at the buoy every hour are used as substitutes for the surface values, and the intermediate values are assumed to take on a

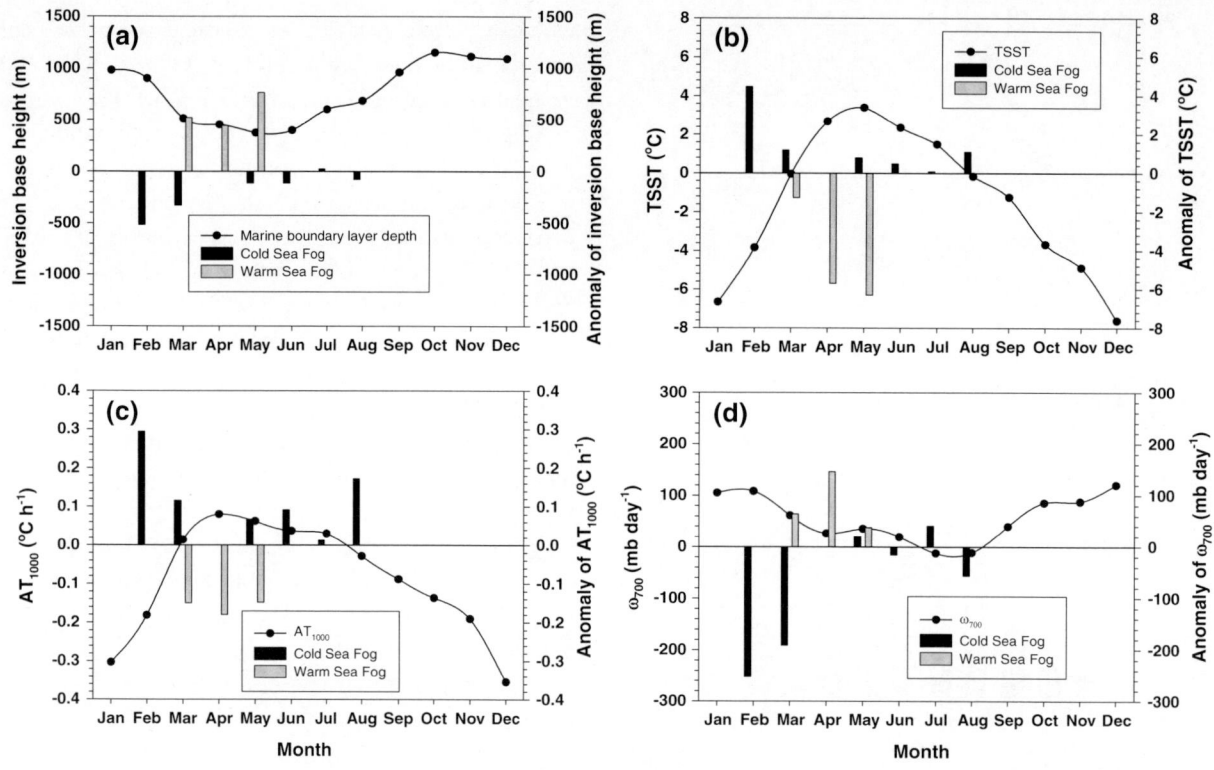

Figure 8
Monthly averages (*line*) and anomalies (*vertical bars*) during cold and warm sea fogs: marine boundary layer depth (**a**); TSST (**b**); $AT_{1,000}$ (**c**); and ω_{700} (**d**)

logarithmic and a linear profile for air temperature and water vapor mixing ratio, respectively. The RTM is then run with a vertical profile synthesized from the 12-hourly rawinsonde soundings and the hourly buoy data. Clear sky is assumed at upper altitude before the formation of fog.

The bulk Richardson number (R_b) is used as a measure of the turbulence:

$$R_b = \frac{g(\theta_r - \theta_S)z_r}{T_S U^2}, \qquad (7)$$

where g is gravitational constant, z_r is the reference level, T_S is the temperature at the surface, θ_r is the potential temperature at the reference level, and θ_S is the potential temperature at the surface. ARYA (1972) suggested the critical value of R_b as 0.25, above which the air is non-turbulent and mesoscale motion is important. Here we apply 0.30 as the R_b threshold for turbulent and non-turbulent regimes since there is a clear separation of R_b values smaller and larger than

0.30 in our dataset as shown in Figs. 9, 10, 11. Table 2 lists the average values of the micrometeorological variables for each regime during cold and warm sea fogs. The CR_T, MR_T, CR_R and R_b, values are based on the averages between the onset of fog formation and 3 h before for each fog case.

4.1. Cold Sea Fogs

It is found that 22 out of 27 cold sea fogs are formed in the turbulent regime. In the turbulent regime, CR is highly correlated with R_b (Fig. 9), implying that cold sea fogs can form even when cooling rate is weak if turbulence is strong (i.e., small R_b) but strong cooling rate is required if turbulence is weak. On the other hand, MR is mostly positive but does not show strong correlation with R_b. In contrast, in the non-turbulent regime, MR is negative or zero, that is, a decrease in water vapor mixing ratio or equivalently T_d (Fig. 9), but CR is strong, which may

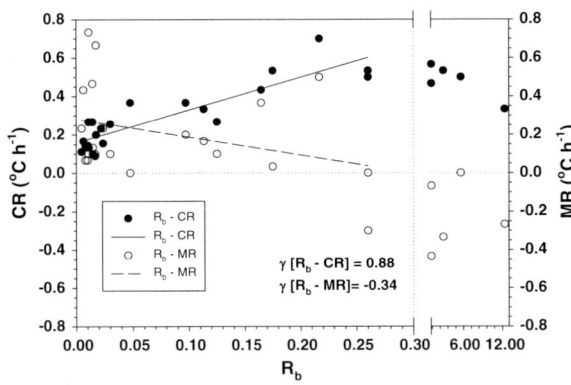

Figure 9
CR (cooling rate) and MR (moistening rate) as a function of R_b (bulk Richardson number) for cold sea fog cases. The *vertical dotted line* indicates the critical R_b, 0.30 that separates the turbulent and non-turbulent regimes

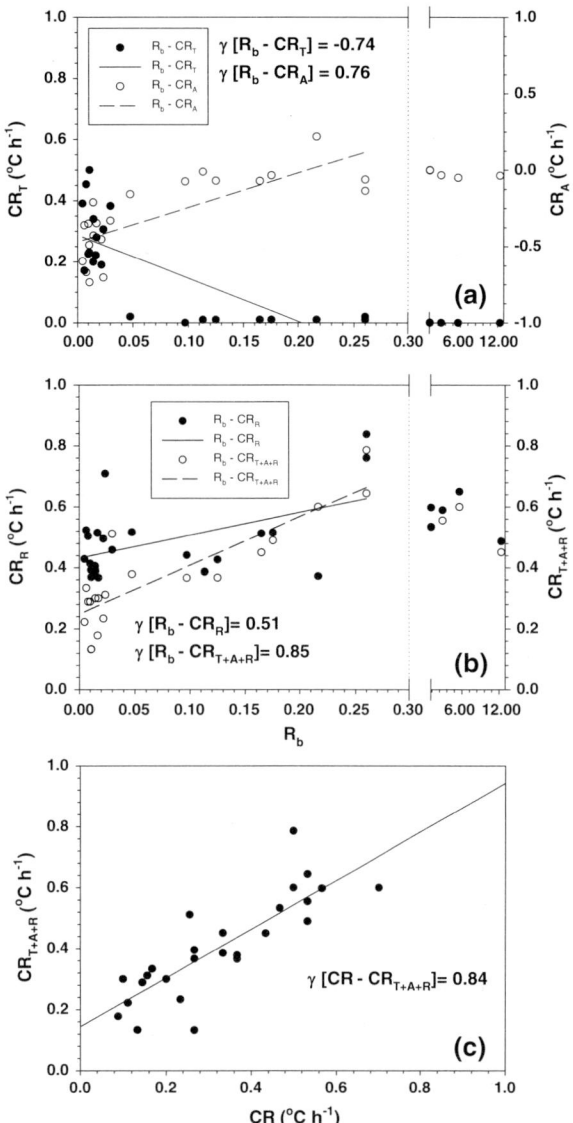

Figure 10
Each component of CR as a function of R_b: R_b versus CR_T and R_b versus CR_A (**a**); R_b versus CR_R and R_b versus CR_{T+A+R} (**b**); and CR versus CR_{T+A+R} (**c**). The meaning of CR_T, CR_A, CR_R, and CR_{T+A+R} is described in the text

compensate for the reduction of water vapor amount and eventually lead to water vapor saturation. In Table 2 the average CR and MR in the non-turbulent regime are $0.48°C\ h^{-1}$ and $-0.22°C\ h^{-1}$, respectively; so on average, T decrease is more than twice faster than T_d decrease.

Each component of CR is examined in Fig. 10. CR_T is large only when turbulence is strong ($R_b < 0.03$) and it is virtually zero even for R_b well within the critical value of 0.30 as well as in the non-turbulent regime (Fig. 10a). This implies that sensible heat loss to the surface can significantly contribute to the cooling of air only for strong turbulence ($R_b < 0.03$). On the other hand, CR_A is negatively very large (i.e., very strong warm advection; note that the scale of CR_A is twice larger than that of CR_T in Fig. 10a) when R_b is small but the magnitude of CR_A decreases as R_b increases. Since both R_b (Eq. 7) and advection are affected by wind speed, this relationship may be expected. Figure 10b shows that CR_R does not show high correlation with R_b but importantly it is always positive and in magnitude it exceeds that of CR_T in most cases in the turbulent and non-turbulent regimes, implying that longwave radiative cooling of the air is critically important during cold sea fog formation. This is consistent with previous studies on the effects of turbulence and radiation in the nocturnal boundary layer over land and water surface (GARRATT and BROST 1981; ANDRE and MAHRT 1982; ESTOURNEL 1986; COANTIC and SEGUIN 1971). The important point

is that when turbulence is strong the warm advection is so large that the strong cooling by turbulence and radiation is largely compensated and when the three effects are combined, a solid positive correlation is established in the turbulent regime (see CR_{T+A+R} in Fig. 10b). Lastly Fig. 10c shows that CR_{T+A+R} matches reasonably well with CR although there are some gaps, which indicates that the assessment of each process is reasonable in this study.

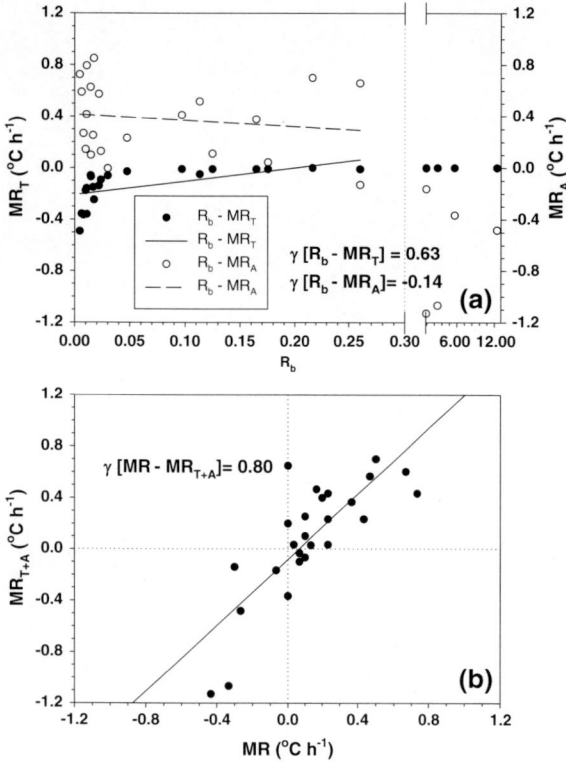

Figure 11

Same as Fig. 10 except for MR. The meaning of MR_T, MR_A and MR_{T+A} is described in the text

Figure 11a shows how the moistening rates by turbulence and advection vary with R_b. MR_T is negatively large only when turbulence is strong ($R_b < 0.03$). This behavior is similar to that of CR_T

(Fig. 10a) but opposite in sign. Over the sea surface, MR_T is usually positive because of evaporation. The negative sign of MR_T in Fig. 11a implies that the effect of strong turbulent mixing is actually drying up the air over the sea surface. This may occur when the moisture content of the air is greater than that at the sea surface, which is assumed to be saturated at SST. Apparently moisture is supplied by the strong moisture advection as the large positive MR_A for small R_b indicates. This is consistent with BOUTLE et al. (2010) numerical simulations that showed that warm and moist air over the cold sea surface can become supersaturated with respect to the sea surface, with moisture eventually condensing out in a manner similar to dew deposition over land. In our cold sea fog cases under strong turbulence, however, the air is not only supersaturated with respect to SST but also saturated or supersaturated with respect to the air temperature to successfully form fog. MR_A is mostly positive in weaker turbulence but in non-turbulent regime it is strongly negative (i.e., meaning very dry advection) and this is responsible for the negative MR in this regime (Fig. 9). MR_{T+A} matches very well with MR (Fig. 11b).

In summary, for cold sea fog cases with strong turbulence ($R_b < 0.03$), turbulence and advection exert opposite contributions to the cooling and moistening of the air but the magnitude is larger for advection: turbulent cooling is more than compensated by warm advection while the moisture

Table 2

Average values of the cooling and moistening rate components

	Cold sea fog		Warm sea fog
	Turbulent regime	Non-turbulent regime	Turbulent regime
Number of cases	22	5	8
CR (°C h^{-1})	0.27 ± 0.17	0.48 ± 0.09	0.05 ± 0.10
MR (°C h^{-1})	0.21 ± 0.23	-0.22 ± 0.18	0.24 ± 0.11
CR_T (°C h^{-1})	0.18 ± 0.17	0.00 ± 0.00	-0.24 ± 0.20
MR_T (°C h^{-1})	-0.13 ± 0.15	0.00 ± 0.00	0.30 ± 0.16
CR_A (°C h^{-1})	-0.30 ± 0.26	-0.02 ± 0.02	0.02 ± 0.01
MR_A (°C h^{-1})	0.38 ± 0.28	-0.64 ± 0.43	-0.02 ± 0.02
CR_R (°C h^{-1})	0.49 ± 0.13	0.57 ± 0.06	0.27 ± 0.06
R_b	0.07 ± 0.09	4.61 ± 4.71	-0.14 ± 0.30

The meaning of each component is described in the text. Positive values of CR (CR_T, CR_A and CR_R) and MR (MR_T and MR_A) indicate the cooling and moistening of air, respectively advection more than compensates for turbulent drying. Consideration of only these two processes implies that the air becomes moister but warmer, which is not a favorable condition for fog formation. In this sense, additional radiative cooling is critical in the successful formation of fog. This is true also for weaker turbulent cases: turbulent cooling is negligible and likewise warm advection is small but radiative cooling is still strong. This is even more critical for non-turbulent regime, where the air gets drier.

4.2. Warm Sea Fogs

All warm sea fog cases exhibit negative R_b, indicating that turbulence is generated by positive buoyancy flux from the warmer sea surface. As listed in Table 2, on average, CR is slightly positive. Turbulence induces warming ($CR_T = -0.24°C \ h^{-1}$) because of warmer sea surface but radiative cooling influence ($CR_R = 0.27°C \ h^{-1}$) that compensated for turbulent warming is slightly larger. Meanwhile, cooling due to thermal advection ($CR_A = 0.02°C \ h^{-1}$) is almost negligible. On the other hand, MR is relatively much larger than CR and this apparently is the main reason for warm sea fog formation. In detail, it is the turbulent moisture supply from the sea surface ($MR_T = 0.30°C \ h^{-1}$) that causes fog formation while moisture advection is almost negligible ($MR_A = -0.02°C \ h^{-1}$). This is consistent with PILIE et al. (1979) that showed that water vapor supply from the warmer sea surface is the crucial factor for formation of sea fogs off the coast of southern California. As in the cold sea fog cases, CR and CR_{T+A+R} match well and so do MR and MR_{T+A} (not shown).

Most of the warm sea fogs are formed during the time of the year where the monthly average TSST is actually positive (see the large negative anomaly of TSST during warm sea fogs in Fig. 8b). This TSST reversal can occur when even colder air moves over the cold sea surface. Moreover, the fact that the effect of advection is small suggests that the wind is calm during warm sea fogs. In summary, the turbulent air-sea interaction supplies the moisture while the turbulent warming is more than compensated by radiative cooling, which leads to warm sea fog formation.

5. Summary and Conclusions

This paper examines the marine boundary layer (MBL) structure for the formation of sea fogs off the west coast of the Korean Peninsula during the investigation period from January 2002 to August 2006. Based on the vertical soundings measured twice daily at a coastal island during the investigation period, the MBL structure is classified as convective boundary layer (CBL; when inversion exists aloft but at altitudes lower than 3 km), stable boundary layer (SBL; when inversion base is at the surface) or near-neutral boundary layer (NNBL; when there is no inversion or inversion base is higher than 3 km altitude). CBL is more likely to be found in colder months when TSST (=T–SST) is negatively large and cold advection is strong at near surface levels. During the CBL condition, the most frequently found lower level cloud is stratocumulus but fogs do form in spring and summer months mostly as warm sea fogs (TSST < 0). On the other hand, SBL is more likely to be found in warmer months when TSST is positive, and warm advection near the sea surface increase the chance of SBL formation. Stratus and cold sea fogs (TSST > 0) are the most frequently found lower level clouds under the SBL condition.

The effects of turbulence, advection and radiation on sea fog formation are found to be clearly different between the cold and warm sea fogs, based on the assessment of the heat and moisture balance equations. For cold sea fog cases, when turbulence is strong (bulk Richardson number, $R_b < 0.03$), turbulent cooling and drying and warm and moist advection exert opposite contributions to the heat and moisture budget of the air but the magnitude is larger for advection, implying that the air becomes moister but warmer when these two effects are combined. Since this is not a favorable condition for fog formation, the additional radiative cooling that is as strong as or even stronger than turbulent cooling turns out to be critical in the successful formation of cold sea fogs. In weak turbulent and non-turbulent ($R_b > 0.30$) regimes, the effects of turbulence decrease dramatically and so do the advection effects but the radiative cooling is still strong, making it the crucial reason for the successful formation of cold sea fogs. On the other hand, the turbulent moisture supply

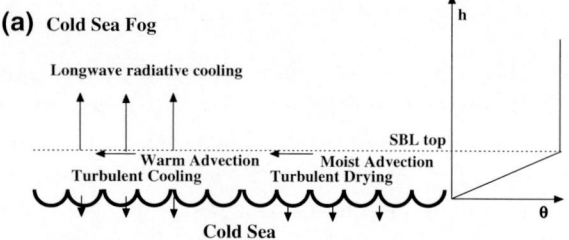

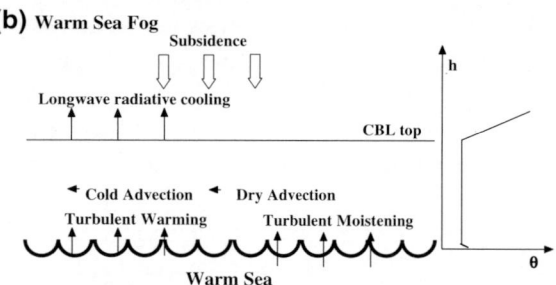

Figure 12
Schematic plots of the conceptual models for cold (**a**) and warm (**b**) sea fog formation off the west coast of the Korean Peninsula. Typical vertical soundings of SBL and CBL are displayed next to each schematic plot. The length of arrow indicates the relative strength of each component of CR and MR, based on Table 2

from the warmer sea surface is the crucial factor for the formation of warm sea fogs while turbulent warming and radiative cooling largely cancel each other out and the advection effects are negligibly small. Figure 12 shows the conceptual model of the typical cold and warm sea fog formation off the west coast of the Korean Peninsula that summarizes the foregoing analyses. Figure 12 is self-explanatory but to note is that the length of the arrows is proportional to the strength of the corresponding effects listed in Table 2.

In this study, we encountered some difficulties in characterizing the marine boundary layer and estimating the turbulent and radiative effects on the sea fog formation off the west coast of the Korean Peninsula, owing to the limitation of measurements over the sea surface. In order to increase the reliability of sea fog analysis, further effort should be made to establish a higher spatial resolution meteorological measurement network in the region, make measurements of vertical thermodynamic and wind structures in a continuous manner (e.g., by using remote sensing instruments such as cloud radar, lidar, and wind profilers), develop a good numerical model for fog

simulation, and carry out intensive field experiments that can draw a concerted effort from the related scientific community and achieve a comprehensive understanding of the relevant processes that lead to sea fog formation.

Acknowledgments

This work was supported by the Mid-Career Researcher Program through an NRF grant funded by the MEST (No. 2010-0000272). The authors would like to express deep thanks to the Korean Meteorological Administration for providing meteorological measurement data.

REFERENCES

ACKERMAN, A. S., P. V. HOBBS, and O. B. TOON, 1995: *A model for particle microphysics, turbulent mixing, and radiative transfer in the stratocumulus-topped marine boundary layer and comparisons with measurements*. J. Atmos. Sci., *52*, 1204–1236.

ALBRECHT, B. A., R. S. PENC, and W. H. SCHUBERT, 1985: *an observational study of cloud-topped mixed layers*. J. Atmos. Sci., *42*, 800–822.

ALBRECHT, B. A., D. A. RANDALL, and S. NICHOLLS, 1988: *Observations of marine stratocumulus clouds during FIRE*. Bull. Amer. Meteor. Soc., *69*, 618–626.

ANDRE, J. C., and L. MAHRT, 1982: *The nocturnal surface inversion and influence of clear-air radiative cooling*. J. Atmos. Sci., *39*, 864–878.

ARYA, S. P. S., 1972: *The critical condition for the maintenance of turbulence in stratified flows*. Quart. J. Roy. Meteor. Soc., *98*, 264–273.

BOUTLE, I., R. BEARE, S. BELCHER, A. BROWN, and R. PLANT, 2010: *The moist boundary layer under a mid-latitude weather system*. Boundary-Layer Meteor., *134*, 367–386.

CHOU, M.-D., M. J. SUAREZ, X.-Z. LIANG, and M.-H. YAN, 2001: *A thermal infrared radiation parameterization for atmospheric studies*. NASA/TM 2001-104606, 65 pp.

COANTIC, M., and B. SEGUIN, 1971: *On the interaction of turbulent and radiative transfers in the surface layer*. Boundary-Layer Meteor., *1*, 245–263.

DRAXLER, R. R., and G. D. ROLPH, 2011: HYSPLIT (HYbrid Single-Particle Lagrangian Integrated Trajectory) Model access via NOAA ARL READY Website [Available online at http://ready.arl.noaa.gov/HYSPLIT.php.].

DUYNKERKE, P. G., 1999: *Turbulence, radiation and fog in Dutch stable boundary layers*. Boundary-Layer Meteor., *90*, 447–477.

ESTOURNEL, C., R. VEHIL, and D. GUEDALIA, 1986: *An observational study of radiative and turbulent cooling in the nocturnal boundary layer (ECLATS experiment)*. Boundary-Layer Meteor., *34*, 55–62.

FINDLATER, J., W. T. ROACH, and B. C. MCHUGH, 1989: *The haar of north-east Scotland*. Quart. J. Roy. Meteor. Soc., *115*, 581–608.

GARRATT, J. R., 1994: The atmospheric boundary layer. Cambridge University Press, 316 pp.

GARRATT, J. R., and R. A. BROST, 1981: *Radiative cooling effects within and above the nocturnal boundary layer.* J. Atmos. Sci., *38*, 2730–2746.

GOPALAKRISHNAN, S. G., M. SHARAN, R. T. MCNIDER, and M. P. SINGH, 1998: *Study of Radiative and Turbulent Processes in the Stable Boundary Layer under Weak Wind Conditions.* J. Atmos. Sci., *55*, 954–960.

JOHNSON, A., and J. J. O'BRIEN, 1973: *A study of an oregon sea breeze event.* J. Appl. Meteor., *12*, 1267–1283.

KAGAN, B. A., 1995: Ocean-atmosphere interaction and climate modelling. Cambridge University Press, 377 pp.

KIM, C. K., and S. S. YUM, 2010: *Local meteorological and synoptic characteristics of fogs formed over Incheon international airport in the west coast of Korea.* Adv. in Atmos. Sci., *27*, 761–776.

KLEIN, S. A., and D. L. HARTMANN, 1993: *The seasonal cycle of low stratiform clouds.* J. Climate, *6*, 1587–1606.

KLEIN, S. A., D. L. HARTMANN, and J. R. NORRIS, 1995: *On the relationships among low-cloud structure, sea surface temperature, and atmospheric circulation in the summertime northeast pacific.* J. Climate, *8*, 1140–1155.

KORACIN, D., L. D. F, and L. J. M, 2005b: *Modeling sea fog on the US California coast during a hot spell event.* Geofizika, *22*, 59–82.

KORACIN, D., J. BUSINGER, C. DORMAN, and J. LEWIS, 2005a: *Formation, evolution, and dissipation of coastal sea fog.* Boundary-Layer Meteor., *117*, 447–478.

KRISHNA, T. B. P. S. R. V., M. SHARAN, S. G. GOPALAKRISHNAN, and ADITI, 2003: *Mean structure of the nocturnal boundary layer under strong and weak wind conditions: epri case study.* J. Appl. Meteor., *42*, 952–969.

LALA, G. G., E. MANDEL, and J. E. JIUSTO, 1975: *A numerical evaluation of radiation fog variables.* J. Atmos. Sci., *32*, 720–728.

LEE, S. H., 1995: *The division of natural seasons in Korea by air pressure patterns in Korean peninsula and its surroundings.* Geographic Res., *26*, 65–78.

LEIPPER, D. F., 1995: *Fog forecasting objectively in the California coastal area using LIBS.* Wea. Forecasting, *10*, 741–762.

NEIBURGER, M., 1960: *The relation of air mass structure to the field of motion over the eastern North Pacific Ocean in summer.* Tellus, *12*, 31–40.

NORRIS, J. R., 1998a: *Low cloud type over the ocean from surface observations. Part I: Relationship to surface meteorology and the vertical distribution of temperature and moisture.* J. Climate, *11*, 369–382.

NORRIS, J. R., 1998b: *Low cloud type over the ocean from surface observations. Part II: Geographical and seasonal variations.* J. Climate, *11*, 383–403.

NORRIS, J. R., and C. B. LEOVY, 1994: *Interannual variability in stratiform cloudiness and sea surface temperature.* J. Climate, *7*, 1915–1925.

PAGOWSKI, M., and G. W. K. MOORE, 2001: A *numerical study of an extreme cold-air outbreak over the labrador sea: sea ice, air–sea interaction, and development of polar lows.* Mon. Wea. Rev., *129*, 47–72.

PILIE, R. J., E. J. MACK, C. W. ROGERS, U. KATZ, and W. C. KOCMOND, 1979: *The formation of marine fog and the development of fog-stratus systems along the California coast.* J. Appl. Meteor., *18*, 1275–1286.

SAVIJARVI, H., 2006: *Radiative and turbulent heating rates in the clear-air boundary layer.* Quart. J. Roy. Meteor. Soc., *132*, 147–161.

SMEDMAN, A.-S., M. TJERNSTROM, and U. HOGSTROM, 1994: *The near-neutral marine atmospheric boundary layer with no surface shearing stress: a case study.* J. Atmos. Sci., *51*, 3399–3411.

SMITH, S. D., 1988: *Coefficients for sea surface wind stress, heat flux, and wind profiles as a function of wind speed and temperature.* J. Geophys. Res., *93*, 15467–15471.

STEVENS, B. et al., 2003: *Dynamics and chemistry of marine stratocumulus-DYCOMS-II,* Bull. Amer. Met. Soc., *84*, 579–593.

STULL, R. B., 1988: An introduction to boundary layer meteorology. Kluwer Academic, 666 pp.

ZHANG, S.-P., S.-P. XIE, Q.-Y. LIU, Y.-Q. YANG, X.-G. WANG, and Z.-P. REN, 2009: *Seasonal variations of yellow sea fog: observations and mechanisms.* J. Climate, *22*, 6758–6772.

(Received October 7, 2010, revised March 1, 2011, accepted April 18, 2011, Published online June 1, 2011)

Pure Appl. Geophys. 169 (2012), 1137–1155
© 2011 Springer Basel AG
DOI 10.1007/s00024-011-0344-9

Summary of a 4-Year Fog Field Study in Northern Nanjing, Part 2: Fog Microphysics

S. J. Niu,[1] D. Y. Liu,[1] L. J. Zhao,[1] C. S. Lu,[1] J. J. Lü,[1] and J. Yang[1]

Abstract—Comprehensive field observations of fog were conducted during the winters of 2006–2009 at the Nanjing University of Information Science & Technology, in order to study macro- and micro-physical structures and physical–chemical processes of dense fogs in northern Nanjing. The fog boundary-layer structures of different types and their corresponding characteristics are presented in Part I of these twin papers. In this second part, microphysical characteristics and droplet spectrum distributions of different types of fogs, microphysical relationships (among fog droplet concentration, liquid water content, and mean diameter), and microphysics of atmospheric aerosols during haze/fog events are discussed. The results show that there are large differences in microphysical parameters among four types of haze/fog. Many interesting phenomena, including fog burst reinforcement, fog droplet spectrum broadening, fog bimodal or multi-modal drop-size distributions, and critical triggering maxD value of fog coagulation growth, were captured during the 4-year field study.

Key words: Fog microphysical process, fog droplet spectrum, spectrum broadening, bimodal and multi-modal peaks.

1. Introduction

A summary of features in fog boundary layer (FBL; Zhou and Ferrier, 2008) based on the 4-year field study in northern Nanjing is presented in Part I. In this second part, the focus is on the fog microphysical characteristics observed during the same four-year field study.

It has been realized that intricate relationships exist between aerosol and fog characteristics since the activation and diffusion growth of droplets depend on the physical–chemical character of the ambient aerosols (Gultepe *et al.*, 2007). Fog condensation nuclei (FCN) have been studied by many researchers (Schumann, 1940; Spencer *et al.*, 1976; Hudson, 1980; Hung and Liaw, 1980; Ogren *et al.*, 1992; Boreux and Guiot, 1993; Podzimek, 1997). Kuroiwa (1951) confirmed that not only hygroscopic but also non-hygroscopic nuclei could act as FCN. The relationships between marine fog droplets and salt nuclei (Woodcock, 1978) and between organic aerosols and fog droplet spectra (Ming and Russell, 2004) have also been studied.

Fog drop-size distributions and droplet growth are important features of fog microphysical process. Many studies have illustrated the phenomenon of droplet growth (Eldridge, 1961, 1966, 1971; Goodman, 1977; Pickering and Jiusto, 1978; Gerber, 1991). Supersaturated environment (Baronti and Elzweig, 1973), radiative cooling (Roach, 1976a, b), and turbulent mixing (Gerber, 1981) were suggested to be the principal agents of droplet growth. However, the study of Choularton *et al.* (1981) indicated that larger drops cannot be produced by radiative cooling but by large supersaturation fluctuations or by convective motions. Spencer *et al.* (1976) suggested that the major effect upon the droplet growth process is from the increasing competition for vapor due to the nucleation of new droplets. Intense droplet growth due to water condensation and droplet coagulation was found to lead to bimodal size distribution (Podzimek, 1997). Similar phenomena like fog bimodal or multi-modal drop-size distributions are discussed in many other studies (Pilie *et al.*, 1975; Goodman, 1977; Garcia *et al.*, 2002).

Vertical structures of fog droplet size distribution and microphysical property have long been researched, but the results by different researchers are quite controversial. Pilie *et al.* (1975) found that fog droplet size distributions became narrower and the mean radius decreased with both increasing altitude and increasing age of fog. However, Goodman (1977)

---
[1] Key laboratory of Meteorological Disaster, Ministry of Education, School of Atmospheric Physics, Nanjing University of Information Science & Technology, Nanjing, China. E-mail: niusj@nuist.edu.cn

and PINNICK et al. (1978) observed opposite features; in all of their cases, the mean droplet diameter and liquid water content (LWC) increased with height.

In the last 30 years, fog droplet size distributions and microphysical characteristics were investigated in many field experiments, e.g., in Albany, New York (MEYER et al., 1980), and in two field campaigns (in 1989 and 1994) in the Po Valley, Italy (FUZZI et al., 1992, 1998; HEINTZENBERG et al., 1998; LAJ et al., 1998; RICCI et al., 1998), during which bimodal drop-size distribution and quasi-periodic oscillations with a period between 10 and 15 min were discussed (WENDISCH et al., 1998). During the recent 10 years, Paris Fog Field Experiment (ELIAS et al., 2009; HAEFFELIN et al., 2010) and Fog Remote Sensing and Modeling (FRAM) project (GULTEPE et al., 2007, 2009; GULTEPE and MILBRANDT, 2010) were conducted in Paris, France and in Canada, respectively. HAEFFELIN et al. (2010) studied turbulent mixing or aerosol number concentration exceeding critical values to explain droplet activation and fog formation. Fog microphysical characteristics for marine and continental fogs were separately parameterized by GULTEPE et al. (2009) to improve numerical weather predictions.

In an effort to gain insight into the triggering, formation, maintenance, and dissipation mechanisms of fog and extremely dense fog [horizontal visibility $L < 50$ m, according to the China Meteorological Administration (CMA) definition (2003)], four field campaigns of the Fog Monitoring & Early Warning and Disaster Damage Assessment Study in the Yangtze River Delta (YRDFOG in short) project were conducted at the meteorological observation station of the Nanjing University of Information Science & Technology (NUIST; 32°12′N and 118°42′E; 25 m above the sea level) in the winters of 2006–2009 (LIU, 2008; LU et al., 2008; PU et al., 2008a, b; YAN et al., 2009; LIU et al., 2010, 2011a, b; LU et al., 2010a, b; NIU et al., 2010a, b; YAN et al., 2010; LI et al., 2011; YANG et al., 2011). These in situ observations include the structures of FBL, microphysical parameters, turbulence, radiation, and thermal equilibrium components. The discussions of the fog field study are divided into two parts. The investigation of the microphysical characteristics of fog in northern Nanjing is covered here as Part II, while the FBL features are presented in a separated paper as Part I (LIU et al., 2011c).

The main objectives of this paper are to summarize the preliminary results of fog microphysical features in the YRDFOG field experiment project: (1) to understand different microphysical features of four types of fogs; (2) to gain insight into the triggering mechanisms of fog and extremely dense fog; and (3) to study droplet spectrum distribution evolution of four types of fogs.

A brief introduction of the instruments and data is given in Sect. 2. Results are presented in Sect. 3, including atmospheric aerosols microphysics during haze/fog events, fog microphysical characteristics and droplet spectrum distribution, dense fog burst reinforcement and droplet spectrum broadening, and fog microphysical parameter relationships. Summary and conclusions are included in Sect. 4.

2. Observations and Data

The observational site is located on the Meteorological Observation Base of the CMA, NUIST station (Fig. 1), a flat ground with no tall buildings or tall trees within 200 m. The instruments include a fog measuring device (FMD; FM-100), a visibility meter, an automatic weather station, and Wide-Range Particle Spectrometer (WPS™, MSP Corporation model 1000XP) (Table 1). The droplet measurement technology (DMT) FMD (FM-100) continuously measured the droplet number concentration and fog droplet spectrum, LWC at 1-Hz sampling rate with a diameter within 2–50 μm; and its maximum number concentration limit is $10^4/cm^3$ (LIU et al., 2011a). The ground weather elements and visibility were collected at 1-min intervals, using an automatic weather station (EnviroStation™ by the ICT, Australia) and a visibility meter (ZQZ-DN by the Jiangsu Province Radio Science Institute, China), respectively (LIU et al., 2011a). The WPS-1000XP was used to measure size distributions in the range of 0.01–10 mm at 5-min intervals (GAO et al., 2009).

The FM-100 has 20 classes, with ranges in: 0–2, 2–4, 4–6, 6–8, 8–10, 10–12, 12–14, 14–16, 16–18, 18–20, 20–23, 23–26, 26–29, 29–32, 32–35, 35–38, 38–41, 41–45, 45–48, and 48–50 μm. Each diameter value is the center of the class.

Figure 1

Meteorological observation base of the China meteorological administration (CMA), Nanjing university of information science & technology (NUIST) station

Table 1

List of instruments used during the YRDFOG project

Instruments	Measurement	Resolution time	Precision
Visibility meter (ZQZ-DN, Jiangsu Province Radio Science Institute, China)	Visibility	1 min	10%, <1,000 m; 20%, >1,000 m;
Fog Measuring Device (FM-100, DMT, USA)	Fog droplet spectrum, LWC, N_d	1 s	Range 2–50 μm;
Automatic weather station (EnviroStation™, ICT, Australia)	T, Pressure, RH, wind	1 min	1–3%;
Wide-range Particle Spectrometer WPS-1000XP(USA, MSP Corporation)	Aerosol particles	5 min	Range 0.010–10 μm;

After LIU *et al.* (2011a)

T temperature, *P* atmospheric pressure, *LWC* liquid water content, N_d droplet number concentration, *RH* relative humidity

From 2006 to 2009, a total of 29 fog events with visibility less than 1,000 m were observed. Fog droplet spectrum data were acquired for 23 of them (Table 2). Based on the fog droplet spectrum data, microphysical characteristics of these fog events were analyzed and are presented in this paper.

3. Results

In Part I, we divide the observed fogs into four types: radiation fog, advection–radiation fog, advection fog, and precipitation fog, according to the mechanisms of fog processes. We will use the same classification here.

3.1. Microphysics of Atmospheric Aerosols During Haze/Fog Events

Using simultaneous measurements of aerosol particle (WPS) and fog droplet size distribution (FM-100) data, YANG *et al.* (2010) investigated the microphysical characteristics of coarse and fine particles under four weather conditions (fog: visibility < 1,000 m, LWC ≥ 0.1 mg/m³; mist: 1,000 m < visibility < 10,000 m, LWC ≥ 0.01 mg/m³; wet haze: visibility < 1,000 m, LWC < 0.1 mg/m³; and haze: 1,000 m < visibility < 10,000 m, LWC < 0.01 mg/m³), to obtain microphysics of atmospheric aerosols during haze/fog events.

Table 2

Fog cases at the NUIST and their microstructure characteristics

No.	Fog types	Data	Formation time (LST)	Dissipation time (LST)	Duration (min)
1	Advection–radiation fog	24–27 Dec. 2006	22:00	14:14	3,840
2	Radiation fog	20 Nov. 2007	02:36	08:02	327
3	Radiation fog	21 Nov. 2007	02:00	07:46	347
4	Radiation fog	23 Nov. 2007	05:50	07:38	109
5	Radiation fog	24 Nov. 2007	00:25	12:14	730
6	Radiation fog	24 Nov. 2007	20:00	22:23	144
7	Precipitation fog	26 Nov. 2007	03:09	11:37	509
8	Radiation fog	10–11 Dec. 2007	22:31	12:30	840
9	Precipitation fog	11 Dec. 2007	19:21	23:10	229
10	Precipitation fog	12 Dec. 2007	08:59	17:07	488
11	Radiation fog	13–14 Dec. 2007	19:50	11:21	930
12	Radiation fog	14–15 Dec. 2007	20:48	12:02	902
13	Precipitation fog	16–17 Dec. 2007	21:07	08:39	693
14	Precipitation fog	17 Dec. 2007	09:39	10:56	78
15	Radiation fog	18 Dec. 2007	02:30	11:30	527
16	Advection–radiation	18–19 Dec. 2007	16:06	12:30	1,230
17	Radiation fog	19–20 Dec. 2007	16:36	16:14	1,419
18	Advection fog	20–21 Dec. 2007	17:47	19:07	1,520
19	Precipitation fog	22 Dec. 2007	02:07	10:46	520
20	Radiation fog	22–23 Dec. 2007	22:10	00:38	149
21	Radiation fog	23 Dec. 2007	01:16	05:30	255
22	Precipitation fog	27 Dec. 2007	09:25	14:36	312
23	Advection fog	01–02 Dec. 2009	19:00	11:20	980

According to these continuous observations, the microphysical characteristics of the particles show a large variation during transition processes from haze, mist, or wet haze to fog (Fig. 2). The number, surface area, and volume concentration of coarse particles with diameters larger than 4.0 μm in fog were much higher than those in the other three weather conditions, and the smallest concentration was observed in haze (Fig. 2a–c). Since during the transition processes from haze, mist, or wet haze to fog, the particles smaller than 4.0 μm would grow by hygroscopic mechanism to larger ones, the more water the particles attached, the larger it would get. The size distributions of surface area and volume concentration exhibited multi-peaks in fog droplets, while they showed single peak for coarse particles in mist, haze, and wet haze.

For the fine particles with diameters larger than 0.010 μm, the spectrum shapes of surface area concentration were similar in the four weather conditions (Fig. 2d, e). The dominant size ranges of the fine particle number concentration were 0.04–0.13 and 0.02–0.14 μm for fog and wet haze,

which were 51 and 75% of the total concentration, respectively, while the dominant size ranges were 0.02–0.06 μm for both mist and haze. During the transition processes from haze, mist, or wet haze to fog, the concentration of smaller particles (less than 0.060–0.090 μm) reduced, and that of larger particles increased. It suggests that the smaller particles would react to gaseous pollutants after hygroscopic growth, which would enlarge the particle size, change particle chemical compositions, and increase hygroscopicity; the concentration during this range would also increase. Temporal variation of aerosol number concentration correlated well negatively with the root mean diameter during the observation period. The aerosol number concentration was the lowest, and the mean diameter was the largest in fog.

The development of fog would decrease the number of fine particles of aerosols and increase the number of larger particles. In general, some of the aerosols activated as a condensation nucleus during fog formation. Then, fog droplets would scavenge on aerosols as fog drop size and LWC increase by nucleation, condensation, and coagulation from

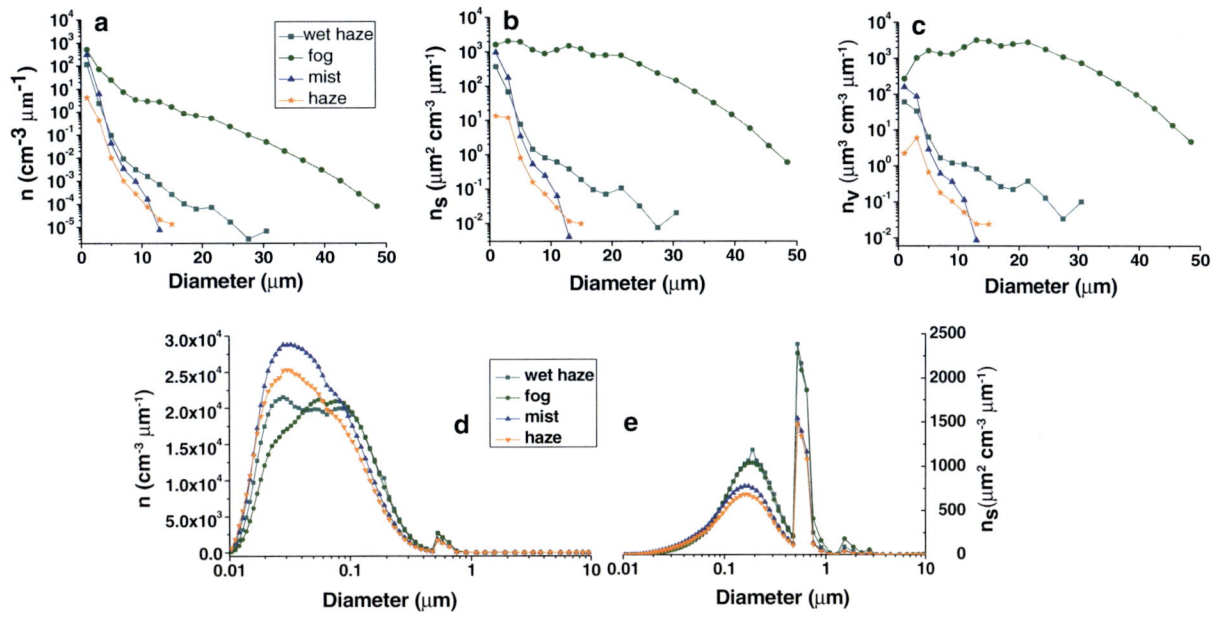

Figure 2
Size distribution of coarse particle (number, surface area and volume concentration, FM-100) and fine particle (number and surface area concentration, WPS) under four kinds of weather conditions

atmosphere to the ground. However, plenty of new hygroscopic aerosols are created by heterogeneous atmosphere reactions in this process, which contributes to the next fog/haze process.

HEINTZENBERG et al. (1998) analyzed the aerosol and fog data from their field experiment. Although the three weather conditions (haze: visibility > 1,000 m; mist:1,000 > visibility > 500 m; fog: visibility < 500 m) were different from what are used here, the concentration of fine particles smaller than 20 nm (diameter) decreased from haze, mist to fog, which was similar with our finding presented here.

The results of this study seem to contradict that of BOTT (1990), who found that the development of fog reduced the number of larger aerosol particles while leaving the smaller ones unchanged. It suggests that aerosol particles, especially larger ones, will be removed from the atmosphere by nucleation scavenging and subsequent gravitational settling of fog droplets. However, during a fog process the intensity of gravitational settling will be lower than that of the condensation, so the concentration of larger particles will remain at a large value. During the dissipative stage or after fog formation, the result is similar to that of BOTT (1990).

3.2. Microphysical Characteristics and Droplet Spectrum Distribution

3.2.1 General Microphysical Characteristics

Table 3 shows the averaged microphysical characteristics of four fog types. There are large differences between any two different types, in terms of microphysical parameters (LIU 2008).

Advection–radiation fog has the largest fog droplet concentration (N) of 211.0 cm^{-3}. It is followed by advection fog and radiation fog, with averaged N of 107.8 and 77.9 cm^{-3}, respectively. The precipitation fog has the smallest N, at 3 cm^{-3}. But the maximum value of each type follows a different sequence, namely in the order of advection fog, advection–radiation fog, radiation fog, and precipitation fog, with the values of 1,914.7, 1,213.8, 993, and 12.5 cm^{-3}, respectively.

The averaged LWC of radiation fog was 23.04 mg/m^3, close to that of advection fog of 21.07 mg/m^3. The averaged LWC of advection–radiation fog was 121.34 mg/m^3, nearly six times that of advection fog or radiation fog. Maximum values were different among them, being 476.24,

Table 3

Microphysical parameters of four types of winter fog

Fog types	Unit	Radiation fog			Precipitation fog			Advection fog			Advection–radiation fog		
		Ave	Max	Min	Ave	Max	Min	Ave	Max	Min	Ave	Max	Min
Vis	m	385	1,729	15	605	1,640	214	292	1,380	50	191	2,005	15
Speed	m/s	0.6	5.2	0.0	1.3	5.7	0.0	1.2	3.9	0.0	0.7	5.3	0.0
T	°C	4.0	14.1	−1.1	7.4	11.3	5.0	6.2	9.7	2.9	3.8	8.6	−0.2
RH	%	99.8	100.0	88.3	99.3	100.0	84.3	97.6	100.0	86.0	99.1	100.0	85.5
N	/cm^3	77.9	993.2	1	2.5	12.3	1	107.8	1,914.7	1	211.0	1,213.8	1
LWC	mg/m^3	23.04	476.24	0.01	0.04	0.21	0.00	21.07	982.03	0.01	121.34	903.03	0.01
maxD	μm	12.69	50.00	4.00	5.09	8.00	0.00	17.39	48.00	4.00	27.13	50.00	4.00
AveD	μm	3.66	8.56	3.00	3.05	3.34	0.00	3.59	5.53	3.00	4.99	9.47	3.00
AveD_S	μm	3.99	11.10	3.00	3.07	3.44	0.00	3.96	7.75	3.00	5.95	12.29	3.00

982.03, and 903.03 mg/m^3 for radiation fog, advection fog, and advection–radiation fog, respectively. The precipitation fog had the smallest value in LWC, with the maximum of just 0.21 mg/m^3.

For fog drop arithmetic mean diameter (AveD), the advection–radiation fog had the largest value, about 4.99 μm on average, with 9.47 μm as its maximum. The radiation fog's AveD was 3.66 μm, with 8.56 μm as its maximum. The averaged values of precipitation fog and advection fog were similar to their maximum values, being 3.05 μm (average)/3.34 μm (max) and 3.59 μm/5.53 μm, respectively, which suggests that there was little change of fog drop size in the whole life time of these two types of fog.

Except for precipitation fog, the other fogs had nearly the same maximum fog droplet spectrum

width (maximum diameter; MaxD for short), which was 50 μm. And the advection–radiation fog had an averaged value that reached 27.13 μm, while the others were no more than 20 μm.

3.2.2 General Droplet Spectra Distribution

Averaged droplet spectrum distributions of the four types of fog are given in Fig. 3a and b. The drop number and mass size distribution are plotted in their logarithmic forms of dN/dlog$D$ and dM/dlog$D$ (log stands for log$_{10}$), with respect to the drop diameter D measured by FM-100. Figure 3 shows that the averaged droplet spectrum distribution had a large difference among the four types. The precipitation fog had a narrow spectrum distribution, with the

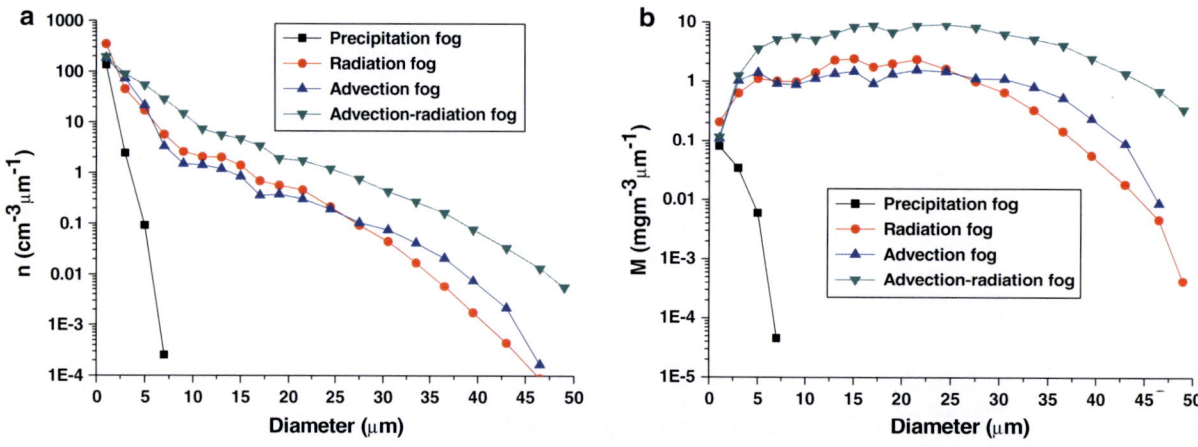

Figure 3

Averaged droplet spectrum distribution of four types of fog. **a** Drop number size distribution in the form of $dN/dlogD$ (cm^{-3} μm^{-1}). **b** Drop mass size distribution in the form of $dM/dlogD$ (mg m^{-3} μm^{-1})

largest droplet of no more than 7 μm; both drop number size distribution and drop mass size distribution curves basically showed exponential declines, which was similar to the other types. Advection–radiation fog had the largest values both in terms of the number and mass size distributions, whose curves were above the other types of fogs. Radiation fog and advection fog had nearly similar distributions of number and mass size distributions, there was little difference between them. The number and mass values of radiation fog were greater (smaller) than those of advection fog when the diameter was smaller (greater) than 24 μm. The drop mass size distribution

had three peaks in advection fog, radiation fog, and advection–radiation fog, with peak values at 9, 16, and 24 μm for advection fog, and at 5, 16, and 22 μm for both radiation fog and advection–radiation fog. Detailed analysis about the three peaks of these three types of fogs will be presented in Sect. 3.3.

3.2.3 Radiation Fog

Of all 23 observed fog cases, there were 12 radiation fogs (Table 4). Observations indicated that the 12 radiation-fog cases formed at different times of the day, and their duration periods also varied: Case X

Table 4

Microphysical parameters of radiation fog

	Data	Formation	Dissipation		Vis (m)	Speed (m/s)	T (°C)	RH (%)	N (/cm^3)	LWC (mg/m^3)	maxD (μm)	AveD (μm)
I	20 Nov. 2007	2:36	8:02	Average	507	0.0	4.2	100.0	2.2	0.059	6.7	3.4
				Max	1,698	0.5	6.3	100.0	20.5	0.767	12.0	3.8
				Min	116	0.0	3.4	100.0	0.8	0.015	6.0	3.1
II	21 Nov. 2007	2:00	7:46	Average	590	0.2	5.7	99.9	2.3	0.053	7.1	3.3
				Max	1,230	1.2	7.2	100.0	8.4	0.274	10.0	3.6
				Min	185	0.0	4.8	98.8	1.3	0.026	6.0	3.1
III	23 Nov. 2007	5:50	7:38	Average	442	0.1	7.6	99.8	4.3	0.185	7.1	3.3
				Max	1,178	0.4	9.0	100.0	172.1	11.069	16.0	4.2
				Min	87	0.0	7.0	98.6	1.0	0.019	6.0	3.2
IV	24 Nov. 2007	00:25	12:14	Average	637	1.4	11.6	96.5	1.7	0.030	5.8	3.1
				Max	1,052	2.7	14.1	100.0	5.9	0.144	10.0	3.4
				Min	240	0.4	9.6	88.3	0.5	0.007	4.0	3.0
V	24 Nov. 2007	20:00	22:23	Average	561	0.2	10.8	98.9	5.0	0.103	7.6	3.2
				Max	1,200	1.9	12.1	100.0	18.6	0.563	12.0	3.5
				Min	152	0.0	9.8	94.3	1.2	0.020	6.0	3.1
VI	10–11 Dec. 2007	22:31	12:30	Average	314	0.4	3.8	99.9	59.1	20.332	10.5	3.6
				Max	1,011	3.3	7.7	100.0	798.6	472.173	38.0	7.0
				Min	15	0.0	1.8	98.9	0.4	0.006	4.0	3.0
VII	13–14 Dec. 2007	19:50	11:21	Average	452	0.2	0.8	99.9	137.8	57.836	14.3	3.9
				Max	1,170	1.0	4.5	100.0	993.2	476.244	45.0	8.0
				Min	15	0.0	−1.1	99.2	0.7	0.009	4.0	3.0
VIII	14–15 Dec. 2007	20:48	12:02	Average	538	0.4	2.2	99.3	20.8	0.996	6.5	3.1
				Max	1,258	2.7	7.2	100.0	723.2	58.786	23.0	4.7
				Min	15	0.0	−0.3	91.6	0.4	0.006	4.0	3.0
IX	18 Dec. 2007	2:30	11:30	Average	317	0.3	4.1	100.0	88.8	22.321	13.9	3.6
				Max	1,190	2.0	6.6	100.0	596.2	210.503	38.0	6.6
				Min	15	0.0	2.9	100.0	0.9	0.013	4.0	3.0
X	19–20 Dec. 2007	16:36	16:14	Average	221	1.8	4.6	100.0	136.7	33.629	19.6	4.0
				Max	1,018	5.2	7.4	100.0	794.2	413.082	50.0	8.6
				Min	15	0.0	2.9	98.9	3.3	0.047	4.0	3.0
XI	22–23 Dec. 2007	22:10	0:38	Average	464	0.2	7.6	100.0	9.2	0.236	6.8	3.2
				Max	1,729	0.6	9.1	100.0	138.0	4.817	14.0	3.7
				Min	66	0.0	6.5	100.0	0.4	0.005	4.0	3.0
XII	23 Dec. 2007	1:16	5:30	Average	149	0.2	5.6	100.0	126.3	34.798	22.6	4.7
				Max	1,230	0.8	7.7	100.0	585.3	208.246	38.0	7.9
				Min	15	0.0	5.1	100.0	1.2	0.020	6.0	3.0

formed in the evening at about 1600 Beijing Standard Time (BST) and dissipated next evening, maintaining for nearly 24 h; five other cases (Cases V, VI, VII, VIII, and XI) formed between 1930 and 2230 BST, two of which lasted only about 2 h, and the others dissipated at noon of the following day; there were four radiation fogs (Cases I, II, IX, and XII) that occurred just after midnight, one of which dissipated at 1100 BST and the other three cases, before or after sunrise; Cases III and IV formed before and after sunrise, respectively, both of which lasted for only 2 h.

Figure 4a and b and Table 4 indicate that there were two types of radiation fog drop spectrum distributions. One of them is called broad spectrum radiation fog, which includes Cases VI, VII, IX, X, and XII because they had a large spectrum width of more than 35 μm, with the second peak at 15 μm in number size distribution, and three peaks at 5, 15, 23 μm in mass size distribution. The other type is called narrow spectrum radiation fog, which has a spectrum width mostly less than 15 μm (Cases I, II, III, IV, V, and XI), with only one case (Case VIII) reached 23 μm. The broad case spectrum lines are all above the narrow ones.

The broad spectrum radiation fogs had large N and LWC at 10^2–10^3 cm^{-3} and 10^2 mg/m^3 in magnitude, with AveD larger than 3.5 μm on average. Extremely dense fogs with visibility less than 50 m also appeared in these cases. In the narrow spectrum radiation fogs, N and LWC were 10 cm^{-3} and 10^{-1}–10^2 mg/m^3 in magnitude, and AveD was less than 3.5 μm on average. These values were all smaller than those of the broad spectrum radiation fogs.

Most broad spectrum radiation fogs lasted more than 8 h, but Case XII lasted for only 4 h. Extremely dense fog also appeared in Case VIII, which was a narrow spectrum radiation fog but lasted no more than 15 min. Radiation fog drop spectrum broadening will be discussed in Sect. 3.3.

3.2.4 Advection Fog and Advection–Radiation Fog

Four fog events related to advection were measured in the 4-year observation project. From Table 5, we can see that all of these cases lasted more than 16 h. Particularly, the event during 24–27 December 2006 lasted for almost 64 h. Extremely dense fog occurred

during both advection–radiation fogs, and continued for 40 h during 24–27 December 2006 and for 7 h during 18–19 December 2007. However, the lowest visibility in the two advection fogs was no less than 50 m.

From Fig. 4c and d and Table 5, the advection-related fogs all had wide spectra of greater than 45 μm, which was similar with the broad spectrum radiation fogs. Only one case during 20–21 December 2007 had a different spectrum distribution from other advection-related fogs and the broad spectrum radiation fogs. Fog droplet number concentrations between 6 and 36 μm (in diameter) were very low, and the LWC was nearly the same between 6 and 36 μm. The microphysical parameters including N, LWC, and AveD were also smaller during 20–21 December 2007 than during the other three cases: the averaged values were 39.4 cm^{-3}, 2.082 mg/m^3, and 3.2 μm, respectively, while the others had 130–250 cm^{-3}, 39.9–150 mg/m^3, and 3.9–5.4 μm, respectively. The advection-related fog drop spectrum broadening will be discussed with radiation fogs in Sect. 3.3.

The persistent time of extremely dense fog was mostly between 1 to 6 h, with the shortest being no more than half an hour and the longest being nearly 40 h during 24–27 December 2006. The appearance of extremely dense fogs was mostly in the morning, which was 1 or 2 h prior to or after sunrise; the others were at midnight.

3.2.5 Precipitation Fogs

Although precipitation fog has been studied by many researchers (TARDIF and RASMUSSEN, 2008, 2009; DONALDSON and STEWART, 1993), the precipitation fog droplet spectrum distribution was never investigated until now.

There are a few studies on the microphysical characteristics and droplet spectrum distributions of precipitation fog (e.g., YAN et al., 2010). Seven precipitation cases were observed, of which one was observed with drizzle. The sustained time dependes on the weather system, from 1 to 12 h. If the weather system changed rapidly, the duration would be short; otherwise, it lasted longer.

The precipitation fog had narrow spectrum width, with the maximum diameter no more than 8 μm, much narrower than that of the narrow spectrum

Table 5

Microphysical parameters of precipitation, advection–radiation, and advection fogs

Data		Formation	Dissipation		Vis (m)	Speed (m/s)	T (°C)	RH (%)	N (/cm³)	LWC (mg/m³)	maxD (µm)	AveD (µm)
Precipitation fogs												
I	26 Nov. 2007	03:09	11:37	Average	590	1.8	10.37	96.7	2.6	0.049	6.1	3.2
				Max	1,035	5.7	11.27	100.0	5.2	0.100	8.0	3.3
				Min	377	0.0	9.60	84.3	0.8	0.011	4.0	3.0
II	11 Dec. 2007	19:21	23:10	Average	787	1.2	7.30	99.1	3.5	0.051	4.7	3.0
				Max	1,176	2.9	8.30	100.0	8.3	0.121	6.0	3.2
				Min	532	0.0	7.09	93.2	0.6	0.008	4.0	3.0
III	12 Dec. 2007	08:59	17:07	Average	677	2.1	5.48	100.0	3.3	0.047	4.0	3.0
				Max	1,243	4.4	5.92	100.0	7.2	0.102	6.0	3.0
				Min	499	0.5	4.97	100.0	0.1	0.001	4.0	3.0
IV	16–17 Dec. 2007	21:07	08:39	Average	466	0.9	5.72	100.0	1.7	0.029	5.5	3.1
				Max	1,092	3.0	6.35	100.0	12.3	0.210	6.0	3.3
				Min	214	0.0	5.39	99.3	0.3	0.004	4.0	3.0
V	17 Dec. 2007	09:39	10:56	Average	740	0.9	6.33	100.0	1.8	0.029	5.5	3.1
				Max	1,640	2.1	6.97	100.0	3.2	0.055	6.0	3.2
				Min	506	0.0	6.03	100.0	0.4	0.005	4.0	3.0
VI	22 Dec. 2007	02:07	10:46	Average	595	1.0	8.87	100.0	3.6	0.057	5.4	3.1
				Max	1,090	3.9	9.58	100.0	7.5	0.133	6.0	3.2
				Min	328	0.0	8.45	100.0	0.6	0.008	4.0	3.0
VII	27Dec. 2007	09:25	14:36	Average	675	0.8	7.58	100.0	0.5	0.006	3.8	2.9
				Max	1,150	3.4	7.82	100.0	1.3	0.018	4.0	3.0
				Min	426	0.0	7.40	100.0	0.0	0.000	0.0	0.0
Advection–radiation fogs												
I	24–27 Dec. 2006	22:00	14:14	Average	116.9	–	5.1	99.8	241.5	150.214	31.3	5.4
				Max	1,629.8	–	7.6	100.0	875.4	903.032	50.0	9.5
				Min	15.0	–	0.6	93.6	0.5	0.008	4.0	3.0
II	18–19 Dec. 2007	16:06	12:30	Average	292.0	0.7	3.8	99.1	130.0	39.926	15.7	3.9
				Max	1,078.0	5.3	8.6	100.0	1213.8	478.507	45.0	7.5
				Min	15	0.0	−0.19	85.5	1.6	0.024	4.0	3.0
Advection fogs												
I	20–21 Dec. 2007	17:47	19:07.	Average	361	0.7	7.77	100.0	39.4	2.082	9.8	3.2
				Max	1,090	3.3	9.72	100.0	360.5	65.576	48.0	4.5
				Min	51	0.0	6.65	99.1	0.7	0.009	4.0	3.0
II	01–02 Dec. 2009	19:00	11:20	Average	179	1.9	3.72	93.7	219.0	51.959	29.8	4.2
				Max	1,380	3.9	5.30	95.0	1914.7	1040.890	48.0	5.5
				Min	51	0.4	2.90	86.0	1.2	0.017	4.0	3.0

radiation fog. They had small N, LWC, and AveD, no more than 4 cm^{-3}, 0.06 mg/m^3, and 3.2 µm, respectively. For these reasons, the minimum visibility was all greater than 200 m, mostly greater than 350 m.

3.3. Burst Reinforcement and Droplet Spectrum Broadening

From the above analysis, there were five radiation fogs, two advection fogs, and two advection–radiation fogs that have broad spectrum width of more

than 30 µm. Different types of fog had different droplet spectrum broadening characteristics.

3.3.1 Radiation Fog Burst Reinforcement and Droplet Spectrum Broadening

Li *et al.* (2011) analyzed radiation fog droplet spectrum broadening, revealing that when radiation fog developed from a dense stage to an extremely stage, the fog displayed explosive reinforcement features, showing a nature of burst broadening in

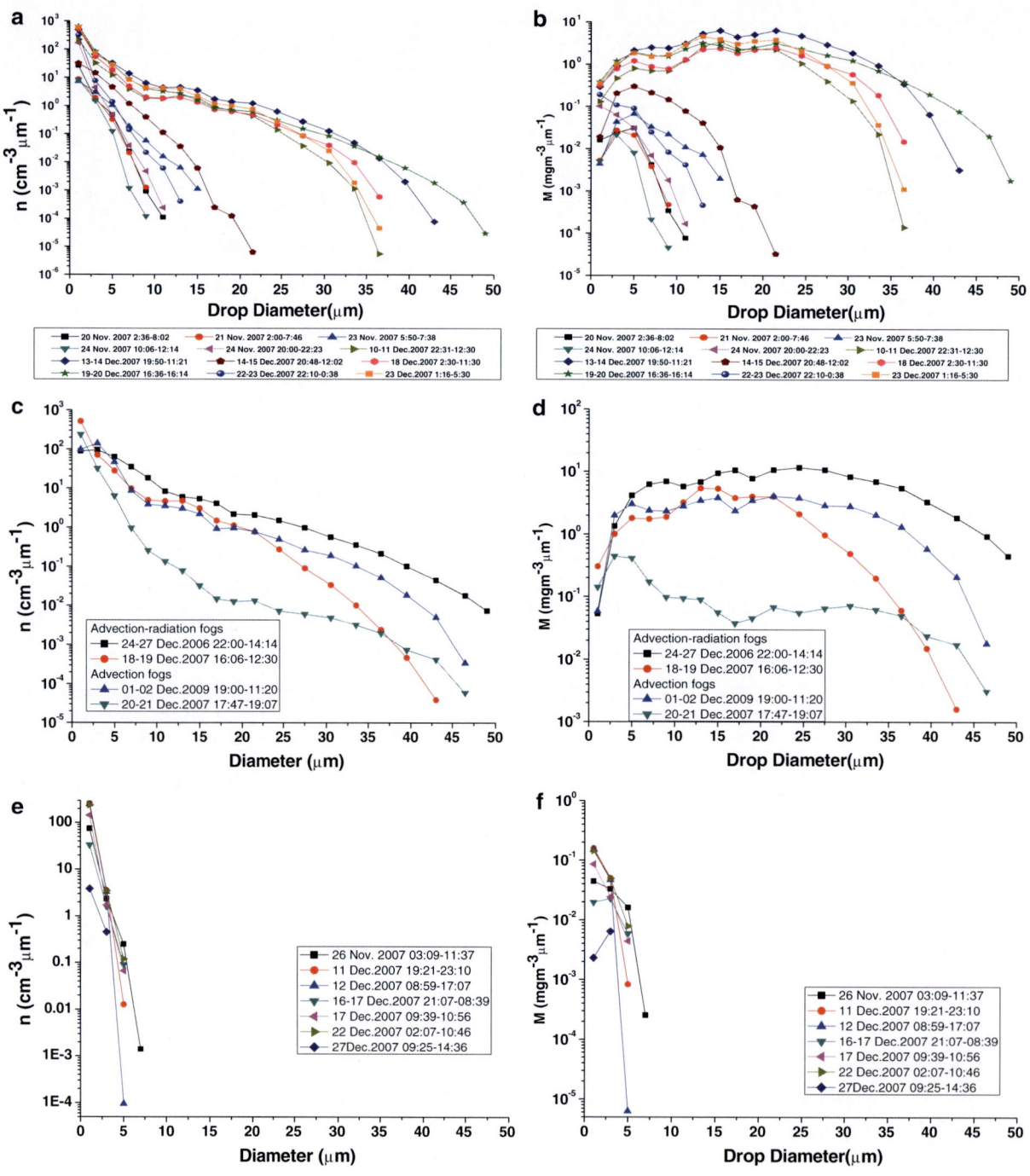

Figure 4

Averaged droplet spectra distribution of fogs: drop number size distribution in the form of *dN/dlogD* (cm^{-3} μm^{-1}) (**a** radiation fog; **c** advection fogs and advection–radiation fogs; **e** precipitation fogs), and drop mass size distribution in the form of *dM/dlogD* (mg m^{-3} μm^{-1}) (**b** radiation fog; **d** advection and advection–radiation fogs; **f** precipitation fog)

the fog droplet spectrum. Radiation fog was also associated with increases in N and LWC, and with decrease in visibility. From Fig. 5 one can see that

the spectrum width increased significantly in a short time (about half an hour), and N and LWC increased sharply in all size classes. For example, the case

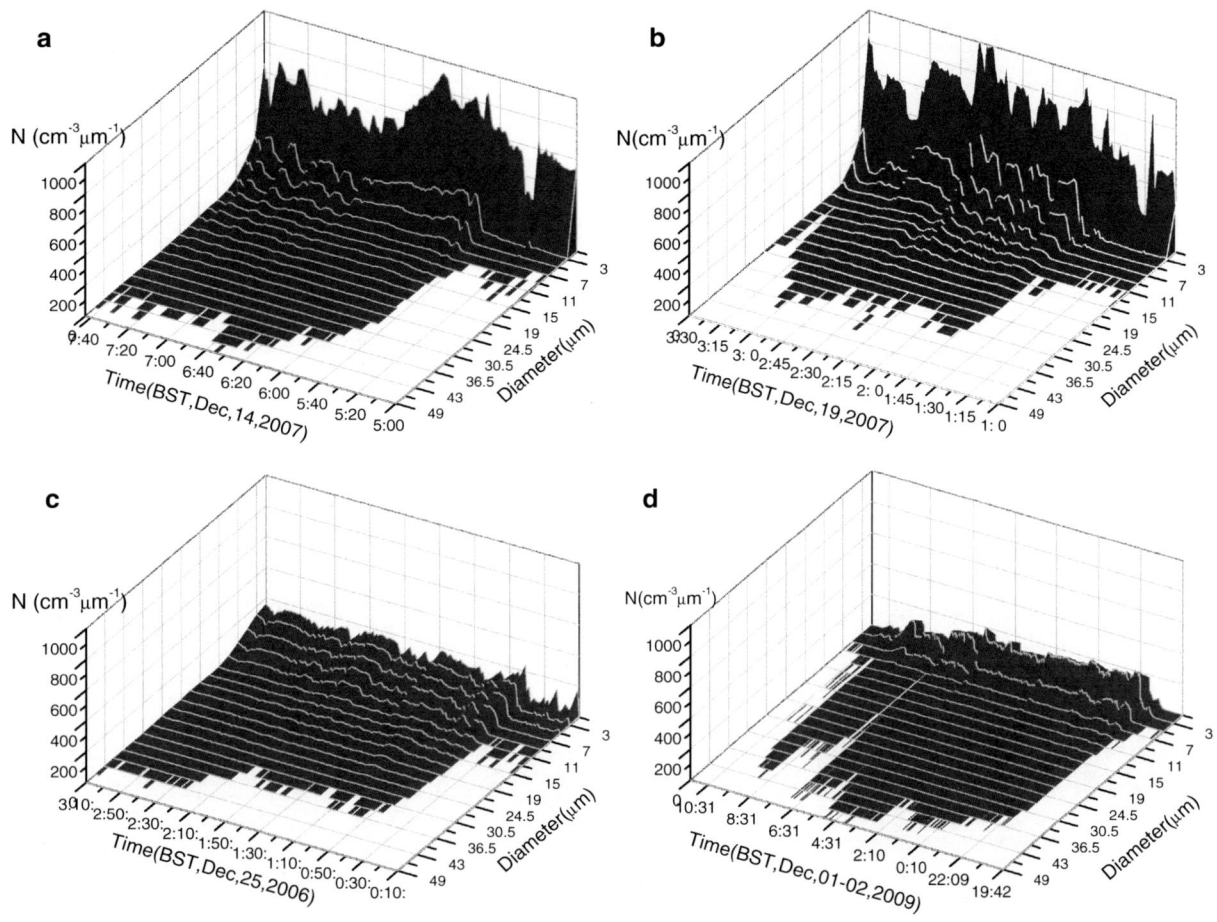

Figure 5
Time evolution of radiation fog and advection-related fog droplet spectra. *BST* Beijing Standard Time

during 13–14 December 2007 had a spectrum width of no more than 10 μm before 0545 BST (Figs. 5a, 6), and it then broadened to 23 μm in 2 min at 0547 BST. In the meantime, the visibility dropped from 200 m to less than 50 m, and the N, LWC, and AveD increased from 30 cm^{-3}, 0.68 mg/m^3, and 3.2 μm to 440 cm^{-3}, 44.9 mg/m^3, and 4.5 μm, respectively.

Li *et al.* (2011) carefully analyzed the macroscopic conditions of the radiation fog droplet spectrum. They pointed out that the radiation fog droplet spectrum burst broadening occurred during the period of rapid temperature decline or significant humidity increase (super-saturation increase), and that turbulence played an important role not only in the vertical transmission of momentum, heat, and water vapor but also in the radiation fog droplet spectrum broadening.

Roach (1976a, b) estimated that radiative cooling was the principal agent of droplet growth in radiation fog, and that supersaturation was very small or even relatively unimportant for droplet growth. Spencer *et al.* (1976) found that the major effect on the droplet growth process was due to increased competition for vapor due to the nucleation of new droplets. However, Choularton *et al.* (1981) argued that large supersaturation fluctuations or convective motions near the fog top produced large drops in radiation fog. Gerber (1991) had a similar opinion about supersaturation. He suggested that nongradient turbulent mixing of saturated air parcels at different temperature and the release of excess vapor by molecular diffusion at the interface between the mixing parcels were the mechanisms causing the large supersaturation.

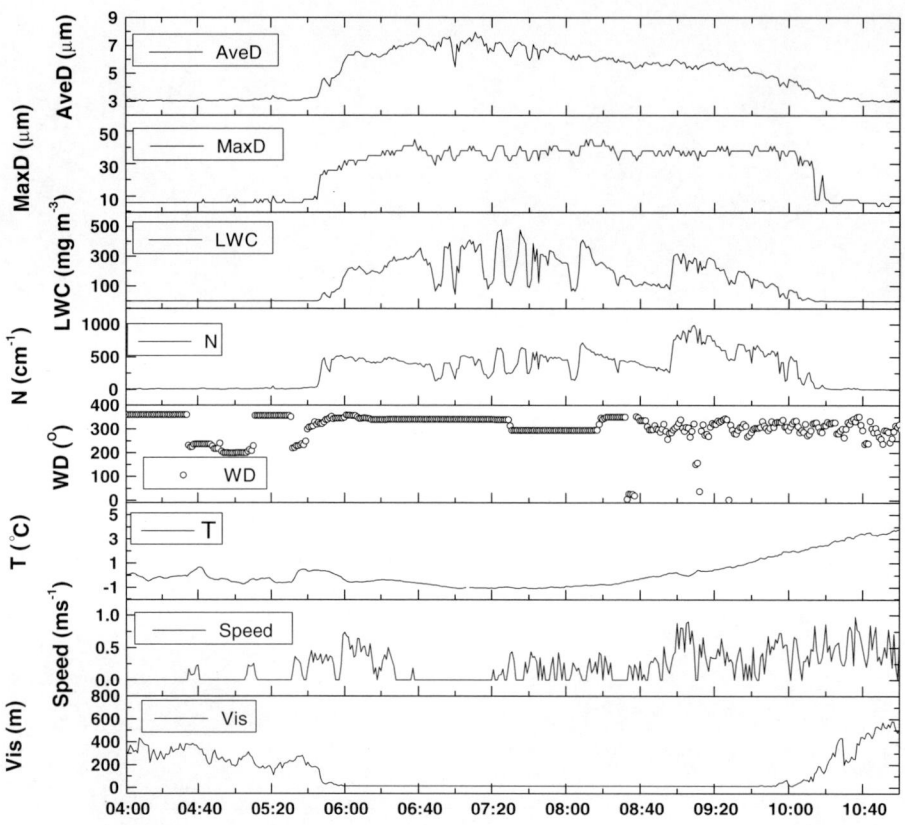

Figure 6

Time evolution of microphysical parameters of extremely dense radiation fog during 13–14 December 2007. *BST* Beijing Standard Time, *WD* wind direction

3.3.2 Advection-Related Fog Burst Reinforcement and Droplet Spectrum Broadening

Advection-related fogs had different droplet spectrum broadening characteristics. The extremely dense fog occurred only in radiation-related cases, namely, radiation fog or advection–radiation fog. However, the lowest visibility of advection fog was more than 50 m (Table 3).

One advection–radiation fog during 18–19 December 2007 had a similar fog droplet spectrum distribution as that of a radiation fog (Fig. 5b). N of small drops of less than 10 μm increased from about 200 cm^{-3} to more than 600 cm^{-3} during the burst reinforcement stage; simultaneously, the droplet spectrum width broadened from no more than 10 μm to more than 30 μm.

Analogous fog droplet spectrum distributions occurred during 24–27 December 2006 and 01–02 December 2009, which were advection–radiation fog and advection fog, respectively (Fig. 5c, d). In these cases, N of small drops increased to no more than 200 cm^{-3}, although the droplet spectrum width broadened similarly with previous radiation fog cases. Moreover, the visibility showed great difference between these two cases, notwithstanding the similar spectrum distribution.

The advection fog during 20–21 December 2007, in which there was no burst reinforcement feature in the fog cycle, had a spectrum distribution different from the advection fog during 01–02 December 2009 (Fig. 7). Although there were large values of N, LWC, and AveD in some periods, most time there were little changes of these values during the whole fog life time, and the visibility changed very little.

3.4. Fog Bimodal or Multi-Modal Drop-Size Distributions

Fog bimodal or multi-modal drop-size distributions were observed during the 4-year field

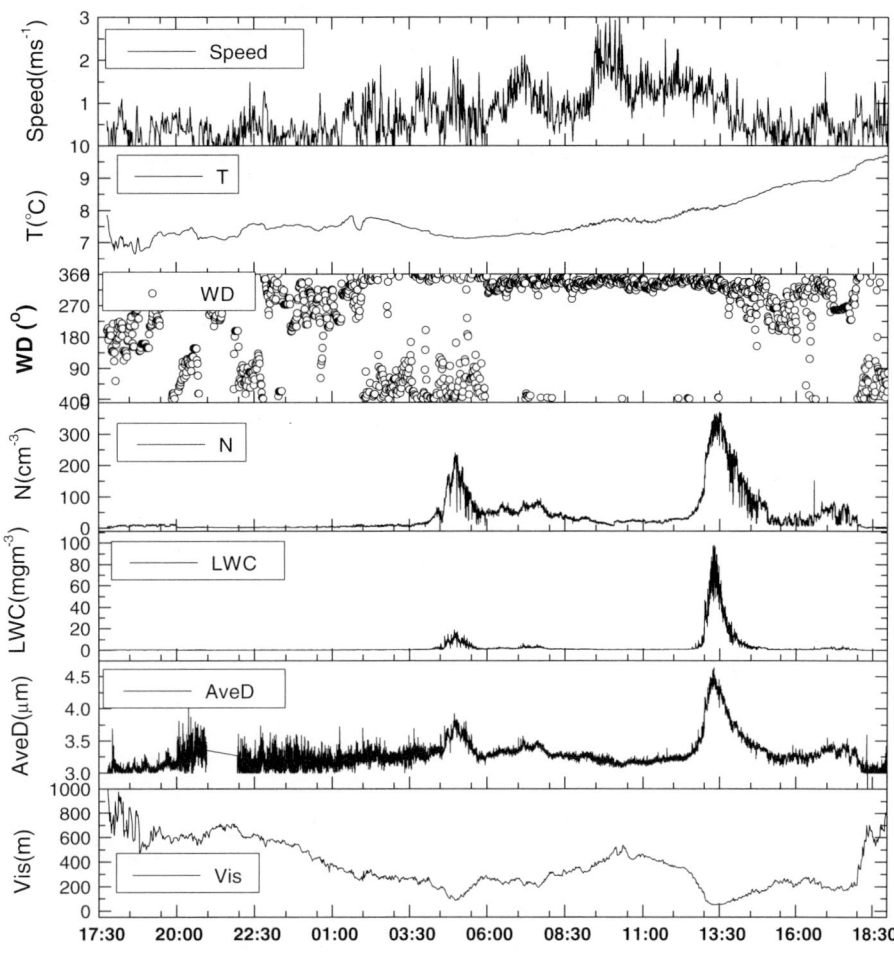

Figure 7

Time evolution of microphysical parameters of advection fog during 20–21 December 2007. *BST* Beijing Standard Time, *WD* wind direction

observation. These phenomena occurred in both number and mass droplet size distributions; however, bimodal/multi-modal number droplet size distributions appeared only in radiation-related fogs.

From Fig. 8, there are six radiation-related cases in which bimodal number droplet size distributions took place (using the FM100 classes in the *y*-axis). All of these bimodal phenomena occurred during extremely dense fogs, in which both *N* and LWC had very large magnitudes, temperature was very low, and visibility was less than 50 m in the process. In all of these cases, peaks were at 1–5 µm and 13–17 µm (the center of the classes) for the first and second mode, respectively. Multi-modal distributions also occurred in radiation-related fogs, but only for their mass droplet size distributions. The advection–radiation fog during 24–27 December 2006, for example,

had four peaks between 1800 and 2400 BST on 25 December 2006 at 8, 14–18, 23–27, and 32–37 µm, respectively (Fig. 9a1, a2).

Bimodal number droplet size distributions also occurred in advection fogs. However, the peaks lay at 2–4 and 20 µm, respectively (Fig. 9b1, b2).

ELDRIDGE (1966) argued that no two distributions were alike in detail, and he divided fog drop-size distributions into four types: stable fog (first type, and second type), selective fog, and evolving fog. The evolving fog drop-size distributions changed from monomodal to bimodal and multi-modal. PILIE *et al.* (1975) showed that bimodal drop size distributions occurred in approximately half of the fogs with one mode at 2–3 µm radius and a second mode between 6 and 12 µm; this study found similar results. MEYER *et al.* (1980) also showed multi-modal distributions

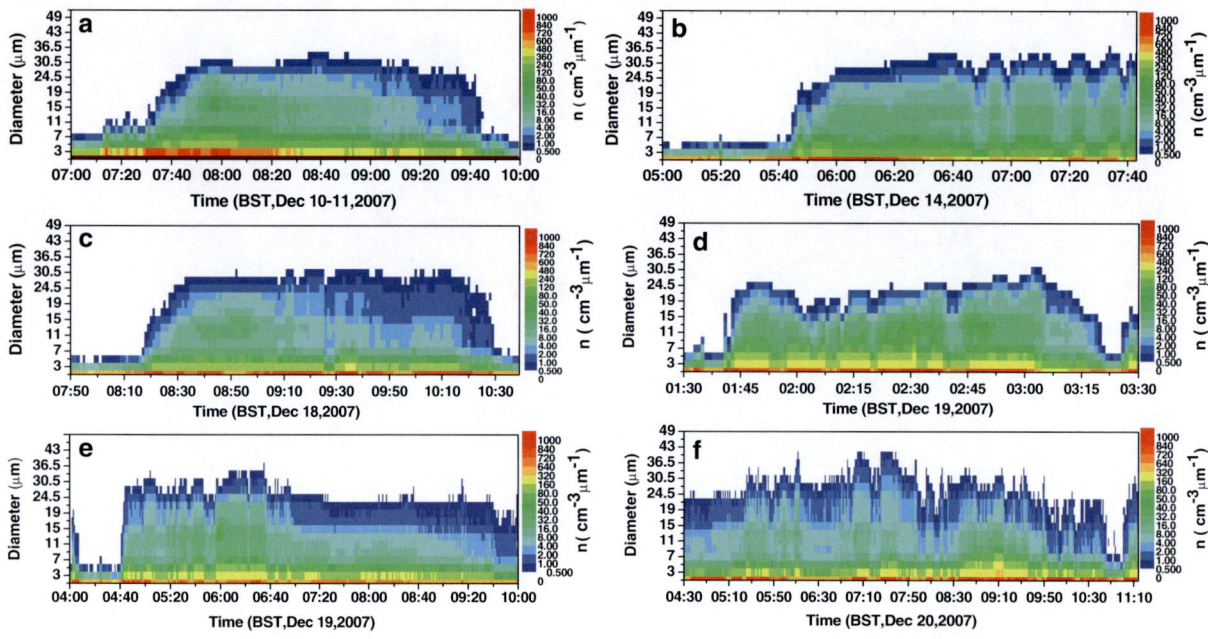

Figure 8
Time evolution of bimodal or multi-modal drop-size distributions of radiation related fogs. *BST* Beijing Standard Time. The *y*-axis is for upper class-size diameter, using FM-100 classes

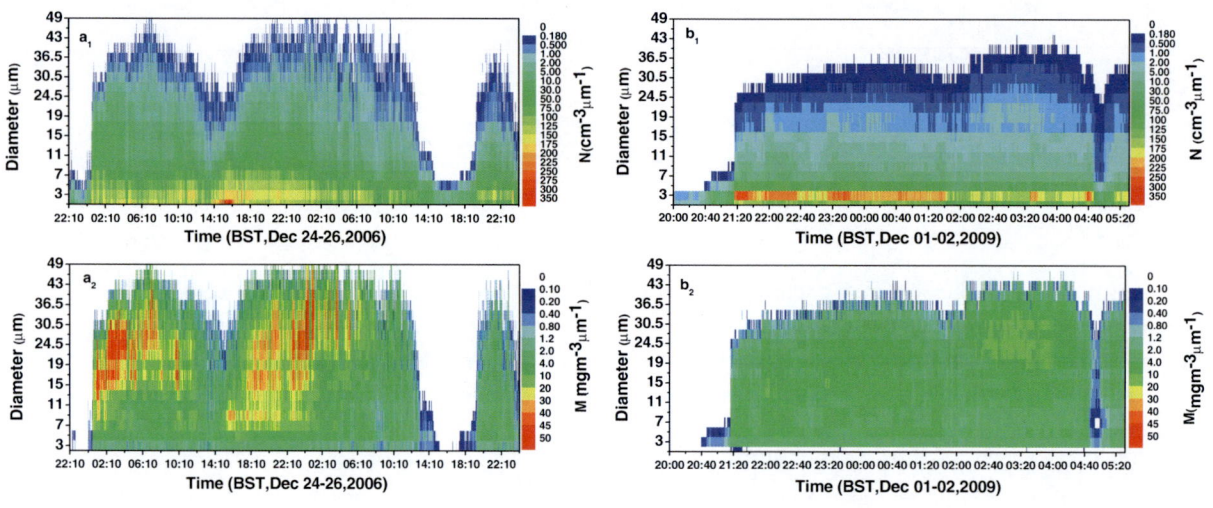

Figure 9
Time evolution of bimodal or multi-modal drop-size distributions of advection related fogs. *BST* Beijing Standard Time. The *y*-axis is for upper class-size diameter, using FM-100 classes

that peaked at the submicron region and 3, 10, and perhaps 20 μm. PODZIMEK (1997) thought that intense droplet growth due to water condensation and droplet coagulation lead to bimodal size distribution. Maybe multi-modal distributions are also due to water condensation and droplet coagulation.

3.5. Microphysical Relationship of Fog

In order to understand the physical processes of these fogs, mutual relationships between any two key microphysical properties, including LWC, *N*, and mean radius, were researched (NIU *et al.*, 2010a, b).

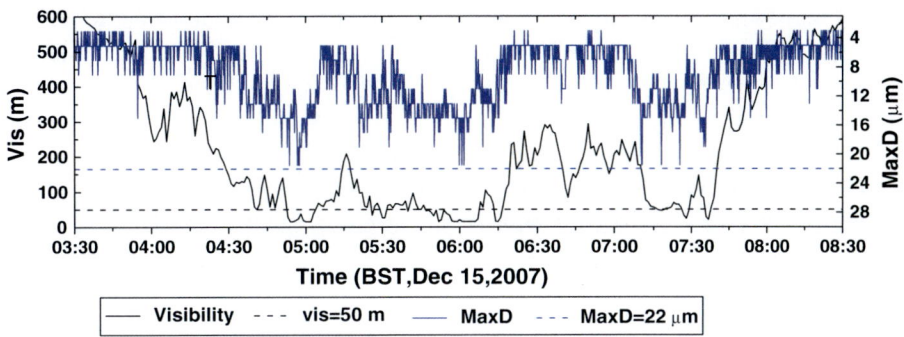

Figure 10
Time evolution of maxD and Vis of narrow radiation fog during 14–15 December 2007. *BST* Beijing Standard Time

An unusual advection–radiation fog during 24–27 December 2006 was selected for the analysis. From their results, positive correlations among N, LWC, mean radius, and their standard deviations were studied, showing that the droplet activation and condensational growth were dominant, leading to strong positive correlations. A stronger coagulation process was suggested to destroy some of the positive correlations, in agreement with theoretical expectation (NIU et al., 2010b).

Moreover, it is anticipated that different microphysical processes act in different stages/periods as different combinations, and that the exact microphysical relationship is determined by the degree of balance of these processes.

During the formation and dissipation stages, droplet activation and condensation should be dominant, leading to strong positive correlations, while in the descending periods, evaporation and turbulent mixing may also be factors. During the development stage, however, droplet activation with subsequent condensational growth, which was promoted by sufficient supplies of water vapor and fog condensation nuclei, and coagulation were all considered to be dominant processes that can influence fog microphysics. Small droplets formed from condensation nuclei may compensate for the loss caused by coagulation.

4. Discussions

Radiation fog droplet spectrum distribution has been comprehensively studied (ELDRIDGE, 1961; GERBER, 1981, 1991; PODZIMEK, 1997), but broad and narrow distributions are presented here for the first time. It is possible that the narrow spectrum distribution fog is the initial stage of the broad spectrum distribution fog (Fig. 8), between which the short duration black area or intermediate is the critical area or the critical stage for the small one to broaden to a larger one. Take the case during 14–15 December 2007, for example (Fig. 10); the visibility and the maxD were in negative correlation, where maxD increased with the decreasing visibility. When the Vis lowered to less than 50 m, the maxD would become greater than 15 μm but no more than 22 μm, so the extremely dense fog occurred, not needing the larger drops to be produced. However, the Vis that was lower than 50 m lasted no more than 20 min in the four periods of time; meanwhile, the burst reinforcement did not occur in these periods and droplet spectrum broadening only reached 22 μm. The explanation may be that the coagulation growth of the fog/cloud in the condition that the extremely dense fog must sustain should be more than 20 min. The bimodal or multi-modal drop-size distribution also occurred after the extremely dense fog lasted more than 30 min. FRANK et al. (1998) discussed the fog droplet formation and growth in polluted fogs; they thought that most of the time the fog consisted of inactive droplets, which were smaller than the critical diameter for activation according to the Köhler equation. So there was a critical maxD for the triggering of fog coagulation growth that the maxD must be greater than 22 μm and lasted more than 20 min.

Regardless of any type, the fog microphysics is related to physical processes, including radiative cooling, decrease of temperature, advection, turbulence, condensation, coagulation, and gravitational settling. In any fog case, parts of the physical processes are playing the leading role. Since the advection-radiaton fog had the similar fog droplet spectrum distribution with radiation fog, there may be analogous microphysical processes in both of them. However, different droplet spectrum distributions occurred among advection fogs and between radiation fogs and advection fogs, the important issue regarding the different physics between them needs to be further investigated. GOODMAN (1977) showed that the mean droplet diameter and LWC of advection fogs increased with height in all observed cases, broader drop size distributions were observed near the inversion interface, and the distribution also broadened throughout the entire layer during the onshore portion of the fog cycle, but without detailed physical processes to explain these phenomena.

Another explanation for advection fog droplet broadening is combustion-related pollutants, which can easily grow and form a dense fog without having the air attaining supersaturation (HUNG and LIAW, 1980). So fog microphysics are also related to chemical processes. By a modeling study, PANDIS et al. (1990) found that acidity and solute concentration in droplets larger than 10 μm in diameter increased with size, and the droplets of diameter 20 μm attained a solute concentration that was a factor of 3.6 larger than that in the 10 μm droplet. After that, the larger droplets could grow larger in a short time.

Fog microphysical characteristics for marine and continental fogs were separately parameterized by GULTEPE et al. (2009) to improve numerical weather predictions. By developing the microphysical parameterizations for operational model applications, interactions of microphysical, dynamical, radiative processes, and surface conditions (e.g., soil moisture and temperature) were simulated (GULTEPE and MILBRANDT, 2007). In the YRDFOG project, modeling studies (such as microphysical parameterizations) need to be done in future, which could lead to improve not only our understanding of fog processes (different fog droplet spectrum distributions and droplet spectrum broadening) at various spatial scales but also the fog forecast.

5. Conclusions

This paper provides an overview on the fog microphysical features of the YRDFOG project in Nanjing. The project's field observations were conducted at the meteorological observation station of the NUIST during the winters of 2006–2009, which was an effort to reveal FBL structures and its macro-/microphysical characteristics and to gain insights into the triggering, formation, maintenance, and dissipation mechanisms of fogs (including extremely dense fog). Detailed FBL observations were obtained using a fog measuring device (FMD; FM-100), a visibility meter, an automatic weather station, an acoustic Doppler radar, a tethersonde system, and an open path eddy covariance system.

The important conclusions we obtained are as follows:

1. There were large differences of microphysical parameters among radiation fog, precipitation fog, advection fog, and advection–radiation fog. Advection–radiation fog had the largest values in N, LWC, and AveD, which were 211.0 cm^{-3}, 121.34 mg/m^3, and 4.99 μm on average, respectively.

2. There were two types of radiation fog drop spectrum distributions. One of them had large spectrum width of more than 30 μm, with the second peak at 15 μm in number size distribution and three peaks at 5, 15, and 23 μm in mass size distribution. It is called the broad spectrum radiation fog. The other type had a spectrum width mostly less than 15 μm, therefore it is called the narrow spectrum radiation fog. The advection-related fogs all had wide spectrum greater than 45 μm, which was similar with the broad spectrum radiation fog. The precipitation fog had a narrow spectrum width, with the maximum diameter no more than 8 μm, narrower than that of the narrow spectrum radiation fog.

3. There were five radiation fogs, two advection fogs, and two advection–radiation fogs that had broad spectrum width of more than 30 μm. Rapid temperature decline or significant humidity increase (supersaturation increase), and turbulence played important roles in radiation fog droplet spectrum broadening.

4. Fog bimodal or multi-modal drop-size distributions occurred in both number and mass droplet

size distributions. Bimodal/multi-modal number droplet size distributions appeared only in radiation-related fogs.

5. The triggering of fog coagulation growth needed a critical maxD and the maxD must be greater than 22 μm and lasted more than 20 min.

Acknowledgments

Funding for this work was jointly provided by the National Natural Science Foundation of China (Grant No.40775012), the Natural Science Fund for Universities in Jiangsu province (Grant No. 08KJA170002), the Meteorology Fund of the Ministry of Science and Technology [Grant Nos. GYHY (QX) 2007-6-26 and GYHY200906012], Jiangsu province Qinglan Project of "Cloud fog precipitation and Aerosol Research," and the Graduate Student Innovation Plan for the Universities of Jiangsu province (Grant No. CX10B_292Z).

REFERENCES

BARONTI, P., and ELZWEIG, S. (1973), *A Study of Droplet Spectra in Fogs,* Journal of the Atmospheric Sciences *30*(5), 903-908.

BOREUX, J. J., and GUIOT J. (1993), *A method of comparison of two close batches data: Application to analysis of fog formation causes,* Geophys. Res. Lett. *20*(12), 1179-1182.

BOTT, A. (1990), *On the influence of the physico-chemical properties of aerosols on the life cycle of radiation fogs,* Boundary-Layer Meteorology *56*(1),1-31.

CHOULARTON, T. W., FULLARTON, G., LATHAM, J., MILL, C. S., SMITH, M. H., and STROMBERG, I. M. (1981), *A field study of radiation fog in meppen, West Germany,* Quarterly Journal of the Royal Meteorological Society *107*(452), 381-394.

DONALDSON, N.R.and STEWART, R. E., (1993), *Fog induced by mixed-phase precipitation,* Atmospheric Research *29*(1-2), 9-25.

ELDRIDGE, R. G. (1961), *A few fog drop-size distributions,* Journal of the Atmospheric Sciences *18*(5), 671-676.

ELDRIDGE, R. G. (1966), *Haze and fog aerosol distributions,* Journal of the Atmospheric Sciences *23*(5), 605-613.

ELDRIDGE, R. G. (1971), *The relationship between visibility and liquid water content in fog,* Journal of the Atmospheric Sciences *28*(7), 1183-1186.

ELIAS, T., HAEFFELIN, M., DROBINSKI, P., GOMES, L., RANGOGNIO, J., BERGOT, T., CHAZETTE, P., RAUT, J.C., and COLOMB, M. (2009), *Particulate contribution to extinction of visible radiation: Pollution, haze, and fog,* Atmospheric Research *92*(4), 443-454.

FRANK, G., MARTINSSON, B., CEDERFELT, S., BERG, O., SWIETLICKI, E., WENDISCH, M., YUSKIEWICZ, B., HEINTZENBERG, J., WIEDENSOHLER, A., ORSINI, D., STRATMANN, F., LAJ, P., and RICCI, L. (1998),

Droplet Formation and Growth in Polluted Fogs, Contributions to Atmosphoric Physics *71*: 65-85.

FUZZI, S., FACCHINI, M. C., ORSI,G., LIND, J. A., WOBROCK, W., KESSEL, M., MASER, R., JAESCHKE, W., ENDERLE, K. H., ARENDS, B. G., BERNER, A., SOLLY, I., KRUISZ, C., REISCHL, G., PAHL, S., KAMINSKI, U., WINKLER, P., OGREN, J. A., NOONE, K. J., HALLBERG, A., FIERLINGEROBERLINNINGER, H., PUXBAUM, H., MARZORATI, A., HANSSON, H. C., WIEDENSOHLER, A., SVENNINGSSON, I. B., MARTINSSON, B. G., SCHELL, D., and GEORGII, H. W. (1992), *The Po Valley fog experiment 1989,* Tellus B *44*(5), 448-468.

FUZZI, S., LAJ, P., RICCI, L., ORSI, G., HEINTZENBERG, J., WENDISCH, M., YUEKIEWICZ, B., MERTES, S., ORSINI, D., SCHWANZ, M., WIEDENSOHLER, A., STRATMANN, F., BERG, O.H., SWIETLICKI, E., FRANK, G., MARTINSSON, B.G., GUNTHER, A., DIERSSEN, J.P., SCHELL, D., JAESCHKE, W., BERNER, A., DUSEK, U., GALAMBOS, Z., KRUISZ, C., MESFIN, N.S., WOBROCK, W., ARENDS, B., and BRINK, H.T.(1998), *Overview of the Po Vally Fog Experiment 1994(CHEMDROP),* Contributions to Atmosphoric Physics *71*: 3-19.

GAO, J.,WANG T., ZHOU X.H., WU W.S., and WANG W.X., (2009), *Measurement of aerosol number size distributions in the Yangtze River delta in China: Formation and growth of particles under polluted conditions,* Atmospheric Environment *43*(4), 829-836.

GARCIA, G. F., VIRAFUENTES, U., and MARTINEZ, G. M. (2002), *Fine-scale measurements of fog-droplet concentrations: a preliminary assessment,* Atmospheric Research *64*(1-4), 179-189.

GERBER, H. E. (1981), *Microstructure of a Radiation Fog,* Journal of the Atmospheric Sciences *38*(2), 454-458.

GERBER, H. E. (1991), *Supersaturation and Droplet Spectral Evolution in Fog,* Journal of the Atmospheric Sciences *48*(24), 2569-2588.

GOODMAN, J. (1977), *The Microstructure of California Coastal Fog and Stratus,* Journal of Applied Meteorology *16*(10), 1056-1067.

GULTEPE, I., and MILBRANDT, J. A. (2007), *Microphysical Observations and Mesoscale Model Simulation of a Warm Fog Case during FRAM Project,* Pure and Applied Geophysics *164*(6), 1161-1178.

GULTEPE, I., TARDIF, R., MICHAELIDES, S. C., CERMAK, J., BOTT, A., BENDIX, J., MULLER, M. D., PAGOWSKI, M., HANSEN, B., ELLROD, G., JACOBS, W., TOTH, G., and COBER, S. G.. (2007), *Fog Research: A Review of Past Achievements and Future Perspectives,* Pure & Applied Geophysics *164*, 1121-1159.

GULTEPE, I., PEARSON, G., MILBRANDT, J. A., HANSEN, B., PLATNICK, S., TAYLOR, P., GORDON, M., OAKLEY, J. P., and COBER, S. (2009), *the Fog Remote Sensing and Modeling Field Project,* Bulletin of the American Meteorological Society *90*, 341-359.

GULTEPE, I., and MILBRANDT, J. A. (2010), *Probabilistic Parameterizations of Visibility Using Observations of Rain Precipitation Rate, Relative Humidity, and Visibility,* Journal of Applied Meteorology and Climatology *49*(1), 36-46.

HEINTZENBERG, J., WENDISCH, M., YUSKIEWICZ, B., ORSINI, D., WIEDENSOHLER, A., STRATMANN, F., FRAND, G., MARTINSSON, B.G., SCHELL, D., FUZZI, S., and ORSI, G. (1998), *Characteristics of Haze, Mist and Fog,* Contributions to Atmosphoric Physics *71*, 21-31.

HAEFFELIN, M., BERGOT, T., ELIAS, T., TARDIF, R., CARRER, D., CHAZETTE, P., COLOMB, M., DROBINSKI P., DUPONT, E., DUPONT, J. C., GOMES, L., MUSSON-GENON, L., PIETRAS, C., PLANA-FATTORI, A., PROTAT, A., RANGOGNIO, J., RAUT, J. C., REMY, S., RICHARD, D., SCIARE, J., and ZHANG, X. (2010), *PARISFOG: Shedding New*

Light on Fog Physical Processes, Bulletin of the American Meteorological Society *91*(6), 767-783.

HUDSON, J. G. (1980), *Relationship Between Fog Condensation Nuclei and Fog Microstructure,* Journal of the Atmospheric Sciences *37*(8), 1854–1867.

HUNG, R. J. and LIAW G. S. (1980), *Advection fog formation associated with atmospheric aerosols due to combustion-related pollutants,* Water, Air, & Soil Pollution V*14*(1), 267-285.

KUROIWA, D. (1951), *Electron-M1croscope Study of Fog Nuclei,* Journal of the Atmospheric Sciences *8*(3), 157-160.

LAJ, P., FUZZI, S., LAZZARI, A., RICCI, L., ORSI, G., BERNER, A., DUSEK, U., SCHELL, D., GUNTHER, A., WENDISCH, M., WOBROCK, W., FRANK, G., MARTINSSON, B.G., and HILLAMO, R. (1998), *The Size Dependent Composition of Fog Droplets,* Contributions to Atmosphoric Physics *71*: 115-130.

LI, Z.H., LIU, D.Y., and YANG, J. (2011), *The microphysical processes and macroscopic conditions of the radiation fog droplet spectrum broadening,* Chinese Journal of Atmospheric Sciences (In Chinese) *35(1),* in press.

LIU, D.Y, PU, M.J, YANG, J., ZHANG, G.Z., YAN, W.L., and LI, Z.H. (2010), *Microphysical Structure and Evolution of a Four-Day Persistent Fog Event in Nanjing in December 2006,* Acta Meteorologica Sinica *24,* 104-115.

LIU D.Y., YANG, J., NIU, S.J., and LI, Z.H. (2011a), *On the Evolution and Structures of a Radiation Fog Event in Nanjing,* Adv. Atmos. Sci., *28*(1), 1-15, doi:10.1007/s00376-010-0017-0.

LIU,D.Y., NIU, S.J., PU, M.J., YANG, J., and LI, Z.H. (2011b), *On the physical process of advection fog influenced by cold & warm advection,* Chinese Journal of Geophysics, *in press* (In Chinese).

LIU D.Y., NIU S.J., YANG J., ZHAO L.J., LÜ J.J., and LU C.S. (2011c) *A Summary of Four-Year Fog Field Study in Northern Nanjing-Part 1: Fog Boundary Layer,* Pure and Applied Geophysics, in submitted.

LIU, D.Y., (2008), *The Micro-physical Characteristic of Winter's Fog in Nanjing Area, Master Thesis, Nanjing University of Information Science &Technology, China.*

LU, C.S., NIU, S.J., YANG, J., and WANG, W.W. (2008), *An observational study on physical mechanism and boundary layer structure of winter advection fog in Nanjing,* Journal of Nanjing Institute of Meteorology *31,* 520-529.(In Chinese).

LU, C.S., NIU, S.J., YANG, J., LIU, X., and ZHAO, L.J. (2010a), *Jump features and causes of macro and microphysical structures fog a winter fog in Nanjing,* Chinese Journal of Atomospheric Sciences *34,* 681-690. (In Chinese).

LU, C.S. NIU, S.J., TANG, L.L., LV, J.J., ZHAO, L.J., and ZHU, B. (2010b). *Chemical composition of fog water in Nanjing area of China and its related fog microphysics.* Atmospheric Research *97*(1-2): 47-69.

MEYER, M. B., JIUSTO, J. E., and LALA, G.G. (1980), Measurements of Visual Range and Radiation-Fog (Haze) Microphysics, Journal of the Atmospheric Sciences *37*(3), 622-629.

MING, Y., and RUSSELL, L. M. (2004), *Organic aerosol effects on fog droplet spectra,* J. Geophys. Res., *109,* D10206, doi: 10.1029/2003JD004427.

NIU, S.J., LU, C.S., YU, H.Y., ZHAO, L.J., and ZHAO, L.J. (2010a), *Fog research in China: an overview,* Adv. Atmos. Sci. *27*(3),639-662.

NIU, S.J., LU, C.S., ZHAO, L.J., LU, J.J, and YANG, J. (2010b), *Analysis of the microphysical structure of heavy fog using a droplet spectrometer: a case study,* Adv. Atmos. Sci. *27*(6), 1259-1275, doi:10.1007/s00376-010-8192-6.

OGREN, J. A., NOONE, K. J., HALLBERG, A., HEINTZENBERG, J., SCHELL, D., BERNER, A., SOLLY, I., KRUISZ, C., REISCHL, G., ARENDS, B.G., and WOBROCK, W. (1992), *Measurements of the size dependence of the concentration of nonvolatile material in fog droplets,* Tellus B *44*(5), 570-580.

PANDIS, S. N., SEINFELD, J. H., and PILINIS, C. (1990),*Chemical composition differences in fog and cloud droplets of different sizes,* Atmospheric Environment. Part A. General Topics *24*(7),1957-1969.

PICKERING, K. E. and JIUSTO J. E. (1978), *Observations of the Relationship Between Dew and Radiation Fog,* J. Geophys. Res. *83,* 2430-2436.

PILIE, R. J., MACK, E. J., EADIE, W.J., and ROGERS, C.W. (1975), *The Life Cycle of Valley Fog. Part II: Fog Microphysics,* Journal of Applied Meteorology *14*(3), 364–374.

PINNICK, R. G., HOIHJELLE, D. L., FERNANDEZ, G., STENMARK, E.B., LINDBERG, J.D., and HOIDALE G.B. (1978), *Vertical Structure in Atmospheric Fog and Haze and Its Effects on Visible and Infrared Extinction,* Journal of the Atmospheric Sciences *35*(10), 2020-2032.

PODZIMEK, J. (1997), *Droplet Concentration and Size Distribution in Haze and Fog,* Studia Geoph. et Geod. *41,* 277-296.

PU, M.J., YAN, W.L., SHANG, Z.T., YANG, J., and LI, Z.H. (2008a), *Study on the physical characteristics of burst reinforcement during the winter fog of Nanjing,* Plateau Meteorology *27,* 1-8. (In Chinese).

PU, M.J., ZHANG, G.Z., YAN, W.L., and LI, Z.H. (2008b), *Features of a rare advection-radiation fog event,* Science in China Series D: Earth Sciences *51,* 1044-1052.

RICCI, L., FUZZI, S., LAJ, P., LAZZARI, A., ORSI, G., BERNER, A., GUNTHER, A., JAESCHKE, W., WENDISCH, M., and ARENDS, B.G. (1998), *Gas-Liquid Equilibria in Polluted Fog,* Contributions to Atmosphoric Physics *71:* 159-170.

ROACH, W. T. (1976a), *On some quasi-periodic oscillations observed during a field investigation of radiation fog,* Quarterly Journal of the Royal Meteorological Society *102*(432), 355-359.

ROACH, W. T. (1976b), *On the effect of radiative exchange on the growth by condensation of a cloud or fog droplet,* Quarterly Journal of the Royal Meteorological Society *102*(432), 361-372.

SCHUMANN, T. E. W. (1940), *Theoretical aspects of the size distribution of fog particles,* Quarterly Journal of the Royal Meteorological Society *66*(285), 195-208.

SPENCER, W. P., JOHNSON, R. A., and VIETTI, M.A. (1976), *On fog formation in a coronal discharge: effects of the discharge on droplet growth,* Journal of Aerosol Science *7*(6), 441-445.

TARDIF, R. and RASMUSSEN, R. M. (2008), *Process-Oriented Analysis of Environmental Conditions Associated with Precipitation Fog Events in the New York City Region,* Journal of Applied Meteorology & Climatology *47*(6), 1681-1703.

TARDIF, R. and RASMUSSEN, R. M. (2009), *Evaporation of non-equilibrium raindrops as a fog formation mechanism,* Journal of the Atmospheric Sciences *49*(6), 1247-1267.

WENDISCH, M., MERTES, S., HEINTZENBERG, J., WIEDENSOHLER, A., SCHELL, D., WOBROCK, W., FRANK, G., MARTINSSON, B.G., FUZZI, S., ORSI, G., KOS, G., and BERNER, A. (1998), *Drop Size Distribution and LWC in Po Vally Fog,* Contributions to Atmosphoric Physics *71:* 87-100.

WOODCOCK, A. H. (1978), *Marine Fog Droplets and Salt Nuclei-Part I,* Journal of the Atmospheric Sciences *35*(4), 657-664.

YAN, W. L., LIU, D.Y., PU, M.J., and LI, Z.H. (2010), *The Formation and Structure Characteristics of Precipitation Fog in Nanjing,* Meteorological Monthly, *in press.* (in Chinese).

YAN, W.L., PU, M.J., WANG, W.W., YANG, J., and LIU, D.Y. (2009), *A study on a rare radiation-advection fog (I): the analysis of physical process of genesis and dissipation,* Scientia Meteorologica Sinica *29,* 9-16. (In Chinese).

YANG, J., NIU, Z.Q., SHI, C.E., LIU, D.Y., LI, Z.H., (2010), *Microphysics of atmospheric aerosols during winter haze/fog events in Nanjing,* Environmental Science *31*(7),17-23.

ZHOU B.B. and FERRIER B.S. (2008), *Asymptotic Analysis of Equilibrium in Radiation Fog,* Journal of Applied Meteorology & Climatology, *47*(6), 1704-1722.

(Received November 14, 2010, revised April 25, 2011, accepted May 7, 2011, Published online June 5, 2011)

Pure Appl. Geophys. 169 (2012), 1157–1163
© 2011 Springer Basel AG
DOI 10.1007/s00024-011-0366-3

The Continued Reduction in Dense Fog in the Southern California Region: Possible Causes

S. LaDochy[1] and M. Witiw[2]

Abstract—Dense fog appears to be decreasing in many parts of the world, especially in western cities. Dense fog (visibility <400 m) is disappearing in the urban southern California area also. There the decrease in dense fog events can be explained mainly by declining particulate levels, Pacific sea surface temperatures (SST), and increased urban warming. Using hourly data from 1948 to the present, we looked at the relationship between fog events in the region and contributing factors and trends over time. Initially a strong relationship was suggested between the occurrence of dense fog and the phases of an atmosphere–ocean cycle: the Pacific Decadal Oscillation (PDO). However, closer analysis revealed the importance to fog variability of an increasing urban heat island and the amount of atmospheric suspended particulate matter. Results show a substantial decrease in the occurrence of very low visibilities (<400 m) at the two airport stations in close proximity to the Pacific Ocean, LAX (Los Angeles International) and LGB (Long Beach International). A downward trend in particulate concentrations, coupled with an upward trend in urban temperatures were associated with the decrease in dense fog occurrence at both LAX and LGB. LAX dense fog that reached over 300 h in 1950 dropped steadily, with 0 h recorded in 1997. Since 1997, there has been a slight recovery with both 2008 and 2009 recording over 30 h of dense fog at both locations. In this study we examine whether the upturn is a temporary reversal of the trend. To remove the urban effect, we also included fog data from Vandenberg Air Force Base (VBG), located in a relatively sparsely populated area approximately 200 km to the north of metropolitan Los Angeles. Particulates, urban heat island, and Pacific SSTs all seem to be contributing factors to the decrease in fog in southern California, along with large-scale atmosphere–ocean interaction cycles. Case studies of local and regional dense fog in southern California point to the importance of strong, low inversions and to a lesser contributor, Santa Ana winds. Both are associated with large-scale atmospheric circulation patterns, which have changed markedly over the period of studied. These changes point to continued decreases in dense fog in the region.

Key words: Coastal fog, fog climatology, fog trends, PDO, climate change.

[1] California State University, Los Angeles, CA 90032, USA. E-mail: sladoch@calstatela.edu
[2] Embry-Riddle Aeronautical University Worldwide, Everett, WA 98203, USA. E-mail: witiw170@erau.edu

1. Introduction

The California coast is one of the foggiest regions in North America (Leipper, 1994). During the warm season, the semi-permanent Pacific high pressure remains off the California coast. The clockwise circulation around this high results in the California Current and upwelling of cold water close to the coast. Although the upwelling is strongest farther north, where summer water temperatures usually remain around 11 to 12°C, temperatures remain relatively cold at latitudes well to the south including the Los Angeles area. Moist, relatively warm air moving over the cold water is chilled to its dew point, resulting in the formation of sea fog. This sea fog typically rises as it moves inland forming a low stratus deck. The density and horizontal coverage of fog and low clouds are negatively correlated with the sea surface temperature (SST) (Norris and Leovy, 1994). Usually, only the immediate coastal regions experience dense fog conditions. Even there, because of the generally large fog droplet size (resulting from the large condensation nuclei available over the ocean) very low visual ranges are rare. The main type of fog in the cool season is advection–radiation fog (Byers, 1959). Typically fog-free marine air moves inland, and frequently after the wind changes direction to an offshore component, advection–radiation fog forms. Because of the smaller fog nuclei available in urban coastal California, this type of fog tends to produce lower visibility than advection fog (sea fog) (Leipper, 1994). October through February are characterized by a relatively high frequency of low visibility (<400 m) at both coastal Los Angeles (LAX) and Long Beach (LGB) International Airports (Baars et al., 2002). This is also the season of lower inversion heights which are necessary for dense fog formation (LaDOCHY and BEHRENS, 1991). However, dense fog events are decreasing along the southern California coast (WITIW and LaDOCHY 2008). Evidence of disappearing dense fog is coming from many parts of the world (WITIW et al., 2003; VAUTARD et al., 2009), even the Arctic (YE 2009), although reports from rapidly developing countries, for example China and India, indicate increases. Often global warming and or air pollution reduction are cited as the cause of fog reductions, whereas increased air pollution often leads to increased fog (WANG et al., 2001; NIU et al. 2010).

In this study, we examine the relationships that control the frequency of dense fog in coastal southern California. Particulates, the urban heat island, and local SSTs are all contributing factors there (WITIW and LaDOCHY 2008). Case studies of local and regional dense fog in southern California point to the importance of strong, low inversions and to a lesser contributor, Santa Ana winds. Both features are controlled by large-scale atmospheric circulation patterns, which in turn may be affected by warming.

2. Data and Methodology

Visibility and significant weather is recorded hourly at the three airport stations in the coastal plains of the Los Angeles Basin, LAX, LGB, and BUR (Burbank International Airport). Dense fog is recorded when visibility is less than ¼ mile (<400 m). Hourly visibilities for both LAX and LGB from 1948 to 2009 were available from the National Climate Data Center, NCDC. From the mid-1960s through June 1999, downtown Los Angeles (CBD) had automatic weather observations without visibility observations, resulting in visibility data being available only for the complete years 1961–1964 and 2000–2004. Burbank Airport had hourly data from 1982 through 2008, with missing data for the first half of 1998. Therefore, 1998 data were omitted. Monthly mean, max and min temperatures for LAX and downtown Los Angeles were also obtained from NCDC for the period of the study. For comparison, hourly dense fog data were collected for Vandenberg Air Force Base (VBG), a coastal site far removed (about 200 km north of Los Angeles) from urban

effects. Complete years included the periods 1958–1960, 1967–1973, and 2006–2008 only.

Monthly and annual particulate air pollution data were recorded as total suspended particulates (TSP) from 1966 to 2008 from the South Coast Air Quality Management District (SCAQMD). Monthly Pacific climatic indices, Pacific Decadal Oscillation (PDO), and Southern Oscillation Index (SOI) were downloaded from the MANTUA (2010) and NOAA (2010) websites, respectively, for the 1948–2009 period. Santa Monica pier sea surface temperatures (SST) were obtained from NOAA National Oceanographic Data Center (NOAA 2010).

Simple linear regressions and Pearson correlations were calculated between annual totals of hourly and daily dense fog occurrences at LAX and LGB and contributing variables—annual PDO and SOI values, downtown Los Angeles annual TSP amounts, mean monthly and annual temperatures at LAX and downtown Los Angeles, and annual sea surface temperatures (SST) at Santa Monica Pier. Multivariate analyses were performed to see which factors were most important in explaining fog variability and how strong the relationships were. Fog frequencies at LAX and LGB were also examined for trends. For comparison, fog frequencies were also examined for BUR and downtown Los Angeles, and for the rural Vandenberg station, VBG (Fig. 1).

As there was a clear decreasing trend in the fog data, both dense fog hours and dense fog days datasets were detrended using a linear regression equation. Pearson correlations were then calculated between annual hourly and daily fog occurrences at LAX and LGB and the above variables of annual TSP, temperatures, SST, SOI, and PDO values to show how well these variables explained interannual fog variability.

3. Results

Table 1 shows the average annual number of hours visibility was less than 400 m at various Los Angeles Basin locations over the study period, 1948–2008. Data for downtown Los Angeles (CBD) were only available for the two periods noted, although the station has been moved since 1999

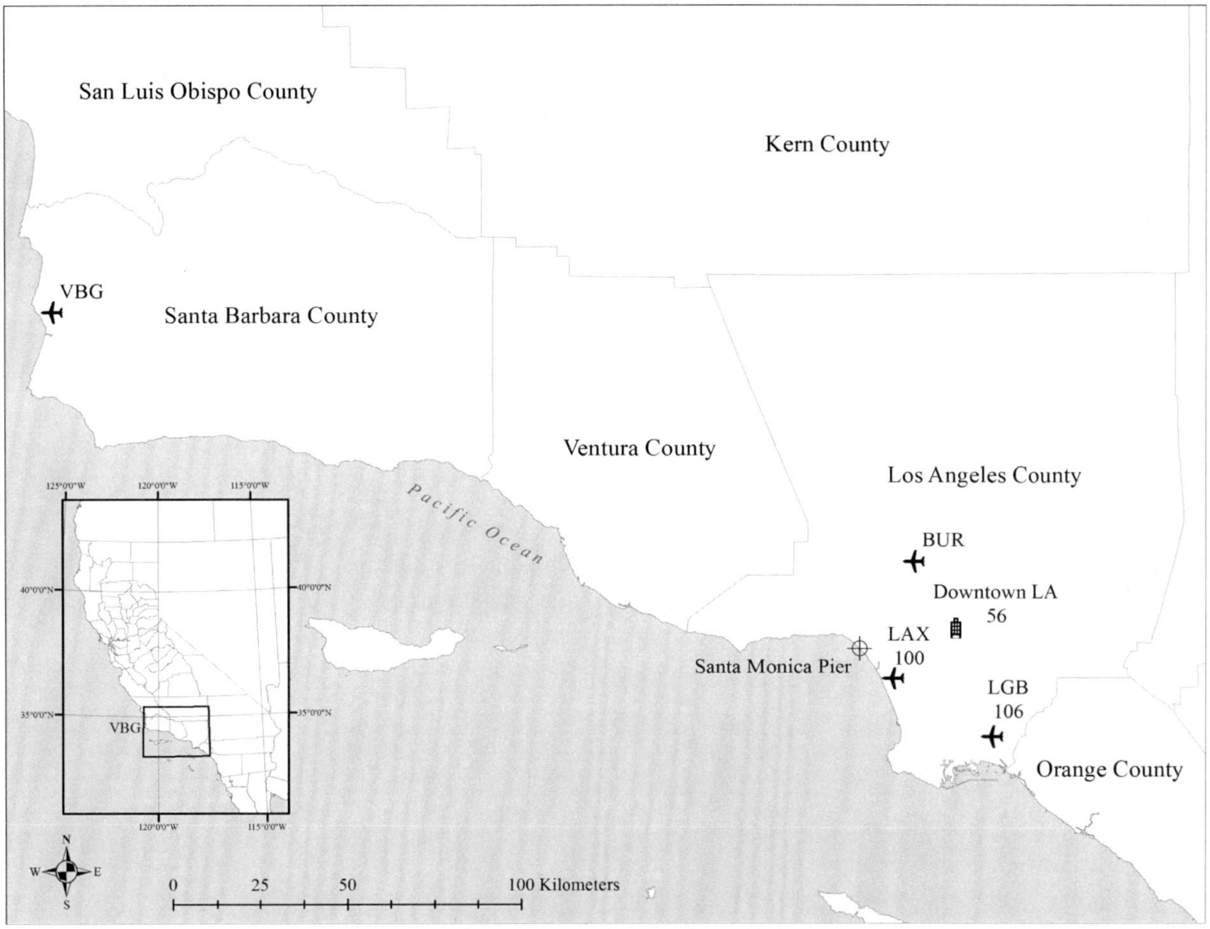

Figure 1

Study area-southern California locations, including annual average number of fog days at selected sites. *Inset* shows study area's location in California. VBG is Vandenberg AFB

Table 1

Average annual hours visibility <400 m (excludes 1998 for BUR)

LAX 1948–2009	73
LGB 1948–2009	107
LAX 1966–2009	44
LGB 1966–2009	59
BUR 1982–2006	7
CBD 1961–1964	10
CBD 2000–2006	2

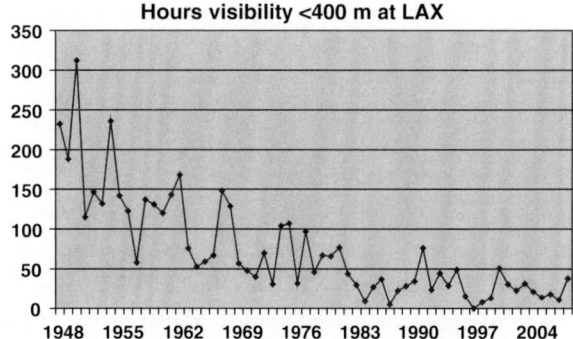

Figure 2

Dense fog trend at Los Angeles International Airport, 1948–2008

closer to the coast. Data for Burbank were not available for the first part of 1998 and therefore 1998 were omitted.

There have been significant decreases in dense fog at all locations studied in the Los Angeles Basin. Most notable are the two coastal airports, LAX and LGB (Figs. 2, 3). Dense fog disappeared completely in 1997 at both stations, but rebounded somewhat in more recent years.

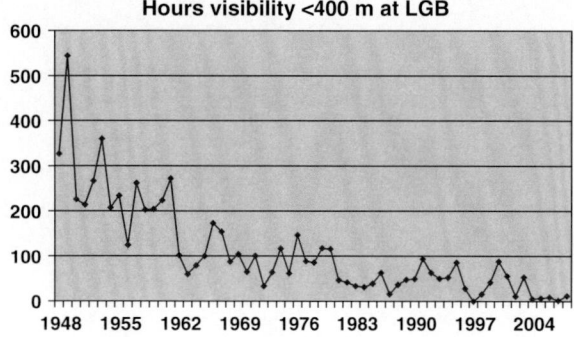

Hours visibility <400 m at LGB

Figure 3
Dense fog trend at Long Beach International Airport, 1948–2008

Table 2

Regression results from use of TSP and PDO to explain the variance of dense fog (visibility <400 m) in 1966–2009

Location	Variable (s)	Variance (R^2)	Significance
LAX	TSP	0.372	$p < 0.001$
	PDO	0.118	$p < 0.05$
	TSP and PDO	0.393	$p < 0.001$
	PDO incremental R^2	0.021	$p > 0.05$
LGB	TSP	0.386	$p < 0.001$
	PDO	0.049	$p > 0.05$
	TSP and PDO	0.386	$p < 0.001$
	PDO incremental R^2	0.000	$p > 0.05$

3.1. Air–sea Interactions and Their Effect on Fog Frequency

Previous studies found that coastal sea surface temperatures and Pacific climatic indices, for example the El Niño-Southern Oscillation (ENSO) and the Pacific Decadal Oscillation (PDO) affect southern California weather and climate, including dense fog frequency. Whereas ENSO, as quantified by the Southern Oscillation Index (SOI), was only weakly correlated with dense fog, annual PDO values explained more than 34% of the variance seen in the amount of dense fog ($p < 0.001$) at LAX and 18% of the variance at LGB ($p < 0.05$) (WITIW and LaDOCHY, 2008). The fact that the study period overlaps with mostly one cycle of the PDO from 1948 to 1997 is not trivial, although more cycles would make the analyses more rigorous. As noted by MANTUA (2010), the PDO shifted from a cool phase to a warm phase in approximately 1977. During cool phases of PDO there are more numerous La Niña episodes and fewer El Niños, whereas the opposite occurs during the warm phase. The frequency of dense fog decreased during El Niño events (LaDOCHY 2005) and disappeared entirely in southern California during the 1997–1998 major El Niño event. In the same study, annual sea surface temperatures recorded along the coastline at Santa Monica Pier were found to be highly significant when explaining the frequencies of dense fog at LAX ($p < 0.001$) and at LGB ($p < 0.001$) for the years 1950–2001. The same is true for the 1950–2008 data (not shown).

3.2. Effect of Particulate Air Pollution

Annual TSP data for downtown Los Angeles were available from 1966 to 2008 (SCAQMD 2009). The trend in TSP indicates a significant decrease throughout the period ($p < 0.001$), dropping to less than half the values of the 1960s by the end of the century. The last decade, 1999–2008, was the cleanest of the record. WITIW and LaDOCHY (2008) found TSP to be highly significant in explaining hours of dense fog variability with $R^2 = 0.343$ for LAX and 0.314 for LGB. Extending the data from 1966 to 2009, TSP was highly significant when explaining dense fog ($p < 0.001$), with $R^2 = 0.372$ for LAX and 0.386 for LGB (Table 2). As TSP levels continue to fall, fewer condensation nuclei are available for fog droplet formation. Using multiple regression analysis, the combined effect of TSP and PDO index increased the variance in fog hours explained to 39.3% at LAX but did not change variance at LGB for the years 1966–2009 (Table 2). Combining local SST values and TSP increased the amount of variance in fog explained at LAX to 40% (higher than with PDO and TSP) while explaining slightly less variance in fog at LGB (Table 3). Particulate levels have continued to decrease, although not as much as in previous decades.

3.3. Effect of Urban Heat Island

The temperature trend for downtown Los Angeles shows marked warming over the study period, more so for T_{min} than T_{max}. Comparing downtown temperatures with annual days of dense fog, LaDOCHY (2005) found correlations between annual average

Table 3

Regression results from use of TSP and SST to explain the variance of dense fog, 1966–2009

Location	Variable (s)	Variance (R^2)	Significance
LAX	TSP	0.372	$p < 0.001$
	SST	0.181	$p < 0.01$
	TSP and SST	0.398	$p < 0.001$
	SST incremental R^2	0.026	$p > 0.05$
LGB	TSP	0.386	$p < 0.001$
	SST	0.091	$p < 0.05$
	TSP and SST	0.386	$p < 0.001$
	SST incremental R^2	0.000	$p > 0.05$

temperatures and coastal dense fog days highly significant. As temperatures increased, dense fog frequency decreased. Pacific Decadal Oscillation values and coastal SSTs also correlated highly with Los Angeles temperatures, and all three variables correlated inversely with TSP. With urban warming, the mixing layer tends to increase, reducing particulate concentrations and minimizing conditions for fog formation. Similarly, with positive PDO values, and higher SSTs along the west coast, the winters tend to be warmer and wetter. TSP values are much lower and better mixing disrupts formation of fog.

In order to test whether factors besides urban heating and particulates are affecting decreasing fog, we also looked at dense fog frequency at Vandenberg Air Force Base, approximately 210 km northwest of downtown Los Angeles near the central California coast. While Vandenberg is far from urban effects, it has a different fog regime, with most dense fog coming in the summer months, whereas the Los Angeles area has a winter maximum. Preliminary data show some slight decreasing tendencies, but not to the same extent as the more drastic declines near Los Angeles. This small decline may, however, be an artifact of how visibilities are calculated using automated sensors versus humans, and should be examined more thoroughly. Automated sensors calculate a mean visibility, whereas human observers determine visibility by identifying objects that are known fixed distances from the observing site. Data were incomplete from the mid-1970s to the mid-1990s when automated observing began. Only 2006 had complete hourly data, showing hourly values similar to the 1960s, but lower in terms of dense fog

days (Table 4). Further north along the California coast, Johnstone and Dawson (2010) found that fog supporting redwoods declined 0.9% per decade over the 1951–2009 period. They based this trend on hourly fog data from central and northern California stations, Monterey (36.6°N) and Arcata (40.9°N), respectively. These areas have not seen urbanization similar to that found in southern California and could be more indicative of background coastal conditions. The warming trend in coastal SSTs may be leading to these slight decreases.

3.4. Detrended Fog Data Analysis

Because fog frequencies have declined at LAX and LGB steadily throughout the study period, as particulate matter has also declined and while urban temperatures and offshore SST values have steadily increased, we removed the trend from both hourly and daily fog frequencies to see how variables would explain the interannual variability at the two airports. We assumed a linear trend for all fog datasets. Relationships between the fog frequencies and tested variables proved to be weaker than when using original annual fog values. However, both Pacific climatic indices, PDO and SOI, had modest strength in explaining detrended fog variability, whereas TSP was poorly related to the detrended fog values. For the 1950–2006 detrended daily fog frequencies, PDO explained 11.3% of fog variability at LAX and 17.6% at LGB. SOI was slightly better with 15.7% explained at LAX and 20.8% at LGB. While these values are not strong, they do point in the correct direction in that fog frequencies (using detrended data) increase during the cool phase (negative) of the PDO and the cool (La Niña) phase of the SOI (Table 5).

Table 4

Average annual hours visibility <800 m (1/2 mile) at Vandenberg Air Force Base

	Hours <0.5 mile	Days <0.5 mile
VBG 1958–1960	660	107
VBG 1967–1969	796	124
VBG Dec 2007–Nov 2008	475	108

December 2007–November 2008—observations were automated. Previous years were manual

Table 5

Regression results for the occurrence of annual dense fog (visibility <400 m) days in 1950–2006 using detrended fog data

Location	Variable (s)	Variance (R^2)	Significance
LAX	SOI	0.157	$p < 0.01$
	PDO	0.113	$p < 0.05$
	TSP	0.010	$p > 0.10$
	SST	0.042	$p < 0.10$
	T	0.040	$p < 0.10$
LGB	SOI	0.208	$p < 0.01$
	PDO	0.176	$p < 0.01$
	TSP	0.033	$p < 0.10$
	SST	0.118	$p < 0.05$
	T	0.079	$p < 0.10$

SST are annual sea surface temperatures at Scripps' pier, *T* are mean annual temperatures in downtown Los Angeles

4. Discussion

In this study we continue to explore the relationships between atmospheric and oceanic variables and the trend in decreasing Los Angeles area's dense fog frequency. While urban warming and decreasing pollution are definite contributors, the data suggest that there are also effects (albeit weaker) from the Pacific. PDO values, which may be shifting back into the cool (negative) phase, may nudge fog frequency higher in the coming years. However, dense fog is associated with strong, low inversions over coastal California (LEIPPER, 1994). Recent trends show that inversion strength has decreased over the last few decades (1960–2007) and inversion frequencies vary, partly in association with SSTs, along the California coast (IACOBELLIS *et al.* 2009). This could be one of the mechanisms linking rising SSTs with decreasing dense fog.

Another factor that may contribute to dense fog events in southern California is Santa Ana winds. HUGHES *et al.*, (2009) found that Santa Ana events declined by more than 30% over the period 1959–2001. They believe that differential warming, that is more rapid warming inland than over the ocean, would make conditions less favorable for these offshore wind events. Santa Ana events are often followed by dense fog, such as on 7 Feb. 2007, when several stations in southern California recorded visibility at 0.0 km. This followed a strong Santa Ana event. Santa Anas provide large amounts of particulates carried from inland out over coastal waters. When winds finally shift to sea breezes, a shallow marine layer laden with dust may form a very dense fog (LEIPPER 1994). Although not significant for all months, there tends to be more Santa Ana events during La Niña winters than El Niño winters (FINLEY and RAPHAEL 2007). During the 1950–2000 period, on which many climate studies are conducted, the PDO shifted from cool to warm phase, with increasing El Niño winters. This would favor fewer Santa Anas, warm, wetter conditions, and higher inversion layers, all leading to less favorable fog conditions. Recently, the PDO has shifted back to the negative phase, with two short switches to positive in the last 10 years. Although dense fog events have risen slightly, there is no discernible return to the higher frequencies of earlier decades.

Several trends in climatic variables, many driven by climate change, point to decreasing dense fog frequency in the Los Angeles region. If the same trend is occurring along the entire west coast, this may become another important positive feedback enhancing global warming. We recommend thorough investigation of the trends in the eastern North Pacific stratus, because possible diminishing low clouds play a similar role as diminishing Arctic sea ice as a positive feedback mechanism in the global radiational balance.

Acknowledgments

The authors are grateful to Dr Richard Medina for his graphics help.

REFERENCES

BAARS, J.A., WITIW, M.R., AL-HABASH, A., and RAMAPRASAD, J.(2002), Determining fog type in the Los Angeles basin using historic surface observation data. Proceedings of the 16th Conf. on Probability and Statistics in the Atmospheric Sciences, Orlando, FL, American Meteorological Society, Boston, pp. J104–107.

BYERS, H.R., General Meteorology, 3rd ed. (McGraw-Hill, New York 1959).

FINLEY, J. AND RAPHAEL, M. (2007), *The relationship between El Niño and the duration and frequency of the Santa Ana winds of southern California.* The Professional Geographer 59:2, 184–192.

HUGHES, M., HALL, A., AND KIM, J. (2009). Anthropogenic Reduction of Santa Ana Winds. CEC Publication # CEC-500-2009-015f. Available at: http://www.energy.ca.gov/2009publications/CEC-500-2009-015-F.PDF.

IACOBELLIS, S.F., NORRIS, J.R., KANAMITSU, M., TYREE, M., CAYAN, D.C. (2009), Climate Variability and California Low Level Temperature Inversions. CEC Publication # CEC-500-2009-020-F. Available at: http://www.energy.ca.gov/2009publications/CEC-500-2009-020-f.pdf.

JOHNSTONE, J.A., AND DAWSON, T.E. (2010), *Climatic context and ecological implications of summer fog decline in the coast redwood region*, PNAS Online access: http://www.pnas.org/content/early/2010/02/09/0915062107.full.pdf+html.

LADOCHY, S. (2005), *The disappearance of dense fog in Los Angeles: Another urban impact?* Physical Geography 26, 177–191.

LADOCHY, S., BEHRENS, D. (1991), Particulate air pollution patterns over metropolitan Los Angles In Air Pollution: Environmental Issues and Health Effects (eds. Majumdar, S.K., Miller, and E.W., Cahir, J.C.) (Pennsylvania Academy of Sciences) pp. 444-459.

LEIPPER, D.F. (1994), *Fog on the U.S. west coast: a review.* Bull. Amer. Meteorol. Soc. 75, 1794–1796.

MANTUA, N.J. (2010), PDO homepage. Available at: http://www.atmos.washington.edu/mantua/pdo.html.

NOAA (2010), ENSO homepage. Available at: http://www.enso.noaa.gov.

National Oceanographic Data Center, NOAA (2010). Santa Monica Pier data available at: http://tidesandcurrents.noaa.gov/station_info.shtml?stn=9410840+Santa+Monica,+CA.

NIU, F. AND LI, Z. AND LI, C. AND LEE, K.-H. AND WANG, M. (2010), *Increase of wintertime fog in {China}: Potential impacts of weakening of the {Eastern Asian} monsoon circulation and increasing aerosol loading.* J. Geophys. Res. *115*, doi: 10.1029/2009JD013484.

NORRIS, J.R., LEOVEY, C.B. (1994), *Interannual variability in stratiform cloudiness and sea surface temperatures.* J. Climate 7, 1915–1925, 1994.

SCAQMD (2009), Air Quality Annual Summaries, 1979–2008. Southern California Air Quality Management District, Diamond Bar, California, 1980–2009.

VAUTARD, R., P. YIOU AND G.J. VAN OLDENBORGH (2009), *The decline of fog, mist and haze in Europe during the last 30 years: a warming amplifier?* Nature Geoscience 2, 115–119, doi: 10.1038/NGEO414.

WANG, Q., B. DENG, H. XU,, X. ZHOU, J. LI, AND Q. ZHANG (2001), The experimental study on Beijing urban fog and its effect on environment. In Proceedings of the 2nd Int. Conf. on Fog and Fog Collection, St. John's, Canada, (eds. Schemenaur, R.S., H. Puxbaum), IDRC, Ottawa, pp. 53-56.

WITIW, M.R., LADOCHY, S. (2008), *Trends in fog frequencies in the Los Angeles Basin.* Atmospheric Research 87, 293–300.

WITIW, M. R., BAARS, J., FISCHER, K. 2003. Urban influences on visibility. In: Proceedings of the 5th International Conference on Urban Climate, Lodz, (Poland). (eds. Klysik, K., Oke, T.R., Fortuniak, K., Grimmond, C.S.B., and Wibig, J.) (University of Lodz) pp. 279-282. Available at http://www.geo.uni.lodz.pl/~icuc5/text/O_32_6.pdf.

YE, H. 2009. *The influence of air temperature and atmospheric circulation on winter fog frequency over Northern Eurasia.* Int. J. of Climatology 29, 729–734.

(Received November 12, 2010, revised June 4, 2011, accepted June 9, 2011, Published online July 22, 2011)

Pure Appl. Geophys. 169 (2012), 1165–1172
© 2011 Springer Basel AG
DOI 10.1007/s00024-011-0326-y

Analyzing the Temporal and Spatial Variation of Fog Days in Iran

Mohammad Rahimi[1]

Abstract—In order to study the temporal and spatial variation of fog days in Iran, the data of 115 synoptic meteorological stations have been analyzed for years 1960–2005. The results revealed that different types of fogs form all over the country, apart from central areas of Iran that are located in the big dessert of Iran. Advection fogs are common in the south coast (Persian Gulf) and north coastal (Caspian Sea) regions. Upslope fogs form in the mountainous areas of the northwest and north parts of Iran. This study shows no height dependence relationship on fog days for all types of fogs in overall. The trend analysis of fog days during the last 20 years shows some significant negative and positive trends. The frequency of advection fogs shows positive trends and most upslope fogs show negative trends. The results show that there are suitable places for fog collection projects in the north and south coastal regions during the year, especially in cold months.

Key words: Fog, Fog days, Fog collection, Persian Gulf, Iran.

1. Introduction

There is an ever-growing need to identify new sources and methods for collecting water in both the developed and developing countries, specially in arid and semi-arid regions, and it is clear that serious consideration needs to be given to unconventional water supplies (Acosta Baladon, 1995). One of these methods is fog collection. A broader look at the meteorological and oceanographic condition on a worldwide basis, as well as the topography, leads us to the conclusion that many countries may have the potential to benefit from fog collection (Schemenauer and Cereceda, 1997). One of the largest projects has provided, since March 1992, an average of 11,000 L of water per day to a village of 330 people in the arid coast desert of North Chile (Al-Fenadi, 2001).

Fog water deposition is not only relevant for single plants, but is also contributing to the hydrological budget of an area and can contribute to the recharge of ground water (Werner Eugster, 2008).

Water can be collected from fogs under favorable climatic conditions. Fog is defined as a mass of water vapor condensed into small water droplets at, or just above, the Earth's surface. The small water droplets present in the fog precipitate when they come in contact with objects (Schemenauer and Cereceda, 1991). Fog collection rates are typically $1–10$ L m^{-2} of vertical collecting surface per day but can reach values of $30–40$ L m^{-2}/day (Schemenauer and Cereceda, 1994). The large fog-water collectors consist of a double layer of mesh made from a 1 mm wide flat polypropylene ribbon. The theoretical collection efficiencies of a 1 mm wide ribbon, for droplets with the observed median volume diameter (MVD), at wind speeds from 2 to 8 m s^{-1}, are 75–95%. Studies show that the mesh shows a marked decrease in droplet collection as the ribbon width increases, while maintaining a constant percentage areal coverage (Schemenauer and Joe, 1989).

Iran is located in southwest Asia and borders the Oman Sea, Persian Gulf, and Caspian Sea. Most of the country is located in regions of subtropical high pressure and more than 75% of its area is arid and semi-arid. The average annual rainfall of all the country is about 250 mm.

Iran consists of rugged, mountainous rims surrounding high interior basins. The main mountain chain is the Zagros Mountains, a series of parallel ridges interspersed with plains that bisect the country from northwest to southeast. Many peaks in the Zagros exceed 3,000 m above sea level, and in the south-central region of the country there are at least five peaks that are over 4,000 m (Fig. 1).

---
[1] Assistant Professor, Faculty of Desert Studies, Semnan University, Semnan, Iran. E-mail: mrahimi@sun.semnan.ac.ir

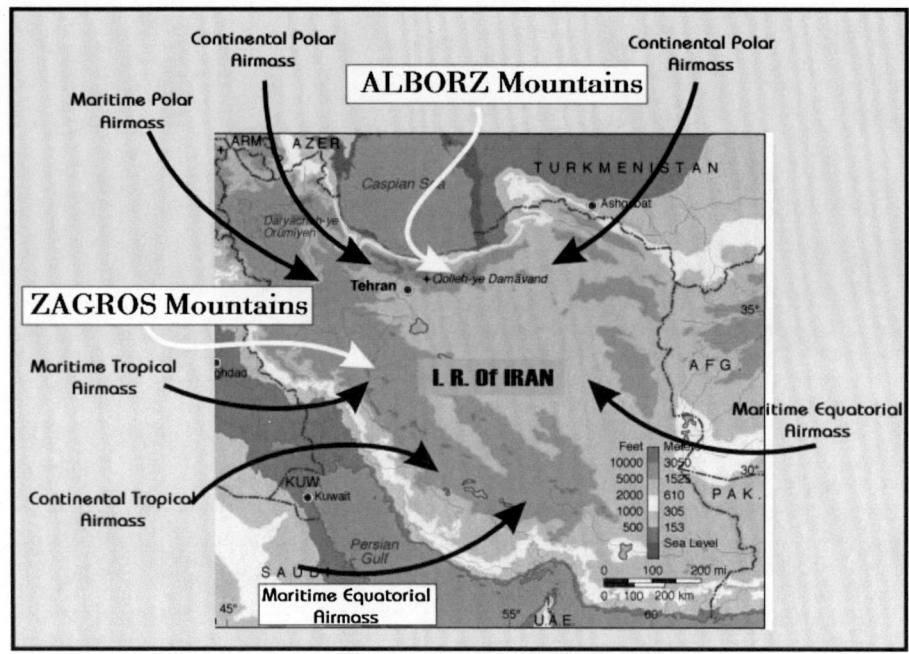

Figure 1
Geographical regions and air masses of Iran

As the Zagros continue into southeastern Iran, the average elevation of the peaks declines dramatically to under 1,500 m. Rimming the Caspian Sea littoral is another chain of mountains, the narrow but high Alborz Mountains. Volcanic Mount Damavand, 5,610 m, located in the center of the Alborz, is not only the country's highest peak but also the highest mountain on the Eurasian landmass west of the Hindu Kush.

The center of Iran consists of several closed basins that collectively are referred to as the Central Plateau. The average elevation of this plateau is about 900 m, but several of the mountains that tower over the plateau exceed 3,000 m. The eastern part of the plateau is covered by two salt deserts, the Dasht-e Kavir (Great Salt Desert) and the Dasht-e Lut. Except for some scattered oases, these deserts are uninhabited.

Iran has only two expanses of lowlands: the Khuzestan Plain in the southwest and the Caspian Sea coastal plain in the north. The former is a roughly triangular-shaped extension of the Mesopotamia plain and averages about 160 km in width. It extends for about 120 km inland, barely rising a few meters above sea level, and then meets abruptly with the first

foothills of the Zagros. Much of the Khuzestan plain is covered with marshes.

Iran has a variable climate. In the northwest, winters are cold with heavy snowfall and subfreezing temperatures during December and January. Spring and fall are relatively mild, while summers are dry and hot. In the south, winters are mild and the summers are very hot, having average daily temperatures in July exceeding 38°C. On the Khuzestan Plain, summer heat is accompanied by high humidity. The main air masses affecting Iran are shown in Fig. 1.

In general, Iran has an arid climate in which most of the relatively scant annual precipitation falls from October through April. In most of the country, yearly precipitation averages 250 mm or less. The major exceptions are the higher mountain valleys of the Zagros and the Caspian coastal plain, where precipitation averages at least 500 mm annually. In the western part of the Caspian, rainfall exceeds 1,000 mm annually and is distributed relatively evenly throughout the year. This contrasts with some basins of the Central Plateau that receive 10 cm or less of precipitation annually.

The Alborz Mountains are elongated from west to east in the north part of the country and capture all humidity of air masses which come from the north and take humidity from the Caspian Sea that creates the central deserts. The Zagros Mountains (northwest to southeast) take the humidity of systems which come from the Persian Gulf. The windward part of these mountains is green and the other parts of Iran are mainly semi-arid and arid.

However, in the northern parts (Caspian Sea coasts) the annual rainfall is about 1,200 mm, whereas in central parts of Iran annual rainfall is about 60 mm. These common conditions, and specific conditions such as droughts, make it necessary to quest for new resources and methods for water supply (Fig. 2).

Research interests related to fog are often closely linked with geographical places where fog occurs frequently and thus are considered a relevant component of the respective climate (WERNER EUGSTER, 2008). Because of the geography of Iran, upslope fogs form in mountainous areas. Proximity to the Persian Gulf in the south and the Caspian Sea in the north of Iran provides suitable conditions to form advection fogs. The wet and warm air from the sea comes over the land and forms fog when it reaches the cold surface of the neighboring land area.

Since advection and upslope fogs form regularly in most parts of Iran, fog collection is a candidate for a suitable operational method of water supply. ESFANDIARNEJAD *et al.* (2010) studied the feasibility of water harvesting from fog and atmospheric moisture in the south coast region of Iran. According to their findings the Persian Gulf coast has high potential for water harvesting from fog and also dew.

2. Materials and Methods

The length of the fog season in a region should be determined. Analyzing the data of fog days could be useful in this step. However, the duration of fog in a day is also important for water collected from fog. A fog day is a day that fog has been observed in an observational point (American Meteorological Society, 2000). There is a difference between water collected from a 1 h fog day and a 20 h fog day. The other point that should be considered is the density of

Figure 2
Precipitation map of Iran

fog. Mist and dense fog have different water collection productivity. In conclusion, the number of fog days, fog duration in a day and the density of fog are important factors of water productivity of a fog event in a given place.

If the region is recognized to have potential fog days, through travel in a region, discussion with the local people, and meetings with government officials and meteorologists, an idea can be obtained as to whether there are high elevation regions with a water requirement and frequent fog.

A more sophisticated program uses standard fog collectors of 1 m² to measure the fog water production rates on specific terrain features. Because fog collection is a non-conventional method of obtaining water, a public education program should be started early in any project. The next step is to design a system of collection, transport and distribution of the fog water.

We have undertaken the first step of a fog collection program. Determining the fog days is necessary to define potential places for fog collection projects.

The study area includes all of Iran, which is about 164 million² km. The data of more than 115 synoptic stations from the I.R. Iran Meteorological Organization (IRIMO) are used to determine fog days (IRIMO, 2007). Locations of these stations are shown in Fig. 3.

A fog day is a day when fog is observed at an observation point. Data included fog days in each month throughout the year. However, since synoptic and airport stations mainly are located in flat areas and low level lands near the cities, the fog days in highlands and mountainous areas are greater than at neighboring synoptic stations. The distribution of stations over the country seems to cover well the fog days. The fog days data is an average of a long term period that is different for each station. However, for most of the stations this period exceeds 20 years. Since there are some trends in temporal variation of fog days during past 10 years because of climate change, we did not consider the WMO suggested base period of time (i.e. 1961–1990) as we need to know the current situation of fog days in the country. Map zoning of fog days was done for all the country. The interpolation method for mapping of fog days

was inversed distance weighting (IDW). In this method the value of each point in the map varies with inverse distance from the known input of near station with given fog days.

Using this method, areas with a high number of fog days were determined for the map. The places with no stations would also be recognized and the fog days interpolated. Also, the graphs of fog days variation during the year were determined for some specific stations. There are four stations, each representative of a topo-climatological area including coastal, island, highland, and low level stations. The last two types are far from the sea.

In order to investigate altitude dependence, a correlation relationship between altitude and fog days was developed. However, in this survey we select just stations that are located inland and far from the sea. Since most fogs in stations near the sea are advections fogs and do not depend on altitude, no significant correlation was anticipated.

3. Results

Table 1 shows the fog days in stations which have more than 50 fog days in a year. As is clear from the table, most of the fog types in island and coastal stations are advection and upslope fogs. Advection fogs especially occur at night because the land becomes colder than the sea and condensation occurs. Since island stations are encircled by water, they have more fog days than coastal stations. Islands include the Kish and Siri Islands. Regarding coastal stations such as Astara, Bandar Abbas, and Abadan, the fog days decrease by distance from the sea. This issue is related to advection fogs and is not relevant for other types of fog.

In land sites far from the sea, such as Parsabad and Ardabil, radiation fogs and upslope fogs frequently form due to low temperature and high altitudes. In some regions all kinds of fog form together and increase the fog days. Astara is a good example of this. Astara is located in the northwest highlands in the west coast of the Caspian Sea. The average length of the fog season in Astara is 220 days/year.

The map of fog days is shown in Fig. 4. The high amount of fog days in the northern and southern

Distribusion of the synoptic stations in Iran

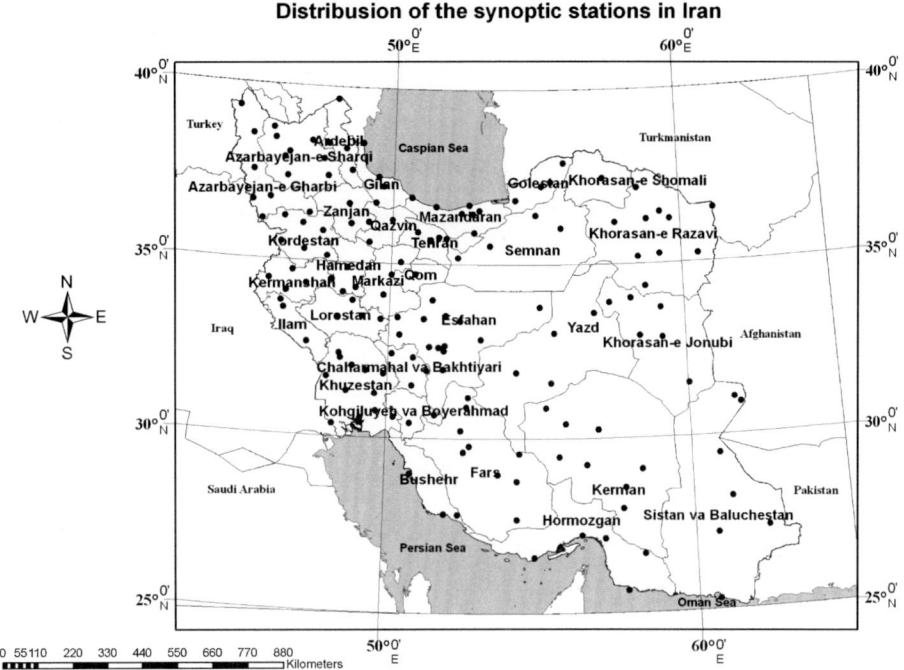

Figure 3
Synoptic meteorological stations

Table 1

Stations with annually fog days more than 50 days

Station	Altitude (MASL)	Fog days	Trend	R^2	Station	Altitude (MASL)	Fog days	Trend	R^2
Astara Coast	−18	220	−12.1	0.69	Dogonbadan	699.5	83	NS	–
Kish Island	30	213	NS	–	Bostan	7.8	83	NS	–
Konarak	12	191	NS	–	Jiroft	601	79	NS	–
Qaem Shahr	14.7	159	NS	–	Masjed soleyman	32.5	78	NS	–
Siri Island	4.9	154	NS	–	Ahar	1390.5	78	−3.2	0.6
Parsabad	31.9	152	−8.6	0.73	Omidieh	34.9	77	2	0.43
Booshehr Coast	8.4	151	NS	–	Mashad	999.2	72	NS	–
Ardabil	1332	146	−4.2	0.47	Lar	792	70	NS	
Rasht	−6.9	143	NS	–	Ramsar	−20	67	−3.79	0.65
Bandar Abbas	9.8	128	2.4	0.36	Hamedan	1741.5	66	NS	–
Jask Coast	4.8	117	2.6	0.37	Abadan	6.6	65	NS	–
Safi Abad	82.9	110	−3.4	0.33	Ahvaz	22.5	63	NS	–
Booshehr Coast	19.6	106	−4.1	0.67	Kangavar	1468	62	NS	–
Ravansar	1379.7	103	−6.6	0.58	Isfahan	1543	61	NS	–
Dezfool	143	103	NS	–	Makoo	1411.3	60	NS	–
Lengeh Port	14.2	100	NS	–	Hamedan	1678.7	58	NS	–
Chabahar	8	99	NS	–	Goochan	1287	55	NS	–
Minab	27	94	NS	–	Golmakan	1176	55	NS	–
Noshahr Coast	−20.9	89	NS	–	Gazvin	1279.2	51	NS	–
Mahshahr Port	6.2	87	NS	–	Myaneh	1110	50	NS	–
Agajari	27	85	−3.2	0.48	Gorveh	1906	50	NS	–

The Annual Number of the Days with Fog During 1960-2005

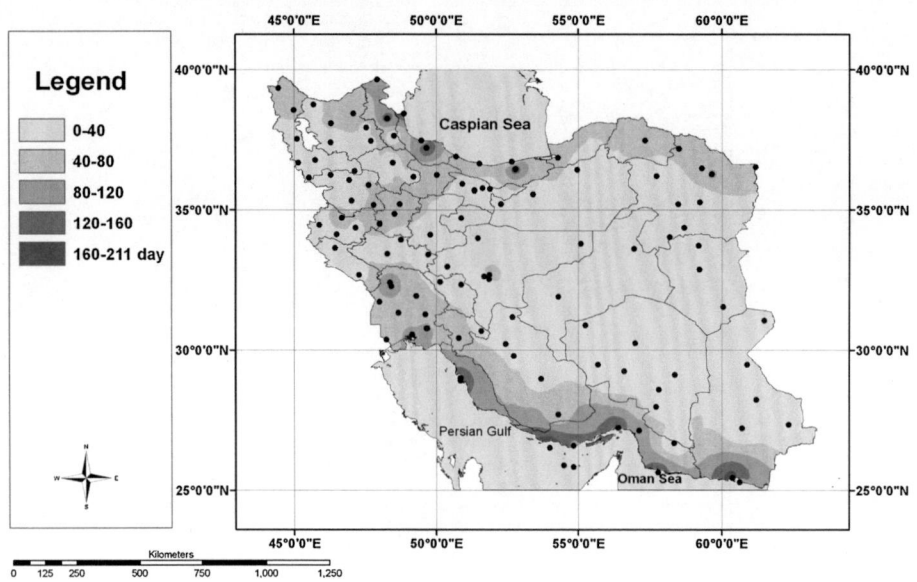

Figure 4
Annual number of fog days (1960–2005)

coastal regions is clear in the map. The farther from the sea, the lower the number of fog days.

There is also a monthly variation in fog days. Figure 5 shows the distribution of fog days in Astara, Kish Island, Anzali, Bandar Abbas, and Qaem Shahr throughout the year. In Astara fog forms during all months of the year but there are more fog days in colds months. In the Kish Islands fog occurs mostly during warm months. On the northern coastline of Iran (Anzali), fog mostly occurs in cold months. On the southern coast of Iran (Bandar Abas), fog occurs mostly in warm months. In some regions (Qaem Shahr) fog forms equally in all months.

The trends of fog days during the study period were also studied. Table 1 shows also the trend of fog days during 1986–2005 with related correlation coefficient (R^2). There are negative and positive trends. However, some trends were non-significant (NS). The trends show the amount of increase or decrease of fog days per year. The significant trends mostly are negative, corresponding to decreasing fog days in these areas. Most of these places are mountainous and upslope fogs are more frequent than other types. This is consistent with global warming effects in this part of world. Upslope fog formation and

frequency decreases as temperature increases. Only three stations, namely Bandar Abbas, Jask, and Omidyeh exhibit positive trends of fog days. This is consistent with advection fogs being common along the Persian Gulf coasts. As temperature increases, the evaporation will also increase, and fog formation conditions will increase.

4. Conclusion

Countries located in arid and semi-arid zones such as Iran can use fog collection to supply water. In this regard fog collection by man-made collectors may be a non-conventional source of water. The water source is sustainable over periods of hundreds and probably thousands of years because the driving forces for the formation of the cloud decks are global in nature and will change only slowly.

Most fogs in Iran are advection fogs and because of proximity to the Persian Gulf in the south and the Caspian Sea in the north of the country. However, a small number of fogs are upslope fogs.

The results show that Iran is a country with special potential for fog occurrences. The northern

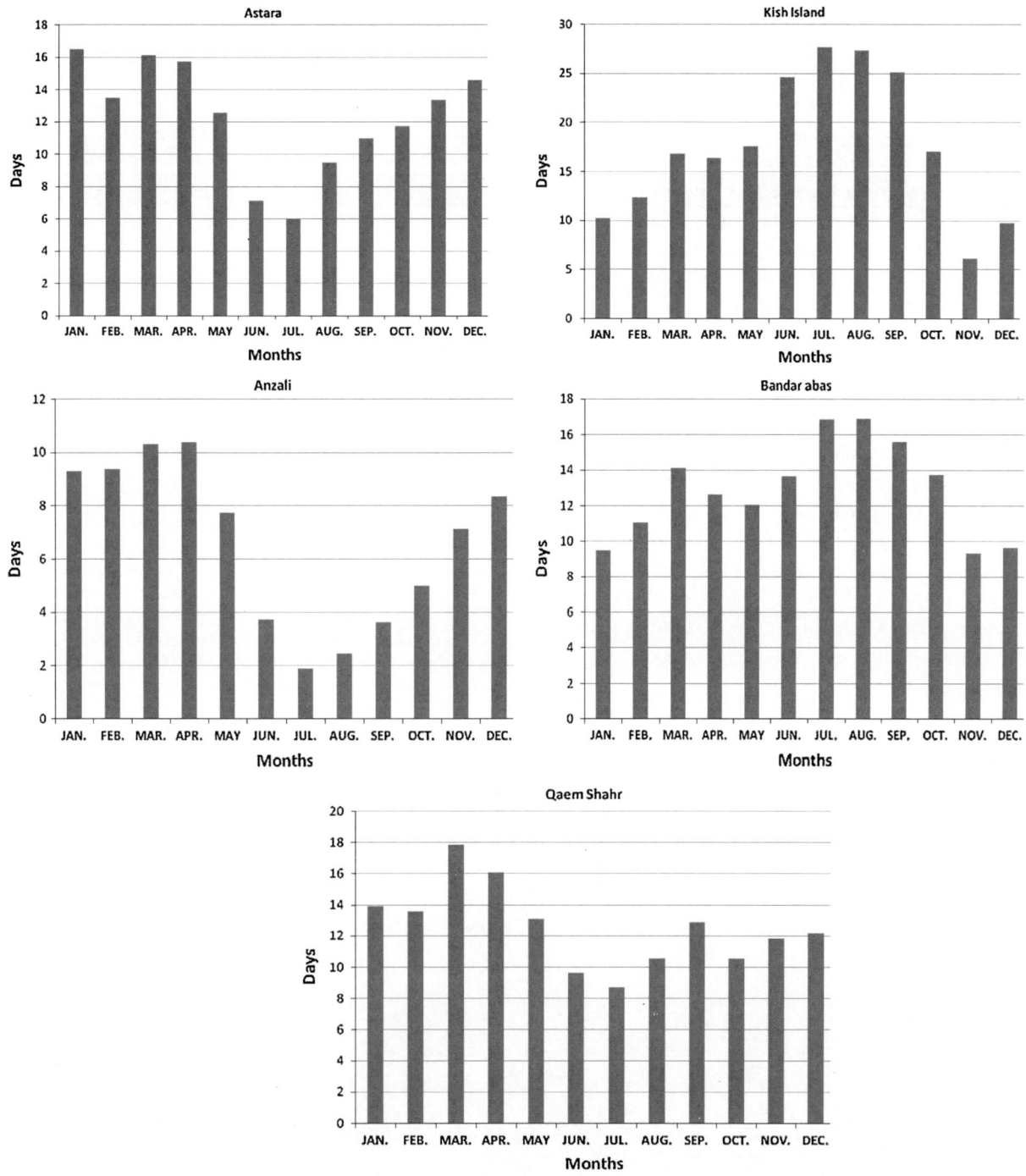

Figure 5
Temporal variation of long term monthly fog days (1960–2005)

coastline with its high relative humidity is suitable for fog collection projects. Also many fog-bearing winds pass over the Caspian Sea where they gain humidity from the sea and increase the number of fog days. The same conditions occur in southern coastal regions. Some fog places are located inside the country, such as Parsabad, and Qaem Shahr are far from the waters but have high elevation.

The north of Iran is suitable for rice cultivation. Rice requires a lot of water during growing season. In spite of high precipitation (Fig. 2), in some years the annual rainfall is not enough for rice production. Fog collected water would be a suitable option to supplement existing rainfall to provide water requirements for rice crops.

In the south coastal region of Iran (Persian Gulf coast) the water shortage is always a big problem for drinking water. The amount of precipitation is low and so water from other resources is necessary. Fog collection projects in this area could be cost effective and necessary.

Because of the unique climate and topography of Iran, many kinds of fogs occur in the northwest, north coast, northeast, and south coast regions of Iran. Defined places would be suitable for implementation of fog collection assessment projects.

The author would like to recommend that water resources planners and managers take this issue into more consideration and initiate fog water collection projects in suitable locations.

Acknowledgments

Here I would like to thank Dr. Robert Schemnauer from Environment Canada for his opinions in preparing this paper, Ian Bell from the Australian Bureau of Meteorology and two anonymous reviewers for their kind improvements in preparing this paper. I also would like to thanks Nooshin Mohammadian from I.R. of Iran Meteorological Organization for preparing maps.

REFERENCES

Acosta, Baladon A.N., Agricultural uses of occult precipitation (Ornex, France 1995).

Al Fenadi, Y.Sh. (2001), fog studies in North Libia, 2nd International conference on fog and fog collection, St. John's, Canada, July 15-20, 2001, pp. 411–412).

American Meteorological Society, Glossary of Meteorology (Boston, 2000).

Esfandiarnejad A., and Ahangar R., and Kamalian U.R., and Sangchouli T.(2010), Feasibility studies for water harvesting from fog and atmospheric moisture in Hormozgan coastal zone (south of Iran), 5th International Conference on Fog, Fog Collection and Dew, Munster, Germany, 25–30 July 2010.

IRIMO, Meteorological Yearbook (Tehran, Iran, 2007).

Schemenauer, R.S. and P. Joe (1989) The collection efficiency of a massive fog collector. Atmos. Res., 24, 53–69.

Schemenauer, R.S. and P. Cereceda (1991), Fog water collection in arid coastal locations. Ambio, 20 (7), 303–308.

Schemenauer, R.S. and P. Cereceda (1994) A proposed standard fog collector for use in high elevation regions. J. Appl. Meteor., 33, 1313–1322.

Schemenauer R. and Cereceda P. (1997) Fog Collection, Tiempo, Issue 26 Dec.

Werner Eugster (2008), fog research, DIE ERDE 139(1-2), Zurich, pp. 1–10.

(Received November 1, 2010, revised March 22, 2011, accepted April 18, 2011, Published online May 19, 2011)